X-RAY DIFFRACTION
PROCEDURES
For Polycrystalline and Amorphous Materials

X-RAY DIFFRACTION PROCEDURES
For Polycrystalline and Amorphous Materials

HAROLD P. KLUG

PROFESSOR OF CHEMISTRY AND STAFF FELLOW, EMERITUS

LEROY E. ALEXANDER

PROFESSOR OF CHEMISTRY AND SENIOR FELLOW

Mellon Institute of Science

Carnegie-Mellon University

SECOND EDITION

A WILEY-INTERSCIENCE PUBLICATION

JOHN WILEY & SONS, New York · **London** · Sydney · Toronto

Copyright © 1954, 1974, by John Wiley & Sons, Inc.

Library of Congress Cataloging in Publication Data

Klug, Harold Philip, 1902–
 X-ray diffraction procedures.

 "A Wiley-Interscience publication."
 First ed. published in 1954 under title: X-ray diffraction procedures for polycrystalline and amorphous materials.
 Includes bibliographical references.
 1. X-rays—Diffraction. I. Alexander, Leroy Elbert, 1910– joint author. II. Title.
 [DNLM: 1. X-ray diffraction. QC482 K66x 1974]
 QC482.D5K55 1974   548′.83   73-21936
 ISBN 0-471-49369-4

Printed in the United States of America

10 9 8 7 6 5 4 3 2

To our patient and understanding wives

HELEN and ELEANOR

PREFACE

Nineteen years have passed since the first edition of this book was published, and great progress and change have appeared in many areas treated therein. Indeed, this revision should have been completed years ago, but one of us (L.E.A.) has been involved in authoring the monograph, *X-ray Diffraction Methods in Polymer Science*, Wiley, New York, 1969. As authors know all too well, one such task at a time is enough. So by 1970, when we both could give serious consideration to the revision, it had become a major rewriting job. Except for the first three chapters, all or extensive portions of most chapters had to be rewritten. To bring the subject matter up to date, more reference material had to be searched than for the original edition. The volume of new material to be considered was such that it has often not been possible to present the detailed treatment we would have preferred. As the best alternative in such instances, we have tried to give generous references to lead the interested reader quickly to an understanding of such topics.

The first edition appeared when the counter diffractometer (then the Geiger-counter spectrometer) was in its infancy. In the intervening years every aspect of the diffractometer has been enormously improved, and it has become the major instrument for precise intensity measurements in all areas of x-ray diffraction. The theory of intensity measurements and of diffractometer line profiles has been extended. The new proportional counters and scintillation counters have been introduced as detectors. Solid-state circuitry has provided pulse-height discrimination and pulse-height analysis, which further increase the accuracy and precision of intensity measurements. Finally, the stability of the instrument and of its x-ray source has been greatly improved, and the instrument has been automated. Accordingly, we have felt it desirable to present as complete a treatment of the diffractometer as space would permit.

Another powerful tool in diffraction studies was just beginning to emerge when the first edition appeared in 1954, the high-speed electronic computer. The applications to powder diffractometry did not begin to appear until the late 1950s. Many of the calculations involving powder diffraction data are readily programmed for a computer: *d* value calculations, correction of film data for various factors, application of the least-squares method to lattice constant calculations, searching powder diffraction data for identification purposes, and, finally, the programming of the diffractometer for automatic collection of data. The Fourier transformation of diffraction data in radial-distribution analysis and the evaluation of crystallite size and strain parameters are also best computerized. In the appropriate places these computer applications are outlined and discussed. The computing facilities of diffractionists vary widely, and it is well-known too that computers and computer programs are constantly in a state of change through improvements, updating, and revision. Hence we have considered the appropriate treatment usually to be a brief outline of the possibilities of the computer application with references to available programs and specific installations.

One chapter in the first edition, Chapter 12, "Small-Angle X-ray Scattering," has been omitted from the new edition. The justification of this omission deserves brief comment. The topic had always seemed to us to be a less coherent part of the original book than the other chapters. Moreover, the monograph by L.E.A. mentioned above contains an up-to-date chapter on small-angle scattering together with a large bibliography on the topic. Thus a single chapter on small-angle scattering prepared for this book would largely be a repetition of the chapter in *X-ray Diffraction Methods in Polymer Science*. Finally, the great amount of new material to be covered in the new edition was already expanding it to a rather large book. Therefore, it seemed desirable to delete the chapter on small-angle scattering, and refer our interested readers to the above monograph. The second edition, however, still lists 12 chapters as a result of splitting former Chapter 10 into two chapters: "Preferred-Orientation Determination" (Chapter 10), and "Stress Measurement in Metals" (Chapter 11). Former Chapter 11, "Diffraction Studies of Noncrystalline Materials," is now Chapter 12, "Radial-Distribution Studies of Noncrystalline Materials." The current heightened interest in line-broadening analysis in relation to both crystallite size and lattice strains is reflected in the rather extensive treatment of Chapter 9.

We hope that this second edition will continue, as did the original book, to serve as textbook, manual, and teacher to plant workers, graduate students, research scientists, and all others who seek to work in or understand the field.

The kindness and courtesy of numerous individuals and organizations in furnishing new illustrative material is appropriately acknowledged in the figure legends. We are indebted to many colleagues and friends in x-ray diffraction who pointed out errors in the first edition or made suggestions to improve it. Hopefully all outright errors have been corrected, and suggestions have received our thoughtful consideration. Our laboratory associates Dr. S. S. Pollack and Mr. Sherman Swisher deserve our thanks for their interest and their assistance in data gathering. Mr. Gilbert Arnold and associates did an outstanding job of the drafting and preparing the figure copy. Highest praise and commendation go to Miss Mary L. Condy, who again has typed the manuscript and given much valuable assistance in other ways. Lastly, the authors are indebted to Dr. A. A. Caretto, Chairman of the Department of Chemistry, for his interest, understanding, and encouragement during the revision, and to Dr. R. M. Cyert, President of Carnegie-Mellon University, for the excellent facilities available for production of the manuscript.

HAROLD P. KLUG
LEROY E. ALEXANDER

Pittsburgh, Pennsylvania
August 1973

PREFACE TO THE FIRST EDITION

In the last decade or two the numerous and varied applications of x-ray diffraction have established it as one of our most fundamental and important research tools. Laboratories by the hundreds, both academic and industrial, have installed such facilities, and others are considering how it fits into their research "picture." The frequent mention of the technique in journal articles, the steadily increasing membership of the diffractionists' national society, the American Crystallographic Association, and the development and wide sale of excellent commerical diffraction units are further evidences of its present standing in the laboratory. Despite all this, many who would benefit greatly by using diffraction methods are still unaware that numerous research applications of the science are actually routine, and require little knowledge of the intricate theory of x-ray diffraction.

The most widely used diffraction procedures are those applied to polycrystalline materials. Yet, at this writing, most previous presentations of the powder method are long out of date, and the Geiger-counter spectrometric techniques are to be found only in the journal literature. Likewise, the small-angle scattering methods and (to a lesser extent) the radial-distribution techniques, so important for amorphous materials, must still be sought out in numerous separate sources. It has thus become imperative to bring together in a single volume the most useful manipulative and interpretative details of these applications, so that the beginner, and others, may learn the various techniques and develop a skill in their use without each new application's requiring a major library and literature search. Obviously, to keep within the compass of a single volume, advanced details and the less widely used applications must be omitted, and single-crystal structure analysis cannot be considered within the book's scope.

We know full well that there is no substitute for a skillful and inspiring teacher. Today, however, many must necessarily acquire certain skills and knowledge largely by their own efforts. This book has been prepared with such individuals in mind; and it is presented with the hope that it may, with some measure of success, serve as both manual and teacher for the plant worker, the graduate student, the research scientist, and others who wish to do experimental work in the field. At the same time it is hoped that the research director and group leader will find in it a handbook and guide to the power and utility of diffraction methods in the laboratory.

The book is the outgrowth of more than twenty-five years of research and graduate teaching in the field of x-ray diffraction, a varied industrial consulting experience, and wide application of the technique to the researches and problems of the eighty Industrial Fellowships at Mellon Institute.

In presenting the most basic techniques, procedures, and applications, we have felt it desirable to err on the side of too much rather than too little detail. Thus simple steps in mathematical derivations are always included; simple as well as more advanced examples are freely used to illustrate concepts and procedures; convenient forms for recording data are presented; and suggestions on equipment maintenance are given in some instances. The book assumes no special knowledge of crystallography and x-rays on the part of the reader, but gives at the beginning sufficient introductions to these sciences that the novice in diffraction may rather quickly understand and apply the simpler techniques. In keeping with the title (and paragraph two above), the fundamental equations of diffraction theory have not as a general rule been developed but have been taken from the previous literature. Those interested in their rigorous derivation must refer to the original sources or to more theoretical books on the subject. Other mathematical details directly connected with the procedures and applications are generally given. The literature cited in the footnotes furnishes the reader a ready reference to original work on a particular topic, while the general list at the ends of the chapters has been carefully selected to provide the most useful and logical approach to the more advanced aspects of the science. Books marked with an asterisk (*) are considered particularly desirable for a small reference library on x-ray diffraction and its applications.

No one can write a book of this type without being greatly indebted to the host of outstanding workers, past and present, in the science, and we gratefully acknowledge this indebtedness to our colleagues throughout the world. As far as possible individual credit has been given throughout, but in many instances we have long since forgotten the source of our information, and, accordingly, have considered it a part of the general body of knowledge in the field.

The kindness of numerous individuals and organizations in furnishing illustrative material is appropriately acknowledged in the legends to the figures. We are indebted also to several of our colleagues in diffraction who have critically read and commented on portions of the manuscript. For this valuable service, we express our gratitude and appreciation to Dr. W. N. Lipscomb (Chapters 1, 2, 3), Dr. M. E. Straumanis (Chapter 8), Dr. L. J. E. Hofer (Chapters 9 and 11), Mr. R. W. Turner (Chapter 10), and to Drs. K. L. Yudowitch and C. G. Shull (Chapter 12). We are similarly obligated to Dr. R. S. Tipson, who kindly read portions of the manuscript for style. Our own laboratory associates, Dr. N. C. Baenziger, Dr. R. L. Collin, Dr. E. S. Hodge, Miss Helen R. Golob, and Mr. D. T. Pitman deserve hearty commendation for their continued interest in the manuscript during preparation, for their criticisms and comments on various parts of it, and for their assistance in gathering data and preparing photographic material. To Dr. E. R. Weidlein, President of Mellon Institute, we are greatly indebted for continued interest and encouragement and for excellent facilities for the production of the manuscript. We are grateful also to Dr. W. A. Hamor, Director of Research, for stimulating discussions relative to the manuscript and to publishing practice. Mrs. Eleanor Alexander gave assistance in the proofreading and indexing that is deeply appreciated. Finally, highest praise and special thanks are due Miss Mary L. Condy, who has typewritten the manuscript and in many other ways given invaluable assistance in its preparation.

HAROLD P. KLUG
LEROY E. ALEXANDER

February 10, 1954
Mellon Institute

CONTENTS

CONTENTS

CHAPTER 1

ELEMENTARY CRYSTALLOGRAPHY

The diffraction of x-rays by matter is the basis of a unique scientific tool. It is most powerful when applied to crystalline materials, but it can yield fundamental and important data when applied to liquids and amorphous solids. The complete realization of its many potentialities requires an elementary knowledge of crystallography and x-ray science. Admittedly, there are certain routine applications of x-ray diffraction that demand little or no familiarity with these fields. Most workers, however, want to go beyond these "push-button" applications and use the various techniques as the powerful aids they can be in both pure and applied research.

Without a clear concept of such common crystallographic terms as habit, form, symmetry element, axial ratios, Miller indices, space lattice, and unit cell, the student is severely handicapped in applying diffraction techniques. A limited knowledge of geometrical crystallography is necessary, of course, for an understanding of the simple geometry of x-ray diffraction by crystals. This chapter provides a working introduction to crystals which will enhance the usefulness of even the simplest diffraction techniques, and which will form the background for more advanced diffraction studies.

1-1 THE CRYSTALLINE STATE

1-1.1 Crystalline and Amorphous Solids

From ancient times crystals have been a fascinating enigma to man. Among the solid bodies in nature they seemed always to be the rarest exception, so that the great bulk of solid matter came to be regarded as *amorphous* (without form). With the advent of modern chemistry and physics, crystals posed a challenging question concerning the internal

1

structure of solid matter. Ultimately, however, their characteristic geometrical form and external symmetry suggested a regular internal arrangement as highly probable. Indeed, such considerations in 1912 led Laue[1] to his epoch-making discovery of x-ray diffraction. Here, in a single brilliant experiment, two of the era's most baffling scientific problems were solved; the regular internal structure of crystals was established, and the wave nature of x-rays was demonstrated. Today, x-ray diffraction is recognized as the most powerful tool available for the study of solids.

The years since Laue's discovery have demonstrated that very few solids are amorphous. Such materials as rocks and soil, metals and alloys, wood, concrete, and even textile fibers are crystalline, at least in part. Rubber becomes crystalline when stretched, and even bones, hair, muscle fibers, and tendons of the animal body are partly crystalline. The surest means of demonstrating this crystalline nature is by the use of x-ray diffraction. Under x-ray examination at least 95 per cent of the solid inorganic chemical substances of the laboratory are found to be crystalline, and among the minerals more than 98 per cent show definite crystalline structure.

The term, *amorphous* solid, must then be reserved for those few substances that show no crystalline nature whatsoever by any of the means available for detecting it. Glasses and certain resins and polymers fall in this class, and opals are probably an example. They give a liquid-type diffraction pattern and are to be regarded simply as supercooled liquids of very great viscosity. The terms *vitreous* and *glassy* are now preferred to *amorphous** by most investigators. The very finely divided precipitates and various carbons, formerly called amorphous, have been shown to be composed of exceedingly fine crystals. Accordingly, the term *solid* has now taken on a meaning synonymous with *crystalline*.

Much of the solid matter we meet in our daily contacts shows little outward evidence of crystalline form because it is composed of an aggregate of tiny crystals in random orientations with respect to each other. Metals are always polycrystalline, except when very special efforts are made to prepare them in the form of single crystals. Most mineral matter occurs in this polycrystalline form and is described by the mineralogist as *massive.* The properties of polycrystalline matter are largely the averaged

---

*The term *amorphous* has also been used by some investigators to describe a third, rather rare, class of solids. Matter deposited by vaporization from a hot filament with subsequent sudden condensation on a cold surface may yield a structureless solid. The deposits on the inside of incandescent lamp bulbs and vacuum tubes are frequently of this nature.

values of those of the individual crystals; hence the proper study of crystal phenomena begins with the single crystal.

1-1.2 Definition of a Crystal

Dana[2] has supplied an excellent definition of a crystal from the viewpoint of morphological crystallography: "A crystal is the regular polyhedral form, bounded by smooth surfaces, which is assumed by a chemical compound, under the action of its interatomic forces, when passing, under suitable conditions, from the state of a liquid or gas to that of a solid." Recalling that the chemist considers the condition of a solute in solution somewhat analogous to the gaseous state, we see that this definition is very complete. A natural consequence of this definition is the inference that a crystal is the normal form of all solid elements and chemical compounds.

A definition by Antonoff[3] supplements Dana's: "A crystal is a homogeneous, anisotropic body having the natural shape of a polyhedron." This second definition implies the internal variation of structure with direction which results in the observed variations of crystal properties with direction.

1-1.3 Characteristics of the Crystalline and Vitreous States

The most outstanding fact x-ray diffraction has revealed about crystals is that they are composed of atoms or groups of atoms arranged in a regular and repeated pattern. The extreme minuteness of the atoms and molecules, however, causes crystals to appear homogeneous macroscopically. A direct consequence of this regular internal arrangement is that, on a microscopic scale, crystals are heterogeneous or *anisotropic*. They are not, in general, the same in nonparallel directions, and certain physical properties which depend upon the structural arrangement show variations with direction. These are known as *directional* or *vectorial properties*.

Vectorial properties are conveniently divided into two classes: (I) external properties and (II) internal properties. Among the external properties are: (1) the production of plane faces during growth, (2) the constancy of interfacial angles, (3) symmetry, and (4) properties on dissolution. A few of the inner properties are: (1) the optical properties, (2) thermal expansion and thermal conductivity, (3) elasticity, (4) hardness, and (5) the electrical and magnetic properties. Some of the foregoing properties may be nonvectorial in certain crystals. Cubic crystals, for example, sodium chloride, are optically isotropic but anisotropic as regards hardness. On the other hand, calcite, a form of calcium carbonate ($CaCO_3$), is anisotropic with respect to both these properties. Other properties, such as the density and specific volume, are obviously independent of direction.

Further characteristics of the crystalline state are the sharp melting point and definite heat of fusion. As the temperature is raised, a point is reached at some definite temperature where the forces holding the atoms in their regular array are overcome, and melting takes place. At this temperature a fixed amount of energy per unit weight is needed to overcome the crystalline forces, and a definite latent heat is observed. For example, the latent heat of fusion of ice is 79.71 cal/gram.

Vitreous substances, in contrast, are isotropic, and their properties do not vary with direction. When heated, they do not melt at a fixed temperature, but gradually soften by imperceptible degrees and become more fluid. They exhibit no latent heat of fusion. Although they appear rigid, they slowly flow under pressure. When broken they yield irregular pieces with curved surfaces, and the fracture is said to be *conchoidal.* They are formed by cooling liquids under such conditions that the melts fail to crystallize. Instead, the viscosity of the melt constantly increases with decrease in temperature until the material sets to a rigid mass- All such properties of vitreous materials point to a random internal arrangement of the atoms, ions, or molecules such as occurs in liquids.

Polycrystalline substances in large masses (large compared to the size of the individual crystals) will generally be isotropic if the crystallites are randomly oriented. Their properties in any given direction represent an average of the vectorial properties of the individual crystals.

1-2   CRYSTAL GEOMETRY

1-2.1 External Form and Habit of Crystals

The smooth plane surfaces bounding a crystal are called *faces* or *planes.** On a well-developed crystal the arrangement of the faces is such as to impart to the crystal as a whole a certain characteristic symmetry of form. It is rare, however, that a crystal grows under conditions so favorable as to yield the ideally perfect polyhedral form. The normal deviations from the ideal form are of two types. Equivalent faces on an individual crystal may vary in size, or some may even be completely missing. It is common, too, for the faces not to be absolutely smooth

---

*The term *plane* is also used in a broader sense in x-ray diffraction to indicate any direction through the crystal that is a possible crystal face, whether or not actually developed as a face on the crystal.

and polished, but instead roughened by minute elevations, depressions, striations, and the like. The crystallographer, nevertheless, in his drawings usually depicts the crystal as an ideal polyhedron.

Crystals of a given chemical compound may show rather great diversity of form. By change in the number of faces, as well as in their relative sizes, the so-called *habit* may vary almost indefinitely. This is a common occurrence among mineral species from different localities, and is frequently observed in laboratory specimens. In Fig. 1-1, the habit of calcite ($CaCO_3$) is illustrated by examples of a few of its observed forms. The internal structure is constant throughout the habit variations of a given species, and, in several samples having identical compositions but different external forms, the constancy of the powder diffraction pattern is one of the simplest means of proving a change in habit.

The variations in crystal habit exhibited by a given species have been demonstrated to result from variations in the conditions under which the substance crystallized. There is considerable experimental evidence that small amounts of foreign material present in the crystallizing solution can affect the habit of the crystals formed. Gaubert[4] reported on the effect of methylene blue on the form of lead nitrate crystals. Normally, they appear as octahedra from pure water, but when the solution is saturated with this dye the crystals have a cubic habit. More recently, Buckley[5] and France and his co-workers[6] studied the effects of both

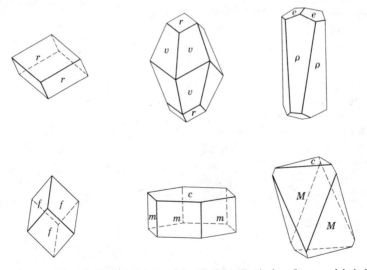

Fig. 1-1. Examples of habit variation in calcite ($CaCO_3$). Equivalent faces are labeled with the same letter. The well-known cleavage rhombohedron is at the upper left. (From E. Dana, *A Textbook of Mineralogy*, 1932; courtesy of Wiley.)

organic and inorganic foreign substances on the crystal habit of many salts, particularly alums and chlorates. Thus from pure water sodium chlorate crystallizes as cubes, but from a solution containing a trace of sodium dithionate its habit is completely changed to tetrahedral.

1-2.2 Constancy of Interfacial Angles

The angle between corresponding faces on crystals of a pure substance is always constant. This observation has been called the *first law of crystallography*. The faces may not be developed to the same size on all its crystals, but the interfacial angles are constant. In fact, these angles not only are constant, but are characteristic of the species and may serve for identification of the substance. Thus, for calcite ($CaCO_3$) the angle between the faces of the cleavage rhombohedron is 74°55′.

The angle commonly given as the interfacial angle is that between the normals to the two adjacent faces. If Fig. 1-2 represents a section through

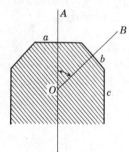

Fig. 1-2. Crystal section depicting the interfacial angle between the face normals.

a crystal perpendicular to the faces *a*, *b*, and *c* the lines *OA* and *OB* represent the face normals to *a* and *b*, and angle *AOB* is the commonly measured interfacial angle. It is seen that these normal angles are the supplements of the actual interfacial angles. In many crystallographic calculations, simpler mathematical relationships exist between these normal angles than between the actual interfacial angles. Furthermore, these angles are usually measured on a reflecting goniometer[7], which gives directly the angle between the face normals.

1-2.3 Symmetry Elements of Crystals

Much of the beauty displayed by fine crystal specimens springs from their symmetry of form, which is by far the most striking external evidence of their regular internal structure. Indeed, a brief study of the distribution of the faces actually reveals the geometrical elements of symmetry possessed by the crystal. The optical properties, and other physical properties, likewise, may be used for symmetry determination. From the examination of many thousands of crystals, it has been demonstrated that these symmetry elements are very simple and few in number. They are of three types: axes, planes, and centers of symmetry. Composite symmetry arises from the combination of these axes, planes, and centers of symmetry. Not every crystal possesses all three types of symmetry elements. In fact, there is one class of crystals with no symmetry whatsoever except for the trivial identity operation (see below) possessed by all objects. It is rather common, however, for a crystal to have all three kinds of symmetry elements present.

A crystal possesses an *axis of symmetry* if, when rotated about an imaginary line through its center, it presents exactly the same appearance more than once during a complete rotation. The imaginary line or direction is the symmetry axis. Although real crystals rarely grow as perfect polyhedra, there is usually no difficulty in recognizing corresponding faces as they are repeated around the crystal, and thereby identifying a symmetry axis. The symmetry axis is twofold, threefold, fourfold, or sixfold, depending on whether the crystal is brought into a similar position every 180, 120, 90, or 60° during rotation. Typical examples of the four types of symmetry axes are illustrated by the crystals depicted in Fig. 1-3. The vertical axes of these crystals represent,

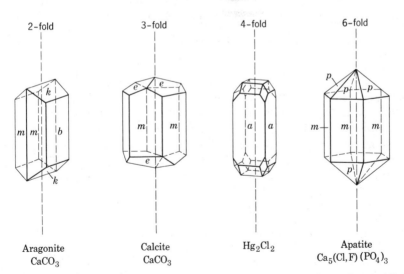

Fig. 1-3. The four types of symmetry axes exhibited by crystals. (After E. Dana, *A Textbook of Mineralogy*, 1932, and A. Winchell, *Microscopic Characters of Artificial Minerals*, 1931; courtesy of Wiley.)

respectively: aragonite ($CaCO_3$) twofold; calcite ($CaCO_3$) threefold; mercurous chloride (Hg_2Cl_2) fourfold; and apatite [$Ca_5(Cl,F)(PO_4)_3$] sixfold axes. Study of the diagrams reveals that each crystal has additional symmetry axes.

Actually, these four rhythms are the only possible ones in crystals. Fivefold, sevenfold, and higher rhythms are never found in crystals, and it can be proved mathematically that they are not possible[8]. All bodies possess an infinite number of onefold axes of symmetry, since a rotation of 360° about any axis brings a body into coincidence with its original position. This is referred to as the *identity operation*.

A symmetry axis is always normal to a possible crystal face and, likewise, always parallel to the edge formed by the intersection of two adjacent faces. A cube possesses three of the four possible kinds of symmetry axes. Figure 1-4 illustrates its six twofold axes *A*, four threefold axes, *B*, and three fourfold axes *C*.

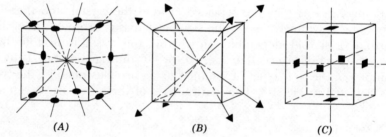

(A)   (B)   (C)

Fig. 1-4. The six twofold axes (*A*), four threefold axes (*B*), and three fourfold axes (*C*) of a cube.

A crystal may show symmetry about a plane, in which case the *symmetry plane* so divides the crystal that each face on one side is matched by a similar face in a reflected or mirror-image position on the other side of the plane. In Fig. 1-5, one of the symmetry planes of the crystal of orthorhombic sulfur is illustrated by the shaded plane. The locations of other symmetry planes are evident on brief study of the drawing. Again it must be remembered that during actual growth crystals rarely develop as

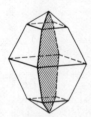

Fig. 1-5.   One of the symmetry planes (shaded) of orthorhombic sulfur.

perfect or ideal geometrical forms. Hence a crystal with a plane of symmetry seldom shows two halves that are the exact mirror images of each other. The symmetry shown by the facial arrangement is merely an expression of the regularity of internal structure, which, in general, is alike in parallel directions. If all like faces have not developed to the same size, or if their distances from the symmetry plane vary, it is an accident of growth and of no consequence; only their angular position is important. Thus, in Fig. 1-6, the crystal of $K_2Zn(SO_4)_2 \cdot 6H_2O$ at A with its

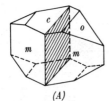   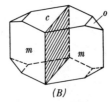

(A)   *(B)*

Fig. 1-6. Nonideal (*A*) and ideal (*B*) development of a crystal of $K_2Zn(SO_4)_2 \cdot 6H_2O$. A vertical symmetry plane is shown by shading. (After A. Tutton, *J. Chem. Soc.*, **63**, 337.)

unequal facial development is crystallographically considered to have a symmetry plane (shaded plane perpendicular to the side-to-side axis) just as truly as if it were the ideal form shown at B. The facial development in both A and B indicates that the internal structure is mirrored about such a vertical plane.

The maximum number of planes of symmetry possible for a crystal is nine. Certain cubic crystals show three planes of symmetry parallel to the cube faces, and six diagonal planes of symmetry passing through opposite cube edges, as shown in Fig. 1-7. Symmetry planes are always parallel to possible crystal faces, and their normals are always parallel to the intersection of two crystal faces.

A crystal possesses a *center of symmetry* when each face is matched by a parallel face in the antipodal position. In this case any straight line passing through the center of the crystal intersects similar points, lines, or faces on opposite sides of the crystal. In Fig. 1-8 the lone center of symmetry in crystals of schizolite, a complex calcium manganese silicate, is delineated by the small black dot. Note that schizolite has neither axes nor planes of symmetry. Many crystals with planes and axes of symmetry also possess a center of symmetry, since a center of symmetry automatically results from the combination of any evenfold axis and a symmetry plane perpendicular to it.

Although there are crystals possessing only a single element of symmetry, such as a plane, axis, or center of symmetry, it has already been hinted that most crystals have several elements of symmetry. Combina-

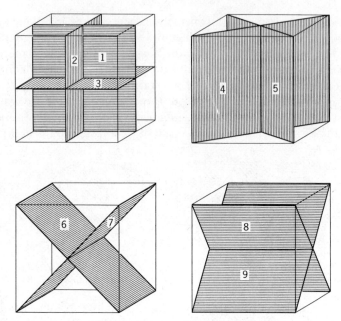

Fig. 1-7.   The nine symmetry planes of a cube.

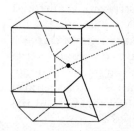

Fig. 1-8.   A crystal with only a symmetry center. Schizolite [HNa(Ca, Mn)$_2$(SiO$_3$)$_3$]. (After J. Dana and E. Dana, *System of Mineralogy*; courtesy of Wiley.)

tion of the symmetry elements in various ways is permissible, as long as the symmetry relations of each element are maintained in the combinaation. An example of such composite symmetry, an evenfold axis with a symmetry plane perpendicular to it, was mentioned above. Another example of composite symmetry is the *axis of rotary inversion* or *inversion axis*. It consists of a rotation through a definite angle, plus an inversion through a point on the axis of rotation. Symbolizing the degree of a regular symmetry axis as 1, 2, 3, 4, or 6, the inversion axis is also conveniently symbolized by its degree with a bar over it, as $\bar{1}, \bar{2}, \bar{3}, \bar{4}$, or $\bar{6}$.

The simple crystals and the geometrical figures which illustrate the inversion axis usually possess additional symmetry elements, but are, nevertheless, most convenient for depicting this special case. Thus the mineral chalcopyrite ($CuFeS_2$) has in its vertical axis (Fig. 1-9) a fourfold axis of rotary inversion ($\overline{4}$). The upper p_1 faces can be brought into coincidence with the lower p_1 faces by a 90° rotation about this axis, plus an inversion across the point at the center. It can be seen that it makes no difference which of these operations is considered to take place first; in either case the final result is the same. The fourfold inversion axis is

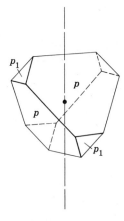

Fig. 1-9.   Crystal with a fourfold axis of rotary inversion. (After E. Dana, *A Textbook of Mineralogy*, 1932; courtesy of Wiley.)

also a regular twofold axis. Study of Fig. 1-9 reveals, however, that the symmetry associated with this axis is more than that of a simple twofold axis but less than that of a twofold axis plus a symmetry center or a horizontal symmetry plane. Brief examination of the calcite rhombohedron (Fig. 1-1) likewise shows its vertical threefold axis to be a threefold inversion axis.

Alternative descriptions of a symmetry distribution are often possible, the description preferred being largely dictated by convenience. Thus the reader will have little difficulty in convincing himself that $\overline{1}$ is equivalent to a center of symmetry, and that $\overline{2}$ is equivalent to a reflection plane, the second description in each case being the one universally used. Another compound symmetry element used in the past, but now largely abandoned in favor of the inversion axis, is the *axis of rotary reflection* or *alternating axis*. This element combines a rotation with reflection in a plane normal to the axis. Consideration of Fig. 1-9 reveals that its vertical axis is likewise a fourfold alternating axis. The vertical axis of the calcite rhombohedron (Fig. 1-1) is, however, a sixfold alternating axis. A simple and useful exercise for the reader is the demonstration of the following relationships between these compound axes.

Degree of Alternating Axis	Equivalent Inversion Axis
1	$\bar{2}$
2	$\bar{1}$
3	$\bar{6}$
4	$\bar{4}$
6	$\bar{3}$

The symmetry properties of all crystals are merely various combinations of these simple or composite axes, planes, and centers of symmetry. If we explore the various distinctive combinations of these symmetry elements, we find that there are just 32 possible. These constitute the 32 *crystal symmetry classes*, to be discussed briefly in Section 1-4.1. Most of the 32 classes are represented by at least one known mineral or synthetic crystalline compound.

The symmetry arrived at from a consideration of the distribution of the faces must be regarded as only preliminary and approximate, and must be used with caution until it is confirmed by additional studies. Such properties as etch figures, electrical properties, and optical properties can be used for additional confirmation, and the best proof of all is obtained by an x-ray structure analysis of the crystal.

Consider a cubic crystal. Geometrically, a cube has three pairs of faces identical in size, and parallel to the three central like planes of symmetry. A cubic crystal, in the crystallographic sense, however, rarely grows as a perfect cube, but as a distorted cube (Fig. 1-10). Thus, to identify such a crystal as cubic, it is necessary to prove by some means (cleavage, optical properties, x-rays, etc.) that all six faces are like faces, that is, that the atomic arrangement is the same in all directions normal to them.

However, a crystal with interfacial angles of 90° is not necessarily cubic. Studies capable of showing the internal structure may reveal that not all the six faces are alike, but that four are alike and the other pair

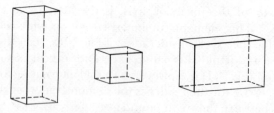

Fig. 1-10. Examples of distorted cubes.

unlike, or, possibly, that there are three pairs of faces, each pair differing from the other pairs. These last examples are not cubic in symmetry. The question has to be settled in every case by the internal structure as as evidenced by those properties which depend upon it.

1-2.4 Pseudosymmetry

In some instances the crystal angles approximate closely those of a higher symmetry system than that to which the crystal belongs. The crystal then appears to possess symmetry higher than its true symmetry, and is said to exhibit *pseudosymmetry*. Pseudosymmetry is common among mica minerals. Cases of pseudosymmetry occasionally occur in twinned crystals as a result of the twin simulating a single crystal of higher symmetry. These cases are further discussed under the subject of twin crystals (Section 1-2.11). Failure to recognize a twinned crystal as such has on several occasions led to erroneous crystal-structure determinations.

1-2.5 Crystallographic Axes

In discussing crystals the need very soon arises for a means of designating the various faces without ambiguity. The crystallographer has devised a way of describing the faces based on crystal axes, which are in reality merely coordinate axes of reference. Any three noncoplanar edges of the crystal, or directions in the crystal, may be chosen for these *crystallographic axes*. If it were entirely an arbitrary matter, an orthogonal set of axes would be chosen because of its simplicity. However, in practice, it is desirable to choose a set of axes which bears as close a relationship to the general symmetry of the crystal as possible, even though at times such a choice results in oblique axes. It is not impossible that different observers may choose different sets of axes, but an attempt is made to choose the set that gives the simplest mathematical relationships between the various faces. The origin of the set of axes is to be thought of as located at the center of the ideal crystal. The unit lengths along the axes are fixed by the symmetry or by the positions of the faces, assuming some face to be the unit face.

There is a certain amount of convention among crystallographers concerning the orientation of the crystal axes. A right-handed set of axes with the positive X axis extending toward the observer, the positive Y to the right, and the positive Z vertical is used. The unique axis of the crystal is normally taken coincident with Z. An exception to this rule is frequently made in the monoclinic system (Section 1-2.7), in which by long established usage and preference the twofold axis is made the side-to-side or Y axis (Table 1-1, second setting). There is also a conventional designation for the angles between various pairs of axes. If the symbol $\wedge$ is used to indicate the angle between two axes, faces, planes, and so on, the axial angles are: $Y \wedge Z = \alpha$, $X \wedge Z = \beta$, and $X \wedge Y = \gamma$.

1-2.6 Axial Ratios

After the choice of axes has been made, a plane intersecting all three axes is chosen as the *unit plane* or *parametral plane*. The choice is arbitrary, and only trial can show the best face to select for the greatest simplicity in later calculations. From angular measurements made on a gonio-meter, the three lengths a, b, and c of the plane's intercepts on the crystal axes, X, Y, and Z, can be calculated. These lengths are called the *parameters* of the crystal. They are merely relative (since they vary with the size of the crystal), and are constant only as regards their ratios to one another and not in absolute value. These ratios of the parameters are known as the *parametral* or *axial ratios*, and are customarily expressed with the value of b equal to unity. Thus for iodine (orthorhombic) the axial ratios are

$$a:b:c = 0.661:1:1.348.$$

If the unit lengths along two of the axes are the same, they are designated a, a, and c. Rutile, TiO_2 (tetragonal), is an example, and its axial ratio is expressed as

$$a:c = 1:0.6442 \quad \text{or, simply,} \quad c = 0.6442.$$

The plane containing any two of the axes is called an *axial plane*, and the three axial planes divide the space about the axes into eight parts or *octants*.

1-2.7 The Six Crystal Symmetry Systems

Although there appears to be a certain arbitrariness in the choice of the crystal axes, nevertheless, if wisely chosen, they bear a very fundamental relationship to the symmetry and internal structure of the crystal. In practice the proper choice of axes inevitably identifies the crystal as belonging to one of six large symmetry groups known as the six crystal *systems*. The names of the six crystal systems and the data necessary for their complete description are tabulated in Table 1-1.

Early crystallographers listed seven crystal systems, the trigonal (rhombohedral) division of the hexagonal system being considered a separate system. However, since rhombohedral crystals can always be described in terms of hexagonal axes [9], it seems more logical to include them as a division of the hexagonal system. The relation between the X and Y hexagonal axes (solid lines) and the X_0 and Y_0 orthohexagonal

Table 1-1. Data Defining the Six Crystal Symmetry Systems

System	Axial Ratios	Angles between Crystal Axes
Triclinic (anorthic)	$a:b:c$	α, β, γ (generally all $\neq 90°$)
Monoclinic (oblique, monosymmetric)		
(A) 1st setting	$a:b:c$	γ ($\alpha = \beta = 90°$)
(B) 2nd setting	$a:b:c$	β ($\alpha = \gamma = 90°$)
Orthorhombic (rhombic, prismatic)	$a:b:c$	All angles 90°
Tetragonal	$a:c$ ($b = a$)	All angles 90°
Hexagonal		
(A) Hexagonal division		
(1) Hexagonal axes	$a:c$ ($b = a$)	$\gamma = 120°$ ($\alpha = \beta = 90°$)
(2) Orthohexagonal axes	$a:b:c$ ($b = a\sqrt{3}$)	All angles 90°
(B) Trigonal division (rhombohedral)		
(1) Rhombohedral axes	$a = b = c$	$\alpha = \beta = \gamma < 120° \neq 90°$
(2) Hexagonal axes	$a:c$ ($b = a$)	$\gamma = 120°$ ($\alpha = \beta = 90°$)
Cubic (regular, isometric)	$a = b = c$	$\alpha = \beta = \gamma = 90°$

axes (dotted lines) is clearly shown in Fig. 1-11. Use of orthohexagonal axes may simplify the interpretation of some kinds of x-ray diffraction patterns and make the calculations easier because of the right angles present.

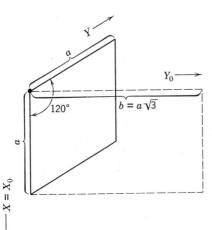

Fig. 1-11. Relation between the orthohexagonal axes X_0 and Y_0 and the regular hexagonal axes X and Y.

1-2.8 Miller Indices

Once the choice of a unit plane has been made, it is possible to express the intercepts of all other planes as ma, nb, and pc, where m, n, and p are small integers or infinity. This is always possible, since a face may be moved parallel to itself without losing its identity or altering the ratios of its intercepts on the axes. These three numbers, m, n, and p, might be used as indices to denote a given face. However, because sometimes an index would have the value infinity, which is troublesome in certain mathematical calculations, it is more convenient to use as indices numbers proportional to the reciprocals of m, n, and p. If

$$h \propto \frac{1}{m}, k \propto \frac{1}{n}, \text{ and } l \propto \frac{1}{p}, \tag{1-1}$$

h, k, and l, expressed as integers without a common divisor,* are known as the *Miller indices* of the face or plane, and are customarily written (hkl). The parametral plane is thus (111). The parameters ma, nb, and pc are known as *Weiss indices*.

In Fig. 1-12, several faces have been indicated on a set of ortho-rhombic axes. Unit distance along each axis is different here, but all axes are at 90° to each other. The derivation of the Miller indices of these faces is tabulated in Table 1-2. Skill in writing Miller indices of planes is readily achieved by sketching various planes on sets of coordinate axes and calculating their indices as illustrated in Table 1-2.

*Miller indices containing a common factor are used by the crystal analyst to indicate higher-order x-ray spectra from a given crystal plane. Thus (222) represents the second-order spectrum from the (111) plane, (333) the third-order, and so on.

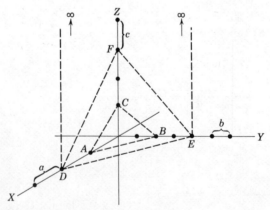

Fig. 1-12.   Several faces intersecting a set of orthorhombic axes. Their Miller indices are readily calculated from the intercepts on the three axes.

Table 1-2. Derivation of Miller Indices from the
Face Intercepts (Parameters)[a]

Face	Intercepts	Reciprocals of Intercept Multiples	Cleared of Fractions	Miller Indices (hkl)
ABC	$1a:2b:1c$	$\dfrac{1}{1}\ \dfrac{1}{2}\ \dfrac{1}{1}$	2 1 2	(212)
DEF	$2a:4b:3c$	$\dfrac{1}{2}\ \dfrac{1}{4}\ \dfrac{1}{3}$	6 3 4	(634)
DE ∞	$2a:4b:\infty c$	$\dfrac{1}{2}\ \dfrac{1}{4}\ \dfrac{1}{\infty}$	2 1 0	(210)

[a](Data from Fig. 1-12).

It must be remembered that crystal axes are axes of reference, hence they extend in negative as well as positive directions from their origin. Faces that intersect an axis on its negative side have a negative intercept on that axis. The Miller index for that particular axis is thus negative, and its sign is indicated by a minus sign placed over the corresponding index: $(1\bar{1}2)$, $(\bar{3}\bar{4}2)$, and so on. Negative indices serve to locate a plane in its particular octant formed by the three axes; thus the eight faces of the general form (hkl) are:

$$(hkl), \ (\bar{h}kl), \ (h\bar{k}l), \ (hk\bar{l}),$$
$$(\bar{h}\bar{k}\bar{l}), \ (h\bar{k}\bar{l}), \ (\bar{h}k\bar{l}), \ (\bar{h}\bar{k}l).$$

Note that the members of each vertical pair above are parallel faces, and that they have the same indices but with complementary signs. Hence to change all the signs of a Miller symbol is to change the plane to its parallel, and therefore equivalent, plane on the opposite side of the crystal.

In describing the planes of hexagonal crystals, four axes are sometimes used. The corresponding Miller symbol then becomes (hkil), known as the *Miller-Bravais indices* of the plane. The additional index, i, is derived from a third coordinate axis lying in the plane of X and Y, and making an angle of 120° with the positive Y. With axes of this sort, $h+k=\bar{i}$, where i is the index of the new axis; that is, $h+k+i=0$.

In Fig. 1-13 the plane containing the three equal horizontal hexagonal axes at 120° angles is depicted, the c axis being perpendicular to the drawing at the center of coordinates. Consider the vertical plane (parallel to the c axis) whose trace is the dashed line tt. Its intercepts are: $2a$, $2a$, $\bar{1}a$, ∞c. Taking reciprocals and clearing of fractions gives

$$\frac{1}{2}\ \frac{1}{2}\ \frac{\bar{1}}{1}\ \frac{1}{\infty} \qquad \text{or} \qquad 11\bar{2}0.$$

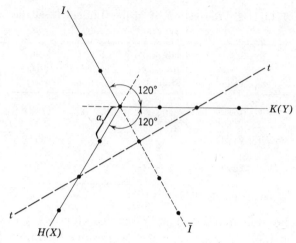

Fig. 1-13. Origin of the third index i in a Miller-Bravais symbol. The trace tt is that of plane (11$\bar{2}$0).

Hence $(hkil) = (11\bar{2}0)$, $h+k = \bar{i} = 2$, and $h+k+i = 1+1-2 = 0$.

In describing the faces of a crystal this additional axis is often of value, but for other purposes it may profitably be omitted. Thus, in the literature, hexagonal indices are frequently found in the form $(hk{\cdot}l)$, when the third index is said to have been suppressed. Obviously, the third index is readily supplied, when desired, because of its simple relationship to h and k.

1-2.9 The Law of Rational Indices

It was discovered very early that for every crystal species a set of axes can be found such that the indices for all crystal faces are small integers or zero. This fact, known as the *law of rational indices*, is of such fundamental importance in crystallography that it is also called the *second law of crystallography*. The Abbé René Just Haüy (1743–1822), principal founder of scientific crystallography, first stated the law.

The law is sometimes known as "the law of simple rational axial sections of the crystal faces." This means that, in the external form of crystals, faces arbitrarily placed do not occur; only those that bear a rational relationship to one another are possible. Thus, if the axial sections on the $A(X)$ and $B(Y)$ axes of a crystal are considered, such ratios as $a:b$, $a:\frac{1}{2}b$, $a:\frac{3}{2}b$, $a:2b$, $a:\infty b$, or any simple rational ratios are possible. But it is not possible for an irrational ratio such as $a:\sqrt{2}b$ to occur. If the plane intersects additional axes, these axial sections must also be rational. That this is true for the intersection of tt on the $\bar{I}$ axis of Fig. 1-13 is easily observed.

We have pointed out that the crystallographic rhythms are two-, three-, four-, and sixfold, but never any other number. A brief consideration shows that the law of rational indices demands this. For example, a fivefold rhythm cannot occur, since a regular pentagon leads to an axial section of $0.809017\cdots$ on one axis. The plane whose trace is tt (Fig. 1-14) intersects axes I and II at unit distance, but its section on $\overline{IV}$ is OM, which by simple trigonometry is seen to be equal to $\cos 36°$. It can be demonstrated that $\cos 36° = (\sqrt{5}+1)/4 = 0.809017\cdots$, and that OM is therefore irrational.

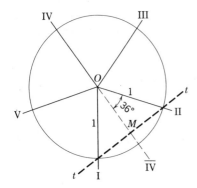

Fig. 1-14. Irrational axial section, $OM = 0.809017\ldots$, which arises with a fivefold axis.

1-2.10 Crystal Forms

On many crystals several of the faces very obviously appear equivalent. Sometimes two, four, eight, or more faces bear a similar relationship to the planes or axes of symmetry. All such faces are said to belong to the same *form*, the term being used in a more specialized sense than heretofore. A form may be defined as the total group of equivalent planes or faces whose presence is required by the symmetry of the crystal when one of them is present. The number of faces making up a complete form thus depends upon the symmetry. The six faces of a cube constitute a form, likewise the eight faces of an octahedron make up a complete form, as do also each of the pairs of faces on certain triclinic crystals. The symbol for a form is enclosed in braces, as {110}, and denotes collectively all the faces of the form, whereas the symbol (110) refers to a particular face belonging to the same form.

Crystal forms are divided into three general classes according to the number of reference axes their planes intersect. (1) *Pinacoids* are composed of planes parallel to two crystal axes and intersecting the third.

The pair of pinacoids intersecting the vertical axis has the general indices {001}, and they are called *basal* pinacoids. The forms {100}* and {010}, intersecting the X and Y axes, respectively, are also pinacoids. (2) *Prisms* and *domes* are forms whose planes intersect two crystal axes but are parallel to the third. The prisms are parallel to the vertical axis, and have the general indices {hk0}, whereas the domes are parallel to one of the remaining two axes and have the indices {h0l} or {0kl}. (3) *Pyramids* are forms whose planes intersect all three crystal axes. Their general indices are of the type {hkl}. In Fig. 1-15 the crystal of chrysolite [(Mg, Fe)$_2$SiO$_4$] exhibits all three form types. The faces *a*, *b*, and *c* are pinacoids, *m* and *s* are prisms, *d*, *h*, and *k* are domes, and *e* and *f* are pyramids.

The largest number of faces constituting a form is 48 for the form {hkl} of the normal class of the cubic system, whereas in several cases a single face constitutes a complete form. If the faces of a form yield an enclosed solid, as in the cube and octahedron, the form is called a *closed form*. Forms such as the pinacoids of the monoclinic system are known as *open forms*.

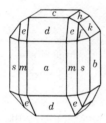

Fig. 1-15. A crystal of chrysolite [(Mg, Fe)$_2$SiO$_4$] exhibiting a variety of forms and several well-defined zones. (From E. Dana, *A Textbook of Mineralogy*; courtesy of Wiley.)

Often crystal forms occur which, in the light of the definition of the completed form given above, appear to be partial. Sometimes only half the faces of a high-symmetry form are present, and the crystal is said to be *hemihedral*. If only a quarter of the form appears, a crystal is termed *tetartohedral*, and a completely symmetrical crystal is termed *holohedral*. Before the advent of geometrical crystallography, observers assumed that these partial forms resulted from an unexplained suppression of part of the faces of the form. Now that mathematical crystallography has developed the entire 32 crystal symmetry classes (Section 1-4.1) on the basis of their specific symmetry elements, these individuals, formerly

---

*In the tetragonal system, however, the form {100} is referred to as the second-order prism, and {101} is called the second-order pyramid. Similarly, in the hexagonal system the form {100} is known as the first-order prism, and {101} is designated a pyramid of the first order.

termed hemihedral or tetartohedral, are seen to be just as perfect individuals as those termed holohedral, inasmuch as they exhibit the complete symmetry of their particular class. The growth of these so-called partial forms is, indeed, a direct result of the demands of internal symmetry. The terms still persist in crystallographic usage, so that an example is worthwhile.

In the normal class of the cubic system the *octahedron* with its eight faces is a holohedral form. The tetrahedral class of the same system, referred to the same crystallographic axes, calls forth but four similar faces, giving rise to a *tetrahedron*, the hemihedral form of this system. The relation between the faces of the tetrahedron and the original octahedron is shown in Fig. 1-16. Note that there is a positive and a negative tetrahedron, which together embrace all the faces of the octahedron.

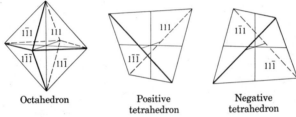

| Octahedron | Positive tetrahedron | Negative tetrahedron |

Fig. 1-16. Holohedral and hemihedral forms in the cubic system. (From E. Dana, *A Textbook of Mineralogy*; courtesy of Wiley.)

One other case, in which only half the expected faces of a form are present, is important. In this case one half the planes of the holohedral form are grouped about one extremity of the principal axis of symmetry. This type of development is termed *hemimorphism*. Hemimorphic crystals thus have one end of the principal axis with different face development than the other. Figure 1-17 illustrates hemimorphic crystals of

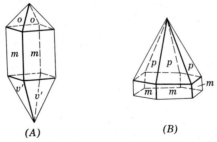

(A) (B)

Fig. 1-17. Hemimorphic crystals. (*A*) Iodosuccinimide [$(CH_2CO)_2NI$]; (*B*) zincite (ZnO). (From F. Rinne, *Crystals and the Fine-Structure of Matter*, and E. Dana, *A Textbook of Mineralogy*; courtesy of Dutton and Wiley.)

iodosuccinimide [$(CH_2CO)_2NI$] and zincite (ZnO). Hemimorphism, too, is the result of internal symmetry requirements, and not of accidental growth. Such crystals possess no center of symmetry and, as a result, show certain peculiar physical properties. They are *polar* crystals and exhibit *piezo-* and *pyroelectric effects* [10].

A few crystals are known that show both a right- and a left-handed form, related to each other as an object and its mirror image. Such forms are said to be *enantiomorphous*. In Fig. 1-18 the enantiomorphism of quartz is shown. Another well-known example is cane sugar. Such crystals possess neither a plane nor a center of symmetry, and cannot be converted into each other by rotation. They bear the same relation to each other as do the right and left hands.

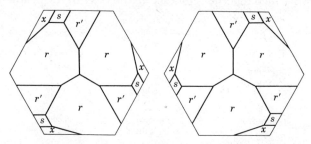

Fig. 1-18.   Left- and right-handed quartz crystals viewed along their threefold axis. (After F. Rinne, *Crystals and the Fine-Structure of Matter*, 1924; courtesy of Dutton.)

1-2.11 Composite Crystals and Twinning

Crystals commonly grow in groups with a haphazard arrangement among the individuals of the group. Less frequently a parallel relationship appears between similar edges, faces, and so on, of the various individuals, and the crystals are said to exhibit *parallel growth*. An excellent example of parallel growths is shown by potassium alum [$KAl(SO_4)_2 \cdot 12H_2O$] (Fig. 1-19).

Fig. 1-19.   Parallel growths in potassium alum [$KAl(SO_4)_2 \cdot 12H_2O$].

Sometimes crystals develop that are built up of two or more individuals having a definite but not parallel arrangement. These are called *twin crystals*. In one type of twinned crystal the components may be thought of as reflections of one another in a plane, the *twinning plane*. In such twins the components are entirely separate from one another except at the plane of meeting (Fig. 1-20), known as the *composition plane*.

Fig. 1-20. Contact twin of iodosuccinimide [$(CH_2CO)_2NI$]. (After F. Rinne, *Crystals and the Fine-Structure of Matter*, 1924; courtesy of Dutton.)

Accordingly, they are known as *contact twins*. Another type of twin has its components related by a rotation of 180° with respect to each other about some line of the crystal called the *twinning axis*. This results in the separate parts closely intergrowing, forming a *penetration twin* (Fig. 1-21). By repeated twinning *triplets*, *fivelings*, *eightlings*, and even much higher-fold twins may be formed.

Twinning sometimes results in elbow-shaped, T-shaped, crosslike, starlike, and so on, individuals. As a rule twinned crystals can be re-

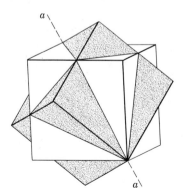

Fig. 1-21. Penetration twin of fluorite (CaF_2). The twinning axis is *aa*.

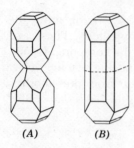

(A)   (B)

Fig. 1-22.   Twinned crystals of calamine. (After E. Dana, *A Textbook of Mineralogy*; courtesy of Wiley.)

cognized by their unusual shape and appearance, and by the presence of reentrant angles (Figs. 1-20, 1-21, 1-22A). Occasionally, however, twinned forms imitate almost exactly a single crystal of greater symmetry, as does the calamine ($H_2Zn_2SiO_5$) twin in Fig. 1-22B. As has already been mentioned, failure to recognize that the specimen under examination was a twin has led to incorrect structure determinations.

1-2.12 Equation for the Plane (hkl)

For a complete understanding of the geometry of x-ray diffraction, it is valuable to be able to write the equation for a plane whose Miller indices are known. Because the cubic case is the simplest, the equation for the plane (hkl) in this system will be derived.

Consider the plane RST in Fig. 1-23, whose intercepts on the coordinate axes X, Y, and Z are OR, OS, and OT. The intercept form for the equation of a plane (from analytical geometry) is $x/a + y/b + z/c = 1$, and in this case the equation becomes

$$\frac{x}{OR} + \frac{y}{OS} + \frac{z}{OT} = 1. \qquad (1\text{-}2)$$

Since these are cubic axes, unit length is the same along each, a, and $OR = ra$, $OS = sa$, and $OT = ta$, where r, s, and t are the intercept

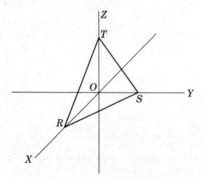

Fig. 1-23.

multiples of a along the axes. Substituting and clearing of fractions gives

$$\frac{x}{ra}+\frac{y}{sa}+\frac{z}{ta}=1 \qquad (1\text{-}3)$$

and

$$stx+rty+rsz=rsta. \qquad (1\text{-}4)$$

The Miller indices of a plane have been defined (Section 1-2.8) as numbers proportional to the reciprocals of the plane's intercept multiples, and are expressed as a triplet of integers containing no common divisor. It is possible then to write the Miller indices of this plane as follows:

$$h\infty\frac{1}{r}, \quad k\infty\frac{1}{s}, \quad \text{and} \quad l\infty\frac{1}{t}. \qquad (1\text{-}5)$$

These may be cleared of fractions by multiplying their numerators by rst; then

$$h\infty st, \quad k\infty rt, \quad \text{and} \quad l\infty rs. \qquad (1\text{-}6)$$

If the right-hand quantities have a greatest common divisor p, division thereby gives the Miller indices (hkl) of the plane,

$$h=\frac{st}{p}, \quad k=\frac{rt}{p}, \quad \text{and} \quad l=\frac{rs}{p}. \qquad (1\text{-}7)$$

Equations 1-4 and 1-7 now lead to

$$hpx+kpy+lpz=rsta \qquad (1\text{-}8)$$

and

$$hx+ky+lz=rsta/p. \qquad (1\text{-}9)$$

The general equation for the plane (hkl), then, is

$$hx+ky+lz-D=0, \qquad (1\text{-}10)$$

where $rsta/p=D$.

There is a general theorem (from analytical geometry) that a plane parallel to $Ax+By+Cz-D=0$, and passing through the point $P(x_1, y_1, z_1)$, is given by

$$Ax+By+Cz-(Ax_1+By_1+Cz_1)=0. \qquad (1\text{-}11)$$

Thus, if the point $P(x_1y_1z_1)=P(000)$, the origin, the equation for the plane with indices (hkl) and passing through the origin is

$$hx+ky+lz=0. \qquad (1\text{-}12)$$

These expressions are readily generalized for axial systems of less symmetry. For instance, with tetragonal axes, the plane (hkl) is expressed by the equation

$$hx + ky + \frac{lza}{c} - D = 0. \qquad (1\text{-}13)$$

where a and c are unit lengths along the axes.

1-2.13 Zones and Zone Relationships

Crystal planes and faces are frequently observed to be arranged in belts around the crystal, with all the faces in a given belt parallel to an imaginary line, or direction, passing through the center of the crystal. These belts of crystal faces are known as *zones*. The imaginary line to which they and their intersections are all parallel is the *zone axis*. In Fig. 1-15 the crystal of chrysolite exhibits several zones; faces c, d, a, d, and so on, belong to the same zone; faces s, m, a, m, s, b, and so on, belong to another zone, and c, h, k, b, and so on, constitute a third zone.

The important thing to know about a zone is the direction of its zone axis, usually expressed as the bracketed symbol [uvw]. This may be obtained from the indices (hkl) and ($h'k'l'$) of any two planes or faces in the zone by means of the relations

$$u = kl' - lk', \qquad v = lh' - hl', \qquad \text{and} \qquad w = hk' - kh'. \qquad (1\text{-}14)$$

Thus all planes ($hk0$) belong to the zone [001], all ($h0l$) planes to the zone [010], and all ($0kl$) planes to the zone [100]. Indeed, in the cubic system and for special zones in most other systems, the zone indices u, v, and w are simply the Miller indices of the plane normal to the zone axis.

In practice the calculation of the zone indices from two planes in the zone is made in a very simple way. The indices of the two planes are written down according to the following scheme:

$$\begin{array}{c|ccccc|c} h & k & l & h & k & l \\ & & & & & \\ h' & k' & l' & h' & k' & l' \end{array} \qquad (1\text{-}15)$$

Then the part between the lines is treated as three second-order determinants, giving in order the values of u, v, and w. It is important, of course, to pay attention to the signs of the indices in these calculations. The following examples illustrate these simple calculations.

1. *Example.* What zone contains the planes (210) and (111)?

$$
\begin{array}{c|ccc|c}
2 & 1 & 0 & 2 & 1 & 0 \\
 & \times & \times & \times & \\
1 & 1 & 1 & 1 & 1 & 1
\end{array}
$$

Thus $u = 1 - 0 = 1$, $v = 0 - 2 = \bar{2}$, $w = 2 - 1 = 1$, and $[uvw] = [1\bar{2}1]$.

2. *Example.* What zone contains the planes $(33\bar{1})$ and (001)?

$$
\begin{array}{c|ccc|c}
\bar{3} & 3 & \bar{1} & \bar{3} & 3 & \bar{1} \\
 & \times & \times & \times & \\
0 & 0 & 1 & 0 & 0 & 1
\end{array}
$$

Thus $u = 3 - 0 = 3$, $v = 0 + 3 = 3$, $w = 0 - 0 = 0$, and $[uvw] = [330]$, which becomes [110] by removal of the common factor.

The reverse calculations can be made in a similar manner. For instance, the indices of any face or plane common to two zones can be obtained from the zone indices of the two zones.

Other important relationships exist between the indices of the faces that lie in a zone. If the indices of two faces in the same zone are added together, the sum will be the indices of a face in the same zone lying between them. In Fig. 1-15, faces $d = (101)$, $e = (111)$, and $f = (121)$ belong to the same zone, and the indices of e can be obtained by adding those of d and f:

$$
\begin{array}{ccc}
1 & 0 & 1 \\
1 & 2 & 1 \\
\hline
2 & 2 & 2, \quad \text{or} \quad (111).
\end{array}
$$

Since there may be several faces or planes intermediate between the two faces, this rule cannot be relied upon to give the indices of a particular face with certainty. Hence the indices should always be checked by cross-multiplication of the two zonal symbols.

Furthermore, every face in a given zone must satisfy the zonal equation. If (hkl) is a face in the zone $[uvw]$,

$$hu + kv + lw = 0. \tag{1-16}$$

3. *Example.* Does face $(\bar{2}11)$ belong to the zone $[5\bar{2}8]$?

$$hu + kv + lw = (-2 \times 5) + (-1 \times -2) + (1 \times 8) = -10 + 2 + 8 = 0$$

Thus face $(\bar{2}11)$ belongs to zone $[5\bar{2}8]$. In this way it can quickly be ascertained whether or not a given face lies in a particular zone.

1-3 SPACE LATTICES

1-3.1 Historical Introduction

The outward form and symmetry of crystals very early led crystallographers to believe in a regular internal arrangement. In the seventeenth century, Huygens[11] explained the outward form and cleavage of calcite ($CaCO_3$), as well as the variation of its hardness and double refraction with direction, by visualizing a regular packing of spheroidal bodies as depicted in Fig. 1-24. Many years later, in the eighteenth

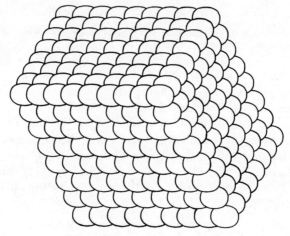

Fig. 1-24.   Fine structure of the calcite cleavage rhombohedron according to Huygens. [From H. P. Klug, *Am. Sci.*, **36**, 377 (1948).]

century, Haüy[12], professor of the humanities at the University of Paris and often termed the "father of crystallography," pointed out that, if we were to cleave a piece of calcite repeatedly, the fragments, which always take the shape of the original crystal, would eventually be reduced to minute parallelepipeds of the same shape as the original crystal and containing one molecule. Obviously, this ultimate unit of structure or "elementary building stone" could not be subdivided without destroying its composition. Haüy, moreover, demonstrated how a structure built up of such elementary parallelepipeds would develop exterior faces that are plane and that are represented by rather simple indices. Such ideas of crystalline fine structure are illustrated schematically by the development of the end faces in Fig. 1-25. Since these building blocks are molecular in size the "stair-step" appearance of the plane faces is not detectable by any visual means.

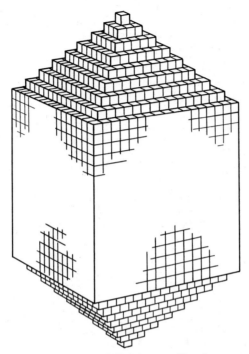

Fig. 1-25. Schematic representation of Haüy's crystalline fine structure. [From H. P. Klug, *Am. Sci.*, **36**, (1948).]

Investigators gradually realized that chemical molecules in crystals could not be represented as solid parallelepipeds closely packed without interspaces, since such phenomena as elastic compression and thermal expansion were difficult to explain on this basis. Particularly, with the advent of atomic and molecular theories, such ideas became untenable, and the crystal came to be looked upon as an orderly assemblage of points, the centers of gravity of individual chemical molecules, or other analogous representative points of molecules.

When a substance crystallizes, the symmetry of the crystal, and therefore its form, is determined by the equilibrium positions which the atoms or molecules take up with respect to one another during the process of crystallization. It is reasonable to expect the magnitude of the interatomic and intermolecular forces to vary with direction, if the molecule or atom group is asymmetric. Furthermore, these forces, as well as the size of the crystal units themselves, are expected to vary widely from compound to compound. Hence the resulting equilibrium groupings of crystal units show great diversity with respect to the distance between units and the angles between the planes in which they lie. These arrangements may be represented by orderly assemblages of points in three dimensions, making a three-dimensional network called a *space lattice* (Fig. 1-26). The points represent the positions of the atoms or molecules, or the centers of similar groups of atoms, and the crystal is then a three-dimensional pattern with an atomic or molecular motif, quite analogous to wallpaper in two dimensions.

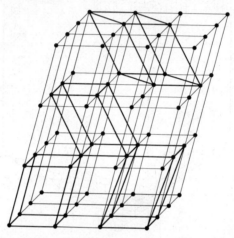

Fig. 1-26.   A space lattice of points with several alternative unit cells outlined.

1-3.2 Definition

Geometrically, a space lattice may be defined as a regular and unlimited distribution of identical points in space. The arrangement is such that a straight line drawn through any two adjacent points of the network and continued intersects at equal intervals the same succession of points. An important characteristic of a space lattice is that the environment about any particular point is the same as that about any other point. The most special case is a lattice in which the points are distributed at equal intervals along three directions at right angles to each other. This net has the symmetry of the normal class of the cubic system. The most general case is that in which the points are distributed at a different repeat interval along each of the three directions, which, furthermore, are oblique to each other. This arrangement has the symmetry of the normal class of the triclinic system. It should always be remembered that the space lattice of a crystal is merely the translational repetition in three dimensions of the positions occupied by its atoms or molecules. Hence the term *lattice* should never be used to mean the actual crystal structure of packed atoms, an incorrect usage frequently encountered in the literature of crystal structure.

1-3.3 The Unit Cell

By joining the points of a space lattice, a series of parallel-sided *unit cells* is produced, each of which in an actual crystal contains a complete unit of the crystal pattern. The whole structure is obtained by packing these cells side by side in space. There are alternative ways of outlining the unit cell (Fig. 1-26), but the volume of each simple cell is the same. In practice the choice of a unit cell for a given space lattice is decided by considerations of convenience in visualizing the symmetry and making mathematical calculations, even though at times the chosen cell is not a simple one. Obviously the space lattice is completely defined by stating the distances between neighboring lattice points along the three directions and the angles between these directions. These distances are referred to as the *primitive* or *unit translations* of the lattice, and are frequently expressed as $2\tau_x$, $2\tau_y$, and $2\tau_z$. They are the same as the edge lengths of the unit cell, and in crystals the proportionality, $a:b:c = 2\tau_x:2\tau_y:2\tau_z$, holds between them.

1-3.4 The 14 Bravais Lattices

Various investigators have studied the possible parallelepipedal networks of points. Outstanding among them was Bravais, who in 1848 demonstrated that there are 14 distinct space lattices. All these space lattices exhibit the property of homogeneity. If the most highly symmetrical unit cell possible is chosen for each network, the symmetry of each lattice is that of some one of the six crystal systems. Their symmetries are as follows: 1 is triclinic; 2 are monoclinic; 4 are orthorhombic; 2 are tetragonal; 1 is hexagonal; 1 is rhombohedral; and 3 are cubic. Unit cells of the 14 space lattices are illustrated in Fig. 1-27.

Inspection of Fig. 1-27 reveals that some of the lattice cells outlined are not simple cells. Thus, 1, 2, 4, 8, 9, 10, and 12 are simple cells, but all others possess additional lattice points. Considering the four orthorhombic lattices as an example, 4 is simple, 5 has an additional point at the centers of the top and bottom faces, 6 has an extra point at its geometrical center, and 7 has extra points at the centers of all faces. The origin of these more complex cells is best understood from a consideration of the cubic lattices. By starting with a lattice in which the distances between lattice points along all three axes are equal, and the angles between the axes are likewise equal, we find that the structure possesses cubic symmetry not only when the angles are 90°, but also when they are 60° or 109°28′. Figure 1-28 illustrates the three cubic lattices. In each diagram the simple cell is outlined with solid lines; then enough additional lattice points are included to permit the cube-shaped cells to be outlined. These are known as the simple cubic lattice, the face-centered cubic

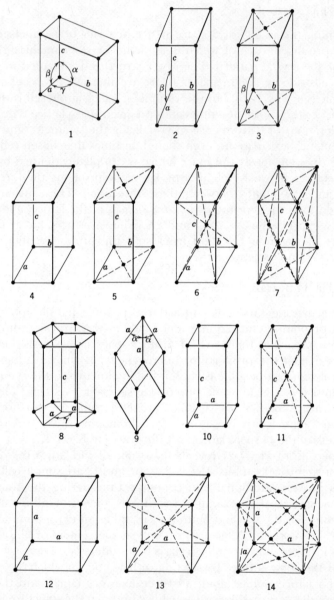

Fig. 1-27.   Unit cells of the 14 space lattices. *1*, Triclinic; *2,3*, monoclinic; *4,5,6,7*, ortho-rhombic; *8*, hexagonal; *9*, rhombohedral; *10,11*, tetragonal; *12,13,14*, cubic.

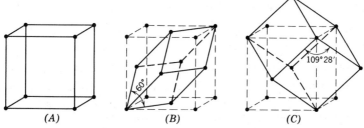

Fig. 1-28. Unit cells of the three cubic space lattices with the simple cells of the face-centered (*B*) and body-centered (*C*) lattices outlined.

lattice, and the body-centered cubic lattice. Trial shows no other lattices are possible with cubic symmetry. Naturally, the cubic aspects of these lattices are far more convenient to handle in mathematical calculations than their simple lattices would be, and this fact, together with the greater ease of picturing their symmetry, makes their cubic cells of prime importance in x-ray diffraction. An exploration of the remaining Bravais lattices discloses several other instances that are better depicted as the more complex cells shown in 3,5,6,7, and 11 of Fig. 1-27.

The lattice points have been defined as the positions occupied by the repeating units of the crystal pattern, atoms, ions, or molecules. A simple cell always contains the equivalent of one complete unit of pattern, and is known as a *primitive* cell. In the two-dimensional lattice of diamonds (Fig. 1-29), the cell *ABCD* is primitive and contains a total of one complete diamond. Usually a primitive cell is chosen, but when it is evident that we can gain geometrical advantages, such as an orthogonal cell, by including additional lattice points, it is convenient to use a nonprimitive cell. Thus, in Fig. 1-29, the cell *CEFG*, containing two complete units of pattern, is designated doubly primitive, and is highly desirable because of its rectangular characteristics. Returning to three dimensions and the 14 space lattices (Fig. 1-27), we see that 3, 5, 6, 11, and 13 are doubly primitive and that 7 and 14 are quadruply primitive.

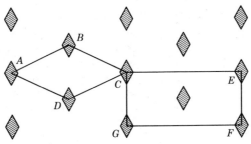

Fig. 1-29. Simple primitive and doubly primitive cells of a two-dimensional lattice of diamonds.

1-3.5 Some Crystallographic Implications of Space Lattices

A consideration of regular molecular networks based on space lattices leads to some important conclusions.

1. Prominently developed crystalline faces are those most thickly populated with lattice points. This consequence is readily represented in two dimensions by a normal orthorhombic network in the *ab* section (Fig. 1-30). Considering faces parallel to the *c* axis, we note that those whose traces are *pp* and *qq* are pinacoids, and that they have the greatest density of lattice points. The prism traces *ee*, *ff*, and *gg* are less densely packed. As the form becomes more and more complex, the distance (in the plane) between neighboring lattice points becomes greater and greater, and the probability that it will appear as a developed face becomes less and less. This observation is verified by the accumulated data on thousands of crystals. The common forms are pinacoids and simple prisms and pyramids. Complex forms, when present, usually develop as small faces on the predominating simple forms.

2. The numerical relations existing between the developed faces of a crystal must be both rational and simple. This statement, of course, is the law of rational indices (Section 1-2.9). The common faces *ee*, *ff*, and *gg* (Fig. 1-30) have the ratios 1:1, 1:2, and 3:2, respectively, in the *a* and *b* directions.

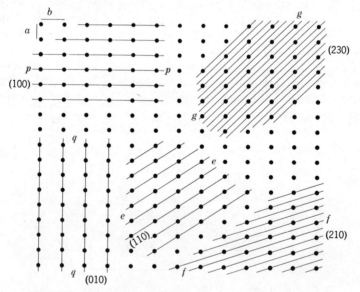

Fig. 1-30.   Plane traces on an *ab* section of an orthorhombic net.

3. When a crystal exhibits the natural easy fracture called *cleavage*, the break usually takes place parallel to important directions in the crystal; for instance, both basal and prismatic cleavage are common. Cleavage is evidence of a minimum interatomic or intermolecular force in the direction perpendicular to the cleavage surface which, in general, is a prominent face relatively crowded with lattice points. This weaker bonding force is partly explained by the fact that planes in which the points are closest together are farthest separated from the next molecular plane. The interplanar spacing and the bond strength are inversely related, the strength being less the greater the spacing. From Fig. 1-30 it is seen that the distance separating adjacent *pp* or *qq* planes is much greater than that separating *ff* or *gg* planes. Other structural features, such as the number and kind of bonds, are also determining factors in cleavage.

1-3.6 Distance between Neighboring Lattice Planes in the Series (*hkl*)

In Fig. 1-30 the traces of successive planes of several pinacoids and prisms have been sketched on an *ab* orthorhombic net of points. Likewise, in a space lattice each Miller index triplet is represented by a series of parallel equispaced planes containing all the lattice points. In actual crystals these planes are the loci of the atomic or molecular units of the crystal pattern, and cleavage, when observed, occurs between successive planes of a series. The planes of a given series are parallel to the corresponding crystal face.

It is frequently convenient to be able to calculate the perpendicular distance between successive planes of a series. Accordingly, the expression for the interplanar spacing d of the series (*hkl*) in a cubic lattice will be developed, after which the expression for d_{hkl} for systems of lower symmetry will be indicated.

From Section 1-2.12 the equation for the plane (*hkl*) in the cubic lattice is

$$hx + ky + lz - D = 0. \tag{1-17}$$

From simple analytical geometry, its distance from the origin is

$$\frac{D}{\sqrt{h^2 + k^2 + l^2}}. \tag{1-18}$$

Now consider a parallel plane passing through the lattice point $P_1(x_1, y_1, z_1)$. Its equation is

$$hx + ky + lz - (hx_1 + ky_1 + lz_1) = 0, \tag{1-19}$$

and its distance from the origin is

$$\frac{hx_1 + ky_1 + lz_1}{\sqrt{h^2 + k^2 + l^2}}. \tag{1-20}$$

The distance between these two planes is simply the difference in their distances from the origin, or

$$\frac{(hx_1 + ky_1 + lz_1) - D}{\sqrt{h^2 + k^2 + l^2}}. \tag{1-21}$$

Likewise, the distance from the first-mentioned plane to the parallel plane passing through the lattice point $P_2(x_2, y_2, z_2)$ is

$$\frac{(hx_2 + ky_2 + lz_2) - D}{\sqrt{h^2 + k^2 + l^2}}. \tag{1-22}$$

The distance between the parallel planes passing through points P_1 and P_2 is the difference in their distances from the first plane, or

$$\frac{h(x_1 - x_2) + k(y_1 - y_2) + l(z_1 - z_2)}{\sqrt{h^2 + k^2 + l^2}}. \tag{1-23}$$

In a cubic lattice the unit distance along all three axes is the same, and is equal to $2\tau_x$ or a. The coordinates of points P_1 and P_2 can thus be expressed in terms of their integral multiples of the unit translation, as $x_1 = r_1a$, $y_1 = s_1a$, $z_1 = t_1a$, and so on, where r_1, s_1, and t_1 are integers. Expression 1-23 can then be rewritten

$$\frac{a\{h(r_1 - r_2) + k(s_1 - s_2) + l(t_1 - t_2)\}}{\sqrt{h^2 + k^2 + l^2}}. \tag{1-24}$$

Since h, k, and l are always integral, the coefficient of a must be integral, and the minimum possible separation for the two planes is given when this coefficient has its smallest possible value, unity. When the coefficient becomes zero the two planes are coincident. Given any three integers h, k, and l, we can always choose three other integers m, n, and p such that

$$hm + kn + lp = 1. \tag{1-25}$$

Since a space lattice is limitless in extent, the integers r, s, and t may assume any desired values, and it is always possible to satisfy the relation

$$hr + ks + lt = 1. \tag{1-26}$$

Thus relation 1-24 gives for the cubic lattice

$$d_{hkl} = \frac{a}{\sqrt{h^2 + k^2 + l^2}}. \tag{1-27}$$

Table 1-3.   Formulas for Calculating Interplanar Spacings $d_{hkl}{}^a$

System	Axial Translations	Axial Angles	d_{hkl}	
1. Cubic	$a = b = c$	$\alpha = \beta = \gamma = 90°$	$a(h^2 + k^2 + l^2)^{-1/2}$	(1-27)
2. Tetragonal	$a = b \neq c$	$\alpha = \beta = \gamma = 90°$	$[(h^2/a^2) + (k^2/a^2) + (l^2/c^2)]^{-1/2}$	(1-28)
3. Orthorhombic	$a \neq b \neq c$	$\alpha = \beta = \gamma = 90°$	$[(h^2/a^2) + (k^2/b^2) + (l^2/c^2)]^{-1/2}$	(1-29)
4. (A) Hexagonal	$a = b \neq c$	$\alpha = \beta = 90°,\ \gamma = 120°$	$[(4/3a^2)\,(h^2 + k^2 + hk) + (l^2/c^2)]^{-1/2}$	(1-30)
(B) Rhombohedral	$a = b = c$	$\alpha = \beta = \gamma \neq 90° < 120°$	$a\left[\dfrac{(h^2 + k^2 + l^2)\sin^2\alpha + 2(hk + hl + kl)(\cos^2\alpha - \cos\alpha)}{1 + 2\cos^3\alpha - 3\cos^2\alpha}\right]^{-1/2}$	(1-31)
5. Monoclinic	$a \neq b \neq c$	$\alpha = \gamma = 90°,\ \beta > 90°$	$\left[\dfrac{(h^2/a^2) + (l^2/c^2) - (2hl/ac)\cos\beta + \dfrac{k^2}{b^2}}{\sin^2\beta}\right]^{-1/2}$	(1-32)
6. Triclinic	$a \neq b \neq c$	$\alpha \neq \beta \neq \gamma \neq 90°$	$\left[\dfrac{h}{a}\begin{vmatrix} h/a & \cos\gamma & \cos\beta \\ k/b & 1 & \cos\alpha \\ l/c & \cos\beta & 1 \end{vmatrix} + \dfrac{k}{b}\begin{vmatrix} 1 & h/a & \cos\beta \\ \cos\gamma & k/b & \cos\alpha \\ \cos\beta & l/c & 1 \end{vmatrix} + \dfrac{l}{c}\begin{vmatrix} 1 & \cos\gamma & h/a \\ \cos\gamma & 1 & k/b \\ \cos\beta & \cos\alpha & l/c \end{vmatrix}\middle/ \begin{vmatrix} 1 & \cos\gamma & \cos\beta \\ \cos\gamma & 1 & \cos\alpha \\ \cos\beta & \cos\alpha & 1 \end{vmatrix}\right]^{-1/2}$	(1-33)

^{a}For simpler expressions involving reciprocal lattice (Section 1-3.7) units see *International Tables for X-Ray Crystallography*, Vol. II, Kynoch Press, Birmingham, England, 1959.

For lattices of lower symmetry, similar reasoning leads to the expressions listed in Table 1-3.

1-3.7 The Reciprocal Lattice

An important aid in interpretative methods and, indeed, in the general understanding of diffraction phenomena in crystals, has developed from the concept of the reciprocal lattice as introduced by Ewald [13]. At the outset it must be pointed out that the reciprocal lattice is a geometrical convenience that has no physical significance whatsoever. It is impossible to show in it the mechanism of crystal diffraction. But, assuming the Bragg law, we can obtain a geometrical picture (Section 3-2.8) of the conditions necessary for diffraction from a plane (*hkl*). The great utility of the concept lies in the ease with which it permits the investigator to visualize the crystal planes, their slopes, and their spacings. Thus it is now one of the most important tools in diffraction studies, largely as a result of Bernal's [14] pioneering application to the interpretation of rotation and oscillation photographs.

From the "real" space lattice, whose primitive translations are *a*, *b*, and *c*, another lattice, an imaginary lattice of points, can be built up, each point of which bears a "reciprocal" relation to a plane in the original lattice. The point *hkl* in the reciprocal lattice representing the plane (*hkl*) in the real lattice lies on the normal from the origin to the plane, and at a distance ρ from the origin, where

$$\rho = k^2/d_{(hkl)}. \tag{1-34}$$

and k is a constant, conveniently taken as unity for the present discussion. The array of points resulting from this operation on a real lattice is a lattice reciprocal to the original one, and the application of the same operation to the new lattice leads again to a replica of the original lattice. The points of the reciprocal lattice represent the slopes of the real lattice planes through the directions of the ρ's (their normals), and the interplanar spacings d_{hkl} through the lengths of the ρ's (their reciprocal spacings).

That this collection of reciprocal points has the properties of a lattice is most readily demonstrated by a consideration of the simpler two-dimensional case of a general parallelogram lattice, such as the *ac* net of a simple monoclinic lattice. In Fig. 1-31 four cells of a simple *ac* monoclinic lattice net are outlined, the lattice points being designated by open circles. Several lattice planes have their traces and Miller indices indicated. Likewise, a portion of the reciprocal lattice net is sketched in with dashed lines, and the reciprocal lattice points are represented by

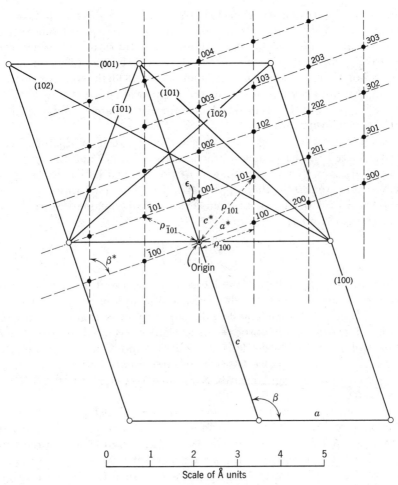

Fig. 1-31.   Projection of a simple monoclinic lattice on (010), and a portion of its reciprocal lattice.

black dots. No parentheses are used on the indices of the reciprocal lattice points in order to distinguish them better from indices of the planes in the real lattice. An arbitrary value of 2 is assumed here for the constant k in order to give ρ a convenient length relative to a and c in the drawing. It is seen that ρ_{100} lies on the perpendicular from the origin to plane (100) and has a length

$$\rho_{100} = a^* = \frac{4}{d_{(100)}}. \qquad (1\text{-}35)$$

Similarly $\rho_{001} = c^* = 4/d_{(001)}$, and the location of reciprocal points 101,

$\bar{1}01$, and others is clearly depicted. In the reciprocal lattice net, points such as 200, 300, 002, 202, 303, and so on, which have no significance as lattice planes in the real lattice, are indicated. Such indices have a very useful significance in terms of the diffracted beams, (200) being the second-order reflection from the (100) planes, (300) the third-order, and so on. An important property of the reciprocal lattice is thus evident; there is a point in the reciprocal lattice for every plane of the real lattice reflecting in every order.

Referring again to Fig. 1-31, we see that the reciprocal lattice net is characterized by the primitive translations a^* and c^* and the angle β,* analogous to a, c, and the angle β of the original lattice net. It is observed, too, that β^* is the supplement of β. The axial ratio of the reciprocal net relative to the original lattice is easily obtained. Since $d_{(100)} = a \sin(180 - \beta)$ and $d_{(001)} = c \sin(180 - \beta)$,

$$\frac{a^*}{c^*} = \frac{4/a \sin(180 - \beta)}{4/c \sin(180 - \beta)} = \frac{c}{a}. \tag{1-36}$$

Thus, for any plane lattice, the lattice reciprocal to it is a similar lattice with an axial ratio equal to the reciprocal of that of the regular lattice and with an axial angle equal to the supplement of that of the regular lattice. The orientation of the reciprocal lattice is rotated 90° from that of the regular lattice. The student can readily demonstrate for himself that the reciprocal to a square lattice is another square lattice, and that the reciprocal to a rectangular lattice is a similar rectangular lattice oriented 90° from the former.

A monoclinic space lattice has a third dimension b perpendicular to the plane of Fig. 1-31. This b dimension in turn leads to $\rho_{010} = b^* = k^2/d_{(010)}$ perpendicular to the plane of ac and also of a^*c^*. Planes of the real lattice which intersect the b axis then give rise to a succession of identical a^*c^* networks at distances nb^* in each direction from the origin, where $n = 1, \bar{1}, 2, \bar{2}, 3, \bar{3}, \ldots$, and so on. The three-dimensional network thus obtained is a reciprocal space lattice. The reciprocal aspects of the space lattices of higher symmetry involve no geometrical difficulties. Some analytical complexities accompany the most general lattice, the triclinic lattice, where the b axis is no longer perpendicular to the plane of ac as it is in the monoclinic system (Fig. 1-31). The primitive translation b^*, however, is perpendicular to the plane ac, but a^* and c^* move out of the ac plane as a result of the obliqueness of the b axis.

The primitive translations of the reciprocal lattice are readily expressed in general form in terms of the cell constants of the real lattice:

$$a^* = \frac{k^2 bc \sin \alpha}{V}, \tag{1-37}$$

$$b* = \frac{k^2 ac \sin \beta}{V}, \tag{1-38}$$

$$c* = \frac{k^2 ab \sin \gamma}{V}, \tag{1-39}$$

where V is the volume of the unit cell. The simplifications for more symmetrical systems are obvious. The reciprocal cell angles, $\alpha*$, $\beta*$, $\gamma*$, are merely the supplements of the corresponding angles of the real cell. In many calculations it is convenient to let $k^2 = \lambda$, the wavelength of the x-rays being used. This point and the geometrical picture of diffraction in reciprocal space are considered in Section 3-2.8. It should be pointed out also that ρ, $a*$, $b*$, $c*$, as well as a, b, and c, are all vector quantities, and that elegant developments of all these relationships by vector methods are available for those who prefer such an approach [15].

1-4 POINT GROUPS AND SPACE GROUPS

The ultimate goal of x-ray crystal analysis is the exact location of all atoms in a crystal, in order to explain completely crystal symmetry, molecular geometry, and other crystallochemical properties. For such considerations we find it important to study the possible ways of arranging points (atoms) in space so that the assemblage possesses symmetry consistent with that observed in crystals. Point-group theory develops the possible arrangements of equivalent points about a single point in space. Its results, in the form of symmetry symbols, furnish a useful notation for describing the symmetry of chemical molecules, and that of a single crystal as well. Space-group theory develops the coordinate positions for unlimited regular arrangements of points (atoms) in space, the limitless repeating patterns of atoms in crystals.

1-4.1 The Point Group or Crystal Symmetry Class

The elements of crystallographic symmetry have already been enumerated (Section 1-2.3) as one-, two-, three-, four-, and sixfold axes; horizontal, vertical, and diagonal symmetry planes; and symmetry centers. It was also pointed out (Section 1-2.3) that composite symmetry arises from the proper combination of these elements, there being 32 distinctive combinations known as the 32 *crystal symmetry classes*. Actually, it is possible to build only 31 different symmetry classes from the simple direct elements just mentioned. The concept of inversion axes introduces one additional class not derived from the simple combination of the above elements:

$\bar{1}$ ≡ a symmetry center,
$\bar{2}$ ≡ a reflection plane,
$\bar{3}$ ≡ threefold axis combined with a center,
$\bar{4}$ (new class),
$\bar{6}$ ≡ threefold axis normal to a symmetry plane.

Modern crystallography therefore finds it more useful to discard the symmetry center as a fundamental symmetry element, and to describe the crystal symmetry in terms of reflection planes, direct rotation axes, and inversion axes. The concept of the symmetry center is still, however, most useful to crystallographers and diffraction workers in many other geometrical aspects of their work.

The same 32 groups are obtained when we consider the arrangement of equivalent points about a single point in space, the points in turn being subject to the operation of the various fundamental crystal symmetry elements and their permissible combinations. Accordingly, *such space point-groups and the crystal symmetry classes are equivalent*, and the two names are used interchangeably.

With 32 different point groups to consider, it is important to have symbols and notations for distinguishing them without ambiguity. Several systems of symbols have been developed and used in the past. The double symbols adopted and used in the *International Tables*[16] embody most of their good features, and are at the same time concise and self-explanatory. The first part of the new symbol is the Schoenflies symbol[17], retained because of the extremely wide usage the great German crystallographer's terminology had acquired. The Schoenflies notation may be described by the following defined symbols and their combinations:

C_n = groups having a single rotation axis of degree n (C for *cyclisch*),
C_{nh} = groups having an n-fold rotation axis with a horizontal symmetry plane normal to it,
C_{nv} = groups having an n-fold rotation axis with vertical planes of symmetry,
D_n = groups possessing an n-fold principal axis and n twofold axes normal to it (D for *Dieder*),
d = diagonal symmetry planes, as in D_{2d},
s = a mirror plane when the normal axis is of degree 1,
S_n = groups having an n-fold rotary-reflection axis (S for *Spiegelung*),
T = tetrahedral groups having the four threefold and the three twofold axes of a regular tetrahedron,

O = octahedral groups possessing three fourfold axes, four three-
    fold axes, and six twofold axes distributed as in a cube,
i = an inversion.

The remainder of the double notation is based on symbols proposed
by Hermann[18] and Mauguin[19] in order that the point-group symbol might present, at a glance, all the essential symmetry elements of the
group. The Hermann-Mauguin symbol utilizes the numbers 1, 2, 3, 4,
and 6 to represent the direct rotation axes, and $\bar{3}$, $\bar{4}$, and $\bar{6}$ for inversion
axes. The symbol $\bar{1}$ is used for a symmetry center, and a plane of symmetry is represented by m (mirror). The symmetry of the principal axis
is given first, and, if there is a mirror plane normal to it, the two are
incorporated in the symbol in the following fashion: $\frac{2}{m}$, meaning a twofold axis with a symmetry plane perpendicular to it. For convenience in
printing it is usually written $2/m$. This is one of the monoclinic point
groups whose complete symbol is $C_{2h} - 2/m$. The secondary axes are next
indicated, followed by any additional symmetry planes. Thus $6/mmm$ has
a symmetry plane normal to a sixfold axis $(6/m)$, and two additional
sets of symmetry planes parallel to the principal axis. The group $\bar{4}2m$
has a principal axis of fourfold inversion, a twofold axis perpendicular
to the principal axis, and a reflection plane parallel to $\bar{4}$.

In writing such symbols it is not necessary to indicate more symmetry
elements than the minimum required to define completely the group's
symmetry. For instance, the symbol 42 completely defines the point
group D_4, which has a fourfold axis and four twofold axes normal to it.
The addition of a single twofold axis normal to a fourfold axis immediately creates a second twofold axis 90° from the first one through the action
of the fourfold symmetry operation of the principal axis. Moreover, this
action simultaneously introduces two more twofold axes in the plane of
the original two but making angles of 45° with them. The *International
Tables*, however, present this symbol in slightly longer form, $D_4 - 422$. It
is frequently possible to abbreviate symbols or give alternative symbols
which completely define the symmetry. Thus,

$$\frac{2}{m}\frac{2}{m}\frac{2}{m} = mmm$$

since three planes of symmetry automatically give rise to twofold axes
normal to each plane. Table 1-4 lists the double symbols of the 32 point
groups as given in the *International Tables* [16].

Point groups may be simply represented by the series of plane diagrams[20] in Fig. 1-32. These are essentially projections of point groups

Table 1-4. Symbols of the 32 Point Groups or Crystal Symmetry Classes

No.	Schoenflies-Hermann-Mauguin Symbol	Number of Equivalent Points or Symmetry Operations	No.	Schoenflies-Hermann-Mauguin Symbol	Number of Equivalent Points or Symmetry Operations
	Triclinic system			Hexagonal system	
1^a	C_1-1	1		(A) Trigonal (rhombohedral) division	
2	$C_i-\bar{1}$	2	16^a	C_3-3	3
			17	$C_{3i}-\bar{3}$	6
	Monoclinic system		18^a	$C_{3v}-3m$	6
3^a	C_s-m	2	19^b	D_3-32	6
4^a	C_2-2	2	20	$D_{3d}-\bar{3}m$	12
5	$C_{2h}-2/m$	4			
				(B) Hexagonal division	
	Orthorhombic system		21^b	$C_{3h}-\bar{6}$	6
6^a	$C_{2v}-mm$	4	22^a	C_6-6	6
7^b	D_2-222	4	23	$C_{6h}-6/m$	12
8	$D_{2h}-mmm$	8	24^b	$D_{3h}-\bar{6}m2$	12
			25^a	$C_{6v}-6mm$	12
	Tetragonal system		26^b	D_6-62	12
9^b	$S_4-\bar{4}$	4	27	$D_{6h}-6/mmm$	24
10^a	C_4-4	4			
11	$C_{4h}-4/m$	8		Cubic system	
12^b	$D_{2d}-\bar{4}2m$	8	28^b	$T-23$	12
13^a	$C_{4v}-4mm$	8	29	T_h-m3	24
14^b	D_4-422	8	30^b	$T_d-\bar{4}3m$	24
15	$D_{4h}-4/mmm$	16	31	$O-43$	24
			32	O_h-m3m	48

[a]These groups show both the piezoelectric effect and the true pyroelectric effect.
[b]These groups show only the piezoelectric effect.

on a plane normal to the group's principal axis, which is to be thought of as located at the center of a large circle. Points shown as solid circles are at a different level along the principal axis from those depicted as open circles. Each type of point is brought into coincidence with other points of its kind by rotation about the principal axis or by reflection across a vertical symmetry plane (not indicated in the drawing). Solid points are brought into coincidence with open points through the action of an

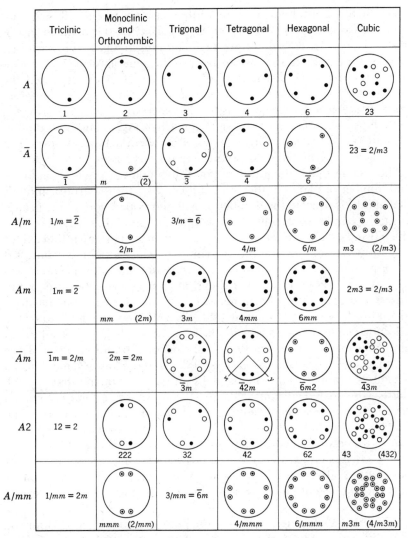

Fig. 1-32. The 32 crystal point groups. (Courtesy of F. Phillips, *An Introduction to Crystallography*, 1946; and Longmans, Green.)

inversion axis, a horizontal reflection plane (cases in which such points are superimposed), or a horizontal twofold axis. The general symmetry operations along a horizontal row of the chart are represented by a Hermann-Mauguin symbol at the left-hand edge, in which the symbol *A* denotes a principal axis of any degree. A most useful exercise for the reader in understanding point groups is to analyze each diagram and symbol equivalence indicated.

An equally useful interpretation of the diagrams in Fig. 1-32 is to consider each one as a tiny *stereographic projection** of the point group or crystal class under consideration. The diagrams in Fig. 1-32 in each case represent the poles of the various faces of the general form $\{hkl\}$, and as such show the same distribution of symmetry elements as would the ideal geometrical crystal when similarly oriented. More elaborate stereograms (stereographic projections) of the 32 crystal classes are provided by Phillips in his detailed discussions of each class[21], and both Phillips [21] and Wyckoff[22] provide a series of excellent perspective drawings of the axes and symmetry elements of the various classes.

The total number of equivalent points in a point group is equal to the total number of planes in the general form $\{hkl\}$ for the group, and this total in turn is equal to the number of symmetry operations of the group. It is to be emphasized again that a symmetry operation is any operation which brings about a point-for-point coincidence. Thus an *n*-fold axis of rotation possesses *n* operations of symmetry, since a point is multiplied into *n* points by its action. Similarly, a reflection plane or a center each produces from a point another identical point, so that they are said to possess two operations of symmetry. Group $4/m$, for example, has four operations as a result of the fourfold axis, and these are doubled by the reflection plane making eight in all. In some symbols we must bear in mind that axes of a certain degree are normal to one or more of the mirror planes. For instance, $m3m \equiv 4/m3m$, and the unwary might be led to total its symmetry operations as 12 on the basis of the first symbol, whereas the second version yields 48, which is correct. Likewise, a symbol frequently lists symmetry elements in excess of those needed to define completely the class symmetry, as in $4/mmm$. The last *m* is not independent of the second *m*, and the group has 16 symmetry operations rather than 32. Inversion axes with *n* even possess *n* symmetry operations; but those with *n* odd have $2n$ operations. The number of symmetry operations and equivalent points associated with each point group is indicated in Table 1-4.

A useful analytical description of the point groups is provided for the x-ray diffractionist by deriving the coordinates of the points arising from a point *xyz* being subjected to the symmetry operations of a given crystal class. Such treatment yields the coordinates of the point group's set of equivalent points. Figure 1-33, for instance, depicts the monoclinic point group $C_{2h} - 2/m$, with *Y* the twofold axis, and the *XZ* plane (normal to *Y*) the mirror plane. It is immaterial which of these two operations is considered to take place first. Suppose therefore that the point *xyz* is first

---

*For a discussion of the stereographic projection, see Section 10-5.2.

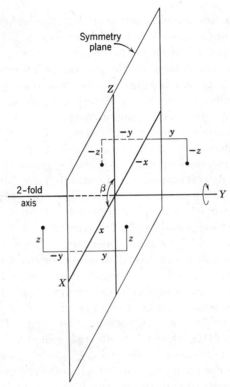

Fig. 1-33. The monoclinic point group C_{2h}–2/m, and its four equivalent points xyz, $\bar{x}y\bar{z}$, $x\bar{y}z$, $\bar{x}\bar{y}\bar{z}$.

acted upon by the twofold axis. From this operation the two points xyz and $\bar{x}y\bar{z}$ result. *The general action of a twofold axis coincident with a crystal axis is to change the sign of the coordinates along the other two axes.* When the two points above are reflected in the XZ plane, they yield two more points, $x\bar{y}z$ and $\bar{x}\bar{y}\bar{z}$. *Reflection in a plane containing two of the crystal axes, and which is normal to the third axis, always changes the sign of the coordinate along the third axis.* The point group $C_{2h}-2/m$ then has the four equivalent points:

$$xyz; \quad \bar{x}y\bar{z}; \quad x\bar{y}z; \quad \bar{x}\bar{y}\bar{z}.$$

Inspection of Fig. 1-33 reveals that the origin is a center of symmetry, and the coordinates of the equivalent points verify this, since the four points can be arranged in pairs in which one member is the negative of the other:

$$
\begin{array}{ccc}
xyz & & x\bar{y}z \\
& \text{and} & \\
\bar{x}\bar{y}\bar{z} & & \bar{x}y\bar{z}
\end{array}
$$

Because of space limitations a similar analytical treatment of the remaining 31 point groups must be omitted here. Wyckoff[22] has, however, carried out and published such a treatment for all the groups.

The problem of determining the correct point group or symmetry class to which a crystal belongs can be very difficult. Symmetrically developed crystals can usually be identified as to symmetry system by visual inspection, goniometric measurements, and/or optical observations with a polarizing microscope. After such preliminary symmetry identification, data of two kinds are used to determine the point group: (1) studies of facial development and other external features, (2) information concerning certain internal physical characteristics.

The early crystallographer customarily based his selection of the point group on the symmetry of face distribution, adopting as the true point group the lowest suggested by the facial development of the specimens observed. This would be a most satisfactory procedure if the general form of each class always occurred on all specimens. Unfortunately, the general form frequently is not present, only special forms occurring, and no special form is uniquely characteristic of one point group. Sometimes low symmetry makes the precise identification of the general form difficult even when it is present. Etch figures[23], if cautiously interpreted, can often be used to advantage in this decision. Another pitfall in the determination of the point group from external symmetry features alone is the frequency with which peculiarities in habit together with other accidents of growth may result in specimens on which only a few faces of a form are developed. Whether or not the absence of faces on a single specimen is to be considered significant is most difficult to decide, and the former incorrect assignments of cuprite (Cu_2O) and low ammonium chloride (NH_4Cl) (now believed to be O_h-m3m), to the point group $O-43$ are celebrated examples of such errors.

Among the internal characteristics which may be used to allocate a crystal to a particular point group are optical activity, piezoelectric and pyroelectric properties, and x-ray diffraction symmetry. X-rays add a center of symmetry to diffraction effects, so that the diffraction-pattern symmetry of a single crystal is that of 1 of the 11 centrosymmetric point groups: numbers 2, 5, 8, 11, 15, 17, 20, 23, 27, 29, and 32 of Table 1-4. Once the centrosymmetric point group has been found from diffraction symmetry, we usually attempt to establish the presence or absence of a center of symmetry by means of piezoelectric and pyroelectric studies. Details of these methods largely involve single-crystal techniques, which are outside the province of this volume.

Thus no one method can be relied upon to establish the point group with certainty. Instead, the investigator must make judicious use of various methods, conservatively interpreting the results in some instances.

1-4.2 The Space Group

Since large and small crystals of the same chemical species have identical symmetry and symmetry-dependent properties, their atomic (or molecular) arrangement must necessarily be based on an infinitely repetitive pattern. For the precise description of such unlimited atomic arrays in crystals, a knowledge of the possible infinite distributions of points in space is needed, such that the distribution as a whole possesses crystallographic symmetry. Indeed, this fundamental problem formed the major interest of several geometrical crystallographers in the latter part of the nineteenth century, and three of them, independently and through different approaches, arrived at a final and complete answer in one of the most remarkable examples of simultaneous discovery in all science. Fedorov's[24] results, published in Russian between the years 1885 and 1890, were not readily available. Schoenflies'[17] work appeared in German in 1891, and the studies of Barlow[25] appeared 3 years later. These investigators proved that there are 230 nonidentical space-point arrangements, designated the 230 space groups.

The 32 point groups define all the possible ways of distributing points about a single point in space such that the arrangement has crystallographic symmetry. If we place a point group at the points of a Bravais lattice of the same symmetry system, an infinitely repetitive point system in space results, which is, moreover, one of the simpler of the space groups (Fig. 1-34). The space groups derivable in this fashion were

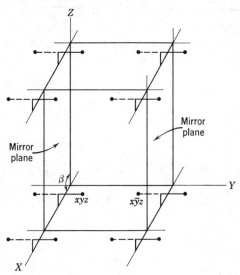

Fig. 1-34. The simple space group, $C_s^1-P_m$, obtained by placing the point group C_s-m at the points of a primitive monoclinic lattice.

the first to be discovered. Sohncke [26], requiring only identity of environment about each lattice point but not necessarily similar orientation, demonstrated the existence of 65 simpler space groups. The remaining space groups, developed later, include two new types of symmetry elements resulting from combinations of simple axes and reflection planes with translations. In the point groups, and the space groups derived directly therefrom, the repeated action of their symmetry elements brings a point back into coincidence with itself or with another point associated with the same origin or lattice point. Repeated action of these new symmetry elements, termed *glide* planes and *screw* axes, never brings an equivalent point back into coincidence with itself, but instead with the corresponding point associated with some neighboring lattice point. Although such symmetry elements have no place in the point groups, in an infinitely extended point system they are just as legitimate as the simpler ones. Because of their action the symmetry associated with a single lattice point is less than that of the whole array. It is to be noted that the elementary lattice translations, possessed by all space groups, are symmetry operations of this same general type, and each has an infinity of translational operations which may be designated by the symbol $2m\tau_x$, $2n\tau_y$, $2p\tau_z$, where m, n, and p are any integers or zero. Translations and operations derived from them are characterized by the property that no amount of repetition of the operation will bring a point back to its original position.

A glide plane of symmetry combines a reflection in a plane with a translation of definite length in the plane. In Fig. 1-35 the XY plane is a glide reflection plane, and point P with coordinates xyz goes over into point P' by its action, which includes a translation of half a lattice length

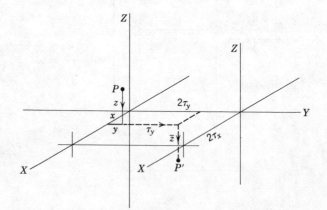

Fig. 1-35.   An XY glide plane with a glide of $\tau_y = b/2$.

parallel to Y. The coordinates of the new point P' are $x, y+\frac{1}{2}, z$. For an XY glide plane several other glides are also possible, for instance, $a/2$, $a/2+b/2$, and occasionally $a/4+b/4$. A glide parallel to the Z axis, however, is meaningless for this particular plane. A crystal possessing such a glide plane gives no externally visible evidence of it, but appears to possess a true mirror reflection plane. The diffraction effects from the crystal, however, definitely reveal the presence of the glide plane (Table 3-3).

A screw axis combines a rotation of a definite angular magnitude with a translation of a definite length parallel to the axis. An n-fold screw axis thus includes a rotation of $2\pi/n$, plus a translation based on $1/n$th of the repeat distance along the axis. Actually, there may be several screw axes of a given degree, since there is one for each of the possible multiples of the basic translation:

$$\frac{1}{n}, \frac{2}{n}, \frac{3}{n}, \ldots, \frac{n-1}{n}.$$

These are indicated by the appropriate subscripts, 1 to $n-1$, placed to the right of the numeral expressing the degree. Accordingly, there is only one twofold screw axis, 2_1; there are two triad types, 3_1 and 3_2; three tetrad types, 4_1, 4_2, and 4_3; and five hexad types, 6_1, 6_2, 6_3, 6_4, and 6_5. Figure 1-36 depicts the various types of screw axes together with the normal and inversion axes for degree 6. It is seen that two pairs of the hexad axes, 6_1 and 6_5 and 6_2 and 6_4, are enantiomorphous pairs related as mirror images. If each pair is viewed by looking upward along the axes, the first member is seen to have a right-handed and the other a left-handed screw direction. Similarly, 3_1 and 3_2, and 4_1 and 4_3 are enantiomorphous pairs among the triad and tetrad axes. Crystals with screw axes are indistinguishable in outward appearance from those with simple

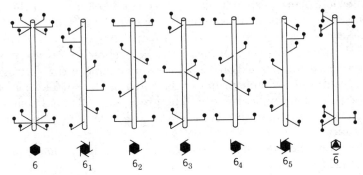

Fig. 1-36. Types of sixfold axes. (Courtesy of C. Bunn, *Chemical Crystallography*, and Clarendon Press, 1945.)

rotation axes, but their x-ray patterns distinguish the two types (Table 3-4). The two individuals of enantiomorphous pairs, however, cannot be differentiated by x-ray means.

A Schoenflies-Hermann-Mauguin symbol notation has been developed to designate each of the 230 space groups without ambiguity. The Schoenflies part is a simple modification of the Schoenflies point-group symbol. For instance, D_{2h}^m represents the mth space group based on the point group D_{2h}. There are actually 28 space groups based on D_{2h}, so that such a system of consecutive numbering gives no clue to the characteristic symmetry elements which distinguish one space group from the others isomorphous with the same point group. The Hermann-Mauguin portion of the symbol, however, takes care of this aspect.

The Hermann-Mauguin notation starts with a symbol denoting the underlying space-lattice type:

P = a simple primitive lattice,
F = completely face-centered lattice,
I = body-centered lattice,
A = (100) face-centered lattice,
B = (010) face-centered lattice,
C = (001) face-centered lattice (also used for primitive hexagonal lattices in the earlier *International Tables* [16],
R = a rhombohedral lattice (a special case of P),
H = a triply primitive hexagonal lattice with axes rotated 30° with respect to the C hexagonal lattice (used only in the earlier *International Tables* [16]

Following the lattice-type symbol is a number denoting the degree of the principal axis, using the simple numerical schemes already described. A reflection plane perpendicular to this axis is indicated in the same manner as for point groups. The remainder of the symbol then lists secondary axes at right angles to the principal axis, or planes (mirror or glide) parallel to the principal axis. Thus the symbol C_{4h}^1–$P4/m$ represents the first space group based on the point-group symmetry C_{4h}, a tetragonal space group with a simple primitive cell and a single fourfold axis with a mirror plane perpendicular to it. The group, C_{3v}^5–$R3m$, has a rhombohedral lattice, a single threefold axis, and a reflection plane parallel to this axis.

When the reflection plane is not a mirror plane, but a glide plane, the following symbols are used:

Translation	Symbol
$a/2$	a
$b/2$	b
$c/2$	c
$\dfrac{a+b}{2}, \dfrac{b+c}{2},$ or $\dfrac{c+a}{2}$	n
$\dfrac{a+b}{4}, \dfrac{b+c}{4},$ or $\dfrac{c+a}{4}$	d

One of the commonest of the orthorhombic space groups is D_{2h}^{16}–$Pnma$, which has a primitive cell, an n glide parallel to (100), that is, a glide $b/2 + c/2$ (since a glide involving $a/2$ would be perpendicular to (100) and therefore meaningless), mirror planes parallel to (010), and an a glide plane parallel to (001). A symbol such as $P4/ncc$ indicates that the principal axis c is fourfold and has an n glide plane normal to it [(001) is an n glide], and that parallel to both (100) and (010) are c glide planes. For further details the reader is referred to the sections on space-group notation and nomenclature in the *International Tables* [27], where a much more detailed account is presented.

Diagrammatic and analytical representation of the space groups is even more necessary than for the point groups. In the *International Tables* [27] pictorial representation of each space group is very satisfactorily achieved in one or two plane diagrams (Fig. 1-37). The point positions to be represented are the general equivalent point positions of the space group, that is, the positions occupied by asymmetric atoms or molecules in building the crystal. It is necessary to distinguish between points which can be brought into coincidence by translations or rotations alone (congruent points), and points which are mirror images of them or related by an axis of rotary inversion. In Fig. 1-37 points represented by open circles are related to those depicted as circles containing a comma (,) by the mirror-image relation (combined in some instances with a translation). Note that all points in a space group represented by open circles are congruent, as are also all points represented by circles containing commas.

The origin of the diagrams lies at the upper left-hand corner with the positive X axis directed toward the bottom of the page, the positive Y axis to the right, and positive Z axis toward the reader (normal to the page for orthogonal cells). Point positions at the same height above the base of the unit cell are indicated by the symbol +, and those at an equal distance below the base bear the symbol (−). Symmetry planes perpendicular to the plane of the diagram are represented by various kinds of lines

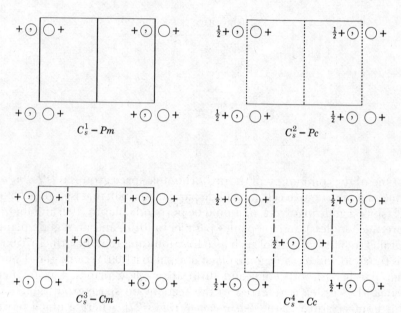

Fig. 1-37.   The four monoclinic space groups isomorphous with point group C_s–m.

m———, $c \bullet \bullet \bullet \bullet \bullet$, a, b— — — — —, and n— $\bullet$ — $\bullet$ — $\bullet$ —. Edges of the unit cell when not coincident with a symmetry element are outlined by a thin line. Additional symbols and diagrams are presented in the *International Tables* [27].

The analytical description of each space group is achieved by deriving the coordinates of its equivalent points referred to a fixed origin with respect to the symmetry elements of the group. Special point positions come into existence when the general coordinates assume the precise values necessary to cause two or more symmetry-related points to coalesce on the symmetry element. If, for instance, in Fig. 1-34, $y = \bar{y} = 0$, the two points xyz and $x\bar{y}z$ coalesce to a single point on the symmetry plane at $x0z$. Also, when $y = \bar{y} = \frac{1}{2}$, the two become a single point at $x\frac{1}{2}z$. These last two positions are said to be special cases of space group Pm. Likewise, points on an n-fold symmetry axis are special positions resulting from n general equivalent points coalescing on the axis. It is to be noted that the complete analytical description of a space group is accomplished in terms of a selected unit of the pattern that is representative of the unlimitedly extended whole. Most commonly this is a unit domain around the origin, with the same shape and dimensions as the unit cell. The coordinates of the general and special positions of all 230 space groups are tabulated in the *International Tables* [27].

Since space-group determination is almost always a single-crystal diffraction problem, only the basis for it is outlined here. The translations of glide planes and screw axes, and the lattice centering present in nonprimitive lattices, lead to the extinguishing of certain classes of x-ray spectra. Each type of extinction is characteristic of the symmetry element or lattice centering producing it, and therefore of the space groups containing that element or lattice. Thus the absence of such characteristic spectra from diffraction data is a major criterion for space-group determination. Actually, each of the above effects causes the absence of a different class of reflections, so that it is a simple procedure to examine single-crystal diffraction data and therefrom write down the Hermann-Mauguin symbol for the space group. Powder diffraction data are, however, rarely sufficient to establish the space group because they are usually too limited in number and often cannot be indexed with certainty. For a consideration of the origin of these extinctions and a tabulation of their significance, the reader is referred to Section 3-3.6.

GENERAL REFERENCES

*1. F. C. Phillips, *An Introduction to Crystallography*, 3rd ed., Longmans, London, 1963; Wiley, New York, 1963.

2. E. S. Dana, *A Textbook of Mineralogy*, 4th ed. (revised by W. E. Ford), Wiley, New York, 1932, Part I, pp. 7–206.

3. A. B. Dale, *Form and Properties of Crystals*, Cambridge University Press, London, 1932. A rather brief but excellent introduction to crystallography.

4. A. E. H. Tutton, *Crystallography and Practical Crystal Measurement*, 2nd ed., 2 vols., Macmillan, London, 1922; Stechert, New York, 1964.

5. H. Hilton, *Mathematical Crystallography*, Oxford University Press, London, 1903; Dover, New York, 1963.

6. M. A. Bravais, *On the Systems Formed by Points Regularly Distributed on a Plane or in Space*, (translated by A. J. Shaler), Crystallographic Society of America Memoir No. 1, 1949; from the original in *J. Ec. Polytech.*, Cahier 33, Tome XIX, pp. 1–128, Paris, 1850.

7. P. P. Ewald, "Historisches und Systematisches zum Gebrauch des Reziproken Gitter in der Kristallstrukturlehre," *Z. Kristallogr.*, **93**, 396 (1936).

8. M. J. Buerger, *X-ray Crystallography*, Wiley, New York, 1942. Chapters 1, 2, 4, 6, 18, 22.

9. M. J. Buerger, *Elementary Crystallography*, rev. ed., Wiley, New York, 1963.

10. C. W. Bunn, *Chemical Crystallography*, 2nd ed., Oxford University Press, London, 1961, Chapters I–IV.

11. W. Kleber, *Einführung in die Kristallographie*, Zehnte, verbesserte Auflage, Verlag Technik, Berlin. *An Introduction to Crystallography*, first English edition (translated by W. A. Wooster and A. M. Wooster), Verlag Technik, Berlin, 1970.

*Books marked with an asterisk are considered particularly desirable for a small reference library on x-ray diffraction and its applications.

*12. *International Tables for X-ray Crystallography*, Vol. I, *Symmetry Groups*, Vol. II, *Mathematical Tables*, Vol. III, *Physical and Chemical Tables*, Kynoch Press, Birmingham, England, Vol. I, 1952, Vol. II, 1959; Vol. III, 1962.

SPECIFIC REFERENCES

[1]   W. Friedrich, P. Knipping, and M. v. Laue, *Ann. Physik*, **41**, 971 (1912).

[2]   E. S. Dana, *A Textbook of Mineralogy*, 4th ed., (revised by W. E. Ford), Wiley, New York, 1932, p. 7.

[3]   G. Antonoff, *J. Phys. Chem.*, **48**, 95 (1944).

[4]   P. Gaubert, *Recherches récentes sur les facies des cristaux*, Publications de la Société de chimie physique II, Hermann, Paris, 1911.

[5]   H. E. Buckley, *Crystal Growth*, Wiley, New York, 1951, Chapter 10, also Appendix, pp. 529–559.

[6]   See W. G. France, "Adsorption and Crystal-Habit Modification," in *Colloid Chemistry*, Vol. V, (J. Alexander, ed.) Van Nostrand Reinhold, New York, 1944, pp. 443–457.

[7]   For a description of the various types of optical goniometers and a discussion of their use in measuring crystal angles, see A. E. H. Tutton, *Crystallography and Practical Crystal Measurement*, 2nd ed., Vol. I, Macmillan, London, 1922; Stechert, New York, 1964. A brief procedure in optical goniometry is presented by F. C. Phillips in *An Introduction to Crystallography*, 3rd ed., Longmans, London, 1963; Wiley, New York, 1963, Chapter V and Appendix I.

[8]   See H. Hilton, *Mathematical Crystallography*, Oxford University Press, 1903, p. 43; Dover, New York, 1963.

[9]   The rhombohedral-hexagonal lattice transformation is simply demonstrated in M. J. Buerger's *X-ray Crystallography*, Wiley, New York, 1942, p. 68.

[10] See W. G. Cady, *Piezoelectricity*, McGraw-Hill, New York, 1946 and 1963; Dover, New York. A very brief discussion is given in reference 2, pp. 335–336.

[11] C. Huygens, *Traité de la lumière*, 1678. Published at Leyden in 1690.

[12] R. J. Haüy, *J. Phys.*, **19**, 366 (May 1782).

[13] P. P. Ewald, *Z. Kristallogr.*, **56**, 129 (1921).

[14] J. D. Bernal, *Proc. Roy. Soc.* (London), **113A**, 117 (1926).

[15] For instance, M. J. Buerger, *X-ray Crystallography*, Wiley, New York, 1942, pp. 122, 347; or R. W. James, *The Crystalline State*, Vol. II, *The Optical Principles of the Diffraction of X-rays*, G. Bell, London, 1948, pp. 598–607; Cornell University Press, Ithaca, 1965.

[16] *International Tables for X-ray Crystallography*, Vol. I, Kynoch Press, Birmingham, England, 1952, Section 3; or *Internationale Tabellen zur Bestimmung von Kristallstrukturen*, Vol. I, Gebrüder Borntraeger, Berlin, 1935, Chapter III.

[17] A. Shoenflies, *Krystallsysteme und Krystallstruktur*, Leipzig, 1891.

[18] C. Hermann, *Z. Kristallogr.*, **68**, 257 (1928); **76**, 559 (1931).

[19] C. Mauguin, *Z. Kristallogr.*, **76**, 542 (1931).

[20] F. C. Phillips, *An Introduction to Crystallography*, 3rd ed., Longmans, London, 1963, p. 106; Wiley, New York, 1963.

[21] Reference 20, pp. 107–150.

[22] R. W. G. Wyckoff, *The Structure of Crystals*, 2nd ed., Chemical Catalog Company, New York, 1931, pp. 25–38.

[23] See reference 2, pp. 211–213; or A. P. Honess, *Nature, Origin, and Interpretation of the Etch Figures of Crystals*, Wiley, New York, 1927.

[24] E. Fedorov, *Z. Kristallogr.*, **24**, 209 (1895). An account of work published in Russian in the years 1885–1890.

[25] W. Barlow, *Z. Kristallogr.*, **23**, 1 (1894).

[26] L. Sohncke, *Entwickelung einer Theorie der Krystallstruktur*, Leipzig, 1879.

[27] Reference 16, (1) Vol. I, Section 4; or (2) Vol. I, Chapters I, II, and V.

CHAPTER 2

THE PRODUCTION AND
PROPERTIES OF X-RAYS

Although modern commercial x-ray generators require little more than the manipulation of a few switches for their control, an introductory knowledge of x-ray tubes, transformers, and generator circuits is most useful to the diffractionist for efficient operation and routine maintenance of the equipment. Similarly, an acquaintance with the nature and properties of x-rays provides the worker with an understanding of the serious absorption and fluorescence problems arising when some substances are being investigated. Choice of the proper wavelengths and of the correct monochromatization filters is likewise greatly aided by a limited knowledge of x-rays. In short, this chapter seeks to present briefly certain aspects of x-ray physics necessary to the most successful application of diffraction techniques. Because of the hazardous nature of x-rays, the safe handling of x-rays is appropriately made the first concern of this discussion.

2-1 X-RAY SAFETY AND PROTECTION

The careful, alert, and informed worker need feel no hesitancy whatsoever in operating x-ray diffraction equipment, but the use of such equipment by careless or uninformed persons is dangerous. The hazards involved come from two sources, the x-rays themselves and the high voltages used in generating them. *Workers must never, knowingly or otherwise, expose themselves to direct or secondary radiation.* The effects of exposure to x-radiation are cumulative, and may lead to serious and permanent injury. The voltages normally used in x-ray diffraction lie in the range 25 to 55 kV peak, and, needless to say, *all contact with equipment at such potentials must be avoided.*

58

Many of the pioneer workers with x-rays became crippled or died from the effects of overexposure to x-radiation. Modern knowledge of the physiological effects of x-rays, however, provides the basis for completely adequate protective measures. Nearly all radiation injuries caused by x-ray diffraction equipment have been to the fingers of operators. The beams emitted through the windows of modern crystal-diffraction tubes are so powerful that even momentary exposure to the direct beam near the tube may result in permanent skin injuries. Dose rates of several thousand roentgen (R) per second have been measured in the useful beam. The roentgen is defined as the quantity of x-radiation such that the associated corpuscular emission per 0.001293 gram of air produces, in air, ions carrying 1 esu of electricity of either sign[1]. The biologically effective dose is the important quantity in connection with radiation injury. The unit of absorbed dose is the rad (roentgen absorbed dose), and the biologically effective unit dose is the rem (roentgen equivalent man). For x-radiations used in x-ray diffraction, the dose in rads or rems is equivalent numerically to the exposure dose in roentgens measured in air.

The National Council on Radiation Protection and Measurements[2] has provided recommendations for the "permissible" dose of radiation to which a person may be occupationally exposed:

Part of Body	Maximum 13-Week Dose	Maximum Yearly Dose
The whole body, gonads, red bone marrow, lens of the eye	3 rems	5 rems
Skin of whole body	10 rems	15 rems
Forearms	10 rems	30 rems
Hands	25 rems	75 rems

Long term accumulation to age N years: $(N - 18) \times 5$ rems

The continuous whole-body dose calculates to be 2.5 mR/hr for a 2000-hr work year. Sufficient protection to keep the worker's exposure far below this permissible dose rate is easily provided. Many laboratories use the film-badge technique to monitor continuously the exposure of their staff.

Modern commercial diffraction generators are usually designed to be suitably rayproof and shockproof for the operator's protection. The safety features should include "fail-safe"-type shutters over the beam ports of the x-ray tube, and labyrinth-type couplings to prevent side scatter where camera collimators meet x-ray tube windows[3]. It is the responsibility of the operator to test his setup for leakage of radiation. Convenient small R meters* and counting rate meters are available for testing x-ray laboratories, and every diffraction laboratory should have such instruments among its facilities. The conversion of x-ray counting rates to roentgens has been discussed by Kohler and Parrish[4]. All special diffraction setups designed and built in the operator's laboratory must likewise meet the foregoing safety requirements. Finally, signs should be prominently posted in diffraction laboratories to remind visitors and workers of the possible radiation hazard.

Protective devices and designs, of course, do not insure that the worker will take adequate precautions, nor can they prevent a careless person from unwisely or unknowingly exposing himself and/or others to radiation. *The responsibility for safe operation rests directly on the individual operator.*

2-2 THE PRODUCTION OF X-RAYS

2-2.1 The Origin of X-rays

X-rays are produced when fast-moving electrons impinge on matter. The phenomena resulting from the deceleration of such electrons are very complex, and x-rays result from two general types of interaction of the electrons with the atoms of the target material. A high-speed electron may strike and displace a tightly bound electron deep in the atom near the nucleus, thereby ionizing the atom. When a certain inner shell of an atom has been ionized in this manner, an electron from an outer shell may fall into the vacant place, with the resulting emission of an x-ray characteristic of the atom involved. Such production of x-rays is a quantum process similar to the origin of optical spectra. The foundations of this theory were originally developed by Kossel[5], on the basis of the Bohr atomic theory[6] and Moseley's celebrated measurements of x-ray spectra[7]. Similar ideas were independently conceived by Barkla[8].

A high-speed electron may be slowed down by another process. Instead of colliding with an inner electron of an atom of the target

---

*For example: Victoreen 440, Victoreen 440 RF, Victoreen 444, Victoreen 1B85 (Victoreen Instrument Division, 10101 Woodland Ave., Cleveland, Ohio); and Eberline E500B with Geiger tube (Eberline Instrument Corporation, Box 2108, Santa Fe, New Mexico). A discussion and comparison of these instruments is presented in general reference 4. Their response to MoKα radiation is satisfactory, but for CuKα radiation they tend to read low. It was suggested that their response to low-energy x-rays might be checked with an iron-55 source.

material, it may simply be slowed down in passing through the strong electric field near the nucleus of an atom. This also is a quantum process, the decrease in energy ΔE of the electron appearing as an x-ray photon of frequency ν as given by Einstein's equation

$$h\nu = \Delta E, \qquad (2\text{-}1)$$

in which h is Planck's constant. X-radiation produced in this manner is independent of the nature of the atoms being bombarded, and appears as a band of continuously varying wavelength whose lower limit is a function of the maximum energy of the bombarding electrons. With this brief introduction to the origin of x-rays, it is appropriate to proceed to a discussion of x-ray tubes. Additional details of the elementary theory of x-ray spectra are presented (Section 2-3.1) in discussing the properties of x-rays.

2-2.2 X-ray Tubes

X-ray tubes are devices for bringing about the interaction of high-speed electrons with matter for the purpose of producing x-rays. The earliest x-ray tubes were merely cathode-ray or Crookes tubes which, although they produced some x-rays, were not highly efficient sources of useful x-radiation. The gas tube or cold-cathode tube, in a much improved design, however, was an important x-ray source in the earlier days of diffraction. Certain operational disadvantages of the gas tube were overcome by the hot-cathode or Coolidge tube [9], which has become the standard x-ray source in many fields of technology and medicine.

A. Gas Tubes. Roentgen's original x-ray tube and the early gas tubes (Fig. 2-1) were essentially glass tubes or bulbs provided with a pair of metal electrodes between which a high-voltage direct current could be passed. When the gas pressure in such a system is reduced to a small fraction of a millimeter of mercury, and a voltage applied, a few electrons and positive ions are supplied by ionization of the rarefied gas. The positive ions impinging on the cathode release additional electrons which are projected perpendicularly from the cathode in straight lines and give rise to x-rays when they strike the walls of the tube, or the other electrode, referred to as the anode, anticathode, or target.

The early cold-cathode tubes were troublesome to operate because the character of the x-rays produced is in part conditioned by the residual gas pressure, which changes as the tube operates unless a means is provided for maintaining it constant. The character of the x-rays produced is also dependent on the voltage applied. If the voltage is changed, the

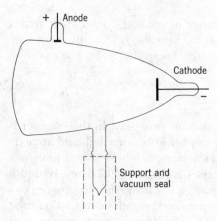

Fig. 2-1.   Diagram of cathode-ray tube with
which Roentgen discovered x-rays.

current through the tube changes, being dependent on the number of
ions formed; and the latter in turn is dependent on the applied voltage
and the gas pressure. This inability to vary voltage and current independently, and the difficulties in maintaining constant residual-gas
pressure, were serious disadvantages in the use of the early gas tube.
Some of the advantages of gas tubes were spectral purity, availability of
several radiations from one tube, and inexpensive servicing in the
laboratory.

B. Hot-Cathode Tubes. The invention of the hot-cathode or electron-
type tube by Coolidge[9] was the result of a direct attempt to provide a
tube free from the operational disadvantages of the gas tube. In this
kind of tube (Fig. 2-2) the bulb is evacuated as thoroughly and completely as possible. With such thorough evacuation there is no appreciable
amount of gas remaining in the tube, so that electrons must be supplied
from another source. They are commonly supplied by application of the

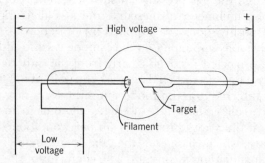

Fig. 2-2.   Early hot-cathode or Coolidge x-ray tube.

Edison effect, that is, by emission from a heated filament which forms the cathode of the tube. The filament is usually a tungsten-wire spiral requiring a current of 1.5 to 5 A at 4 to 12 V to heat it to incandescence. Correspondingly, the range of filament temperature is 1800 to 2600° absolute. A potential gradient applied between the cathode and the target draws the thermal electrons across the tube to the target, and all the current through the tube is carried by these electrons.

The most outstanding advantage of the hot-cathode tube is the independence of tube current and voltage. Either may be altered without appreciably affecting the other, in sharp contrast to the behavior of the gas tube, in which they are interdependent. At higher voltages the current i in a Coolidge tube is solely a function of the number of thermal electrons n, which is controlled by the temperature of the filament in accordance with the Richardson relationship*[10],

$$i = ne = CT^2 \epsilon^{-d/T}, \tag{2-2}$$

where e is the charge on the electron, T is the absolute temperature, ϵ is the Napierian base, and C and d are constants of the filament metal (approximately 1.80×10^{11} and 5.2×10^4, respectively, for tungsten). At lower voltages, less than about 3 kV, a space-charge effect develops, and the tube current responds only feebly to changes in filament temperature.

The evacuation and complete removal of occluded gases from the metal and glass parts of these tubes require a very special factory technique involving pumping and baking, with simultaneous operation under increasing potential during part of these operations. Failure of electron tubes frequently results from sudden outbursts of traces of occluded gas, leading to a gassy condition and the passage of large currents. Other types of failure are due to puncture resulting from accidentally exceeding the voltage limit, to high-tension surges induced by sparking, or to gassiness. Even when operation proceeds normally, there is a limit to tube life. Tungsten vaporizes from the glowing filament of the cathode, so that the filament gradually grows thinner and thinner until it finally burns through. On the average, the life of a Coolidge tube is that time over which the resistance of the filament increases by 10 per cent, usually about 2000 hr of service. Operation at slightly under the tube's maximum rating is conducive to longer life.

The generation of x-rays is a very inefficient process. The x-ray energy obtained from a Coolidge tube is directly proportional to the tube current and to the square of the voltage applied. However, the x-ray output at

---

*Richardson's earlier development of this expression contained the factor $T^{1/2}$ instead of T^2, but the latter is now generally considered the more correct form.

100 kV, even under the best operating conditions, is only about 0.2 per cent of the energy supplied to the tube. This small output spreads out from the target face in a hemisphere, so that only a small fraction of it passes through the tiny slits or pinholes of the diffraction apparatus. In addition, filters inserted in the beam further reduce its intensity by one-half or more. The small proportion finally available at the crystal sample for diffraction is but a minute fraction of the total energy input to the tube.

The operational convenience of the hot-cathode type of tube has made it the favorite among diffraction workers. Two or three disadvantages should, however, be mentioned. One of the most serious is the deposition of tungsten vapor on the surface of the target, resulting in contaminated radiation. In fact, such tubes frequently become so contaminated that they must be discarded, although they may be operating entirely satisfactorily otherwise. Unsuspected use of impure radiation has several times led to the publication of erroneous results. Procedures for determination of spectral contamination of x-ray tubes have been described by Ladell and Parrish[11]. Another disadvantage, where several different radiations are needed, is that a separate tube is required for each kind of radiation. Finally, installation and maintenance costs are generally higher for these tubes.

C. Modern Diffraction Tube Design. In the earliest days of crystal analysis and x-ray diffraction, discarded medical and dental x-ray tubes, and other makeshift sources of x-rays, were used. Soon, however, specially designed diffraction tubes began to be available commercially, and some of them are now very sophisticated devices. The principal objects sought in the design of diffraction tubes are to: (1) increase the intensity of the x-rays; (2) minimize absorption of the beam by the tube windows; (3) avoid contamination of the radiation, as from tungsten deposited on the target face; and (4) incorporate features of shockproof and rayproof design.

Increased beam intensity is achieved in a variety of ways. The linear focal spot is an important improvement in this respect. The electron beam is brought to a line focus instead of to a point focus on the target face which is perpendicular to the axis of the tube. Typical dimensions of a line focus are 1.2 mm width and 12 mm length. When such a focal spot is viewed along its length at an angle of 6° to the target face, it gives a projected focal spot 1.2 mm square and thus provides two powerful beams approximately 180° apart. When the linear focal spot is viewed perpendicularly to its length, it is projected as a very narrow line focus which is useful with linear slits but is rather weak for pinhole slits. Some tubes have four windows 90° apart to take advantage of these four beams; others provide only two windows, end-on to the focal spot. The linear focal spot is achieved in electron tubes through the use of a linear filament and a focusing cup. In gas tubes the cathode was rather early made concave to bring the electrons to a point focus on the target; then Corey, Lagsdin, and Wyckoff showed that a vertical slot milled in the concave cathode would bring the electron beam to a line focus in the horizontal plane [12].

More powerful beams are obtained by designing tubes to pass greater amounts of energy. This is, however, one of the most difficult of design problems, since about 98 per cent of the energy of the high-speed electrons unavoidably goes into the generation of heat in the target. In the early tubes this heat was dissipated by making the target a massive block of metal, which frequently operated at almost a cherry-red temperature. Sometimes the heat is conducted to radiating fins outside the tube, but the most satisfactory cooling is obtained by circulating a stream of cold water through the target. Even with water cooling, the energy which the tube will safely carry is usually limited by the efficiency of conduction of heat from the front to the back of the target. The face may become hot even though the copper back of the target is kept cold by an adequate flow of water. If the face becomes too hot, pitting or melting will ensue, or the target metal may separate from the copper back. Still greater energy input is achievable in tubes with oscillating [13] or rotating targets [14–24] that provide a greater focal area, which by constantly changing brings relatively cool metal into the path of the electron beam. Such targets are the basic feature of high-energy diffraction tubes (Section 2-2.2E).

A further method for achieving greater beam intensity is by decreasing the absorption of the x-rays in passing through the walls of the tube. This is accomplished by providing modern tubes with especially transparent windows through which the beams are led. Such windows are particularly necessary for the wavelengths (> 1.50 Å) generated by copper and elements of smaller atomic number than copper, which are so necessary in many diffraction studies today. The most suitable window materials are composed of elements with as low an atomic number as possible. Lindemann glass, containing boron, lithium, and beryllium instead of the silicon, sodium, and calcium common in ordinary glass, was formerly widely used as a window material, but it tends to devitrify in a moist atmosphere and has too low a transmission for wavelengths longer than those of copper. Mica windows of 0.0005-in. thickness, backed up by thin beryllium foil to remove charges which may produce punctures, are presently used. Actually, the most satisfactory window material is thin sheet beryllium metal[25], which can be produced in malleable form through the introduction of 0.2 per cent of titanium. Since beryllium is a fair conductor of heat and electricity, these windows may be placed very close to the focal spot without becoming overheated and charged by secondary bombardment, resulting in an additional gain in intensity by permitting equipment to be brought closer to the x-ray source (the focal spot). The curves of Fig. 2-3 compare the transmission characteristics of these three window materials for the thinnest practicable windows of each material.

The elimination of contamination by vaporized tungsten in hot-cathode tubes is a difficult problem. Little can be done to prevent the slow vaporization of the tungsten filament, but the effect can be minimized by increasing the anode-cathode distance. The beryllium win-

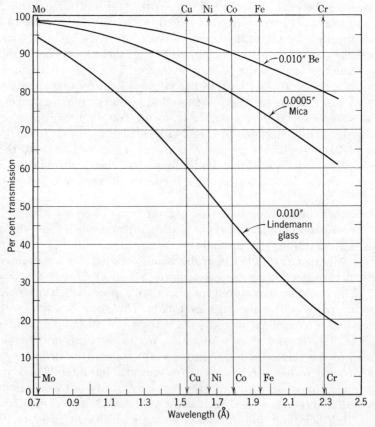

Fig. 2-3. Transmission of x-ray window materials for $K\alpha$ radiation.

dows, fortunately, can be recessed in the walls of the discharge chamber and thus be completely shielded from tungsten vapor.

Most manufacturers of sealed-off tubes provide them either with or without grounded shockproof shields with lead linings, and with shockproof cable leads. In many commercial diffraction units, rayproofing and shockproofing are achieved partly in the tube design and partly in the design of the tube housing and transformer cabinet of the diffraction unit itself. It is imperative, of course, that the individual who assembles his own diffraction setup provide the unit with adequate rayproofing and shockproofing. Several commercial diffraction tubes are depicted in Fig. 2-4, and various useful data on commercial tubes are provided in Table 2-1.

D. Cold-Cathode Diffraction Tubes.

The cold-cathode (gas) tube today is used chiefly in very specialized applications. These tubes were favored in the early days of diffraction because several radiations could be provided from the same tube, by interchange of targets, and because they could be serviced in the laboratory [26–30]. The modern cold-cathode tube is generally an all-metal tube, except for a glass or porcelain insulator between the cathode and anode ends. Since diffraction tubes are almost universally operated with the target at ground potential, the entire anode end of a gas tube, as well as the target, is provided with cooling water channels. A series of interchangeable targets is usually provided. The cathode typically extends well into the metal anode end of the tube and terminates in a replaceable concave aluminum cathode tip. The cathode rod usually requires some cooling by means of a stream of compressed air passed through its interior, or by a series of external fins. Windows of thin sheet beryllium are standard. The designing of modern gas tubes to be completely rayproof (except for the useful beams through the windows) is rather simple, since the cathode tip ordinarily extends well into the metal anode chamber. In a good design all radiation is effectively baffled by metal parts of the anode compartment and of the cathode rod.

Such gas tubes operate under a vacuum of approximately 0.01 mm of mercury, which is best maintained by a high-speed mechanical oil pump pumping against a controlled air leak. Fine-leak controls with day-to-day reproducible settings are available, and operational stability good enough to permit all-night operation without supervision is not difficult to attain. Many gas tubes are self-rectifying in operation, just as are the electron-type tubes. The mechanism of this rather unexpected property is, however, not entirely clear, so that designs for achieving it are more or less empirical. If the cold cathode of a gas tube is replaced

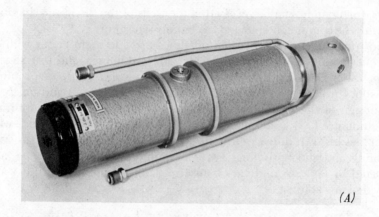

(A)

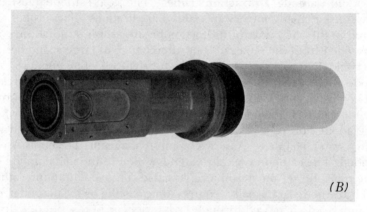

(B)

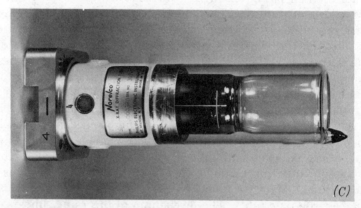

(C)

Fig. 2-4. Examples of commercial diffraction tubes: (A) Siemens type AG . . . 3ö Tube. (Courtesy of Siemens Corporation.) (*B*) Dunlee-Picker DZ-1B Tube. (Courtesy of Picker Nuclear.) (*C*) Norelco Diffraction Tube. (Courtesy of Philips Electronic Instruments.)

Table 2-1. Tabulated Characteristics for Several Commercial Diffraction Tubes

Make	Number of Windows	Window Material	Focal Spot (mm)	Filament Current (A)	Filament Voltage (V)	Maximum Voltage (peak kV)	Maximum Current at Maximum Voltage (mA)	Cooling Water
DZ-1B, Dunlee-Picker	Four (two spot foci, two line foci)	Be	0.75 × 15.0	2.1–2.9	3.2–8.5	50	Mo, 22 Cu, 20 Fe, 10 Cr, 16	5 pt/min
Y613-618 series, Enraf-Nonius	Four (two spot foci, two line foci)	Mica–Be	1 × 10	2.8–3.8	6.5–11.0	50	Mo, 24 Cu, 24 Fe, 10 Cr, 8	3.5 l/min
CA-8S, General Electric	Three (two spot foci, one line focus)	Be	0.8 × 12.5	2.5–4.5	3.0–9.0	50	Mo, 25 Cu, 20 Fe, 10 Cr, 16	8 pt/min
Type 140-00X-00, Norelco	Four (two spot foci, two line foci)	Mica–Be	1 × 10	3.3–3.7	6.9	60	Mo, 29 Cu, 29 Fe, 19 Cr, 29	3 qt/min
Type A Gq-3Ö, Siemens	Three (two spot foci, one line focus)	Be	0.75 × 10	4.1	5.6–7.3	50	Mo, 15 Cu, 15 Fe, 9 Cr, 9	4 l/min
AEG-F50, Syntex	Four (two spot foci, two line foci)	Be	1 × 10	3.8	11.0	60 55 55 55	Mo, 16.5 Cu, 18 Fe, 6.5 Cr, 5.5	3.5 l/min

with a suitable hot filament of tungsten mounted in a focusing cup, and a better vacuum is maintained by means of an oil diffusion pump, the cold-cathode tube becomes a demountable hot-cathode tube. Such an arrangement incorporates essentially all the advantages of both the gas tube and the electron-type tube [31–33].

Cold-cathode tubes must usually be built in a laboratory's own shop, hence, the interested reader is referred to the designs published in the literature [26–30]. Many additional details on cold-cathode tubes were presented in the earlier edition of this book [34].

E. High-Intensity Diffraction Tubes. Conventional sealed-off diffraction tubes provide an output in the general neighborhood of 15 to 40 mA at 40 kV. Exposure periods with such tubes in photographic recording of diffraction effects vary with the nature of the sample and the type of camera used, and may be anywhere between 15 min and 24 hr, with the average exposure probably in the range of 2 to 4 hr. Such prolonged exposures, however, may be unacceptable in certain types of studies. Metallic specimens being examined at elevated temperatures may recrystallize, oxidize, or volatilize. Proteinlike materials may be subject to dehydration or deterioration over relatively short periods of time. For instance, in the preparation of useful small-angle diffraction patterns of muscle, it is necessary to keep the exposures as short as possible so that the sample remains essentially in the *live* condition. One may also wish to follow a reaction in the solid state. Thus for many investigations it may be desirable, indeed even mandatory, to cut conventional exposure times by a factor of 25 or more. The answer to such problems is a high-intensity diffraction tube.

The simplest means for greatly increasing energy input, hence x-ray output, of an x-ray tube is to provide it with a rotating target [14–24]. Such a target (Fig. 2-5) provides a much greater focal spot area which, as a result of the rotation, constantly brings cool metal into the path of the electron beam. Whereas a stationary water-cooled copper target in in a conventional tube may permit a focal spot loading of 100 W/mm², a rotating copper-target tube with efficient water cooling may allow a loading of several thousand watts per square millimeter. The diameter of the rotating anode in these tubes may range from 89 to 400 mm, and the speed of rotation from 1000 to 3000 rpm.

In theory developed by Müller [35], the performance of a tube is raised by target rotation by a factor approximately proportional to $a\sqrt{bv}$, where $2a$ is the focal spot length, $2b$ is the focal spot width, and v is the speed of the target surface. For a constant focal spot size, increased loading is achieved only by increased target-surface speed. The latter is

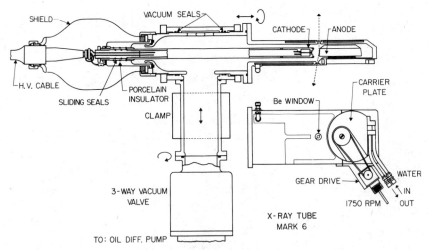

Fig. 2-5. Diagram of a rotating-target diffraction tube. (Courtesy of A. Taylor, *Advances in X-Ray Analysis*, Vol. 9, Plenum Press, 1966, and Westinghouse Electric Corporation.)

attained by increasing rotational speed of the target or by increasing the target diameter, or both. Obviously, mechanical difficulties limit the ultimate practical maximum target-surface speed attainable. Focal spot drift of a few microns is reported for some of these tubes. Characteristics of several rotating-target diffraction tubes are presented in Table 2-2.

F. Microfocus Diffraction Tubes. Special techniques often require extremely small and intense x-ray sources. Such small, intense sources, available from microfocus tubes, are used to study lattice inhomogeneities and dislocation networks in single and polycrystalline samples, to obtain back-reflection Laue patterns from single grains in a polycrystalline body, and to achieve high resolution at very small diffraction angles close to the central beam, as in protein crystallography, polymer chemistry, particle size, or precipitate studies. A microfocus tube is also a good x-ray source for small-angle scattering investigations, and can give improved line resolution in normal Debye-Scherrer powder patterns.

A typical microfocus tube is a hot-cathode continuously evacuated tube, often with replaceable targets of various metals. Focal spot sizes may vary from 10 to 50-μ circular focal areas, or be linear in ranges such as 100×1000 to 300×3000 μ. Focal spot loading can vary widely: 30,000 W/mm² (0.7 mA at 50 kV) on a 40-μ spot focus; 2000 W/mm² (4 mA at 50 kV) on a 100×1000 μ line focus. The focal spot-to-window distance may be as short as 10 mm, making it possible to place cameras close to the x-ray source and reduce exposure times. Operation and maintenance are

Table 2.2. Characteristics of Several Rotating-Target Diffraction Tubes[a]

	Maximum Voltage (kV)	Maximum Current (mA)	Focal Spot (mm)	Focal Spot Loading (W/mm²)	Target Diameter (mm)	Target Speed (rpm)	Cooling Water (l/min)
Delft[19]	40	500	1 × 10	2000	300	3000	15
Elliott[20]	50	80	0.2 × 2.0	4500	89	3000	9
General Electric-Rigaku[21]	60	100	0.5 × 5	2400	99	2500	> 6
Rigaku-Denki[22]	60	500	0.5 × 10	6000	400	1000	
Taylor (Westinghouse[23, 24])	30	275	1 × 10	825	127	1750	8.2

[a]All have copper targets.

simple—time to change a filament and return to full operation, as little as 10 min. Most purveyors of x-ray equipment list one or more well-engineered microfocus tubes.

2-2.3 Power Equipment for the Production of X-rays

Since most diffraction tubes operate self-rectifying, the power requirements for routine photographic diffraction work can be met with relatively simple electrical equipment. Precise intensity measurements, however, require more elaborate electrical installations varying with the method used. The electrical power source may be either a 100/130-V or a 200/250-V, 50/60-cycle, ac line, and the total load for most units is not over 6.0 kVA. The voltages used in x-ray diffraction are all under 60 kV peak. Hence a high-voltage transformer delivering 50 to 60 kV peak continuously at 30 to 40 mA will serve for most setups. The transformer should have one side of the secondary grounded, since most x-ray tubes are designed to be operated with the anode end at ground potential and with the target connected directly to laboratory water lines for cooling. The voltage applied to the primary of the high-tension transformer is best selected by a rheostat, stepped autotransformer, or continuously variable transformer of the Variac type. A voltmeter in the primary circuit can be calibrated to read directly in peak kilovolts. The x-ray tube filament can be heated by low-voltage current from a suitably insulated storage battery, but a more desirable method is to use a small stepdown filament transformer, often known as a Coolidge transformer, which has its secondary insulated for high voltage against the primary. This may be a small separate transformer, or it may be mounted in the same tank and oil as the high-voltage transformer. A rheostat or inductive type of control in the primary circuit of the filament transformer controls the current through the filament, and in turn the current through the x-ray tube. A milliammeter in the ground side of the tube circuit gives the current through the tube, and (with the necessary switches, fuses, etc.) completes the list of essential parts of the electrical setup. A simple circuit for a tube operating by self-rectification is diagramed in Fig. 2-6.

In addition to the electrical parts just listed, there are several very useful and desirable accessories which might be included in the control unit. For instance, a red tell-tale warning light wired into the circuit so as to be on at all times when x-rays are being produced is a most useful means for continuously warning the operator of the x-ray danger. An overload relay set to open the circuits when the tube current exceeds normal operating values by a predetermined amount is a useful protective device for parts of the equipment. There should always be a

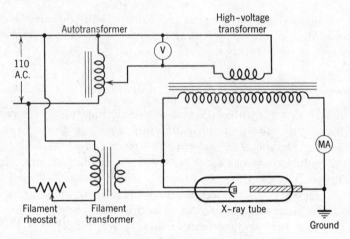

Fig. 2-6.   Simple circuit for tube operating by self-rectification.

pressure switch in the water line from the tube to the drain to open the circuits when the flow of water to the tube is insufficient to cool the tube. Such a switch prevents the tube from being started until after the cooling water is turned on and flowing at the proper rate. An electric clock or exposure timer and a time totalizer in the x-ray circuit are very useful. The latter makes it easy to keep an accurate record of tube life, and if it reads to 0.1 hr can be used as an exposure timer. Exposure timing devices which will shut off the unit at the end of an exposure are very handy. They permit the operator to leave exposures running at night, to be shut off automatically when completed. A magnetic switch, which energizes the equipment except for impressing the high potential on the tube is desirable. Any failure of the power supply, or the opening of the water-pressure switch, opens the magnetic switch which may be of the type that must be manually closed again for further operation. Modern commercial assemblies provide most of the foregoing (as well as other) features, sometimes with time-delay relays to keep the number of control switches a minimum.

Another feature which is nearly universal in commercial assemblies, and which is most desirable in any installation is line voltage stabilization ahead of the unit. Line voltage variations, of course, lead to corresponding variations in the high voltage applied to the x-ray tube. More serious, however, is the effect of such fluctuations on the tube current. Tube current is dependent upon the output of electrons from the hot tungsten filament, which output is strongly influenced by slight changes in the filament temperature. Thus relatively small voltage variations causing small filament temperature changes can bring about somewhat more serious tube current changes, which may endanger the life of a tube being operated at near full rating. A stabilized power source, moreover, makes it easier to set up a uniform exposure technique.

The stabilizer may be of the constant-voltage transformer type which usually provides a constant output voltage to ±1 per cent for input variation up to ±15 per cent. More desirable, however, is a solid-state electronic type of stabilizer which may control the output voltage to ±0.1 per cent and operate satisfactorily over a range of 0.1 to full load. The stabilizer must be large enough to handle the maximum power demand of the unit. For the most precise control, an isolation transformer is often added to the generator components. When placed ahead of the unit, it filters out high-frequency transient pulses which the line voltage stabilizer cannot handle.

A hot-cathode x-ray tube operates unidirectionally (self-rectifying) on an alternating high potential, because the target is cold and produces no electrons to carry the current across the tube during the half cycle when it is negative and the cathode positive. During the next half cycle, when the cathode is negative and the target positive, electrons from the hot filament are propelled across to the target and generate x-rays. The tube thus recifies the alternating high voltage applied to it, resulting in a pulsating direct current which records as a steady current on the dc milliammeter because of the milliammeter's sluggish action. The little-understood self-rectifying property of many gas tubes has already been mentioned.

The efficiency of tube operation in self-rectification, and in half-wave rectification, is not particularly good, as evidenced by the schematic waveform curve in Fig. 2-7. Actually, only the shaded portion of the useful half cycle produces x-rays, while the clear portions simply pass current at low voltage and contribute to the heating of the target. The

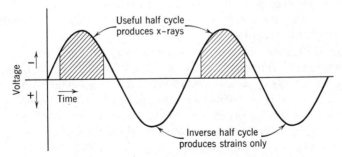

Fig. 2-7. Efficiency of tube operation in self-rectification (schematic).

inverse half cycle not only produces no x-rays, but also subjects the tube to undesirable strains. For the moderate voltages used in diffraction techniques, these strains are not so serious but that self-rectification is a satisfactory setup where economy is important. Rectification, however, is generally considered to contribute to the performance and stability of diffraction tubes, and most commercial units are provided with full-wave rectification.

Before the development of solid-state electronics, rectification was achieved through the use of rectifier or valve tubes (formerly known as Kenotrons). A rectifier tube is an electron tube operating on the same principle as an x-ray tube. Current passes in one direction, since its hot filament is the only source of electrons. The filament is larger and more rugged than an x-ray tube filament, and capable of passing several hundred milliamperes. The potential drop across the tube is normally only a few hundred volts, but if too low a filament current is applied the voltage drop may increase to a value causing the tube to emit x-rays. Such improper operation is to be avoided, both because of possible danger to the operator and the resulting shortened valve tube life. The rectifier tube during the inverse cycle protects the x-ray tube by sharing the inverse potential, or by assuming the entire load, depending on the particular circuit involved, and in some circuits the valve tube may have to withstand an inverse voltage of twice the x-ray tube voltage. It is common practice to place valve tubes in oil inside the case with the high-tension transformer, thus providing protection from any x-rays generated, and eliminating trouble from dust and moisture.

Figure 2-8 illustrates a full-wave target-grounded circuit suitable for camera x-ray diffraction work. In the circuit shown the valve tubes (or rectifiers) must withstand the full transformer voltage, which is twice the x-ray tube voltage. Without a condenser the full-wave potential pulsates, but with the proper-sized condenser (often mounted in the tank with the transformer, *B* in Fig. 2-9) the potential to the x-ray tube is maintained at close to its maximum value with only a small "ripple." The condenser indicated is helpful but not absolutely necessary for camera diffraction studies.

Many diffraction investigations, however, require highly accurate intensity measurements. Such work demands constant-potential direct current (cpdc). The circuit of Fig. 2-8 with good line voltage stabilization and a condenser of sufficient capacity to reduce the ripple to less than 0.1 per cent may then be entirely satisfactory. For a given tube current, the condenser can be smaller if the frequency is increased. Thus for severe cpdc demands it may be best to use the line voltage to drive a synchronous motor coupled to a generator. This will provide

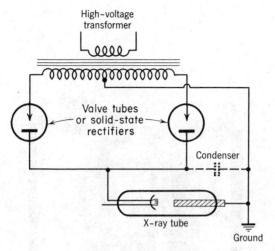

Fig. 2-8. Full-wave circuit with two valve tubes or solid-state rectifiers.

any primary voltage stabilization necessary, and the generator may be designed to furnish its output at 600 to 2000 cycles. Full-wave rectification in the high-tension circuit with suitable high-tension condensers will then provide a very constant output. The waveform for cpdc units may look something like Fig. 2-10.

It is not at all difficult to assemble a reliable x-ray generator from reconditioned medical or dental x-ray transformers and a suitable diffraction tube, and where strict economy is necessary this is the only solution. For further information on x-ray tubes and circuits the reader is referred to x-ray equipment manufacturers' pamphlets and to the numerous publications on the subject [36].

2-2.4 Commercial X-ray Generators for Diffraction

Commercial assemblies offer an immediately available setup, with the advantages of years of engineering in the design and construction of diffraction units. For instance, modern x-ray generators for diffraction (Fig. 2-9) frequently achieve rectification and high stability through solid-state circuitry. Among the acknowledged advantages of solid-state devices are high reliability, low temperature in critical control circuitry, fast warm-up, integrated circuits giving fast stabilization response, and elimination of radiation often present with tube-type rectifiers. A typical high-voltage generator may use a Graetz-bridge circuit with four silicon rectifiers, or a voltage-doubling circuit with silicon or selenium diodes. The high voltage and tube current may each be stabil-

Fig. 2-9. Simple, modern high-voltage power supply (removed from tank). *A*, Solid-state (selenium) rectifiers; *B*, capacitors to reduce ripple; *C*, high-voltage transformer coils. (Courtesy of Picker Nuclear.)

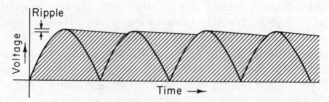

Fig. 2-10. Waveform for full-wave rectified voltage with ripple suppression (schematic).

78

ized to 0.03 per cent, or even to 0.01 per cent, by means of transducers in transistorized circuits. An excess-high-voltage safety circuit is often incorporated to protect the generator and x-ray tube.

Such commercial units are usually built around a two-, three-, or four-window tube which is mounted on, or in, a table in such a manner that the table serves as a base for camera supports in front of the various windows. The table is a planed iron surface in order to prevent loss of alignment of the cameras with humidity changes. Suitable tube housings and shields, including safety covers for the tube windows, are built into the assembly to make it rayproof. Shockproof design may be achieved by placing the high-tension transformer and other control equipment in the space beneath the table, which forms a completely enclosed cabinet with convenient access doors provided with safety switches. Alternatively, the transformer and controls may be in a separate cabinet with shockproof cables to the x-ray tube. A control panel with switches, control knobs, and instruments is arranged in one wall of the cabinet. Electrical lines, water connections, and similar services may be brought into the cabinet through the floor and thus prevent the laboratory from being cluttered with wires and tubes. Some of the additional features commonly available in a commercial unit are: stabilization of input power; water cooling of the high-tension trans-former; convenient tube interchange — completed in as little as 5 min; stepless kilovolt and milliampere controls with direct-reading meters for same; relay control to simplify operation; a main switch operated by a key; an instrument-panel light and indicator lights to show the condition of circuits and rectifier tubes and to warn of x-rays; filters in the water line, and pressure regulator; an automatic solenoid valve in the water line; a pressure gauge and a pressure switch for protecting the tube against too little or too great flow of cooling water; filter selectors on beam ports; exposure timers; and power outlets for auxiliary equip-ment. Typical commercial x-ray diffraction generators are depicted in Figs. 2-11, 5-5C, and 5-5D.

2-2.5 Isotopic X-ray Sources

Useful x-radiation for diffraction purposes can be obtained from the decay of radioactive isotopes. For instance, iron 55 decays by electron capture to the ground state of the stable isotope manganese 55. At the same time, essentially the characteristic spectrum of manganese is produced with $K\alpha_1 = 2.1$ Å. In comparison with the usual x-ray generator, such a source has the advantages of small weight (a few grams) and size (a few millimeters), no power consumption, no need for cooling, extreme

Fig. 2-11. Picker constant-potential two-tube generator (with accessories). (Courtesy of Picker Nuclear.)

stability, and of course unlimited portability. One almost crucial difficulty is the much lower intensity available from isotopic sources. Iron 55 sources of sufficient activity have been developed to produce a diffraction pattern with lithium fluoride (LiF) powder in 20 hr in a 2-in.-diameter Debye-Scherrer camera. Other serious problems are poor resolution and poor ratio of line intensity to background. For very special investigations, however, isotopic sources are worthy of consideration. Toothacker and Preuss[87] have reviewed the subject and present a list of 18 references.

2-3 PROPERTIES OF X-RAYS AND THEIR MEASUREMENT

2-3.1 The X-ray Spectrum of an Element

From the brief discussion of the origin of x-rays (Section 2-2.1) it is evident that they are to be considered electromagnetic in nature, and thus a part of the *electromagnetic spectrum*. Like ultraviolet, visible, and infrared radiations, x-rays exhibit a distinctly dual nature, behaving under some conditions as waves and under others as particles. The major interest of this book, the phenomenon of x-ray diffraction, is largely concerned with their wavelike character.

Crystals act as diffraction gratings (Chapter 3) for analyzing x-ray spectra and measuring their wavelengths. If one uses as a crystal grating a face of a cubic crystal of rock salt, natural sodium chloride (NaCl), and obtains the spectrum of the radiation from a tungsten-target tube operating at about 55 kV peak, the x-ray spectrogram will resemble the photograph in Fig. 2-12. The spectrum is seen to consist of two parts, a

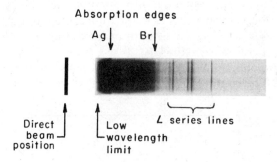

Fig. 2-12. X-ray spectrum from a tungsten-target tube operating at about 55 kV peak.

continuous spectrum composed of a wide band of wavelengths, and a line spectrum of intense single wavelengths superimposed on the continuous background. The former is analogous to white light in the visible region, and is frequently referred to as *white* or *general radiation*. Each line of the line spectrum corresponds to monochromatic light; and since the various wavelengths are characteristic of the target material, it is termed the *characteristic radiation*. Usually both types of radiation are present in the spectrum from a tube target, but at lower voltages the characteristic radiation excited may be of such long wavelengths as to be completely absorbed by the walls and windows of the tube.

The unit of wavelength commonly used in x-ray measurements is the ångström (abbreviated Å), defined as 10^{-8} cm. The x-rays of interest to diffraction workers occur in the range 0.2 to 2.5 Å. The x-ray region of the spectrum, however, covers the range 0.10 Å ($K\beta_2$ for U $= 0.1086$ Å) to nearly 700 Å ($M_{II,III}N_I$ for K $= 692$ Å). For purposes of orientation it is appropriate to indicate the overlap of the long-wavelength end of the x-ray region with the short-wavelength end of the ultraviolet region whose range is commonly given as 100 to 4000 Å. The visible region in turn extends from 4000 to 7700 Å. The terms *soft* and *hard* are sometimes used to describe the penetrability of x-rays. Actually, *hardness* is a relative measure of wavelength and, other things being equal, the harder the x-rays the shorter the wavelength.

A. The Continuous X-ray Spectrum. The distribution of energy in the continuous spectrum of an element can be studied by measuring the x-ray intensity at several wavelengths. When this is done at several applied voltages, a series of curves similar to those for tungsten (Fig. 2-13) results, and several important features of such spectra are immediately evident. The wavelength limits of the radiation and the distribution of intensity within are determined by the magnitude of the applied voltage. With increasing potential impressed on the tube, the minimum wavelength progressively shifts to lower values, the intensity at all wavelengths increases, and the wavelength of maximum intensity decreases (peaks of the curves move to the left).

The definite short-wavelength limit is in agreement with quantum expectations. The continuous spectrum results from the deceleration of

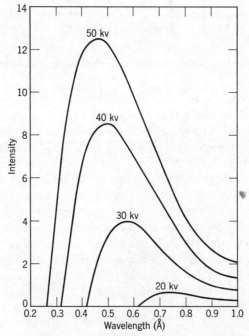

Fig. 2-13. Intensity distribution with wavelength in the continuous x-ray spectrum of tungsten at several voltages. [Courtesy of C. Ulrey, *Phys. Rev.*, **11**, 401 (1918).]

the high-energy electrons by the electric fields of the target atoms, each successive decrease in an electron's energy ΔE appearing as an x-ray photon of frequency v, according to the relation $\Delta E = hv$. There is a very small, but finite, probability that an electron in a single encounter with a target atom may lose all of its kinetic energy. When this happens, the energy of the photon released is equal to the energy E of the electron. The energy of the fastest-moving electrons striking the target is

$$E = Ve, \qquad (2\text{-}3)$$

where V is the applied voltage and e is the charge on the electron $(4.803 \times 10^{-10} \, \text{esu} = 1.602 \times 10^{-20} \, \text{emu})$. The frequency of the excited rays is then related to the energy by

$$E = Ve = hv, \qquad (2\text{-}4)$$

and since

$$v = \frac{c}{\lambda}, \qquad (2\text{-}5)$$

where c is the speed of light $(2.998 \times 10^{10} \, \text{cm/sec})$ and λ is the wavelength,

$$\lambda = \frac{hc}{Ve}. \qquad (2\text{-}6)$$

In calculating λ by this expression, V and e must both be expressed in similar units. If e is in electrostatic units, V must be in the same units $(1 \, \text{esu of potential} = 300 \, \text{V})$, and λ will be given in centimeters. If V is expressed in volts and e in electromagnetic units, the value of λ will be given in ångström units. The latest value of Planck's constant h is $6.6255 \times 10^{-27} \, \text{erg-sec}$. For convenience in calculations the values of the various constants in the equation may be combined into a single numerical value. Thus

$$\lambda = \frac{6.6255 \times 10^{-27} \times 2.998 \times 10^{10}}{V \times 1.602 \times 10^{-20}} = \frac{12{,}399}{V}, \qquad (2\text{-}7)$$

in which λ is given in ångström units when V is expressed in volts. If the tube voltage is pulsating between limits, the peak value and not the root-mean-square value must be used in this expression. Thus the short-wavelength limit, known also as the "Duane-Hunt[37] limit," for a tube operating at 40 kV peak is

$$\lambda_{min} = \frac{12{,}399}{40{,}000} = 0.31 \, \text{Å}.$$

The data of Fig. 2-13, as well as accurate measurements of the short-wavelength limit on spectral photographs, confirm this result.

B. The Characteristic X-ray Spectrum. If the voltage applied to the tube target is high enough to excite its harder characteristic radiation, the lines will be found superposed on the curve of the continuous radiation, as in Fig. 2-14 for molybdenum. Except for the intense peaks at 0.63 and 0.71 Å, this curve is a typical continuous radiation curve, although its intensity at all wavelengths is decreased as compared with tungsten under similar conditions (Fig. 2-13). The two powerful peaks at the positions indicated are characteristic of the element molybdenum. Similar peaks characteristic of tungsten appear on the tungsten spectrum at $\lambda = 0.18$ and 0.21 Å if the voltage is raised above 70 kV. This accounts for their absence in Fig. 2-13. Moreover, the molybdenum peaks disappear at voltages below 20 kV, just as they do with tungsten below 70 kV. This series of x-ray lines and also a series of longer wavelength were discovered by Barkla[38] who named them the "K series" and the "L series," respectively, by analogy to optical spectra. Actually, Barkla's identification of these radiations relied upon absorption measurements

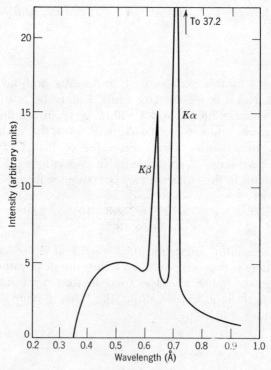

Fig. 2-14. Intensity curve for x-rays from a molybdenum target operated at 35 kV peak. [Courtesy of C. Ulrey, *Phys. Rev.*, **11**, 401 (1918).]

which naturally did not resolve the individual wavelengths, and the discovery of the monochromatic components of the characteristic x-ray spectrum was made by Bragg, using rhodium and platinum radiations, a sodium chloride crystal, and an ionization spectrometer[39]. Shortly thereafter, Moseley[40] made the first systematic investigation of these two series of spectra. Later two additional series, N and M, with wavelengths so long that they can be studied only with a vacuum spectrograph, were discovered. Thus for a given element the wavelengths of its four series of spectra vary as follows: $N > M > L > K$.

Using successively 38 different elements as the target of an x-ray tube, Moseley studied their K and L series spectra. He observed that each series consisted of relatively few lines. In each series the same lines occurred for the various elements with progressively altered wavelengths. Figure 2-15 reproduces Moseley's photographs of the K series

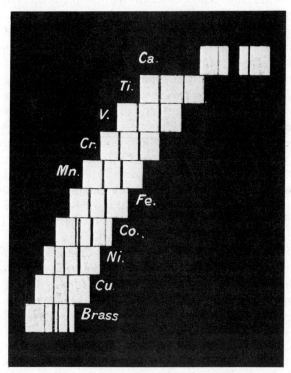

⟶ Increasing Wave Length

Fig. 2-15.   Moseley's photographs of the K series spectra of the elements calcium to copper in the periodic table. (G. Kaye, *X-rays*, 1923; courtesy of Longmans, Green.)

spectra of the elements calcium to copper in the periodic arrangement. The regularity of the diagram is most striking. We can consider the wavelengths of the various lines to be nearly proportional to their distances from the left-hand edge of the figure, thus qualitatively establishing the progressive change in wavelength with position in the periodic table. Although scandium was unavailable for Moseley's studies, its position between calcium and titanium could hardly be more strongly confirmed than by the large gap between the spectra of the last two.

Careful examination of these spectra revealed that the square root of the frequency of either of the two lines is nearly proportional to the atomic number of the target element. The exact relationship, known as Moseley's law, is

$$\nu^{1/2} = K(Z - \sigma) \tag{2-8}$$

where K is a universal constant for all elements, Z is the atomic number, and σ is another universal constant. With appropriate changes in K and σ, the expression also holds for the L, M, and N series lines.

Actually, the two lines of Moseley's K spectra are both doublets, so close that his spectrograph did not resolve them. It is customary to designate the various lines of a series by means of Greek letters with numerical subscripts. The more intense and longer-wavelength K series line is the $K\alpha$ line, and the two components of the doublet are designated $K\alpha_1$ and $K\alpha_2$. The intensity ratio of these two lines for all elements is very close to $K\alpha_2 : K\alpha_1 = 1:2$. The less intense and lower-wavelength line of Moseley's spectra is the $K\beta$ line, whose two components are $K\beta_3$ and $K\beta_1$. A weaker line, now known as $K\beta_2$, was not present on Moseley's photographs. It likewise is a doublet, usually unresolved, and it appears in the K spectrum of elements with atomic number above about 29. For molybdenum the intensity ratio of the unresolved doublets is $K\beta : K\alpha = 1:5.4$, and for targets of high atomic number the intensity ratios $K\alpha_2 : K\alpha_1 : (K\beta_3 + K\beta_1) : K\beta_2$ approximate $50:100:35:15$.

The quantum origin of the characteristic radiation is evident from the fact that the lines of a series appear only when the impressed voltage exceeds a definite minimum value, for the K series 20.0 kV for a molybdenum target and 69.5 kV for a tungsten target. These minimum voltages correspond to the energy required to eject an electron from the K shells of the respective atoms, leaving them in an excited state. The wavelength corresponding to this minimum voltage is frequently designated the *quantum wavelength*, and it is always slightly shorter than the wavelength of the shortest line of the series. After excitation of the K shell, an electron from an outer shell may fall into the vacancy in the K shell, accompanied by the simultaneous emission of a characteristic x-ray, in this case one of the lines of the K series. If an L-shell electron fills the K-shell vacancy, the emitted line will be $K\alpha$ radiation; and if a K-shell vacancy is filled by an M electron, the $K\beta$ line is emitted. Similarly, such electronic transitions from outer shells to vacancies in the L shell of atoms lead to the production of the lines of the L series.

An atom may also be put into an excited state by the absorption of an x-ray photon (quantum). If the energy of the photon is sufficiently high, it will eject an electron from one of the shells of the absorbing atom, and the same characteristic radiation is emitted as in excitation by high-speed electrons. For a given shell to be excited, the energy of the x-ray quantum, as calculated by the relation $E = h\nu_{x\text{-ray}}$, must equal or exceed that corresponding to the quantum wavelength for the shell and series. Accordingly, K-shell absorption results in K-shell excitation and the subsequent emission of K radiation. An abrupt change in absorption occurs when the wavelength of the incident radiation equals the quantum wavelength, and this discontinuity in absorption is called the *absorption edge* or *critical absorption wavelength*. The latter terms are used synonymously with the term quantum wavelength in the field of x-ray spectra.

These quantized absorption and emission processes are readily shown schematically in energy-level diagrams (Fig. 2-16). Modern quantum mechanics has revealed the finer details of atomic structure, and has

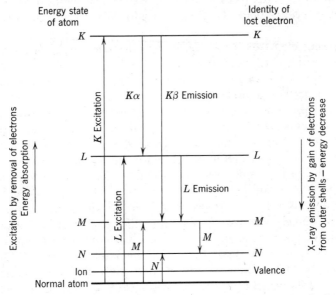

Fig. 2-16. Schematic energy-level diagram for a many-electron atom, indicating (by arrows) the processes of excitation and emission.

therefore evolved more detailed and complete energy-level diagrams such as Fig. 2-17 for uranium. All shells outside the K shell are seen to possess several energy levels, resulting in the more complex spectra of the L and M series. Since the critical excitation potentials for the energy levels of the shells decrease with increasing distance of the electrons from the nucleus, the various series appear in the following order with the application of increasing potential to the tube:

1. N series (in several sections)
2. M series (in five sections)
3. L series (in three sections)
4. K series (all lines appear simultaneously)

It is evident of course that atoms which possess no electrons in the M and outer shells cannot have an L series spectrum, but only a K series; and similarly no one of the x-ray series can be present unless the atom possesses electrons in shells beyond the ionized shell.

This simple Bohr-Kossel-Barkla electronic shell explanation, all developed before 1916, is still considered a fundamentally correct picture of the origin of $K\alpha$ radiation. Complete discussion of the more complex details of x-ray spectra, however, requires the introduction of the modern quantum mechanics, and is beyond the purpose of this book. The interested investigator may pursue these details further in the excellent reference books by Compton and Allison[41] and by Sproull [36].

The target elements most useful for diffraction purposes are those whose $K\alpha$ radiation falls in the wavelength range from 0.56 to 2.29 Å; these are silver, molybdenum, copper, nickel, cobalt, iron, and chromium. The K radiation of elements of lower atomic number is too readily absorbed by the tube windows, and even in air, to be useful, whereas the heavier elements produce such intense continuous radiation as to render their line radiation of little value. Table 2-3 lists the wavelengths of several K lines of the targets must useful for diffraction work.

Although the excitation potentials listed in Table 2-3 are sufficient to excite the K spectra of the respective elements, in practice higher voltages are always used. It was found empirically that, for optimum intensity of the characteristic radiation relative to that of the continuous radiation, an applied voltage about 3.5 times the critical excitation voltage voltage is required. A theoretical study by Witty and Wood[42] concludes that this factor should be 4 for dc potentials and 5 for ac potentials. Actually, the shapes of their curves are such as to indicate that any factor between 3.5 and 5 that is practicable for use will be suitable. Furthermore, the intensity of the characteristic radiation increases with the

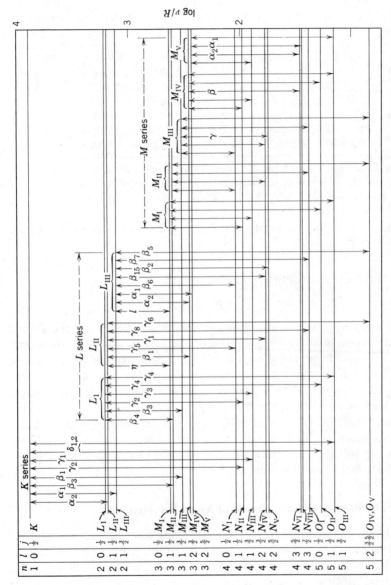

Fig. 2-17. X-ray energy-level diagram for uranium 92. (By permission, from F. Richtmyer and E. Kennard, *Introduction to Modern Physics*, Copyright, 1947, McGraw-Hill.)

89

Table 2-3.   X-ray Wavelengths Most Useful in Diffraction Studies[a]

Element	$K\alpha_1$ (Å)	$K\alpha_2$ (Å)	Un-resolved[b] $K\alpha$ (Å)	$K\beta_1$ (Å)	K Absorption Edge (Å)	Excitation Potential (kV)
Ag	0.55941	0.56380	0.56084	0.49707	0.4859	25.52
Mo	0.70930	0.71359	0.71073	0.63229	0.6198	20.00
Cu	1.54056	1.54439	1.54184	1.39222	1.3806	8.98
Ni	1.65791	1.66175	1.65919	1.50014	1.4881	8.33
Co	1.78897	1.79285	1.79026	1.62079	1.6082	7.71
Fe	1.93604	1.93998	1.93735	1.75661	1.7435	7.11
Cr	2.28970	2.29361	2.29100	2.08487	2.0702	5.99

[a]These values are taken from Bearden[43] in which they were listed on a re-adjusted scale of Å* units based on $\lambda(WK\alpha_1) = 0.2090100$ Å*. Since 1 Å* = 1 Å to ±5 ppm (probable error), values in this table are designated as being in ångström units.

[b]These values are the customary weighted mean of $K\alpha_1$ and $K\alpha_2$, $K\alpha_1$ being given twice the weight of $K\alpha_2$.

applied voltage according to the relation[44]

$$I = Ai(V - V_k)^{1.5},$$ (2-9)

where i is the tube current, V is the voltage applied, V_k is the critical excitation potential for the material of the target, and A is a constant. Thus we should operate as near to $5V_k$ as possible, and at as high a tube current as is consistent with the energy rating of the tube. The tube load (in watts) is readily calculated for comparison with its rating (commonly expressed in watts) by taking the product of the applied voltage in kilovolts and the tube current in milliamperes. Operation at slightly under the maximum rating is usually conducive to longer tube life. Suitable operating voltages are: molybdenum, 50 to 55 kV; copper, 35 to 40 kV; iron, 25 to 30 kV; and chromium, 25 kV.

2-3.2 The Precise Determination of X-ray Wavelengths

Shortly after Laue's discovery, Bragg[45] gave a very simple geometrical interpretation of diffraction by a crystal grating. Using an analogy to specular reflection, he showed (Section 3-2.7) that the conditions for a "reflected" (diffracted) beam are given by the relation

$$n\lambda = 2d \sin \theta,$$ (2-10)

where n is an integer, the "order of the reflection," λ is the wavelength of the x-rays, d the interplanar spacing between successive atomic planes in the crystal, and θ the angle between the atomic plane and both the incident and reflected beams. This fundamental relation is known as the Bragg equation or Bragg law.

If a crystal is mounted as in Fig. 2-18 with a face on the axis of a rotating or oscillating table, and a narrow beam of monochromatic x-rays, defined by the lead slits S_1 and S_2, strikes the face, the conditions of equation 2-10 will be met at certain definite values of the angle θ, and the

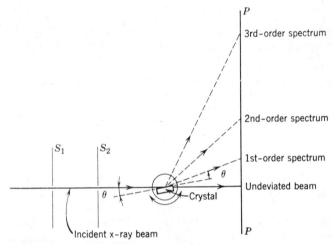

Fig. 2-18.   Production of a spectrum from a face of a single crystal.

beam will produce a spectrum of several orders on a photographic plate P. An ionization chamber was substituted for the photographic plate by Bragg, and with such an ionization spectrometer he discovered the monochromatic characteristic x-ray spectrum[39]. Measurements of the positions of these diffracted beams yield data on only n and θ of equation 2-10. Not until a structure determination had been made on one crystal was it possible to evaluate d and make a precise absolute determination of x-ray wavelengths. Structure determinations of the alkali halides, sodium chloride (NaCl) and potassium chloride (KCl), however, were soon accomplished by Bragg[46] through a brilliant consideration of their symmetry and diffraction intensities. This structure analysis provided the value of 2.81 Å for the d spacing of the cube face of rock salt, and permitted Bragg to make the first absolute x-ray wavelength determination, 1.10 Å for the strongest line in his platinum spectrum. This was apparently the entire $L\beta$ group whose most prominent lines, $L\beta_1$ and $L\beta_2$, now have the values 1.11984 and 1.10196 Å, respectively.

Soon it became evident that relative measurements of x-ray wavelengths could be made with a precision much greater than the best measurements of Avogadro's number N and the crystal density ρ, both of which enter into the calculation of the $d_{(200)}$ spacing of NaCl from symmetry considerations alone. At this point, therefore, a new unit of length, designated the "X unit" or "X.U.," was adopted, which was defined precisely by assuming the grating space of NaCl at 18°C to be *exactly* 2814.00 X.U.* The unit was thus *very nearly* equal to 1×10^{-11} cm. Shortly thereafter, it was realized that calcite (CaCO$_3$) formed much more perfect crystals than NaCl, and was, accordingly, a more desirable crystal for precise wavelength measurements. Its grating space and that of others used in x-ray spectroscopy are listed in Table 2-4.

Table 2-4.   Grating Spaces of the Principal Crystals Used in X-ray Spectroscopy[a]

	Grating Space Siegbahn, X.U.	d at 18°C, Corrected, (Å)	Change in d per °C (Å)
Sodium chloride (NaCl)	2814.00	2.81968	0.00011
Calcite (CaCO$_3$)	3029.45	3.03557	0.00003
Quartz (SiO$_2$)	4246.02	4.25460	0.00004
Gypsum (CaSO$_4$·2H$_2$O)	7584.70	7.60002	0.00029
Mica	9942.72	9.96280	0.00015

[a]The Siegbahn values are taken from reference [47].

Foremost in the field of x-ray spectroscopy for many years was Siegbahn [47], whose grating-space values are reported in Table 2-4. Details of his precision wavelength determinations are not presented here, but the interested reader is referred to his outstanding book on the subject. The immense amount of precise experimental work and critical study that has since gone into our present wavelength values is evident from Bearden's review [43] and its accompanying bibliography.

The Bragg equation, 2-10, limits the maximum wavelength that can be measured with a given crystal to $2d$, since sin θ cannot exceed unity. To overcome this difficulty, organic crystals of large grating space (sugar, $d = 10.57$ Å; lead melissate, $d = 87.5$ Å) have occasionally been used for very long-wavelength measurements. Absolute x-ray wavelength measurements, however, can also be made by means of ruled reflection gratings similar to those used in the visible region of the

*A larger unit, the kX.U. = 1000 X.U., has also been widely used.

spectrum, and there is no upper limit to the wavelengths measurable by such gratings. Early experiments of this type failed, because it was not realized that the x-rays must graze the surface at a very sharp angle. Within an angle of 0.5° or less, it is possible to use a ruled grating. At first investigators thought such gratings would have to have a prohibitively large number of lines (approximately 100,000,000/cm) in order to produce spectra by diffraction. Actually, just the opposite is true, and the grating space is relatively coarse (approximately 200 lines/ cm). The diffraction formula for such a grating is

$$n\lambda = d(\cos\theta - \cos\phi), \tag{2-11}$$

where n is the order of the spectrum, d is the grating space, θ is the glancing angle of incidence, and ϕ is the glancing angle of diffraction. With a grating of 212 lines/cm, $d = 0.0043$ cm, the first order of $MoK\alpha$ with $\lambda = 0.71$ Å will have values of θ and ϕ equal to 5 and 8 minutes, respectively. The first measurements of this type were made by Compton and Doan [48]. Since then, refinements in technique have made possible grating wavelength measurements of five-figure accuracy. For example, Bearden's[49] determination of the $K\alpha_1$ line of copper from ruled grating measurements yielded $\lambda = 1.5406$ Å. The present accepted value (Table 2-3) is 1.54056 Å.

With increasing precision of grating wavelengths, discrepancies began to appear between them and the crystal-determined wavelengths of Siegbahn. These discrepancies were a source of great concern to diffraction workers, and pointed to a very definite nonequivalence of the ångström unit and the kX unit, the difference being about 0.2 per cent. The discrepancy was finally traced to a small error in Avogadro's number. Accordingly, values of wavelengths and lattice constants expressed in kX units should be multiplied by a factor 1.002056 to convert them to absolute metric or ångström units[43]. Although this correction is important only when measurements have been made with a precision of a few parts in 100,000, confusion may be avoided by expressing all such data in true absolute units. The corrected grating-space values of Table 2-4, all precise wavelengths (Table 2-3), and lattice constant data throughout this book are expressed in true ångström units.*

---

*It should be pointed out here that attempts to attain the very highest precision in x-ray diffraction measurements have directed attention to measuring the centroid of the diffraction line rather than the peak of the line. Thus for complete consistency centroid wavelengths rather than peak wavelengths should be used with such techniques. Almost all wavelength measurements, however, have been based on peak measurements. The reader is referred to Section 5-2.2C for a consideration of centroid measurements and the efforts in this direction.

2-3.3 Absorption of X-rays

The ability of x-rays to penetrate opaque substances is their most striking property. Nevertheless, all x-rays are absorbed to some extent in passing through matter. Neglecting for the present the nature of this absorption, we note that x-ray absorption follows the same law as ordinary light traversing an imperfectly transparent medium; that is, for an incident beam of monochromatic x-rays, the fraction absorbed is the same for equal thicknesses of the absorbing material. This behavior is expressed by the well-known equation,

$$I = I_0 e^{-\mu x}, \tag{2-12}$$

where I is the intensity transmitted, I_0 is the original beam intensity, e is the Napierian base, χ is the thickness of the absorbing layer in centimeters, and μ is the linear absorption coefficient of the material for x-rays, a constant for a given wavelength. Equation 2-12 can be put into a more convenient form, in terms of mass traversed rather than thickness, by multiplying and dividing the exponent by the density ρ of the material, thus:

$$I = I_0 e^{-(\mu/\rho)\rho x}. \tag{2-13}$$

The quantity μ/ρ is another constant of the absorber, the mass absorption coefficient, and it possesses the very useful property of being independent of the physical and chemical state of the material, in contrast to μ, which is not. Thus the linear absorption coefficient for a given x-ray beam is much greater in water than in steam or in a stoichiometric mixture of oxygen and hydrogen, whereas the mass absorption coefficient is the same for all three. This remarkable property of the mass absorption coefficient sharply distinguishes x-rays from visible light. For example, diamonds are very transparent to light, whereas carbon as graphite is a strong absorber; both, however, have the same mass absorption coefficient for x-rays. Liquid and solid mercury are opaque to light, but mercury vapor is almost perfectly transparent; all three forms of course have the same μ/ρ.

When the absorbing material is a chemical compound, alloy, or solid solution instead of a single element, its mass absorption coefficient is easily calculated from those of its constituent elements and the composition. A calculation of μ/ρ for the specific case of cupric oxide (CuO) serves as an illustration. For $\lambda = 0.71$ Å (Mo$K\alpha$) the mass absorption coefficients of copper and oxygen are 50.9 and 1.31, respectively (see Appendix V). The atomic weights of copper and oxygen are 63.57 and 16.00. Thus CuO is $63.57/(63.57 + 16.00) = 0.80$ Cu, and $16.00/(63.57 + 16.00) = 0.20$ O. Therefore,

$$\left(\frac{\mu}{\rho}\right)_{CuO} = 0.80\left(\frac{\mu}{\rho}\right)_{Cu} + 0.20\left(\frac{\mu}{\rho}\right)_{O} = (0.8 \times 50.9) + (0.2 \times 1.31) = 40.98.$$

Thus for any material the mass absorption coefficient is equal to the sum of the mass absorption coefficients of each constituent element multiplied by the corresponding mass fraction of that element present. The linear absorption coefficient μ of the material is then obtained by multiplying the mass absorption coefficient by the density ρ of the material.

If the mass absorption coefficient of an element is determined* at a large number of wavelengths and a plot of μ/ρ versus λ is prepared, the resulting graph is similar to that for platinum shown in Fig. 2-19. The absorption coefficient decreases rapidly with decrease in wavelength,

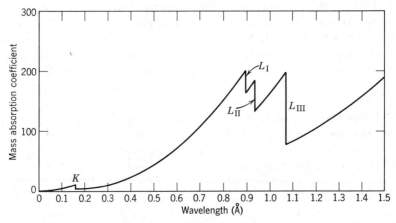

Fig. 2-19. Plot of the mass absorption coefficient for platinum versus wavelength, showing positions of the K and L absorption edges.

except for several sharp discontinuities. Thus for platinum the absorption decreases rapidly from $\lambda = 1.5$ Å to about $\lambda = 1.07$ Å. At this point there is a sharp increase in μ/ρ, followed again by a decrease and, shortly, two more similar increases. Then, from a wavelength of approximately 0.9 Å it begins another long decrease to $\lambda = 0.16$ Å, where again there is a sharp increase followed by the usual decrease. The sharp discontinuities in the absorption are referred to as the critical absorption edges of the material. These absorption edges mark the point on the frequency scale where the x-rays possess sufficient energy to eject an

*For references on the determination of x-ray absorption coefficients see the *International Tables* [84]. Ergun and Tiensuu have described a method particularly applicable to inhomogeneous materials [85].

electron from one of the shells. In Fig. 2-19 the edge at $\lambda = 0.16$ Å is the K absorption edge for platinum, and all wavelengths shorter than this value (wavelengths with higher frequency) will possess sufficient energy to eject K electrons from platinum, and those immediately shorter will be strongly absorbed. As the wavelength decreases and the energy of the photons increases farther above the critical excitation value, there is less likelihood of ionization and more chance that the photon will simply pass through unchanged and therefore unabsorbed. Absorption then decreases rather rapidly on the short-wavelength side of the absorption edge. On the long-wavelength side of an absorption edge, the photons do not have sufficient energy to eject an electron from the shell concerned; hence there is little absorption, and the value of μ/ρ is low. The group of three edges between $\lambda = 0.9$ and $\lambda = 1.1$ Å in Fig. 2-19 are the L absorption edges associated with the three discrete energy levels of the L shell, as depicted in Fig 2-17. It has been determined that μ varies as the third power of the wavelength in the regions between absorption edges (Fig. 2-19). These marked variations of μ and μ/ρ with wavelength make it most evident that accurate calculations can be made with equations 2-12 and 2-13 only when these quantities are known for the specific wavelength concerned.

An important application of these equations in diffraction techniques is in calculating the proper thickness of filters for use in monochromatizing an x-ray beam (Section 2-3.6A, B). The absorption law likewise enters into the calculation of the necessary thickness of shielding materials for x-ray protection. The heavy elements are the most efficient absorbers, since μ/ρ increases as the fourth power of the atomic number Z. Sheet lead is thus one of the most commonly used shielding materials, because of its high atomic number, $Z = 82$, and its relatively low cost. For high-voltage setups the cost of the necessary lead shielding sometimes becomes prohibitive, in which case concrete is frequently used. Concrete has a much lower absorption (at 100 kV peak, 7.5 cm concrete $\cong$ 1 mm lead) but is inexpensive, so that walls several feet in thickness may be used if required.

The fraction of radiation of $\lambda = 0.71$ Å (Mo$K\alpha$) transmitted by a 1-mm lead shield is calculated as follows. Equation 2-12 is first expressed in the form

$$\text{Fraction transmitted} = \frac{I}{I_0} = e^{-\mu x}. \qquad (2\text{-}14)$$

At this wavelength (Appendix V) lead has $\mu/\rho = 120$, $\rho = 11.35$, and $\mu = 1362$. Accordingly,

$$\frac{I}{I_0} = e^{-1362 \times 0.1} = e^{-136.2} = 7.1 \times 10^{-60}.$$

and the transmitted radiation is an infinitesimal fraction of the original beam intensity. If this molybdenum-target tube is operating at 40 kV peak, the low-wavelength limit will be at about 0.3 Å, and the peak of the continuous radiation will be at about 0.5 Å. For these more penetrating wavelengths μ/ρ for lead is approximately 12.8 and 50.6, respectively. Similar calculations for these wavelengths then give for I/I_0 for a 1-mm shield the values 4.9×10^{-7} and 1.1×10^{-25}, indicating that such a shield is entirely ample for these operating conditions. For a tungsten-target tube operating at 75 kV peak and 20 mA, however, 1 mm of lead is not sufficient shielding. According to the National Bureau of Standards *Handbook* 76[50], such a barrier around this tube would pass radiation at the rate of approximately 400 mR/hr at 1 m from the target. Thus, even at a 1-m distance, this is nearly 160 times the permissible dosage rate (2.50 mR/hr) for continuous bodily exposure. A barrier of about 2.6 mm of lead is required to permit continuous exposure (five 8-hr days/ week) at 1 m under the specified operating conditions.

Single-crystal diffraction photographs taken with radiation containing an appreciable portion of the continuous radiation possess a distinctive appearance as a result of selective absorption by the elemental constituents of the photographic emulsion, essentially silver and bromine. The K absorption edges of silver and bromine are at 0.486 and 0.920 Å, respectively. These wavelengths are in the region of intense continuous radiation from targets operating at 30 to 50 kV. Hence the wavelengths immediately shorter than these are strongly absorbed, leading to strong blackening of the film and to a peculiar banded appearance in spectral photographs (Fig. 2-12).

2-3.4 Secondary Fluorescent and Scattered X-rays

The foregoing discussion of x-ray absorption largely neglects the nature of the absorption process. The loss of intensity as a beam traverses matter is actually due to two processes, "true absorption" and "scatter." True absorption involves the transformation of x-rays into kinetic energy of ejected electrons, whereas scattering is a process in which there is a transfer of radiant energy from the primary beam to scattered beams originating in the atoms of the absorber. The absorption coefficient is thus made up of two separate terms,

$$\frac{\mu}{\rho} = \frac{\tau}{\rho} + \frac{\sigma}{\rho}, \tag{2-15}$$

where τ/ρ is the true absorption coefficient, and σ/ρ is the scattering coefficient. The latter term is small for elements beyond iron ($Z = 26$), and varies only slightly with changes in wavelength or atomic number.

The true absorption coefficient τ/ρ is related to the marked variation of absorption with wavelength and atomic number mentioned in the previous section. When the energy of the x-ray photons is such that they are efficient in ejecting electrons and ionizing inner shells of the atoms of the absorber, appreciable amounts of x-radiation characteristic of the absorber are generated by outer electrons falling into the vacancies of the inner shells as described earlier (Section 2-3.1). This secondary radiation is commonly called "fluorescent" radiation. It can never have a wavelength shorter than that of the primary rays producing it, so that Stokes' law for the production of fluorescent radiation in the visible range also holds for the x-ray range, and indeed fluorescence is a great deal more important in the x-ray range than in the visible.

In diffraction studies it is important to choose the x-ray wavelengths used so that no appreciable fluorescent radiation is produced by the crystal sample. There are no phase relations between such secondary radiations, or between them and the incident and diffracted beams, but they produce an undesired distributed background which can be intense enough to obscure the diffraction effects being sought. For instance, chromium compounds fluoresce strongly when irradiated with $\text{Fe}K\alpha = 1.937$ Å, which is just shorter than the K absorption edge for chromium at 2.07 Å. Thus $\text{Fe}K\alpha$ radiation is not very suitable for studying chromium compounds. Chromium radiation, however, may be used, since no element is capable of exciting its own fluorescence, or $\text{Cu}K\alpha$ radiation might be used, since it is far enough below the chromium absorption edge not to excite strong $\text{Cr}K$ radiation. Similarly, cobalt and its compounds (K_{abs} edge $= 1.61$ Å) fluoresce strongly with $\text{Cu}K\alpha = 1.54$ Å, and molybdenum compounds (K_{abs} edge $= 0.62$ Å) cannot well be studied with $\text{Pd}K\alpha = 0.58$ Å. L fluorescent radiation is less frequently a problem in diffraction work. However, $\text{Mo}K\alpha = 0.71$ Å can excite the entire L spectrum of lead ($L_{\text{I}_{\text{abs}}}$ edge $= 0.78$ Å). The rule is to choose an incident-beam wavelength as far removed as possible from the K or L absorption edges of the prominent metallic elements in the sample under investigation, or to use the K radiation from a target of the same metallic element as that in the sample.

Most of the loss in intensity as x-rays traverse a substance results from the fluorescent absorption discussed above. The smaller fraction of the beam intensity lost through scatter, the σ/ρ quantity of equation 2-15, results from a deviation of some of the primary rays by the atoms of the absorber so that they emerge in directions different from that of the incident beam. Accordingly, a measurement in the direction of the emergent primary beam makes it appear that they too have been absorbed. A part of these scattered x-rays has a wavelength identical with that of the primary beam; another part has a wavelength slightly longer than this. The former are referred to as "unmodified" or "coherently" scattered x-rays; the latter are called the "modified" or "incoherently" scattered rays. The incoherent scattering is also called "Compton scattering" or the "Compton effect," from A. H. Compton, who measured and explained the effect[51]. A portion of the coherently scattered radiation comprises the phenomenon of x-ray diffraction or "Bragg scattering," which is the major interest of this book, and which is discussed in detail in Chapter 3.

An intensity curve of the x-rays (say, MoKα) scattered by powdered graphite exhibits two interesting features. First, there are the expected maxima of the powdered-crystal diffraction pattern. Second, the background between the lines does not fall to zero intensity as required in a crystal with perfect regularity. In the latter case diffraction theory requires that rays at any angle other than near $\sin^{-1}(n\lambda/2d)$ undergo complete destructive interference. Debye[52], however, showed that diffuse scattering from a crystal should appear as a consequence of the thermal motions of the atoms in the crystal. A wavelength determination on the background radiation scattered at any given angle with the incident beam reveals a maximum with a wavelength exactly that of the primary beam, and a second maximum of slightly longer wavelength resulting from Compton scattering. This diffuse scattering and Compton scattering thus contribute a background on which the crystalline diffraction lines are superposed, and these two factors have been demonstrated to be completely adequate to account for the phenomenon.

The Compton (incoherently) scattered radiation[51] is always of slightly longer wavelength than that of the primary incident beam. It is much more efficiently produced by shorter wavelengths, and by lighter elements, such as lithium, carbon, and aluminum. Furthermore, the increase in wavelength is a function of the angle ϕ between the primary beam and the direction of scatter in accordance with the equation

$$\Delta\lambda = \frac{h}{mc}(1-\cos\phi) \qquad\qquad (2\text{-}16)$$

$$= 0.0243(1-\cos\phi), \qquad (\text{in Å})$$

where h is Planck's constant, m is the mass of the electron, and c is the velocity of light. Compton explained the shift in wavelength as the result of an encounter of an x-ray photon with a loosely bound or free electron in the absorber. As a result of the impact the electron recoils slightly, and in doing so absorbs a small amount of the energy of the x-ray quantum, which thereafter continues on its way as a quantum of slightly longer wavelength. The laws of conservation of energy and momentum are completely preserved in the encounter. This same theory also accounts for the presence of the unmodified (coherent) radiation in the scattered beam. For a more tightly bound electron, the same photon may not have sufficient energy actually to eject the electron from the atom. Hence it merely bounces off the electron, with its energy $h\nu$ the same, and its wavelength unchanged.

2-3.5 Refraction of X-rays

With x-rays exhibiting so many of the properties of visible light, we might expect that they would be refracted in passing from one medium to another. Roentgen, soon after his discovery, tried refraction experiments with prisms of materials such as ebonite, aluminum, and water, but with no success. Nearly 30 years later refraction of x-rays by a prism was first accomplished by Larsson, Siegbahn, and Waller[53]. This, however, was not the first direct evidence that x-rays are refracted slightly. A measurable refractive index for x-rays should make itself evident by deviations from Bragg's law, $n\lambda = 2d \sin \theta$, which requires that the rays have the same velocity inside the crystal as outside for its exact fulfillment. In a series of refined experiments, Stenström[54] showed that, for x-rays of wavelength greater than about 3 Å, diffracted from crystals of gypsum and sugar, the Bragg equation was not accurately obeyed. The discrepancy, he found, could be explained if an index of refraction of slightly less than unity was assumed for x-rays. Since then it has been shown that these same discrepancies, though somewhat smaller, occur when the common wavelengths used in diffraction studies are reflected from calcite.

If the refractive index is designated by μ, classical optical theory permits the derivation of the following expression[55]:

$$\mu = 1 - \frac{Ne^2}{2\pi m\nu^2} = 1 - \frac{Ne^2\lambda^2}{2\pi mc^2} = 1 - \delta, \qquad (2\text{-}17)$$

where

$$\delta = \frac{Ne^2}{2\pi m\nu^2} = \frac{Ne^2\lambda^2}{2\pi mc^2}. \qquad (2\text{-}18)$$

Here N is the number of electrons per cubic centimeter of the refracting material, e and m are the charge (in electrostatic units) and mass of the electron, respectively, ν is the frequency of the refracted x-rays, and λ and c have their usual significance. It is seen that the refractive index of x-rays is expected to be less than unity by a quantity δ, which is of the order of 10^{-6}. Experimental determinations of δ are in good agreement with values calculated from equation 2-18.

For the most precise work, such as precision lattice constant measurement to 1 part in 50,000 to 100,000, the Bragg law must be corrected for refraction, becoming

$$n\lambda = 2d\left(1 - \frac{\delta}{\sin^2 \theta}\right) \sin \theta. \qquad (2\text{-}19)$$

When μ or δ is not known from experiment, it may be calculated by means of equation 2-17. The effect of the refractive index is to cause the measured angle θ_{obs} to be slightly larger than the angle θ_{calc}, which we would obtain from the usual Bragg equation, $\theta_{calc} = \sin^{-1}(n\lambda/2d)$, by inserting the true values of λ and d. In turn, values of d_{obs} or a_{obs} calculated from θ_{obs} would be too small. Strictly, relation 2-19 holds only for a crystal whose face is parallel to the reflecting planes. Wilson[56], however, demonstrated that it is also a good approximation for powdered crystals of a highly absorbing substance, and that the lattice spacings of cubic crystals may be corrected for refraction merely by multiplying the observed values by $1 + \delta$. Thus,

$$a_{corr} = a_{obs} \times (1 + \delta). \qquad (2\text{-}20)$$

Accordingly, when the precision of the measurements warrants it, the refraction correction may be made by using equation 2-19 and the original data, or equation 2-20 and the results from the usual Bragg equation.

The correction of some precision measurements of the lattice constant of lead[57] illustrates the application. Using a lead sample of 99.999+ per cent purity, six separate determinations at $25 \pm 0.1°$ (Cu$K\alpha$ radiation used) yielded an average value of $a_{obs} = 4.95064 \pm 0.00006$ Å, where the error indicated is the probable error of the mean. To calculate δ for lead N, the number of electrons per cubic centimeter of lead, must be calculated. For any element,

$$\text{Number of electrons per cubic centimeter} = N = \frac{N_0 Z \rho}{A}, \qquad (2\text{-}21)$$

where N_0 is Avogradro's number, 6.023×10^{23}, Z is the atomic number, ρ is the density, and A is the atomic weight. Thus

$$N(\text{Pb}) = \frac{6.023 \times 10^{23} \times 82 \times 11.35}{207.19} = 27 \times 10^{23},$$

and

$$\delta(\text{Pb}) = \frac{Ne^2\lambda^2}{2\pi mc^2} = \frac{27 \times 10^{23} \times 23.07 \times 10^{-20} \times 2.3771 \times 10^{-16}}{2 \times 3.1416 \times 0.9109 \times 10^{-27} \times 8.987 \times 10^{20}}$$

$$= \frac{1480.7 \times 10^{-13}}{51.436 \times 10^{-7}} = 28.8 \times 10^{-6}.$$

Then, using equation 2-20, we have

$$a_{\text{corr}} = 4.95064 \times 1.0000288 = 4.95079 \text{ Å}.$$

Considering the probable error of the uncorrected value, we conclude that the final value for the lattice constant of lead at $25 \pm 0.1°$C should be reported as 4.9508 ± 0.0001 Å.

2-3.6 Monochromatization of X-radiation

Most diffraction techniques require x-radiation that is essentially mono chromatic; a few actually demand a strictly monochromatic beam. The source of such radiation is usually the $K\alpha$ doublet from one of the targets listed in Table 2-3. The beam from an x-ray tube operating at the potential required to produce radiation suitable for diffraction studies (Section 2-3.1) not only contains the complete K line spectrum but also a continuous spectrum ranging over a wide band of wavelengths. Each individual wavelength of this beam is capable of being diffracted by the various crystal planes, giving rise with each plane to a spectrum such as Fig. 2-12. It is evident, therefore, that a separation of a suitable wavelength from this complex spectrum must be effected, and an intensity versus wavelength curve, such as Fig. 2-14 for molybdenum, suggests the $K\alpha$ line as most useful because of its great intensity. The $K\alpha$ line is actually a close doublet, with the α_1 line having twice the intensity of the α_2 line. For most requirements the doublet itself is entirely satisfactory, but it is possible to separate the $K\alpha_1$ line if such rigid monochromatization is required. For a large part of the diffraction work of most laboratories, the monochromatization may be achieved through the use of filters, whereas the stricter requirements are met by means of crystal monochromators. A detailed review of the subject is presented in the *International Tables*[58], and a very extensive bibliography appeared in 1967[59].

   A. Single-Filter Technique. Monochromatization by use of a single filter is simple, being based on the strong absorption exhibited by an element immediately below its K critical absorption edge. Figure 2-20 depicts the output of a molybdenum-target tube at 35 kV, with the absorption curve for zirconium superposed. The K edge, $\lambda = 0.689$ Å, for zirconium is seen to fall between the $K\beta$ and $K\alpha$ wavelengths of molybdenum, and just on the short-wavelength side of the latter. Accordingly, wavelengths shorter than $\lambda = 0.689$ Å, including Mo$K\beta$, $\lambda = 0.63$ Å, are strongly absorbed in passing through zirconium, but wavelengths immediately longer than $\lambda = 0.689$, such as Mo$K\alpha$, 0.71 Å, are only slightly absorbed by the metal. The proper thickness of zirconium in the

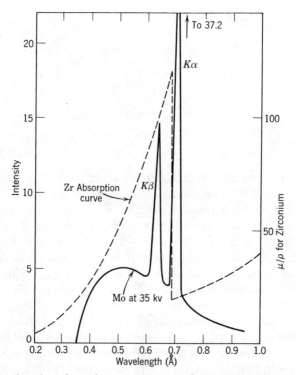

Fig. 2-20.   The zirconium absorption curve superposed on 35-kV molybdenum radiation.

form of thin metal foil is therefore a suitable filter for monochromatization of molybdenum radiation to give essentially pure MoKα. Niobium
metal foil (K edge, $\lambda = 0.653$ Å) is also a satisfactory filter material for
monochromatizing MoKα radiation. The rule for choosing a suitable
filter is: *Choose for the filter an element whose K absorption edge is just to the
short-wavelength side of the Kα line of the target material.*

Calculation of the proper thickness of zirconium foil* for use as a
filter for molybdenum radiation is readily carried out by means of equation 2-12. The problem is to use sufficient thickness to cut down the fairly
intense $K\beta$ line to negligible intensity and yet not reduce the intensity of
the $K\alpha$ radiation to the point where exposures are unduly increased. For
purposes of illustration, the effects of a 0.005-cm (approximately 0.002-
in.) thickness of zirconium foil on the $K\beta$ and $K\alpha$ lines of molybdenum
are calculated. For this purpose the following absorption data (Appendix

---

*Zirconium foil, nickel foil, and other sheet materials for filters are obtainable from the
various manufacturers of diffraction equipment.

V) for zirconium are needed:

Wavelength	μ/ρ	ρ	$\mu = (\mu/\rho) \cdot \rho$
0.63 Å	79.0	6.4	506
0.71 Å	15.9	6.4	102

Then the amounts of radiation transmitted by the 0.005-cm foil are:

β radiation transmitted $= I/I_0 = e^{-(506 \times 0.005)} = e^{-2.53} = 0.08 = 8$ per cent,

α radiation transmitted $= I/I_0 = e^{-(102 \times 0.005)} = e^{-0.51} = 0.60 = 60$ per cent.

Originally, the intensity ratio $K\beta:K\alpha$ was 18.5:100, but after the foregoing filtration it is 1.5:60, or approximately 2.5:100. By this procedure the objectionable $K\beta$ is reduced to about one-eight of its original intensity, and to approximately one-fortieth of the intensity of the $K\alpha$ radiation, which results in only the very strongest diffraction lines or spots showing any $K\beta$ component. The less intense continuous radiation of neighboring wavelengths likewise is absorbed to a comparable extent. This too is important, since there is less of it left to contribute to air scatter and to the continuous bands and streaks on the diffraction photograph. This cleaner film, uncluttered by the $K\beta$ pattern, is achieved with an increased exposure time not quite double that with unfiltered radiation. Moreover, this is just about the right thickness of zirconium filter to use on molybdenum radiation because, although a thicker one would remove even a greater portion of the $K\beta$ radiation, it would also absorb more $K\alpha$ radiation and unduly increase the exposure time required.

The correct materials for filtering out $K\beta$ radiation from the commonly used targets are listed in Table 2-5. Also listed is the thickness of filter which reduces the $K\alpha$ radiation to one-half its original value, and the per cent absorption of the $K\beta$ radiation by this thickness of filter. These filters require a doubling of the exposure time. For use in making up filters, the weight of the element in grams per square centimeter has been calculated. Approximately half the thickness indicated requires only a 50 per cent increase in exposure time, but absorbs on the average 15 per cent less of the $K\beta$ radiation. Satisfactory filtering is obtained with filter thicknesses which result in an exposure-increase factor of about 1.5 to 2.

Thin ductile foils of the metals make the most suitable materials for filters, when it is possible to prepare or obtain such. Wood [60] has prepared a manganese filter for use with iron radiation by electrolytic deposition on aluminum foil. When it is not possible to obtain the element in thin sheet form by rolling or plating techniques, it is necessary to prepare filters by dispersing the required weight of some compound of

Table 2-5. Filters for the Production of Monochromatic X-rays[a]

Target	λ of Kα Doublet (Å)	Filter	Thickness in Millimeters	Mils[b]	Grams/cm³	Kβ Absorbed (per cent)
Ag	0.561	Pd	0.0461	1.8	0.056	96.0
		Rh	0.0480	1.9	0.060	96.4
Mo	0.711	Nb	0.0481	1.9	0.040	96.5
		Zr	0.0678	2.7	0.044	96.8
Cu	1.542	Ni	0.0170	0.7	0.015	98.4
Ni	1.659	Co	0.0158	0.6	0.014	98.7
Co	1.790	Fe	0.0166	0.7	0.013	98.9
Fe	1.937	Mn	0.0168	0.7	0.012	99.2
Cr	2.291	V	0.0169	0.7	0.010	99.4

[a]Thickness given reduces Kα radiation by 50 per cent.
[b]1 Mil = 0.001 in.

the element, usually the oxide, in a suitable wax or plastic base such that when formed into a thin sheet it has the requisite weight of filtering element per square centimeter to provide the desired absorptive power [61].

B. Balanced-Filter Technique. The use of a single filter provides a beam in which the relative intensity of the Kα component is strongly enhanced with respect to the Kβ radiation. In the filtered beam the Kβ line and other shorter wavelengths are considerably decreased in intensity but by no means eliminated. Such a beam is very useful for many situations, but it is not monochromatic. By the use of two filters, it is possible to achieve a much more nearly monochromatic beam, as first proposed by Ross[62] in the balanced-filter technique. The experimental procedure consists in making intensity measurements while alternately interposing in the x-ray beam two filters whose absorption edges lie just above and just below the wavelength of the Kα radiation sought. The proper pairs of filters (and suitable thicknesses) are presented in Table 2-6 for the radiations commonly used in x-ray diffraction.

Consider, for example, balanced filters for CuKα radiation, $\lambda = 1.542$ Å. A nickel filter (NiK edge, 1.488 Å) is chosen for the low-wavelength side, and a cobalt filter (CoK edge, 1.608 Å) for the high-wavelength side. Figure 2-21A presents the absorption coefficient μ versus wavelength for these two metals relative to the CuKα line. It is necessary to adjust very carefully the thicknesses of these two filters, metal foils or their finely divided oxides, until the transmission of the general radiation

Table 2-6. Calculated Thickness of Ross-Filter Components for Commonly Used Radiations[a]

Target Element	Filter Pair (A)	Filter Pair (B)	(A) Thickness Milli-meters	(A) Thickness Mils[b]	(A) Thickness Grams/cm²	(B) Thickness Milli-meters	(B) Thickness Mils[b]	(B) Thickness Grams/cm²
Ag	Pd	Mo	0.0275	1.08	0.033	0.039	1.53	0.040
Mo	Zr	Sr	0.0392	1.54	0.026	0.104	4.09	0.027
Mo	Zr	Y	0.0392	1.54	0.026	0.063	2.49	0.028
Cu	Ni	Co	0.0100	0.38	0.0089	0.0108	0.42	0.0095
Ni	Co	Fe	0.0094	0.37	0.0083	0.0113	0.45	0.0089
Co	Fe	Mn	0.0098	0.38	0.0077	0.0111	0.44	0.0083
Fe	Mn	Cr	0.0095	0.37	0.0071	0.0107	0.42	0.0077
Cr	V	Ti	0.0097	0.38	0.0059	0.0146	0.58	0.0066

[a]Reprinted by special permission from *International Tables for X-ray Crystallography*, Vol. III. Kynoch Press, Birmingham, England, 1962, p. 79.
[b]1 Mil = 0.001 in.

at a wavelength just shorter than the nickel absorption edge is the same (balanced) for both filters. Ideally, a similar balance should simultaneously be achieved at a wavelength just longer than the cobalt absorption edge. Under these circumstances the situation is represented by Fig. 2-21B in which the absorption exponent μt (where $t =$ filter thickness) is plotted against wavelength. The difference between the intensities with the two filters is due only to the narrow *pass band* of wavelengths between the two absorption edges. This band of radiation evidently consists almost entirely of the very intense $CuK\alpha$ characteristic line. The difference between intensities observed on two x-ray photographs, or two counter diffractometer traces taken alternately with such a pair of filters, can thus be attributed to the narrow pass band of wavelengths, essentially, that is, to the $K\alpha_1\alpha_2$ wavelengths. The technique is also useful for obtaining a desired narrow band of wavelengths from regions of the x-ray spectrum where characteristic lines are not present.

In actual practice it is virtually impossible to achieve a perfect balance of the two filters over the entire general-radiation spectrum of the x-ray tube, even though balance may be rather good in the vicinity of the pass band. In counter diffractometry, however, by utilizing pulse-height discrimination together with balanced filters, the exclusion of wavelengths outside the pass band can be made practically complete. Kirkpatrick [63] made a careful analysis of the Ross balanced-filter technique, and

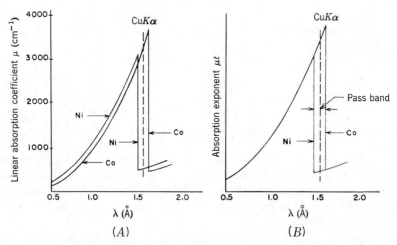

Fig. 2-21.   Balanced filters of nickel and cobalt for copper radiation. (*A*) Absorption coefficient μ versus wavelength. (*B*) Absorption exponent μt, after balancing. (From L. Alexander, *X-ray Diffraction Methods in Polymer Science*, Wiley, New York, 1969.)

Kirkpatrick and Chang[64] have suggested a four-filter procedure to eliminate certain errors. Wollan[65] used a three-filter scheme. Both Soules, Gordon, and Shaw[66] and Ruland[67] have discussed the design of balanced filters and methods for achieving optimum filter thicknesses. Ruland concluded that the most satisfactory balance of nickel and colbalt for Cu$K\alpha$ occurs at thicknesses of 0.0068 and 0.00755 mm, respectively. Ruland then calculated that the nickel and cobalt thicknesses indicated for Cu$K\alpha$ in Table 2-6 indicate an imbalance of a maximum of 3 per cent. In addition to filter thickness, Bol[68] investigated the difficulties caused by incoherently scattered x-rays, and a method of eliminating fluorescent radiation. When balanced filters are used with photographic recording, a study of the monochromatization error as a function of thickness[86] suggests that thinner filters are better monochromators than thicker ones.

   C. Crystal Monochromator Techniques. Crystal monochromators provide the strictly monochromatic radiation required in special investigations, particularly in radial distribution studies of amorphous materials (Chapter 12), in small-angle scattering investigations, certain diffractometric studies, and in all cases in which complete freedom from background is important. The plane-crystal monochromator is simply a device whereby the $K\alpha$ line from a target is separated by diffraction from a suitable crystal, the diffracted beam becoming the primary incident beam for the particular study at hand. This procedure yields a beam made up of the two wavelengths of the $K\alpha$ doublet, plus a very small intensity of harmonics. These harmonics are to be expected because the wavelength λ, the second order of $\lambda/2$, third order of $\lambda/3$, and so on, are all diffracted at the same angle by the monochromatizing crystal. This "di- or tri-chromatic" beam usually serves the most rigid requirements for monochromatic radiation. It is possible, however, using a plane crystal and a lead wedge or a straightedge at a suitable distance from the monochromator, to adjust them so as to pass only the $K\alpha_1$ component of the doublet. If the harmonics are objectionable, they too can be removed. The $\lambda/2$ harmonic can be eliminated only through the use of a crystal with a negligible second-order reflection, such as 111 of fluorite, whereas higher-order harmonics can usually be eliminated by a suitable choice of the tube voltage so that they fall below the low-wavelength limit of the beam.

Because the efficiency of diffraction is so low, the intensity of the diffracted $K\alpha$ doublet is feeble compared to that of the original beam. It might be anticipated therefore that studies with crystal monochromatized radiation require unduly long exposures. Although this is true to some extent, the handicap is not as great as would be expected. The freedom from background usually makes it unnecessary to expose to the same intensity in order to have lines that stand out and are readily measured with ease and precision. Every means must be used, however, to provide as intense a beam as possible. This may be accomplished by choosing a crystal which provides as intense reflection as possible, by adaptation of the crystal to give an increased intensity, and by the most advantageous mounting of the crystal.

The choice of the monochromating crystal depends on whether it is to produce radiation for use with ordinary pinholes or vertical slits, or is itself to act as the collimator. The former requires a strong beam of sufficient area to cover the slit and bathe the specimen. Such a beam is obtained from a mosaic crystal (Section 3-3.1). If the crystal's mosaic character is not sufficiently developed, it can often be increased by dipping the crystal in liquid air or by lightly grinding the reflecting face. For use as a collimator to produce a very narrow beam, the crystal should be as perfect as possible, that is, nonmosaic. The crystal must have a reflecting face about 3- to 5-mm square and a thickness of about 1.5 mm. It must also be mechanically strong and not appreciably affected by prolonged exposure to x-rays and the atmosphere. Obviously, it should not be polycrystalline or have gross imperfections, as these lead to multiple reflections and nonuniformity of the diffracted beam. Finally, the strong reflection used must have a low Bragg angle in order to keep polarization (Section 3-3.4A) at a minimum. No one crystal, of course, meets all these requirements. Pentaerythritol is an excellent general-purpose monochromator except for its poor stability in the beam. Generally, about a 100-hr exposure to a collimated direct x-ray beam causes a breakdown of the crystal, with broadening and loss of intensity and parallelism of the reflected beam. Workers have also experienced difficulties in growing pentaerythritol crystals, but an elaborate technique for this is described by Whetstone[69]. Mica crystals as monochromators are reported to have about a 200-hr life[70]. Fluorite and rock salt have less intense reflections but much better stability. Gypsum and calcite are the most suitable crystals for collimation, although the former is reported to have poor stability. Table 2-7, taken from the *International Tables*[58], summarizes the characteristics and properties of various crystals used as monochromators.

Under the topic, adaptation of the crystal to provide increased intensity, might be listed the techniques mentioned earlier for increasing mosaic character. These broaden the reflection and give it a greater integrated intensity, although they may not always result in a greater effective intensity. A condensing monochromator technique described by Fankuchen[71] is of interest in this respect. By grinding the crystal so that the surface makes an angle of $\theta/2$ with the reflecting planes, a concentration of the x-ray beam of approximately 2:1 should be effected, as depicted in Fig. 2-22. Fankuchen found the technique successful with pentaerythritol crystals and CuKα radiation, for which the Bragg angle is $\theta = 10°$ and $\theta/2 = 5°$. We have used this procedure for increasing the intensity from synthetic rock salt.

The most advantageous mounting of a crystal is achieved by building into the monochromator assembly the necessary adjustments to permit precise alignment in the beam. The monochromator assembly should be designed to provide: (1) rotation of the crystal about an axis lying in or near the reflecting face, (2) translation of the whole assembly parallel to

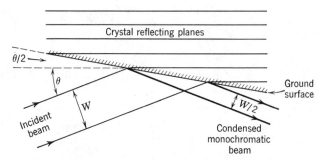

Fig. 2-22. Fankuchen's condensing monochromator.

Table 2-7. Reflection Characteristics and Properties of Various Monochromator Crystals (Not Bent)[a,b]

Crystal	Reflection	Spacing (Å)	Properties of Reflection		Properties of Crystal			Special Uses
			Peak Intensity	Breadth	Crystal Imperfection	Stability	Mechanical Properties	
β alumina	0002 0004	11.24 5.62	Weak Weak to medium	Moderate	Great	Perfect	Hard, brittle	For long wavelengths, but usable crystals hard to obtain
Mica	001 004	10.1 2.53	Weak	Small	Negligible for selected specimens	Fair	Flexible, easily cleaved	For point-focusing devices; exhibits irradiation effects
Gypsum	020	7.60	Medium to strong	Very small	Good specimens hard to find	Poor	Soft, flexible	For small-angle scattering; focusing long wavelengths
Pentaerythritol	002	4.40	Very strong	Moderate	Great	Poor	Soft, easily deformed	General purposes; exhibits irradiation effects
Quartz	$10\bar{1}1$	3.35	Weak to medium	Very small	Negligible	Perfect	Can be elastically bent	For small-angle scattering; focusing
Potassium bromide	200	3.29	Medium to strong	Moderate	Negligible	Slightly deliquescent		
Fluorite	111 220	3.16 1.94	Medium to strong Very strong	Moderate	Small	Perfect	Moderately hard	For eliminating harmonics; general purposes; short wavelengths
Urea nitrate	002	3.14	Strong	Very large	Very great	Very poor	Very easily deformed	For large specimens; soon decays
Calcite	200	3.04	Medium	Small	Negligible	Perfect	Moderately soft	For small-angle scattering; isolation of α_1 or α_2

Rock salt	200	2.82	Medium to strong	Large	Great	Slightly deliquescent	Can be plastically bent in warm supersaturated saline	For focusing
Aluminum	111	2.33	Very strong	Moderate to large		Good	Soft, can be seeded and grown to shape, then plastically shaped at room temperature	For focusing; diffuse scattering
Diamond	111	2.05	Weak	Very small	Negligible	Perfect	Very hard	For eliminating harmonics
Lithium fluoride	200	2.01	Very strong	Small to moderate	Negligible	Perfect	Hard; can be plastically bent at high temperature	For focusing; diffuse scattering; general purposes

[a]Reprinted by special permission from *International Tables for X-ray Crystallography*, Vol. III, Kynoch Press, Birmingham, England, 1962, p. 80.
[b]For germanium and silicon as monochromator crystals, see Maloof[73].

the axis of the tube, (3) an adjustable slit between the crystal and target, and (4) the shortest possible distance between the camera and focal spot.* Since the machine work is relatively simple, the building of a monochromator is not a difficult matter. Lipson, Nelson, and Riley[72] have published a good design, and other simple ones have been described[74]. All crystal monochromatized radiation is polarized. Hence a more complicated polarization correction is needed for the resulting twice-diffracted beam when such radiation is used in diffraction studies (see Section 3–3.4A).

In addition to plane-crystal monochromators, there are bent-crystal monochromators. Such monochromators produce radiation converging to a line focus, and find their chief use with focusing diffractometers and focusing-type powder cameras. In addition to the production of monochromatic radiation with low background and the elimination of spurious effects on powder patterns, they furnish a high intensity compared to plane-crystal monochromators, and the focusing properties provide high resolving power. The bent crystal may be elastically bent mica, gypsum, or quartz, used either as a transmission grating[75] or as a reflection grating[76].

The increase in intensity provided by a bent focusing crystal is obtained at the expense of increased divergence of the monochromatic beam. Thus bent crystals do not provide perfect focusing, but it can be achieved if the crystal is both cut and bent[77]. First the crystal is bent so that the parallel atomic planes become concentric circles of radius R; then a surface of radius $R/2$ is ground into the bent crystal (Fig. 2-23). Plastically, bent rock salt crystals may be used for this purpose[78a, b]. When immersed in hot, concentrated salt solution, rock salt is readily deformed without fracture. Smith[78-b] concluded that lukewarm water was even more effective, immersion of 15 sec being sufficient, and solution effects were negligible in this short time. Fracture occurs about once in five bending attempts, and the bending is usually not perfect. Of 20 crystals bent and ground by Smith, 6 were satisfactory in all respects, and he concluded that synthetic optical rock salt† is far superior to natural rock salt in its integrated reflection and in its cleavage and bending properties. His bent crystals had exposure factors as small as $\frac{1}{10}$, compared with plane-crystal monochromators. For further details on single-crystal monochromators the reader should consult the literature cited [58, 59, 69–79].

---

*Suitable monochromator assemblies are available commercially from the manufacturers of diffraction equipment and accessories.
†Such as prepared by Harshaw Chemical Company, Cleveland, Ohio.

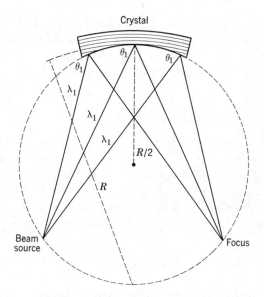

Fig. 2-23.   Accurate focusing by a bent-ground monochromatizing crystal.

D. Graphite Monochromators. In 1956, Renninger[80] reported the very high x-ray intensity diffracted from a natural single crystal of Ceylon graphite, an observation which suggested its desirability as a monochromator if available in usable-sized pieces. Encouraged by C. J. Sparks of Oak Ridge National Laboratories, the Union Carbide Corporation,* was led to discover how to force graphite into this highly oriented configuration. Such material with its hexagonal basal planes aligned parallel to within $\pm 0.3°$ can diffract x-rays with an efficiency exceeding that of other known materials. Although the material is only a pseudo single crystal, it simulates a single crystal, and as a monochromator it has four to six times the reflectivity of LiF[81] and gives a more uniformly distributed diffracted beam. Used in the incident beam, it lowers the background for film-recorded powder patterns and enhances the quality of rotating crystal photographs. It is even possible to use these graphite pseudo crystals in the form of bent-crystal monochromators[82].

Although the more common position for the insertion of a monochromator is in the incident x-ray beam, there are situations in which it is desirable to place it in the diffracted beam[82] (also see Section 5-3.4). Indeed, the latter arrangement has two distinct advantages over the

*Union Carbide Corporation, Carbon Products Division, 270 Park Avenue, New York, New York.

former. Primarily, it eliminates background originating in the specimen, whether from secondary fluorescence or radioactivity. Second, it suppresses incoherent radiation. Installation of the monochromator in the diffracted beam need not be anymore complicated than in the incident beam.

2-3.7 The Photographic Effects of X-rays

One of the first properties of x-rays to be observed was their effect on a photographic plate. The rays initiate the photochemical change in a silver bromide emulsion, which results in a deposit of silver grains upon development. The degree of blackening or density D of this silver deposit is defined in terms of the fraction T of an incident light beam which it transmits. The reciprocal of the transmission is known as the opacity O. Thus

$$O = \frac{1}{T}. \tag{2-22}$$

The density is then defined either as the common logarithm of the opacity, or as the logarithm of the reciprocal of the transparency,

$$D = \log O = \log \frac{1}{T} = \log \frac{\text{incident light}}{\text{transmitted light}}. \tag{2-23}$$

Thus the blackening corresponding to a film density of 1.0 is such that it permits only one-tenth of an incident light beam to pass through, and a film density of 2.0 transmits only one-hundredth of the light.

Film darkening by x-rays obeys the *reciprocity law* that blackening is proportional to the exposure E defined as the product of the beam intensity I and the time of exposure t. That is,

$$E = It. \tag{2-24}$$

In other words, a 50-kV beam operating at 10 mA gives the same exposure and blackening in 1 hr as it would in 10 hr at 1 mA. Furthermore, it makes no difference whether the exposure is continuous or intermittent. Actually, this law fails when carried to extremes, but it holds with good accuracy over an intensity ratio of 100:1, and with fair accuracy over a ratio of 1000:1[83]. In diffraction studies the latter ratio is seldom exceeded.

When the density of blackening D is plotted against the logarithm of the exposure E, the so-called *characteristic curve* of the photographic material is obtained. Typical forms of characteristic curves for x-ray films are presented in Fig. 2-24. Curve 1 is similar to that of many negative photographic materials on exposure to ordinary light. There is

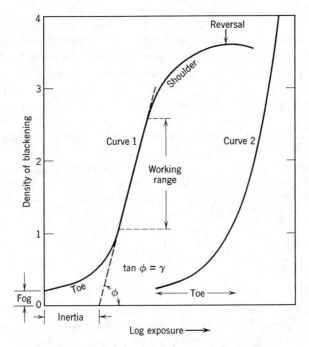

Fig. 2-24.   Typical characteristic curves of x-ray films.

present on development a certain minimum density representing "fog" in the emulsion, even when no exposure has been made. The initial part of the curve is referred to as the "toe." Above this is a straight-line portion, called the "working range of densities," and, finally, an upper curved portion, the "shoulder," where overexposure and reversal take place. Curve 2, typical of some of the x-ray film most commonly used in diffraction studies,* has no straight-line portion or shoulder in the range of usable densities. The upper limit of the toe is likewise poorly defined, but its working range may be said to start at about a density of 1.0 and continue upward along the slightly curved main portion of the curve. The eye observes a density of 2.0 as very black, but with double-coated x-ray films densities as high as 3.0 are useful. If the straight-line portion is projected to the abscissa axis, the intercept is called the "inertia." The inertia is an index of film speed; "fast" films have a short inertia range, and "slow" films a much longer one. The slope of the working range (tan ϕ in Fig. 2-24) is called the "gamma" of the film, and is a measure of

---

*Typical of Kodak industrial x-ray film, types A and K, and Eastman No-Screen x-ray film.

film contrast. When the slope is steep, the film has a high-gamma value and is referred to as "hard" or "contrasty." Similarly, low-gamma film is "soft." The characteristic curve not only varies with the type of emulsion, but even with different samples of the same film type and with variations in the developing technique. When a film is used with an intensifying screen of zinc sulfide or calcium tungstate, to increase the film blackening and thereby shorten the exposure, the film-screen combination has a characteristic curve entirely different from that of the film alone.

The practical aspects of measuring x-ray diffraction intensities photographically are treated elsewhere in this volume (Section 6-7), and a detailed discussion of x-ray film, including its handling and processing, is presented in Appendix II.

GENERAL REFERENCES

X-ray Safety

1. *Code of Practice for the Protection of Persons Exposed to Ionizing Radiations in Research and Teaching*, 2nd ed., Her Majesty's Stationery Office, London, 1968.

2. International Union of Crystallography, Commission on Crystallographic Apparatus, "Radiation Hazards Associated With X-ray Diffraction Techniques," *Acta Crystallogr.*, **16**, 324 (1963).

3. J. E. Cook and W. J. Oosterkamp, "Protection Against Radiation Injury," *International Tables for Crystallography*, Vol. III, Section 6, Kynoch Press, Birmingham, England, 1962.

*4. T. M. Moore, W. E. Gundaker, and J. W. Thomas, Eds., *Radiation Safety in X-ray Diffraction and Spectroscopy*, 1971. Available from the Superintendent of Documents, U.S. Government Printing Office, Washington, D.C. Price $2.00. An excellent book on radiation safety.

5. C. Faloci, A. Susanna, and F. Zampini, "Radiation Hazards from X-ray Diffraction Equipment," CNEN-Divisione di Protezione Sanitaria e Controlli, Rome, Italy; *J. Appl. Crystallogr.*, **5**, 375 (1972).

X-ray Physics

*1. E. F. Kaelble (ed.), *Handbook of X-rays*, McGraw-Hill, New York, 1967, Chapters 1, 2, 3.

2. M. A. Blokhin, *The Physics of X-rays*, 2nd ed. (translated from the Russian by the U.S. Joint Publications Research Service, New York), Office of Technical Services, Washington, 1961.

3. F. Jaundrell-Thompson and W. J. Ashworth, *X-ray Physics and Equipment*, Blackwell, Oxford, 1965, especially Chapters 11, 12, 18.

4. J. A. Bearden, *X-ray Wavelengths and X-ray Atomic Energy Levels*, U.S. Government Printing Office, 1967. [Reprinted from *Rev. Mod. Phys.*, **39**, 78 (1967)].

5. W. T. Sproull, *X-rays in Practice*, McGraw-Hill, New York, 1946.

6. A. H. Compton and S. K. Allison, *X-rays in Theory and Experiment*, 2nd ed., Van Nostrand Reinhold, New York, 1935.

7. G. L. Clark, *Applied X-rays*, 4th ed., McGraw-Hill, New York, 1955.

8. H. M. Terrill and C. T. Ulrey, *X-ray Technology*, Van Nostrand Reinhold, New York, 1930.

9. M. Siegbahn, *Spectroskopie der Röntgenstrahlen*, 2nd ed., Springer, Berlin, 1931. An English translation of the first edition was published by Oxford University Press, London, in 1925.

10. A. Guinier, *La radiocristallographie*, Dunod, Paris, 1945, Chapters 1 and 2. An English translation, *X-ray Crystallographic Technology* (T. L. Tippell, translator), was published by Hilger and Watts, London, in 1952.

SPECIFIC REFERENCES

[1] National Bureau of Standards, *Handbook HB93*. Procurable from the Superintendent of Documents, Washington, D.C. Price $0.30.

[2] National Council on Radiation Protection and Measurements, *Basic Radiation Protection Criteria*, NCRP Report No. 39, 1971, NCRP Publications, P. O. Box 30175, Washington, D. C. Price $2.00.

[3] See for instance: *Safety Considerations in the Design of X-Ray Tube and Collimator Couplings on X-Ray Diffraction Equipment*, K. E. Beu, Report No. 1, Apparatus and Standards Committee, American Crystallographic Association, 1962; *Safety Devices for X-Ray Diffraction*, S. Samson, Proceedings of the "Conference on Radiation Safety in X-Ray Diffraction and Spectroscopy", Philadelphia, Pennsylvania, January 6–7, 1970; J. A. Dunne, *Norelco Rep.*, **17**, 7 (1970).

[4] T. R. Kohler and W. Parrish, *Rev. Sci. Instrum.*, **27**, 705 (1956).

[5] W. Kossel, *Verh. deut. phys. Ges.*, **14**, 953 (1914); **18**, 339, 396 (1916); *Phys. Z.*, **18**, 240 (1917); *Z. Phys.*, **1**, 119 (1920); **2**, 470 (1920).

[6] N. Bohr, *Phil. Mag.*, (6)**26**, 1 (1913).

[7] H. G. J. Moseley, *Phil. Mag.*, (6)**26**, 1024 (1913); **27**, 703 (1914).

[8] C. G. Barkla, *Nature*, **95**, 7 (1915).

[9] W. D. Coolidge, *Phys. Rev.*, **2**, 409 (1913).

[10] O. W. Richardson, *Phil. Mag.*, (6)**28**, 633 (1914); S. Dushman, *Phys. Rev.*, **21**, 623 (1923).

[11] J. Ladell and W. Parrish, *X-ray Analysis Papers*, Centrex Publishing Company, Eindhoven, The Netherlands, 1965, Chapter 4.

[12] R. B. Corey, J. B. Lagsdin, and R. W. G. Wyckoff, *Rev. Sci. Instrum.*, **7**, 193 (1936).

[13] W. T. Astbury and I. MacArthur, *Nature*, **155**, 108 (1945).

[14] A. Muller and R. E. Clay, *J. Inst. Elec. Eng.*, **84**, 261 (1939).

[15] F. Fournier, *Bull. soc. fr. elec.*, **9**, 531 (1939).

[16] J. Beck, *Phys. Z.*, **40**, 474 (1939).

[17] A. Taylor, *Proc. Phys. Soc.* (London), **61**, 86 (1948).

[18] P. Gay, P. B. Hirsch, J. S. Thorp, and J. N. Kellar, *Proc. Phys. Soc.*, (London), **64B**, 374 (1951).

[19] G. M. Fraase Storm and J. A. Prins, *Ned. Tijdschr. Natuurk.*, **24**, 99 (1958).

[20] Descriptive literature from Elliot Automation Radar Systems, Ltd., Elstree Way, Borehamwood, Herts., England.

[21] Sales literature on General Electric Rotating Anode X-ray Generator, Diano

Corporation, Industrial X-ray Division, 2 Lowell Avenue, Winchester, Massachusetts. Diano Corporation acquired the x-ray diffraction and fluorescence instrument lines from the General Electric Company in 1972.

[22] From a pamphlet by Y. Shimura, M. Mizunuma, and K. Nakamura, Rigaku Denki Co., Ltd., 9-8, 2-chome, Sotokanda, Chiyoda-ku, Tokyo, Japan.

[23] A. Taylor, X-Ray Metallography, Wiley, New York, 1961, p. 227.

[24] A. Taylor, Advances in X-Ray Analysis, Vol. 9, Plenum Press, New York, 1966, p. 194.

[25] R. R. Machlett, J. Appl. Phys., 13, 398 (1942); H. Brackney and Z. J. Atlee, Rev. Sci. Instrum., 14, 59 (1943); F. C. Kelley, Iron Age, 156, 68 (1945); H. Smith, J. Sci. Instrum., 26, 378 (1949).

[26] R. W. G. Wyckoff and J. B. Lagsdin, Radiology, 15, 42 (1930); Rev. Sci. Instrum., 7, 35 (1936).

[27] C. J. Ksanda, Rev. Sci. Instrum., 3, 531 (1932).

[28] I. Fankuchen, Rev. Sci. Instrum., 4, 593 (1933).

[29] G. Hägg, Rev. Sci. Instrum., 5, 117 (1934).

[30] F. G. Chesley, Rev. Sci. Instrum., 14, 3 (1943).

[31] P. M. Harris, Rev. Sci. Instrum., 8, 478 (1937).

[32] T. B. Rymer and P. G. Hambling, J. Sci. Instrum., 29, 192 (1952).

[33] M. J. Buerger, X-ray Crystallography, Wiley, New York, 1942, pp. 459–460.

[34] H. P. Klug and L. E. Alexander, X-Ray Diffraction Procedures, 1st ed., Wiley, New York, 1954, Chapter 2.

[35] A. Müller, Proc. Roy. Soc., (London), 132A, 646 (1931).

[36] For instance, W. T. Sproull, X-rays in Practice, McGraw-Hill, New York, 1946; G. L. Clark, Applied X-rays, McGraw-Hill, New York, 1955; H. M. Terrill and C. T. Ulrey, X-ray Technology, Van Nostrand Reinhold, New York, 1930; E. F. Kaelble (ed.), Handbook of X-rays, McGraw-Hill, New York, 1967.

[37] W. Duane and F. L. Hunt, Phys. Rev., 6, 166 (1915).

[38] C. G. Barkla, Proc. Camb. Phil. Soc., (May 1909); Phil. Mag., (6)22, 396 (1911).

[39] W. H. Bragg, Nature, 91, 477 (1913).

[40] H. G. J. Moseley, Phil. Mag., (6) 26, 1024 (1913); 27, 703 (1914).

[41] A. H. Compton and S. K. Allison, X-rays in Theory and Experiment, 2nd ed., Van Nostrand Reinhold, New York, 1935.

[42] R. Witty and P. Wood, Nature, 163, 323 (1949).

[43] J. A. Bearden, Rev. Mod. Phys., 39, 78 (1967).

[44] A. Guinier, La radiocristallographie, Dunod, Paris, 1945, p. 6.

[45] W. L. Bragg, Proc. Camb. Phil. Soc., 17, 43 (1912).

[46] W. L. Bragg, Proc. Roy. Soc. (London), 89A, 248 (1913).

[47] M. Siegbahn, Spektroskopie der Röntgenstrahlen, 2nd ed., Springer, Berlin, 1931. An English translation of the first edition was published by Oxford University Press, London, in 1925.

[48] A. H. Compton and R. L. Doan, Proc. Nat. Acad. Sci. U.S., 11, 598 (1925).

[49] J. A. Bearden, Phys. Rev., 48, 385 (1935).

[50] National Bureau of Standards, Handbook HB76. Issued 1961, and procurable from the Superintendent of Documents, Washington, D. C. Price, $0.25.

[51] A. H. Compton, Phys. Rev., 21, 715 (1923); 22, 409 (1923).

[52] P. Debye, Ann. Phys., 43, 49 (1914).

[53] A. Larsson, M. Siegbahn, and I. Waller, Naturwissenschaften, 12, 1212 (1924); Phys. Rev., 25, 235 (1925).

[54] W. Stenström, Dissertation, University of Lund (1919).

[55] W. T. Sproull, X-rays in Practice, McGraw-Hill, New York, 1946, p. 98.

[56] A. J. C. Wilson, *Proc. Camb. Phil. Soc.*, **36**, 485 (1940).

[57] H. P. Klug, *J. Am. Chem. Soc.*, **68**, 1493 (1946).

[58] *International Tables for X-ray Crystallography*, Vol. III, Kynoch Press, Birmingham, England 1962, Section 2.3.

[59] *Methods of Obtaining Monochromatic X-rays and Neutrons*, Bibliography 3, International Union of Crystallography Commission on Crystallographic Apparatus, 1967. Orders can be placed with Polycrystal Book Service (P. O. Box 11567, Pittsburgh, Pa.) or with any book seller. Price $3.00.

[60] W. A. Wood, *Proc. Phys. Soc. (London)*, **43**, 275 (1931).

[61] H. Kersten and J. Maas, *Rev. Sci. Instrum.*, **4**, 14 (1933); S. S. Sidhu, *Rev. Sci. Instrum.*, **8**, 308 (1937); O. Kratky, *Naturwissenschaften*, **31**, 325 (1943); E. A. Wood and L. M. Towsley, *Rev. Sci. Instrum.*, **24**, 547 (1953).

[62] P. A. Ross, *Phys. Rev.*, **28**, 425 (1926); *J. Opt. Soc. Am.*, **16**, 375, 433 (1928); O. Kratky, *Naturwissenschaften*, **31**, 325, 442 (1943).

[63] P. Kirkpatrick, *Rev. Sci. Instrum.*, **10**, 186 (1939); **15**, 223 (1944).

[64] P. Kirkpatrick and C. K. Chang, *Phys. Rev.*, **66**, 159 (1944).

[65] E. O. Wollan, *Phys. Rev.*, **43**, 955 (1933).

[66] J. A. Soules, W. L. Gordon and C. H. Shaw, *Rev. Sci. Instrum.*, **27**, 12 (1956).

[67] W. Ruland, *Acta Crystallogr.*, **14**, 1180 (1961).

[68] W. Bol, *J. Sci. Instrum.*, **44**, 736 (1967).

[69] J. Whetstone, *Research*, **2**, 194 (1949).

[70] T. C. Furnas, *Rev. Sci. Instrum.*, **28**, 1042 (1957).

[71] I. Fankuchen, *Nature*, **139**, 193 (1937); *Phys. Rev.*, **53**, 910 (1938); R. C. Evans, P. B. Hirsch, and J. N. Kellar, *Acta Crystallogr.*, **1**, 124 (1948).

[72] H. Lipson, J. B. Nelson, and D. P. Riley, *J. Sci. Instrum.*, **22**, 184 (1945).

[73] S. R. Maloof, *Rev. Sci. Instrum.*, **27**, 146 (1956).

[74] J. D. H. Donnay and I. Fankuchen, *Rev. Sci. Instrum.*, **15**, 128 (1944); F. M. Wrightson and I. Fankuchen, *Rev. Sci. Instrum.*, **22**, 212 (1951).

[75] Y. Cauchois, *J. phys. radium* (7)**3**, 320 (1932); J. W. M. DuMond and B. B. Watson, *Phys. Rev.*, **46**, 316 (1934).

[76] H. H. Johann, *Z. Phys.*, **69**, 185 (1931); B. E. Warren, *Rev. Sci. Instrum.*, **21**, 102 (1950).

[77] J. W. M. DuMond and H. A. Kirkpatrick, *Rev. Sci. Instrum.*, **1**, 88 (1930); T. Johansson, *Naturwissenschaften*, **20**, 159 (1932); *Z. Phys.*, **82**, 507 (1933).

[78a] R. M. Bozorth and F. E. Haworth, *Phys. Rev.*, **53**, 538 (1938) [78b] C. S. Smith, *Rev. Sci. Instrum.*, **12**, 312 (1941).

[79] J. Witz, *Acta Crystallogr.*, **A25**, 30 (1969).

[80] M. Renninger, *Z. Kristallogr.*, **107**, 464 (1956).

[81] C. J. Sparks, Jr., *Highly Oriented Graphite as an Incident and Diffracted Beam Monochromator*, a paper presented at the Pittsburgh Diffraction Conference, Pittsburgh, Pa, November, 4–6, 1970.

[82] E. M. Proctor, T. C. Furnas, and W. F. Loranger, *Advances in X-ray Analysis*, Vol. 14, Plenum Press, New York, 1971, p. 38.

[83] G. E. Bell, *Brit. J. Radiol.*, **9**, 578 (1936).

[84] *International Tables for X-ray Crystallography*, Vol. III, Kynoch Press, Birmingham, England, 1962, Section 3.2.1.4, p. 160.

[85] S. Ergun and V. H. Tiensuu, *J. Appl. Phys.*, **29**, 946 (1958).

[86] G. Becherer, G. Hermes, and V. Motzfeld, *Z. Kristallogr.*, **128**, 85 (1969).

[87] W. S. Toothacker and L. E. Preuss, *Advances in X-ray Analysis*, Vol. 14, Plenum Press, New York, 1971, p. 139.

CHAPTER 3

FUNDAMENTAL PRINCIPLES OF
X-RAY DIFFRACTION

3-1 KINEMATICAL AND DYNAMICAL DIFFRACTION THEORY

Within 2 years after the discovery of x-ray diffraction, Darwin[1] had produced several papers discussing the intensity of diffraction from a perfect crystal. His approach treated the phenomenon as a problem of Fraunhofer diffraction in three dimensions and became known as the kinematical theory of x-ray diffraction. By this theory the scattering from each volume element of a crystal was considered independent of that of other volume elements, and was considered to pass out of the crystal without further scattering. Such treament gives the directions of the diffracted beams, and often their intensities, with quite adequate accuracy for many purposes. The underlying assumptions of the kinematical theory, however, are not completely justifiable. For instance, the scattered beams within the crystal may be rescattered, and may combine again with the primary beam and with one another. Darwin soon realized this, and made allowance for multiple scatter in his theory.

Ewald[2] also treated the same problem, independently and in greater scope. His theory, called the dynamical theory, took into account all wave interactions within the crystal. The incident and diffracted beams are coherently coupled within the crystal and swap energy back and forth. Thus the dynamical theory considers the total electromagnetic field in a crystal. For fine crystal powders the same intensity expression results from either theory. However, when diffraction from large perfect crystals is considered, dynamical theory must be used. A detailed treatment of dynamical diffraction theory is inappropriate to the immediate purpose of this book. Accordingly, the interested reader is referred to two excellent reviews of the subject[3,4] and additional material in the original literature[5–8].

120

3-2 THE GEOMETRY OF DIFFRACTION

The phenomenon of x-ray diffraction by crystals results from a scattering process in which x-rays are scattered by the electrons of the atoms without change in wavelength (coherent or Bragg scattering, Section 2-3.4). A diffracted beam is produced by such scattering only when certain geometrical conditions are satisfied, which may be expressed in either of two forms, the *Bragg* law or the *Laue* equations. The resulting diffraction pattern of a crystal, comprising both the positions and intensities of the diffraction effects, is a fundamental physical property of the substance, serving not only for its speedy identification (Chapter 7) but also for the complete elucidation of its structure. Analysis of the positions of the diffraction effect leads immediately to a knowledge of the size, shape, and orientation of the unit cell. To locate the positions of the individual atoms in the cell, the intensities must be measured and analyzed. Most important in relating the positions of the atoms to the diffraction intensities is the *structure factor* equation.

3-2.1 Scattering of X-rays by Electrons and Atoms

X-rays are electromagnetic waves, and as such are accompanied by a periodically changing electric field as they proceed outward from their source. An electron in the path of such a wave is excited to periodic vibrations by the changing field, and itself becomes a source of electromagnetic waves of the same frequency and wavelength. There thus arises from this interaction a new spherical wave front of x-rays, with the electron as its origin, deriving its energy from the impinging beam. By this process the electron is said to *scatter* the original beam.

An atom is made up of a positively charged nucleus surrounded by a cloud of electrons, one for each increment of nuclear charge, the number being equal to the atomic number of the element in question. The scattered waves from the several electrons in an atom combine, so that the scattering effect from an atom may be regarded as essentially that of a point source of scattered x-rays. The intensity of the scattering of course is dependent on the number of electrons in the atom, but because the electrons are distributed throughout the volume of the atom rather than concentrated at a point, the intensity varies with direction (Section 3-3.4D). For the present, however, in treating the geometry of diffraction, the atom is considered a point scattering source.

3-2.2 Scattering by a Regularly Spaced Row of Atoms

Interference phenomena with water waves and light are well known. In a similar fashion constructive and destructive interference can arise between the scattered x-ray waves from atoms. Suppose an x-ray beam encounters a row of regularly spaced atoms, as in Fig. 3-1. The parallel wave fronts cause each atom to become a source of a set of spherical scattered waves of the same frequency and wavelength. In Fig. 3-1 we consider the succession of wave crests and troughs about two neighboring atoms at some instant in time. It is necessary only to consider the scattering about a pair of neighboring atoms, as the interatomic distance and the wavelength of the x-rays determine the geometry of the diffraction effects. Scattering from atoms more distant in the row merely contributes (at the same angular directions) to the scattered beams depicted in Fig. 3-1. All points of intersection of the two sets of concentric arcs are points at which the crests of the waves from both atoms coincide and their amplitudes add, leading to constructive interference and a diffraction maximum. At points between the intersections the waves are more or less out of phase and lead to various degrees of destructive interference or extinction.

An obvious direction of reinforcement is that perpendicular to the

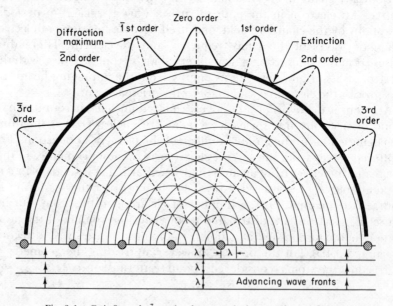

Fig. 3-1.   Reinforced scattering by a regularly spaced row of atoms.

original wave front. Here the wave crest difference between the waves scattered from the two atoms is zero, and gives rise to the *zero-order* diffracted beam. To the right of the zero-order beam is a prominent direction of crest intersections characterized by a wave crest or phase difference of one. This is the *first-order* diffracted beam. Similarly, farther to the right, *second-order*, *third-order*, and so on to the *nth-order*, diffracted beams represent 2, 3, 4, . . . , *n* wave crest (wavelength) differences in phase between the wavelets from neighboring atoms. Corresponding negative orders of diffraction (minus first-order, minus second-order, etc.) arise on the opposite side of the direction of the zero-order beam. Although Fig. 3-1 represents the special case of a beam impinging at right angles on a row of atoms, the general case of a beam making any angle with the row is entirely analogous.

3-2.3 Conditions for Diffraction by a Linear Lattice of Atoms

A straight row of regularly spaced atoms constitutes a linear lattice. Consider that a parallel beam of x-rays meets such a row of atoms at an angle Δ (Fig. 3-2), a being the constant spacing between the atoms. All

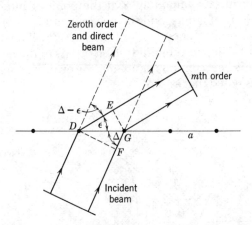

Fig. 3-2. Conditions for diffraction from a row of atoms.

atoms of the row act as centers for series of spherically spreading scattered waves, and reinforcement leading to zero-, first-, second-, and higher-order diffracted beams occurs in certain directions. Suppose one of these directions of constructive interference makes an angle ϵ with the row axis. Then, since x-rays scattered at D must be in phase with those scattered at G, the paths DE and FG must differ by a whole number of wavelengths. That is

$$DE - FG = m\lambda, \tag{3-1}$$

where m is an integer, and λ is the wavelength of the x-ray beam. From simple trigonometry

$$DE = a \cos \epsilon, \quad \text{and} \quad FG = a \cos \Delta. \tag{3-2}$$

Hence the path difference is

$$a \cos \epsilon - a \cos \Delta = a\,(\cos \epsilon - \cos \Delta) = m\lambda, \tag{3-3}$$

and equation 3-3 is the condition to be satisfied by the several discrete orders of diffracted beams from this lattice row. The direction of any given order of diffracted beam is obtained by solving for $\cos \epsilon$ and substituting the appropriate value of m,

$$\cos \epsilon_m = \cos \Delta + \frac{m\lambda}{a}. \tag{3-4}$$

Obviously, the incident beam could have met the lattice row at the angle Δ by impinging on the row from any direction that is a generator of a cone (the incident beam cone) concentric with the row and of semi-apex angle Δ (Fig. 3-3). The locus of all zero-order beams is, then, an identical cone with a common apex at the point of intersection of the beam and the row of atoms; and the incident and zero-order beams are diametrically opposite generators of the two cones. The directions satisfying equation 3-4 for the other diffracted orders of the beam lie on other cones with the same common apex and appropriate semiapex angles ϵ (Fig. 3-3). It is noted that $\epsilon_0 = \Delta$.

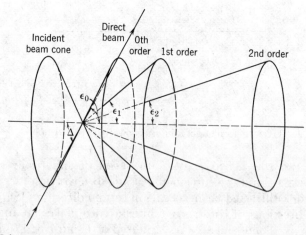

Fig. 3-3.   Schematic representation of the positive diffraction cones from a line of atoms.

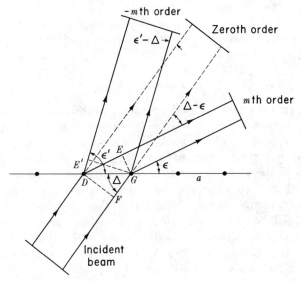

Fig. 3-4.   Geometrical relations between the mth and $-m$th orders of diffraction from a row of atoms.

To the left of the zero-order beam (Fig. 3-4) are negative orders of the diffracted beam, the $-m$th order making an angle $\epsilon' - \Delta$ on the left with the zero-order beam, and an angle ϵ' with the lattice row, where

$$\epsilon'_{-m} - \Delta \leqq \Delta - \epsilon_m. \tag{3-5}$$

Note that angle ϵ' is measured from the positive end of the lattice row. Angle $\epsilon'_{-m} - \Delta$ is always smaller than $\Delta - \epsilon_m$ when angle $\Delta < 90°$, and the two angles are equal in the special case $\Delta = 90°$. The locus of the directions of the negative orders of the diffracted beam is thus a series of cones to the left of the zero-order cone and having the same common apex as the positive orders. The inner semiapex cone angle ϵ' becomes the external apical angle when $\epsilon' > 90°$. A trigonometrical construction and treatment similar to that leading to equation 3-4 gives

$$FG - DE' = a\,(\cos \Delta - \cos \epsilon') = m\lambda, \tag{3-6}$$

and

$$\cos \epsilon'_{-m} = \cos \Delta - \frac{m\lambda}{a}. \tag{3-7}$$

In the special case in which the incident beam is perpendicular to the row of atoms, $\Delta = 90°$, the zero-order cone degenerates into a disk perpendicular to the linear lattice, and the cones of each positive and negative order of diffraction become symmetrical about the zero order. For a given λ and interatomic spacing a, only a limited number of diffraction orders is possible, since when m is such as to cause the right member of equation 3-4 to exceed unity (or -1 in equation 3-7) no solution for ϵ or ϵ' is possible.

3-2.4 Diffraction by a Simple Cubic Lattice

The geometrical analysis based on cones of diffraction can readily be extended to the two- and three-dimensional cases [9]. In the plane lattice there are two axial directions, each giving rise to a coaxial series of diffraction cones; in a space lattice there are three such directions. The condition for in-phase scattering from the diffraction cones about the several axial directions is that the nests of cones intersect each other; and, for simultaneous diffraction in a single direction from *every* atom in the lattice, special beam orientation is required such as to cause the two cone intersections of the plane-lattice array, and three of the six intersections of the three-dimensional case, to coalesce to a single defining direction. Further details of this approach are, however, purposely omitted here, as they are considered to add little to the beginner's understanding of the geometry of diffraction in space.

Analytical treatment of diffraction was used most successfully by Wyckoff many years ago in his monumental contribution to the literature of x-ray diffraction [10]. For present purposes it is likewise deemed most fruitful to apply the analytical approach, and to pass directly, as Wyckoff did, to a consideration of the simplest space lattice, the simple cubic lattice. A treatment of diffraction in the most general three-dimensional lattice, although desirable perhaps on purely logical grounds, is accompanied by mathematical gymnastics which contribute little to, if they do not actually hinder, a speedy understanding of diffraction geometry. Moreover, after this simplest case is analyzed, it is sufficient merely to point out the modifications of the basic diffraction equation needed for crystals of less symmetry. Laue [11], of course, after his discovery of x-ray diffraction, immediately developed the mathematical theory of the positions of diffraction effects from crystals.

In Fig. 3-5 a portion of a simple cubic lattice of atoms is depicted with a beam of parallel x-rays incident upon its three orthogonal lattice directions OX, OY, and OZ at angles Δ_1, Δ_2, and Δ_3, respectively. The corresponding diffracted beam makes angles ϵ_1, ϵ_2, and ϵ_3 with these directions. The conditions to be satisfied for in-phase scattering are analogous to the linear lattice case, but now three directions in space are involved and there are three equations similar to equation 3-3:

$$a\,(\cos \epsilon_1 - \cos \Delta_1) = m\lambda,$$

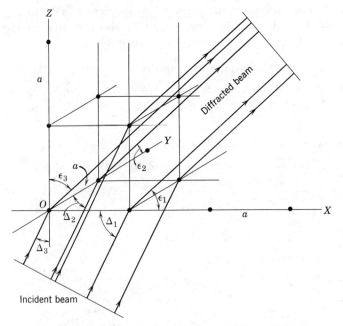

Fig. 3-5. Diffraction by a portion of a simple cubic lattice.

$$a\ (\cos \epsilon_2 - \cos \Delta_2) = p\lambda, \tag{3-8}$$
$$a\ (\cos \epsilon_3 - \cos \Delta_3) = q\lambda.$$

These equations with m, p, and q integral represent the conditions for constructive interference leading to the mth-order diffraction from row OX, the pth-order from row OY, and the qth-order from row OZ. They are known as the *Laue* equations of x-ray diffraction, and the condition for x-ray diffraction from the lattice as a whole is that they be simultaneously satisfied. For a simultaneous solution to the Laue equations, it is necessary that the directions defined by each of the three equations be the same. In general this is not the case, and no diffraction occurs. The angles Δ between the incident beam and the lattice rows may, however, be varied at will, and it is possible to choose them so that the three equations (3-8) define an identical direction.

The three cosines Δ are merely the direction cosines of the incident beam direction, whereas the cosines ϵ are the direction cosines of the diffracted beam. The most desirable solution of the Laue equations is one in terms of the angle between the diffracted beam and the undeviated direct beam, since this is an angle readily measured experimentally. This angle, to be designated 2θ, is the angle of deviation of the diffracted beam, and is measured in the plane of the incident and diffracted beams (Fig. 3–6). Cos 2θ is readily expressed by the formula for the angle between two lines in terms of their direction cosines:

$$\cos 2\theta = \cos \epsilon_1 \cos \Delta_1 + \cos \epsilon_2 \cos \Delta_2 + \cos \epsilon_3 \cos \Delta_3. \qquad (3\text{-}9)$$

The solution of equations 3-8 in terms of the angle 2θ may now proceed as follows. First square the equations and add them together, obtaining

$$a^2(\cos^2 \epsilon_1 - 2 \cos \epsilon_1 \cos \Delta_1 + \cos^2 \Delta_1 + \cos^2 \epsilon_2 - 2 \cos \epsilon_2 \cos \Delta_2 + \cos^2 \Delta_2$$

$$+ \cos^2 \epsilon_3 - 2 \cos \epsilon_3 \cos \Delta_3 + \cos^2 \Delta_3) = \lambda^2(m^2 + p^2 + q^2). \qquad (3\text{-}10)$$

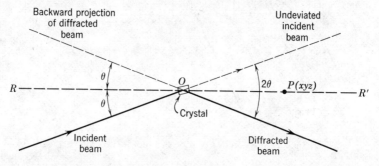

Fig. 3-6. Relation between the incident and diffracted beams and the "diffracting plane."

Recalling that the sum of the squares of the direction cosines of a line is equal to unity, we have

$$\cos^2 \epsilon_1 + \cos^2 \epsilon_2 + \cos^2 \epsilon_3 = 1, \qquad (3\text{-}11)$$

and

$$\cos^2 \Delta_1 + \cos^2 \Delta_2 + \cos^2 \Delta_3 = 1.$$

Then, substituting equations 3-9 and 3-11 in 3-10, we obtain

$$a^2(2 - 2 \cos 2\theta) = \lambda^2(m^2 + p^2 + q^2). \qquad (3\text{-}12)$$

From the trigonometric relation,

$$1 - \cos 2\theta = 2 \sin^2 \theta,$$

equation 3-12 simplifies to

$$4a^2 \sin^2 \theta = \lambda^2(m^2 + p^2 + q^2), \qquad (3\text{-}13)$$

which, when solved for λ, gives

$$\lambda = \frac{2a}{\sqrt{m^2 + p^2 + q^2}} \sin \theta.$$

If the integers m, p, and q possess a greatest common divisor n, such that $m = nm_0$, $p = np_0$, and $q = nq_0$, then

$$n\lambda = \frac{2a}{\sqrt{m_0^2 + p_0^2 + q_0^2}} \sin \theta. \tag{3-14}$$

Equation 3-14 thus relates the wavelength of the x-rays to the primitive translation of the lattice, to half the angle of deviation of the diffracted beam, and to the integers m_0, p_0, and q_0, derived from the orders of the linear-lattice diffraction cones. It is immediately evident too, that a more satisfying interpretation of equation 3-14, and especially of the integers m_0, p_0, and q_0, is greatly to be desired. Fortunately, an elegant interpretation is provided by the simple demonstration that the diffracted beam behaves as if it were specularly reflected from a lattice plane of atoms in the crystal, an interpretation first pointed out by Bragg[12] in 1913. Following an analytical proof that the "diffracting plane" is a lattice plane, a simple derivation of the Bragg equation from the "reflection" analogy is presented.

3-2.5 Proof That the "Diffracting Plane" Is a Lattice Plane

Referring again to Fig. 3-6, consider the line RR' to be the trace of a plane perpendicular to the plane of the incident and diffracted beams, and bisecting the angle 2θ. This plane is the "diffracting plane," a plane from which the diffracted beam appears to have been specularly reflected. Consider $P(xyz)$ to be the projection of a point P with coordinates (xyz) lying in the plane. The radius vector OP of the line to P makes equal angles with the diffracted beam and the undeviated beam, and x/OP, y/OP, and z/OP are its direction cosines. Designating the angle between OP and the undeviated beam δ, we have

$$\cos \delta = \frac{x}{OP} \cos \Delta_1 + \frac{y}{OP} \cos \Delta_2 + \frac{z}{OP} \cos \Delta_3. \tag{3-15}$$

The angle between OP and the diffracted beam is also δ, hence

$$\cos \delta = \frac{x}{OP} \cos \epsilon_1 + \frac{y}{OP} \cos \epsilon_2 + \frac{z}{OP} \cos \epsilon_3. \tag{3-16}$$

Equating equations 3-15 and 3-16, and collecting terms, gives

$$x(\cos \epsilon_1 - \cos \Delta_1) + y(\cos \epsilon_2 - \cos \Delta_2) + z(\cos \epsilon_3 - \cos \Delta_3) = 0. \tag{3-17}$$

Then substituting from equation 3-8, we obtain

$$\frac{m\lambda x}{a} + \frac{p\lambda y}{a} + \frac{q\lambda z}{a} = 0,$$

or

$$m_0x + p_0y + q_0z = 0. \tag{3-18}$$

This is seen to be the equation for the lattice plane (Section 1-2.12) with Miller indices $(hkl) = (m_0p_0q_0)$ and passing through the origin. The "diffracting plane" is thus demonstrated to be the lattice plane (hkl), and it is customary to write equation 3-14 as

$$n\lambda = \frac{2a}{\sqrt{h^2 + k^2 + l^2}} \sin \theta, \tag{3-19}$$

and to identify each diffracted beam by its corresponding Miller index triplet (hkl). This is the basis and justification for the Bragg "reflection" analogy and for our wide use of the term "reflections" in speaking of the diffracted beams. The integer n is also seen to be merely the order of the diffracted beam or reflection, and it may be left within the Miller symbol (Section 1-2.8) or expressed separately, as in equation 3-19.

3-2.6 The Bragg Equation

Equation 3-19 is the *Bragg* equation for the cubic system. Reference to Table 1-3 (Section 1-3.6) reveals that the factor $a/\sqrt{h^2 + k^2 + l^2}$ in equation 3-19 is simply the interplanar spacing d for the plane (hkl). The Bragg equation in its general form is then written

$$n\lambda = 2d \sin \theta. \tag{3-20}$$

An analytical derivation of the Bragg equation when carried through for a crystal system of lower symmetry leads to an expression identical with equation 3-19 or 3-20, except with a more complicated d term. These d's are in every case the interplanar spacing for the reflecting plane. Thus to obtain the equation in the special form for calculations in a given crystal system, it is necessary only to substitute in place of d the expression (Table 1-3) for $d_{(hkl)}$ for the appropriate system.

3-2.7 Derivation of the Bragg Equation from the "Reflection" Analogy

Bragg's explanation of x-ray diffraction effects in term of "reflection" from a stack of parallel atomic planes deserves brief consideration both because of its simplicity and its historical interest. In the discussion of space lattices in Chapter 1, it was indicated that the atomic or molecular units in a crystal lie at the intersections of a space lattice, that the prominent crystal faces are those most thickly populated with lattice points (atoms or molecules), and that parallel to every possible crystal face or plane is a series of equispaced identical planes. When a beam of x-rays strikes an extended crystal face and is reflected in the Bragg sense the phenomenon is not a surface reflection, as with ordinary light. Parallel to the face is an effectively infinite series of equispaced atomic planes which the x-rays penetrate to a depth of several million layers before being appreciably absorbed. At each atomic plane a minute portion of the beam may be considered to be reflected. For these tiny reflected beams to emerge as a single beam of appreciably intensity, they must not be absorbed in passing through layers nearer the surface as they emerge, and, far more important, the beams from successive layers must not interfere and destroy each other. If conditions can be arranged so that reinforcement rather than destruction occurs, all planes in the series that are not too deep in the crystal will contribute to the reflection. Bragg demonstrated these conditions in the following manner.

Consider the lines pp, p_1p_1, p_2p_2, and so on, of Fig. 3-7 to represent the traces of a series of atomic planes of constant interplanar spacing d parallel to a crystal face. AB, $A'B'$ is a train of incident x-rays of wavelength λ impinging on the planes and reflecting off in the direction CD. For the reflected wavelet from B' to reinforce the one reflected at C, it must arrive at C in phase with the wave ABC. This will be the case if the path difference is a whole number of wavelengths, that is, if

$$B'C - BC = n\lambda. \tag{3-21}$$

By simple trigonometry,

$$B'C = \frac{d}{\sin\theta}, \tag{3-22}$$

and

$$BC = B'C\cos 2\theta = \frac{d(\cos 2\theta)}{\sin\theta}. \tag{3-23}$$

Substituting in equation 3-21,

$$\frac{d}{\sin\theta}(1-\cos 2\theta) = \frac{d}{\sin\theta}2\sin^2\theta = n\lambda. \tag{3-24}$$

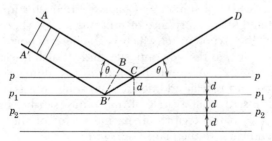

Fig. 3-7. Geometry of the Bragg "reflection" analogy.

and

$$n\lambda = 2d \sin \theta. \tag{3-25}$$

This is the Bragg equation, also known as the *Bragg* law. For a crystal of a given d spacing, and for a given wavelength λ, the various orders n of reflection occur *only* at the precise values of angle θ which satisfy equation 3-25. At other angles there is no reflected beam because of interference. This is in marked contrast to reflection of a beam of light from a polished metal surface, which may take place over a large continuous angular range. The reflection point of view thus provides a very satisfying picture of diffraction in crystals, and it has been widely used from the beginning.

3-2.8 The Geometrical Picture of Diffraction in Reciprocal Space

Although the mechanism of x-ray diffraction cannot be explained or demonstrated in *reciprocal space* (the space which the reciprocal lattice (Section 1-3.7) is imagined to occupy), a useful geometrical picture [13] of diffraction can be obtained therein on the basis of Bragg's law. By including the n in the Miller symbol of the reflection indices and re-arranging, equation 3-25 can be written in the form

$$\sin \theta_{(hkl)} = \frac{\lambda}{2d_{(hkl)}}. \tag{3-26}$$

An interesting result is now obtained by substituting in equation 3-26 the value of $d_{(hkl)}$ from equation 1-34, namely, k^2/ρ. At this time also, the proportionality constant k^2 is assigned the value λ, the wavelength of the x-rays used. Making these substitutions gives

$$\sin \theta_{(hkl)} = \frac{\lambda}{2(\lambda/\rho)} = \frac{\rho_{(hkl)}}{2}. \tag{3-27}$$

Accordingly, for each plane (hkl) the Bragg angle θ is an angle whose sine is equal to one-half its ρ value. It is recalled, too, that all plane triangles inscribed in a sphere and having a diameter for one side are right triangles. Thus, in a sphere of unit radius, θ is the angle between one side and the hypotenuse (which is a sphere diameter) of such an inscribed triangle, ρ is the side opposite, and all such inscribed triangles satisfy equation 3-27. A simple diagram in two dimensions illustrates these relations and reveals additional interesting aspects. In Fig. 3-8 the triangle AOP inscribed in a circle of unit radius satisfies the above relation. Since $\rho_{(hkl)}$ is the reciprocal spacing vector of the plane (hkl), it stands perpendicular to the plane (hkl), and tt in the drawing is the trace of this plane. The angle between the trace tt and AO is θ of triangle AOP.

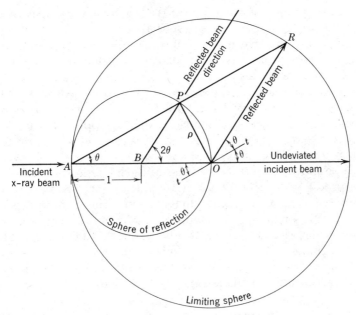

Fig. 3-8. The geometrical conditions for reflection in reciprocal space. [Courtesy of J. Bernal, *Proc. Roy. Soc.* (London), **113A**, 117 (1926).]

If now the direction AO is the direction of an incident x-ray beam, this beam makes the correct Bragg angle with (hkl) for reflection, and the direction of the reflected beam is OR, which is the same as BP.

Corresponding relations hold in space for all similarly inscribed triangles about a diameter of a sphere of radius 1. Thus, if an x-ray beam passes along a diameter of this sphere of unit radius, all points on the surface of the sphere satisfy the condition for Bragg reflection of the beam, and it is logical to term it the *sphere of reflection*. This sphere is now considered to be placed in the reciprocal lattice of some crystal with the point O of Fig. 3-8 at the origin of the reciprocal lattice and the beam direction AO coincident with a reciprocal lattice primitive translation. For all orthogonal cases this is equivalent to an incident beam perpendicular to one of the crystal axes and parallel to another. For this position of the sphere, all points of the reciprocal lattice which lie on its surface are in position to reflect the beam. Obviously, few reciprocal points can be expected to lie exactly on the surface of a sphere so placed in a lattice, and therefore few reflections can be expected. This is in line with experience, as we obtain few reflections by irradiating a stationary crystal of average cell size with a monochromatic x-ray beam. As cell size increases, the reciprocal lattice constants decrease, the reciprocal lattice points become closer together, and there is greater probability of their lying on the surface of the sphere of reflection.

One way to insure that many reciprocal lattice points will lie on the sphere of reflection, at least momentarily, and thus be in position to reflect the x-ray beam for a short time, is to rotate or oscillate the reciprocal lattice about an axis through its origin and tangent to the sphere at O, at the same time holding the sphere of reflection stationary. With such rotation or oscillation of the lattice, many reciprocal lattice points are caused to intersect the surface of the sphere of reflection, and at the particular instant of intersection the point is on the sphere of reflection and in a position to reflect the x-ray beam. Such rotation or oscillation of the reciprocal lattice is achieved by rotation or oscillation of the individual crystal in the x-ray beam, and is the basis of most of the important single-crystal techniques. In a complete rotation of the reciprocal lattice, the intersection with the sphere of reflection cuts from it a tore shaped like a doughnut with an infinitely small hole. Sections of this *tore of reflection* are shown in Fig. 3-9. With the x-ray beam inclined to the axis of rotation, the surface is an interlacing tore (Fig. 3-9*B*).

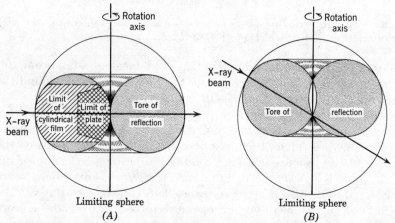

Fig. 3-9.   Sections of the tore of reflection. [Courtesy of J. Bernal, *Proc. Roy. Soc.* (London), **113A**, 117 (1926).]

For a given x-ray wavelength and position of the axis of rotation with respect to the incident beam, no reciprocal lattice points lying outside the tore can reflect. All points within the tore reflect the beam, but the reflected beams proceed outward in discrete directions in all eight quadrants of the sphere of reflection, so that a spherical photographic film is required to record all of them. In practice of course we are limited to flat or cylindrical films which register only a limited portion of the tore

(Fig. 3-9A), a cylindrical film being considerably better than a flat film. One rotation photograph evidently does not register very much of the reciprocal lattice, which actually extends to infinity. A brief consideration makes it clear, however, that the tores of reflection for rotations about three mutually perpendicular axes contain most of the points out to a distance of radius 2 around the origin, and cover those near the origin at least twice.

It is important now to consider briefly the infinity of reciprocal lattice points farther from the origin. From equation 3-27,

$$\rho_{(hkl)} = 2 \sin \theta_{(hkl)}. \tag{3-28}$$

Thus ρ cannot exceed the value 2, since the maximum value of $\sin \theta$ is 1. This means that no point outside a sphere of radius 2 around the origin can ever give a reflection with a given wavelength, hence the term *limiting sphere* (Figs. 3-8 and 3-9). More reciprocal lattice points will fall within the limiting sphere, and thus be potential reflecting spots, if a shorter x-ray wavelength is used. From equation 1-34,

$$\rho = \frac{k^2}{d} = \frac{\lambda}{d}. \tag{3-29}$$

Thus ρ for a given plane is directly proportional to λ, and the number of reciprocal lattice points along a radius of the limiting sphere is inversely proportional to the λ used. In turn, the total number of points within the limiting sphere is inversely proportional to λ^3, so that to obtain data from as many planes as possible, it may be important to give careful consideration to the choice of the most suitable wavelength for a study.

3-3 THE INTENSITY OF DIFFRACTION

3-3.1 Perfect and Imperfect Crystals

Both Darwin[1] and Ewald[2], in their studies of the intensity of diffraction, postulated an ideally perfect crystal. Such a crystal is one whose atoms throughout its entire extent are placed accurately at the points of an undistorted space lattice. Moreover, in these treatments the crystal is considered to have negligible absorption for the x-ray beam. For such a crystal with real atoms at the lattice points, the reflected wavelets at the various atomic planes interact with each other and with the incoming beam, and the observed intensity of the reflected beam is the result of this interaction. Up to a certain angle θ' (Fig. 3-10), only a few seconds of arc from the angle θ which satisfies the Bragg equation for the plane under consideration, the resultant intensity is zero. Then,

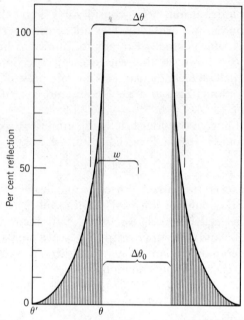

Fig. 3-10.   Reflection from a perfect crystal (schematic).

within the small range θ' to θ, the intensity rises sharply to total reflection. The reflection remains total over a small angular range $\Delta\theta_0$ and then drops to zero again. Darwin concluded that the partial reflection (shaded area of Fig. 3-10) amounted to 25 per cent of the total area, whereas Ewald found 20 per cent. Both treatments, however, yield essentially identical expressions for the *effective* breadth of reflection $\Delta\theta$, that of Darwin being

$$\Delta\theta = \frac{4}{3}\,\Delta\theta_0 = \frac{4}{3}\,\frac{N''e^2\lambda^2}{\pi mc^2}\cdot F\cdot\frac{1+|\cos 2\theta|}{\sin 2\theta}, \tag{3-30}$$

where N'' is the number of atoms in unit volume, F is a structure factor term depending on the positions of the electrons and atoms relative to the reflecting plane, and the other terms have their usual significance.

To test expression 3-30 experimentally, we must make a measurement which compares the total amount of reflection from the crystal plane with the amount of radiation incident upon it. A crystal with an extended face is mounted with the face on the axis of a counter diffractometer* and the counter adjusted to the angle 2θ at which reflection

*Earliest measurements of this kind were made with an ionization spectrometer (see Section 2-3.2).

occurs. The crystal is then rotated in the x-ray beam at constant angular velocity ω from a position on the low-angle side of θ to a similar point on the high-angle side sufficient to reflect all radiation for the particular order. If E is the integrated radiation entering the counter, and I_0 is the original incident intensity per unit time, then $E = tI_0$, where t is the time required for the crystal to pass through the region of total reflection. It follows too that $\omega t = \Delta\theta$, the effective angular breadth of reflection. With these arrangements the so-called integrated reflection is measured:

$$\frac{E\omega}{I_0} = \omega t = \Delta\theta. \tag{3-31}$$

The integrated reflection, also called the "coefficient of reflection," is a quantity having the physical dimensions of a pure number.

The experimental test results with real crystals were very confusing at first. Compton[14] found for fine-quality calcite using $\lambda = 0.708$ Å a measured $\Delta\theta = 8.7 \times 10^{-5}$ radian, whereas theory leads to a calculated value of 2.1×10^{-5} rad (3.6″). The results of Allison[15] and Parratt [16] have shown that calcite surfaces can be obtained which for not too short incident wavelengths give results very close to those predicted for perfect crystals. Diamond crystals likewise seem to have comparable perfection. The great majority of crystals, however, when tested in this manner give integrated reflections many times greater than theory predicts. Typical rock salt crystals may give values of 300 to 900″ for the half width w (Fig. 3-10) at half-maximum intensity ($2w \cong 1.06 \Delta\theta_0$ for Darwin's theoretical curve), although Kirkpatrick and Ross[17] reported a value of $w = 87″$ for the finest specimen of rock salt they could find, and Renninger[18] measured a few crystals on which small areas (approximately 1 mm²) gave values with $CuK\alpha$ of the order of 7.1″ wide, the predicted value being about 5″. It has also been observed that grinding or roughening of a crystal face leads to still larger values of the integrated reflection. Since such operations would be expected to destroy the lattice regularity to some extent in the surface layers of the crystal, we are led to the conclusion that even fine specimens of most crystals deviate considerably from the perfection postulated for the ideally perfect crystal.

After consideration of such results, a second kind of ideal crystal was postulated, the *ideally imperfect* crystal. This crystal is visualized as being made up of tiny crystal fragments in completely random arrangement; in other words, it is a fine crystalline powder. The fragments are so small that each one extracts only a negligible amount of energy from the beam. In this situation no phase relationships exist between the waves reflected from different fragments, and the complete reflection is the sum of the contributions from the individual fragments. Darwin[1, 19] and others have calculated the integrated reflection from one of these tiny crystal fragments to be

$$\frac{E\omega}{I_0} = \frac{N''^2 e^4 \lambda^3}{2m^2 c^4} \cdot F^2 \cdot \frac{1 + \cos^2 2\theta}{\sin 2\theta} \cdot \delta V = Q \delta V. \tag{3-32}$$

For diffraction from a sample of powder of volume V, sufficiently large to contain enough tiny fragments for their orientations to be considered to have a continuous random distribution, and yet small enough for absorption to be neglected, the expression becomes

$$\frac{P}{I_0} = \tfrac{1}{2} Q V \cos \theta, \tag{3-33}$$

where P is the total energy reflected into the powder diffraction halo for the particular set of planes considered.

Between the extremes just considered is the imperfect single crystal, one made up of these tiny crystal fragments or blocks arranged in a nearly but not quite parallel arrangement to one another, the units being in the neighborhood of 10^{-5} cm on a side and their deviation from parallel alignment several minutes or seconds or arc. Such a crystal would be a *mosaic* crystal. Only a portion of its blocks would reflect a beam at a given orientation, but slight rotation of the crystal would bring other blocks into reflecting position, apparently increasing the angular range of total reflection. Whether the imperfection of real crystals results from such a mosaic block structure is not entirely certain; imperfections due to warping of the lattice, or to other causes, are possible. The behavior of other possible types of irregularity toward x-rays, however, is similar to the behavior of the mosaic structure, provided that the regions of exact regularity do not extend over distances of more than a few thousand atomic planes without a dislocation.

If we also include the linear absorption coefficient μ_l of the radiation in the crystal, the integrated reflection from an extended face of a mosaic crystal is

$$\frac{E\omega}{I_0} = \frac{Q}{2\mu_l}. \tag{3-34}$$

In similar fashion the integrated reflection by transmission through a thin parallel-sided slip of crystal (the reflecting planes being perpendicular to the sides of the crystal section) is

$$\frac{E\omega}{I_0} = Qt \sec \theta e^{-\mu_l t \sec \theta} \tag{3-35}$$

where again μ_l is the linear absorption coefficient, and t is the thickness of the crystal slip. Comparing equations 3-34 and 3-35 with equations 3-30 through 3-31 and 3-32, we see that the important quantities leading to differences in the integrated reflection from mosaic and perfect crystals are the number of atoms per unit volume N'' and the structure factor F, both of which enter in as the square in a mosaic crystal and as the first power in a perfect crystal. These differences in turn cause decided changes in the numerical values for the same reflection from the two crystal types. For rigorous derivations and other theoretical details on absolute intensities, the reader is referred to the original papers cited, and to the excellent references on x-ray diffraction theory at the end of the chapter.

3-3.2 Primary and Secondary Extinction

Darwin[19] imagined the differences between the intensities from real single crystals and those expected from perfect crystal theory to arise from two factors, which he called *primary* and *secondary extinction*,* both of which act as an additional absorption coefficient when at the exact angle for reflection. A perfect crystal, not at the critical position for a Bragg reflection, exhibits a small but finite absorption for an x-ray beam passing through it (equation 2-12). During a Bragg reflection each successive plane extracts and reflects a tiny amount of energy from the incident beam, with the result that the beam becomes increasingly weaker as it passes to lower and lower planes. Furthermore, some of the reflected wavelets may also undergo a second reflection from the undersides of the atomic planes, sending them back in the direction of the incident beam with which they interfere, resulting in an additional decrease in direct beam intensity. These two effects act during reflection, in addition to the ordinary absorption coefficient μ, and combine in rather perfect crystals of calcite to produce an absorption coefficient 70-fold larger at the position for Bragg reflection than at neighboring positions. This abnormal absorption effect is referred to as primary extinction. Its effects are less for the higher-angle reflections, with the result that the measured values of the strongest reflections are usually less strong than they should be in comparison with the weaker reflections.

*The term extinction, introduced by Darwin for these effects, is unfortunate because it suggests complete obliteration of the beam, which is not at all the case. Moreover, the term is also used in connection with the so-called space-group extinctions (Section 3-3.6) wherein certain reflections have zero intensity as a result of subperiodicities introduced by non-primitive lattices, glide planes, and screw axes. The reader must carefully distinguish between these two usages.

Most real crystals, however, are more or less imperfect, and approximate the mosaic type of structure. In such a crystal, when suitably oriented, an incident beam is reflected from a large number of crystal blocks (perhaps several thousand) that are precisely oriented at the Bragg reflection angle. On rotation of the crystal a few seconds of arc, this first group of blocks takes on an incorrect orientation, but a similar new group then assumes the right orientation. The total diffracted beam is the sum of these successive contributions throughout the angular range of the Bragg reflection; and because of the slight misorientation of the blocks, it carries more energy than if the crystal were perfect. During such diffraction reflecting blocks near the surface reduce the beam intensity for deeper blocks simultaneously oriented for reflection. This same thing can happen on the outward path of the beam as well, resulting in an increased absorption coefficient over that of the crystal's regular μ value. This effect is referred to as secondary extinction. The little blocks may show primary extinction too, but even if it is high only a small fraction of the blocks are oriented for reflection at any one beam direction, and all those not at the reflection angle show only the ordinary absorption μ. Again, secondary extinction is much more pronounced for strong reflections than for weak ones. Fine powders (below 10^{-2} to 10^{-3} cm) do not show the effect because of the relatively few crystal particles in the sample that are critically oriented to reflect the incident radiation. Extinction is a dynamical diffraction effect, and it is best treated by dynamical theory. Correcting for extinction is a problem of great difficulty, but, fortunately, much diffraction work, even single-crystal studies, can be done without considering the question of extinction.

3-3.3 Relative and Absolute Intensities

Before we consider in detail the factors affecting diffraction intensities, it is desirable to inquire as to the quality of intensity data required in diffraction studies. Actually, the type of study being pursued determines the quality of intensity data needed. Simple relative intensities from visual estimation of diffraction maxima against a calibrated scale suffice for much of the more routine diffraction work. Indeed, much excellent structure work in the early days was done with only the characterization of the spots or lines as *very very strong, very strong, strong, medium strong, medium, weak medium, weak, very weak,* and *very very weak.* Certain studies, however, demand the most precise absolute intensity data obtainable. At this point only a general discussion of intensity data is presented, to be augmented later by specific details for intensity measurements in each of the special techniques.

A measurement of the integrated reflection from a crystal plane, when carried out so that an actual comparison is made between the total energy in the reflected beam and the intensity I_0 of the incident beam, provides the absolute reflection intensity, the most accurate intensity information obtainable. From equation 3-31 the absolute reflection intensity is $E = tI_0 = \Delta\theta I_0/\omega$. Such measurements are usually made on a counter diffractometer, with the crystal moving through the reflection position at constant angular velocity ω. Comparison with the primary beam is difficult to make because of the great difference in intensity between it and the reflected beam, although an absorber of known thickness may be used to reduce its intensity to a convenient value. The measurement is most simply made on large crystal faces, and is complicated for internal planes not present as faces. These factors make it very tedious to obtain such data from more than a few simple planes.

Since many crystals are unobtainable as large single crystals, Robinson [20] devised methods for accurately measuring the integrated reflections from minute single crystals when they are completely bathed in a monochromatic x-ray beam. Working with tiny anthracene crystals, he established the proportionality between integrated intensity and mass for crystals up to approximately 0.25 mg in weight, and investigated questions of extinction and absorption. This was the first absolute intensity measurement on an organic crystal, and served as a standard for converting relative intensity measurements to the absolute scale in a long series of precise studies by Robertson and coworkers [21]. Azaroff has described a method for measuring integrated intensities photographically [82].

Such absolute intensity data are extremely desirable, if not indispensable, in the determination of atomic positions in complex crystals; and studies dealing with bond nature, bond type, and the contributions of various bond structures to the normal state of a molecule depend on small variations in interatomic distance which can be reliably established only by complete and accurate intensity data.

Visual intensity data, however, are universally used in routine identification techniques and in most other diffraction procedures presented in this volume. The human eye, which is a remarkably good photometer, can estimate the relative order of blackness of a series of spots with exceptional accuracy, but not on a true numerical scale. With a calibrated step wedge of intensities, preferably placed on a diffraction photograph at the time of exposure, a highly satisfactory set of visually estimated relative intensities can be obtained. When necessary, such relative intensity data can be put on an absolute scale by calibrating the films with a few reflections from a standard crystal or powder whose absolute intensities are known. Details of the actual intensity measurements for the various techniques are provided at the appropriate points in later chapters.

3-3.4 Factors Affecting the Diffraction Intensities

Earlier in the chapter expressions for the diffracted intensity from perfect and imperfect crystals were presented, and the general effect of crystal perfection as it is evidenced by extinction was discussed. It is important now to discuss the origin of the individual factors in the intensity formulas for the various diffraction methods, and to consider their role in intensity calculations. Such detailed considerations of intensity calculations rarely enter into the more routine applications of diffraction, but are important for single-crystal analysis and other specialized applications. Various more elaborate treatments of these factors can be found in the references at the end of the chapter.

A. The Polarization Factor. The characteristic radiation from an x-ray tube is considered to be unpolarized [22], but such radiation after being scattered or diffracted is polarized, the amount of polarization depending upon the angle through which it is scattered or diffracted. Various workers have found good agreement between their experimental results and the classical theory, first worked out by Thomson [23], whereby the total energy of the scattered beam is proportional to the factor $(1 + \cos^2 2\theta)/2$.

Consider an unpolarized primary x-ray beam incident upon a crystal at the proper angle θ for diffraction (Fig. 3-11). In such a beam the electric vector vibrates in all directions perpendicular to the beam direction. It is possible, however, to consider this randomly oriented vector as resolved into two equal unit vectors at right angles to each other, and each of amplitude proportional to $\sqrt{2}/2$. Diffracting electrons in the crystal are set into forced vibrations parallel to these two vectors in the plane perpendicular to the incident beam. Thus in the direction of the undeviated beam there is no polarization. Along a diffracted beam, as at point P, the horizontal component is acting with its full effect, but the vertical component is decreased since the vibrating electrons radiate no energy parallel to their vibration directions. Accordingly, at P the effective component of the vertical vector is $(\sqrt{2}/2) \cos 2\theta$. For a diffracted beam to point P', where $2\theta = 90°$, the effective component of the vertical vector becomes zero, and only the horizontal component

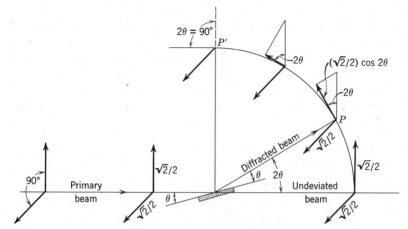

Fig. 3-11. Polarization of a diffracted x-ray beam. [Courtesy of F. Blake, *Rev. Mod. Phys.*, **5**, 169 (1933).]

acts. Since the intensity at P is proportional to the square of the amplitudes of these vectors,

$$I \propto \left(\frac{\sqrt{2}}{2}\right)^2 + \left(\frac{\sqrt{2}}{2}\right)^2 \cos^2 2\theta, \tag{3-36}$$

hence

$$I \propto \tfrac{2}{4} + \tfrac{2}{4}\cos^2 2\theta = \tfrac{1}{2}(1 + \cos^2 2\theta). \tag{3-37}$$

The polarization factor in this form is common to all the methods of x-ray diffraction, except when a crystal-monochromatized primary beam is used. For convenience, values of the polarization factor (also combined with the Lorentz factor) have been tabulated as a function of θ and $\sin \theta$ [24].

Since all crystal-monochromatized radiation is polarized, its use as a primary beam results in reflected beams from the sample which have been twice diffracted, and which require a more complicated polarization factor. The final expression depends on the experimental arrangement used. For the usual powdered crystalline technique it is $(1 + \cos^2 2\theta_1 \cos^2 2\theta_2)/(1 + \cos^2 2\theta_1)$, where $2\theta_1$ refers to the diffraction angle of the monochromator crystal. Both Whittaker [80] and Azaroff [81] investigated the appropriate twice-diffracted polarization factor for other experimental arrangements.

B. The Lorentz and "Velocity" Factors. Unless we use a crystal-monochromatized source of radiation as a primary beam, the beam will not be strictly monochromatic; and in the Laue method a polychromatic beam is actually required. Furthermore, the beam is generally not strictly parallel but more or less divergent. These features, together with any motion imparted to the crystal, contribute to a plane's "opportunity" to reflect by virtue of its orientation or the length of time it is in position to reflect. Thus in the Laue method the intensity is proportional to $1/\sin^2 \theta$, the so-called polychromatic Lorentz factor, which Lorentz originally presented directly to his classes. It appeared in the literature first as an addendum to Debye's paper on the temperature factor[25]. The monochromatic Lorentz factor as it applies to the powder method was first given by Debye and Scherrer[26] as

$$\frac{1}{(2 \sin^2 \theta \cos \theta)} = \frac{1}{(\sin 2\theta \sin \theta)} = \csc 2\theta \csc \theta. \qquad (3\text{-}38)$$

This expression holds for both flat cakes and cylindrical rods of powdered crystal, provided the crystallite orientation is perfectly random. For the theoretical development of these two factors the student is referred to treatments by Blake[27] or Schleede and Schneider[28].

In the various rotating- and oscillating-crystal methods, the Lorentz factor is essentially a reflection-time factor. Darwin[19] first gave this factor for a single crystal for the "equatorial" reflections, reflections from planes parallel to the axis of rotation. Here it takes the form

$$2 \csc 2\theta = \frac{2}{(\sin 2\theta)} = \frac{1}{(\sin \theta \cos \theta)}. \qquad (3\text{-}39)$$

Later it was realized that the planes that make an angle with the rotation axis are in reflecting position a longer time than those parallel to the axis, hence with photographic recording of the reflections they show abnormally strong intensities relative to the equatorial reflections. The problem is purely a geometrical one resulting from the angular velocity of these planes being effectively decreased the greater the inclination of the plane to the rotation axis. Details of this correction, so important in single-crystal studies, have been presented in successively more convenient form by several investigators[29].

Values of the Lorentz factor for ready use in calculations are tabulated in the *International Tables*[30] and in Appendix VIII.

C. The Temperature Factor. From the long-established concepts of the kinetic theory of matter, it is obvious that heat motions of atoms and molecules persist in a crystal, the atoms describing small vibrations about their equilibrium positions on the space lattice. At all finite temperatures above absolute zero, the amplitudes of these vibrations increase, so that the higher the temperature the weaker the diffraction intensities until at a sufficiently high temperature the intensity becomes zero. For high-melting substances at room temperature or substances far from their melting points, the heat motions may not be enough to affect noticeably the sharpness of the regular diffraction maxima, but as the temperature is raised and the melting point of the material is approached the reflections become progressively more diffuse and finally merge with the background.

The influence of temperature upon the diffraction intensities was first investigated theoretically by Debye[31], but later Laue[32], Darwin [1], Schrödinger[33], Faxen[34], Brillouin[35], Waller[36], and Zener and Jauncey[37] contributed to the theory. In his first treatment of the effect, Debye made the assumption that all the displacements of the atoms in their heat motions were independent. This obviously is not true, for the atoms are bound to one another by both attractive and repulsive forces, rather than merely to fixed equilibrium positions, so that a displacement of an atom necessarily disturbs its neighbors. Debye then recalculated the effect, and Waller[36] made a further correction. Thus, with zero-point energy assumed, the influence of heat motion is to introduce the factor e^{-2M} into the intensity expression. The quantity M, for convenience in tabulations, is commonly expressed as $M = B (\sin^2 \theta)/\lambda^2$, in which

$$B = \frac{6h^2}{m_a k\Theta} \left[\frac{\phi(x)}{x} + \frac{1}{4} \right], \qquad (3\text{-}40)$$

where m_a is the mass of the atom, h is Planck's constant, k is Boltzmann's constant, and $x = \Theta/T$, where Θ is the so-called characteristic temperature of the crystal expressed in degrees absolute. The term $\phi(x)$ is a Debye function given by the expression

$$\phi(x) = \frac{1}{x} \int_0^x \frac{\xi d\xi}{e^\xi - 1}. \qquad (3\text{-}41)$$

The temperature factor in the Debye form actually applies strictly only to cubic crystals containing a single kind of atom. In a crystal with more than one atomic species, such as sodium chloride (NaCl), each species has its own temperature factor, thus complicating the application of the factor. The Debye-Waller formula, however, has been found to represent fairly accurately the effects of temperature on simple cubic crystals at not too high temperatures, as evidenced by James and Firth's [38] results on NaCl from liquid air temperature to 500° abs, James and Brindley's[39] study of sylvine, potassium chloride, and the work of James, Brindley, and Wood[40] on aluminum at liquid-air temperature.

For crystals of less than cubic symmetry, and even for some cubic crystals, it has been found necessary to apply an anisotropic temperature factor as illustrated by Helmholz[41] in his study of silver phosphate (cubic), by Hughes[42] in the study of melamine (monoclinic), by Hughes and Lipscomb[43] in the investigation of methylammonium chloride (tetragonal), and by Shaffer[44] in his work on hexamethylenetetramine (cubic). Indeed, in modern single-crystal diffraction studies an anisotropic temperature factor is routinely applied in a form such as:

$$M = \tfrac{1}{4}(h^2a^{*2}B_{11} + k^2b^{*2}B_{22} + l^2c^{*2}B_{33} + 2hka^*b^*B_{12} + 2hla^*c^*B_{13} + 2klb^*c^*B_{23}).$$
$$(3\text{-}42)$$

In a great deal of work, particularly in the powder diffraction field, the temperature factor may be neglected without serious inconvenience to the study. The effect of such neglect is merely to cause the calculated intensities or structure factors to fall off less rapidly than the observed values in the high-θ region. Bearing this in mind, the diffractionist may not be especially handicapped in his data comparisons. It is frequently possible, however, to estimate a suitable value for $2M$, and even to use an isotropic factor with crystals of less than cubic symmetry. The reader may consult Appendix IX for tabulated values of the Debye-Waller temperature factor, or these data, as well as values of the function $\phi(x)$ and the characteristic temperature, can be obtained from the *International Tables*[45].

After noting the attenuation of the crystal reflection by the temperature factor e^{-2M}, we may inquire what happens to the lost intensity. Rigorous treatment of this point is outside the scope of this book, but suffice it to say that the time-averaged intensity of a reflection (following Azaroff[83]), expressed in electron units, can be written in the form

$$I_{eu}^t = Nf^2(1 - e^{-2M}) + (CR)\,e^{-2M}, \qquad (3\text{-}43)$$

where CR stands for the crystal reflection, the quantities on the right side of equations 3-75, 3-77, and 3-78 (except for the e^{-2M} factor). A schematic plot of equation 3-43, after dividing both sides by f^2, is presented in Fig. 3-12. The temperature factor e^{-2M} decreases rapidly with increasing angle θ [or its function $(\sin\theta)/\lambda$], and the reflection peaks are successively weakened relative to their values (dashed peaks) without a temperature effect. The term in $1 - e^{-2M}$ is known as the temperature diffuse scattering (TDS), and it appears to vary monotonically, thus superposing a background on the weakened reflections.

Actually, equation 3-43 represents the original Debye theory, and the situation with Faxen-Waller corrections involves a more complicated expression for the term in $1 - e^{-2M}$. Then the TDS, especially for simple inorganic substances, no longer varies monotonically but possesses maxima at the reciprocal lattice points of the regular reflections, Fig. 3-13. Each sharp reflection maximum is surrounded by a broad diffuse

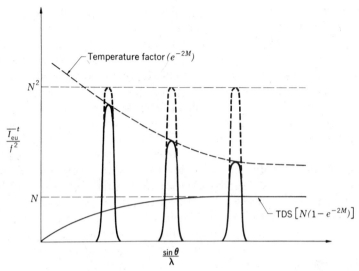

Fig. 3-12. Schematic representation of the effect of the temperature factor e^{-2M} on crystal reflection peaks (simple Debye theory). The dashed lines indicate the absence of the temperature effect. The TDS varies monotonically. (Used by permission; from L. Azaroff, *Elements of X-ray Crystallography*, McGraw-Hill, 1968.)

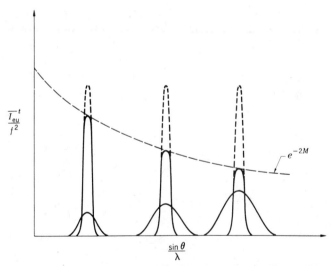

Fig. 3-13. Schematic representation of the temperature factor effects based on Debye-Faxen-Waller theory. The sharp maxima represent the regular crystal reflections, and the broad maxima are the result of TDS. (Used by permission; from L. Azaroff, *Elements of X-ray Crystallography*, McGraw-Hill, 1968.)

147

base which adds to the difficulty of reflection intensity measurements. In more complex inorganic crystals, however, and especially in molecular crystals, the TDS may be distributed over relatively large volumes of reciprocal space[91]. Moreover, these TDS volumes often appear to have little relation to the reciprocal lattice points corresponding to the strong Bragg reflections. Diffraction photographs of anthracene crystals show pronounced broad bands of TDS blackening in some reciprocal lattice directions[84]. Warren[85], who first drew attention to the fact that TDS contributes a nonlinear background to powder diffraction patterns, calculated an approximate TDS curve for the powder pattern of a face-centered cubic element. Herbstein and Averbach[86] made additional calculations on cubic crystals, and Mitra and Misra[87] have calculated the TDS curve for certain hexagonal crystal reflections. For rigorous TDS theory the reader is referred to James[88], Wooster[89], and Cochran and Pawley[90]. In addition to TDS there is also structural diffuse scattering, which arises from structural imperfections such as stacking faults, plastic deformation, dislocations, disorder, and so on [89].

D. The Atomic Scattering Factor. Atoms are not mere mathematical points in space, but possess finite sizes which are of the same magnitude as the x-ray wavelengths used in diffraction studies. Moreover, the electrons are spread throughout the volume of the atom, with the result that not all of them can be expected to scatter in phase (Fig. 3-14). These phase differences lead to partial interference and a net decrease in the scattered amplitude for an atom, so that the scattering efficiency is a function of the Bragg angle θ and falls off with $(\sin \theta)/\lambda$ (Fig. 3-15). The efficiency of cooperation among the scattering electrons of an atom is expressed by the atomic scattering factor f_0 which may be defined as the

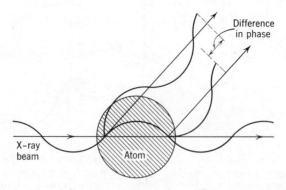

Fig. 3-14.   Phase difference in scattering from different parts of an atom.

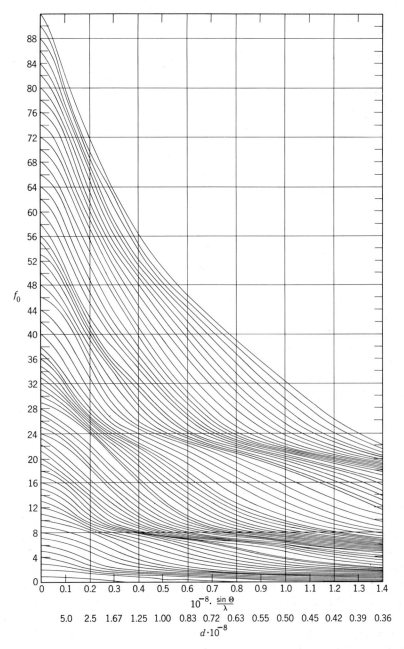

Fig. 3-15. Values of f_0 for neutral atoms. [Courtesy of L. Pauling and J. Sherman, Z. *Kristallogr.*, **81**, 1 (1932).]

ratio of the amplitude scattered by an atom at rest to that scattered by a single electron. At $\theta = 0$ the value of f_0 is equal to the atomic number Z of the atom, but it rapidly decreases with increasing $(\sin \theta)/\lambda$ (Fig. 3-15). With the atoms vibrating in the crystal as a result of heat motion, f_0 must be multiplied by the temperature factor e^{-M} before use in crystal calculations.

Mathematically, the relation of the atomic scattering factor to the electron distribution in the atom is expressed by the formula,

$$f_0 = \int_0^\infty U(r) \frac{\sin kr}{kr} dr, \qquad (3\text{-}44)$$

where $k = 4\pi(\sin \theta)/\lambda$, and λ is the wavelength of the x-rays. The quantity $U(r) dr$ on the basis of the classical theory is interpreted as the probability of finding an electron between r and $r + dr$ from the center of the spherical atom, so that

$$\int_0^\infty U(r) dr = Z. \qquad (3\text{-}45)$$

A more satisfactory interpretation may be had on the basis of wave mechanics, whereupon it becomes the density of electronic scattering matter between radii r and $r + dr$, and it can be demonstrated that $U(r) = 4\pi r^2 |\psi|^2$ in terms of Schrödinger's wave function ψ. Indeed, wave mechanical models of the atom serve for the prediction of f_0, and experimentally determined f_0 values in turn provide convenient data for testing atomic models.

For diffraction studies f_0 data are most conveniently tabulated or plotted as a function of $(\sin \theta)/\lambda$ (Fig. 3-15), since they can then be readily used with any wavelength. Present tables of f_0 values have been obtained largely through theoretical calculations, but enough values have been obtained experimentally to serve as satisfactory checks on the theory. Early work of Hartree[46], with modifications by Fock[47] and Slater [48], led to the well-known self-consistent field or HFS (Hartree-Fock-Slater) wave functions for computing f_0. The HFS method yields the best scattering factors for light atoms, but has been difficult to apply to heavier atoms because of the enormous labor involved. Heavy atoms have been treated by a statistical approach developed independently by Thomas[49] and Fermi[50]. Based on Fermi-Dirac statistics, it is known as the TFD (Thomas-Fermi-Dirac) method. Extensive tables of HFS and TFD scattering factors are to be found in the *International Tables*[51]. Liberman, Waber, and Cromer[52] have developed self-consistent field calculations for heavy atoms by using the Dirac equations and Slater's[53] approximate exchange correction. These DS (Dirac-

Slater) wave functions yield scattering factors[54] for heavy atoms, which Cromer[55] believes are to be preferred to the TFD values.

The earliest attempts to determine f_0 values experimentally were by crystalline scattering, although scattering measurements from crystals are less satisfactory for theory confirmation than gaseous measurements because their scattered beams are limited as to direction, and the further fact that the crystal atoms are not free like those of a gas. Bragg, James, and Bosanquet[56] used their careful absolute determinations of the integrated reflection from rock salt for a determination of f_{Cl} and f_{Na}. Some of the spectra measured involve scattering of the type $f_{Cl} + f_{Na}$; others are of the type $f_{Cl} - f_{Na}$. It was a simple matter, therefore, to plot f_{Cl} and f_{Na} against $\sin \theta$ by taking as ordinates half the sums and half the differences, respectively (Fig. 3-16). Their original values tended to approach 14 and 8 at low values of $\sin \theta$, rather than 18 and 10, the expected limiting values at $\theta = 0$. A correction for secondary extinction, however, greatly improved the trend of the curves in the expected directions (Fig. 3-16). Excellent measurements of f_0 for aluminum have been made by James, Brindley, and Wood[40], using single crystals, and by Bearden[57], using powdered crystals. A similar outstanding study of carbon (diamond) has been made by Brill[58].

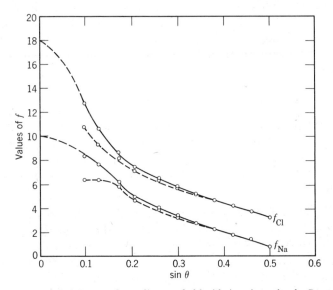

Fig. 3-16. Experimental f curves for sodium and chloride ions in rock salt. Corrected for secondary extinction, ———; uncorrected, -----. [Courtesy of Bragg, James, and Bosanquet, *Phil. Mag.*, (6)**42**, 1 (1921).]

In the use of tables of scattering factors, it is well to emphasize again that the temperature correction has not been included and that its effect can be large. For instance, the f_0 values for (10.0.0) for NaCl are reduced to one-third of their values for atoms at rest as a result of the thermal vibrations at room temperature. Indeed, it is frequently worth the additional effort in special studies to prepare an empirical f_0 curve from the experimental data in order better to take into account the trouble-some temperature correction and eliminate other errors in the f_0 data [59]. We should also keep in mind that the f_0 value is dependent upon the atom's state of chemical combination, since the latter partly determines the number and distribution of its electrons [58]. Thus, S^{+6}, S, and S^{-2} show appreciable differences in their f_0 curves at low values of $10^{-8} \times$ (sin θ)/λ. Calculations by McWeeny [60] of the effective scattering factors for bonded atoms show significant differences from earlier results based on spherical atoms.

There was early evidence that the f_0 value of an atom shows sharp fluctuations for wavelengths in the vicinity of one of its critical absorption edges [61]. This is the phenomenon now known as anomalous dispersion. The atomic scattering factors have been calculated on the assumption that the binding energy of the electron is small compared with the energy of the x-ray photon scattered, that is, that the electron is essentially a free electron. When the atom has an absorption edge, say a K edge, close to the wavelength of the x-rays, this is no longer true, and the scattering amplitude shows a sharp change. This effect may be expressed by representing the atomic scattering factor f as a complex number:

$$f = f_0 + \Delta f' + i\Delta f'' \tag{3-46}$$

where $\Delta f'$ and $\Delta f''$ are the real and imaginary dispersion corrections. The effect is of considerable importance in single-crystal studies, but of much less importance in powder diffraction work. Data on the real and imaginary dispersion corrections of the elements for the wavelengths commonly used in diffraction studies are presented in the *International Tables* [62].

E. The Structure Factor. Only an insignificant number of the known crystalline substances consists of a single kind of atom at the points of a simple space lattice. All other real crystals contain more than one kind of atom and/or have atom groups, such as molecules or complex ions, repeated at the lattice points. This group of atoms making up the repeating unit or motif of pattern may or may not have its atoms related by one or more symmetry operations such as a center, plane, or axes of symmetry. Repetition of this pattern motif by the lattice translations builds the crystal structure. Any atom of the motif may be chosen as the origin of a lattice array of points all of which are identical, so that the crystal may be considered a series of interpenetrating space lattices, one for each spatially different atom in the pattern. This decomposition of a crystal into lattice arrays is readily visualized by the simple two-dimensional case of Fig. 3-17A, in which the pattern unit is a group of two atoms. The two slightly displaced lattice arrays of Fig. 3-17B result from passing lines through each kind of atom, and it is to be noted that the lattices are similarly oriented and identical in dimensions.

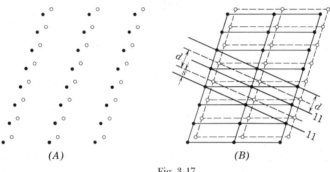

(A) $\qquad$ (B)

Fig. 3-17.

Further consideration of Fig. 3-17B demonstrates that any series of planes, such as the (11) planes of the solid atoms, is interleaved by a corresponding series from the ringed atoms. The interplanar spacing of the series is the same for the solid atoms as for the ringed atoms, and the reflection angles, which depend only on the lattice dimensions, are also the same for each. Reflection takes place from each set simultaneously, with all the solid-atom planes cooperating and all the ringed-atom planes likewise cooperating. Diffraction from the two sets of planes is not in phase, however, by an amount equal to $2\pi s/d$, considering the spacing d to represent an angular phase difference of a complete period, 2π or $360°$. This period of 2π corresponds to a path difference of exactly one wavelength for a first-order reflection. For an nth-order reflection the path difference is n wavelengths, and the spacing d corresponds to a phase difference of $2\pi n$. The observed reflection is then a composite of the two reflected waves with the same wavelength and period but with a different amplitude resulting from their partial interference.

The general situation may be summarized by saying that, no matter how complicated a crystal, it is always possible to picture it as a series of interpenetrating simple lattices, one for each structurally different kind of atom present. Each reflection plane (hkl) then is no longer essentially a geometrical plane, but becomes a series of interleaved atomic planes (one for each atom type) with constant d spacing. The reflection geometry is determined entirely by the lattice dimensions and not by the complexity of the pattern motif. The latter, through the displacements of the several lattice arrays, however, controls the phase relations of the reflections from the interleaving atomic planes parallel to (hkl), and thereby determines the amplitudes of the reflections.

These effects of the pattern motif or atomic arrangement on the diffraction intensity from a plane are taken account by the so-called structure factor $F(hkl)$ which enters into the intensity expression as the square F^2. It may be defined as the ratio of the amplitude scattered by the plane (hkl) relative to the amplitude scattered by a single electron. The problem of evaluating F is seen to be that of compounding several wave motions, all of the same period, but differing in both phase and amplitude. Before we begin a detailed discussion of the structure factor, it is well to have a look at one or two very simple examples.

Figure 3.18A depicts two adjacent unit cells of a cesium metal crystal which possesses body-centered cubic structure. If we remember that these two cells are merely a tiny section from an unlimited contiguous packing of similarly oriented cells, it is evident that the cube planes C are interleaved exactly halfway by the identical C' planes. For diffraction from either the C or C' planes, the phase difference must be an integral

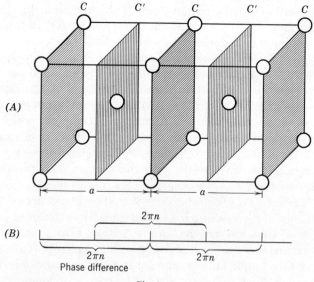

Fig. 3-18.

multiple of 2π as indicated in Fig. 3-18B. No first-order reflection is expected, because when $n = 1$ the C and C' planes are exactly halfway out of phase (one half wavelength apart) and exactly cancel each other since their amplitudes are identical and equal to f_{Cs}. There is a strong second-order reflection, however, because then the distance CC corresponds to a phase difference of 4π (two wavelengths), and CC' is half this distance or 2π (one wavelength); hence the two reflected waves are exactly in phase and reinforce each other. Similar reasoning discloses that all odd-order reflections from these planes are missing (have zero intensity), whereas all the even orders are present, their relative intensities of course being subject to the various other factors that affect diffraction intensities.

If we again consider Fig. 3-18A but now assume the C' planes to contain chlorine instead of cesium, the structure represented is that of cesium chloride. The situation is still geometrically the same as before, but now the amplitudes of the two waves are different, being f_{Cs^+} for the C planes and f_{Cl^-} for the chlorine planes C'. There is a weak first-order reflection in this instance because, although the cesium and chlorine planes are exactly out of phase, the smaller amplitude of the chlorine wave only partially destroys the amplitude of the cesium wave. The second-order reflection is again strong, since it results from addition of the amplitudes from the two reflected waves. Higher odd-order reflections are likewise expected, but each is weaker than the next higher even-order reflection.

After these simple examples the manner of compounding the structure factor of the resultant wave must be considered for the general case. There are actually several equivalent mathematical approaches to the problem, which above was seen to be a matter of adding amplitudes in accordance with their phases. The amplitudes are the appropriate atomic scattering factors for the atoms concerned, and the phase factors express the distances of the atomic planes from a suitably chosen origin. Consider then that Fig. 3-19 represents the succession of parallel atomic planes producing a reflection (hkl), that f_1, f_2, f_3, and so on, are the amplitudes of the respective atomic planes, and that ϕ_1, ϕ_2, ϕ_3, and so on, are their phase factors referred to the origin indicated. A general expression for $F(hkl)$ is

$$F(hkl) = \sum_N (f_N, \phi_N), \tag{3-47}$$

where f_N is the f value of the Nth kind of atom in the cell, and ϕ_N is its phase factor. The summation is made over all atoms in the unit cell, even though in special cases certain atoms may be equivalent.

One of the simplest expressions in optics for such a composite wave

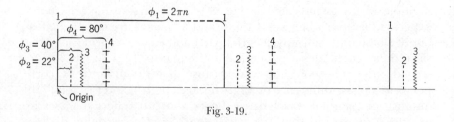

Fig. 3-19.

takes the form of an exponential sum [63], which for Fig. 3-19 becomes

$$F(hkl) = f_1 e^{i\phi_1} + f_2 e^{i\phi_2} + f_3 e^{i\phi_3} + \cdots + f_N e^{i\phi_N}, \tag{3-48}$$

$$= \sum_N f_N e^{i\phi_N}. \tag{3-49}$$

This exponential relation is not only very concise, but it also leads directly to the simple compounding of structure amplitudes and phases by means of vectors in the complex plane. From a well-known identity,

$$e^{i\phi} = \cos\phi + i\sin\phi = A + iB, \tag{3-50}$$

where $A = \cos\phi$ and $B = \sin\phi$. When this relation is plotted, as in Fig. 3-20A it is evident that $e^{i\phi}$ represents a unit vector of phase angle ϕ,

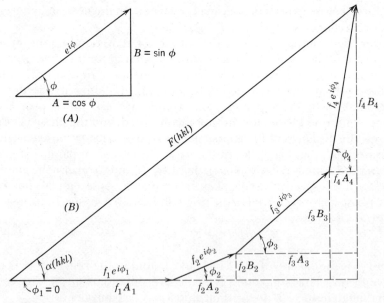

Fig. 3-20.   (A) Representation of a unit vector of phase ϕ in the complex plane. (B) Vector compounding of amplitudes of waves with different phases.

whose real component is $A = \cos \phi$ and whose complex component is $B = \sin \phi$. Thus the phase and magnitude of an atomic amplitude vector are both expressed in the relation,

$$fe^{i\phi} = f\cos\phi + fi\sin\phi = fA + fiB. \tag{3-51}$$

Similarly, the composite amplitude $F(hkl)$ corresponding to Fig. 3-19 and equation 3-48 is represented by the Argand diagram of Fig. 3-20B. Note that phase ϕ_1 may be considered as zero or 2π, since in either case $\cos \phi_1 = 1$ and $\sin \phi_1 = 0$, with the result that $f_1e^{i\phi_1}$ has no complex component. In actual practice F values are usually not calculated by this vector method, since it becomes very cumbersome and laborious with a complex crystal. Nor is the exponential expression 3-48 convenient for routine work.

In diffraction studies the measured intensities yield only values of F^2 and ultimately of the absolute magnitude of $|F|$, the phase factor [angle $\alpha(hkl)$, Fig. 3-20B] of the complete structure factor being indeterminate directly from intensities. A satisfactory trigonometric expression for calculating $|F|$ is evident, however, from Fig. 3-20B, that is,

$$|F| = [(f_1A_1 + f_2A_2 + \cdots + f_NA_N)^2 + (f_1B_1 + f_2B_2 + \cdots + f_NB_N)^2]^{1/2} \tag{3-52}$$

$$= \left[\left(\sum_N f_NA_N\right)^2 + \left(\sum_N f_NB_N\right)^2\right]^{1/2} \tag{3-53}$$

$$= \left[\left(\sum_N f_N\cos\phi_N\right)^2 + \left(\sum_N f_N\sin\phi_N\right)^2\right]^{1/2}. \tag{3-54}$$

It is also observed that

$$\alpha(hkl) = \tan^{-1}\frac{\sum_N f_NB_N}{\sum_N f_NA_N}, \tag{3-55}$$

thus making it possible to calculate $\alpha(hkl)$ when the individual atomic phases ϕ are known. These phases ϕ are related to the positions of the atoms in the unit cell, as is now demonstrated.

Consider first the two-dimensional case delineated in Fig. 3-21. A unit cell with solid atoms S at the corners contains a single ringed atom R at the point (xa, yb), where x and y are fractions of the cell edges. Any series of planes whose Miller indices are (hk) divides a into h equal parts and b into k equal parts, as demonstrated here for the (23) planes. Hence for the (hk) planes a/h is equivalent to a 2π phase change in the a direction. The phase component ϕ_a of the ringed atom along the a direction is equivalent to xa. From these considerations the following proportionality can be written:

$$\phi_a : 2\pi :: xa : \frac{a}{h}. \tag{3-56}$$

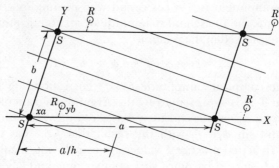

Fig. 3-21.

which, on being solved, leads to

$$\phi_a = 2\pi hx. \tag{3-57}$$

Similar reasoning gives for the phase component of atom R along the b direction

$$\phi_b = 2\pi ky. \tag{3-58}$$

The phase factor ϕ_R for the ringed atoms relative to the origin chosen is the sum of these two expressions:

$$\phi_R = \phi_a + \phi_b = 2\pi hx + 2\pi ky. \tag{3-59}$$

Generalization of equation 3-59 to three dimensions merely adds an additional term for a component ϕ_c along the c direction:

$$\phi_R = 2\pi hx + 2\pi ky + 2\pi lz$$
$$= 2\pi(hx + ky + lz). \tag{3-60}$$

With the phase factor in this form, equation 3-49 becomes

$$F(hkl) = \sum_N f_N e^{\,2\pi i(hx_N + ky_N + lz_N)}, \tag{3-61}$$

where x_N, y_N, z_N are the coordinates of the Nth atom in the cell expressed as fractions of the cell edge lengths. Similarly, expression 3-54 for $|F|$ may be written

$$\left| F(hkl) \right| = \left\{ \left[\sum \sum f_N \cos 2\pi(hx_N + ky_N + lz_N) \right]^2 \right.$$
$$\left. + \left[\sum \sum f_N \sin 2\pi(hx_N + ky_N + lz_N) \right]^2 \right\}^{1/2}. \tag{3-62}$$

Double summation signs have been inserted here to remind the reader that two summations are really involved, one over the positions (the trigonometric factor) of all the atoms, and the other over the kinds of atoms (the f factors). The trigonometric part is often referred to as the *geometrical structure factor* and written separately as

$$A = \sum \cos 2\pi(hx + ky + lz)$$

and   (3-63)

$$B = \sum \sin 2\pi(hx + ky + lz).$$

A and B are entirely a function of the coordinates of the positions of the atoms in the cell, and therefore are characteristic of the space group to which the crystal belongs.

When the crystal structure possesses a center of symmetry, and the origin of coordinates is placed at that center, a most important simplification of the structure factor results. In this instance an atom at the point xyz is matched by an identical atom at the point $\bar{x}\bar{y}\bar{z}$. The phases of these two atoms are:

$$\phi_{xyz} = 2\pi (hx + ky + lz),\tag{3-64}$$

$$\phi_{\bar{x}\bar{y}\bar{z}} = 2\pi (-hx - ky - lz)\tag{3-65}$$

$$= -2\pi (hx + ky + lz)\tag{3-66}$$

$$= -\phi_{xyz}.\tag{3-67}$$

Indeed, in the presence of a symmetry center at the origin, atoms always occur in identical pairs whose phases are equal in magnitude but opposite in sign. Since $\cos -\phi = \cos \phi$ for all values of ϕ, this simplifies the cosine summation by reducing it to half as many terms, it being necessary to consider only the positive phases of the pairs and then multiply the sum by two. Far more important, however, is the fact that $\sin -\phi = -\sin \phi$ for all values of ϕ, with the result that the sine contributions from the atoms of each pair cancel each other, and the whole sine summation term goes to zero. In the presence of a center of symmetry, the structure factor may then be expressed as

$$F(hkl) = \sum_N f_N \cos 2\pi (hx_N + ky_N + lz_N).\tag{3-68}$$

Figure 3-22 illustrates vectorially the disappearance of the sine terms in the composite wave as a result of the imaginary components neutralizing each other. The phase angle $\alpha(hkl)$ then becomes either 0 or π, depending on whether the cosine sum is positive or negative.

The use of expressions 3-62 and 3-68 for computing structure factors can conveniently be demonstrated by the simple example of face-centered cubic NaCl whose structure is depicted in Fig. 3-23. Both the sodium ions and the chloride ions are at symmetry centers, and it makes no difference which is chosen as the origin. Assuming, however, that the

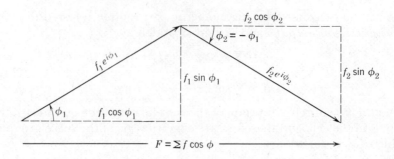

Fig. 3-22.

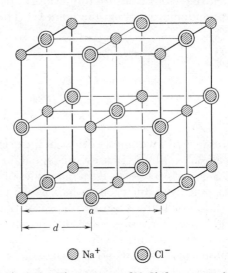

⊘ Na⁺   ◎ Cl⁻

Fig. 3-23.   The structure of NaCl (face-centered cubic).

origin is at a sodium ion, the coordinates of the $4Na^+$ and $4Cl^-$ are

$$Na^+ \text{ at: } 000; \tfrac{1}{2}\tfrac{1}{2}0; \tfrac{1}{2}0\tfrac{1}{2}; 0\tfrac{1}{2}\tfrac{1}{2};$$

$$Cl^- \text{ at: } \tfrac{1}{2}\tfrac{1}{2}\tfrac{1}{2}; 00\tfrac{1}{2}; 0\tfrac{1}{2}0; \tfrac{1}{2}00.$$

According to equation 3-62 the structure factor is

$$|F(hkl)| = [(f_{Na}A + f_{Cl}A)^2 + (f_{Na}B + f_{Cl}B)^2]^{1/2} = [A'^2 + B'^2]^{1/2},$$

where A and B have the significance of equations 3-63. Although this structure has a center of symmetry and the sine terms are therefore expected to go to zero, this is not assumed but will be demonstrated in the calculations. Substituting the atomic coordinates in the expression above, we obtain

$$A' = f_{Na} [\cos 2\pi(0) + \cos \pi(h+k) + \cos \pi(h+l) + \cos \pi(k+l)]$$
$$+ f_{Cl} [\cos \pi(h+k+l) + \cos \pi l + \cos \pi k + \cos \pi h],$$

$$B' = f_{Na} [\sin 2\pi(0) + \sin \pi(h+k) + \sin \pi(h+l) + \sin \pi(k+l)]$$
$$+ f_{Cl} [\sin \pi(h+k+l) + \sin \pi l + \sin \pi k + \sin \pi h].$$

It is immediately seen that the B' term will go to zero, since the sine of 0 or of any multiple of π is zero, and, accordingly,

$$F(hkl) = A'.$$

For further consideration of F, it is convenient to group the various index triplets (hkl) into several reflection types:

1. h, k, and l all odd:

$$F = A' = 4f_{Na} - 4f_{Cl} \quad \text{(reflections of this type are weak).}$$

2. h, k, and l all even:

$$F = A' = 4f_{Na} + 4f_{Cl} \quad \text{(reflections of this type are strong).}$$

3. h, k, and l, two odd and one even:

$$F = A' = 0f_{Na} + 0f_{Cl} = 0 \quad \text{(reflections of this type are missing).}$$

4. h, k, and l, two even and one odd:

$$F = A' = 0f_{Na} + 0f_{Cl} = 0 \quad \text{(reflections of this type are missing).}$$

Thus the reflections on a NaCl single-crystal or powder photograph have either all odd or all even indices, but no mixed indices. Indeed, all crystals based on a face-centered lattice are found to follow this same pattern with regard to absent reflection types, so that the absence of mixed indices is a criterion for identification of the presence of a face-centered lattice. It is to be noted, too, that all planes of a given reflection type, that is, all odd indices or all even indices, do not have identical F's because of the dependence of f on the $(\sin \theta)\lambda$ value for the plane concerned, and the effect of the temperature factor provided it has been included.

Since space does not permit additional sample calculations at this point, the interested reader is encouraged to carry out similar calculations with the two sets of data below and thereby determine the values of F for the various reflection types:

Cesium chloride (1-molecule cell):

$$\text{Cs at: } 0\,0\,0; \quad \text{Cl at: } \tfrac{1}{2}\tfrac{1}{2}\tfrac{1}{2}.$$

Zinc sulfide (4-molecule cell):

$$Zn\ at:\ 0\,0\,0;\ \tfrac{1}{2}\tfrac{1}{2}0;\ \tfrac{1}{2}0\tfrac{1}{2};\ 0\tfrac{1}{2}\tfrac{1}{2};$$

$$S\ at:\ \tfrac{1}{4}\tfrac{1}{4}\tfrac{1}{4};\ \tfrac{1}{4}\tfrac{3}{4}\tfrac{3}{4};\ \tfrac{3}{4}\tfrac{1}{4}\tfrac{3}{4};\ \tfrac{3}{4}\tfrac{3}{4}\tfrac{1}{4}.$$

The same procedure is followed in calculating the structure factor for a crystal of any degree of complexity. In many instances the coordinates of the atoms are not the convenient fractions of the cell edges observed in the crystals above, but are instead the decimal fractions x, y, z. Tables of $\cos 2\pi hx$ and $\sin 2\pi hx$ for small increments of x and a range of values of h then become a necessity for routine work. The excellent tables by Buerger[64] provide three-figure values of these two terms for values of the argument h (or k or l) from 1 to 30, and for all values of the argument x (or y or z) from 0 to 1 in increments of 0.001. As a further convenience for the investigator, the general expressions for the geometrical parts of the structure factor for each of the 230 space groups are listed in the *International Tables* [65a,b]. The new *International Tables* also contain Lonsdale's[66] simplified structure factors for special reflection cases[65a].

F. The Multiplicity Factor. In some diffraction techniques the reflected beams from several planes superpose to give a single spot or line on the photograph. Accordingly, such spots or lines show a stronger intensity than if they resulted from a single plane, and their increased intensity is taken into account by the introduction of the so-called multiplicity factor into the intensity formula. This factor, to be designated j, has been variously represented in the literature by such letters as n, j, p, and z. Its value is determined not only by the experimental method and conditions being used, but also by the symmetry of the crystal. In the Laue and counter diffractometer methods with stationary single crystals, the value is unity. The factor is likewise unity in the moving-film techniques such as the precession and Weissenberg methods. In the oscillated-crystal method the factor may or may not be greater than one, whereas in the rotating-crystal and powder techniques it usually enters as an integer greater than unity.

In the powder diffraction techniques the value of j depends only on the symmetry of the crystalline material. Consideration of the Bragg equation,

$$\lambda = 2d \sin \theta, \tag{3-69}$$

makes it immediately evident that with monochromatic radiation all planes of the same interplanar spacing d reflect at the same angle θ. This means that they all contribute to the same powder diffraction halo on the photograph (Chapter 4). That more than one plane in a crystal can have the same spacing is most readily demonstrated by examining the expression for d in the cubic system (Table 1-3):

$$d(hkl) = \frac{a}{\sqrt{h^2 + k^2 + l^2}}. \tag{3-70}$$

It is at once clear that (111) and $(\bar{1}\bar{1}\bar{1})$ have the same spacing, as also do all other combinations of positive and negative unit indices: $(1\bar{1}\bar{1})$, $(\bar{1}11)$, $(1\bar{1}1)$, $(\bar{1}1\bar{1})$, $(11\bar{1})$, and $(\bar{1}\bar{1}1)$. Indeed, all eight of these planes contribute to the cubic powder line designated (111), and the line is eight times as strong as it would be if the (111) plane alone were responsible for the reflection.

Further study of formula 3-70 reveals that a general plane such as (213) has the same spacing as (123), (321), and the rest of the six permutations of this index triplet. Moreover, each of these six permutations may have eight permutations of sign, making in all 48 reflections that contribute to this cubic powder line. In addition, planes with different index triplets, such as the pairs (300) and (221) or (333) and (511), which lead to the same value of the sum $h^2 + k^2 + l^2$, have the same spacing in the cubic system. It is thus seen that the total number of reflections contributing to a powder diffraction line may be rather large, but is completely dependent upon the symmetry of the crystalline material. For any crystal symmetry class the number of contributing reflections for any given form is equal to the total number of faces in the holohedral representative of that form. For some low-symmetry crystals (certain types of hemihedrism, hemimorphism, and tetartohedrism) account must be taken of the fact that part of the faces of the form may have structure factors different from the others. Thus for pyrite (FeS_2), belonging to the cubic hemihedral class T_h, the planes (210) and (120) have different structure factors. Table 3-1 lists the multiplicity factors for all forms in the 32 crystal classes for use in the powder method.

G. The Absorption Factor. In all diffraction techniques the incident and diffracted beams in passing through the crystal are partially absorbed, so that the diffracted beams are less intense than they would be with a completely nonabsorbing crystal. Rather early this absorption was taken account of in reflection from extended faces of single crystals and in transmission through thin crystal slips (equations 3-34 and 3-35). In other situations, however, the problem of evaluating the absorption factor was so complex that it was largely neglected, its use being attempted only in special cases[67–69]. Indeed, since the absorption factor and the temperature factor oppose and thus tend to cancel each other, the

Table 3-1.   Multiplicity Factor j for the Powder Method

Symmetry[a]	Reflection Type and Multiplicity[b]						
Cubic system	(hkl)	(hhl)	$(hk0)$	$(hh0)$	(hhh)	$(h00)$	
O_h, O, T_d	48	24	24	12	8	6	
T_h, T	2(24)	24	2(12)	12	8	6	
Hexagonal and trigonal							
system	(hkl)	(hhl)	$(h0l)$	$(hk0)$	$(hh0)$	$(h00)$	$(00l)$
$D_{6h}, D_6, C_{6v}, D_{3h}$	24	12	12	12	6	6	2
C_{6h}, C_6, C_{3h}	2(12)	12	12	2(6)	6	6	2
D_{3d}, D_3, C_{3v} with hex-	2(12)	2(6)	12	12	6	6	2
C_{3i}, C_3 agonal	4(6)	2(6)	2(6)	2(6)	6	6	2
lattice							
Tetragonal system	(hkl)	(hhl)	$(h0l)$	$(hk0)$	$(hh0)$	$(h00)$	$(00l)$
$D_{4h}, D_4, C_{4v}, D_{2d}$	16	8	8	8	4	4	2
C_{4h}, C_4, S_4	2(8)	8	8	2(4)	4	4	2
Orthorhombic system	(hkl)	$(h0l)$	$(hk0)$	$(0kl)$	$(h00)$	$(0k0)$	$(00l)$
D_{2h}, D_2, C_{2v}	8	4	4	4	2	2	2
Monoclinic system	(hkl)	$(h0l)$	$(h00)$				
C_{2h}, C_2, C_s	4	2	2				
Triclinic system							
C_1, C_i	All types: 2						

[a]For the meaning of these symmetry symbols see Chapter 1.
[b]A multiplicity expressed as 2(12) indicates that there are two sets of reflections at the same angle but having different structure factors.

complete neglect of these two factors has in frequent instances led to better agreement between observed and calculated intensities than when allowance was made for the factors. This is obviously no solution to the problem, and any precise consideration of intensities demands that both factors be considered. In 1930, Claassen [70] devised a graphical method for the Debye-Scherrer powder technique which is of general application. Shortly thereafter, Rusterholz [71] showed that for fairly highly absorbing substances the absorption factor is a rather simple function of the glancing angle θ. Superseding both was a treatment by Bradley [72] which makes it possible to calculate the absorption factor for cylindrical powder samples, or for the equatorial reflections from a single-crystal specimen that is cylindrical about its rotation axis, with an accuracy of at least 1 per cent.

The Bradley method assumes an essentially parallel incident beam of x-rays which completely bathes the specimen. The absorption factor (fraction transmitted) $A(\theta)$ for a cylindrical specimen of radius r and height h is given by

$$A(\theta) = \frac{1}{\pi r^2} \int \int e^{-\mu a} d\sigma = \int \int e^{-\mu r x}\, ds, \qquad (3\text{-}71)$$

where a is the total path length of the ray in the specimen before and after reflection from a small fragment, μ is the linear absorption coefficient of the whole specimen, $d\sigma$ is a small element of cross section, $x = a/r$, and $ds = d\sigma/\pi r^2$. The integration is carried out over the whole cross section of the specimen. In the evaluation of this expression, it is necessary to calculate the value of s, that is, the proportion of the cross section of the sample for which the absorbing path xr is less than a given value. This gives s as a function of x, from which ds/dx may be obtained by differentiation. The problem is complicated by the fact that the same formulas are not applicable to all portions of the cross section. In Fig. 3-24, the front area A_F is by far the most important for all angles θ except for small absorption, and the center and back areas are of such minor importance as to be merely correction terms. Moreover, the center area cannot be represented directly by a formula but must be estimated. The general absorption factor $A(\theta)$ is the sum of the contributions from these three areas:

$$A(\theta) = A_F + A_C + A_B. \qquad (3\text{-}72)$$

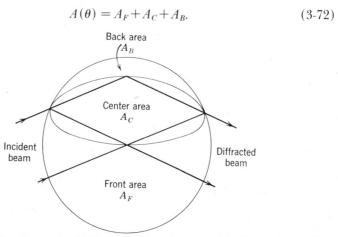

Fig. 3-24. Relative sizes at $\theta = 22.5°$ of the front, center, and back areas in Bradley's absorption correction for the powder method. [From A. Bradley, *Proc. Phys. Soc.* (London), **47**, 879 (1935).]

The reader is referred to the original article[72] for the mathematical details and the formulas for the A's, but a few special cases are of interest. Thus, when $\theta = 0°$, $A(\theta) = 2A_F$. Also the quantity A_B is operative only in the range $0° < \theta < 30°$. Finally, when $\mu r > 5$, the terms A_B and A_C may be neglected except when $\theta = 0$ or $90°$. Using Claassen's graphical method for the range below $\mu r = 2$, and his own formulas above $\mu r = 2$, Bradley calculated two tables of absorption factors[72], a table of absorption factors $100\,A(\theta)$ for values of $\mu r < 5$ at selected angles θ, and a table of relative values of $A(\theta)$ for large values of μr. To obtain the absolute values at high μr, it is only necessary to calculate the value of $A(\theta = 90°)$ from the equation

$$A(\theta = 90°) = \frac{1}{\pi\mu r} - \frac{1}{16\pi\mu^3 r^3}. \tag{3-73}$$

To correct observed intensities one multiplies by the reciprocal of $A(\theta)$,

$$A^* = \frac{1}{A(\theta)}. \tag{3-74}$$

In some instances the Debye-Scherrer powder sample is not a cylinder of powder, but takes the form of a layer of powder on the outside of a hair or solid glass fiber. Møller and Jensen[92] have presented a method for calculating absorption factors for such rod-mounted specimens, and tables of $100\,A(\theta)$ are given for three different rod (core) materials: hair, Lindemann glass, and common glass.

The problem of absorption has always been very troublesome in single-crystal studies, because of the great variations in shape they exhibit. In the rare instance in which a crystal approximates a cylinder, or can be ground to cylindrical form[73], Bradley's absorption factor is directly applicable to the reflections on the equatorial layer line. A spherical crystal is an even more symmetrical shape for application of an absorption correction, and grinding to a spherical shape is often possible[74]. Thus Bond[75] was led to evaluate A^* for cylindrical crystals by more elegant mathematical methods, and likewise to treat the problem of crystal spheres. His A^* absorption corrections are tabulated in the *International Tables*[76]. Although Bond developed his corrections for cylindrical single crystals, they apply equally to cylindrical powdered crystal specimens whose powder (crystallites plus voids) has a given μr value. Indeed, Bond's values are almost identical with those of Bradley. Weber[77] has recalculated the cylindrical A^* absorption corrections to a reported accuracy of ± 0.001. For the calculation of absorption corrections for a crystal of any shape (no reentrant angles between faces), see Coppens, Leiserowitz, and Rabinovich[78].

3-3.5 Expressions for the Relative Intensity of Diffraction by the Various Techniques

The different factors affecting diffraction intensities can readily be grouped into single expressions for use in calculating the relative intensities of reflections by the commoner techniques. The various factors in the expressions are easily identified.

1. *Polychromatic Laue method:*

$$I \cong j \cdot \frac{1 + \cos^2 2\theta}{2} \cdot \frac{1}{\sin^2 \theta} \cdot F^2 \cdot A(\theta), \tag{3-75}$$

where $j = 1$, and

$$F = \sum_N f_N e^{-M_N} e^{2\pi i(hx_N + ky_N + lz_N)}. \tag{3-76}$$

This is the proper method for applying the temperature factor e^{-2M}, since in a complex crystal the mean-square vibration of each kind of atom must be considered separately unless it can be shown that a single temperature factor can be applied. Where the structure factor can be multiplied directly by the temperature factor Fe^{-M}, the unit cell may be considered as vibrating as a whole. That this behavior is reasonably frequent is apparent from the number of successful applications of this type.

2. *Powder method:*

$$I \cong j \cdot \frac{1 + \cos^2 2\theta}{2} \cdot \frac{1}{2 \sin^2 \theta \cos \theta} \cdot F^2 \cdot A(\theta), \tag{3-77}$$

with F having the significance of equation 3-76.

3. *Oscillating- or rotating-crystal method, counter diffractometer, and Weissenberg method (equatorial layer line using a single crystal and filtered monochromatic radiation):*

$$I \cong j \cdot \frac{1 + \cos^2 2\theta}{2} \cdot \frac{1}{\sin \theta \cos \theta} \cdot F^2 \cdot A(\theta). \tag{3-78}$$

Here, again, F has the form of equation 3-76. With the proper form of the Lorentz-velocity factor [29, 30], equation 3-78 is readily adapted to nonequatorial layer lines.

3-3.6 Lattice-Centering and Space-Group Extinctions

A most important aspect of intensity distribution among reflections is that relating to the extinctions produced by glide planes, screw axes, and the lattice centering of nonprimitive lattices. The translations involved in these symmetry elements and centered lattices add a new periodicity to the patterns, which shows itself by extinguishing certain classes of x-ray spectra. Each type of extinction is characteristic of the translational element producing it, and therefore of the space groups or nonprimitive lattices containing that pattern element. Accordingly, the absence of such characteristic spectra from diffraction data is a major criterion for determination of the lattice type and space group. Since each of the foregoing elements causes the extinction of a different class of reflections, a tabulation of the types of absences and their significance will be very useful for the ready recognition of lattice type and space group. Indeed, only 37 space groups, those possessing simple symmetry elements and based on a primitive cell, out of the entire 230 show all expected classes of spectra. The remaining 193 groups have one or more systematic reflection-type absences.

The additional periodicities introduced by glide planes, screw axes, and lattice centering cause the structure factors, the F's, of certain characteristic reflections to go to zero, and in turn the reflection is extinguished. This was strikingly demonstrated (Section 3-3.4E) in the calculation of structure factors for NaCl crystals which are based on the face-centered cubic lattice. By these calculations it was established that NaCl patterns show only reflections with all odd or all even indices,

Table 3-2.   Reflection Extinctions Characteristic of Lattice Type

Condition for Nonextinction of (hkl)	Lattice Type Indicated	Lattice Symbol
All (hkl) present	Simple primitive	P, C (hex.), R
$h+k+l = 2n$	Body-centered	I
$h+k = 2n$	(001) faces centered	C
$h+l = 2n$	(010) faces centered	B
$k+l = 2n$	(100) faces centered	A
$\left.\begin{array}{l}h+k = 2n\\h+l = 2n\\k+l = 2n\end{array}\right\}$	Face-centered	F
$h-k = 3n$	Triply primitive hexagonal	H
$\left.\begin{array}{l}-h+k+l = 3n\\ \text{or}\\ h-k+l = 3n\end{array}\right\}$	Trigonal (rhombohedral) lattice indexed on hexagonal axes	R
$h+k+l = 3n$	Primitive hexagonal lattice indexed on trigonal (rhombohedral) axes	P, C (hex.)

Table 3-3. Space-Group Extinctions Characteristic of Glide Planes

Reflection Type	Condition for Non-extinction	Glide Plane Type Indicated	Glide Plane Symbol
$(hk0)$	$h = 2n$	Glide plane normal to c, component $\dfrac{a}{2}$	a
	$k = 2n$	Glide plane normal to c, component $\dfrac{b}{2}$	b
	$h + k = 2n$	Glide plane normal to c, component $\dfrac{a}{2} + \dfrac{b}{2}$	n
	$h + k = 4n$	Glide plane normal to c, component $\dfrac{a}{4} + \dfrac{b}{4}$	d
$(h0l)$	$h = 2n$	Glide plane normal to b, component $\dfrac{a}{2}$	a
	$l = 2n$	Glide plane normal to b, component $\dfrac{c}{2}$	c
	$h + l = 2n$	Glide plane normal to b, component $\dfrac{a}{2} + \dfrac{c}{2}$	n
	$h + l = 4n$	Glide plane normal to b, component $\dfrac{a}{4} + \dfrac{c}{4}$	d
$(0kl)$	$k = 2n$	Glide plane normal to a, component $\dfrac{b}{2}$	b
	$l = 2n$	Glide plane normal to a, component $\dfrac{c}{2}$	c
	$k + l = 2n$	Glide plane normal to a, component $\dfrac{b}{2} + \dfrac{c}{2}$	n
	$k + l = 4n$	Glide plane normal to a, component $\dfrac{b}{4} + \dfrac{c}{4}$	d
(hhl)	$l = 2n$	Glide plane normal to [110], component $\dfrac{c}{2}$	c
	$h = 2n$	Glide plane normal to [110], component $\dfrac{a}{4} + \dfrac{b}{4}$	b
	$h + l = 2n$	Glide plane normal to [110], component $\dfrac{a}{4} + \dfrac{b}{4} + \dfrac{c}{2}$	n
	$2h + l = 4n$	Glide plane normal to [110], component $\dfrac{a}{4} + \dfrac{b}{4} + \dfrac{c}{4}$	d

169

never with mixed indices. The absence of mixed indices is, indeed, the criterion for the identification of a face-centered lattice in any symmetry system. Similar calculations for crystals based on other centered lattices reveal the criteria for these lattices. Table 3-2 lists the reflection extinctions characteristic of the various lattice types. These absences are the only systematic absences found among the pyramid planes (hkl).

The presence of glide planes in a space group gives rise to systematic absences in the prism planes $(hk0)$, $(h0l)$, and $(0kl)$, and in planes of the type (hhl). In an entirely similar fashion, these extinctions can be analytically demonstrated by means of the structure factor equation and the general equivalent points of the space group. For ready reference glide-plane extinctions and their significance have been listed in Table 3-3.

The translations characteristic of a screw axis result in systematic absences in the planes normal to the screw axis, whereas normal rotation axes give rise to no absences in the corresponding sets of planes. Specifically, the period of the screw axis leads to a periodicity along the axis

Table 3-4. Space-Group Extinctions Characteristic of Screw Axes

Reflection Type	Condition for Non-extinction	Screw Axis Type Indicated	Screw Axis Symbol
$(h00)$	$h = 2n$	Screw axis parallel to [100], component $\dfrac{a}{2}$	$2_1, 4_2$
	$h = 4n$	Screw axis parallel to [100], component $\dfrac{a}{4}$	$4_1, 4_3$
$(0k0)$	$k = 2n$	Screw axis parallel to [010], component $\dfrac{b}{2}$	$2_1, 4_2$
	$k = 4n$	Screw axis parallel to [010], component $\dfrac{b}{4}$	$4_1, 4_3$
$(00l)$	$l = 2n$	Screw axis parallel to [001], component $\dfrac{c}{2}$	$2_1, 4_2, 6_3$
	$l = 3n$	Screw axis parallel to [001], component $\dfrac{c}{3}$	$3_1, 3_2, 6_2, 6_4$
	$l = 4n$	Screw axis parallel to [001], component $\dfrac{c}{4}$	$4_1, 4_3$
	$l = 6n$	Screw axis parallel to [001], component $\dfrac{c}{6}$	$6_1, 6_5$
$(hh0)$	$h = 2n$	Screw axis parallel to [110], component $\dfrac{a}{4}+\dfrac{b}{4}$	2_1

which is always a submultiple of the axial periodicity along a normal axis of like degree. Analytical treatment of the various cases leads to the screw-axis extinctions conveniently tabulated in Table 3-4.

A logically planned structure analysis proceeds through a series of rather definite steps. The first of these steps is usually the determination of the symmetry system and point group of the crystal, as outlined earlier (Section 1-4.1). This step is followed by measurement of the size and shape of the unit cell, and calculation of the number of molecules (formula weights) in the cell. It must be emphasized that this second step is possible from polycrystalline diffraction data only for substances of high symmetry, and it is usually best attacked by single-crystal methods. The analysis next proceeds to the deduction of the space group from a consideration of reflection absences. The investigator critically inspects the reflection data for missing reflection types and lists those found. It is convenient to do this systematically, looking first for pyramid absences which give evidence of the lattice type, next examining prism reflections for glide plane absences, and finally hunting for the absences characteristic of screw axes. These data together with the symmetry data of step one permit the writing of a Hermann-Mauguin symbol representing the data. Reference to the *International Tables* [79], may then make it possible to identify the space group unequivocally.

GENERAL REFERENCES

*1. R. W. James, *The Crystalline State*, Vol. II, *The Optical Principles of the Diffraction of X-rays*, G. Bell, London, 1948; Cornell University Press, Ithaca, 1965.

2. W. H. Zachariasen, *Theory of X-ray Diffraction in Crystals*, Wiley, New York, 1945; Dover, New York, 1968.

3. M. v. Laue, *Röntgenstrahl-Interferenzen*, Akademische Verlagsgesellschaft, Leipzig, 1941; photo-lithoprint reproduction by Edwards Brothers, Ann Arbor, 1943.

4. F. C. Blake, "On the Factors Affecting the Reflection Intensities by the Several Methods of Analysis of Crystal Structures," *Rev. Mod. Phys.*, **5**, 169 (1933).

5. A. J. C. Wilson, *X-ray Optics*, Methuen, London, 1949.

6. P. P. Ewald, "Der Aufbau der festen Materie und seine Erforschung durch Röntgenstrahlen," in *Handbuch der Physik*, Vol. XXIV, Springer, Berlin, 1927, pp. 191–369.

*7. L. V. Azaroff, *Elements of X-ray Crystallography*, McGraw-Hill, New York, 1968, especially Chapters 7–11.

SPECIFIC REFERENCES

[1]   C. G. Darwin, *Phil. Mag.*, (6) **27**, 315, 675 (1914).

[2]   P. P. Ewald, *Ann. Phys.*, **49**, 1, 117 (1916); **54**, 519, 557 (1917); *Z. Phys.*, **2**, 323 (1920); **30**, 1 (1924); *Phys. Z.*, **21**, 617 (1921); **26**, 29 (1925); *Acta Crystallogr.*, **11**, 888 (1958).

[3]   R. W. James, "The Dynamical Theory of X-ray Diffraction," in *Solid State Physics (F. Seitz and D. Turnbull,* eds.), Vol. 15, Academic Press, New York, 1963, p. 53; also general reference 1, p. 66.

[4]   B. W. Batterman and H. Cole. *Rev. Mod. Phys.*, **36**, 681 (1964).

[5]   M. v. Laue, *Ergeb. Exakt. Naturwiss.*, **10**, 133 (1931); *Acta Crystallogr.*, **2**, 106 (1949).

[6]   P. B. Hirsch and G. N. Ramachandran, *Acta Crystallogr.*, **3**, 187 (1950).

[7]   P. B. Hirsch. *Acta Crystallogr.*, **5**, 176 (1952).

[8]   N. Kato, *J. Phys. Soc. Jap.*, **7**, 397 (1952); *Acta Crystallogr.*, **11**, 885 (1958).

[9]   M. J. Buerger, *X-ray Crystallography*, Wiley, New York, 1942.

[10]   R. W. G. Wyckoff, *The Structure of Crystals*, 2nd ed., Chemical Catalog Company, New York, 1931.

[11]   M. v. Laue, *Sitz. math. phys. Klasse bayer, Akad. Wiss.*, p. 303 (1912); *Ann. Phys.* **41**, 971 (1913); *Enzyklopädie math. Wiss.*, **24**, 359 (1915).

[12]   W. L. Bragg, *Proc. Cambridge Phil. Soc.*, **17**, 43 (1913).

[13]   J. D. Bernal, *Proc. Roy. Soc.* (London), **113A**, 117 (1926).

[14]   A. H. Compton, *Phys. Rev.*, **10**, 95 (1917).

[15]   S. K. Allison, *Phys. Rev.*, **41**, 13, 688 (1932).

[16]   L. G. Parratt, *Phys. Rev.*, **41**, 561 (1932).

[17]   P. Kirkpatrick and P. A. Ross *Phys. Rev.*, **43**, 596 (1933).

[18]   M. Renninger, *Naturwissenschaften*, **21**, 334 (1934).

[19]   C. G. Darwin, *Phil. Mag.*, (6) **43**, 800 (1922).

[20]   B. W. Robinson, *Proc. Roy. Soc.* (London), **142A**, 422 (1933).

[21]   See the brilliant series of papers on the structures of aromatic hydrocarbons, starting with the quantitative x-ray investigation of anthracene: J. M. Robertson. *Proc. Roy. Soc.* (London), **140A**, 79 (1933).

[22]   J. A. Bearden, *Proc. Nat. Acad. Sci. U.S.*, **14**, 539 (1928); E. O. Wollan, *Proc. Nat. Acad. Sci. U.S.*, **14**, 864 (1928); H. Haas, *Ann. Phys.*, **85**, 470 (1928); H. Mark and K. Wolf, *Z. Phys.*, **52**, 1 (1928).

[23]   J. J. Thomson, *Conduction of Electricity through Gases*, 2nd ed., Cambridge University Press, London, 1906, p. 325.

[24]   (1) *International Tables for X-ray Crystallography*, Vol. II, Kynoch Press, Birmingham, England, 1959, pp. 238–240, 270, 271; (2) or *Internationale Tabellen zur Bestimmung von Kristallstrukturen*, Vol. II, Gebrüder Borntraeger, Berlin, 1935, pp. 567–568; also Appendix VIII.

[25]   P. Debye, *Ann. Phys.*, **43**, 93 (1913–1914).

[26]   P. Debye and P. Scherrer, *Phys. Z.*, **19**, 481 (1918).

[27]   F. C. Blake, *Rev. Mod. Phys.*, **5**, 169 (1933).

[28]   A. Schleede and E. Schneider, *Röntgenspektroscopie und Kristallstrukturanalyse*, Vol. II, de Gruyter, Berlin, 1929, pp. 240–244.

[29]   H. Ott, *Z. Phys.*, **22**, 201 (1924); **88**, 699 (1934); H. Hoffmann and H. Mark, *Z. phys. Chem.*, **111A**, 321 (1924); E. G. Cox and W. F. B. Shaw, *Proc. Roy. Soc.* (London), **127A**, 71 (1930); A. Hettich, *Z. Kristallogr.*, **90**, 479 (1935); M. J. Buerger. *Proc. Nat. Acad. Sci. U.S.*, **26**, 637 (1940); M. J. Buerger and G. E. Klein, *J. Appl. Phys.*, **16**, 408 (1945).

[30]   Reference 24-1, pp. 266–290.

[31] P. Debye, *Verh. deut. phys. Ges.*, **15**, 678, 738, 857 (1913); *Ann. Phys.*, **43**, 49 (1913–1914).

[32] M. v. Laue, *Ann. Phys.*, **42**, 1561 (1913); **81**, 877 (1926).

[33] E. Schrödinger, *Phys. Z.*, **15**, 79, 497 (1914).

[34] H. Faxen, *Ann. Phys.*, **17**, 615 (1918); *Z. Phys.*, **17**, 266 (1923).

[35] L. Brillouin, *Ann. Phys.*, **17**, 88 (1922).

[36] I. Waller, *Z. Phys.*, **17**, 398 (1923); **51**, 213 (1928); dissertation, University of Uppsala (1925); *Ann. Phys.*, **79**, 261 (1926); **83**, 154 (1927).

[37] C. Zener and G. E. M. Jauncey, *Phys. Rev.*, **49**, 17 (1936); C. Zener, *Phys. Rev.*, **49**, 122 (1936).

[38] R. W. James and E. M. Firth, *Proc. Roy. Soc.* (London), **117A**, 62 (1927).

[39] R. W. James and G. W. Brindley, *Proc. Roy. Soc.* (London), **121A**, 155 (1928).

[40] R. W. James, G. W. Brindley, and R. G. Wood, *Proc. Roy. Soc.* (London), **125A**, 401 (1929).

[41] L. Helmholz, *J. Chem. Phys.*, **4**, 316 (1936).

[42] E. W. Hughes, *J. Am. Chem. Soc.*, **63**, 1737 (1941).

[43] E. W. Hughes and W. N. Lipscomb, *J. Am. Chem. Soc.*, **68**, 1970 (1946).

[44] P. A. Shaffer, Jr., *J. Am. Chem. Soc.*, **69**, 1557 (1947).

[45] *International Tables for X-ray Crystallography*, Kynoch Press. Birmingham, England, Vol. II, pp. 241–265, 327, Vol. III, pp. 232–244, 1959 and 1962.

[46] D. R. Hartree, *Proc. Camb. Phil. Soc.*, **24**, 89, 111 (1928).

[47] V. Fock, *Z. Phys.*, **61**, 126 (1930).

[48] J. C. Slater, *Phys. Rev.*, **34**, 1293 (1929); **35**, 210 (1929).

[49] L. H. Thomas, *Proc. Camb. Phil. Soc.*, **23**, 542 (1926).

[50] E. Fermi, *Z. Phys.*, **48**, 73 (1928).

[51] *International Tables for X-ray Crystallography*, Vol. III, Kynoch Press, Birmingham, England, 1962, pp. 201–212.

[52] D. Liberman, J. T. Waber, and D. T. Cromer, *Phys. Rev.*, **137**, A27 (1965).

[53] J. C. Slater, *Phys. Rev.*, **81**, 385 (1951).

[54] D. T. Cromer and J. T. Waber, *Acta Crystallogr.*, **18**, 104 (1965).

[55] D. T. Cromer, *Acta Crystallogr.*, **19**, 224 (1965).

[56] W. L. Bragg, R. W. James, and C. H. Bosanquet, *Phil. Mag.*, (6) **41**, 309; **42**, 1 (1921).

[57] J. A. Bearden, *Phys. Rev.*, **29**, 20 (1927).

[58] R. Brill, *Acta Crystallogr.*, **3**, 333 (1950).

[59] J. M. Robertson and I. Woodward, *J. Chem. Soc.*, p. 36 (1940); S. C. Abrahams, J. M. Robertson, and J. G. White, *Acta Crystallogr.*, **2**, 233 (1949).

[60] R. McWeeny, *Acta Crystallogr.*, **4**, 513 (1951).

[61] For instance, R. W. G. Wyckoff, *Phys. Rev.*, **36**, 1116 (1930).

[62] Reference 51, pp. 213–216.

[63] A. Schuster and J. W. Nicholson, *An Introduction to the Theory of Optics*, 3rd ed, Edward Arnold, London, 1924, p. 17.

[64] M. J. Buerger, "Numerical Structure Factor Tables," *Geol. Soc. Am. Special Paper 33*, 1941, 119 pp.

[65a] *International Tables for X-ray Crystallography*, Vol. I, Kynoch Press, Birmingham, England, 1952, Sections 4.5 and 4.7, pp. 353–366, 373–525.

[65b] *Internationale Tabellen zur Bestimmung von Kristallstrukturen*, Vol. I, Gebrüder Borntraeger, Berlin, 1935, Chapter V, pp. 80–377.

[66] K. Lonsdale, *Structure Factor Tables*, G. Bell, London, 1936.

[67] L. W. McKeehan, *J. Franklin Inst.*, **193**, 231 (1922).

[68] G. Greenwood, *Phil. Mag.*, (7) **3**, 963 (1927).

[69] H. Möller and A. Reis, *Z. phys. Chem.*, **139A**, 425 (1928).

[70]   A. Claassen, *Phil. Mag.*, (7) **9**, 57 (1930).

[71]   A. Rusterholz, *Helv. Phys. Acta*, **4**, 68 (1931).

[72]   A. J. Bradley, *Proc. Phys. Soc.* (London), **47**, 879 (1935).

[73]   H. Kersten and W. Lange, *Rev. Sci. Instrum.*, **3**, 790 (1932); R. Pepinsky, *Rev. Sci. Instrum.*, **24**, 403 (1953); F. Barbieri and J. Durand, *Rev. Sci. Instrum.*, **27**, 871 (1956); *Bell Lab. Rec.*, **35**, 428 (1957).

[74]   W. L. Bond, *Rev. Sci. Instrum.*, **22**, 344 (1951); K. S. Revell and R. W. H. Small, *J. Sci. Instrum.*, **35**, 73 (1958); J. Durand, *Rev. Sci. Instrum.*, **30**, 840 (1959); R. F. Giese, Jr., *Rev. Sci. Instrum.*, **34**, 185 (1963); R. C. Linares, *Rev. Sci. Instrum.*, **35**, 1610 (1964).

[75]   W. L. Bond, *Acta Crystallogr.*, **12**, 375 (1959).

[76]   Reference 24-1, pp. 291–305.

[77]   K. Weber, *Acta Crystallogr.*, **23**, 720 (1967).

[78]   P. Coppens, L. Leiserowitz, and D. Rabinovich, *Acta Crystallogr.*, **18**, 1035 (1965).

[79]   Reference 65a, (1) Sections 4.3 and 4.4, pp. 73–352; (2) reference 65b, chapters V and VI, pp. 80–404.

[80]   E. J. W. Whittaker, *Acta Crystallogr.*, **6**, 222 (1953).

[81]   L. V. Azaroff, *Acta Crystallogr.*, **8**, 701 (1955).

[82]   L. V. Azaroff, *Acta Crystallogr.*, **10**, 413 (1957).

[83]   General reference 7, pp. 234–243.

[84]   E. Sándor and W. A. Wooster, *Brit. J. Appl. Phys.*, **14**, 506, 515 (1963).

[85]   B. E. Warren, *Acta Crystallogr.*, **6**, 803 (1953).

[86]   F. H. Herbstein and B. L. Averbach, *Acta Crystallogr.*, **8**, 843 (1955).

[87]   G. B. Mitra and N. K. Misra, *J. Phys.*, **D2**, 27 (1969).

[88]   General reference 1, Chapter V.

[89]   W. A. Wooster, *Diffuse X-ray Reflections from Crystals*, Oxford University Press, London, 1962.

[90]   W. Cochran and G. S. Pawley, *Proc. Roy. Soc.* (London), **A280**, 1 (1964).

[91]   J. L. Amorós, M. L. Canut, and A. DeAcha, *Z. Kristallogr.*, **114**, 39 (1960).

[92]   E. Møller and E. Jensen, *Acta Crystallogr.*, **5**, 345 (1952).

CHAPTER 4

PHOTOGRAPHIC POWDER TECHNIQUES

Not long after Laue's discovery of the diffraction of x-rays by single crystals, the first diffraction halos of the "powder" type were observed by Friedrich[1] upon passing x-rays through certain waxes, and by Keene [2] when x-rays were allowed to penetrate thin metal sheets. However, because they regarded these materials as being essentially noncrystalline, neither Friedrich nor Keene arrived at a proper interpretation of their observations.

The reciprocal-lattice concept can aid us in understanding the origin of powder diffraction diagrams. Referring to Fig. 3-8, a discrete diffracted ray is emitted in the direction BP (equivalent to OR) when a single crystal is tilted so that a reciprocal-lattice point or node P_{hkl} falls on the sphere of reflection. In the case of a polycrystalline specimen characterized by random crystallite orientations, however, the reciprocal-lattice vector $\boldsymbol{\rho}$ corresponding to the set of planes (hkl) assumes all orientations with equal probability, with the consequence that the point P is dispersed over all positions on the surface of a sphere of radius $|\boldsymbol{\rho}_{hkl}| = 1/d_{hkl}$, as indicated in Fig. 4-1. This sphere intersects the sphere of reflection in the circular zone shown in perspective as a dotted ellipse. Reflection from planes (hkl) therefore takes place in all directions defined by this zone in reciprocal space, and it can be seen that the diffracted rays, thought of as originating at the point B, lie on a cone of semiapex angle 2θ concentric with the undeviated beam. A film placed perpendicular to the undeviated beam records the diffracted rays as a circle. In the same way another reciprocal-lattice node with a different index triplet $(h'k'l')$ generates a diffraction cone of a different angle, leading to a different diffraction circle on the film.

The essential features of the powder diffraction technique thus include a narrow beam of monochromatic x-rays impinging upon a crystalline powder composed of fine, randomly oriented particles. As shown in Fig. 4-2, under these conditions all the diffracted rays from sets of planes of

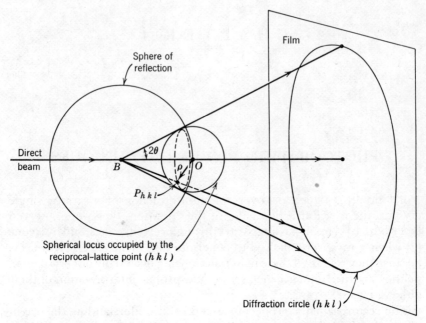

Fig. 4-1. Reciprocal-lattice model of the origin of a powder reflection. (Courtesy of Alexander [174].)

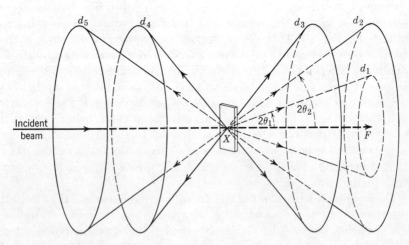

Fig. 4-2. Diffraction of x-rays by a flat powder cake.

176

spacing d_1 generate a cone of semiapex angle $2\theta_1$, planes of spacing d_2 generate a cone of angle $2\theta_2$, and so on. A pattern of concentric rings is produced by those cones of diffracted rays that intersect a film placed perpendicular to the undeviated beam XF at the point F.

If for any reason the total number of orientations of the crystallites irradiated by the x-ray beam is too small, the diffraction rings will be discontinuous and spotty rather than continuous and uniform in intensity. This difficulty can be remedied either by increasing the number of crystallites irradiated by the beam, or by moving the sample in such a way as to cause each crystallite to assume a variety of orientations with respect to the x-ray beam. The necessity of a very large number of crystal orientations is evident when we realize that under ordinary experimental conditions any given diffraction halo corresponding to planes of some spacing d is produced by only a small fraction of all the crystallites, that is, only those that chance to be oriented so that planes of spacing d are in position to satisfy the Bragg reflection condition for x-rays of wavelength λ:

$$n\lambda = 2d \sin \theta.$$

Calculations for a stationary specimen under typical experimental conditions show that only 0.1 to 10 per cent of irradiated crystallites contribute to any given diffraction halo. When the sample is rotated, as in the Debye-Scherrer method, the situation is much more favorable, but still the proportion of crystallites contributing to the useful portion of the halo which can be recorded on the film is usually less than 35 and may be as little as 2 per cent.

It is evident, then, that generation of a powder reflection is a relatively inefficient process and that, furthermore, the fineness of subdivision of the sample is of critical importance from the standpoint of the quality and reliability of the reflections obtained. Specific recommendations for the preparation of satisfactory powder specimens are included in the following sections.

4-1 THE DEBYE-SCHERRER METHOD

4-1.1 Introduction

This method of powder analysis is also referred to as the Hull-Debye-Scherrer method because it was independently devised by Hull[3] in the United States and Debye and Scherrer[4] in Germany between 1915 and 1917. The characteristic geometrical features are portrayed in Fig. 4–3. The incident beam is usually passed through a filter F to eliminate the characteristic $K\beta$ radiation, after which the $K\alpha$ rays are collimated by the

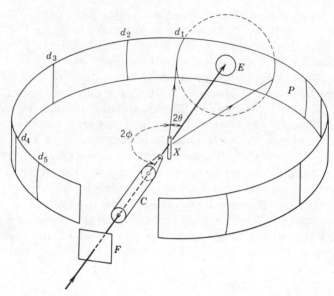

Fig. 4-3.   Geometrical features of the Debye-Scherrer technique.

pinhole system C. The resulting narrow pencil of rays bathes the small cylindrical sample X, and sections of the various diffraction cones are intercepted by the cylindrically disposed strip of photographic film P. By comparing this figure with Fig. 4-2, the reader can appreciate how this arrangement permits us to record not only sections of the halos diffracted in the forward direction, but also most of those in the back-reflection region (for which $2\theta > 90°$). The powder specimen is customarily rotated about its cylindrical axis in order to increase the number of particles contributing to each reflection. The features shown in Fig. 4-3 are typical of the Debye-Scherrer technique, but some, such as location of the filter, nature of the collimator, and positions of the ends of the film, may be varied in order to achieve particular advantages. The invariant features of the method include a collimated beam of characteristic x-rays impinging upon a small powder sample located at the center of a cylindrically arranged strip of film.

The chief merits of the Debye-Scherrer method are (1) the small amount of sample powder required (as little as 0.1 mg can be used), (2) the practically complete coverage of all the reflections produced by the specimen, and (3) the relative simplicity of the apparatus and technique required.

4-1.2 Camera Design

The most satisfactory general-purpose camera is one providing high-quality diffraction patterns with short exposures. More specifically, the goals of good camera design may be given as: (1) maximum line intensity, (2) optimum resolution of lines, (3) minimum background, (4) completeness of pattern including weak lines, and (5) good line profile and quality (in particular, smoothness rather than spottiness). Of these goals 3, 4, and 5 and either 1 or 2 can to a large extent be achieved in a single design. To a considerable extent goals 1 and 2 are incompatible, so that we cannot expect both highest-quality patterns and minimum exposure times from the same camera. However, it is possible to design a single camera with interchangeable beam collimators and exit tubes by which either condition 1 or 2 can be rather well satisfied. Such a modifiable camera comes close to fulfilling the need for a general-purpose instrument in a diffraction laboratory.

In the following treatment of camera design it is assumed that the sample is in the form of a small cylinder mounted perpendicular to the incident beam, which is sufficiently large to irradiate its entire cross section. This is the arrangement in most general use. After a brief discussion of the basic principles of Debye-Scherrer camera design in terms of this particular model, certain modifications are described which present special advantages.

A. General Geometry. The geometry of the standard Debye-Scherrer camera has evolved over the years, and has been presented in detail by Klug and Alexander [5], Buerger [6,7], and others. In turn, the purveyors of diffraction equipment have incorporated the best design features into a wide selection of well-engineered, popularly priced cameras, which the diffractionist can buy with confidence. Some of the basic features of good camera design are depicted in Fig. 4-4, which represents a section through the camera perpendicular to the camera and specimen axis and containing the axis of the collimator-exit port system. The x-ray source, which is the focal spot of the x-ray tube, is ordinarily rectangular in shape and "viewed" by the camera longitudinally ("end on") at a small angle (the so-called take-off angle) to the target surface, so that its apparent shape approximates a square. A typical commercial tube might have a focal spot of 1.2×12.0 mm, which appears as a 1.2-mm square at a take-off angle of 6°. The line ff' in Fig. 4-4 is the projected length of the x-ray focal spot; S_1, S_2, and S_3 are the three circular apertures of a pinhole collimating system; and X is the sample.

The sample diameter is less than the projected width of the focus, and the second aperture S_2 is just large enough to permit the entire sample to

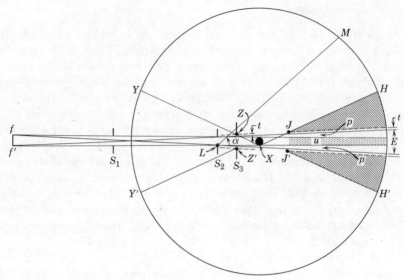

Fig. 4-4.   Illustrating geometrical features of good camera design. (For clarity dimensions and angles are exaggerated relative to camera diameter.)

be immersed in the zone of maximum illumination (dotted zone, denoted u). The rest of the undiffracted beam is less intense than u and has an overall width of E at the exit port. For maximum intensity aperture S_1 must be large enough to accommodate the entire beam defined by S_2. In ordinary practice these apertures are made of equal size, in which case S_1 actually limits the incident beam to some extent. S_1 is usually located on or slightly outside the periphery of the camera, while S_2 is commonly placed about halfway between the center and the periphery. Narrower collimation is more simply achieved (but with probably greater loss of intensity) by the use of a pair of smaller apertures, S_1 and S_2, than by reduction of the target take-off angle. The take-off angle is seldom made less than 5 or 6°, because of the rapid fall-off of intensity with decreasing angle (Fig. 4-5).

The requirement of minimum film background is met in part by minimizing the amount of air-scattered radiation reaching the film. This can be accomplished by shielding the direct beam from the film over a large part of its path through the camera. The collimator tube serves this purpose between the entrance port and the sample, and this new function of the collimator dictates that S_2 be brought as near as possible to the sample without any desired portion of the diffraction pattern being lost. Between the sample and the exit port, a second tubular fixture, which may be designated the "exit tube," should be

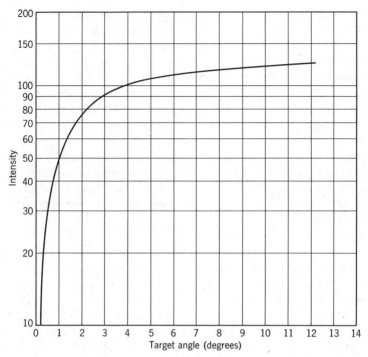

Fig. 4-5. Dependence of intensity upon angle at which target surface is viewed for a typical diffraction tube. (Courtesy of H. W. Pickett, and General Electric Company.)

placed to perform the same shielding function in the low-angle portion of the camera.

Collimator aperture S_3 must now be considered. Its primary purpose [6,7] is to prevent x-rays scattered by the periphery of S_2 from reaching the film. Scatter from S_2 occurs only at angles greater than some minimum angle α which is determined by the material composing aperture S_2 and the lower-wavelength limit (λ_m) of the beam. Suppose that aperture S_2 is fabricated from lead and that the x-ray tube is operated at such a voltage that $\lambda_m = 0.35$ Å; then $\alpha = 2\theta_m = 7.0°$. The optimum position of S_3 is then a function only of the size and position of S_2. As can be seen from Fig. 4-4, the inner edge of S_3, designated Z, should be placed outside the extreme limits of the x-ray beam by a narrow safety interval t, and it should be placed slightly to the right of the line LM (defined by the angle α) so as to intercept the lowest wavelength reflections from S_2. The two apertures S_2 and S_3 are thus interdependent. In order to minimize air scatter, the two apertures should be moved as near to the sample as possible without sacrificing a significant portion of the back-reflection pattern. Since the collimator tube unavoidably eliminates the zone YY' of the diffraction pattern, some compromise may be needed here to keep YY' as small as possible (not over 10 to 12° of arc).

The optimum dimensions of the exit tube can also be determined diagrammatically, as illustrated in Fig. 4-4. Suppose the zone HH' represents the low-angle portion of the diffraction pattern that can be sacrificed for the contemplated applications of the camera. Commonly, HH' is the diameter of the hole punched in the film to accommodate the base of the exit tube, and it should normally not eliminate more than the portion of the diffraction pattern below $2\theta = 6°$. From the geometry of the diagram, it can be seen that the walls of the exit tube may be placed anywhere within the hatched zones without the desired portion of the diffraction pattern being lost. To minimize air scatter the optimum position of the forward rim of the exit tube is then J, allowing a reasonable tolerance zone of thickness t between the outer periphery of the beam and the inner wall of the tube. It is of utmost importance that sufficient tolerance be allowed so that there is no chance for the undiffracted beam to strike the tip of the exit tube, thereby producing undesired diffraction effects from the metal of which it is composed.

For practical reasons S_1 is located as far as possible from S_2. This means that it is close to the extreme outer end of the collimating fixture and approaches the x-ray tube window very closely when the camera is in operating position. This arrangement is conducive to optimum resolution of lines. Parrish and Cisney[8] have provided a rigorous mathematical treatment for obtaining the optimum position of S_2 and the optimum positions and diameters of the apertures S_3 and JJ'.

Next to be considered is the problem of attaining good resolution, which is equivalent to achieving minimum angular width of lines. Two principal cases must be distinguished according to whether the width of the projected focal spot is (I) considerably larger than the diameter of the first collimator pinhole, or (II) of approximately the same size. Case I is usually applicable when very small pinholes, for instance, 0.5 mm or less in diameter, are employed to view a typical rectangular focal spot 8 to 12 mm in length longitudinally at an angle of 6 or 7° to the target surface. Case II is applicable when pinholes approximately 0.8 mm or more in diameter are employed with a similar take-off angle.

Mathematical descriptions of both of the above cases have been presented by Klug and Alexander[5]. It suffices here to point out that, in case I in the forward-reflection region ($2\theta < 90°$), the line width decreases with decreasing diameter of the first pinhole S_1 (Fig. 4-4), decreasing radius of the sample X, increasing distance S_1 to X, increasing film radius, and increasing Bragg angle. In the back-reflection region, the line width is approximately equal to the width of the beam-defining aperture S_1, a fact that has been pointed out by Buerger [7]. For a vanishingly small aperture S_1, this leads to an interesting and useful focusing effect at back-reflection angles, where line widths approach zero. In practice further sharpening of the lines may be achieved by making the distance S_1 to X somewhat greater than the film radius. In case II the size of S_1 is large enough that its position becomes irrelevant. The resolution now improves with decreasing focal spot width ff' and increasing source-to-sample distance, ff' to X. A smaller take-off angle is a practical way to decrease the focal spot width ff'. If the projected area of the source is reduced by one-half as a result of halving the take-off angle, the energy loss is much less than one-half, except at very small take-off angles, whereas if the aperture areas are reduced by one-half, the transmitted energy is likewise diminished by one-half. Adjustment of the take-off angle, however, is usually a somewhat more time-consuming operation than substituting a pair of smaller pinhole apertures.

Increased intensity and shortened exposure time are promoted by decreasing the camera radius, reducing to a minimum the source-to-specimen distance, and simultaneously increasing the aperture sizes and the specimen radius in conformity with the geometrical principles portrayed in Fig. 4-4. These desiderata are seen to be opposed to the conditions promoting line resolution.

For a laboratory performing diversified powder diffraction work, the best solution to the problem of camera design is to construct or obtain two cameras, or one camera with interchangeable collimators and exit tubes, one designed to provide high intensity and/or short exposures for work of a routine nature, and the other designed for good resolution and line quality for tasks of a more demanding nature. In many cases three designs are warranted, the intermediate one being suitable for general-purpose work and providing both reasonably good resolution and intensity. Table 4-1 lists suitable dimensions for such a set of three cameras, together with the approximate exposure times expected when used with any of the modern commercial Coolidge tubes operated at recommended wattage ratings. The apparently strange camera diameters of 5.73 and 11.46 cm have the effect of making 1 mm on the film equal 2 and $1°\,2\theta$, respectively. This permits direct measurement of Bragg angles with a millimeter scale to an accuracy sufficient for many purposes. If the camera diameters are made larger than these figures by an amount equal to the thickness of the film used (for double-emulsion film), the resulting effective film diameters will approach these ideal values even more closely.

For special purposes Debye-Scherrer cameras of unusual dimensions

Table 4-1. **Suggestive Debye-Scherrer Camera Dimensions for a Laboratory Carrying out Diversified Work**

Purpose	Inside Camera Diameter (cm)	Diameter of Pinholes S_1 and S_2 (cm)	Specimen Diameter (cm)	Exposure Time (hr)
Rapid work of a routine nature	5.73	0.08–0.10	0.06–0.08	$\frac{1}{4}$–$\frac{1}{2}$
General purpose	11.46	0.08–0.10	0.06–0.08	1–2
Exacting work (precision measurements, crystallite-size measurements, and so on)	11.46	0.04–0.05	0.03	6–10

may be required. Thus the need for an unusual degree of resolution can be satisfied either by using very fine specimens (perhaps as small as 0.02 cm in diameter) in conjunction with a camera diameter of 11.46 cm, or by increasing the camera diameter beyond this size. In either case a longer exposure is required. Conversely, still greater intensity can be obtained by employing a camera smaller than 5.73 cm in diameter, but for most purposes a limit is soon reached beyond which the resolution suffers excessively. Parrish and Cisney[8] describe cameras modified somewhat so as to record the diffraction pattern to either unusually low or unusually high angles ($2\theta = 4.5$ and $175.5°$, respectively), but in each case a small portion of the pattern elsewhere is sacrificed.

The preceding discussion of camera design has presupposed the use of a pinhole collimator. This type of collimator is most generally satisfactory for Debye-Scherrer work of a varied nature and, in particular, it is conducive to highest-quality line profiles and minimum film background. When these considerations are outweighed by the need for greater intensity, it may be of advantage to substitute slits for the pinholes S_1, S_2, and S_3 of Fig. 4-4. Figures 4-6A and 4-6B show, respectively, the effects upon the diffraction halos of substituting one and two slits for the pinholes defining the beam. With the slits disposed parallel to the axis of the cylindrical sample, the extreme rays a and c give rise, respectively, to the halos a' and c', which are displaced from each other by the maximum amount AC. All other rays from the target are intermediate between a and c in direction, and they produce halos which are intermediate in position between a' and c'. The overlapping halos result in fan-shaped lines on the film F. The broadening is least at the center and greatest at the edges of the film, and the extent of distortion is greatest

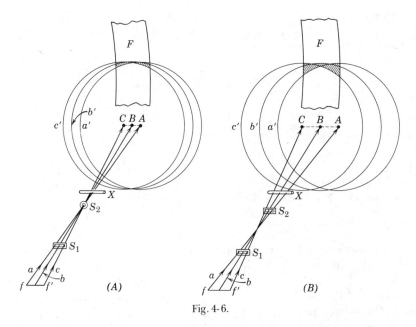

Fig. 4-6.

for 2θ near 0 and 180° and least near 90°. However, greater intensity and/ or shorter exposures are gained by such slit arrangements because of (1) the larger amount of sample contributing to the diffraction lines, and (2) the greater intensity of illumination of the sample. It should also be noted that no advantage is gained in making the slit S_1 longer than necessary to permit rays from the extreme ends of the focal spot, f and f', to illuminate the sample.

The relative intensities to be expected with different collimator and source arrangements have been predicted by Alexander[9] by means of a geometrical and algebraic analysis. The results of such an analysis are summarized in Table 4-2 for four different sets of experimental conditions, including viewing the rectangular focal spot through (1) two 0.75-mm pinholes longitudinally, (2) two 0.75-mm pinholes laterally, (3) two 0.75×5.5-mm slits longitudinally, and (4) two 0.75×5.5-mm slits laterally. Here *longitudinally* signifies viewing the focal spot "end on" at a small angle, and *laterally* means viewing the focal spot at right angles to its long dimension at the same small angle, ordinarily 6 or 7°.

These results demonstrate clearly that a pair of pinholes is greatly inferior to a pair of slits of moderate length from the standpoint of irradiation intensity. Furthermore, the gain in intensity on substituting slits for pinholes is found to be about the same whether the source is viewed longitudinally or laterally. Experimental measurements[10–12]

Table 4-2.   Computed Integrated Intensities of Sample Irradiation
for Four Typical Experimental Cases[a]

Case	Apertures S_1 and S_2	Aperture Dimensions	View of Focal Spot	Calculated Relative Intensity
1	Pinholes	0.75-mm diam	Longitudinal	10.5
2	Pinholes	0.75-mm diam	Lateral	2.5
3	Slits	0.75 × 5.5 mm	Longitudinal	100.0
4	Slits	0.75 × 5.5 mm	Lateral	97.5

[a]Source is a 1 × 8-mm rectangular focal spot viewed at 7°; source-to-S_1 distance (FS_1), 60 mm; source-to-S_2 distance (FS_2), 100 mm; source-to-sample distance (FX), 150 mm.

confirm that the order of magnitude of the calculated intensity increase is correct. The added intensity is obtained at the expense of decreased resolution and a shift in line positions, but both these effects are very small for moderate slit lengths, whereas the gain in intensity is several-fold. Consequently, such an arrangement is entirely satisfactory for the numerous applications of powder diffraction for which optimum line quality is not a prime consideration, and furthermore it permits a very substantial saving in exposure time and increase in x-ray tube life.

It is particularly important that the gain in intensity on replacing pin-holes with slits is approximately the same whether the focal spot is viewed longitudinally or transversely, thus making it possible for the "side" windows of an x-ray tube, commonly neglected by diffractionists, to be put to good use. The reason for this neglect of side windows when employing pinhole apertures is illustrated by the results obtained for case 2 in Table 4-2. Here a marked loss of intensity is indicated in viewing a rectangular focus laterally through a pair of pinholes with diameters much smaller than the length of the focus.

B. Details of Camera Construction. A completely satisfactory Debye-Scherrer camera must not only incorporate sound geometrical principles of design, but must also be constructed with regard to convenience and precision of manipulation, particularly if it is to be suitable for more routine applications. In general, the trend in American design has been toward convenience and speed of manipulation, whereas European efforts, particularly British, have to a greater degree accented precision features. If a newcomer to the x-ray diffraction field is fortunate enough to acquire a camera manufactured by one of the several commercial firms in the field, he may reasonably expect that it will not only incorporate the minimum features of good design but that it will also provide a number of features conducive to speed and ease of use. If, however, he builds his own camera, he will be immediately faced with a need to understand the rudiments of the theory and practice of camera construction even though he follows a detailed set of plans representing the results of experience of earlier workers. In either event he will become a more proficient and versatile diffractionist if he understands the fundamentals of camera construction, because he cannot work seriously in this field for long without having to fall back on his own resources and ingenuity in constructing cameras or modifying cameras of standard design to fill special requirements. For instance, Becherer and Herms [54] modified a Debye-Scherrer camera for the balanced-filter technique or the comparison of two samples.

Debye-Scherrer cameras are of several types. Perhaps the most widely used in this country is typified by the Buerger design [6, 7] shown in Fig. 4-7. The film is held tightly against the inner cylindrical surface of the camera and protected from light during the exposure by a well-fitted cover. This kind of camera does not permit access to the specimen or collimating fixtures after the camera has been loaded. A second type of camera is distinguished by the film's being held against the outside of a supporting cylinder, which is slotted over a large angular range to permit the diffracted rays to reach a central strip of the film. A camera of this design [13] evolved in the laboratories of W. L. Bragg. In such a camera the film is inserted in a light-tight holder, or cassette, in the darkroom, the loaded cassette then being placed in position in the camera, which does not offer protection from light. A minor advantage of this design is that the camera is not disturbed after the sample- and beam-aligning operations have been completed.*

It is important that the metal surface against which the film is held be precisely cylindrical in shape, and even more important that the axis of the sample-supporting spindle be accurately centered in this cylinder. Perhaps the most satisfactory method of accomplishing these aims is to turn the film-supporting surface and bore the bearing for the spindle from a solid metal casting with the same lathe setting [6]. Because of its strength and ease of machinability, brass is the most commonly employed material for fabricating camera bodies and other parts not subject to an unusual amount of wear. For cassette cameras or other designs in

*The casette-type camera was available from the General Electric Company, whose diffraction instrument line was acquired in 1972 by the Diano Corporation, Industrial X-ray Division, 2 Lowell Avenue, Winchester, Massachusetts; and from Pye Unicam Ltd., whose former line of diffraction equipment is now manufactured and marketed by Crystal Structures, Ltd., Swaffham Road, Bottisham, Cambridge, England.

Fig.4-7.   Small and large Debye-Scherrer powder cameras based on the design of Buerger [6, 7]. (Courtesy of Philips Electronic Instruments and Alexander [174].)

which the spindle bearing and film-supporting surface cannot be constructed as parts of a single metal block, it is necessary to achieve accurate centering of the spindle bearing either by very precise machining of the respective parts to permit a close-tolerance assembly, or by making the spindle bearing fixture adjustable so that the spindle can be empirically centered once and for all at the time the camera is put into service. The latter solution of the problem is not recommended unless the first alternative is impracticable for some reason.

Figure 4-8 illustrates some desirable features of beam-collimator and exit-tube construction. Both tubes for convenience are inserted into the camera from the outside through cylindrical housings, the tubes remaining in place because of the close tolerance of the sliding friction fit. Utmost care must be exercised so that the axes of these two housings coincide with a diameter of the film cylinder. A consequence of this is that the common axis of the housings H_1 and H_2 intersects the perpendicular axis of the sample spindle bearing at the center of the camera. An

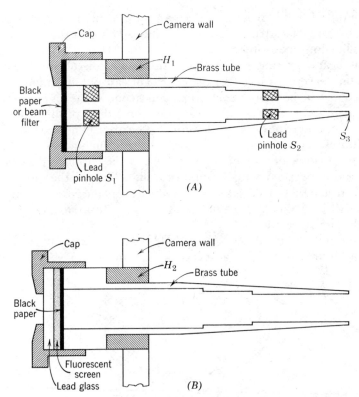

Fig. 4-8.   Typical features of good collimator (*A*) and exit-tube (*B*) construction (vertical dimension exaggerated somewhat).

alternative plan is to thread H_1 and H_2 so that both tubes are screwed into place, but it is difficult to accomplish this with the precision required to keep the axes of both housings on the camera diameter. The problem of accurately aligning the collimator and exit tubes on the camera diameter can be solved in a home-built camera by boring the tube housings H_1 and H_2 in small circular plates which are later adjusted to the proper positions by utilizing four loose holes through which the plates are screwed to the inner camera wall [6].

The pinhole diaphragms S_1 and S_2 (Fig. 4-8A) are made of lead, or alternatively of another element of sufficiently high atomic number that its K fluorescence spectrum is not excited by the shortest wavelengths in the general radiation component of the primary beam. Lead is most commonly employed, but because of its softness it must be protected from damage. Diaphragm S_1 is inserted a short distance within the collimator tube for this reason. If desired, a cap containing a β-radiation filter disk or a black paper disk may be slipped on over the outer end of the collimator tube as shown.

The beam-exit tube B is similar to the collimator tube in general dimensions, except for its larger diameter. For ease of fabrication its inner surface may be bored out to approximate a cone by means of a series of steps as indicated. An exit-port cap containing disks of the three kinds shown is found to be of particular utility. The black paper excludes light, the fluorescent screen gives an image of the beam and usually also reveals the shadow of the sample, and the lead glass protects the observer from the undeviated x-ray beam. For a slit collimator the tube may be oval or rectangular in cross section, in which event the cross section of the beam-exit tube should be rectangular so as to accommodate the undeviated beam efficiently.

Three devices are described for holding the film firmly against the inner cylindrical surface of a Buerger-type camera. In Fig. 4-9 the film is

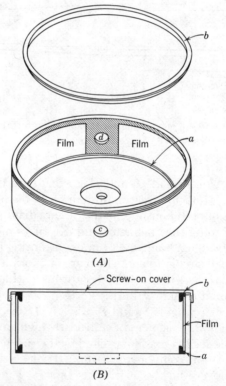

Fig. 4-9. Beveled-ring arrangement for holding film in position.

held in place by means of two beveled rings *a* and *b*, shown in perspective in *A* and cross-sectionally and in position in *B*. Ring *a* may be soldered in position, whereas *b* is removable. The film strip is inserted behind
the beveled edge of ring *a*, where it is held snugly in position by *b* when
the light-tight cover is screwed down. This technique has the decided
advantage that the film may be arranged in the camera in any desired
orientation, provided only that holes are first punched in the film to
coincide with the collimator and exit-tube apertures *c* and *d*. A second
method of holding the film is shown in Fig. 4-10. One end of the film is
held by a "stationary finger" *a*, and the other end is held by a "movable
finger" *b* attached to a slide *s* which can be moved from outside the
camera. The film is compressed tangentially by moving *b* until it fits
snugly against the camera wall. Then the slider is clamped in position
by turning a milled head *m* which acts against a brake [7]. When the film
is arranged in this way, commonly referred to as the asymmetric position,
holes must be accurately punched at *C* and *E* to accommodate the collimator and exit tubes. This film arrangement records the diffraction
pattern through nearly 360° of arc and is of particular value in the precision measurement of lattice constants (see Chapter 8). Because of its
several merits, this film position may justifiably be adopted as the stand-

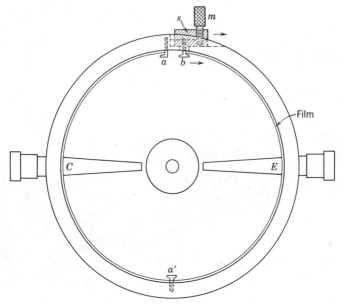

Fig. 4-10.   Arrangement involving a movable finger for compressing film against camera
wall.

ard one for recording all powder patterns. If for reasons of film economy only half the diffraction pattern is desired, a fixed stop can be plugged into a receptacle at a', and the film held between a' and b (centered about E). Alternatively, a can be removed and the film held between a' and b, but centered about C. It is necessary of course that the slide fixture s be so designed that no light can enter the camera through the slot in the camera wall through which b projects. Straumanis has described a pair of internal expansion rings for pressing the film tightly against the inside surface of a cylindrical camera. Two identical rings, an upper and a lower, are used[14].

In cameras of the second type in which the film is placed against the outside of a slotted cylindrical surface, the film can be held snugly in position by means of a felt-lined strip of spring steel which is placed outside the film and is provided with a tightening device. A similar arrangement is commonly used in circular cassettes, except that in addition the inner exposed strip of film is protected from light with black paper, and the film and steel-strip holder fit into a channel in such a way as to exclude light from the outside surface of the film.

No screen x-ray film is currently available in 35-mm roll form which fits most commercial cameras. It remains only to cut the film to length and punch holes to accommodate the collimator and exit tubes. Figure 4-11 shows an instrument for both punching roll film and cutting it to length. Strips of film to fit 57.3-mm and 114.6-mm cameras can also be cut from 5×7 in. and 11×14 in. sheets, respectively, with little or no wastage provided about 340° of the angular range is to be recorded (as in

Fig. 4-11.   Combination film punch and end cutter. (Courtesy of Philips Electronic Instruments.)

the asymmetric film position). When completeness of the pattern at high and low angles is a prime requisite, a simple cutter can be made that produces holes of precisely the right size with the film in place in the camera [15].

A very important feature of a satisfactory camera is a means for bringing the specimen axis into coincidence with the axis of the spindle by means of which it is rotated. Figure 4-12 shows schematically three devices which have been successfully used for centering the sample holder. In the scheme shown in the two drawings A, the spindle S is rigidly attached to the cylindrical member t [6]. This part contains a short cylindrical channel n in which the sample holder r is free to slide in two dimensions. The position of r is adjusted by the two pushers a and b, which hold r tightly against the spring-actuated plug p. The sample itself is mounted on a small cylindrical brass base of the proper diameter to fit comfortably in the cavity x, where it is held in place by the set screw c. This device is somewhat tedious to work, because the operation of the pushers a and b does not cause r to move in perpendicular directions. This defect is remedied in the design shown in drawings B [13]. Here a and b either push or pull the sample holder r along mutually perpendicular directions. This result is accomplished by providing a and b with endpieces a' and b' which fit snugly but without binding in the perpendicular channels l and m.

A particularly satisfactory scheme for centering the sample is shown in drawings C [8]. The basic design is approximately the inverse of that shown in A. A disk t is rigidly fixed to the spindle S and rests in a channel n in the cylindrical member r which holds the sample. The spring s holds t firmly against the lower ledge of r and permits r to be moved in two dimensions with respect to t, provided that sufficient force is exerted to overcome the friction induced by s. The sample is centered by means of the threaded pusher p which extends through the camera wall w to the outside and is operated by a knurled knob. When the centering has been accomplished, the pusher is retracted into w so as not to collide with r when the latter is rotated.

Recently, manufacturers have employed Alnico magnets in the form of disks to support the sample (filling the function of member r in Fig. 4-12, A, B, and C). The member r then adheres magnetically to the surface of a second disk-shaped part which performs the role of t in Fig. 4-12. The sample can be easily centered by translating r as required over the surface of t.

For reasons given earlier it is necessary that means be provided for rotating or oscillating the powder specimen during an exposure. This is commonly done with a synchronous motor which is either directly

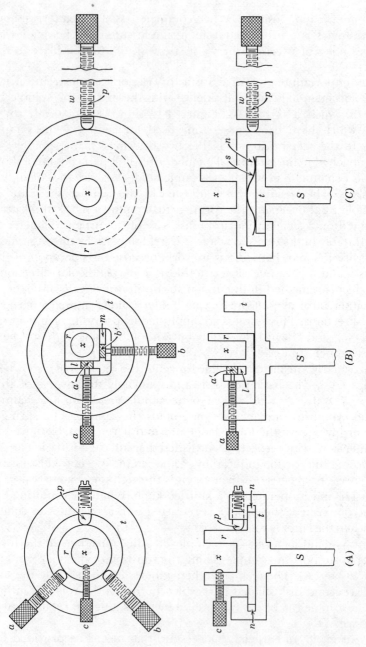

Fig. 4-12.   Three devices for centering a powder specimen.

194

geared to the sample spindle or is coupled to it by a pulley-and-belt drive. The second alternative is somewhat simpler to incorporate in a home-built camera. A slow speed of sample rotation, for example, 1 rpm or less, is desirable. When a wedge specimen mount is used, or when a photograph is to be made of the edge of a massive sample of noncylindrical shape, it is necessary to oscillate the specimen through a small angular range rather than rotate it. This can be done by replacing the gear drive with a cam assembly of the right dimensions. In order to avoid warming of the camera as a result of heat generated by the motor, it is good practice to allow a minimum of metallic contact between the camera and the motor. In particular, the motor should not be contained in a housing that is an integral part of the camera body or, if this is unavoidable, provision should be made for adequate ventilation of the motor. These considerations are important when the camera is to be employed in precisely determining the interplanar spacings of a specimen at some fixed temperature.

C. Camera Support and Alignment. In order that the camera may be brought into the most favorable orientation with respect to the x-ray beam, the means of support must provide four kinds of motion, as shown in Fig. 4-13: (1) translational motion approximately perpendicular to, and directed toward or away from, the axis of the x-ray tube, (2) translational motion parallel to the axis of the x-ray tube, (3) rotational motion about the axis of the sample spindle, and (4) translational motion parallel to the axis of the sample spindle. The process of aligning the camera will then consist of the following fundamental steps:

a. Utilizing all the degrees of freedom, place the camera in approximately the right position by direct inspection. In particular, by means of motion (1) bring the camera forward with the collimator assembly in

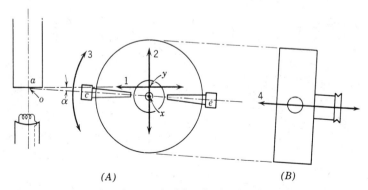

(A)   (B)

Fig. 4-13.   Types of motion required for aligning a Debye-Scherrer camera.

place until it is as close to the x-ray tube window as possible without interfering with further adjustments.

b. While observing the beam through the collimator with a fluorescent screen, translate along direction (2) while simultaneously rocking with motion (3) so as to keep the beam in view. At some time during this operation, translate along (4) to make the beam of maximum brightness, which insures that the collimator-exit tube axis *ce* actually intersects the center of the focal spot *o.* Continue motions (2) and (3) until the beam fades from view. When this happens, the sample axis *x* has reached the point *y* and the target surface is being viewed at an angle of $\alpha = 0^0$.

c. By means of motion (2) move the camera in the reverse direction until the distance $xy = oy \times \tan \alpha$, α being the desired angle of view as computed from the length of the focal spot and also in some cases the size of the collimator apertures (Section 4-1.2A). For example, for two 1-mm pinhole apertures and a focal spot 1×10 mm in size, the optimum angle of view is $\alpha = 6°$, whereas for two 0.5-mm pinholes the angle should be reduced to about 3° for maximum intensity.

d. Utilize motions (3) and (4) simultaneously (principally the former) until the observed beam intensity appears to be a maximum.

We have used two simple accessories for improving the ease and precision of the aligning process. The collimating tube is replaced by a short cylindrical plug *c* (Fig. 4-14) containing a single accurately centered

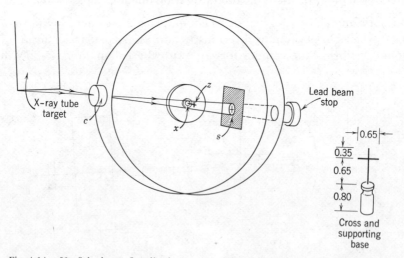

Fig. 4-14.   Useful scheme for aligning camera. Dimensions of cross are in centimeters.

pinhole of the same diameter as that of the two pinholes of the collimator which is to be used in the contemplated program of work. A small metal cross z is placed in the sample holder x. The insert shows satisfactory dimensions of such a cross for 1-mm pinholes. The arms of the cross are cut from no. 26 wire and soldered together and to the small cylindrical brass base.

The alignment procedure is now slightly modified, the first step being to view the cross by sighting down the pinhole system of the regular collimator with the aid of a suitable magnifying lens or long-focus microscope and adjusting its position in the sample holder until it is accurately centered in the circular field of view. In a properly constructed camera, the center of the cross then also lies on the axis of rotation of the sample spindle, hence it will stay in the same position when the cross is rotated, The line-up pinhole then replaces the collimator, and the beam-exit tube is removed so that the beam can be viewed on a sheet of fluorescent screen s placed between the cross and the exit port of the camera. The alignment process is then carried out as described above, except that the camera motions are always made with a view to centering the image of the focal spot about the intersection of the cross arms. This insures that the axis of the collimator pinholes accurately intersects both the center of the x-ray tube focus and the axis of rotation of the sample spindle, which is in effect equivalent to assuring that the camera alignment will provide an x-ray beam of greatest intensity. A final test of the beam should be made before using the camera by placing the regular collimator and exit tube in place, removing the cross, and noting whether the beam transmitted by the collimator is approximately round in cross section and of good intensity.

The nature of the supporting and orienting mechanisms depends partly upon the position of the x-ray tube, that is, whether the tube axis is vertical or horizontal, since the best relative positions of tube and camera are the ones shown in Fig. 4-13. Thus, if the x-ray tube is vertical, the camera-spindle axis is horizontal, and if the x-ray tube is horizontal, the spindle axis becomes vertical. If the focal spot is always to be viewed longitudinally and at such an angle as to appear square in shape, it becomes immaterial whether the camera is supported with the spindle axis perpendicular to the tube axis or nearly parallel to it (actually at an angle of α degrees to it). However, the more flexible arrangement and the one almost universally employed is the former (shown in Fig. 4-13), and it is henceforth assumed in the present discussion that this arrangement is to be utilized. It should be pointed out that if slit collimators are to be employed in conjunction with the "side" windows so that the focus is viewed laterally, it will become mandatory that the x-ray tube be in the vertical position, at least if simultaneous use is to be made of the "end" windows.

Besides providing means for adjusting the camera position with respect to the beam, a prime requisite of a satisfactory camera support is that it permit the camera to be removed and later replaced in precisely the same position without the necessity of repeating any adjustments. A useful plan for accomplishing this, particularly for home-built apparatus, is to provide the camera with a base having three leveling screws which rest on a "point-slot-plane" type of supporting platform (see Fig. 4-15). The diagram illustrates in a simplified manner the arrangement employed when the x-ray tube is vertical. The triangular base plate g is rigidly attached to the camera body by means of the block f. The base plate is supported near its three corners by the three leveling screws a', b', and d', which rest, respectively, in a concave hole a, in a slot b of triangular cross section, and on the plane surface of the plate h at d. The entire assembly is supported above the table top by the vertical tube j, which may merely be a section of 2-in. pipe of the proper length. The pipe is attached to the table top and the plate h by means of brackets or pipe flanges. By making the screw holes k in the bracket considerably oversize as indicated, a degree of freedom of motion is permitted in directions 1 and 4 of Fig. 4-13. Vertical translation (2) and rotational motion (3) can be effected by proper operation of the three base leveling screws. Because of the limited degree and rather arbitrary types of motion permitted by the adjustable features just described, it is necessary that the initial installation of the vertical support j be performed with sufficient care to place the camera in nearly the desired operating position. Since the remaining necessary movements are very limited in extent, they can be executed within the comparatively small degrees of freedom remaining. Once the adjustments have been completed, they will need to be changed only infrequently by small amounts if and when the relative positions of the x-ray tube and camera support change slightly with the passage of time. The camera can be removed at any time (for example, to load with film in the darkroom), and its original position will be perfectly restored when it is again returned to its point-slot-plane support. This method of supporting the camera has the merits of simplicity and cheapness of construction, but the limited nature and comparative crudity of the adjustments are serious disadvantages, particularly if diffraction work of a varied nature is contemplated, which usually requires flexibility of instrumentation. Two additional simple designs of camera supports are those of Karkhanavala[16] and Lindsay [17].

Figure 4-16 is a schematic diagram of a camera-supporting mechanism

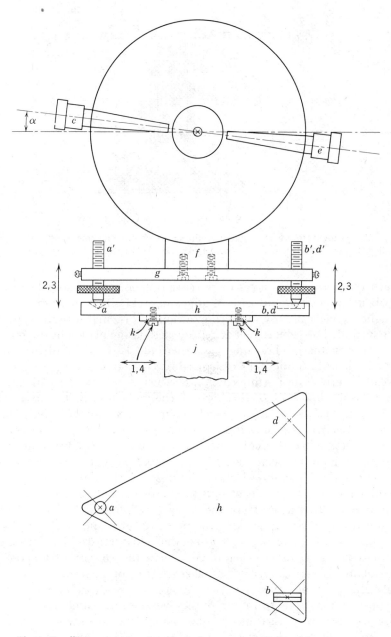

Fig. 4-15.   "Point-slot-plane" method of supporting a Debye-Scherrer camera.

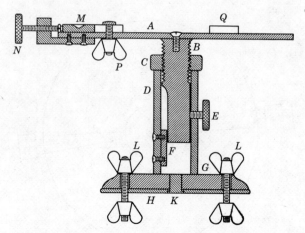

Fig. 4-16. Camera-supporting table permitting precision adjustments. [Courtesy of Bradley, Lipson, and Petch, *J. Sci. Instrum.*, **18**, 216 (1941); copyright, The Institute of Physics.]

which offers the advantages of point-slot-plane support together with flexibility, precision, and ease of adjustment[13]. This mechanism was designed specifically for supporting a camera oriented with the spindle axis vertical. The camera feet are supported on the plate *A* by the point-slot-plane method. The camera is raised and lowered by turning the large kurled nut *C* (motion 4 of Fig. 4-13). This elevates or depresses the threaded vertical supporting column *B*, which is kept from rotating by a key *F* on the inside of *D* working in a channel cut in *B*. The tube *D* is rigidly attached to the base *G*, which rotates on a spindle *K* fastened to a plate *H*. The rotation of *D* and *G* (motion 3 of Fig. 4-13) is permitted by curved slots cut in *G* through which pass two screws fixed in *H*. When the desired position has been reached, wing nuts *L* are tightened, locking *G* and *H* together. The screws also serve to attach the supporting base to the tabletop, and by means of slots in *H* they permit the assembly to be moved toward or away from the x-ray tube (motion 1 of Fig. 4-13). Motion 2 is provided by making the hole *M* of the hole, slot, and plane support movable. The plate containing the hole is translated by means of the screw *N*, and can be locked in position by the wing nut *P*. Of course the slot on the plate *Q* must be parallel to the axis of the screw *N*.

By attaching the camera-supporting device directly to the x-ray tube support, it is possible to minimize play between the camera and x-ray tube, with a greater assurance of good alignment. A camera bracket of this kind, which was first described by Buerger[7], is now incorporated into one of the commercial x-ray diffraction units. Most manufacturers in the United States have now conformed to the practice of supporting the camera by means of a special base which slides on a track fixed to the camera-mounting bracket. A typical modern arrangement of this sort is shown in Fig. 4-17. Figure 4-7 shows more clearly the construction of the camera-supporting base, and Fig. 4-17 shows two cameras in position on the tracks. In this design the various necessary motions for adjusting the camera are provided for as follows. The camera-mounting bracket fastens to the x-ray tube housing on a dovetail rail and is translated vertically (motion 2 of Fig. 4-13) by turning a coarsely threaded screw which is actuated by a large knurled knob resting against the top of the cabinet of the x-ray unit (see Fig. 4-17). Motion 1 is accomplished by sliding the camera on the track. It is important that the fit between the camera base and the track be precise, so that the position of the camera is recoverable. Motions 3 and 4 are provided in a particularly convenient way. As can be seen from Fig. 4-17, the camera is attached to its base by means of a horizontal shaftlike extension of the body which is coaxial with the sample spindle and acts as an extended housing for it. This cylindrical extension of the body fits into a horizontal bearing in the top of the upright member of the camera base. This bearing is split so that, when a tightening screw is loosened, the camera can be rotated or trans-

Fig. 4-17.   Modern bracket and track arrangement for supporting cameras. (Courtesy of Philips Electronic Instruments.)

lated in the bearing (motions 3 and 4, respectively). When the screw is tightened, motion of the camera is prevented, and the camera and its supporting base become a rigid unit. A further convenient feature of this design is the mounting of the synchronous motor in a case which forms an integral part of the camera bracket. The method of coupling the motor and sample spindle by means of a belt and pulleys can be seen in Fig. 4-17.

4-1.3 Preparation of the Powder

In order to yield a satisfactory diffraction pattern, a crystalline powder must meet certain specifications. The sample examined should be representative of the particular lot of material being investigated, the size of the crystallites composing the powder must fall within acceptable limits, and the method of preparing the powder for analysis must not distort the lattice and thereby impair the quality of the diffraction lines.

The most common type of sample is that which is either procurable directly in a finely divided state or sufficiently brittle to be easily reduced to a fine state by crushing and grinding. Materials of this class include most chemical precipitates, smokes and dusts, soil minerals, most rocks, most inorganic single crystals whether synthetic or naturally occurring, organic single crystals (unless very soft), metal corrosion products, and so on. When the material occurs in such a fine state as to be very smooth to the "feel," we can be reasonably certain that the ultimate crystallite size is small enough to give smooth diffraction lines, in which event grinding of the sample is not needed. When the powder grains are course, it is good general practice to reduce the material to a fine powder by grinding in a small agate mortar. This procedure should normally be followed, despite the fact that coarse particles frequently consist of much smaller crystallites, thereby generating smooth diffraction lines.

The necessary reduction in crystallite size depends upon the amount of powder irradiated by the beam, the crystalline symmetry, and the degree of motion of the specimen during the exposure. For ordinary exposure conditions, including 360° rotation of the specimen, most materials give smooth lines if the mean crystallite dimension is less than 45 μ. This is equivalent to a size small enough to pass a 325-mesh screen. For crystals of high symmetry (cubic, hexagonal, and tetragonal) 270- or even 200-mesh is often fine enough because the multiplicity factor is large (see Section 3-3.4F). Conversely, materials of lower symmetry (orthorhombic, or particularly monoclinic and triclinic crystals) may give somewhat spotty lines even after passing a 325-mesh screen. However, a degree of irregularity or spottiness is tolerable in powder diffraction work of a less exacting nature. A safe standard procedure for preparing sample powder from materials of this class is to grind the material in a small mullite or agate mortar until practically all of it passes a 325-mesh screen. Particles too coarse to pass the screen should not be used, because of the possibility of their contributing spots to the diffraction lines. It should be pointed out that screening may change the composition of a sample that consists of constituents varying in brittleness and hardness. For this reason such heterogeneous samples should always be ground until practically all the material passes the screen. However, if a degree of spottiness of the lines can be tolerated, it is still better (and much simpler) to omit screening the sample and simply grind long enough to achieve a reasonably small average particle size. Devices have been described for superposing a translatory motion upon the usual rotational motions of a Debye-Scherrer specimen in order to reduce line spottiness by exposing a larger number of particles to the x-ray beam [7, 18–20].

When the sample is oscillated through a small angular range, rather than rotated, the number of orientations of the crystallites contributing to the pattern is greatly reduced, and consequently line irregularity and spottiness are more common and set in at a smaller average crystallite dimension. Thus for wedge-type mounts (see below), which can only be oscillated, it is especially important that the particle size be small. Stationary samples of course are the worst offenders in this respect and tend to give spotty lines down to particle dimensions as small as 5 to 10 μ. Samples of sufficiently fine size can be prepared, if necessary, by sedimentation from a liquid of suitable density and viscosity [21], but it must be borne in mind that the composition of a heterogeneous powder is usually changed in the process.

Some inorganic and many organic crystals are so soft that they cannot be ground to fine powders under ordinary conditions without plastically deforming the lattice, so that broad, diffuse diffraction lines are produced, especially in the back-reflection region. Examples of such inorganic substances are iodine and the iodides of some heavy metals. Often such materials can be reduced to strain-free powders by first chilling to liquid-air or Dry Ice temperatures and then grinding or crushing in the usual way.

Metallic samples require special preparatory techniques. Metals may be regarded as texturally analogous to rock minerals, which are composed of an assemblage of variously oriented crystallites and are termed *massive* (see Section 1-1.1). If the crystallite size is small and the orientations highly random, a rock specimen of the proper dimensions is actually an excellent Debye-Scherrer "powder" sample. In fact, the diffractionist may occasionally find it to his advantage to prepare a powder diagram of a rock mineral by mounting a tiny particle of the material on the end of a fine supporting fiber and rotating it in the x-ray beam. Other materials of a fine polycrystalline texture that cannot be reduced to a powdered state can also be photographed directly if specimens of satisfactory size and shape can be prepared. Examples of such material are waxes and soaps. In a similar way metals and alloys fill the role of more or less perfect "powder" specimens, since they are also assemblages of variously oriented crystallites, customarily referred to as grains. Short pieces of wire of the proper diameter make the most satisfactory samples of this kind. However, a small metal block can serve as a sample if it is placed in the camera so that one edge splits the x-ray beam. Unless the grain size is very fine, the specimen must be oscillated through a short angular range when using this technique. Debye-Scherrer cameras designed to accommodate massive samples of this sort and also provide for the necessary degree of oscillation are procurable* Most of the powder reflections on one side of the camera are lost because of shadowing by the sample when this method is employed, but a few lines appear on the "sample" side of the camera in the low- and high-angle regions, permitting reasonably accurate determination of interplanar spacings when needed.

For the most dependable results, metal and alloy specimens should be reduced to fine filings for x-ray diffraction analysis. Hume-Rothery and Reynolds[22] have described a procedure for the preparation of filings of high purity, which is applicable to most metals and alloys. First, sufficient filings must be prepared, and they must be taken from appropriate parts of the specimen under study to insure that they are truly representative. The filings may be prepared with a clean jeweler's file, using light strokes so as to avoid heating of the sample. It is best to use a new file for each sample. The cold work due to the filing introduces lattice strains which reduce the sharpness of the diffraction lines, so that if accurate measurements are contemplated the strains must be relieved by annealing. An alloy containing only elements of low volatility may be satisfactorily annealed at a temperature slightly below its melting point for a short period of time, whereas when a highly volatile constituent is involved a longer annealing period at a lower temperature is preferable. Filings of lead, aluminum, and some other metals self-anneal to varying degrees upon standing at room temperature. Hume-Rothery[23] has

---

*Cameras of this type are available from the Diano Corporation, Industrial X-ray Division, 2 Lowell Avenue, Winchester, Massachusetts, which acquired the General Electric Company's x-ray equipment line in 1972; from Siemens Aktiengesellschaft, Schöneberger Strasse 2-4, Berlin, Germany; and from Siemens Corporation, 186 Wood Avenue South, Iselin, New Jersey.

also devised special apparatus and techniques for preparing pure filings of metals and alloys which are very reactive or which alloy with nitrogen and oxygen. Brittle alloys can often be crushed to a coarse powder in a hardened-steel mortar of the Plattner "diamond" type, after which further reduction can be accomplished in an agate mortar.

Figure 4-18 illustrates the effects on the diffraction pattern of movement or lack of movement of the sample and of differences in the nature of the powder. All the patterns were prepared in a Debye-Scherrer camera of 57.3-mm radius with $CuK\alpha$ radiation, the powder being contained inside special Parlodion specimen tubes of 0.6-mm inside diameter. The films, 14 in. in length, were inserted in the camera in the asymmetric position. The beam entered through the right hole and emerged through the left hole punched in the film. Patterns A and B show the effect of differences in crystallite size upon line quality, the sample being stationary. The 5-μ and 15- to 50-μ fractions were pre-

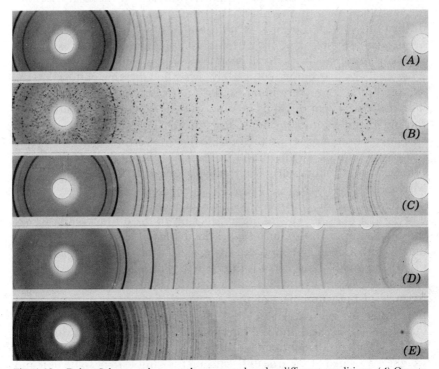

Fig. 4-18.   Debye-Scherrer photographs prepared under different conditions. (A) Quartz powder, <5 μ, specimen stationary. (B) Quartz powder, 15–50 μ, specimen stationary. (C) Quartz powder, 15–50 μ, specimen rotating. (D) NaCl, <325 mesh, specimen rotating. (E) Feldspar, <325 mesh, specimen rotating.

pared by crushing a large quartz crystal in a Plattner mortar and then fractionating the powder by sedimentation from methanol. By comparing B and C we can appreciate the value of rotating the sample in smoothing out spotty lines. Patterns D and E illustrate the effect of crystal symmetry upon line smoothness. The highly symmetrical sodium chloride (cubic) gives comparatively smooth lines because of the many sets of planes contributing to each reflection, whereas the poorly symmetrical feldspar (monoclinic) gives more spotty lines.

4-1.4 Mounting the Powder

Ideally, the powder should be mounted in such a way that no foreign material is exposed to the x-ray beam. In practice this situation is usually not realizable, and indeed for much routine work we are not justified in attempting it. As pointed out above, polycrystalline specimens of suitable dimensions can sometimes be supported directly in the beam without any extraneous mounting materials being irradiated. For most powder diffraction work, however, it is sufficient to support the powder in the beam with a limited amount of noncrystalline material such as a glass, polymer, or fine fiber which generates only a faint, diffuse pattern. These effects are either completely swamped by the pattern of the sample itself, or are very minor and do not cause interpretative errors. The three principal methods of supporting a cylindrical specimen with the aid of an extraneous substance include (a) forming a cylindrical specimen of the powder on the outside of a supporting fiber, (b) holding the powder inside a thin-walled cylindrical tube, and (c) molding the powder into a cylindrical specimen by mixing a little adhesive material with it.

A hair or very fine thread constitues an excellent support for method a because it does not contribute appreciably to the diffraction pattern if the final sample is 0.2 mm or more in diameter. The powder can be mixed with Canada balsam, gum tragacanth, or other suitable amorphous binder to the state of a thick paste, which is then coated on the fiber and rolled into the form of a cylinder of the required diameter. Because of the suppleness of the specimen, this type of mount is most practicable for cameras of the kind described earlier[13], in which the sample axis is vertical and the sample itself projects downward from its holder. One end of the fiber is clamped in the holder, and a small lead weight is attached to the free end, causing the sample to become taut and assume a precisely vertical position. It then becomes easy to align the sample and spindle axes, provided the camera has been carefully leveled.

When the camera axis is horizontal, it is necessary to use a rigid supporting fiber of some kind. Suitable ones 0.2 to 0.4 mm in diameter can be drawn from Pyrex rod or small tubing. If it is required that the fiber have very low absorption for the x-ray beam, a lithium borate (Linde-mann) glass may be used, but it has the disadvantage of devitrifying rapidly in moist atmospheres. A section of fiber about 15 mm long (Fig. 4-19A) is held at one end between the thumb and forefinger, and a por-tion near the opposite end is dipped in a small drop of adhesive on a glass microscope slide. The coated fiber is rotated against a clean portion of glass surface to smooth the adhesive into a thin, uniform film, after which the filmed area is rotated in a tiny pile of the sample powder until a thick layer adheres. The crudely cylindrical sample is then made more cylindrical by rolling it on a clean plane surface. If the sample is too small, the foregoing operations can be repeated once or twice until the diameter is large enough (Fig. 4-19B). The required speed of these manipulations is governed by the rate at which the adhesive dries. The adhesive must be properly selected so that the sample powder is in-soluble or, at most, sparingly soluble in it. Glyptal, collodion, and Duco cement are examples of satisfactory adhesives for water-soluble powders, whereas water-base mucilage or glue is suitable for many organic materials and most minerals.

Handy bases for mounting rigid sample supports can be made by cutting 0.125 in. brass rod into 0.375 in. lengths and drilling them longitudinally with a 0.07 in. drill to accommodate the fibers (Fig. 4-19C). A small lump of picein, beeswax, Plasteline, modeling clay, or similar material is softened and molded into a cap over one end of the brass base (D). The long end of the glass fiber mount (B) is then inserted through the softened cap into the drill hole so as to be approximately co-axial with the cylindrical brass base (E). When the wax has stiffened, the the specimen is transferred to the sample holder of the camera, where the brass base fits in the cylindrical cavity x of Fig. 4-12 and is held tightly by means of a set screw or spring. Photographs made with glass-fiber

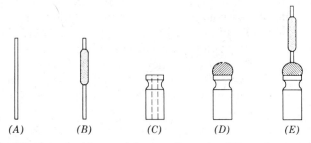

(A)   (B)   (C)   (D)   (E)

Fig.4-19.   Simple method for mounting sample powder. (Dimensions exaggerated.)

mounts show, in addition to the pattern of the powder, one or two diffuse glass halos of an intensity that depends upon the degree of absorption of the x-rays by the sample powder. For moderately or highly absorbing substances, the glass halos are faint and sometimes unnoticeable.

A highly satisfactory mounting technique consists in placing the powder inside thin-walled, low-absorbing tubes. The earliest sample tubes were prepared by drawing out Pyrex test tubes or 1- to 2-cm Pyrex tubing in an oxygen-gas flame to thin-walled tubes of the proper diameter. The wall thickness must be very thin (of the order of 0.01 mm) to avoid excessive absorption of the x-ray beam. Glass containing lead cannot be used. Pyrex tubes are particularly suitable for hermetically sealing off powder samples that are unstable in the atmosphere. Lindemann glass sample tubes are very useful when high transparency to x-rays is especially needed, but tubing of this type has not been generally availably in the United States in recent years. Glass capillaries with internal diameters from 0.1 to 2.00 mm and quartz capillaries with internal diameters from 0.2 to 2.00 mm are available commercially.*

Tallman and Margrave [31] have described a technique for pulling polystyrene and polyethylene tubing (while heating in a coil of Nichrome wire) to produce capillary sample tubes. Pulling these materials is quite different from pulling glass, and it must be done very slowly. The empty capillary tubes show some weak diffraction lines characteristic of the partly crystallized plastics, but these are readily identified and do not interfere with the patterns of the powders. Tallman and Margrave found these tubes very useful in working with fluorine compounds which react with glass and which also must be protected from air, water vapor, and carbon dioxide.

Pulled glass or polyethylene capillaries frequently do not provide sufficiently reproducible dimensions from tube to tube for some studies. Fortunately, several techniques have been described [24–30] for preparing plastic tubes of very uniform diameter. Most of these methods involve the dipping of 22- to 28-gage annealed copper wire in a plastic solution of the proper concentration. Suitable solutions are 12 grams of cellulose acetate in 90 ml of chloroform and 10 ml of 95 per cent ethanol [26], and a 10 per cent solution of Parlodion in amyl acetate [25]. After allowing the coated wires to dry, they are stretched to loosen the coating which is cut into tubes of the requisite length. Some investigators felt the tubes could be removed more easily if the wires were initially rubbed

*Among the purveyors of such capillaries are: Syntex Analytical Instruments, Stanford Industrial Park, Palo Alto, California; and Unimex-Caine Corporation 8133 Maple Avenue, Gary, Indiana.

with a little lubricant, such as anhydrous sodium stearate[26] or graphite powder[25].

Thomas and Alexander[28], considering the use of a lubricant objectionable because of the possible production of unwanted diffraction lines, studied several of these procedures and developed a technique giving consistently good results in the production of 0.6-mm Parlodion tubes. Straightened 6-in. pieces of 22-gage bare copper wire are annealed in water-pumped commercial grade Linde nitrogen at 700°C. The softened wires are dipped once rather rapidly in an 18 per cent solution of Parlodion in amyl acetate, and dried in a vertical position over night. The Parlodion sheath is then scraped off the wire at each end and the wire stretched to free the tube. The ready loosening of the plastic tube by this process is dependent upon the production of a cuprous oxide film on the wire surface during annealing, apparently by the action of water vapor (an impurity in the nitrogen) upon the hot copper surface. Adhering cuprous oxide must be removed from the inner surface of the tubes by dipping them in 6 N hydrochloric acid solution until the reddish brown color disappears. The tubes are then rinsed in water, residual water blown out, cut to length, and stored in a desiccator until used. Additional details are presented in reference[28].

Williams[30] prepared capillary tubes from gelatin solution by twice dipping 28-gage General Electric copper magnet wire. The wires required no annealing, no acid cleaning, no coating with lubricants, and no other previous treatment, except wiping with lint-free tissue. After the gelatin coating dried the wires were stretched 20 to 23 per cent, and the capillary tubes removed with forceps. Gelatin capillaries are reported to remain straight indefinitely, but should be handled with forceps when loading because they take up moisture from the fingers.

For average powders these tubes can be loaded in a simple manner. A piece of the wire from which the tubes were removed is dipped in the forming solution and inserted 1 or 2 mm into one end of a section of tubing. This seals the tube in position on the wire, which now serves as a support for the tube in the subsequent loading operations. By taking special pains the tube can be sealed to the wire tightly enough to exclude moisture from samples requiring a degree of protection. The mounted tube is cut off to the proper length (about 1 cm) with sharp scissors and filled with the powder by pressing the open end into a small pile of the powder repeatedly, tamping it in if desired with a wire tamper of slightly smaller diameter than that of the tube. The open end is then sealed off with a drop of the forming solution.

Gibson and Bicek[32] have devised a modified filter-stick technique for filling capillaries with powdered materials. The method is simple and may be used with dry powders or with liquid or gaseous suspensions. Figure 4-20 is an "exploded" diagram of the device. A rubber sleeve, B, fits over a 10- to 15-cm length of capillary tubing, A, with about 5 mm projecting. Into this cup fit a filter paper disk, C, and a short piece of capillary tubing, D, on the outer end of which is cemented a disk of soft rubber, E, perforated in the center. To assemble the parts, the sample capillary is pushed through D and the hole in E from the left until flush with the surface of D. The filter disk and short capillary section are then inserted snugly in the 5-mm cavity at the right end of B. The sample capillary is filled by inserting it in the sample material, while the left end of A is connected to the vacuum line. The vacuum should be broken before removing the loaded sample capillary. This technique can be adapted to the collection of a sample under a controlled atmosphere by passing the main capillary tube A through a rubber stopper which is sealed into the wall of the controlled-atmosphere chamber with a flexible joint. In a similar fashion, and rather simpler, Mathews[33] placed a tiny plug of glass wool in his sample capillaries. Then a vacuum was applied through the plug to suck the powders into the tubes.

Hagelston and Dunn[34] describe a simple device (Fig. 4-21A) by means of which a capillary may be filled very efficiently, a prime requirement when the available sample material is very limited. In addition, the method prevents contamination of the specimen during the filling operation, and it facilitates filling and sealing off the tube while the specimen is kept in a controlled atmosphere. The open end of the sample capillary tube is first sealed accurately in place over a small hole in the female member of the filling jig, using collodion, an operation best performed with a special assembly jig shown in Fig. 4-21B. The two members of the filling jig are then separated and inverted, and the sample powder placed in the female member. The two members are next refitted, and the powder made to fall into the capillary by drawing a fine-cut file across the assembly. When the capillary is filled to within $\frac{1}{4}$ in. of the collodion joint, it can be cut off from the assembly and simultane-

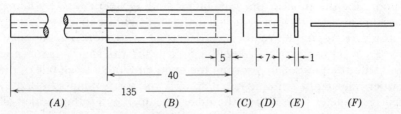

Fig. 4-20. Filter-stick assembly for filling capillary tubes with powder (dimensions in millimeters). [Courtesy of Gibson and Bicek, *Anal. Chem.*, **20**, 884 (1948); by permission of the American Chemical Society.]

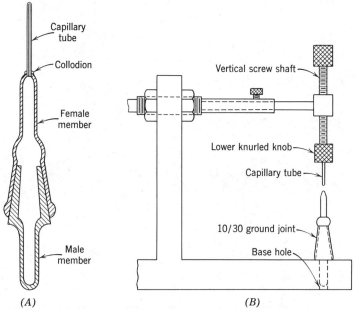

Fig. 4-21. Device for efficiently filling capillary tubes with sample powder. [Courtesy of Hagelston and Dunn, *Rev. Sci. Instrum.*, **20**, 373 (1949).]

ously sealed with a small flame (glass capillaries). Another manipulative device for loading capillaries, particularly for the "dry-box" technique in which handling with bulky gloves is often necessary, has been described by Larsen and Leddy [35].

By mixing the sample powder with an amorphous binder, such as library paste, collodion, or gum tragacanth, it is possible to extrude a self-supporting cylindrical sample with very simple equipment [36]. Even if the use of a binder is undesirable, it is usually possible to produce extruded specimens [37–40]. Kossenberg [37] has described the extrusion press shown in Fig. 4-22. The press itself consists of three separate parts, *a*, *b*, and *c*, which fit one on another. Block *c* has a funnel-shaped cavity, and a specimen holder *d* fits in the central opening in *b* and *c*. Specimen holders of different internal diameter are available, each fitted with its corresponding punch *e*. With the parts *a*, *b*, and *c* assembled as in Fig. 4-22, a specimen holder is placed in position. Some of the powdered specimen is next placed in the funnel. The punch *e* is then used to press the powder into the holder until it is about half full. Part *a* is then removed, and about 3 mm of the specimen is pressed out with the punch. On removal of part *c*, the specimen holder with its pencil of specimen *f* is carefully removed. The specimen holders are prepared with an outside

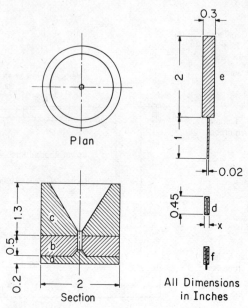

Plan

Section

All Dimensions
in Inches

Fig. 4-22. Simple press for extruded powder specimens. [Courtesy of Kossenberg, *J. Sci. Instrum.*, **32**, 117 (1955); copyright, The Institute of Physics.]

diameter to fit into *x* of the centering device of the camera, Fig. 4-12. A minor objection to the extrusion technique is that it may cause some powders to orient preferentially (see Sections 7-2.3B and 10-1).

Another type of support for powder specimens, the wedge mount, offers several advantages. Figure 4-23 shows the details of such an arrangement. The powder is pressed into the wedge-shaped cavity in the holder *w* (see perspective view *B*), which is then inserted in a hole in the block *b* and held firmly in place by tightening the knurled knob *k*. *C* is a schematic drawing of the mounting assembly looking along the axis of the collimator-exit-tube system *ce*, and *A* shows the assembly in place in the camera. The edge of the powder wedge *p* is moved by turning *k* until it splits the x-ray beam. The spindle *t* projects outside the camera and bears a gear wheel or cam by means of which the sample-mounting assembly can be oscillated through a small angular range, *α*. This method of mounting the powder is generally satisfactory only when the camera spindle axis is horizontal, so that the powder wedge can be set up with its base downward. Furthermore, the most satisfactory film arrangement is the asymmetric one, with the ends of the film located at *s* and *s'* at the bottom of the camera. The outstanding advantage of the wedge mount is that in most cases the powder can be supported in the beam

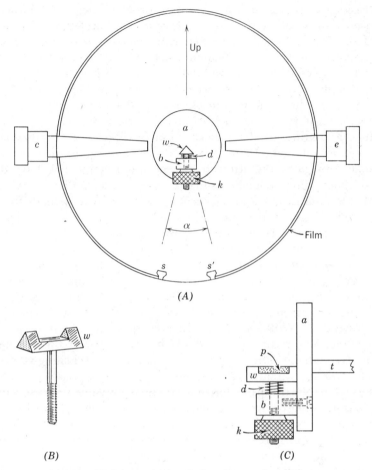

Fig. 4-23.   Wedge mounting technique for powder specimens.

without the aid of any binding agents or other foreign materials. In addition, the mounting process is convenient and very speedy. The disadvantages are (1) the loss of most of the diffraction pattern in the lower half of the film, (2) the need for a somewhat larger amount of powder, and (3) the unsuitability of the method for highly precise interplanar-spacing measurements. Some manufacturers supply cameras with provision for mounting either cylindrical or wedge-shaped specimens. Christ and Champaygne [41] have devised a wedge mount with 0.001-in. cellophane windows by means of which diffraction patterns have been prepared of materials as unstable in the atmosphere as calcium carbide and phosphorus pentoxide.

Occasionally, special sample-mounting problems arise. For instance, radioactive materials require special care in handling[42, 43]. Sherman and Keller[43] modified Debye-Scherrer cameras so that patterns of compounds containing protactinium-231 (α and γ emitter, half-life 3.26×10^4 yr) could be taken without hazard to personnel. The fragile sample capillary tube was protected by a beryllium cup to prevent its being broken while loading the film in the darkroom. It must be remembered that the α, β and γ radiations from radioactive samples produce a continuous dark background on film. The γ radiation is usually more energetic than the x-rays used, and cannot be selectively eliminated by filtration. Runyan, Olsen, and Kempter[99] have described camera filter attachments to cope with α and β radiations. Moore, Wright, and Martin[44] and Barker[45] have described methods for the protection of highly active and toxic metals during exposure in an x-ray diffraction camera. Techniques for investigating thin films and small specimens have been presented by Wilkinson and Calvert[46]. Eeles[47] has given useful details for the preparation of metallic specimens.

Taylor[48] modified a standard Debye-Scherrer camera to accommodate large-diameter specimens. Eccentrically mounted specimens up to $\frac{1}{2}$ in. in diameter may be used. An oscillating flat-specimen holder was designed by Wilkinson and Calvert[49] for a 114.6-mm camera. With it one can record one half of the pattern with the specimen stationary, and the other half with the specimen oscillating, thereby accentuating any variations in crystallite size between phases present in the specimen. Northrop[50] has attacked the spotty-line problem with a sample-translating device which translates the specimen $\frac{5}{8}$ in. while it is simultaneously being rotated. With 4 translations and 60 rotations per hour, samples normally (without translation) producing very spotty lines yield lines sufficiently uniform for good measurements. Northrop lists references for eight other translating devices. Gandolfi[172] has designed a camera for taking normal Debye-Scherrer photographs or for producing powder photographs from single crystals as small as 30 μ.*

The intensity of x-rays diffracted from a specimen at small angles is influenced greatly by the total thickness of the material through which the beam must pass. Assume for the moment that the powder is in the form of a flat sheet of thickness t', and that its linear absorption coefficient for the x-rays is μ'. Consider the diffraction at some small angle 2θ from an element of thickness dx at depth x in the specimen. For a given reflection the intensity of the x-rays diffracted from the element

*Produced by Officina Elettrotecnica di Tenno, Trento, Italy. Available in the United States from Blake Industries, Inc., P.O. Box 464, 379 Morris Avenue, Springfield, New Jersey.

dx is proportional to the incident beam intensity I_0 and to the distance dx, and in addition it is reduced by the passage of the incident and diffracted x-rays through the powder for a distance of approximately t':

$$dI = kI_0e^{-\mu't'}\,dx.$$

The total intensity of the diffracted x-rays is then

$$I = \int_{x=0}^{x=t'} dI = kI_0e^{-\mu't'}\int_0^{t'} dx$$

$$= kI_0t'e^{-\mu't'}.$$

The thickness corresponding to maximum diffracted intensity at small angles can be found by differentiating with respect to t' and equating the resulting expression to zero:

$$\frac{dI}{dt'} = kI_0e^{-\mu't'} - kI_0t'\mu'e^{-\mu't'}.$$

$$kI_0e^{-\mu't'_m} = \mu'kI_0t'_me^{-\mu't'_m}.$$

$$t'_m = \frac{1}{\mu'}. \tag{4-1}$$

This result, which was first obtained by Hull [51], shows that the optimum thickness varies with the nature of the material comprising the powder and with the wavelength of the incident x-rays. Denoting the density by ρ, and referring to the powder, including interstices, by primed quantities, and to the solid material by unprimed quantities, we may write

$$\frac{\mu}{\mu'} = \frac{\rho}{\rho'} = \frac{t'}{t} \tag{4-2}$$

In actual practice it is often inconvenient or impossible to vary the thickness of each specimen to satisfy expression 4-1. More usually, a fixed sample thickness is required. In this event it is still possible to satisfy equation 4-1 more or less closely by diluting heavily absorbing substances with a material that absorbs x-rays only slightly and which generates only a weak and diffuse diffraction pattern. Materials that have been used for this purpose are finely powdered gum tragacanth, flour, and cornstarch. It can be shown that the optimum relative weights of sample powder and diluent, w_s and w_d, respectively, are given with sufficient accuracy by

$$\frac{w_s}{w_d} = \frac{\rho'_s}{\rho'_d(\mu'_st'-1)}. \tag{4-3}$$

Formula 4-3 can be applied directly to cylindrical Debye-Scherrer samples with sufficient accuracy by letting t' be the diameter of the sample, although strictly the diameter should be slightly greater than the calculated value of t'.

For radiologically dense materials it may not be possible to dilute to the theoretical degree without reducing the number of powder particles below the lower limit required to give an effectively statistical distribution of orientations. hence the lines may become spotty. Davey[52] worked out a dilution scheme using flour, gum tragacanth, or cornstarch for Mo$K\alpha$ radiation. Pogainis and Shaw[53] found lampblack distinctly superior to the above diluents for Cu$K\alpha$ radiation. The volume of lampblack to be added to 1 vol of substance may be estimated as $V = (\mu - 18)/16.8$, where μ is the density of the loosely packed powdered material and is usually about half of the μ for the solid crystal.

4-1.5 Making the Exposure

The mounted sample (here assumed to be cylindrical) is placed in the specimen-holding device of the camera, and its axis brought into alignment with the rotation axis by means of the centering mechanism. With the mechanisms described in Section 4-1.2B, this can be done while observing the specimen through the beam-entrance or exit-port opening. The sample is rotated into the position corresponding to its maximum lateral displacement from the rotation axis, after which it is brought into approximate alignment with the aid of the centering device. The sample is then rotated again into the new position corresponding to maximum displacement and recentered, and so on, this process being repeated until no displacement is observed when the sample is rotated. An arrangement of the kind shown in Fig. 4-24 greatly facilitates the aligning process. The collimator and beam-exit tubes are removed and the sample is viewed through a low-power lens l of focal

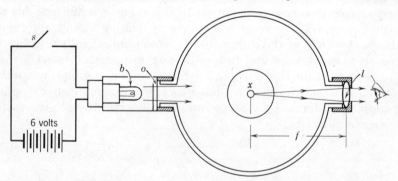

Fig. 4-24.   Simple arrangement for aligning the powder specimen.

length f. The sample periphery is sharply silhouetted against the milky glass disk o illuminated by the low-voltage lamp b. A toggle switch s in the circuit is a convenience. If desired, the batteries may be replaced by a low-voltage transformer operating on a 110-V circuit. It is a convenience if the illuminator case and lens frame are designed to fit directly in or over the collimator and exit-port openings in some manner, as suggested by the figure.

For highly exacting work in which the sample diameter is small, for instance, 0.1 to 0.3 mm, and the sample alignment must be very good, the lens of Fig. 4-24 may be replaced by a long-focus microscope with cross hairs. Eyepieces giving magnifications between 8× and 24× prove satisfactory. The alignment operation is facilitated by utilizing the principle of the optical bench in supporting the camera and telescope; that is to say, they can both be mounted on similar bases which slide on the camera track of the x-ray unit. After the optical axis of the microscope has once been aligned with the collimator-exit tube axis of the camera, the microscope automatically assumes the correct position when it is placed on the track with the camera.

The proper choice of radiation depends both upon the nature of the specimen and the purpose for which the diffraction pattern is to be used. The most generally useful radiation is $CuK\alpha$ ($\lambda = 1.542\,\text{Å}$), rendered free of β radiation by transmission through a nickel filter (see Table 2-5). A first consideration is to make sure that the sample does not contain appreciable amounts of elements which generate secondary fluorescent x-rays when irradiated with the wavelength in question. Thus samples containing much iron or cobalt cannot be studied with $CuK\alpha$ radiation. For further information on x-ray fluorescence the reader is referred to Section 2-3.4. Assuming no difficulty on this score, for a general investigation of the diffraction pattern the choice of radiation is made principally on the basis of the unit-cell dimensions of the material to be investigated. For most solid elements and for simple inorganic compounds and minerals, unit-cell dimensions are small. As a result, a short wavelength such as that of molybdenum is especially suitable, although radiations of moderate wavelength, such as copper, cobalt, or iron, are usually satisfactory. The larger unit cells of more complex inorganic compounds and minerals produce the best diffraction patterns, as a rule, with these radiations of moderate wavelength. Many organic substances, as well as a few complex minerals (certain clays in particular), have unit cells with one or more very large dimensions. For such materials the long wavelength of chromium is advantageous, because the closely spaced lines of the many large interplanar spacings are thereby more highly dispersed and better resolved.

For special applications the choice of radiation is affected by other considerations. For example, in the precision determination of lattice constants, the lines at Bragg angles between 60 and 90° are of the most importance, and it is desirable to select a radiation that produces the best distribution of lines in this region. This problem is discussed at length in Chapter 8. In the measurement of stresses in metals, a wavelength must be employed that yields a diffraction ring at a very high Bragg angle, preferably above 80°, a matter that is considered in Chapter 11. In certain cases it is important to reduce absorption of the x-ray beam of elements of high atomic weight in the sample. This can be accomplished by employing a very penetrating radiation like that of molybdenum or siliver, which gives a more intense pattern and a more accurate relationship between the intensities of lines at low and high Bragg angles.

The proper β filters for the various radiations are listed in Table 2-5. In the investigation of materials of partially or entirely unknown composition, it is advisable to use a β filter because of the added interpretative difficulties introduced by a mixture of lines due to $K\alpha$ and $K\beta$ wavelengths. It is sometimes advantageous to dispense with the β filter disk at the entrance port of the camera and instead cover a strip of the film with a narrow piece of the filter material (in the form of a sheet or foil). Figure 4-25 shows powder patterns prepared in this way. The β lines are easily identified by their presence in the unfiltered part and absence in the filtered part of the pattern, and the important advantage of a reduced exposure time is gained. Completely unfiltered exposures are of course economical of time and tube life and should be used whenever possible. It should be pointed out that when an especially clear

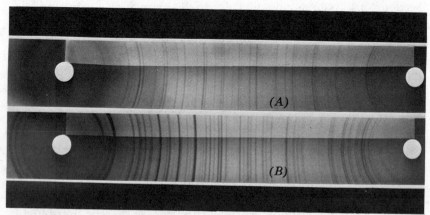

Fig. 4-25.   Powder patterns prepared with unfiltered CuKα radiation, the upper portion of the film being covered with 0.0007-in. nickel foil. (A) Lead nitrate; (B) Quartz.

background is desired there is some advantage in placing the β filter in front of the film rather than at the beam entrance to the camera. When this technique is followed, the filter sheet or foil should be separated from the film by a space of $\frac{1}{8}$ in. or more. With the filter placed in front of the film, the general background is reduced to a greater or smaller degree because of preferential absorption of fluorescent and incoherently scattered x-radiation from the sample.

After the sample has been centered and the proper radiation selected, the alignment of the camera with respect to the beam should be checked. It is assumed here that the camera has previously been placed in good alignment, as explained in Section 4-1.2C, so that any adjustments needed will be minor. The beam is turned on with the camera in position, and any necessary fine adjustments are made to cause the image of the beam on a fluorescent screen (placed behind the sample) to appear symmetrical about the shadow of the specimen. If the exit-port cap contains a fluorescent-screen disk (see Section 4-1.2B), this final alignment check may be made equally well after the camera has been loaded with film. In the event that the sample is too weakly absorbing to cast a perceptible shadow on the screen, it is doubly important that the prealignment of the camera be good.

The camera (or cylindrical cassette) is then loaded with film in the darkroom. Workers the world over have a large selection of excellent films to choose from. In 1956 the Commission on Crystallographic Apparatus of the International Union of Crystallography published a comparative study of 41 commercially available x-ray films [55]. A second comparison of commercial x-ray films was reported by Morimoto and Uyeda [56] in 1963. These studies determined, in particular, the film speeds for $CuK\alpha$ and $MoK\alpha$ radiation, the granularity, the fog density, and the transmission for $CuK\alpha$ radiation. Usually, the films are composed of a transparent, blue-tinted, cellulose acetate base coated on both sides with a layer of sensitive silver halide emulsion about 0.001 in. thick. Table 4-3 lists the characteristics of several commerical x-ray films suitable for diffraction purposes. It is noted that a film with very low fog density is a slow-speed film. Also, graininess increases with speed. For most diffraction applications the faster films are preferred. In special situations fine-grained film may be advantageous, as in microcameras to permit satisfactory enlargement.

Care must be exercised in handling film to avoid abrasion of the emulsion and deposition of fingerprints during the process of cutting and punching it, marking it, and placing it in position in the camera. The best policy is to cut and punch the film while it is kept protected on both sides with the original wrapping paper. Identification symbols can

Table 4-3. Characteristics of Several Commercially Available X-ray Films [56]

Country	Name of Film	Speed[a] for CuKα	MoKα	Granularity[a] (Graininess)	Fog Density[a]	Per Cent of CuKα Transmitted
Australia	Kodak, Industrial	1.49	0.73	3.0	0.15	31.2
France	Kodak, Kodirex	1.14	0.56	3.2	0.30	37.9
Germany	Agfa Leverkusen, Röntgen-Sino	1.84	0.89	4.3	0.25	31.2
Japan	Fuji, Type 200	1.68	0.86	2.8	0.20	33.8
	Konishiroku, Sakura N	2.6	1.34	3.5	0.15	29.4
United Kingdom	Ilford, Industrial G	1.08	0.54	2.3	0.15	34.1
	Kodak, Kodirex	1.38	0.64	2.9	0.15	34.6
United States of America	Ansco, Non Screen	1.44	0.72	3.4	0.20	34.5
	DuPont, Type 504	0.50	0.21	2.0	0.15	54.9
	Eastman Kodak, No Screen	1.59	0.82	2.7	0.20	25.7
	Eastman Kodak, Type M	0.10	0.05	0.96	0.00	46.7

[a]These quantities are explained in detail by Morimoto and Uyeda [56]. For general comparative purposes the larger the numerical value the higher the speed, or the greater the graininess or the fog density.

be easily put on the film by writing with a rather blunt stylus (an eight-penny nail will serve), using firm pressure. After the film has been processed, the legend will be clear and permanent.

The loaded camera is then placed in position before the x-ray tube window, and rotation or oscillation of the sample is started. The exposure is begun, using the appropriate tube voltage (refer to Section 2-3.1B) and a value of the tube current that is within the maximum rating of the tube. For example, a commercial copper-target tube rated at 1000 W might be operated at 40 kV and 20 mA, or at 35 kV and 25 mA, allowing a reasonable margin of safety.

The optimum length of an exposure depends on a considerable number of variables, some of which cannot be accurately determined or controlled. They include: (1) the absorbing power of the sample powder and sample holder for the x-ray wavelength used; (2) the density

Table 4-4. Representative Exposures for Various Experimental Conditions, Using Debye-Scherrer Cameras[a,b]

Camera Diameter (mm)	Pinhole Diameter (mm)	β Filter	Metal Filings	Simple Salts and Minerals	Organic Compounds, Complex Inorganic Compounds, and Minerals
57.3	1.0	No	5 min	10 min	20 min
57.3	1.0	Yes	10 min	20 min	40 min
57.3 114.6	0.5 1.0	No	15 min	30 min	1 hr
57.3 114.6	0.5 1.0	Yes	30 min	1 hr	2 hr
114.6	0.5	No	1 hr	2 hr	4 hr
114.6	0.5	Yes	2 hr	4 hr	8 hr

[a]Conditions assumed: Rectangular x-ray tube focus about 1×10 mm in size, viewed longitudinally at an angle of 6°. Tube voltages as suggested in Chapter 2. Operational energy 800 to 1000 W. Target may be Mo, Ag, Cu, Fe, or Co. Sample in form of cylinder with powder either mounted on the outside of a fine fiber or inside a plastic tube of low absorbing power for x-rays. Sample diameter appropriate for the pinholes used. Collimator has two equal pinholes. Double-coated nonscreen x-ray film used.

[b]For wattages smaller than 800 to 1000, the exposures must be increased inversely. For Cr radiation the exposures should be increased by an additional factor of about 1.5 to 2.0.

of packing of the powder; (3) the scattering factors of the various chemical elements composing the sample; (4) the simplicity or complexity of the crystal structures of the sample materials; (5) the wavelength of the x-rays (radiations of long wavelength, such as Cr$K\alpha$, are more highly absorbed by most materials, including the x-ray tube windows and the air in the camera); (6) the size of the collimating apertures; (7) the size of the sample; (8) the camera radius; (9) The distance from the x-ray source to the sample; (10) the type of photographic film used; (11) the dimensions of the x-ray focal spot and the angle at which it is viewed; (12) the wattage at which the tube is operated; and (13) the presence or absence of a β filter. From the nature of these many factors, it is evident that no definite rules for exposure time can be proposed that will be applicable to all experimental situations. However, with experience the diffractionist learns to make good estimates in most cases. The schedule of exposure times given in Table 4-4 is intended as a guide to help the beginner choose exposure periods of the right order of magnitude. Sometimes exposures are reported in terms of the product of the tube current in milliamperes and the time in hours; thus, 2 hr at 20 mA equals 40 mA hr, and so on. However, since the beam intensity depends not only upon the tube current but also upon the voltage, as well as other factors, this notation has no general significance except possibly in the records of a particular laboratory in which the same x-ray tubes are used repeatedly and the operational voltages have become standardized.

4-1.6 Processing the Film

After the exposure has been completed, the film is removed from the camera in the darkroom and processed. It is scarcely possible to overemphasize the importance of good darkroom practice in successful photographic diffraction work. Many of the sought-after results of painstaking exposure techniques can be nullified in a few minutes by by careless or indifferent handling of the exposed film in the processes of development, fixing, washing, and drying. For this reason the beginning diffractionist must very soon develop a satisfactory standardized procedure for the processing of films. For a detailed discussion of good equipment and procedures, the reader is referred to Appendix II.

4-2 PARAFOCUSING METHODS

By making use of so-called focusing geometry, cameras can be constructed that give much greater resolution without increased exposure time when compared with Debye-Scherrer cameras of the same radius.

The basic arrangements for this technique were first devised independently by Seemann [57] and Bohlin [58], for which reason cameras of this
type are frequently called Seemann-Bohlin cameras. It should be stated
at the outset that focusing cameras are not an indispensable adjunct to a
diversified diffration laboratory. Debye-Scherrer cameras can be used
almost as effectively in their two most important applications, the
precision measurement of lattice constants and the study of alloy phase
diagrams.

A geometrical arrangement of source, sample powder, and film that
causes the diffracted rays from a set of planes (*hkl*) to converge to a
single point on the film is properly designated a *parafocusing* rather than
a *focusing* arrangement [59]. Figure 4-26 shows the geometrical principles utilized in Seemann-Bohlin parafocusing cameras. Theoretically,
x-rays diverging from the point source S, upon impinging on the powder
sample disposed along the arc $\overset{\frown}{ABC}$, generate diffracted rays from planes
(*hkl*) which come to a precise focus at the point F. This result follows
because the several angles 2ϕ are all supplementary to the angle of
deviation 2θ corresponding to the planes (*hkl*), hence equal, and they

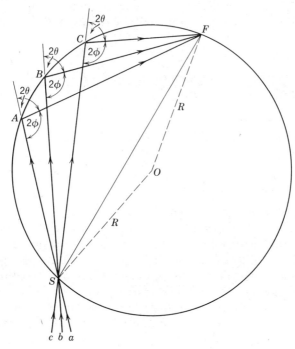

Fig.4-26.   Geometrical features of parafocusing cameras.

must therefore be inscribed in the same arc $\overset{\frown}{SABCF}$ of the circle. In practice the two-dimensional geometry of Fig. 4-26 must be replaced by a three-dimensional camera, in which case rays from S will diffract precisely to a common point F only if the powder is distributed on a segment of the toroidal surface generated by rotating the arc $\overset{\frown}{SABCF}$ about the chord SF [59]. Such a surface is difficult to realize in practice, and in addition a point source is not provided by modern diffraction tubes. Consequently, S is usually a narrow slit of rather limited height, which transmits convergent x-rays (e.g., rays a, b, and c) from a broad source. Within the camera, which is a short cylinder of radius R, these rays diverge and impinge upon the sample powder, the latter being distributed over a short section of the inner cylindrical surface of the camera. Since the sample surface is tangent to the ideal toroidal surface along the arc $\overset{\frown}{ABC}$, a rather good line focus (actually an image of the slit) will be obtained at F if the slit S is not too high, and if the x-ray source dimension in this same direction (perpendicular to the plane of the drawing) is small.

The value of the Bragg angle θ corresponding to an arc length $\overset{\frown}{SABCF}$ can be determined as follows. The angles BSF and BFS subtend the arcs $\overset{\frown}{BCF}$ and $\overset{\frown}{SAB}$, respectively, hence the sum of these two angles subtends the arc $\overset{\frown}{SABCF}$ and is equal to one-half the corresponding central angle SOF:

$$\angle BSF + \angle BFS = \tfrac{1}{2} \times \angle SOF.$$

But in the triangle SBF the sum of the angles BSF and BFS equals the exterior angle 2θ. Hence we find

$$2\theta = \tfrac{1}{2} \times \angle SOF$$
$$= \frac{1}{2} \times \frac{\overset{\frown}{SABCF}}{R}.$$
$$\theta = \frac{l}{4R},$$

where $l = \overset{\frown}{SABCF}$, a measurable distance on the film.

Figure 4-27 shows schematically (A) asymmetric and (B) symmetric parafocusing arrangements. The former is most valuable for providing a highly resolved diffraction pattern at intermediate Bragg angles, as in the study of complex alloy systems; the latter finds its greatest utility at large angles, as in the precision measurement of lattice constants. Cameras of both kinds suffer from certain disadvantages which must be recognised (1) If the sample is not moved during the exposure and, in particular, if the amount of powder irradiated is too small, the diffraction lines will tend to be spotty unless the ultimate crystallite dimension

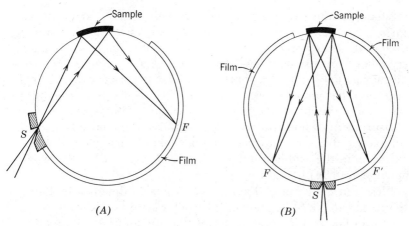

Fig.4-27.   (A) Asymmetric and (B) symmetric parafocusing geometry.

happens to be very small, approximately 5 μ or less. The difficulty can be overcome by doing two things. First, the focal spot should be viewed laterally (see Section 4-1.2), resulting in a wide divergence of the beam and a correspondingly large area of irradiation. Figure 4-28 illustrates right and wrong ways of utilizing the x-ray tube focus. Second, provision should be made for either oscillating the specimen around the axis of the camera, or translating it back and forth parallel to this axis. Such

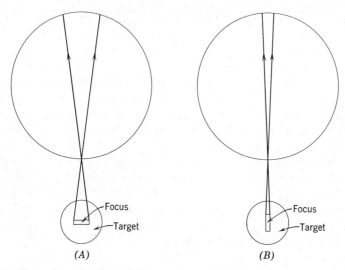

Fig.4-28.   Illustrating (A) right and (B) wrong ways of viewing the focal spot for obtaining best line quality with a parafocusing camera.

mechanisms must be precision constructed, so as to keep the sample surface accurately tangent to the focusing circle at all times. Even though these two precautions are observed, it is still found that a mean crystallite dimension of less than about $10\,\mu$ is required for smooth diffraction lines. (2) The sharpness of focus varies with the angle of reflection, becoming poorer as the angle of incidence of the diffracted rays upon the film becomes more oblique. For the most accurate measurement of line positions, single-emulsion film must be used. (3) A complete diffraction pattern cannot be obtained in any one camera. For example, in the symmetric design only reflections at Bragg angles well above 45° can be photographed.

Westgren[60] employed asymmetric cameras designed by Phragmen to investigate the constituion of many alloy systems. In order to provide optimum line quality at all angles within the limits $\theta = 16$ to 82°, three cameras with overlapping ranges were used. Because of the fine-grained texture of the materials studied, good-quality lines were obtained without oscillating the specimen. It should be emphasized that the superior line resolution afforded by parafocusing cameras can be largely nullified by the line broadening resulting from either lattice strains or inhomogeneities in composition. The state of the sample is of cardinal importance in diffraction work of this kind.

Jette and Foote[61] have published one of the best analyses of the use of symmetrical parafocusing cameras for the precision determination of lattice constants. Since the lattices of most substances expand by detectable amounts for a temperature rise exceeding 1°C, it is essential that the camera temperature be maintained constant to within ±1° during the course of an exposure. A satisfactory way of accomplishing this is to provide channels in the body of the camera through which fluid from a thermostatted bath is circulated. It is essential that the sample oscillating mechanism work very freely; otherwise friction in the moving parts may cause a very appreciable rise in temperature, especially if a circulating fluid is not employed. Figure 4-29 shows a camera (60-mm radius) of this type, which can be either partially evacuated or filled with hydrogen or helium to eliminate background due to air scatter. A specimen oscillation mechanism with motor drive is incorporated into the camera body. The camera accommodates standard 35-mm x-ray film and records reflections in the range $\theta = 59$ to 88.7°. Calibration notches are provided at 10° intervals on the film housing drum to aid in film comparison and to help in correcting film shrinkage errors. Lattice constant measurements to 1 part in 50,000 are claimed possible with it. For the application of the symmetrical parafocusing camera to the precise measurement of lattice parameters, the reader is referred to Chapter 8.

Fig. 4-29. Precision symmetrical back-reflection focusing camera. (Courtesy of Philips Electronic Instruments.)

Patterns of particularly high quality can be obtained by using parafocusing geometry in conjunction with a curved-crystal monochromator (Section 2-3.6C). Cauchois[62] used the converging beam from a Cauchois-bent crystal in a symmetrical transmission arrangement, as depicted in Fig. 4-30C. The powder specimen is placed tangentially to the camera surface, and the transmitted beam comes to a focus on the diametrically opposite edge of the cylindrical camera. This arrangement gives sharp forward reflections in the range $\theta = 0$ to $30°$. Using a Johannson-cut crystal, Guinier[63] described camera arrangements (also including Fig. 4-30A and B) for powder studies. The arrangement in A

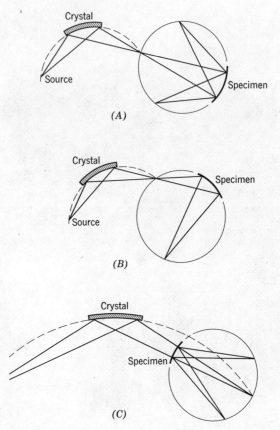

Fig. 4-30.   Focusing arrangements involving a curved-crystal monochromator.

gives the back-reflection lines from $\theta = 55$ to $90°$. When detailed studies near $\theta = 30$ to $60°$ are required, the asymmetric arrangement B is necessary. Guinier-type cameras yield patterns with very clear backgrounds, because of the elimination of wavelengths other than the $K\alpha$ doublet, and with exposure times comparable to the normal Debye-Scherrer method using a β filter. Other advantages are exceptional resolution and dispersion of patterns, and large specimen volume, yielding smooth diffraction lines.

de Wolff[64], an early enthusiast of Guinier cameras, recognized additional unique features. The focusing property largely eliminates the influence of the thickness of the specimen. Also, pairs of diffraction lines corresponding to the two wavelengths of the $K\alpha$ doublet can be made to coincide for any desired value of θ. For an extended region on either side of the chosen value of θ, the doublet separation is much reduced. Making use of this possibility, sharp and intense single lines are obtained, even with the largest cameras. de Wolff also recognized that a singly bent Johannson crystal with a broad x-ray source gives rise to a line focus. Hence a monochromating crystal with a large vertical dimension permits two or even four Guinier cameras to be stacked one above the other. Such an arrangment makes possible the taking of simultaneous exposures of several samples for comparison purposes or serial work.

Both the double and quadruple Guinier-de Wolff cameras engineered by de Wolff[64] have become widely used. The Guinier-de Wolff camera has a diameter of 114.6 mm, and the quadruple model permits four exposures simultaneously on a single 45×180 mm sheet of film. The large dispersion of 2 mm per degree of 2θ is twice that of a Debye-Scherrer camera of the same radius. The curvature of the monochromator crystal can be continuously and precisely varied so as to focus x-rays of any wavelength between 0.7 and 2.5 Å, and the entire camera body can be evacuated to eliminate air-scattered radiation. Figure 4-31 shows a quadruple Guinier-de Wolff camera which has been made available

Fig. 4-31. Guinier-de Wolff quadruple focusing camera. Cover removed. (Courtesy of N. V. Nederlandsche Instrumentfabriek (Nonius) and Alexander [174].)

Fig. 4-32. Guinier-de Wolff quadruple exposure: (*A*) ammonium alum; (*B*) alum admixed with quartz; (*C*) quartz; and (*D*) quartz. (Courtesy of Enraf-Nonius.)

commercially.* Figure 4-32 shows a quadruple photograph of four specimens prepared in this camera.

Hofmann and Jagodzinski[65] have designed an arrangement of two Guinier cameras in which the camera closest to the x-ray tube and monochromating crystal operates in transmission, while the second camera operates in reflection (Fig. 4-33). This arrangement (known commercially as the Guinier camera according to Jagodzinski†) permits simultaneous recording of both the $\theta = 0$ to 45° and $\theta = 45$ to 90° regions of the diffraction pattern. Moreover, each camera permits three samples to be exposed, so that as many as six specimens may be studied at the same time. In the symmetrical scheme of Fig. 4-33*A* (angle of incidence of radiation $= 0°$), the reflection ranges that can be evaluated accurately are $2\theta = 0$ to 60° and $2\theta = 120$ to 180°. With an asymmetrical situation and an angle of incidence of the x-radiation of 45°, reflections in the ranges $2\theta = 0$ to 105° and $2\theta = 76$ to 180° can be evaluated accurately. A vacuum shield, special diaphragms to reduce air scattering, and a linear oscillating device for the specimen are additional features. The precision monochromator provided makes it possible to work with monochromatic $K\alpha_1$ radiation.

With the well-deserved popularity of the Guinier camera, a library of useful technical information and theory is building up which can only be briefly mentioned here. Hellner[66] has given a detailed treatment of intensity measurements from Guinier camera photographs, and Sas and de Wolff[67] have also discussed intensity corrections. Fischmeister [68] listed several adjustment conveniences deemed necessary in a Guinier camera, and then built a camera incorporating these features.

*N. V. Nederlandsche Instrumentfabrik (Nonius), Delft, Netherlands. Available in the United States through Enraf-Nonius, Inc., 130 County Courthouse Road, Garden City Park, New York.
†Available from Siemens Aktiengesellschaft, Schöneberger Strasse 2-4, Berlin, Germany; and in the United States from Siemens Corporation, 186 Wood Avenue South, Iselin, New Jersey.

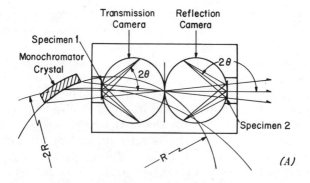

Fig. 4-33.   Guinier camera according to Jagodzinski. (*A*) Geometry. (*B*) Commercial unit. (Courtesy Siemens Aktiengesellschaft.)

Zsoldos[69], however, claimed that all of Fischmeister's requirements (except the vertical adjustment) could easily be met for a de Wolff-type camera with a simple base which he has described. Hägg and Ersson[70] have given a brief description of an easily adjustable Guinier camera of highest precision. Details of the calibration of the Guinier-de Wolff camera have been given by Fisher[71]. Since best results are obtained in focusing cameras by the use of single-emulsion film, Polk[72] has presented a method for quickly removing the emulstion backing on double-emulsion film used in Guinier cameras.

By way of an example of the great utility of the Guinier camera for identification purposes, a study of 60 crystalline sodium phosphates made by Corbridge and Tromans[73] is cited. The greater resolution and increased definition of the diffraction lines made the Guinier-de Wolff camera the choice over the Debye-Scherrer camera for this class of compounds. At that time the powder diffraction file (see Section 7-1.1) contained only about 20 cards on sodium phosphates, and several of these patterns were in error. Corbridge and Tromans in their publication reproduce the 60 photographs and present line data on each.

4-3   MONOCHROMATIC-PINHOLE TECHNIQUES

4-3.1   Forward-Reflection Method

This method is also referred to as the *transmission* or, somewhat ambiguously, *flat-film* technique. Typical geometrical features are diagramed in Fig. 4-34. Characteristic x-rays from the x-ray tube are freed of β radiation by transmission through the filter M, after which they are collimated by the small pinholes S_1, S_2, and S_3 and penetrate perpen-

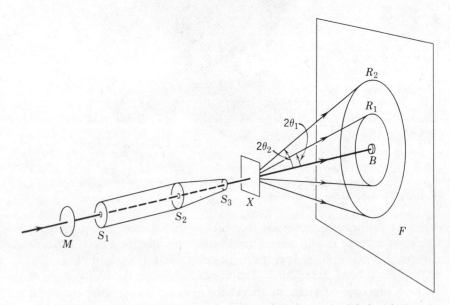

Fig. 4-34.   Forward-reflection arrangement of the monochromatic-pinhole technique.

dicularly through the flat sample X. The undeviated beam is absorbed by the lead beam stop B, or passes out through a small hole punched in the film, and the diffracted rays from different sets of planes produce rings of different diameters on the film F. Normally, both the sample and the film are kept stationary during the exposure. The experimental arrangement is the same as that employed for Laue photographs of single crystals, but the latter method utilizes the general, or white, x-radiation, whereas the monochromatic-pinhole technique requires monochromatic radiation.

A forward-reflection camera of this type is a basic item of equipment in any diffraction laboratory. It ranks next to the Debye-Scherrer camera in utility for the investigation of polycrystalline materials. Its chief limitations are that, first, only the reflections at angles considerably less than $2\theta = 90°$ are recorded and, second, the conversion of film measurements to Bragg angles and d spacings is more difficult than for Debye-Scherrer patterns. However, this technique possesses important advantages. Entire diffraction circles are recorded, making it easy to detect circumferential variations in intensity which would escape observation on a Debye-Scherrer photograph. For this reason preferred orientation and lattice distortions in thin metal sheets are advantageously studied by this method. It is also well suited to the investigation of amorphous and vitreous materials, which give only one or more diffuse halos and these at low angles. This is also the best experimental arrangement for recording the low-angle reflections from large interplanar spacings, such as are generated by certain clay minerals and many organic substances like waxes, soaps, and fibrous proteins.

A monochromatic-pinhole collimator should be designed so that the third, or guard, pinhole prevents radiation scattered by the second pinhole from reaching the film (see Section 4-1.2A). The use of a beam-exit tube is of little advantage in this method, because there is no film in the high-angle region to intercept backscatter from a beam stop. Hence the undeviated beam may be intercepted by a small, concave lead button which is most conveniently located immediately in front of the film. However, background due to air scatter is lessened by placing the beam stop much closer to the sample, a consideration that is important for samples producing rather faint reflections at small angles. Figure 4-35 shows a typical commercially available camera of this type. In this particular design the lead beam stop is supported on a wire which is sufficiently flexible to permit the momentary displacement of the stop from the path of the beam in order to mark on the film the location of the undeviated-beam image. Alternatively, the beam stop button can be mounted with a drop of glue directly on the black paper sheet (or other

Fig. 4-35. Flat camera for preparing monochromatic-pinhole patterns. (Courtesy of General Electric Company.)

material) that excludes light from the film. The light-tight cassette is loaded in the darkroom and then returned precisely to its original position in the supporting bracket on the x-ray unit.

The sample in the form of a thin sheet may be affixed directly to the end of the collimator by a suitable holder. The thickness of the sample, so far as consonant with other objectives of the experiment, should be regulated so as to satisfy the optimum intensity requirements explained in Section 4-1.4. Most finely divided materials can be easily mounted by pressing them into a small hole drilled through a metal plate of the proper thickness. If the material tends to fall out, a thin sheet of cellophane may be placed over both sides of the sample plate to hold the sample in position and act as windows for the x-ray beam, or a little gum tragacanth, collodion, or ethyl cellulose solution may be mixed with the powder as a binding agent. Collimator pinholes 0.5 or 1.0 mm in diameter are suitable for most work, but a diameter of 0.25 mm or even less may be required to resolve reflections from the undeviated-beam image at unusually small angles. For such measurements the beam stop button must be as small as possible without permitting spillover of the main beam, and of course it must be very accurately aligned. Ordinary photographs obtained with 1 mm pinholes and a sample-to-film distance of 3 to 5 cm require exposures of 30 min to 2 hr, depending upon the nature of the sample and radiation employed.

For clarity and uniformity of film background, it is usually advantageous to replace the black paper in front of the film with a thin sheet of aluminum foil or, better yet, with a sheet of the β-filter material if obtainable. Thus 0.016 mm of nickel foil may be very conveniently used to exclude light from the film and eliminate β radiation when a copper-target tube is employed. The filter sheet should be placed at least $\frac{1}{8}$ in. from the film in order to avoid fogging due to the excitation of the (fluorescent) K spectrum of the filter material by the absorbed β wavelength of the incident beam.

4-3.2 Back-Reflection Method[74]

In this monochromatic-pinhole technique, the beam is defined by a pin-hole system which passes through the center of the film as indicated in Fig. 4-36. The beam is incident upon the sample surface at X, and a limited number of diffraction circles at large Bragg angles are recorded on the flat film, C and C' indicating such a typical circle by its points of intersection with the plane of the diagram. It is seen that the Bragg angle θ is related to the measurable angle ϕ as follows:

$$\theta = \frac{\pi}{2} - \phi.$$

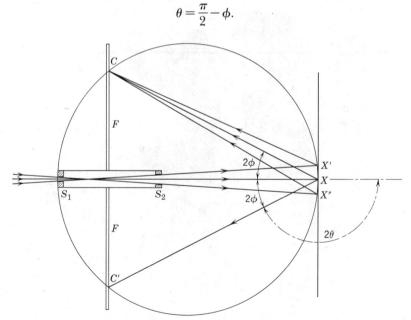

Fig. 4-36.   Geometry of the back-reflection method.

In common practice the pinholes S_1 and S_2 are of equal size and relatively small, in which case the beam has little divergence and irradiates only a very small area about the point X. Under these conditions the positions of S and S_2 with respect to the film and sample are not critical, and the sharpness of all the recorded reflections is equal except for the small increase in line breadth arising from increasing obliqueness of incidence upon the film as 2ϕ increases. The line XC' represents a narrow diffracted beam of this kind. The back-reflection arrangement is especially suited to the measurement of small changes in the interplanar spacings of metals and alloys, because the reflections obtained are those at high angles where the sensitivity of line position to changes in lattice spacings is at a maximum. The most valuable application of such cameras is to the measurement of lattice stress and strains in metals. Figure 4-37 shows a commercial back-reflection cassette which contains a 1 mm pinhole collimator.

Under certain circumstances it is advantageous to employ one small "source" pinhole S, and a larger pinhole S_2 which acts to limit the divergence of the beam. This technique is indicated in Fig. 4-36, where S_2 is sufficiently large to permit the beam to irradiate a circular patch of the sample surface of diameter $X'X''$. When this is done, a sharply

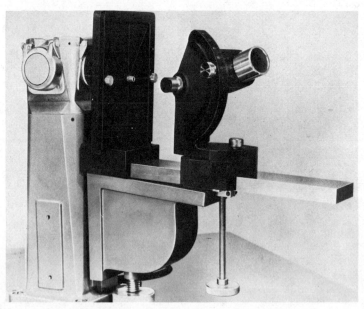

Fig. 4-37. Simple back-reflection cassette and a two-arc goniometer specimen holder. (Courtesy of Philips Electronic Instruments.)

focused reflection can be obtained only if S_1, X, and C and C' lie on the circumference of a common circle so as to satisfy the parafocusing condition (see Section 4-2). Reflections at all other angles are more or less diffuse when the focused reflection CC' is sharp. If it is desired to photograph other reflections, the sample-to-film distance or the collimator position must be changed to satisfy the new parafocusing requirement. Two advantages are gained by employing this focusing technique. (a) The exposure time is reduced manyfold. (b) The larger area of sample surface irradiated is conductive to smoother diffraction lines, because of the larger number of grains contributing to each reflection [75 a, b].

Successful precision measurements depend upon several factors, including (a) choice of the correct radiation to produce a high-angle reflection from the particular specimen under study, (b) proper condition of the specimen to result in reasonably sharp diffraction circles (lattice distortions due to cold work must not be excessive, and the grain size must not be so large as to give very discontinuous lines), and (c) an effective method for determining the sample-to-film distance with an accuracy at least as great as that limiting the measurement of the diameter of the diffraction halo. Several methods for determination of the sample-to-film distance have been used. The most common is to paint the specimen surface with a material that furnishes sharp reflections of known spacing in the back-reflection region. When $CoK\alpha$ radiation is used, as in the study of iron and its alloys, gold and silver powders are suitable calibrating substances. The principal disadvantage of this method is that the exposure time must be approximately doubled. In addition, another significant source of experimental error is introduced in the measurement of the calibrating reflection, unless it is much sharper than the reflection from the specimen itself.

A method that is capable of higher accuracy consists in mechanically measuring the sample-to-film distance with the aid of a specially constructed gage block or template [75 a, b, 76]. In the method of Thomas, the x-ray tube itself is mounted on a support which can be made to approach or recede from the specimen by means of a finely threaded screw. The collimator and cassette in Fig. 4-38 are rigidly fastened to the x-ray tube (not shown). At the start of an experiment, a sleeve L, housing a bullet-shaped member Q, is screwed on the end of the collimator. The x-ray tube and the attached collimator and cassette are than moved forward (by the fine screw) until the tip of the bullet just touches the specimen, as determined by means of a feeler gage 0.0015 in. thick. The x-ray tube is now locked in this position, after which the sleeve is unscrewed, permitting both the sleeve and the bullet to fall clear. The front pinhole P is then screwed on and the exposure made. This technique per-

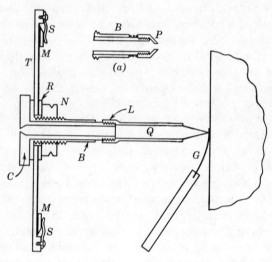

Fig. 4-38.   Gage for setting the specimen at a reproducible distance from the film. [Courtesy of Thomas, *J. Sci. Instrum.*, **18**, 135 (1941); copyright, The Institute of Physics.]

mits the same sample-to-film distance to be reproduced time after time with a precision greater than that with which a sharp diffraction halo can be measured. Another method [100] uses the knife edges of slip gages to cast sharp, straight shadows on the film. This method gives an accurate specimen-to-film distance, and at the same time minimizes film-shrinkage errors.

In certain cases it is desirable to be able to irradiate a precisely selected spot on the sample. For this purpose an adjustable specimen table has been described which can be transferred from a microscope to a back-reflection camera so that the same area viewed through the microscope is irradiated with the x-ray beam [77].

With the advent of high-speed Polaroid film, diffractionists soon investigated its possibilities in both forward- and back-reflection methods. Smith [78] used Polaroid film (ASA 3000) in a Polaroid film holder (type 500) with a high-speed x-ray phosphor. The method proved to be about twice as fast as conventional wet-film x-ray methods, and the development time was only 10 sec. Herglotz [79] used a DuPont type D (blue fluorescent) miniature radiographic screen in close contact with ASA-3000 Polaroid film and found exposure times as short as 3 sec suitable and a processing time of 10 sec. Speed of this kind makes possible observation of transient phenomena in crystalline materials. Herglotz

[79] also investigated special fluorescent screens for use with mono-chromatic x-radiation, since most commercial screens are designed for polychromatic radiation of wavelength less than 1 Å.

Spakowski[80] experimented with the very fast Polaroid film type 410 (ASA 10,000), and found it made possible the elimination of the fluorescent screen. Schmidt and Spencer[81] described their camera and technique for use with Polaroid film, and Dumbleton and Bowles[82] reported on a heating and cooling unit for use with a Polaroid camera.

Fig. 4-39.   Polaroid XR-7 Land diffraction cassette on track mount. (Courtesy of Polaroid Corporation.)

Such interest finally stimulated the development of the Polaroid XR-7 Land diffraction cassette* as a standard piece of diffraction equipment (Fig. 4-39). The cassette permits placing X and Y coordinate fiducial marks on each photograph. Processing is accomplished by merely pulling the exposed film packet through two stainless-steel rollers.

Peters and Kulin[83] carried out a study to evaluate technically the Polaroid film method in diffraction. Their study concludes that type 57 (ASA 3000) film provides the best compromise between exposure time and pattern quality. Primary beam transmission through the negative (as in back-reflection techniques) causes no adverse effects on pattern quality. Development time can be varied between 8 sec and 2 min with no discernible differences in the pattern. Peters and Kulin also studied the effect of x-ray tube target material, tube voltage and current, collimator size, and specimen-to-film distance on pattern quality.

4-4 MICROCAMERAS AND MICROBEAM TECHNIQUES

The Bragg relation reveals that a d spacing of 50 Å has a 2θ angle of $1°46'$ for Cu$K\alpha$ radiation. Hence the measurement of a 50-Å spacing requires a rather special camera with a very finely collimated x-ray beam and a very small beam stop. Indeed, the measurement of spacings of 50 Å and longer (2θ angles $\lesssim 2°$) is one aspect of a rather specialized technique, *small-angle scattering*.† Many other problems of the diffractionist also require special small cameras, microcameras, and finely collimated beams, microbeams. Such problems are: examination of single fibers for orientation of crystallites; studies of the mean particle size, misorientation, and degree of deformation in cold-worked metal grains; back-reflection Laue studies of orientation of single grains in metals; preparation of powder patterns from exceedingly small samples; microanalytical identification problems; and a host of others.

The term "microbeam" generally refers to an x-ray beam collimated by pinholes smaller than 0.1 mm (100 μ) in diameter. Fine-bore lead glass capillary tubes make excellent microbeam collimators (e.g., marine barometer tubing[84]). Figure 4-40 presents a longitudinal section through the collimator system of the Chesley[84] microcamera. The collimators are $\frac{3}{8}$-in. lengths of capillary tubing with bore diameters from 25 to 100 μ. A rubber gasket and a small piece of cellophane over the

*Available from the Polaroid Corporation, Cambridge, Massachusetts; and also from most purveyors of diffraction equipment.
†For an introduction to small-angle scattering the reader is referred to Chapter 12 of reference[5] and, particularly, to Chapter 5 of reference[174] with its accompanying bibliography.

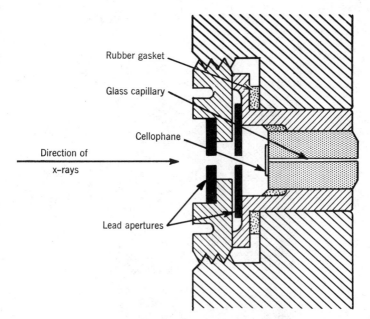

Fig. 4-40.   Longitudinal section of collimator system used in the microcamera designed by Chesley[84]. (Courtesy of Philips Electronic Instruments, and Alexander[174].)

forward (outer) end of the capillary seal the camera and permit evacuation. Lead diaphragms with openings slightly larger than the bore of the capillary shield the collimator from unwanted x-rays and light. In this camera a fiber specimen may be either affixed directly to the back (inner) end of the collimator tube, or mounted on a suitable holder just behind the end of the collimator. A minute powder sample mounted on a stiff hair or plastic fiber, or on very thin sheet plastic, may be supported on the specimen holder. Two motions built into the specimen holder permit selection of the precise region of the specimen to be placed in the x-ray beam.

Figure 4-41 depicts the Chesley camera assembled and aligned on the camera track. On the lower right is the outlet for evacuating the camera. A microdiffraction pattern of a nylon-56 fiber with preferred orientation (obtained with a Chesley camera) is shown in Fig. 4-42. This camera has been available commercially for several years.* Microcameras incorporating a semicylindrical film arrangement and mounted on brackets

---

*Central Research Laboratories, Inc., Red Wing, Minnesota; Philips Electronic Instruments, 750 South Fulton Avenue, Mount Vernon, New York.

Fig. 4-41. Assembled microcamera of Chesley design mounted on x-ray tube housing. (Courtesy of Philips Electronic Instruments, and Alexander [174].)

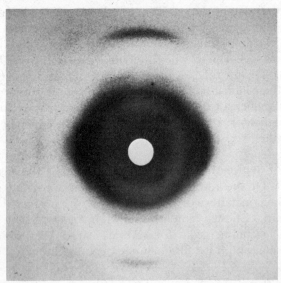

Fig. 4-42. Microdiffraction pattern of a nylon-56 fiber. Fiber axis vertical. Magnification 4.4 ×. (Courtesy of J. H. Magill and S. S. Pollack, and Alexander [174].)

which attach directly to the x-ray-tube housing for improved stability and alignment are also available commercially.†

Oron and Minkoff[85] have described a microcamera similar to the Chesley camera but which employs a conical film for back-reflection techniques. The conical geometry allows a wide range of diffraction angles to be recorded, permits reduced exposure times, and reduces distortion in the shape of diffraction spots. A feature of an evacuable microcamera built by Glas[86] (also based on the Chesley design) is a series of three interchangeable specimen holders, each designed for a different purpose: (a) a holder for orientation texture studies, (b) a holder for single microcrystal analysis, and (c) a holder for ordinary microdiffraction work. Clawson[87] modified the Philips (Norelco) microcamera to make back-reflection studies with it.

Alignment is not difficult when a microcamera is used with an ordinary x-ray tube with a 1×10 mm focal spot, but beam intensity is very low since only a small portion of the focal spot can transmit x-rays through the fine aperture of the collimator. The exposures required under these conditions, even with specimens of optimal thickness, are very long, and small-diameter fibers do not yield useful diffraction patterns with exposures of many hours. The practical solution is to mate the microcamera to a microfocus x-ray tube. The microbeam thus supplied is much more intense, and even small specimens with small collimators yield satisfactory diffraction patterns with practicable exposure times. Although alignment of a microcollimator with the microsource is a more critical and difficult operation, it is well worth the greater effort and time involved.

Increased attention is being given to the diffraction analysis of very minute powder specimens. Indeed, x-ray diffraction has important advantages over chemical and most physical methods when applied to microanalytical problems of this kind. Manipulative difficulties in mounting a few micrograms of powder in the x-ray beam, rather than photographic problems, appear to set the lower limit of sample size. Fried and and Davidson[88] have described effective techniques for mounting microamounts of solid neptunium compounds for x-ray analysis. Rooksby[89] has described an unconventional semicylindrical powder camera of 6-cm radius with which good patterns can be obtained from as little as 10 μg of material. The powder adheres in a narrow band to a sheet of tissue paper and is irradiated by means of a slit collimator.

†Rank Precision Industries, Ltd. (successor to Hilger and Watts), 31 Camden Road, London, N.W. 1, England. In the United States: Rank Precision Industries, Inc., 411 Jarvis Avenue, Des Plaines, Illinois.

Wilkinson and Calvert[46] have presented techniques for investigating thin films and small specimens.

Almost any laboratory doing serious x-ray metallographic studies finds a microbeam back-reflection camera with arrangement for positioning a preselected area a necessity. Often the problem requires positioning a small x-ray beam of the order of 20 to 100 μ in diameter on a preselected area (say, 0.1 to 0.5 mm in diameter) for studies of cold-work effects on grains, orientation of precipitates, identification of phases, and so on, by back-reflection methods. In some cases, however, the desired area for x-raying may be only 10 to 20 μ in diameter or width. Microfocus x-ray sources (Section 2-2.2F) and appropriate collimators readily supply the needed microbeams. For instance, Otte and Cahn[94] used a Hilger microfocus x-ray tube with a focal spot $\sim 40 \mu$ in diameter, and placed a single collimating pinhole in a platinum foil close to the tube window. Hole sizes of 10, 25, and 50 μ in diameter were used. Beam divergence with such pinholes at 3 to 5 cm from the focal spot was slight. For orientation studies with white radiation from a tungsten target operating at 50 kV, 400 μA and a specimen-to-film distance of 10 mm, exposure times of about 2 to 3 hr were required with the 50-μ pinhole, and of the order of 50 hr using the 10-μ hole. Hirsch and Kellar[90] have described collimators made from lead glass capillaries for microbeams down to approximately 1 μ in diameter.

The arrangement for precise alignment of the preselected area with the beam is the crucial part of such equipment, and many cameras, positioning devices, and techniques have been described[91–98]. In the schemes of Lewis[92], Holmes and Fochs[93], and Rudman and Weiss [95], the x-ray beam is first located on fluorescence material with a low-power microscope and focused on the crosshairs. The sample is then brought into position so that the selected area is on the crosshairs. Taylor and Moore[97] located the beam by a few seconds' exposure of a slow photographic plate placed in the position normally occupied by the specimen. After exposure the plate was developed and fixed without being removed from the mount by applying the appropriate solutions with a paintbrush. These four methods are simple, but not very precise.

Otte and Cahn[94] developed an improved method for aligning a selected area of a flat metallic specimen for back-reflection diffraction. The specimen is mounted coplanar with a set of fine (130-μ) tungsten wires in the form of a "crosshair," and both are mounted on a micrometer stage with x and y movements relative to the beam direction which is z. The relative positions of the crosshairs and desired specimen area are located by means of an optical microscope. The micrometer stage is transferred to the x-ray track, and the beam brought on the crosshairs by following it with a Geiger counter and a rate meter. The specimen area is then moved the necessary distance to place it in the x-ray beam. Otte and Cahn claim that areas 10 to $20\,\mu$ in diameter can be reliably located and examined in a beam of approximately the same diameter.

Michels and Wayman[96] sought to improve the Otte and Cahn technique by substituting a light beam for the x-ray beam and using an etched reticule on an optical flat. The light (x-ray) beam is located on a reference mark of the reticule, and the relative positions of the selected specimen area and the reference mark are determined by means of an auxiliary microscope. Finally, with an accurate XY positioner, the desired area is brought in front of the collimator. The positioning accuracy is described as $10\,\mu$ or better. An instrument described by ter Avest and Klostermann[98] probably provides the highest accuracy (better than $2\,\mu$ is claimed).

4-5   HIGH-TEMPERATURE TECHNIQUES

At elevated temperatures no one camera is best for all samples and all temperatures. Generally, the higher the temperatures involved, the more elaborate the camera required. Temperatures only slightly above room temperature ($< 100°C$) can be readily attained by enclosing the camera in a thermostatically regulated chamber [101, 102], or by flowing fluid from a thermostatted reservoir through channels in the camera [102]. Unless some provision is made for cooling the photographic film, such methods are unsuitable above about 70°C. because of deterioration of the emulsion and the consequent formation of an excessively dense background. This objection has been eliminated in a design by Vand[103] in which a Frevel parafocusing arrangement[104] is employed and the sample powder is mounted on the surface of a curved metal holder containing channels through which temperature-controlled water is circulated. In this method the film is not heated appreciably, and any desired temperature between 0 and 100°C can be obtained. Another method of heating a powder sample to elevated temperatures is to mount it on a wire which is heated by the passage of an electric current[105]. In both this method and in Vand's technique, it is important to remember that the sample temperature tends to be less than that of the supporting material; hence, although the temperature of the support can be measured directly with a chromel–alumel or other suitable thermocouple, it is necessary to calibrate the thermocouple emf's in terms of the actual sample temperature. This can be done by utilizing accurately known transition and melting points of solid materials, as discussed later.

For temperatures much above 100°C, the best results cannot be had without resorting to more elaborate equipment. A Debye-Scherrer camera that gives good results up to about 600°C, and at the same time is relatively easy to manipulate, was described several years ago [106] and was formerly produced commercially.* Figure 4-43 shows this camera both assembled (at left) and disassembled. There are several very attractive features. The cylindrical and light-tight film-supporting member (at right of Fig. 4-43) can be removed without disturbing the adjustment or thermal state of the specimen. The furnace unit with its electrical and cooling leads can also be removed without affecting the specimen alignment. The heating element is a coil of chromel wire on the inside of a slotted porcelain tube which surrounds the specimen coaxially. The temperature is maintained by radiation barriers in the form of thin aluminum cylinders of low heat capacity rather than the customary more massive thermal insulating barriers, and as a result the specimen temperature can be varied rapidly, a definite advantage in certain applications such as the investigation of the phase diagrams of minerals. Modifications of this basic design have been worked out [107], which

*Otto von der Heyde, Inc., R. R. 1, Box 187A, Thompson, Connecticut. Unfortunately, von der Heyde no longer has instruments in production.

Fig. 4-43.   High-temperature camera for x-ray diffraction work up to 600°C. (Courtesy of O. von der Heyde.)

permit the use of either a 57.3- or 114.6-mm camera "shell" with the same camera base, specimen holder, and furnace unit. In this way the equipment can be adapted for both "survey" work and applications requiring higher resolution. Hanak and Daane[108] have also described modifications of this camera, which permit studies on reactive metals to considerably higher temperatures. One method is used to 1000°C and another method from 1000 to 2200°C, the differences between the two being chiefly in the arrangements for heating the sample.

For precision measurements at temperatures up to 1000°C and higher, British cameras developed from an earlier design of Hume-Rothery and Reynolds[109] are especially suitable. Figure 4-44 shows the improved design of A. J. C. Wilson[110]. The furnace consists of two bobbins K and L, of oxidation-resistant steel, bearing windings of platinum insulated from the bobbins by mica. The platinum-rhodium–platinum thermocouple is of the ring type and is supported by the lower clamp N. The specimen, if oxidizable or volatile at the temperatures of the experiment, is held in a thin-walled silica tube and rotated. Blackening of the

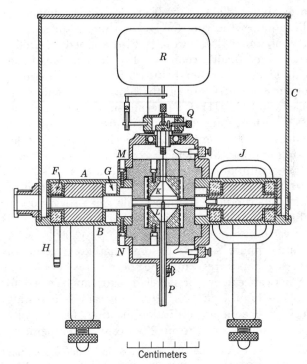

Fig. 4-44.   Improved 19-cm high-temperature camera for x-ray diffraction work up to 1000°C. [Courtesy of Wilson, *Proc. Phys. Soc.* (London), **53**, 235 (1941).]

film by furnace radiations is prevented by screening it with a sheet of aluminum foil. The body of the camera, including the film-supporting member, is cooled by water flowing through special channels F, G, and J. The heating currents in the two bobbins can be varied independently, and in practice they are adjusted so that the temperature distribution about the sample is symmetrical as determined by exploration with the thermocouple junction. The remaining features of design resemble those of the 19-cm room-temperature camera of Bradley, Lipson, and Petch[13]. Owen has designed a somewhat similar camera[111] which possesses two additional features: (a) evacuation of the entire furnace unit, and (b) magnetic coupling mechanism for rotating the specimen without a mechanical linkage between the specimen holder and the motor. Most of the best features just described, as well as further improvements, were combined in another 19-cm camera designed to operate up to at least 1400°C. This camera was manufactured after a design of the British Small Arms Group Research Centre.*

For diffraction studies above about 1400°C, the investigator generally faces major camera construction. Fortunately, several cameras for temperatures above 1400°C have been described. Hanak and Daane's [108] modification of the von der Heyde camera for studies to 2200°C has been mentioned. Matuyama[112] designed a Debye-Scherrer camera fitted with a small carbon-tube furnace capable of attaining temperatures of 1800°C. Another feature of the camera provides for recording a comparison pattern on the film. In a technique designed by Aruja, Welch, and Gutt[113], a thermocouple combines the functions of sample support, heating, and temperature measurement. The device is described as simple, compact, inexpensive, and very reliable. Maximum working temperature depends on the thermocouple material. A 5/20 rhodium–platinum couple provides 1750°C, and a 20/40 rhodium–platinum couple yields 1850°C. Gutt[114] later found that an iridium–60% iridium/40% rhodium thermocouple can provide an upper temperature limit of 2150°C. Austin, Richard, and Schwarz[115] have described a vacuum camera for operation up to 2000°C, which permits controlled atmosphere studies at somewhat lower temperatures and allows successive exposures while the specimen is held at temperature. Another American camera of novel design can be used for work at temperatures as high as 2500°K[116]. This camera is distinguished by (a) 0.5-Mc, high-frequency induction heating, (b) specially designed temperature regulator, good to ±0.2 per cent, (c) temperature measure-

*The producer was Pye Unicam, Ltd., whose former line of diffraction equipment is now manufactured and marketed by Crystal Structures, Ltd. Swaffham Road, Bottisham, Cambridge, England.

ment above 1100°K by means of a disappearing-filament optical pyrometer, and (d) operation with the specimen exposed to either high vacuum or an inert-gas atmosphere. Finally, Revcolevshi, Hubert, and Collongues[117] have described a camera that permits samples to be studied at temperatures up to 3000°C in vacuum or in oxidizing or reducing atmospheres. Its novel heating device consists of two elliptical mirrors, a flat mirror, and a 6.5 kV short-arc xenon lamp in an image-furnace arrangement.

Of special interest is the Guinier-Lenne high-temperature camera.* This device permits the continuous film recording of such thermal phenomena as phase changes, recrystallization, thermal expansion, and various chemical transformations and reactions. The instrument (Fig. 4-45) makes a detailed two-dimensional film record, as presented in Fig. 4-46. Among the features of the camera are focusing geometry, a built-in monochromator, complete enclosure of the sample in the furnace, continuous or discontinuous movement of the film, alignment by the de Wolff method, temperatures to 1200°C, a thermocouple for temperature measurement, controlled atmosphere or vacuum, film marks to indicate reference temperatures, and a platinum sample holder.

Often high-temperature diffraction problems demand the preparation of special heating units for standard cameras or, indeed, the construction of a new camera of unusual design. To assist the reader in such instances, the following references are mentioned. Farrow and Preston[118] prepared a simple camera for studies of polymer specimens to 270°C. It was designed to be used with a Unicam S.25 standard x-ray goniometer.† For small-angle diffraction studies of soap–water systems, Marsden and Sanjana[119] devised an inexpensive camera for heating (up to 250°) samples containing a volatile component. Several cameras with a working range to 1000 to 1200°C have been described. Two are heating-unit adaptations for the General Electric powder camera or cameras of similar design[120, 121]. A unique high-temperature high-vacuum camera for examination of reactive metals up to 1000°C was constructed by Hatt, Kent, and Williams[122]. The camera is constructed of glass without demountable seals, and uses a radiant

*Manufactured by N. V. Nederlandsche Instrumentfabrik (Nonius), Delft, Netherlands. Available in the United States through Enraf-Nonius, Inc., 130 County Courthouse Road, Garden City Park, New York; or through Diano Corporation, Industrial X-ray Division, 2 Lowell Avenue, Winchester, Massachusetts, which acquired General Electric Company's x-ray diffraction instrument line in 1972.

†The Unicam line of cameras and accessories is now manufactured by Crystal Structures, Ltd., Swaffham Road, Bottisham, Cambridge, England.

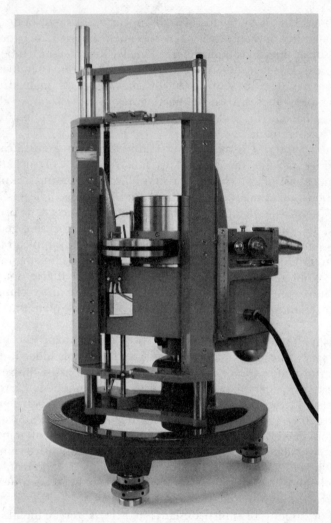

Fig. 4-45.   Nonius Guinier-Lenne high-temperature camera. (Courtesy of Enraf-Nonius.)

heater in place of the conventional furnace assembly. For temperatures up to 1000 to 1200°C, the cameras described by Intrater and Smith[123] and by Zubenko, Kvitka, and Umanskii[124] also deserve to be studied.

The accurate measurement of specimen temperature, regardless of camera design, requires care in the placement of the thermocouples and their calibration.* Large thermal gradients have been observed in

*As a background for such measurements, the reader is strongly urged to read Gray and Finch's[131] excellent article on the attainment of accuracy in temperature measurements.

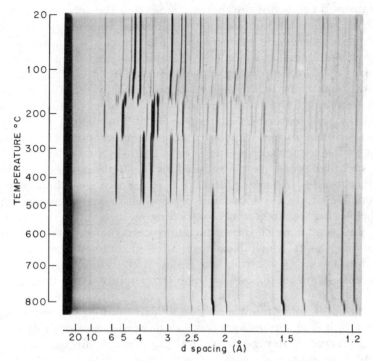

Fig. 4-46. Guinier-Lenne camera high-temperature film scan showing decomposition of magnesium perchlorate. (Courtesy of Enraf-Nonius.)

the region occupied by the specimen, and the difference between the reading of the measuring thermocouple and the actual mean specimen temperature can be as large as 30 to 50°C at 1000°C. Basinski, Pearson, and Christian[125] used a pair of hemispherical furnaces and measured the temperatures by means of a pair of ring thermocouples, one mounted at the level of the mouth of the top furnace, and one at the level of the mouth of the bottom furnace. The leads in both cases were taken out through the furnace gap. The currents through the two furnaces were adjusted until the two thermocouples gave the same emf. By measurements with silver specimens, the mean specimen temperature was indicated to be within 3°C of that indicated by the thermocouples. Pease[126] considered the problem of specimen temperature measurement by thermocouples from both the theoretical and experimental point of view. Errors arising from conduction down the thermocouple leads can be minimized and kept constant by proper design. Studies by Brand and Goldschmidt[127] led to the same conclusion about thermocouple lead losses.

Improved contact of the thermocouple with the specimen obviously is highly important, and Johnson[128] devised four ways of mounting specimens on fine-wire thermocouples: (A) nonreactive powder samples, (B) machined metal specimens, (C) nonmachinable specimens, (D) samples sealed off in capillaries with couple (Fig. 4-47). The mounting of such specimens in the camera is discussed by these investigators. Wilkinson and Calvert[129] have described a direct melting-point cali-bration of the Unicam S150 camera* to 800°C. The melting points of suit-able specimens in silica capillaries were observed optically in polarized light with an internal consistency of ±1°C. Thermal gradients within the specimen were greatly reduced by a simple radiation shield. A simpler, but less reliable, procedure is to prepare several photographs of suitable substances above and below their known transition points, or below their melting points, bracketing by successively smaller inter-vals the point at which the diffraction pattern changes. Some standard reference points which have been used for camera calibration are listed in Table 4-5. Ostertag and Fischer[130] studied the temperature gradients in metal ribbon-type high-temperature x-ray furnaces and the calibration of such furnaces. Heat drainage from the furnace was minimized by using thin thermocouple wire, and eight thermocouples were attached at various locations. Conclusions from this study suggest the preferred heater geometry and the area on the specimen stage where the temperature is nearly constant, and discuss briefly the use of microoptical pyrometers. Table 4-6 lists substances used by Ostertag and Fischer in calibrating high-temperature ribbon heaters.

*The Unicam line of cameras and accessories is now manufactured and marketed by Crystal Structures, Ltd., Swaffham Road, Bottisham, Cambridge, England.

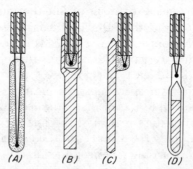

(A)   (B)   (C)   (D)

Fig. 4-47. Suggested methods for mounting specimens on thermocouples: (A) nonreactive powders; (B) machined metal specimens; (C) nonmachinable specimens; (D) samples sealed off in capillaries with couple. [Courtesy of Johnson, *J. Sci. Instrum.*, **38**, 373 (1961); copy-right, The Institute of Physics.]

Table 4-5. Transition- and Melting-Point Temperatures Useful in the Calibration of Thermocouples and High-Temperature Cameras

Transition Point	Melting Point	Temperature (°C)	Reference
	H_2O (triple point)	0.01^a	[131]
	Diphenylmethane	27	[129]
NH_4NO_3		85.4	[101]
	Benzoic acid	122.0 ± 0.5	[129]
NH_4NO_3		126.2	[101]
	Anthracene	216.3 ± 0.1	[148]
	Sn	231.91^a	[131]
$KClO_4$		295.7	[101]
	Pb	327.5	[148]
	KNO_3	333.6 ± 0.2	[129]
	Zn	419.58^a	[131]
Quartz		573.0	[101]
	Sb	630.5	[106]
	Al	660.2	[148]
	KCl	776	[129]
	NaCl	804 ± 3.0	[129]
	Ag	961.93^a	[131]
	Au	1064.43^a	[131]
	Cu	1083.0 ± 0.1	[148]
	Fe	1535	[109]
	Pt	1773.5	[109]

^{a}Assigned temperatures, International Practical Temperature Scale, 1968 [131].

Table 4-6. Melting Points Used in Calibrating High-Temperature Ribbon Heater [130]

	Melting Temperature (°C)	
Substance	Thermocouple Reading	Literature Value^a
CuCl	427	430
NaCl	800	801
Bi_2O_3	817	820
NaF	984	988
K_2SO_4	1064	1069
CaF_2	1355	1360

^{a}*Handbook of Chemistry and Physics*, 46th ed., Chemical Rubber Company, Cleveland, Ohio, 1965.

For additional references on high-temperature diffraction techniques, the reader is referred to the elaborate bibliography by Goldschmidt [132].

4-6   LOW-TEMPERATURE TECHNIQUES

Techniques and equipment for high-temperature work are better developed than those for measurements below room temperature. A simple but effective method of attaining very low temperatures is simply to bathe the sample, sealed in a suitable holder, in a slow stream of liquid nitrogen or oxygen[133a, b]. The flow of liquid prevents the accumulation of frost on the sample surface. Cameras have been constructed that attain low temperatures by keeping the specimen in close thermal contact with the metallic inner bottom member of a specially constructed Dewar vessel[134].

More complicated equipment is necessary for controlling the sample temperature accurately. Barnes and Hampton[135] have obtained any desired temperature between -70 and $0°C$ by mounting the powder on a copper block cooled by the circulation of a fluid kept at some fixed temperature. The liquid flows in a closed system successively through a a pump, a Dry Ice–acetone cooling mixture, electrical resistance heating coils, and the sample-mounting block. By varying the heating current, any desired temperature is attained and kept constant to $\pm0.2°C$ for several hours. The specimen temperature can be measured with a copper–constantan thermocouple, due precautions being observed in its calibration so that the actual sample temperature is obtained. This method of cooling the sample is adaptable to either the Debye-Scherrer or the monochromatic pinhole technique. By selecting an appropriate circulating liquid, the same apparatus can be used to warm the sample to temperatures considerably above the temperature of the laboratory air.

A more generally useful technique consists in flowing a cooled gas around the specimen tube itself, or around a chamber of small dimensions containing the specimen[136–139]. In one version of this technique, which has been successfully applied to single-crystal analysis[137], compressed air from the laboratory line is first dried by passing it through concentrated sulfuric acid and Drierite, after which it is chilled by circulating through metal coils immersed in a suitable refrigerating bath (Dry Ice–acetone, or liquid nitrogen or oxygen) and then directed against the sample through a vacuum-jacketed nozzle. A cylindrical cellophane chimney surrounds the nozzle and sample to

prevent frosting of the sample as the result of diffusion of moist laboratory air to the specimen, and deposition of frost on the outside of the
cellophane cylinder is eliminated by flowing a stream of warm, dry air
over it.

Hume-Rothery and Strawbridge[140] constructed a Debye-Scherrer
camera which has been used at controlled temperatures down to −110°C.
The popular Unicam 19-cm powder camera has often been adapted for
low-temperature work. Tombs[133b] modified it to permit cooling by a
streaming cold liquid. From the sharpness of the diffraction lines obtained, it was evident that the variation in specimen temperature during
an exposure was not more than a few degrees. The simple modifications
of Thewlis and Davey[141] used streaming cold gas as a coolant and
permitted regulation to within ±1°C; temperatures as low as −150°C
could be obtained.

Francombe[142] also modified the 19-cm Unicam camera with special
attention to prevent vaporized coolant from coming into contact with
the film and to conserve coolant. This was achieved by a robust cooling
chamber which prevents coolant gas from escaping into the camera
body. Figure 4-48 is a sectional view of the cooling chamber positioned
in the camera. The coolant gas is obtained in the usual way by boiling
liquid nitrogen with an immersion heater in a Dewar flask. The cold

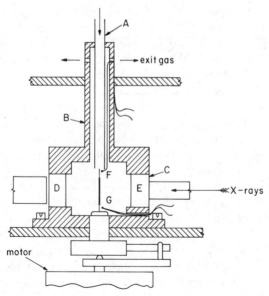

Fig. 4-48.   Section of a cooling chamber for use with a 19-cm x-ray powder camera.
[Courtesy of Francombe, *J. Sci. Instrum.*, **34**, 35 (1957); copyright, The Institute of Physics.]

vapor is led into the cooling chamber through the brass tube A and cools the specimen. The cold gas then passes up through the annular space between the tube A and the cylinder B and exits through vent holes above the camera lid. The upper B and lower parts of the cooling chamber are made of Tufnol, a resin-bonded-paper laminate, and are connected by two pairs of brass shields at D and E, which are placed in line with the ends of the beam trap and collimator. A window C made of film base completely encloses the specimen space. The specimen temperature is taken as the mean of the readings of the two copper–constantan thermocouples at F and G. A liquid-nitrogen consumption of about 1.25 l/hr easily permits temperatures of $-140°C$ to be reached. Lower temperatures require increased coolant consumption.

Other workers have modified the standard 114.59-mm powder camera for low-temperature studies. The conversion unit of Kramer, Venturino, and Mazelsky[143] takes the form of an alternate cover for the camera. The low-temperature adaptor designed by Crandall[144] also takes the form of a camera cover, but is a more elaborate double-chamber device. When assembled with the original camera case, two chambers are formed (Fig. 4-49). The outer chamber A contains the film. A baffle C diverts a continuous flow of warm, dry nitrogen gas around the film chamber via the light-trap ports D. The central chamber B contains the specimen mount E (shown in detail in Fig. 4-50). The two chambers are separated by a layer of lightproof paper G. Additional details are given in Fig. 4-49. The range, room temperature to liquid-nitrogen temperature, is readily covered with an accuracy of $\pm 5°C$.

Owen and Williams[145] have described a camera suitable for examining small metal plates at low temperatures by the back-reflection method. The specimen may be rotated in its own plane, while simultaneously being oscillated through a small angle about a vertical axis in its surface. The surface may also be oriented at different angles to the incident beam during exposure. The lowest temperature accessible (by dropping liquid air on the sample) with the camera is $-194°C$. The parafocusing, symmetrical Seemann-Bohlin (Section 4-2) camera was modified for precision measurement of lattice constants at low temperatures by Mascarenhas and Mascarenhas[146]. A small, low-temperature Dewar vessel with a curved neck is attached at the normal position of the sample holder. A sample holder is located on the end of the Dewar neck, and is accurately machined to the curvature of the parafocusing circle. There is excellent thermal contact between sample and coolant, and thermocouples are provided for monitoring the temperature.

These examples suffice to illustrate the types of cameras and low-temperature techniques commonly used down to liquid-nitrogen

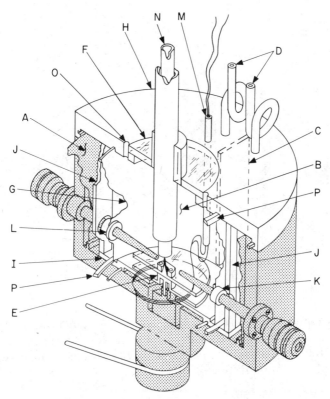

Fig. 4-49. Isometric drawing of camera with low-temperature adapter in place. (Original camera parts are shaded.) The cooling device N is inserted through the acrylic window F to a position close to the specimen yet clear of the reflections. A, Outer film chamber; B, specimen chamber; C, baffle; D, gas ports; E, specimen mount; G, paper divider; H, adapter cover; I, adapter base plate; J, connecting posts; K, brass backing sleeves; L, rubber light seals; M, thermocouple exit tube; O, gas escape vents, and P, locking rings. [Courtesy of Crandall, *Rev. Sci. Instrum.*, **40**, 954 (1969).]

temperature, $-198°C$. Additional references on low-temperature x-ray diffraction are to be had in the elaborate bibliography by Post[147]. Only a few manufacturers of diffraction cameras can provide low-temperature cameras or attachments for work in this range. The interested investigator should make inquiry stating his needs.

Low-temperature diffraction studies in the range from liquid-nitrogen temperature down to liquid-helium temperature can be carried out. Even the rare gases of the atmosphere have had their lattice constants measured with precision, and their densities and expansivities determined (see, for example, references [149–151]). Work in this region

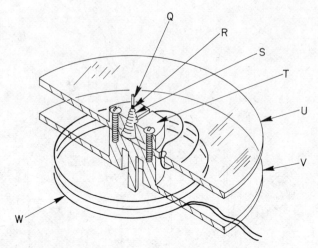

Fig. 4-50. Detail of specimen mount (E in Fig. 4-49). Q, Specimen fiber; R, thermocouple junction; S, thermocouple gradient barrier cone; T, nylon block; U, acrylic flange; V, flange of nylon specimen mount; and W, thermocouple wire loops. [Courtesy of Crandall, *Rev. Sci. Instrum.*, **40**, 954 (1969).]

requires liquid helium as a coolant, and a very much more elaborate cooling arrangement. Although certain powder cameras might be adapted for work in this region, most studies have been made with the flat-specimen counter diffractometer technique (Section 5-4).

Table 4-7. Suitable Materials for Low-Temperature Calibration of Thermo couples

Material	Melting Point (°C)	Material	Melting Point (°C)
$Na_2SO_4 \cdot 10H_2O$	32.38	Butyl chloride	−123.1
Benzene	5.5	Ethyl chloride	−136.4
Water	0.01[a](triple point)	Ethylene	−169.15
Mercury	−38.87	Oxygen	−182.96[a] (bp)
Chloroform	−63.5	Nitrogen	−195.8 (bp)
Ethyl acetate	−83.58	Oxygen	−218.79[a] (triple point)
Methyl chloride	−97.73	Neon	−246.05[a] (triple point)
		Helium	−268.9 (bp)

[a]Assigned temperatures, International Practical Temperature Scale, 1968 [131].

Accurate determination of the specimen temperature is just as important in low-temperature studies as in high-temperature studies as in high-temperature investigations. Moreover, the same care must be used to insure that the thermocouples are placed so that they truly measure the specimen temperature. Thermocouples of copper–constantan wire, 30 gage (0.0095 in.), lacquered and cotton insulated, with welded-bead junctions, are suitable for temperature measurements. Some materials suitable for their low-temperature calibration are listed in Table 4-7.

4-7 HIGH-PRESSURE TECHNIQUES

Many laboratories are making high-pressure diffraction studies today. This interest stems from a growing need to know such properties as structure, density, compressibility, and so on, of high-pressure phases of materials. The pioneering work of Bridgman[152] demonstrated the frequent occurrence of high-pressure polymorphs (allotropes) of the elements. Compounds, likewise, commonly exhibit high-pressure polymorphs. Only rarely can the high-pressure phase be metastably preserved by "quenching in" at room temperature or below. Kamb [153], however, was successful in studying certain high-pressure forms of ice by this method, and Simons and Dachille[154] investigated a high-pressure phase of TiO_2.

To be most useful, a method must produce the diffraction pattern from a substance *while it is under pressure*. Early studies by Cohn[155] and Frevel[156] accomplished this, but not very satisfactorily. Essentially, the first real high-pressure, x-ray diffraction work was that of Jacobs [157] on silver iodide. Jacobs placed the sample and film in a 5-cm-diameter Debye-Scherrer camera which in turn was placed inside a large steel pressure bomb with a beryllium window to admit the x-rays. The bomb was then filled with compressed helium gas, and the x-ray exposure was made. He achieved pressures up to approximately 5 kbar (1 bar = 10^6 dyn cm^{-2} = 0.987 atm), and photographs comparable to those obtained at 1 atm. There were problems involving absorption of the x-rays by the compressed helium, impurities in the helium, and the degassing of the x-ray film before development, which had to be overcome. This method, however, finds special applicability in the study of the structure of metals exposed to controlled gas pressures at elevated temperatures. Previously, studies were made by quenching the sample to room temperature, assuming that the structure of the metal in the presence of the gas did not change during the quenching process.

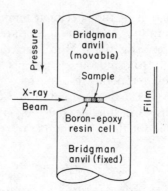

Fig. 4-51.   Schematic diagram of a pressure cell with opposed Bridgman anvils.

Thus Goon, Mason, and Gibb[158] followed a technique similar to Jacobs' in a study of the uranium–uranium hydride–hydrogen gas system. Powder patterns of the specimen under hydrogen gas pressures up to 600 psi and temperatures up to 500°C were obtained.

Presently, three basic designs dominate high-pressure x-ray diffraction equipment. Studies in the 100-kbar range can be made by compressing the sample between Carboloy Bridgman anvils[159–162] (Fig. 4-51). The sample cell is a disk of "amorphous" boron in epoxy resin.* The sample is mounted in a small hole in the center of the disk. Typical dimensions of the disk are: 2.38-mm diameter; 0.64-mm thickness; 0.25-mm hole. The disk collapses under strong compression and, after releasing the pressure, may have a thickness of ~0.28 mm. With a beam (usually Mo$K\alpha$) collimated to 0.25 mm, exposure times may be of the order of 1.5 hr for a sample under zero pressure, and 5 hr for patterns taken under high pressures. It will be noted that with this geometry the beam passes through the sample parallel to the anvil faces and perpendicular to the direction of the compressing force (Fig. 4-51). The camera depicted in Fig. 4-52 is a simple example based on the opposed Bridgman anvil technique.†

Pressure is applied to the anvils usually by means of a hand pump connected to the anvil pistons through a hydraulic system. The geo-

---

*In a private communication, Dr. J. C. Jamieson, Professor of Geophysics, University of Chicago, informs us that his student, Philip Halleck, has successfully replaced the boron-epoxy disk with a beryllium annulus which permits studies to 90 kbar and the use of Cu$K\alpha$ radiation.
†D. B. McWhan (private communication) suggests that this camera would be improved if adjustments were provided to move the press with respect to the cassette, thereby eliminating eccentricity problems.

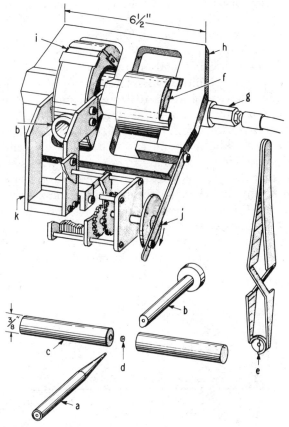

Fig. 4-52. A simple high-pressure x-ray powder camera which permits oscillation of the sample during the exposure. [Courtesy of McWhan and Bond, *Rev. Sci. Instrum.*, **35**, 626 (1964).]

metry of the hydraulic multiplication may provide a pressure factor of the order of 120× to 500× that of the hand pump. While a very high pressure is easily achieved by this scheme, it is difficult to know what sample pressure has been obtained. The nominal hydrostatic pressure on the anvil piston gives only a crude estimate of the pressure on the sample. The sample pressure depends also on such unknown quantities as the filling factor, the compressibility of the sample relative to that of the boron, the stress distribution within the compressed boron wafer, and the friction at the upper piston walls. Hence it is customary to mix an internal standard, such as NaCl, with the sample. The equation of state for NaCl is then used to convert the observed lattice parameter of NaCl to the pressure [163].

An alternative approach is the diamond cell, which consists of a pair of diamonds in the configuration of Bridgman anvils[164–166]. The x-ray beam passes through the diamonds perpendicular to the anvil faces and the sample plane (Fig. 4-53). The diamonds F used are small, brilliant-cut, gem-quality diamonds of approximately 0.15 carat (30 mg). Their culet faces are ground off to enlarge them to about 1.8-mm diameter for the fixed anvil E and 0.2- to 0.6-mm diameter for the movable anvil G, H. A 50-μ collimated beam K (usually from a microfocus x-ray tube) is arranged to pass along the axis of the anvils. A simple support can be arranged for the diamonds, so that the whole camera can be mounted on a modified precession camera. A disadvantage of the diamond cell is that the parallel alignment of the diamond faces is quite critical. Piermarini and Weir[164] found that improper alignment decreased the useful life of the diamonds. A set of diamonds properly aligned, however, may be used daily for a month or more, at pressures up to 30 kbar, before developing small, shallow cracks. Even with these cracks, the diamonds continue to perform satisfactorily. To prepare the

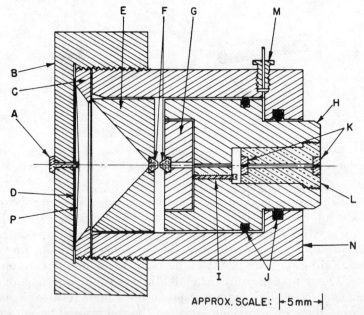

APPROX. SCALE: |–5mm–|

Fig. 4-53. Cross section through a simple diamond cell high-pressure diffraction camera. Some of its parts are: A, beam stop; B, cap; C, thrust washer; D, x-ray film; E, entabulature, F, diamond anvils; G, insert; H, piston; I, alignment screw; J, O rings; K, collimating pinholes; M, pressure fitting for introducing pressure; N, camera body; and P, light-protective paper. [Courtesy of Piermarini and Weir, $J. Res. Nat. Bur. Stand.$, **66A**, 325 (1962).]

sample about 5 mg of powdered material is placed on the fixed diamond and tamped lightly with a spatula. Some of the powder adheres as a small patch of material on the diamond face, and the remaining material is removed. Application of pressure to the movable diamond compresses the patch of material into the desired sample.

Pressure is applied to some diamond cells manually by mechanical means[165, 166]. Piermarini and Weir[164] applied pressure hydraulically through a fitting at M. Measurement of the sample pressure is again a problem. There are large pressure gradients across the face of the anvil, with the pressure varying from 1 atm on the edge to a high pressure at the center. Figure 4-54 depicts the gradients in diamond anvil geometry as mapped by Lippincott and Duecker[167]. Thus in the diamond cell the finely collimated beam must intersect only that part of the sample where the pressure is a maximum, and an internal standard, usually NaCl, must be admixed with the sample. Exposure times may be an order of magnitude greater with diamond anvil cells than with Bridgman anvils, but the quality of the diffraction pattern is usually considerably better. We have been urged,* however, to stress the gener-

*Private communication from D. B. McWhan.

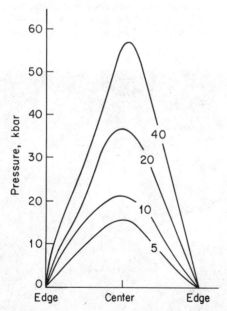

Fig. 4-54. Pressure gradients across a diamond anvil face as a function of the applied load. [Courtesy of Lippincott and Duecker, *Science*, **144**, 1119 (1964); copyright 1964, American Association for the Advancement of Science.]

ally poor quality of most high-pressure diffraction data as compared to normal data, and also the possible lack of uniqueness in the interpretation of many of the data.

A very promising new procedure for quantitative measurement of pressure in a diamond anvil cell was developed by Piermarini, Block, and Barnett[176] in 1971. They demonstrated that the sharp R lines in the fluorescence of ruby shift to lower energies as the pressure is increased. The fluorescence system utilizes standard facilities, and the method is adaptable to any pressure system that has optical access. With ruby as the sensor, the precision of the pressure measurement in a hydrostatic environment up to 100 kbar is 0.5 kbar, which is better than the accuracy of the present (1972) pressure scale above 40 kbar.

The popularity of diamond cameras has resulted in their commercial production. William Bassett of the University of Rochester is producing a model designed after the camera described in reference [165]. Another commercial diamond-type high-pressure camera is that shown in Fig. 4-55, which utilizes gas pressure to actuate the piston diamond.* Class, Iannucci, and Nesor[173] have presented experimental details and observations on the use of the parallel-axis diamond camera, and Weir,

*Available from Materials Research Corporation, Orangeburg, New York, and from Diano Corporation, Industrial X-ray Division, 2 Lowell Avenue, Winchester, Massachusetts, which acquired General Electric Company's x-ray diffraction instrument line in 1972.

Fig. 4-55.   Commercial high-pressure (diamond cell) x-ray diffraction camera with curved film cassette for recording pattern. (Courtesy of Materials Research Corporation.)

Block, and Piermarini[175] have described a small furnace for high-pressure studies at elevated temperatures (up to 650°C) with spring-loaded diamond cells.

A third design utilizes a tetrahedral anvil device to generate the pressure (Fig. 4-56). The x-ray beam passes in and out of the high-pressure volume through compressible gaskets. For high-pressure studies at elevated temperatures (> 500°C), the tetrahedral device is much more suitable than the Bridgman anvil and diamond cell designs which are largely limited to external heating arrangements. The tetrahedral design permits internal heating to 1000°C at 100 kbar. Barnett and Hall [168] have developed such equipment, and have discussed the geometry for matching the Debye-Scherrer camera to the geometry of the tetrahedral anvil press. They demonstrate, however, that counting technique (see Section 5-4) provides advantages over film technique in this design. Pressure control is better than with opposed anvil devices. Unfortunately the cost of the tetrahedral design may prove prohibitive for many laboratories.

Freud and Sclar[169] found that they could enlarge the anvil geometry pressure cell by introducing a split belt or girdle around the anvils to accommodate an internal heater. The apparatus is compact, may be positioned on conventional x-ray sources, and in principle can achieve simultaneous temperatures and pressures of 1000°C and ~ 100 kbar, respectively.

Two other developments in this field deserve mention. Freud and and La Mori[170] have developed a nondispersive, high-pressure, high-temperature x-ray diffraction technique. This technique uses the belted cell described by Freud and Sclar[169]; the x-rays are measured by a

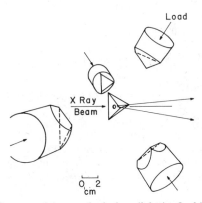

Fig. 4-56. Schematic diagram of the tetrahedral anvil device for high-pressure diffraction studies. [Courtesy of McWhan, *Trans. Am. Crystallogr. Assoc.*, **5**, 39 (1969).]

solid-state detector system (Section 5-3.2E) whereby continuous x-radiation is analyzed at a constant angle with a multichannel analyzer. A great advantage of this scheme is that with constant-angle geometry it is necessary only to provide one small x-ray port, thus simplifying the design of the high-pressure die. As the energy resolution obtainable with solid-state detectors improves, the nondispersive technique may become important. Finally, Johnson, Mitchell, and Evans[171] have succeeded in recording photographically x-ray diffraction patterns of the (111) and (200) reflections of shock-wave-compressed lithium fluoride (LiF). A shock wave of 180-kbar pressure was generated by a conventional high-explosive, plane-wave generator. Whereas the uncompressed LiF unit cell has $a = 4.027$ Å, the shock-wave-compressed LiF has a unit cell dimension of $a = 3.80$ Å. Shock-wave techniques may ultimately permit studies to pressures of 1000 kbar.

GENERAL REFERENCES

*1. L. V. Azároff and M. J. Buerger, *The Powder Method in X-ray Crystallography*, McGraw-Hill, New York, 1958.

2. R. W. M. D'Eye and E. Wait, *X-ray Powder Photography in Inorganic Chemistry*, Butterworths, London; Academic Press, New York, 1960.

*3. H. Lipson and H. Steeple, *Interpretation of X-ray Powder Diffraction Patterns*, Macmillan, London, 1970.

4. E. F. Kaelble (ed.), *Handbook of X-rays*, McGraw-Hill, New York, 1967, Chapters 7, 8, 14, 15.

5. H. S. Peiser, H. P. Rooksby, and A. J. C. Wilson, *X-ray Diffraction by Polycrystalline Materials*, Institute of Physics, London, 1955, especially Chapters 3, 5, 6, 9, 10, 11.

6. A. Guinier, *Théorie et Technique de la Radiocrystallographie*, 2nd ed., Dunod, Paris, 1956, Chapter 4. An English translation of the first edition, *X-ray Crystallographic Technology* (T. L. Tippell, translator), was published by Hilger and Watts, London, in 1952.

7. C. S. Barrett and T. B. Massalski, *Structure of Metals*, 3rd ed., McGraw-Hill, New York, 1966, Chapter 7.

8. A. Taylor, *X-ray Metallography*, Wiley, New York, 1961, Chapter 6.

SPECIFIC REFERENCES

[1] W. Friedrich, *Phys. Z.*, **14**, 317 (1913).

[2] H. B. Keene, *Nature*, **91**, 607 (1913).

[3] A. W. Hull, *Phys. Rev.*, **9**, 84, 564 (1917); **10**, 661 (1917).

[4] P. Debye and P. Scherrer, *Phys. Z.*, **17**, 277 (1916); **18**, 291 (1917).

[5] H. P. Klug and L. E. Alexander, *X-ray Diffraction Procedures*, 1st ed., Wiley, New York, 1954, pp. 165–177.

[6] M. J. Buerger, *Am. Mineral.*, **21**, 11 (1936).

[7] M. J. Buerger, *J. Appl. Phys.*, **16**, 501 (1945).

[8]   W. Parrish and E. Cisney, *Philips Tech. Rev.*, **10**, 157 (1948).

[9]   L. Alexander, *J. Appl. Phys.*, **21**, 779 (1950).

[10] A. Unmack and A. T. Jensen, *J. Sci. Instrum.*, **25**, 399 (1948).

[11] A. Taylor, *J. Sci. Instrum.*, **26**, 61 (1949).

[12] R. I. Garrod, *J. Sci. Instrum.*, **26**, 162 (1949).

[13] A. J. Bradley, H. Lipson, and N. J. Petch, *J. Sci. Instrum.*, **18**, 216 (1941).

[14] M. E. Straumanis, *J. Appl. Phys.*, **20**, 726 (1949).

[15] E. J. Grill and A. H. Weber, *Rev. Sci. Instrum.*, **20**, 532 (1949).

[16] M. D. Karkhanavala, *J. Sci. Instrum.*, **33**, 82 (1956).

[17] K. Lindsay, *J. Sci. Instrum.*, **41**, 477 (1964).

[18] C. S. Barrett and A. G. Guy, *Rev. Sci. Instrum.*, **15**, 13 (1944).

[19] J. Thewlis and A. R. Pollock, *J. Sci. Instrum.*, **27**, 72 (1950).

[20] M. E. Straumanis and E. Z. Aka, *Rev. Sci. Instrum.*, **22**, 843 (1952).

[21] J. M. DallaValle, *Micromeritics*, 2nd ed., Pitman Publishing Corporation, New York, 1948, pp. 73–86; *Symposium on New Methods for Particle Size Determination in the Subsieve Range,* American Society for Testing Materials, Philadelphia, 1941, pp. 66–89.

[22] W. Hume-Rothery and P. W. Reynolds, *J. Inst. Met.*, **60**, 303 (1937).

[23] W. Hume-Rothery, *J. Sci. Instrum.*, **24**, 75 (1947).

[24] K. Lonsdale and H. Smith, *J. Sci. Instrum.*, **18**, 133 (1941); R. Fricke, O. Lohrmann, W. Schröder, G. Weitbrecht, and R. Sammet, *Z. Elektrochem.*, **47**, 374 (1941); A. A. Burr, *Rev. Sci. Instrum.*, **13**, 127 (1942).

[25] J. M. Waite, *Rev. Sci. Instrum.*, **17**, 557 (1946).

[26] K. E. Beu and H. H. Claassen, *Rev. Sci. Instrum.*, **19**, 179 (1948).

[27] K. E. Beu, *Rev. Sci. Instrum.*, **22**, 62 (1951).

[28] L. R. Thomas and L. E. Alexander, *Rev. Sci. Instrum.*, **27**, 174 (1956).

[29] W. E. Armstrong and R. J. Davis, *J. Sci. Instrum.*, **35**, 36 (1958).

[30] J. M. Williams, *Rev. Sci. Instrum.*, **34**, 1430 (1963).

[31] R. L. Tallman and J. L. Margrave, *Rev. Sci. Instrum.*, **31**, 574 (1960).

[32] G. Gibson and E. J. Bicek, *Anal. Chem.*, **20**, 884 (1948).

[33] F. W. Mathews, *Anal. Chem.*, **26**, 619 (1954).

[34] P. J. Hagelston and H. W. Dunn, *Rev. Sci. Instrum.*, **20**, 373 (1949).

[35] E. M. Larsen and J. L. Leddy, *Rev. Sci. Instrum.*, **29**, 736 (1958).

[36] J. S. Lukesh, *Rev. Sci. Instrum.*, **11**, 200 (1940).

[37] M. Kossenberg, *J. Sci. Instrum.*, **32**, 117 (1955).

[38] J. K. Morse, *J. Opt. Soc. Am.* and *Rev. Sci. Instrum.*, **16**, 360 (1928).

[39] L. J. E. Hofer, W. C. Peebles, and P. G. Guest, *Anal. Chem.*, **22**, 1218 (1950).

[40] L. J. E. Hofer, A. Damick, A. F. Headrick, F. Fauth, E. H. Bean, and P. L. Golden, *Anal. Chem.*, **29**, 1563 (1957).

[41] C. L. Christ and E. F. Champaygne, *Rev. Sci. Instrum.*, **19**, 117 (1948).

[42] R. S. Pease, *Rev. Sci. Instrum.*, **26**, 1204 (1955).

[43] R. L. Sherman and O. L. Keller, Jr., *Rev. Sci. Instrum.*, **37**, 240 (1966) .

[44] A. Moore, D. B. Wright, and A. J. Martin, *J. Sci. Instrum.*, **35**, 301 (1958).

[45] M. G. Barker, *J. Phys.*, **E2**, 755 (1969).

[46] J. D. Wilkinson and L. D. Calvert, *J. Sci. Instrum.*, **39**, 87 (1962).

[47] E. G. Eeles, *Rev. Sci. Instrum.*, **28**, 1096 (1957).

[48] A. Taylor, *J. Sci. Instrum.*, **29**, 236 (1952).

[49] J. D. Wilkinson and L. D. Calvert, *J. Sci. Instrum.*, **37**, 399 (1960).

[50] D. A. Northrop, *Rev. Sci. Instrum.*, **31**, 1160 (1960).

[51] A . W. Hull, *Phys. Rev.*, **9**, 84, 564 (1917); **10**, 661 (1917).

[52]   W. P. Davey, *A Study of Crystal Structure and Its Applications*, McGraw-Hill, New York, 1934, p. 118.

[53]   E. M. Pogainis and E. H. Shaw, Jr., *Rev. Sci. Instrum.*, **25**, 524 (1954).

[54]   G. Becherer and G. Herms, *J. Phys.*, **E2**, 1130 (1969).

[55]   Commission on Crystallographic Apparatus of the International Union of Crystallography, *Acta Crystallogr.*, **9**, 520 (1956).

[56]   H. Morimoto and R. Uyeda, *Acta Crystallogr.*, **16**, 1107 (1963).

[57]   H. Seemann, *Ann. Phys.*, **59**, 455 (1919).

[58]   H. Bohlin, *Ann. Phys.*, **61**, 421 (1920).

[59]   J. C. M. Brentano, *Arch. sci. phys. nat.*, (4)**44**, 66 (1917); *Proc. Phys. Soc.* (London), **37**, 184 (1925); *J. Appl. Phys.*, **17**, 420 (1946).

[60]   A. F. Westgren, *Trans. AIME*, **93**, 13 (1931).

[61]   E. R. Jette and F. Foote, *J. Chem. Phys.*, **3**, 605 (1935).

[62]   Y. Cauchois, *C.R. Acad. Sci., Paris*, **195**, 298 (1932).

[63]   A. Guinier, *C.R. Acad. Sci., Paris*, **204**, 1115 (1937); *Ann. Phys.*, (Paris), **12**, 161 (1939).

[64]   P. M. de Wolff, *Acta Crystallogr.*, **1**, 207 (1948).

[65]   E-G. Hofmann and H. Jagodzinski, *Z. Metallk.*, **46**, 601 (1955) .

[66]   E. Hellner, *Z. Kristallogr.*, **106**, 122 (1954).

[67]   W. H. Sas and P. M. de Wolff, *Acta Crystallogr.*, **21**, 826 (1966).

[68]   H. F. Fischmeister, *Acta Crystallogr.*, **14**, 113 (1961).

[69]   L. Zsoldos, *J. Sci. Instrum.*, **38**, 410 (1961).

[70]   G. Hägg and N-O. Ersson, *Acta Crystallogr.*, **A25**, S64 (1969).

[71]   D. J. Fisher, *Z. Kristallogr.*, **109**, 73 (1957).

[72]   M. L. Polk, *Rev. Sci. Instrum.*, **40**, 1645 (1969).

[73]   D. E. C. Corbridge and F. R. Tromans, *Anal. Chem.*, **30**, 1101 (1958).

[74]   G. Sachs and J. Weerts, *Z. Phys.*, **60**, 481 (1930); G. Sachs, *Trans. AIME*, **93**, 39 (1931); F. Regler, *Z. Phys.*, **71**, 371 (1931).

[75a]   D. E. Thomas, *J. Sci. Instrum.*, **18**, 135 (1941).

[75b]   D. E. Thomas, *J. Appl. Phys.*, **19**, 190 (1948); *The Measurement of Stress and Strain in Solids*, Institute of Physics, London, 1948, pp. 75–76.

[76]   J. A. Bennett and H. C. Vacher, *J. Res. Nat. Bur. Stand.*, **40**, 285 (1948).

[77]   C. Wainwright, *J. Sci. Instrum.*, **19**, 165 (1942).

[78]   H. G. Smith, *Rev. Sci. Instrum.*, **33**, 128 (1962).

[79]   H. K. Herglotz, *Rev. Sci. Instrum.*, **34**, 708 (1963); **39**, 1658 (1968).

[80]   A. E. Spakowski, *Rev. Sci. Instrum.*, **34**, 930 (1963).

[81]   P. H. Schmidt and E. G. Spencer, *Rev. Sci. Instrum.*, **35**, 957 (1964).

[82]   J. H. Dumbleton and B. B. Bowles, *Rev. Sci. Instrum.*, **37**, 1613 (1966).

[83]   E. T. Peters and S. A. Kulin, *Rev. Sci. Instrum.*, **37**, 1726 (1966).

[84]   F. G. Chesley, *Rev. Sci. Instrum.*, **18**, 422 (1947).

[85]   M. Oron and I. Minkoff, *J. Sci. Instrum.*, **42**, 337 (1965).

[86]   J-E. Glas, *J. Sci. Instrum.*, **39**, 60 (1962).

[87]   A. R. Clawson, *Rev. Sci. Instrum.*, **39**, 597 (1968).

[88]   S. Fried and N. Davidson, *J. Am. Chem. Soc.*, **70**, 3539 (1948).

[89]   H. P. Rooksby, *Analyst*, **73**, 326 (1948).

[90]   P. B. Hirsch and J. N. Keller, *Proc. Phys. Soc.* (London), **B64**, 369 (1951).

[91]   R. W. Cahn, *J. Sci. Instrum.*, **30**, 201 (1953).

[92]   D. Lewis, *J. Sci. Instrum.*, **32**, 467 (1955).

[93]   P. J. Holmes and P. D. Fochs, *J. Sci. Instrum.*, **33**, 239 (1956).

[94]   H. M. Otte and R. W. Cahn, *J. Sci. Instrum.*, **36**, 463 (1959).

[95]   P. S. Rudman and B. Weiss, *Rev. Sci. Instrum.*, **30**, 1129 (1959).

[96] L. C. Michels and C. M. Wayman, *Rev. Sci. Instrum.*, **33**, 572 (1962).
[97] W. Taylor and A. Moore, *J. Sci. Instrum.*, **40**, 46 (1963).
[98] F. J. ter Avest and J. A. Klostermann, *J. Phys.*, **E2**, 950 (1969).
[99] J. E. Runyan, C. E. Olsen, and C. P. Kempter, *Norelco Rep.*, **17**, 16 (1970).
[100] R. D. Arnell, *J. Sci. Instrum.*, **44**, 67 (1967).
[101] M. Straumanis, A. Ievinš, and K. Karlsons, *Z. anorg. allg. Chem.*, **238**, 175 (1938).
[102] M. E. Straumanis, *J. Appl. Phys.*, **20**, 726 (1949).
[103] V. Vand, *J. Appl. Phys.*, **19**, 852 (1948).
[104] L. K. Frevel, *Rev. Sci. Instrum.*, **8**, 475 (1937).
[105] N. W. Taylor, *Rev. Sci. Instrum.*, **2**, 751 (1931).
[106] M. J. Buerger, N. W. Buerger, and F. G. Chesley, *Am. Mineral.*, **28**, 285 (1943).
[107] F. G. Chesley, *Rev. Sci. Instrum.*, **17**, 558 (1946); L. F. Connell, *Rev. Sci. Instrum.*, **18**, 367 (1947).
[108] J. J. Hanak and A. H. Daane, *Rev. Sci. Instrum.*, **32**, 712 (1961).
[109] W. Hume-Rothery and P. W. Reynolds, *Proc. Roy. Soc.* (London), **167A**, 25 (1938).
[110] A. J. C. Wilson, *Proc. Phys. Soc.* (London), **53**, 235 (1941).
[111] E. A. Owen, *J. Sci. Instrum.*, **20**, 190 (1943).
[112] E. Matuyama, *J. Sci. Instrum.*, **32**, 229 (1955).
[113] E. Aruja, J. H. Welch, and W. Gutt, *J. Sci. Instrum.*, **36**, 16 (1959).
[114] W. Gutt, *J. Sci. Instrum.*, **41**, 393 (1964).
[115] A. E. Austin, N. A. Richard, and C. M. Schwarz, *Rev. Sci. Instrum.*, **27**, 860 (1956).
[116] J. W. Edwards, R. Speiser, and H. L. Johnston, *Rev. Sci. Instrum.*, **20**, 343 (1949).
[117] A. Revcolevschi, J. Hubert, and R. Collongues, *C. R. Acad. Sci., Ser. C*, **269**, 265 (1969).
[118] G. Farrow and D. Preston, *J. Sci. Instrum.*, **37**, 305 (1960).
[119] S. S. Marsden and N. R. Sanjana, *J. Sci. Instrum.*, **30**, 427 (1953).
[120] J. Fridrichsons, *Rev. Sci. Instrum.*, **27**, 1015 (1956).
[121] W. L. Bond, *Rev. Sci. Instrum.*, **29**, 654 (1958).
[122] B. A. Hatt, P. J. C. Kent, and G. I. Williams, *J. Sci. Instrum.*, **37**, 273 (1960).
[123] J. Intrater and D. K. Smith, *Rev. Sci. Instrum.*, **39**, 1491 (1968).
[124] V. V. Zubenko, S. S. Kvitka, and M. M. Umanskii, *Soviet Phys.—Crystallogr.*, **4**, 222 (1959).
[125] Z. S. Basinski, W. B. Pearson, and J. W. Christian, *J. Sci. Instrum.*, **29**, 154 (1952).
[126] R. S. Pease, *J. Sci. Instrum.*, **32**, 476 (1955).
[127] J. A. Brand and H. J. Goldschmidt, *J. Sci. Instrum.*, **33**, 41 (1956).
[128] W. Johnson, *J. Sci. Instrum.*, **38**, 373 (1961).
[129] J. D. Wilkinson and L. D. Calvert, *Rev. Sci. Instrum.*, **34**, 545 (1963).
[130] W. Ostertag and G. R. Fischer, *Rev. Sci. Instrum.*, **39**, 888 (1968).
[131] W. T. Gray and D. I. Finch, *Phys. Today*, **24**, No. 9, 32 (1971).
[132] H. J. Goldschmidt, *High-Temperature X-ray diffraction Techniques*, Bibliography 1, International Union of Crystallography Commission on Crystallographic Apparatus, 1964. Orders can be placed with Polycrystal Book Service (P.O. Box 11567, Pittsburgh, Pennsylvania) or with any book seller. Price $3.00.
[133a] K. Lonsdale and H. Smith, *J. Sci. Instrum.*, **18**, 133 (1941).
[133b] N. C. Tombs, *J. Sci. Instrum.*, **29**, 364 (1952).
[134] E. Pohland, *Z. phys. Chem.*, **B26**, 238 (1934).
[135] W. H. Barnes and W. F. Hampton, *Rev. Sci. Instrum.*, **6**, 342 (1935).
[136] H. S. Kaufman and I. Fankuchen, *Rev. Sci. Instrum.*, **20**, 733 (1949).
[137] S. C. Abrahams, R. L. Collin, W. N. Lipscomb, and T. B. Reed, *Rev. Sci. Instrum.*, **21**, 396 (1950).
[138] B. Post, R. S. Schwartz, and I. Fankuchen, *Rev. Sci. Instrum.*, **22**, 218 (1951).

[139] E. A. Wood, *Rev. Sci. Instrum.*, **24**, 325 (1952).

[140] W. Hume-Rothery and D. J. Strawbridge, *J. Sci. Instrum.*, **24**, 89 (1947).

[141] J. Thewlis and A. R. Davey, *J. Sci. Instrum.*, **32**, 79 (1955).

[142] M. H. Francombe, *J. Sci. Instrum.*, **34**, 35 (1957).

[143] W. E. Kramer, A. S. Venturino, and R. Mazelsky, *Rev. Sci. Instrum.*, **34**, 933 (1963).

[144] P. B. Crandall, *Rev. Sci. Instrum.*, **40**, 954 (1969).

[145] E. A. Owen and G. I. Williams, *J. Sci. Instrum.*, **31**, 49 (1954).

[146] Y. Mascarenhas and S. Mascarenhas, *Rev. Sci. Instrum.*, **38**, 141 (1967).

[147] B. Post, *Low-Temperature X-ray Diffraction*, Bibliography 2, International Union of Crystallography Commission on Crystallographic Apparatus, 1964. Orders can be placed with Polycrystal Book Service (P.O. Box 11567, Pittsburgh, Pennsylvania) or with any book seller. Price $3.00.

[148] *Handbook of Chemistry and Physics*, 50th ed., Chemical Rubber Company, Cleveland, Ohio, 1969.

[149] B. F. Figgins and B. L. Smith, *Phil. Mag.*, **5**, 186 (1960).

[150] L. H. Bolz and F. A. Mauer, *Proceedings of the Eleventh Annual Conference on Application of X-ray Analysis*, **6**, 242 (1962).

[151] D. R. Sears and H. P. Klug, *J. Chem. Phys.*, **37**, 3002 (1962).

[152] P. W. Bridgman, *The Physics of High Pressure*, Bell, London, 1952. Chapter VIII.

[153] B. Kamb, *Trans. Am. Crystallogr. Assoc.*, **5**, 61 (1969).

[154] P. Y. Simons and F. Dachille, *Acta Crystallogr.*, **23**, 334 (1967).

[155] W. M. Cohn, *Proc. Am. Phys. Soc.*, Abstract 63 of the Chicago meeting, June 1933.

[156] L. K. Frevel, *Rev. Sci. Instrum.*, **6**, 214 (1935).

[157] R. B. Jacobs, *Phys. Rev.*, **54**, 325 (1938).

[158] E. J. Goon, J. T. Mason, and T. R. P. Gibb, Jr., *Rev. Sci. Instrum.*, **28**, 342 (1957).

[159] J. C. Jamieson and A. W. Lawson, *J. Appl. Phys.*, **33**, 776 (1962).

[160] D. B. McWhan and W. L. Bond, *Rev. Sci. Instrum.*, **35**, 626 (1964).

[161] E. A. Perez-Albuerne, K. F. Forsgren, and H. G. Drickhamer, *Rev. Sci. Instrum.*, **35**, 29 (1964).

[162] S. Saito and Y. Ozaki, *Jap. J. Appl. Phys.*, **8**, 797 (1969).

[163] D. L. Decker, *J. Appl. Phys.*, **42**, 3239 (1971).

[164] G. J. Piermarini and C. E. Weir, *J. Res. Nat. Bur. Stand.*, **66A**, 325 (1962).

[165] W. A. Bassett, T. Takahashi, and P. W. Stook, *Rev. Sci. Instrum.*, **38**, 37 (1967).

[166] D. B. McWhan, *Trans. Am. Crystallogr. Assoc.*, **5**, 39 (1969).

[167] E. R. Lippincott and H. C. Duecker, *Science*, **144**, 1119 (1964).

[168] J. D. Barnett and H. T. Hall, *Rev. Sci. Instrum.*, **35**, 175 (1964).

[169] P. J. Freud and C. B. Sclar, *Rev. Sci. Instrum.*, **40**, 434 (1969).

[170] P. J. Freud and P. N. LaMori, *Trans. Am. Crystallogr. Assoc.*, **5**, 155 (1969).

[171] Q. Johnson, A. Mitchell, and L. Evans, *Nature*, **231**, 310 (1971).

[172] G. Gandolfi, *Mineral Petrogr. Acta*, **13**, 67 (1967); but see also E. J. Graeber and D. A. Jelinek, *Norelco Rep.*, **13**, 91 (1966).

[173] W. Class, A. Iannucci, and H. Nesor, *Norelco Rep.*, **13**, 87 (1966).

[174] L. E. Alexander, *X-ray Diffraction Methods in Polymer Science*, Wiley, New York, 1969.

[175] C. Weir, S. Block, and G. Piermarini, *J. Res. Nat. Bur. Stand.*, **69C**, 275 (1965).

[176] Private communication to L. E. Alexander from S. Block.

CHAPTER 5

DIFFRACTOMETRIC POWDER TECHNIQUE

The x-ray powder diffractometer is an outgrowth of the Bragg ionization spectrometer[1], which was applied as early as 1913 to the measurement of reflections from single crystals. Unlike the ionization spectrometer, which disperses a spectrum of x-ray wavelengths by means of a crystal grating of some fixed spacing d_{hkl}, the diffractometer is designed to disperse x-rays of a single wavelength by diffracting them from planes of different spacings. Furthermore, an x-ray powder diffractometer is characterized by the use of a local intensity-measuring receiver (quantum counter) rather than a photographic film, and a parafocusing arrangement is usually employed to increase the intensity and resolution.

Because of its small sensitivity, the ionization spectrometer had but very limited success in measuring the relatively feeble reflections from powders[2]. Geiger and Müller first described the use of a Geiger-Müller counter to detect x-rays[3]. More refined experiments of this kind were reported by Huppertsberg[4]. Later, LeGalley[5] devised an instrument incorporating many of the basic features of the x-ray counter diffractometer. The principal weakness of his apparatus was its lack of parafocusing geometry, which resulted in a weak signal. The first counter diffractometer for powder studies was developed at the United States Naval Research Laboratory by Friedman[6], and later engineered and marketed widely by the North American Philips Company. This instrument was in part an outgrowth of an earlier Geiger-counter apparatus employed during World War II for the cutting of quartz oscillator plates[7]. The subsequent development of the diffractometer into the precision instrument of today may be largely credited to Parrish and coworkers[8].

5-1 GEOMETRY OF THE POWDER DIFFRACTOMETER

5-1.1 General Features

Figures 5-1 and 5-2 illustrate how Seemann-Bohlin parafocusing geometry (see Section 4-2) is modified for use in the so-called Bragg-Brentano x-ray powder diffractometer. In the Seemann-Bohlin arrangement (Fig. 5-1), the curved sample S produces reflections from planes of spacings d_1, d_2, and d_3, which focus, respectively, at G_1, G_2, and G_3 on the photographic film. The radius of the focusing circle is constant, but the distances SG_1, SG_2, SG_3, and so on, for the various reflections differ. In the parafocusing diffractometer (Fig. 5-2), however, the receiver G pivots about the sample S, thus maintaining a constant sample-to-receiver distance SG. In order to accomplish this end, the sample surface is made flat rather than curved, and it is caused to rotate with one-half the angular velocity with which the receiver revolves ($2:1$ following, or $2\theta:\theta$ motion), so that the sample surface remains tangent to the focusing circle at all times. As G pivots about S toward larger angles, 2θ, the radius of the focusing circle decreases. Thus, as shown in Fig. 5-2, the focusing circle has the respective radii r_1, r_2, and r_3 for reflections at the deviation angles $2\theta_1$, $2\theta_2$, and $2\theta_3$. For $2\theta = 0°$, $r = \infty$, whereas at $2\theta = 180°$, r reaches a minimum value of $SF/2 = SG/2$.

At this point it is appropriate to denote the plane of Figs. 5-1, 5-2, and 5-3 as the *equatorial* or *focusing* plane, and the perpendicular direction,

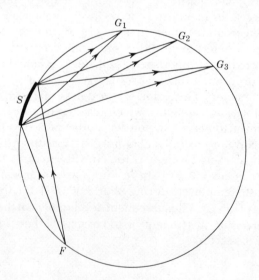

Fig. 5-1.   Seemann-Bohlin parafocusing arrangement.

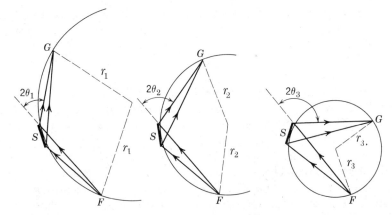

Fig. 5-2. Parafocusing arrangement used in the powder diffractometer.

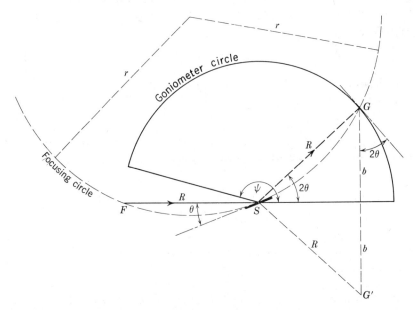

Fig. 5-3. Two-dimensional features of the x-ray diffractometer.

which is parallel to the goniometer axis, as the *axial direction.*[*] The focus at G (Fig. 5-2) is not perfect for several reasons. In particular, parafocusing errors result from the use of a flat rather than a curved sample, from sample "transparency" (low absorption coefficient), and from the

*In the earlier literature the equatorial plane and axial direction were commonly designated the *horizontal* plane and the *vertical* direction, respectively.

third-dimensional properties of the system (finite dimensions of the sample, source, slits, etc., in the axial direction). However, by limiting the equatorial divergence of the primary beam, and by restricting the axial dimensions of the x-ray optical system in various respects, the aberrations can be kept small.

Figure 5-3 depicts in greater detail some of the idealized two-dimensional features of the parafocusing arrangement utilized in a standard powder diffractometer. The flat sample S is tangent to the focusing circle of radius r. The sample-to-source distance FS and sample-to-receiver distance SG are the same and equal to R, the radius of the goniometer circle. The triangle SGG' is an axial section through a diffraction cone of semiapex angle 2θ diverging from S; evidently, this triangle represents the intersection of this cone with the plane of the goniometer circle. The diffracted rays come to an approximate focus at G, the position of the receiving slit, and diverge again as they enter the counter window. An inspection of Fig. 5-3 shows r to possess the angular dependence

$$r = \frac{R}{2 \sin \theta}. \tag{5-1}$$

The base of the diffracted cone intercepted by the receiving slit is a circle of radius

$$b = R \sin 2\theta. \tag{5-2}$$

The angle ψ represents the limiting range of the goniometer arc, which is about 165° in most instruments.

5-1.2 Details of the Optical Arrangement

Figure 5-4A is a schematic perspective of the optical arrangement that was employed in early-model diffractometers and which still finds occasional application. The focal spot F is viewed longitudinally, as is the common practice in Debye-Scherrer photographic techniques when employing a pinhole collimator (see Section 4-1.2), but the angle of view (take-off angle) is made as small as 1.5 or 2.0° to improve the resolving power. The apparent source width is then $h \sin \alpha$. The equatorial (γ) and axial (δ) divergences of the primary beam are limited by the dimensions x and y of the direct-beam aperture X to respective values of about 1.5 and 3.0°, defined by the geometry of Fig. 5-4A. The portion of the specimen surface illuminated by the beam consists of a zone u of maximum intensity and two flanking zones p of partial intensity. Because of the appreciable projected width of the x-ray source, as well as the considerable axial divergence, the focus at the receiving aperture G is not very sharp. Nevertheless, the high intensity obtainable with this geometry renders it of some current usefulness on occasions when ultrasharp focusing is not required.

Figure 5-4B shows the optical properties of a typical present-day diffractometer. The focal spot is viewed laterally (see Section 4-1.2), so that the apparent source width is only $w \sin \alpha$, where α is normally 3 to 6°. An x-ray source of such narrow projected width is often referred to as a line source. The equatorial divergence is limited by the dimension x of the aperture X, whereas the axial divergence, instead of being prescribed by the heights h of the source and y of the divergence slit, is more drastically limited by employing two Soller slit collimators[9] s_1 and s_2 for the incident and diffracted beams, respectively. Such a collimator consists of a number of closely spaced, parallel, and highly absorbing metal (e.g., molybdenum or tantalum) plates which have the effect of dividing the x-ray beam into a number of parallel slices, each with very restricted axial divergence. In this way the radiation from an extended line source can be utilized without the generation of serious aberrations in the focus due to axial divergence, and greatly improved resolving power is achieved at the price of a substantial, but not prohibitive, intensity loss of the order of 50 per cent. In a representative commercial diffractometer, the Soller slit collimators consist of plates of length $l = 12.7$ mm and spacing $s = 0.5$ mm, stacked to a height of 10 mm in the axial dimension. The basic angular aperture of any pair of adjacent plates may then be defined as

$$\Delta = \tan^{-1} \frac{s}{l} \qquad (5\text{-}3)$$

(refer to Fig. 5-4B, right insert).* For $s = 0.5$ mm and $l = 12.7$ mm, $\Delta = 2.25°$. The design of Soller slits has been discussed in detail by Parrish[10].

The background is improved by placing an antiscatter slit at either position M or M' (Fig. 5-4), which excludes from the receiver all x-rays except those emanating from the specimen. Although the slits and Soller plates can be constructed of brass or nickel without serious consequences, it is preferable to make use of some metal of high atomic number in order to minimize the possibility of K fluorescence, which tends to increase the background level. Figures 5-5A and B show two counter goniometers, the one in B being mounted with its axis horizontal

---

*In some of the literature, for example, in articles by Parrish and associates, the divergence of a Soller collimator is specified as $2 \tan^{-1}(s/l)$, leading to an angle of 4.5° rather than 2.25° in the example cited. This represents the limiting angular "cross-fire" permitted by the plate length and spacing.

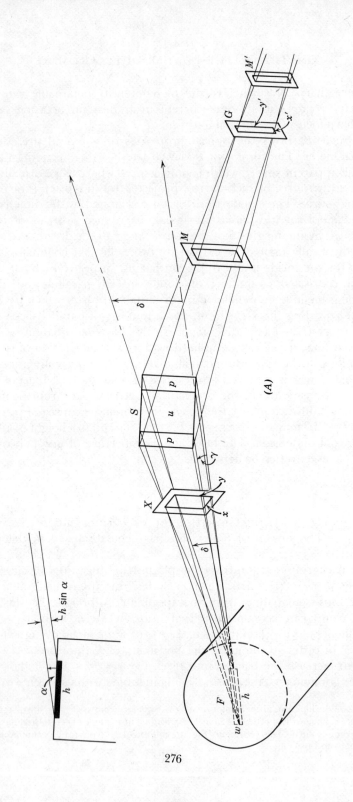

(A)

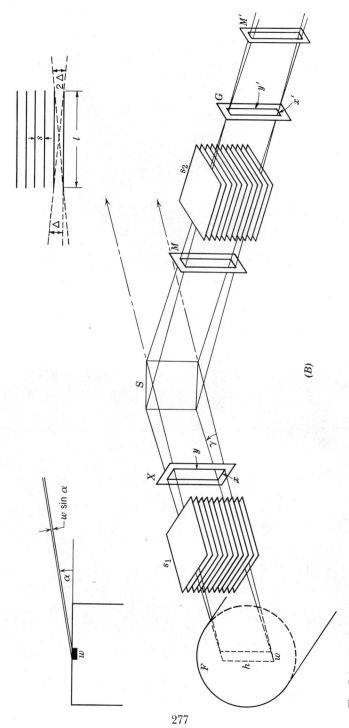

Fig. 5-4. (A) Optical arrangement used in early x-ray diffractometer. (B) Present-day optical arrangement with a "line" x-ray source and Soller slits to limit axial divergence.

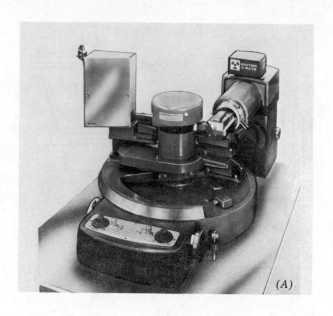

(A)

(B)

Fig. 5-5. (*A* and *B*). Close-ups of two modern counter goniometers. (Courtesy of Diano Corporation, *A*, and Philips Electronic Instruments, *B*).

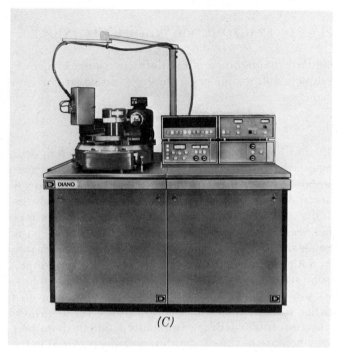

(C)

(D)

Fig. 5-5. (C and D) Complete diffractometers including x-ray generators and counting components. (Courtesy of Diano Corporation, C, and Philips Electronic Instruments, D.)

on a standard photographic x-ray diffraction unit. Figures 5-5C and D show complete diffractometers, including x-ray generators and detector-circuit panels.

5-1.3 The Seemann-Bohlin Diffractometer

Since the mid-1960s renewed interest has been shown in designing diffractometers that utilize true Seemann-Bohlin parafocusing geometry (see Section 4-2), which comprises a focusing circle of fixed radius, a stationary specimen with curved surface coincident with the focusing circle, and a variable specimen-to-detector distance. These investigations show that Seemann-Bohlin geometry has certain important advantages for 2θ larger than about 40°, including (1) higher intensity without sacrifice of resolution, (2) simplification of mechanical design permitted by the use of a stationary specimen, and (3) feasibility of employing several detectors simultaneously for investigating phase changes in specimens as a function of time when subjected to various treatments. Below about 40° 2θ the conventional Bragg-Brentano geometry is to be preferred, principally because of the extraordinary pains required in the preparation of a specimen sufficiently free of surface roughness for a Seemann-Bohlin diffractometer. This requirement becomes increasingly critical as the Bragg angle decreases toward 0°. Other general disadvantages of Seemann-Bohlin geometry are the somewhat limited accessible angular range (in particular, the exclusion of the 0° 2θ position) and the consequent necessity of resorting to standard substances for the calibration of the 2θ scale. Space limitations preclude the treatment of this topic in greater detail, for which the reader is referred to selected literature [11–14a].

5-1.4 Alignment and Angular Calibration of the Diffractometer

The manuals issued by manufacturers contain detailed instructions for the alignment and adjustment of their respective instruments. Because of the very considerable variation in design and engineering, the details of the recommended alignment procedures differ from one to another. Moreover, commercial diffractometers are customarily supplied with special jigs or templates for making various adjustments, and these are peculiar to each instrument. Thus it is not possible to present a universally applicable set of alignment instructions. However, it is worthwhile to describe in somewhat general terms the features of alignment that must be accomplished in order that optimum geometrical efficiency may be realized.

Correct alignment is essential to the attainment of (a) optimum resolution, (b) maximum intensity consistent with this degree of resolution, and (c) correct angular readings. Several essential operations comprise the overall alignment process. Although there is considerable flexibility in the order in which these steps may be performed, the following is a suggestive sequence. The reader may refer to Fig. 5-6.

A. Operations Appropriately Performed in Advance by the Manufacturer.

1. The specimen-positioning holder is adjusted so that its plane surface precisely contains the goniometer axis aa'.

2. The Soller slits s_1 and s_2 are mounted with their flat foils parallel to the equatorial plane of the goniometer and their long axes parallel to the median line cc' when the goniometer is set to read $0°$.

B. Further Internal Alignment of the Goniometer.
In some of the following steps it is necessary to view the direct x-ray beam with a fluorescent screen or to measure its intensity with the radiation detector. To protect the detector from damage, it must be shielded from the direct beam with a metal-absorbing sheet (nickel, brass, etc.) of the necessary thickness. Of course, it is also essential that the investigator at all times avoid exposure to direct or strongly scattered x-rays.

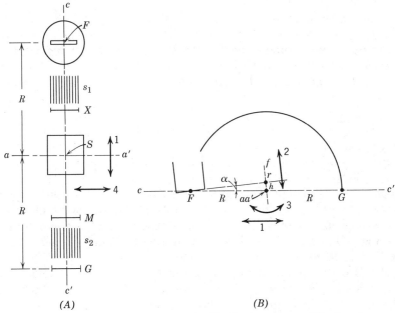

(A)   (B)

Fig. 5-6.   The principal motions required in aligning an x-ray diffractometer.

3. With the goniometer set at $0°$, the surface of the specimen S and the midpoints of the collimating slits X, G, s_1, and s_2 must lie on the median line cc'. In some instruments the antiscatter slit M is located between the Soller slit s_2 and the sample (as shown in Fig. 5-6), and in others between G and the detector (M' in Fig. 5-4).

It is very important that the plane surface of the specimen holder be accurately adjusted so as to contain both the axis aa' and the median line cc' when the goniometer is set at $0°$. Given that the specimen-holder surface has been prealigned to contain the axis aa', the cc' alignment is achieved by disengaging the θ drive of the specimen post and rotating it until the surface of the specimen holder precisely bisects the uncollimated beam. Manufacturers commonly furnish a special collimating slit or bar to facilitate this adjustment. This static $2\theta\!:\!\theta$ setting can be rechecked and finalized after the alignment of the goniometer with respect to the x-ray tube has been completed (see step 11).

4. With the foregoing conditions maintained, it must be made certain that the long axes of the slits X, G, and M are parallel to aa'. Most manufacturers prealign the collimating system in the factory; nevertheless, the user must check the alignment for possible need of refinement.

C. Alignment of the Goniometer with Respect to the X-ray Tube.

5. The entire goniometer must be moved toward or away from the x-ray tube (Fig. 5-6, motion 1), or vice versa, until the distance from the center of the x-ray tube focus F to the axis aa' equals the desired goniometer radius R. Without special templates it is difficult to adjust this distance to an accuracy much better than ± 1 mm, although this setting is not critical.

6. The goniometer as a unit must then be rotated about F in the equatorial plane until the axis aa' "views" the focal spot at the desired take-off angle α. Some diffractometers are provided with special facilities for accurately positioning the focal spot directly over the pivot about which α is varied. A reasonably satisfactory approximation to this rotation about F is a translation of the goniometer in the tangential direction 2. Either means of setting α can be accomplished with the aid of a fluorescent screen in a manner similar to that described in Section 4-1.2 for aligning a Debye-Scherrer camera. With the slits X and s_1 removed and the screen placed in the specimen holder and oriented parallel to the long axis of the x-ray tube (position f in Fig. 5-6B), the goniometer is rotated about F or translated in direction 2 until the axis aa' is a distance h from r, the diffuse edge of the beam produced by the shadow of the target surface (or the x-ray tube may be moved with respect to the

goniometer in a reciprocal manner). If a take-off angle of 3° is desired and R is 17 cm, for example, the correct value of h is given by

$$h = R\alpha = 17 \times 0.0523 = 0.89 \text{ cm.}$$

Two sharp parallel lines drawn 0.89 cm apart on the fluorescent screen are an aid in achieving the right position. For appreciably higher intensity with only slight sacrifice in resolution, α may be set at 6°.

7. The entire goniometer is next rotated about the axis aa' (motion 3) and, if necessary, translated in direction 4 until the midpoint of the focal spot also lies on the line cc'. This may also be accomplished by an equivalent reciprocal motion of the x-ray tube.

8. While condition 7 is maintained, it may now be necessary to rotate the goniometer slightly about the beam axis cc' to bring the long axis of the focal spot more precisely into parallelism with the goniometer axis aa' and the slits X and G. Some instruments permit this orientation to be performed more simply by rotation of the x-ray tube about the axis cc'.

The parallelism of the x-ray line focus and axis aa' can be checked by moving the detector slightly off the $0°2\theta$ position, inserting a strongly absorbing metal filter, and measuring the intensities received with the upper and lower halves of the slit G alternately blocked off. Any tilt of the focus relative to aa' produces a difference in these two measures of the primary beam intensity, and also indicates the direction in which a corrective tilt is required to equalize the two measurements [15].

9. The receiving-slit assembly, as well as the detector if these components are integrally mounted, is then translated toward or away from the axis aa' until SG equals SF.

10. It must be ascertained that the antiscatter slit M is properly centered on the line cc' (with goniometer set at 0°) and parallel to slits X and G. A satisfactory empirical test is to scan over a diffraction line and observe whether or not the peak intensity is diminished when M is inserted. There should be no significant reduction in intensity.

11. Similarly, final empirical tests of the correctness of the 2 : 1 following of detector and specimen established in step 3 and of the equality of distances SG and SF (step 9) can be performed by scanning selected diffraction lines at different angles and noting whether or not small refinements of the relevant alignment motions result in increased peak intensity.

If the foregoing sequence of operations has been precisely executed, very nearly optimum resolution and intensity will be obtained. In practice, however, it is usually necessary to repeat the entire procedure, or at least the more critical steps, once or twice. A test of the resolving power of the diffractometer is to note the smallest angle at which the $K\alpha$ doublet is just resolved when employing $CuK\alpha$ radiation, a take-off angle between 3 and 6°, a receiving slit of about 0.02° width, and an x-ray slit limiting the equatorial divergence to 1°. When strain-free powders prepared from well-crystallized substances are examined under these conditions, resolution of the α_1 and α_2 peaks should be observed to begin at about 30 or 35° 2θ. In making this test it is important that the $K\alpha$ doublet be step-scanned instrumentally (see Section 5-4.5), or scanned manually by counting the intensities at regular angular intervals, or continuously scanned at a very slow speed such as $\frac{1}{8}$°/min with an electronic time constant (RC) not exceeding 4 sec (Section 5-4.4). At larger values of the scanning speed and time constant, the inherent resolving power of the goniometer and counting circuits is impaired.

With copper radiation the appearance of the $K\alpha_1\alpha_2$ reflection of tungsten (wolfram) at 40° can be used as a criterion of the resolution to be anticipated. The specimen should be either strain-free powder passed through a 325-mesh screen or a piece of metal the plane surface of which has been etched to remove lattice strains. Figure 5-7 shows the extent to

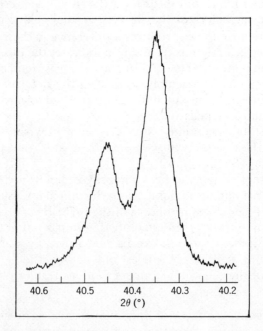

Fig. 5-7.   Slow trace of the 110 $K\alpha$ doublet of tungsten with diffractometer adjusted for high resolution. $CuK\alpha$ radiation, pulse-integrating circuit with linear response.

which the doublet should be resolved. The minimum intensity between the two peaks should drop to a level not much greater than one-half the intensity of the α_2 peak. If this is not the case, step 9 should be repeated with great care, and steps 3 and 4 may possibly require closer attention. If the intensity is definitely below par, the most critical adjustments to be rechecked are those of steps 2, 4 and 8. Furthermore, the intensity will be unnecessarily low if the target take-off angle α is less than 3° (step 6).

D. Calibration of the 0° 2θ Position. The usual practice is to determine the 0° 2θ angle with great care, and then proceed on the assumption that the gear train has been manufactured with the requisite accuracy to provide goniometer readings between 0 and 180° that can be relied on to within ±0.005°. Two methods of locating the 0° angle may be used.

In the method of Parrish and Lowitzsch[16] a flat plate containing a fine pinhole approximately 0.008 in. in diameter is placed in the specimen holder, the specimen post is disengaged from its θ control, and the plate is set at right angles to the direct beam axis cc' with the pinhole very nearly, but not precisely, coincident with the axis aa'. With a narrow receiving slit, the profile of the direct beam is measured with good statistical accuracy, a metal filter of sufficient thickness being used to prevent the maximum counting rate from exceeding 4000 cps. The plate is next rotated through 180°, and the direct-beam profile measured again. Parrish and Lowitzsch employed the following experimental conditions: copper-target tube, full-wave rectification, 15 kV peak, 7 mA, $\alpha = 6°$, 0.0025-in. nickel filter, 1° divergence slit and 0.006-in. receiving slit (goniometer radius 17 cm), scintillation counter. Figure 5-8 shows the intensity plot obtained. The two profiles of the focal spot are symmetrical, and the vertical median line (at a goniometer reading of 0.872° in the illustration) is easily found and can serve to locate the true 0° position to a precision of ±0.001°.

King and Vassamillet[17] have very successfully employed a double-scanning method, which requires a diffractometer capable of measuring reflections from the specimen on both positive and negative sides of the direct beam up to at least moderate angles, for example, 60 to 90°. Before the measurements in the negative-2θ range are made, the specimen must be rotated through 180°, and then the 2θ:θ relationship reestablished. The flat powder specimen used by King and Vassamillet was prepared from filings of 99.999 per cent silver annealed at 600°C and passed through a 325-mesh screen. The filings were dusted evenly onto a Mylar film stretched over the opening in a specimen mount, and 1 or 2 drops of collodion were added. When the specimen was dry, the Mylar film was

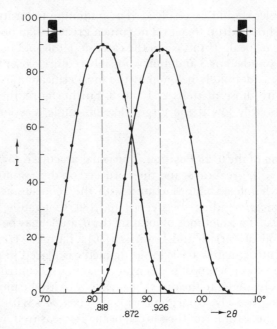

Fig. 5-8.   Plot of data to determine 0° angle using the pinhole method of Parrish and Lowitzsch. (*Am. Mineral.*, **44**, 765 (1959), Fig. 5.)

stripped away, leaving a specimen with a flat surface flush with that of the specimen mount.

Figure 5-9 shows diffraction profiles of the 111 silver reflection scanned toward increasing angles on both sides (+ and −) of the direct beam. Each reflection was double-scanned (in both clockwise and counterclockwise directions), and the mean peak position ascertained to an accuracy of $0.005°2\theta$ or better. By means of this technique the effects of dynamic scanning errors and any backlash in the gear train are eliminated. Figure 5-9A shows the excellent agreement between the positive and negative 2θ positions of the 111 reflection after the position of the gonimeter had been refined by a small translation in the equatorial plane perpendicular to the median line cc' (nearly parallel to movement 2, Fig. 5-6) so as to bring about coincidence of the + and − profiles (alignment 6). Figure 5-9B shows the relative positions of the + and − profiles after the diffractometer was deliberately misaligned by about $0.02°2\theta$ (motion 2). By averaging the peak positions of eight silver reflections below 90° measured as described above, the mean zero correction for alignment A was found to be $-0.00_{03}°$ and for B $-0.01_{25}°$. The foregoing procedure is equivalent to measuring 4θ in powder camera calibrations. It has been termed the Cornu method by Neff[18].

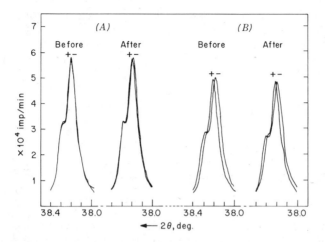

Fig. 5-9. Diffraction patterns of the silver 111 reflection scanned toward increasing angles on both sides of the direct beam. Recorded for alignment conditions 1 and 2 before and after scanning experiments to determine the lattice constant a. (Courtesy of H. W. King and L. F. Vassamillet, *Advances in X-Ray Analysis*, Vol. 5, Plenum Press, 1962, p. 78.)

E. Calibration of Angular Registration between 0 and 180° 2θ. If there is cause for concern relative to the precision of the gear train and the consequent accuracy of registration of angles on the gonimeter between 0 and 180° 2θ, the angular accuracy throughout the range can be optically checked to within 0.001° by the use of a precision-constructed reflecting polygon [19–21]. With the aid of such a regular 12-sided optical polygon and an autocollimating system, the angular accuracies of several kinds of commercial diffractometers in the Pittsburgh area were determined [21], each being studied in its as-currently-used condition. The form of the angular error was found to be a sinusoidal function of 2θ with a period of 360° (see Fig. 5-10 for representative curves) and maximum deviations of the order of 0.019 to 0.045° 2θ. Vassamillet and King point out that the effective magnitude and practical consequences of such errors depend upon the location of the 0° 2θ point on the sinusoidal curve, and also on whether or not the diffractometer is used to scan both positive and negative directions from the 0° point. Thus the error may be symmetrical about 0°, or it may be positive on one side and negative on the other. These considerations can be understood more readily from an inspection of Fig. 5-10.

The investigation of Vassamillet and King [21] also revealed backlash errors originating in the gear train, which fell in the range 0.003 to 0.020° 2θ for the several diffractometers examined but revealed an average scatter of about ± 0.003° for a typical instrument. Optical calibration of the

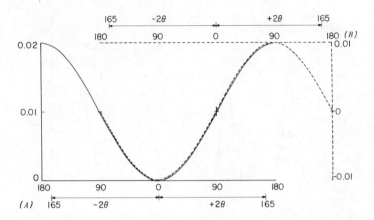

Fig. 5-10.   Typical goniometer errors in the registration of 2θ angles, illustrating two extreme positions of the 0° 2θ position. (A) ——, effective $\Delta 2\theta$ error in high-angle region of same size as the maximum $\Delta 2\theta$ (taken to be 0.02°). (B)- - -, effective $\Delta 2\theta$ error one half the maximum $\Delta 2\theta$. (Courtesy of L. F. Vassamillet and H. W. King, *Advances in X-Ray Analysis*, Vol. 6, Plenum Press, 1963, p. 142.)

2:1 following function ($2\theta:\theta$ setting) disclosed errors ranging from 0.01 to 0.05°. If backlash errors exceed about 0.02°, it may be concluded that the instrument needs reconditioning, provided one is first careful to ascertain that the spring loading of the worm drive to the main 2θ gear is functioning properly.

A convenient but less accurate method of calibrating the goniometer scale between 0 and 180° 2θ is to compare the observed readings with the angles calculated for a standard substance possessing accurately known lattice constants. However, as is explained in Section 5-2.2, certain geometrical properties of the diffractometer, several physical factors, and the thickness and transparency of the specimen to x-rays all cause displacements of the reflections from their calculated positions. Furthermore, these aberrations exhibit various systematic dependences upon 2θ, in general becoming smaller as 2θ approaches 180°. Hence it is not possible to calibrate the entire angular scale by matching the positions of both low- and high-angle lines with the theoretical values for a given reference substance. Furthermore, the purity of the specimen is of critical importance when very accurate lattice constants are required. For these reasons it is doubtful that the employment of a calibrating substance can provide d spacings of an "unknown" to an accuracy of better than 0.013 per cent [16].

Despite the objections cited above, standard substances are useful in appraising the performance of a newly aligned diffractometer both with respect to the resolution attained, as discussed earlier, and, to a first approximation, with respect to the correctness of the angular readings. Reflection angles for diamond, silicon, tungsten (woltram), and other substances for several wavelengths have been published by Parrish[22]. Finely powdered quartz or a massive fine-grained polycrystalline quartz specimen, such as the natural hydrothermal mineral Novaculite, or "Permaquartz"*, is particularly useful because the pattern contains numerous and well-distributed reflections at angles larger than 21° 2θ when CuKα radiation is employed. Table 5-1 lists some of the more useful quartz reflections for the purpose of preliminary or approximate angular calibration of the goniometer scale.

For additional details, as well as alternate procedures for the alignment of diffractometers, the reader is referred to the manufacturers'

*Obtainable from Ward's Natural Science Establishment, Inc., P.O. Box 1712, Rochester, New York.

Table 5-1. Useful Quartz Reflections for the Preliminary Angular Calibration of an X-ray Diffractometer[a,b]

Peak No.	$hk \cdot l$	2θ (°) CuKα	CuKα₁
1	10·0	20.88	
2	10·1	26.66	
3	11·0	36.58	36.55
4	20·0	42.49	42.45
5	11·2	50.19	50.14
6	12·1	60.02	59.96
7	30·2	75.74	75.66
8	13·2		90.83
9	23·1		106.61
10	13·4		120.13
11	11·6		131.23
12	14·3		137.90
13	24·0		146.63
14	23·4		153.54

[a] CuKα radiation.
[b] The data in this table have been calculated from the lattice constants of quartz at 18°C as reported by H. Lipson and A. J. C. Wilson, *Proc. Phys. Soc.* (London), **53**, 245 (1941).

manuals and to publications by Vassamillet and King[23], Parrish and Lowitzsch[16], Delf[24], and Azaroff[15]. Azaroff's procedure utilizes visible light in conjunction with plane mirrors, a knife edge, and an adjustable-focus telescope, thus avoiding the hazard of accidental exposure to x-radiation. Likewise, the specific suggestions of Delf are based on the use of optical devices. Some of the features of optical alignment in the exceptionally precise procedure outlined by Samson[25] for single-crystal diffractometers may be applied to powder diffractometers as well.

5-2 PROFILES AND POSITIONS OF DIFFRACTION MAXIMA

The features of direct counter recording and sharp parafocusing geometry employed in present-day conventional powder diffractometry render this technique well suited to the measurement of the intensity profiles of diffraction maxima (lines of a powder pattern). However, it must be borne in mind that the geometrical properties of the diffractometer introduce aberrations into the pure diffraction profile which cause it to be (1) rendered more or less asymmetrical, (2) broadened, and (3) displaced from its theoretical angle 2θ on the goniometer scale. The breadth and shape of a diffraction line are important in the measurement of crystallite size and lattice distortions (Chapter 9), while the displacement of the line is significant in the determination of accurate interplanar spacings and lattice constants (Chapter 8). We now (1) present a considerably simplified discussion of the origin of experimental line profiles using the convolution principle, and (2) treat in a more detailed and quantitative way the specific instrumental and physical aberrations that broaden and displace lines from their theoretical positions. Methods for the derivation of the pure diffraction profile from the experimental line profile may be appropriately deferred to Chapter 9.

5-2.1 Convolution Synthesis of Line Profiles[26-28]

A pure diffraction maximum produced by a crystalline powder (unmodified by the effect of the measuring apparatus) has a natural profile which is determined largely by the crystallite-size distribution, the nature and magnitude of lattice distortions, and the spectral distribution of energy in the incident radiation (lack of true monochromatism, or *dispersion* [29]), as modified by the Lorentz and polarization factors. In the absence of crystallite-size and lattice-distortion broadening, then, spectral dispersion places an ultimate limit on the sharpness of maxima, especially at large Bragg angles.

The effect of any x-ray diffraction apparatus in modifying a pure diffraction maximum can be analyzed by employing the superposition theorem [30]. According to this theorem, the profile of the observed maximum $h(\epsilon)$ is the *convolution*, or *fold*, of the pure diffraction profile $f(\epsilon)$ and the weight function of the apparatus $g(\epsilon)$:

$$h(\epsilon) = \int_{-\infty}^{+\infty} g(\eta) f(\epsilon - \eta)\, d\eta. \qquad (5\text{-}4a)$$

This convolution may be expressed more succinctly as

$$h(\epsilon) = \widehat{g(\epsilon) f(\epsilon)} \qquad (5\text{-}4b)$$

The function g expresses the sum total of the apparatus effects upon the pure function being measured. The variable ϵ measures the angular deviation of any point from the theoretical scattering angle $2\theta_0$; in equation 5-4a ϵ and the auxiliary variable η have the same dimensions as 2θ.

The geometrical weight function $g(\epsilon)$ of a well-aligned parafocusing diffractometer may reasonably be regarded as consisting of six specific instrumental functions, $g_I, g_{II}, \ldots, g_{VI}$. The total instrumental profile is then the multiple convolution of these six functions [27,28]:

$$g(\epsilon) = \widehat{g_I \widehat{g_{II} \widehat{g_{III} \widehat{g_{IV} \widehat{g_V g_{VI}}}}}}. \qquad (5\text{-}5)$$

Equation 5-5 may be evaluated by performing the five constituent folds in any desired order. In the following treatment of the six instrumental functions, the reader may refer to Fig. 5-11 for representative profiles of each.

A. X-ray Source, g_I. For the present semiquantitative purpose, the projected focal spot profile may be approximated by a Gaussian intensity function,

$$g_I = \exp\left(-k_1^2 \epsilon^2\right), \qquad (5\text{-}6)$$

in which $k_1 = 1.67\, w_I$, w_I being the half-maximum breadth. When the focal spot is viewed horizontally at a take-off angle of 1.5° (see Fig. 5-4A), w_I is about 0.15 or 0.20° for typical commercial diffraction tubes. With this low-resolution geometry, g_I is the most important of the instrumental weight functions.

When the focus is viewed laterally at 3° (Fig. 5-4B), the commonly employed high-resolution arrangement, the half-maximum breadth is only about 0.02°, so small as to be dominated by some of the other weight functions. An examination of Fig. 5-11 will enable the reader to judge the relative importance of g_I in low- and high-resolution geometries.

B. Flat Specimen Surface, g_{II}. This function arises from the varying displacements of different portions of the flat specimen surface from the

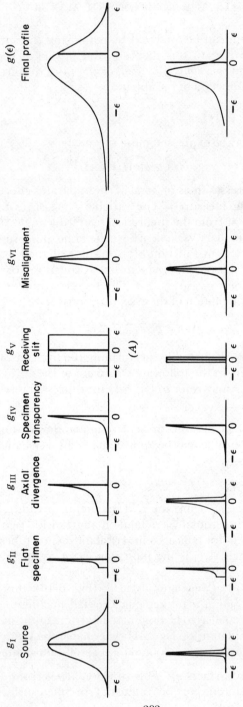

Fig. 5-11. The six instrumental weight functions for diffractometers employing (A) low-resolution, and (B) high-resolution geometry computed for $2\theta = 24°$, $\mu = 34 \text{ cm}^{-1}$, and other representative instrumental parameters. (Courtesy of L. E. Alexander, *J. Appl. Phys.*, **25**, 155 (1954).)

292

focusing circle, to which it is tangent. The profile assumes the form

$$g_{\mathrm{II}} = |\epsilon|^{-1/2}, \tag{5-7}$$

with the angular limits $\epsilon = 0$ and

$$\epsilon_m = -\frac{\gamma^2 \cot \theta}{114.6} \text{ degrees.} \tag{5-8}$$

It can be seen that g_{II} has the effect of displacing the line profile in the $-\epsilon$ direction, that is, to a smaller 2θ value. Compared with the other instrumental factors, the flat-specimen effect is minor except at small Bragg angles or when the horizontal beam divergence γ is large. In the foregoing it has been assumed that the specimen is long enough to intercept the entire incident beam (length $> \gamma R \csc \theta$ for moderately to highly absorbing specimens).

C. Axial Divergence, g_{III}. The profile of this function is in principle determinate, but in actual practice the complexity of the calculations is so great that profiles corresponding to a wide range of useful experimental conditions have not been determined (Chapter 5 of reference[31], and reference[32]). However, profiles have been calculated for certain simplified situations; thus Fig. 5-11 shows representative axial divergence profiles corresponding to $2\theta = 24°$, and a linear absorption coefficient μ of the specimen of 34 Å^{-1}. For diffractometers without Soller slits in which the illuminated axial length of the specimen is large with respect to the lengths of the source and receiving slit, the axial divergence function assumes the form [32,33]:

$$g_{\mathrm{III}} = |2\epsilon \cot \theta|^{-1/2}. \tag{5-9}$$

The limits of equation 5-9 are from $\epsilon = 0$ to

$$\epsilon_m = -\frac{\delta^2 \cot \theta}{4 \times 57.3} \text{ degrees.} \tag{5-10}$$

Figure 5-11B shows the corresponding axial divergence profile calculated by an approximate method for a diffractometer with identical sets of Soller slits in the direct and diffracted beams. The instrumental parameters are: $2\theta = 24°$, $\mu = 34$ Å^{-1}, $R = 17$ cm, lengths of focal spot and slits X and G (h, y and y' of Fig. 5-4B) all 11 mm, Soller collimators having 19 plates spaced 0.6 mm apart and with an angular aperture of $\Delta = 2.25°$.

Although the form of g_{III} is different for diffractometers with and without Soller slits, in either situation the magnitude of the line distortion is greatest at very small values of 2θ, shrinks with increasing 2θ until it is near zero in the general range $90° < 2\theta < 120°$, and increases again at still larger angles. Below $90°$ the displacement of the line profile is in the $-\epsilon$ direction, whereas in the back-reflection region it is in the $+\epsilon$ direction[32, 34]. At low values of 2θ, the distortion of the line profile caused by axial divergence is conspicuous, because of its large magnitude in relation to the distortions resulting from specimen transparency (g_{IV}) and the natural spectral breadth of the characteristic radiation employed. However, in the region $120° < 2\theta < 180°$, the distortion due to axial divergence is difficult to detect not only because of its smaller magnitude but even more because of the overriding influence on the line profile of the natural spectral breadth (dispersion) and the Lorentz and polarization factors.

D. Specimen Transparency, g_{IV} [27,28,31,35]. This weight function arises from the penetration of the specimen of finite absorption coefficient μ by the beam. When the specimen is sufficiently long to intercept the entire direct beam at the particular incident angle concerned, and when the thickness t of the specimen satisfies the condition[36]

$$t \geq \frac{3.2}{\mu} \frac{\rho}{\rho'} \sin \theta, \tag{5-11}$$

that generates a diffracted beam of maximum intensity, the transparency function g_{IV} assumes the form

$$g_{IV} = \exp (k_{IV}\epsilon), \tag{5-12}$$

where

$$k_{IV} = \frac{4\mu R}{114.6} \sin 2\theta \tag{5-13}$$

for ϵ expressed in degrees of arc. In equation 5-12, ϵ varies between the limits 0 and $-\infty$. The quantities ρ and ρ' in equation 5-11 are, respectively, the densities of the solid material composing the powder and that of the powder including interstices; μ is the linear absorption coefficient of the solid material.

The influence of this asymmetrical function upon the final line profile is greatest for thick specimens with small absorption coefficients for x-rays. Thus organic compounds and porous carbons display appreciable broadening and asymmetry of the line profile as a result of sample transparency. Keating and Warren[37], Milberg[38], and Bragg and Packer [39] have given mathematical procedures for dealing with line profile distortions resulting from low absorption. Transparency errors arising from the use of specimens of insufficient length, including the effect of interference by the specimen container itself, have been discussed by Wilson[31,35], Vonk[40], and Langford and Wilson[41]. Specimens prepared by coating a thin layer of powder on a flat surface are too thin to permit significant transparency effects; hence such mounting techniques may be resorted to when optimum resolution is required. However, the penalty of reduced intensity must be paid for this advantage. Flat, fine-grained metal specimens constitute excellent diffractometer subjects from the standpoint of penetration by the x-ray beam, because the thickness required by equation 5-11 is small except for light elements.

It should be emphasized that the three asymmetrical factors g_{II}, g_{III}, and g_{IV} render the final line profile $g(\epsilon)$ more asymmetrical for high-resolution diffractometers than for low-resolution instruments, as can be noted in Fig. 5-11. This is a direct consequence of the much smaller relative weight of the symmetrical factors g_I and g_V for the high-resolution geometry.

E. Receiving Slit, g_V. As shown in Fig. 5-11, this function is evidently a rectangle in profile with a width w_s equal to the angle subtended by the receiving slit at the center of the goniometer arc. For any given angular position ϵ, it covers the range from $\epsilon - (w_s/2)$ to $\epsilon + (w_s/2)$. This function is ordinarily the second most important component of the total instrumental weight function of a diffractometer with a broad source, whereas in diffractometers with narrow sources much narrower receiving slits can be used, making this factor relatively less important. It can be demonstrated analytically or graphically that the broadening of a Gaussian profile of half-maximum breadth b by a receiving slit of width b is only 25 per cent, and the results are not greatly different for profiles of other shapes. This shows that for diffractometers with broad sources a good rule to follow in order to preserve satisfactory resolution without an undue loss of intensity is to make the receiving slit width approximately equal to the source width at half-maximum intensity. For diffractometers with narrow sources (of the order of 0.02°), several instrumental factors of approximately equal weight are involved, yielding a net profile breadth of 0.08 or 0.10°, so that the resolution is not much affected by employing a slit width somewhat greater than the source width, for instance, as wide as 0.05°.

F. Comparison of Calculated and Experimental Line Profiles. The foregoing analyses show that all five instrumental weight functions contribute to the breadth of the diffraction profile, but that two of them, I and V, broaden it *symmetrically*, whereas the other three broaden it *asymmetrically* and shift the peak from its theoretical position. When the sample transparency function IV is small, as is usually the case, the three asymmetric factors combine to produce increasing asymmetry as 2θ

decreases from 90 to 0°. When function IV is weighty, as in thick, very low-absorbing specimens, the same effect is observed at small angles, but in addition the asymmetry becomes larger again in the neighborhood of 90°, because of penetration of the specimen by the x-ray beam.

The instrumental profiles $g(\epsilon)$ synthesized in Fig. 5-11 apply to a moderately small Bragg angle, $2\theta = 24°$, where asymmetry is conspicuous. When the convolution of functions g_{I} to g_{V} for new diffractometers was compared with the observed profile of the $10\bar{1}1$ line of quartz obtained with $\mathrm{Cu}K\beta$ radiation, the observed profile was found to be somewhat too wide. Repeated efforts to sharpen the peaks by refining the geometrical alignment failed to give significant improvement, and it was concluded that a certain residual degree of misalignment must remain that was impossible or very difficult to remove. It is probable that any well-adjusted diffractometer exhibits a similar residual deficiency in focus, arising from either the limiting mechanical imperfections of the instrument or the extreme difficulty in synchronizing the numerous variables of the alignment process. It was found empirically that a misalignment function g_{VI} of the form

$$g_{\mathrm{VI}} = \frac{1}{(1 + k_{\mathrm{VI}}^2 \epsilon^2)} \tag{5-14}$$

led to good agreement with the experimental profile when combined with the five specific instrumental functions discussed above. Such misalignment profiles are shown in Fig. 5-11 for both low- and high-resolution diffractometers, and in both instances they have been included in the synthesis of the final theoretical profile $g(\epsilon)$.

Figure 5-12 compares the $10\bar{1}1$ β peak of quartz powder with the $g(\epsilon)$ profile of Fig. 5-11B, the misalignment function g_{VI} having been assigned a half-maximum breadth of 0.04°. The calculated profile is for the following experimental conditions:

Source: $w_{\mathrm{I}} = 0.02°$ and $k_{\mathrm{I}} = 1.67/0.02 = 83.5$
Flat specimen surface: $\epsilon_m = -0.04°$
Axial divergence: computed for Soller slits and $2\theta = 24°$
Specimen transparency: $k_{\mathrm{IV}} = 50$
Receiving slit: $w_s = 0.025°$
Misalignment function: $w_{\mathrm{VI}} = 0.04°$ and $k_{\mathrm{VI}} = 2/0.04 = 50$.

The inclusion of a misalignment function of this sort has in some instances been found essential to good agreement between observed and theoretical line profiles, and in other cases not. The forms of certain specific misalignment functions is given in Section 5-2.2 (see especially Table 5-3).

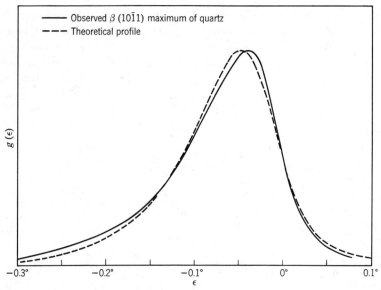

Fig. 5-12. Comparison of theoretical and observed diffractometer line profiles for $2\theta = 24°$. (Courtesy of L. E. Alexander, *J. Appl. Phys.*, **25**, 155 (1954).)

It should be noted that a seventh fundamental weight function, essential to a more rigorous treatment, has been neglected in the foregoing line synthesis. This is the natural spectral dispersion of the characteristic x-radiation employed (see first paragraph of Section 5-2.1). For a "pure" spectral line such as $K\alpha_1$ or $K\alpha_2$, or for the unresolved $K\alpha$ doublet at low Bragg angles, this function may be assigned the Cauchy form,

$$g_{VII} = \frac{1}{(1 + k_{VII}^2 \epsilon^2)}, \qquad (5\text{-}15)$$

to a first approximation [29, 42]. In equation 5-15, $k_{VII} = 2/w_{VII}$, w_{VII} being the angular width of the spectral profile at half-maximum intensity. The width is proportional to $\tan \theta$ and is therefore small at low Bragg angles. Thus for $CuK\alpha$ radiation, Ekstein and Siegel [29] give $w_{VII} = 0.0424 \tan \theta$. A more exact evaluation of the spectral dispersion profile is a very complicated matter which has not yet been satisfactorily resolved [43, 44].

5-2.2 Displacement and Breadth of Diffraction Maxima

In the preceding section an effort was made to explain the origin and principal features of diffractometer line profiles by the method of convolution synthesis. The object of this section is to summarize the present state of understanding of the quantitative aspects of *line position*, which finds direct application in the accurate determination of interplanar spacings, and of *line breadth*, which is relevant to the measurement of crystallite size and lattice strains (distortions). In this effort we rely principally on the admirable and authoritative treatment of Wilson[31], which is based both upon his own researches[35,45,47,49] and those of his associates and other investigators, including in particular Eastabrook [32], Pike[34,46,47], Delf[48], Parrish[49,50], Ladell[50], Taylor[44,50], and Langford[51].

A. Line Position. The two measures of line position most commonly employed in powder diffraction are the *peak*, or *mode*, and the *centroid*, or *center of gravity*, the latter defined by

$$\langle 2\theta \rangle = \frac{\int 2\theta I(2\theta) d(2\theta)}{\int I(2\theta) d(2\theta)}. \tag{5-16}$$

Figure 5-13 shows the peak and centroid of a partially resolved $CuK\alpha_1\alpha_2$ line profile. The peak can be more easily determined experimentally, and is an adequate measure for many practical applications of powder diffractometry. For mathematical evaluation and more precise experimental work, however, the centroid has important advantages. For example, if the centroid of a pure spectral line profile $f(2\theta)$ is F, and that of an instrumental aberration function $g(2\theta)$ is G, both measured

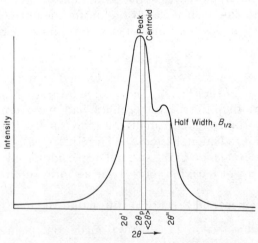

Fig. 5-13. An observed line profile (111 reflection of aluminum, $CuK\alpha$ radiation). Shown are the peak, centroid, and half width (width at half-maximum intensity $B_{1/2}$). (Courtesy of A. J. C. Wilson and Philips Technical Library[31].)

from some particular origin $2\theta_0$, it can be shown by convoluting the two functions that the centroid of the resulting profile is simply

$$H = F + G.$$

It follows that the centroid C of a line profile generated by the effect of N aberrations or physical functions $C_1, C_2, \ldots, C_N$ is their sum

$$C = \sum_{i=1}^{N} C_i. \tag{5-17}$$

A special difficulty is encountered in the experimental determination of the centroids of line profiles, a consequence of the slow asymptotic approach of the tails of the profiles to the underlying background. This feature is contributed to the line profile very largely by the spectral distribution. Without incurring significant error it has been found practicable to solve this problem by truncation of the profile at some arbitrary level above the background[47,49,50,52–56]. Nevertheless, the calculation of line centroids is at best a rather tedious and exacting process not conducive to rapid or routine application.

B. Line Breadth. Three measures of line *breadth*, or geometric dispersion, have been employed in powder diffraction. The most widely used, though the one least amenable to mathematical interpretation, is the so-called *half width*, that is, the overall width of the line profile at half-maximum intensity measured above the background, as illustrated in Fig. 5-13:

$$B_{1/2} = 2\theta'' - 2\theta'. \tag{5-18}$$

A second commonly used measure is the *integral breadth*, which is defined as the integrated intensity of the line profile above background, divided by the peak height:

$$B_{1/2} = \frac{1}{I_p} \int I(2\theta) \, d(2\theta). \tag{5-19}$$

The integral breadth possesses special mathematical significance; thus equation 5-19 shows it to be the width of a rectangle having the same area and peak height as the actual line profile.

However, mathematically most useful as a measure of line breadth is the *variance*, or squared standard deviation of the elements comprising the profile[57–61]. It is directly associated with the *centroid* $\langle 2\theta \rangle$ as a measure of line location, and is defined by

$$W_{2\theta} = \langle (2\theta - \langle 2\theta \rangle)^2 \rangle = \frac{\int (2\theta - \langle 2\theta \rangle)^2 I(2\theta) d(2\theta)}{\int I(2\theta) d(2\theta)}. \tag{5-20}$$

Unlike the first two measures of line breadth cited above, the variance, like the centroid, possesses the advantage of mathematical additivity. Thus by convolution synthesis it can be shown that the mean-square breadth of the final line profile is simply equal to the mean-square breadth of the original "pure" line profile plus the mean-square breadths of all the constituent aberration profiles:

$$W_{2\theta} = \sum_{i=1}^{N} (W_{2\theta})_i. \tag{5-21}$$

Like the centroid $\langle 2\theta \rangle$, the variance W of an observed line profile in actual practice can be determined only after truncation of the profile over a suitable angular range [47,49,50,52–56]. In the treatment that follows, emphasis is placed on *centroids* and *variances*, respectively, as the primary criteria of precise line *positions* and *breadths*. Where appropriate, comments are also made relative to the employment of peak positions and integral breadths.

C. The Practical Determination of the Centroid and Variance. Figure 5-14 and Table 5-2 illustrate the numerical computation of the centroid and variance of a partially resolved $K\alpha_1\alpha_2$ line profile which extends over an angular range of 0.64° after appropriate truncation.

Table 5-3, reproduced from Wilson [31], presents a summary of the analytical expressions derived to date that appear to express most accurately the centroids and variances of the important geometrical aberrations, including the principal misalignment errors. It should be observed that the symbolism adopted in this table, as well as Table 5-4, is that of Wilson and does not necessarily agree with the conventions presented above in Section 5-1.2 and in Figs. 5-3 and 5-4. In particular we note the following correspondences:

	This Text	Wilson [31]
Specimen thickness	t	t
Diffractometer radius	R	R
Angular aperture of Soller slits	Δ	Δ
Axial beam divergence (without Soller slits)	δ	
Equatorial beam divergence	γ	α
Focal line width	w	f_2
Focal line height	h	$2h$
Receiving slit width	x'	r_1

Table 5-2. Calculation of Centroid $\langle 2\theta \rangle$ and Variance W from 11·2 α-Quartz Profile Shown in Fig. 5-14[a]

2θ increment number[b]	$2\theta(°)$	$I(2\theta)^c$ $= I - B$	$2\theta \cdot I(2\theta)$	$(2\theta - \langle 2\theta \rangle)^2$	$(2\theta - \langle 2\theta \rangle)^2 \cdot I(2\theta)$
1	49.866	0.30	14.96	0.09382	0.028146
2	49.879	0.34	16.96	0.08614	0.029288
3	49.892	0.40	19.96	0.07879	0.031517
4	49.905	0.44	21.96	0.07177	0.031579
5	49.918	0.50	24.96	0.06508	0.032538
6	49.930	0.55	27.46	0.05871	0.032290
7	49.943	0.63	31.46	0.05267	0.033182
8	49.956	0.74	36.97	0.04696	0.034750
...	...	...	...	...	...
22	50.135	12.98	650.75	0.00141	0.018253
23	50.148	13.66	685.02	0.00061	0.008334
24	50.161	12.99	651.59	0.00014	0.001840
25	50.174	11.22	562.95	0.00008	0.000909
26	50.186	8.61	432.10	0.00019	0.001616
...	...	...	...	...	...
50	50.494	0.26	13.13	0.10298	0.026774
Σ		193.02	9684.34		1.802474

[a]Calculation of centroid and variance:

$$\langle 2\theta \rangle = \frac{\Sigma \, 2\theta \cdot I(2\theta) \cdot \Delta(2\theta)}{\Sigma I(2\theta) \cdot \Delta(2\theta)} = \frac{9684.34}{193.02} = 50.173°.$$

$$W = \frac{\Sigma \, (2\theta - \langle 2\theta \rangle)^2 \cdot I(2\theta) \cdot \Delta(2\theta)}{\Sigma I(2\theta) \cdot \Delta(2\theta)} = \frac{1.802474}{193.02} = 0.00934°.$$

[b]2θ range from 49.860 to 50.500° by 50 increments of 0.0128°. This angular range corresponds to the wavelength range 1.54178 ± 0.009 Å symmetrical about the preliminary centroid angle $\langle 2\theta \rangle = 50.190°$ 2θ.

[c]$I(2\theta)$ is the overall intensity I less the background B, all in counts per second.

In Table 5-3 expressions for the centroid $\langle \Delta(2\theta) \rangle$ and variance W not followed by a literature reference may be presumed attributable to Wilson, and are derived or explained in his book [31]. In addition to the various geometrical aberrations given in Table 5-3, several physical factors also influence the location and breadth of experimental line profiles. Table 5-4 specifies six physical factors, including *refraction*,

Table 5-3. Centroid and Variance of Certain Geometrical Aberrations[a,b]

Aberration	$\langle \Delta(2\theta) \rangle$	W	Reference
Zero-angle calibration	Constant	Zero	[35]
Specimen displacement	$-2s(\cos\theta)/R$	Zero	[35,62]
Specimen transparency			
Thick specimen	$-(\sin 2\theta)/2\mu R$	$(\sin^2 2\theta)/(2\mu R)^2$	
Thin specimen	$-t(\cos\theta)/R$	$t^2(\cos^2\theta)/3R^2$	
2:1 missetting		$\frac{1}{3}\alpha^2\beta^2\cot^2\theta$	
Inclination of plane of specimen to axis of rotation	Zero if centroid of illuminated area of specimen centered on axis of rotation	$4\gamma^2 h^2(\cos^2\theta)/3R^2$ for uniform illumination	
Flat specimen	Zero if centroid of illuminated area on equator of specimen	$\frac{1}{45}\alpha^4\cot^2\theta$	
	$-\frac{1}{6}\alpha^2\cot\theta$		
Focal-line width	Zero	$\sim f_1^2/12R^2$, but depends on intensity variation across focal line	
Receiving-slit width	Zero	$r_1^2/12R^2$	
Cross term between equatorial extensions of focal line and specimen	Zero if adjustment reasonable	$\sim \alpha^2 f_2^2/144R^2$ but depends on intensity variation across focal line	
Cross term between equatorial extension of specimen and nonequality of source-specimen and specimen-receiving-slit distances	Proportional to r_2 and distance between axis of rotation and centroid of illuminated area, and thus vanishes if the latter is centered	$\alpha^2 r_2^2/12R^2$	

302

Axial divergence

	Second order	Fourth order	Reference
No Soller slits			
Source and receiver small in comparison with specimen	$-h^2(\cot\theta)/3R^2$	$4h^4(\cot^2\theta)/45R^2$	[32,63]
Source, specimen, and receiver equal	$-h^2(2\cot 2\theta + \csc 2\theta)/3R^2$	$h^4(16\cot^2 2\theta + 28\cot 2\theta\,\csc 2\theta + 19\csc^2 2\theta)/45R^4$	[34-1]
One set of Soller slits			
Narrow slits (q small)	$-h^2[Q_1\cot 2\theta + Q_2\csc 2\theta]/3R^2$	Complex $7h^4(\cot^2 2\theta)/45R^4$	[34-1]
$q < 2$	$Q_1 = 1,\ Q_2 = 0$ $Q_1 = (1 - \tfrac{3}{4}q + \tfrac{3}{8}q^2 - \tfrac{1}{10}q^3)/(1 - \tfrac{1}{6}q)$ $Q_2 = (\tfrac{1}{4}q^2 - \tfrac{3}{40}q^3)/(1 - \tfrac{1}{6}q)$	Complex	[34-1]
Wide slits ($q > 2$)	$Q_1 = \left(2 - \dfrac{29}{15q}\right)\bigg/\left(1 - \dfrac{2}{3q}\right)$ $Q_2 = \left(1 - \dfrac{6}{5q}\right)\bigg/\left(1 - \dfrac{2}{3q}\right)$	Complex	
Two sets of Soller slits			
Narrow slits	$-\tfrac{1}{6}\Delta^2\cot 2\theta$	$\tfrac{1}{360}\Delta^4(10 + 17\cot^2 2\theta)$	[32]
Wide slits	See Section 7.6 of reference [31]		

303

[a]Reproduced by permission of A. J. C. Wilson and Philips Technical Library from reference [31].

[b]Notation: α = equatorial divergence of the beam; β = angle of missetting of 2:1 ratio; γ = angle of inclination of plane of specimen to axis of rotation; Δ = angular aperture of Soller slits; μ = linear absorption coefficient; r_1 = width of receiving slit; r_2 = distance of receiving slit from focusing circle; R = diffractometer radius; t = specimen thickness; s = specimen-surface displacement; f_1 = projected width of focal line; f_2 = width of focal line; h = half height of focal line, specimen and receiving slit, taken as equal; $q = R\Delta/h$; Q_1 and Q_2 defined in table.

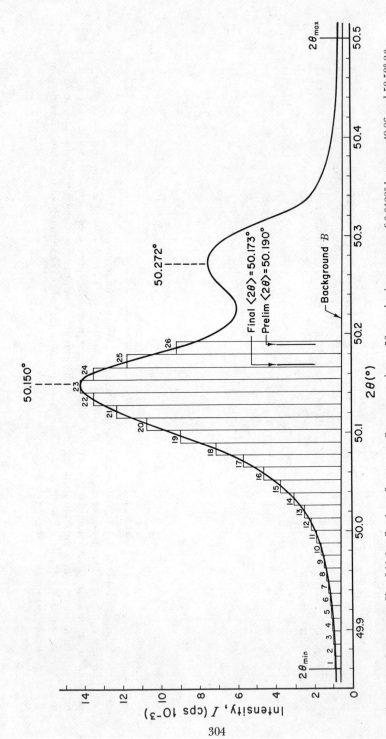

Fig. 5-14. CuKα₁,₂ profile of 11·2 reflection of α-quartz. Step-counted over 50 angular increments of 0.0128° between 49.86 and 50.50° 2θ. Preliminary ⟨2θ⟩ = 50.190°, the weighted mean of Kα₁ and Kα₂ peak positions. Final ⟨2θ⟩ = 50.173°; final W = 0.00934°.

spectral dispersion, and four which may be classified as *response variations*, that is, those that modify the true distribution of intensity as a function of wavelength. The centroid displacements given in Table 5-4 have been deduced to at least a fair approximation. It should be emphasized that four of the centroid displacements are expressed as functions of W, which here designates the variance of the specific range of wavelengths used in determining the centroid displacements of the several physical aberrations under consideration. Although investigations of the spectral distributions actually employed in x-ray diffraction have been under way for some years, there are numerous experimental and interpretative complexities that make very difficult the calculation of the W's at the desired level of accuracy. Among the earlier semiquantitative estimates of the wavelength dispersion profiles of $K\alpha$ doublets [44, 46, 62, 63], the quality of the results obtained by Ladell and co-workers [44] deserves special mention. More recently, Taylor, Mack, and Parrish [55, 69, 70] have increased the accuracy of the analysis of diffraction line profiles by employing two-crystal spectrometer profiles of the $CuK\alpha$ and $FeK\alpha$ spectral lines supplied by J. A. Bearden. Wilson has suggested that for order-of-magnitude estimates the variance of a $K\alpha$ line may be set equal to one-third of the square of the doublet separation [71].

For the physical aberrations (Table 5-4), it may be remarked that the effect of *dispersion* is to stretch out the high-angle portion of the wavelength distribution on the 2θ scale relative to the low-angle portion, thereby displacing the centroid. It may also be noted that the angular shift of the centroid caused by the Lorentz factor is of the same form as that produced by dispersion, only it is numerically twice as large. Thus

Table 5-4. Centroid Displacements Arising From Certain Physical Aberrations[a]

Aberration	$\langle \Delta(2\theta) \rangle$	Reference
1. Refraction	$-2\delta \tan \theta$	[64]
2. Response variations		
Absorption in filter	$(-6\mu t W \tan \theta)/\lambda^2$	[65]
Quantum-counting efficiency		[66]
Polarization	$(2W \tan \theta)/d^2$ approximately	
Lorentz factor	$(2W \tan^3 \theta)/\lambda^2$ approximately	[67, 68]
3. Dispersion	$(W \tan^3 \theta)/\lambda^2$ approximately	[44, 63]

[a]Reproduced by permission of A. J. C. Wilson and Philips Technical Library from reference [31].

the overall effect of both factors is to shift the centroid on the 2θ scale to higher angles by the amount

$$\Delta 2\theta = \left(\frac{3W}{\lambda_0{}^2}\right)\tan^3\theta_0. \tag{5-22}$$

5-2.3   Accurate Determination of Interplanar (d) Spacings [72]

From the Bragg equation in the form

$$d = \frac{\lambda}{2\sin\theta}, \tag{5-23}$$

it follows that the relative error in d is related to the error in θ by

$$\frac{\Delta d}{d} = -\cot\theta\,\Delta\theta. \tag{5-24}$$

Equation 5-24 shows that the precision with which interplanar spacings, and therefore also lattice constants, can be measured increases rapidly at large Bragg angles according to the function $\cot\theta$. Thus, in relation to the accurate determination of d spacings, it is the behavior of the various aberrations at large angles that is of most interest.

Table 5-5 classifies the principal aberrations in two ways, first, according to whether they are functions of the specimen (column i) or of the instrument and radiation employed (column ii) and, second, according to their $\tan\theta$ dependence. The fact that most of the aberrations fall into class ii is of great advantage, inasmuch as the calculations need to be made only once for any given radiation and set of instrumental conditions, after which the results can be applied to a succession of specimens. Classes i and ii are in turn divided into subclasses according to whether their dependence on $\tan\theta$ is to the power -1, 0 (independent of θ), 1, or 3. Bearing in mind equation 5-24, we see that aberrations of type A caused by the specimen, then, result in an error in d which decreases to zero with the functionality $\cot^2\theta$ as θ approaches $90°$, whereas type-C aberrations produce an error in d which is independent of angle.

The foregoing observations prompted Wilson[72] to suggest a procedure, that, in principle at least, should permit the derivation of accurate lattice constants from diffractometric data. For the simplest case, cubic symmetry, the steps are: (1) Obtain the centroids of all high-angle lines; (2) correct each centroid for all aberrations of class (ii) that can be quantitatively assessed; (3) calculate the lattice constant a for each line; and (4) plot the several values of a versus $\cot^2\theta$, which should be linear, and the ordinate intercept of which at $\cot^2\theta = 0$ yields a free of the effects of aberrations of type A. Since some of the errors of type i are proportional to $\cos\theta$ rather than $\cot\theta$, it is possible that extrapolation against

Table 5-5. Aberrations Classified According to Behavior at High Angles[a]

Angular dependence	(i) Function of specimen	(ii) Function of instrument and radiation
A. $(\tan \theta)^{-1}$	Specimen transparency Variable part of specimen-surface displacement	Flat-specimen error Axial divergence, source and receiver small Part of axial divergence with Soller slits proportional to $\cot \phi$[b] Constant part of specimen-surface displacement Missetting of 2:1 ratio gears
B. $(\tan \theta)^{0}$		Zero-angle setting Cross term between equatorial extensions of focal spot and specimen
C. $\tan \theta$	Refraction	Part of axial divergence with Soller slits proportional to $\cot 2\phi$[b] Absorption in filter, and so on Variation of quantum-counting efficiency
D. $(\tan \theta)^{3}$		Polarization Lorentz factor Dispersion

[a]Reproduced by permission of A. J. C. Wilson and Philips Technical Library from reference [31].
[b]As defined by Wilson[31] the angle ϕ is in general very nearly equal to the Bragg angle θ.

$\cos \theta \cot \theta$ rather than $\cot^2 \theta$ may yield a more linear plot. For crystalline substances of less than cubic symmetry, a solution for a, b, c, α, β, and γ can be obtained, provided sufficient high-angle reflections are available, by means of least-squares analytic extrapolation, which not only is less subjective than graphical extrapolation but also permits each aberration to be assigned its particular angular functionality.

We must reemphasize that the ultimate limit on the *absolute* accuracy with which d spacings can be measured is imposed by the accuracy with which the effective wavelength is known. At present, the centroid wavelengths required in the above approach to the determination of lattice constants are known in absolute metric units (true ångström units) to an accuracy of only about 1 part in 20,000. However, the diffractometer is certainly capable of providing *relative* d spacings with precisions in the range of 1/50,000 to 1/100,000. It should also be emphasized that at the largest accessible angles, in the range $150° < 2\theta < 165°$, the overriding distortions introduced by the Lorentz factor, spectral dispersion, and axial divergence tend to preclude practical use of reflections by the centroid method.

Delf[53] and Taylor, Mack, and Parrish[55] made notable inde-pendent efforts to demonstrate the feasibility of determining lattice con-stants with high accuracy by the painstaking procedure of correcting the measured centroid of each reflection for all known aberrations, thereby providing directly values of the lattice constants without the need to resort to extrapolation methods. Their procedures differed principally in the wavelength values employed and in that Delf used lines at both low and medium angles, whereas Taylor and co-workers utilized only lines at medium and moderately high angles. In both investi-gations reflections above 155° were avoided for the reason cited in the preceding paragraph. Delf used the weighted mean peak wavelength, while Taylor and co-workers went to extraordinary efforts to ascertain the proper centroid value to employ. The results of the two studies are clearly of high quality, and the numerical differences in the *a* of tungsten obtained in the two investigations appear to stem mainly from the different wavelengths employed for Cu$K\alpha$.

5-2.4 "Routine" Determination of Interplanar (d) Spacings

In this section we refer exclusively to the use of the *peak*, or *mode*, as the measure of line position. In truth, the vast majority of the applications of powder diffractometry, such as those of an analytical laboratory, require absolute accuracies in the *d* spacings that fall well within the limit of approximately 1/20,000, with which the centroid wavelengths of the x-ray lines are currently known on the metric scale. Furthermore, the more routine diffractometric measurements tend to be concentrated in the low- and intermediate-angle region ($10° < 2\theta < 120°$) and are in-herently incapable of yielding *d* spacings with precisions of the order of 1 part in 50,000 or better, such as is true, in principle at least, of reflec-tions at higher angles ($120° < 2\theta < 165°$). On the other hand, it is pertinent at this point to note that the *d*-spacing measurements under considera-tion in this section are derived by applying the Bragg equation to the *peak positions of the reflections* using the *peak wavelength values* of the x-ray spectral lines, which, unlike the centroids, are known on an absolute (metric) scale with high accuracy (see Section 2-3.2 and especially Table 2-3).

   Like the centroids of the reflections, the peaks are also displaced by
the numerous geometrical and physical aberrations discussed in Section
5-2.2, although in general to a lesser extent. However, the peak displacements cannot be assessed with the quantitative accuracy obtained
for centroid displacements. This results in part from the variety of ways
in which the peak position can be ascertained, and in part from the lack
of a demonstrable systematic relationship between the peak position as
determined by any particular method and the centroid position. Nevertheless, any particular criterion of the peak position that is consistently
adhered to throughout a program of measurements suffices to yield d
values of a quality entirely adequate for the general purposes presently
being considered.

   Three common methods of determining the peak position are illustrated in Fig. 5-15. P_0 is the apparent maximum point of the line profile.
It has the drawback that statistical uncertainty in the counts recorded at
and in the immediate vicinity of the maximum makes its precise location
uncertain. A commonly employed device for circumventing this problem
is to locate the peak at P_x, the point of intersection of two straight lines
projected from the most nearly linear portions of the sloping sides of
the profile. Another method is to assign the peak position to the mid-

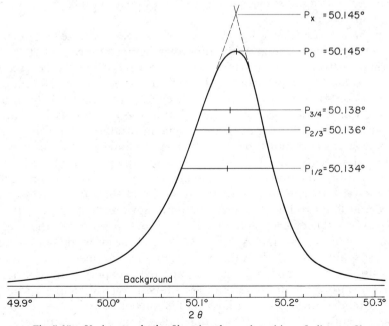

Fig. 5-15.   Various methods of locating the peak position of a line profile.

point of a chord drawn parallel to the background and located at one-half the maximum peak intensity $P_{1/2}$, or better yet, at higher levels such as $P_{2/3}$ or $P_{3/4}$. Clearly, despite the relative sharpness of the line profile shown in Fig. 5-15, these methods all give somewhat different results. Not shown in Fig. 5-15 is Bearden's "center line" peak technique for the measurement of spectroscopic lines (see Section 8-1 and Table 2-3), which may also be applied to diffraction line profiles.

Wilson [73, 74] has discussed at considerable length the problems associated with the location of the peak position and recommends as a more objective criterion the maximum point of a parabola fitted by least squares to a portion of the reflection profile encompassing the peak. This method also has the advantage of affording an objective estimate of the variance of the derived peak position. However, the number of experimental measurements involved coupled with the mathematical operations required to fix the peak location with acceptable statistical accuracy hardly lend this method to routine use. Thus, for the measurement of a large number of lines with acceptable accuracy, we may perhaps first recommend the method of the chord midpoint $P_{x/y}$, with x/y equal to or larger than $\frac{3}{4}$; somewhat less preferable is the P_x method. However, it must be remembered that neither of these procedures, although providing a convenient and systematic means for locating the peak, permits a nonsubjective (unbiased) estimate of the accuracy of its position to be made.

In the measurement of peak positions with the purpose of determining interplanar spacings with somewhat higher accuracy (1/20,000 or perhaps better), we cite two approaches: (1) A set of empirical correction curves can be prepared from peak shifts measured under different conditions for substances of known lattice spacings. (2) It can be shown theoretically, as well as empirically, that in the moderately large angle range $90° < 2\theta < 150°$ the peak displacements produced by the important geometrical aberrations are to a first approximation the same as the centroid shifts [35, 45]. It has also been shown theoretically [75] that a second approximation can be made in terms of the variance of the given geometrical aberration and of the second and third derivatives of the observed line profile at its peak. To date, however, experimental work to confirm the efficacy and practicability of carrying out the second approximation is lacking. At high 2θ's ($> 150°$) the evaluation of the peak shifts resulting from the physical aberrations becomes very complex, and thus far has not been successfully applied to experimental data [75].

By means of the double-scanning technique of King and Vassamillet [17] extended to 2θ angles as large as $163°$ on both the positive and negative sides of the direct beam by King and associates [76, 77], the zero calibration error and the error caused by displacement of the specimen surface from the goniometer axis are both eliminated experimentally, leaving specimen transparency and axial divergence as the principal remaining aberrations, provided the diffractometer is otherwise well aligned and the horizontal divergence of the direct beam is limited to 2°. The effects of these two aberrations on $\Delta d/d$ (or the lattice constants) have the respective functionalities $\cos^2 \theta$ and $\cos^2 \theta$ plus constant. Thus under these experimental circumstances the correct extrapolation function to be applied to individual values of the lattice constants derived from the peak positions of reflections at large 2θ values is $\cos^2 \theta$ [76].

A less favorable reaction to the use of peak measurements in conjunction with extrapolation functions such as $\cos^2 \theta$ or $\cos \theta \cot \theta$ has been expressed by Parrish, Taylor, and Mack [69] after the application of very precise experimental techniques, including use of the resolved $K\alpha_1$ and $K\alpha_2$ lines of both copper and iron radiations, to the determination of the lattice constants of silicon and tungsten. However, it is important to note that the techniques of zero-angle location, calibration of the angular scale, and correction for specimen displacement employed by these investigators differed from those used by King and co-workers. In particular, Parrish and co-workers did not make use of the important double-scanning technique from low to high angles on both sides of the 0° position.

5-3 ELECTRICAL CHARACTERISTICS OF THE DIFFRACTOMETER

5-3.1 General Arrangement of Components [23, 78–80]

The electronic complexity and sophisication of the circuitry of a modern diffractometer are such as to preclude in this chapter anything but a relatively general discussion of the subject. For further details and circuit diagrams, the reader is referred to the manufacturers' manuals.

Figure 5-16 is a block diagram showing a typical arrangement of components. The unregulated line voltage is passed through an isolation transformer and then to an electronic line-voltage regulator providing an output stabilized to within ± 0.1 per cent (see Section 2-2.3). Although ordinary full-wave rectification may be emplyed in diffractometry, considerably higher counting rates with an improvement in linearity (see Section 5-3.3) can be achieved by applying to the x-ray tube almost ripple-free ("constant") dc voltage (refer to Fig. 2-10 and related text).

The relatively low-voltage pulses from the proportional or scintillation counter are first preamplified at the counter locale and then transmitted to the linear amplifier and pulse shaper, which emit pulses of

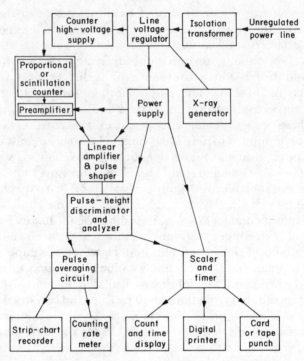

Fig. 5-16.   Typical arrangement of components in an x-ray diffractometer.

square shape and proportional in amplitude to the energy of the incident x-ray photons. Depending upon the particular radiation employed,
the nature of the specimen, and the quality of experimental data
being sought, the pulses may be either subjected to pulse-height discrimination and analysis or sent directly to the pulse-averaging and
scaling circuits. The pulse-averaging, or "tank," circuit possesses an
adjustable capacitance, permitting the RC time constant* to be varied.
This circuit smooths out the pulses it receives into a fluctuating current
which may be fed to a counting-rate meter (microammeter) or a strip-

---

*For a circuit containing only a resistance R and a capicitance C in series, the time
constant is RC. This is the time in seconds required for the charge q on the capacitor to
reach 0.632 of its final value Q when charged through the resistance R by a constant emf.
For such a circuit, with zero initial charge, the charge at time t is

$$q = Q(1 - e^{-t/RC}),$$

and, when $t = RC$,

$$q = Q\left(1 - \frac{1}{e}\right) = 0.632Q.$$

chart recorder. Large values of the RC constant have a greater smoothing-out effect upon the pulse current, which results in reduced differentiation of the details of current variation with time, whereas low RC's increase the resolution of these details. Thus the choice of the time constant has an important bearing upon the authenticity and proper interpretation of both strip-chart records and visual observations of the counting-rate meter. The counting-rate meter plays the useful role of continuously monitoring the output current, while the strip-chart recorder provides a permanent record. It should be mentioned that if a Geiger-Müller counter is used, the pulse voltages are sufficiently high that they do not necessarily require preamplification but may be transmitted directly to a linear amplifier and scaler of appropriate characteristics.

Typical scaling circuits have very high response speeds (resolving times of about 1 μsec) and encompass seven decades, thus permitting up to 9,999,999 counts to be registered. With the aid of an electronic timer, which is also usually of the decade type, the scaler may be operated so as to count for a preset time or to a preset count. The output of the scaler-timer circuit may be utilized in three ways. First, the accumulated counts and time are continuously displayed, thus providing for visual monitoring and permitting scalar data to be manually recorded when desired. Second, in conjunction with a step-scanning mode of operation, the digital printer may be employed to furnish a permanent numerical record extensive sequence of measurements at small angular increments embracing considerable ranges of 2θ. Third, some manufacturers provide the option of a card or paper-tape punch, by means of which the step-scanned scaler data may be registered in a form suitable for direct processing with a computer.

We shall now discuss in greater detail (1) characteristics of the various radiation detectors available at present, (2) nonlinearity of detector response, and (3) monochromatizing techniques including pulse-height discrimination and analysis. Some aspects of these topics are also further clarified in Section 5-4.

5-3.2 Radiation Detectors (Quantum Counters) [80–85]

X-radiation detectors other than photographic film may be classified into three types: gas ionization counters, scintillation counters, and semiconductor detectors. The early ionization counters used in x-ray diffractometry were of the Geiger-Müller type; by the early 1950s these began to be replaced by proportional gas ionization counters, which afforded the important advantages of small resolving time, higher inherent wavelength selectivity, and applicability of pulse-height analysis. Somewhat later, scintillation counters came to be preferred for many applications, and today are predominantly used in diffractometry, possessing as they do practically uniform response (sensitivity) over a broad wavelength band, small resolving time, and amenability to pulse-height analysis. At the present writing, semiconductor (solid-state) detectors are in the process of rapid development and offer the exciting prospect of future application as convenient energy-dispersive, rather than wavelength-dispersive, x-ray receivers.

We now discuss the characteristics of the detectors in widespread use at the present time, with some emphasis on the limitations and advantages of each. At the outset we may observe that the detectors in current use are all *counters*, which produce *discrete pulses* of electrical discharge and consequently have the important advantage of expressing intensity measurements in digital form suitable for data processing.

A. Gas Ionization Counters. An *ionization chamber*, or *Geiger-Müller* or *proportional counter*, consists of a gas-filled cylindrical envelope J (Fig. 5-17) and two electrodes, a central wire W (the anode) maintained at an appropriate positive potential, and a grounded coaxial conducting cylinder C (the cathode). X-rays enter the counter through a thin window of mica or beryllium (Z in Fig. 5-17), which in Geiger-Müller counters is located at the end, and in proportional counters usually at the side, of the cylinder (see also Fig. 5-23). The filling gas is one of the noble gases, argon, krypton or xenon, at or somewhat below atmospheric pressure, mixed with about 1 per cent of a quenching gas, commonly chlorine. When x-ray photons impinge upon the cylinder or gas molecules within, they eject photoelectrons which are accelerated toward the anodic collecting wire by its high positive potential. Basically, the ionization chamber, Geiger-Müller counter, and proportional counter differ from one another in the potential difference between the elec-

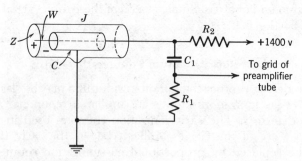

Fig. 5-17.   Geiger-Müller counter for detecting x-rays.

trodes. The following paragraphs are based largely on the excellent treatment of Arndt and Willis [83].

If the electrons generated in the original ionizing event are accelerated sufficiently, they collide with one or more gas molecules, producing additional ion pairs, and these secondary released electrons in turn acquire enough kinetic energy to ionize other molecules. For example, the energy of a CuKα photon is 8×10^3 eV, and the energy required to ionize an argon atom is 29 eV, from which it is seen that about 275 ion pairs are produced for each absorbed x-ray photon. The net result is a cumulative discharge of electrons to the central wire, known as a Townsend avalanche, also referred to as gas amplification. If A is the gas amplification factor and n the number of *initial* ion pairs formed during the absorption of the entering photon, the number of ion pairs collected is An. For CuKα quanta and an argon-filled counter, the voltage of such a pulse is at most less than $1.45 A \times 10^{-6}$ V. Since the amplitude of the noise pulses produced by conventional amplifiers is of the order of 10×10^{-6} V, the gas amplification factor A must be at least 50 to provide an acceptable signal-to-noise ratio. Commonly, A is about 10^3, so that the requisite amplifier gain is between 10^3 and 10^5.

In the range 10 to 10^4, A varies approximately as the exponential of the applied voltage. Whereas at lower voltages (V_0 to V_1, Fig. 5-18) gas amplification is confined to a region very near the anode wire (as in ionization chambers), the region extends over an increasing portion of the counter volume as the voltage is further increased. Nevertheless, in proportional counters the total discharge remains confined to the immediate vicinity of the plane, perpendicular to the counter axis, in which the original ionizing event occurred. As long as the voltage remains low enough (V_1 to V_2, Fig. 5-18) to keep the discharge avalanche localized to this plane, A is independent of the number n of primary ion pairs formed, and the amplitude of the final pulse is proportional to n and consequently also to the energy of the incident x-ray quantum. The detector functioning under these circumstances is said to be a *proportional counter*.

As the voltage across the electrodes is raised further, A continues to increase but is no longer independent of n, and when it reaches the range 10^8 to 10^9 (V_4 to V_5, Fig. 5-18), all pulses generated by the counter are of equal size. Now the discharge is no longer localized as described in the preceding paragraph but is propagated throughout the length of the counter by ultraviolet photons formed in the primary avalanche [83], and the positive ions reaching the cathode can produce further avalanches. Under these conditions the counter functions as a Geiger-Müller counter, and the presence of a quenching agent, such as chlorine, is

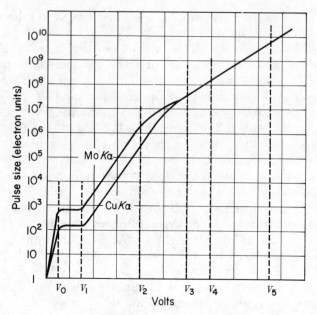

Fig. 5-18. Counter pulse size as a function of applied voltage. Below V_0, partial recombination; V_0 to V_1, ionization-chamber region, $A = 1$; V_1 to V_2, proportional region, A independent of energy; V_2 to V_3, region of limited proportionality; V_4, Geiger threshold voltage; above V_5, continuous discharge. (Courtesy of U. W. Arndt and B. T. M. Willis, *Single Crystal Diffractometry*, Cambridge University Press, 1966.)

required to prevent the discharge from becoming continuous. At still higher voltages (above V_5, Fig. 5-18), the counter goes into a state of continuous discharge, even when a quenching agent is present.

B. Geiger-Müller Counters. Figure 5-19 shows a typical response curve of counting rate versus applied voltage for a Geiger-Müller counter. As the voltage V applied across the electrodes is increased, no counts occur until the threshold voltage V_s is reached. In the narrow interval V_s to V_p (proportional region), the counting rate increases very rapidly and the pulse amplitude is proportional to the number of initial ion pairs formed and therefore to the energy of the ionizing photon. For a counter of given design, the location of V_s and the shape of the initial part of the curve depend only upon the characteristics of the amplifier, because V_s is simply the lowest voltage at which the pulses are large enough to be counted.

Between V_p and V_f, the *plateau*, the counting rate increases only very slowly and rather linearly with the applied voltage, and, furthermore, the pulse height is constant and independent of the energy of the inci-

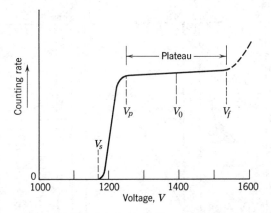

Fig. 5-19. Typical response curve of a Geiger-Müller counter.

dent photon. The plateau extends over a range of 200 to 300 V, and the most satisfactory operating voltage V_0 is near its midpoint, commonly about 1400 V. It is the usual practice to couple the counter to a quenching circuit, which renders the plateau flatter and the performance more stable. Above V_f the counting rate increases more rapidly again, and the counter goes into a state of continuous discharge, a condition that can damage a Geiger-Müller counter in even a short time and should therefore be avoided whenever possible.

A proper combination of counter-tube dimensions and filling gas makes it possible to achieve high quantum-counting efficiencies, that is, approaching one count for every x-ray photon entering the tube. The principal requirements are that the gas pressure be relatively high and the geometry of the counter such as to promote absorption of the x-rays by the gas rather than by the cathode cylinder. Figure 5-20 shows the absorptive properties of argon and krypton for x-rays of the commonly used characteristic wavelengths. It is seen that 10 cm of argon at atmospheric pressure absorbs the $K\alpha$ radiations of copper, nickel, cobalt, iron, and chromium with efficiencies of 85 to 100 per cent. The more penetrating wavelengths of $MoK\alpha$ and $AgK\alpha$, however, are much more efficiently absorbed by krypton than by argon. The low efficiency of an argon-filled counter for absorbing x-rays of short wavelengths, as shown in Fig. 5-21 from the work of Taylor and Parrish [86] and Fig. 5-20, has a useful monochromatizing effect when one of the longer characteristic wavelengths is being measured, because the general-radiation spectrum, which is responsible for much of the background in an x-ray diffraction pattern, is concentrated in the shorter wavelengths and hence inefficiently detected.

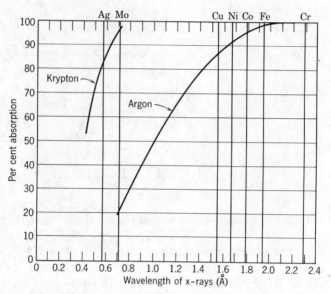

Fig. 5-20. Absorption of x-rays by argon and krypton at 76 cm pressure for a path length of 10 cm.

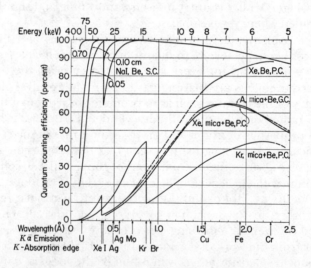

Fig. 5-21. Calculated quantum-counting efficiency versus wavelength of typical x-ray counters. G.C., Geiger-Müller counter; P.C., proportional counter; NaI, sodium iodide crystal; S.C., scintillation counter. Window compositions, beryllium (Be) and mica + Be. Filling gases, argon, krypton, and xenon. (After J. Taylor and W. Parrish, *Rev. Sci. Instrum.*, **26** (4), 367 (1955).)

The overall quantum-counting efficiency of a Geiger-Müller counter depends not only on the geometrical design and choice of filling gas and pressure, but also upon the transmission of the window for the particular wavelength being measured and the extent of any insensitive regions in the tube volume. Currently, beryllium or mica windows are almost universally used because of their high transparency to the longer characteristic x-rays commonly employed. Thus, as can be seen in Fig. 5-21, an argon-filled Geiger-Müller counter with mica–beryllium windows possesses an overall quantum-counting efficiency for longer wavelengths of the order of 60 per cent.

While the positive ions produced in one discharge are traveling outward toward the cathode (expanding positive-ion sheath), the field near the anode wire falls below the threshold potential and no incident photons can generate pulses until the potential once more rises to V_g. This so-called *dead time* is typically about 1.0 to 1.5×10^{-4} sec for Geiger-Müller counters. Moreover, full-size pulses will not be generated until the positive ions have been actually collected by the cathode. This time interval between the initiation of a given pulse and the recovery of the full working potential is considerably longer than the dead time, and is termed the *recovery time*. The recovery time is characteristic of a given counter design and the associated circuitry, and generally lies between 4 and 10×10^{-4} sec.

Figure 5-22 is a diagram of the type of pulse pattern obtained from a Geiger-Müller counter and portrayed on an oscilloscope with the aid of a one-shot sweep circuit [87,88]. Pulses from the preamplifier of the Geiger-Müller tube trigger the sweep circuit and are also fed to the vertical plates of the oscilloscope. The output of the sweep circuit is supplied to the horizontal plates. Since each horizontal sweep is synchronized with the application of a pulse to the vertical plates, a strong, persistent pulse profile appears on the screen (heavy line in Fig. 5-22). At

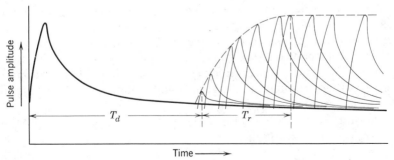

Fig. 5-22. Oscillographic pulse pattern obtained with a one-shot sweep circuit.

moderately high counting rates, fainter pulse profiles of variable amplitude flash into view momentarily near the end of the sweep (represented by thin-lined profiles at right in Fig. 5-22). The distances T_d and $T_d + T_r$ correspond, respectively, to the dead time and recovery time as defined above. We may note that the foregoing technique not only reveals the shape of a pulse on the time axis, but also furnishes a direct means of measuring to a fair approximation the inherent dead time and recovery time of a Geiger-Müller counter.

In its practical application to x-ray diffraction measurements, the major drawback of the Geiger-Müller counter is its large dead time and consequent nonlinear response at relatively low counting rates. In quantitative considerations of this nonlinearity problem, it is preferable to characterize the *resolving time* τ, rather than the dead or recovery time, of the counter. The *resolving time* τ is slightly larger than the dead time, and is defined as the time that must elapse after a given pulse occurs before a succeeding pulse recovers to the necessary amplitude to trip the first stage of the scaling circuit. Because of the large resolving times of Geiger-Müller counters, a significant fraction of the incident x-ray quanta fail to be counted even at counting rates as small as a few hundred counts per second. Under such conditions the response of the counter is said to be nonlinear. In numerical terms the response of a single-chamber Geiger-Müller counter is perceptibly nonlinear above 100 cps. This limitation and means of dealing with it are discussed in more detail in Section 5-3.3.

Further specific limitations of the Geiger-Müller counter are the somewhat restricted angular aperture imposed by its end-window construction and nonuniform radial sensitivity of the window (see Figs. 5-17 and 5-23A). Because of the sharp reduction in sensitivity at the central anode wire, it is essential that the counter window be aligned with respect to the receiving slit so that one of the lateral zones of relatively uniform sensitivity intersects the entire diffracted beam. Despite the limitations just cited, Geiger-Müller counters can be very satisfactorily employed in the measurement of weakly scattered x-rays (as in many small-angle-scattering investigations, for example) and in certain situations in which high levels of stability and reliability are of paramount importance. Low initial cost, simplicity of associated circuitry, and long life are additional features that commend Geiger-Müller counters for use under appropriate conditions.

Following is a suitable series of steps for putting a Geiger-Müller counter into operation [80]:

1. If the detection circuitry has a pulse-height discriminator and analyzer circuit, set it in the pulse-height discrimination mode.

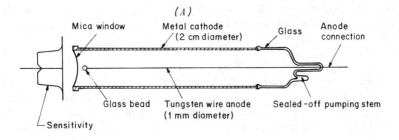

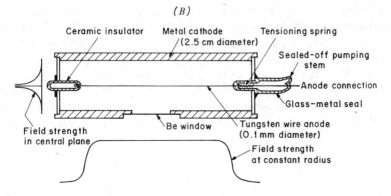

Fig. 5-23.   (*A*) Geiger-Müller counter, showing sensitivity across the window. (*B*) Proportional counter, showing variation of field strength along its length and across its diameter. (Courtesy of U. W. Arndt and B. T. M. Willis, *Single Crystal Diffractometry*, Cambridge University Press, 1966.)

2. Set the detector angle at an appropriate value of 2θ to receive a diffracted beam of moderate intensity (not exceeding about 300 cps).

3. With x-rays off and fixed minimum amplifier gain, prepare a noise curve by plotting counting rate against applied voltage.

4. With x-rays on and the diffracted beam entering the counter, and with a minimal usable amplifier gain, raise the voltage by steps and plot the response curve of counting rate versus applied voltage (the noise from the amplifier is normally greater than that produced by the Geiger-Müller tube).

5. Select as the operating voltage a value 50 to 75 V higher than the knee (point V_p in Fig. 5-19).

The useful life of a Geiger-Müller counter ends when the quenching gas has been consumed. Organic vapors such as methanol are consumed rather rapidly, whereas chlorine is consumed only very slowly, resulting in a relatively long life time. As the supply of quenching gas diminishes with use, the point V_f (Fig. 5-19) shifts to progressively lower voltages, shortening the plateau.

   C. Proportional Counters. Because of fluctuations in the gas absorption and multiplication processes, the production of a Townsend avalanche is a statistical process, so that when monochromatic x-rays are detected by a proportional counter, the pulse heights are not identical but exhibit a Gaussian-like distribution about the mean value [82–84]. Figure 5-24 is a plot of number of pulses versus pulse energy, in kiloelectron volts, as registered by a xenon-filled proportional counter receiving a crystal-monochromatized beam of CuKα x-rays [83]. The principal peak with a mean energy of 8 keV, which we may designate $\bar{V}$, is due to CuKα x-rays of wavelength 1.54 Å. Under the conditions of the experiment the full width of this peak at half-maximum height W was 1.2 keV. Thus the ratio of the width to the mean energy is 0.15, or 15 per cent, which is a measure of the energy resolution of the detector system for CuKα radiation. It is worthwhile to point out that W is proportional to the square root of the energy of the incident x-ray photons,

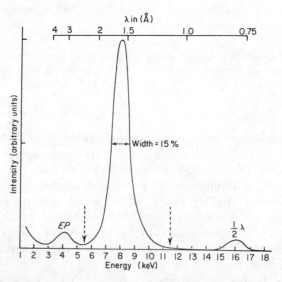

Fig. 5-24.   Pulse-height distribution curve for a xenon-filled proportional counter with crystal-reflected CuKα radiation, showing the escape peak EP and the ½λ component. The normal setting of the "window" is indicated by the arrows. (Courtesy of U. W. Arndt and B. T. M. Willis, *Single Crystal Diffractometry*, Cambridge University Press, 1966.)

whereas $\bar{V}$ is directly proportional to the photon energy, from which it follows that $W/\bar{V}$ varies inversely as the square root of the energy of the incident x-rays[82]. For this reason the energy resolution for $K\alpha$ x-rays, as measured by the ratio $W/\bar{V}$, varies in a regular manner with the atomic number of the target element.

In Fig. 5-24 the small peak at 16 keV is caused by the first harmonic of the $CuK\alpha$ radiation with a wavelength of 0.77 Å. The small peak at 4 keV is typical of a proportional counter, and is known as the *escape peak*. It originates as follows[82, 84]. If the energy of the incident x-ray photon exceeds that of a critical absorption edge of the atoms comprising the filling gas, fluorescent x-ray photons are produced, which may either be totally absorbed in the pulse-producing process within the counter volume or escape from the active counter volume, thereby producing the escape peak. Escape pulses have an average amplitude V_2 proportional to the difference between the energies of the incident photon E_1 and the fluorescent photon E_2:

$$V_2 = k(E_1 - E_2). \tag{5-25}$$

It should be noted at this point that when $MoK\alpha$ x-rays (wavelength 0.711 Å) are measured with a krypton-filled counter (KrK absorption edge at 0.865 Å), the escape peak generated by the fluorescent Kr$K\alpha$ radiation is actually larger than that of the primary $MoK\alpha$ peak. For this reason krypton-filled counters cannot be satisfactorily used for measuring $MoK\alpha$ x-rays. Lang[84] and Parrish[89] have discussed escape peaks in x-ray diffractometry in some detail.

Unlike Geiger-Müller counters, proportional counters are designed with side windows (Fig. 5-23B) and are filled with a heavier gas (usually xenon), necessitated by the short absorbing path length. The side-window construction ensures that entering x-ray photons will be absorbed within an active volume characterized by a highly uniform electric field, which is prerequisite to good proportionality between the incident photon energy and the resulting pulse size. The extremely narrow spatial zone ionized during a given avalanche leaves the remainder of the gas volume active for the detection of a second photon arriving as soon as 0.2 μsec after the first. Thus, in principle, counting rates as high as 10^7 cps can be attained without loss of linearity of response[79, 83]. However, a proportional counter is used in x-ray diffraction principally because of its wavelength- (energy-) discriminating capability, and for this purpose it is necessary that the gas amplification factor A be kept smaller than 500. This permits effective pulse-height discrimination (Section 5-3.4), but only up to counting rates not exceeding 5×10^4 cps[83]. Furthermore, the resolving time imposed by the electronic circuitry involved is of the order of 1 μsec, which sets an effective upper limit of about 10^4 cps for overall linearity of response.

With monochromatic x-rays a plot of counting rate against applied voltage for a proportional counter shows a plateau resembling that of a Geiger-Müller counter (see Fig. 5-25B). However, the response curve of counting rate versus applied voltage originates in a different way. As the voltage is increased, the threshold is reached when the highest-energy pulses in the pulse-amplitude distribution are amplified just enough to exceed the discriminator level at which the counter circuit is set. As the voltage is further increased, the plateau commences at the point where all the pulses in the amplitude distribution generated by the characteristic x-ray wavelength being used (see again Fig. 5-24 and related text) exceed the discriminator level and are counted. For a proportional counter the location of the plateau is a function not only of the applied voltage, as in Geiger-Müller counters, but also the x-ray quantum energy (wavelength) and the amplifier gain.

Figure 5-25B, from the work of Parrish and Kohler[82], shows typical plateaus for three monochromatic x-ray wavelengths as obtained with a discriminator (base) level of 5 V and an amplifier gain of about 2000. This discriminator setting eliminates the noise pulses generated by the counting circuitry, which have maximum amplitudes of only a few volts. From Fig. 5-25B it is seen that at any given voltage a wide spectral range is detected; for example, a voltage of 1500 registers all Ag$K\alpha$ x-rays but no Cu$K\alpha$ x-rays or other longer wavelengths. It is worth noting that this kind of discrimination, which is inherent in a proportional counter, decreases the general background of the diffraction pattern by eliminating all wavelengths larger than a given cutoff value without the necessity of employing a special pulse-height analysis circuit (see Section 5-3.4). From the pronounced flatness of the plateau, it is evident that when the operating voltage is fixed at some point on it, the intensity increases only very slowly with voltage, and therefore a degree of regulation of ± 0.1 per cent is highly satisfactory.

Figures 5-21 and 5-26A (broken line) show that the net quantum-counting efficiency of a xenon-filled proportional counter with a mica-plus-beryllium window reaches a broad maximum in the wavelength range 1.5 to 2.0 Å, thus being favorable for the detection of Cu$K\alpha$ radiation. However, the much lower efficiency in the neighborhood of 0.7 Å renders it unsuitable for measuring Mo$K\alpha$ x-rays.

The following procedure may be utilized in order to put a xenon-filled proportional counter into operation[80]:

1. Set the analyzer circuit for pulse-height discrimination (PHD), permitting a minimum discriminator voltage to be selected.

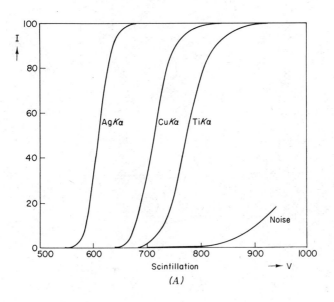

(A)

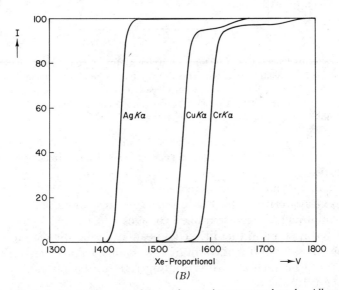

(B)

Fig. 5-25. Typical plateaus for several monochromatic x-ray wavelengths. All curves normalized to same intensity at highest voltage. The noise curve of the photomultiplier is plotted in counts per second. (Courtesy of W. Parrish and T. R. Kohler, *Rev. Sci. Instrum.*, **27**(10), 795 (1956).)

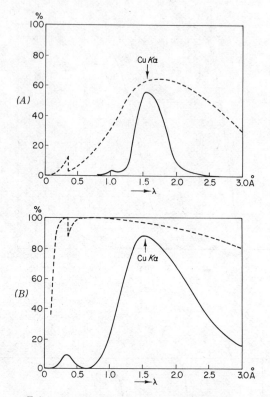

Fig. 5-26. Detector efficiency as a function of wavelength for (A) xenon-filled proportional counter, and (B) scintillation counter. The dashed curves show the efficiencies of the counters themselves; the solid curves show the efficiencies for CuKα radiation with a pulse-height analyzer set to accept 90 per cent of the distribution curve. (Courtesy of W. Parrish and T. R. Kohler, *Rev. Sci. Instrum.*, **27** (10), 795 (1956).)

2. Set the detector at a 2θ angle to receive a diffracted beam of moderate intensity.

3. Select an intermediate amplifier gain.

4. With x-rays off and a fixed amplifier gain, prepare a noise curve by plotting counting rate against applied voltage.

5. With x-rays on and the diffracted beam entering the counter, raise the voltage by steps and plot counting rate against voltage. The beam intensity should be such that the counting rate on the plateau does not exceed about 10^4 cps.

6. Select as the operating voltage the value on the plateau at which the increase in counting rate with voltage is, at least approximately, a minimum.

D. Scintillation Counters [79,80,83]. A scintillation counter consists of two basic elements, (1) a crystal that fluoresces visible light (scintillates) when struck by x-ray photons, and (2) a photomultiplier which converts the light to electrical pulses. Figure 5-27, from Arndt and Willis[83], shows a type of construction in which the sealed scintillator assembly is removable. The entire counter is enclosed in a light-tight container, which may also exert a shielding influence against magnetic fields. Table 5-6 lists the characteristics of four scintillators for detecting x-rays, of which the first, a thin waferlike single crystal of thallium-activated sodium iodide [abbreviated NaI(Tl)] is currently in most general use. Samson[90] has shown that thorough moisture-proofing of the crystal surface is essential to optimal working life, and has described an effective method of encapsulation employing a resin seal.

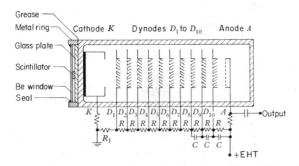

Fig. 5-27. Scintillation counter. In the type of construction shown, the sealed scintillator assembly is removable. The entire counter must be surrounded by a light-tight can, which may also provide shielding against magnetic fields. The resistor R_1 is chosen in relation to other resistors of the chain to provide the recommended $K–D_1$ voltage. The decoupling condensers C are required only when the output pulses are large. (Courtesy of U. W. Arndt and B. T. M. Willis, *Single Crystal Diffractometry*, Cambridge University Press, 1966.)

Table 5-6. Properties of Scintillators for X-rays[83]

Material	Fluorescence Efficiency	Maximum Emission (Å)	Remarks
NaI(Tl)	0.08	4200	Single crystal, very deliquescent
CsI(Tl)	0.04	4200–5700	Single crystal, nondeliquescent
CaI$_2$(Eu)	0.16	4700	Single crystal, very deliquescent
ZnS(Ag)	0.28	4500	Powder only[a]

[a]The amount of light collected is generally less than with monocrystalline scintillators.

An incident x-ray photon is transmitted by the beryllium window and absorbed by an atom of the crystal, which produces a photoelectron and commonly one or more Auger electrons that activate a large number of the fluorescent centers located at the thallium ion sites[79]. In this way the energy of the x-ray photon is converted into the energy of a considerable number of fluorescent photons in the optical region. For about 10 light photons, 1 electron is liberated from the cathode of the photomultiplier tube. The light-collection factor can be made nearly unity by coating the outside faces of the crystal with a reflecting layer, and by optically coupling the inner face to the front of the photomultiplier tube with a thin layer of transparent grease. The photoelectrons ejected from the cathode are accelerated and strike the first photomultiplier electrode, called a dynode, which is maintained at a positive potential with respect to the cathode. At the first dynode additional electrons are produced by a secondary emission process with a multiplication factor of about 4. These electrons are accelerated to the second dynode, where the absorption, reemission, and multiplication processes are repeated, after which the electron stream continues to be amplified at a succession of dynodes, commonly numbering 10 or 11 in all, before final collection takes place at the anode of the photomultiplier. The overall result is an amplification factor as high as 10^6 to 10^8, the specific value depending on the dynode voltage. The output pulse is fed to an amplifier, which activates the counting circuits. Normally, the photomultiplier is connected as shown in Fig. 5-27, with the chain of resistors R providing the necessary accelerating potential differences between the dynodes.

In four respects the performance of a scintillation counter closely resembles that of a proportional counter:

1. The pulse amplitudes produced by monochromatic x-radiation are subject to statistical fluctuations from their mean, resulting in a Gaussian-type distribution. The functional spread in pulse heights is proportional to $n^{1/2}$, where n is the number of initial photoelectrons produced by one incident x-ray photon[83].

2. The mean pulse amplitude is proportional to the energy of the incident x-ray photon.

3. Because of the very short decay time of the scintillation crystal (phosphor), about 10^{-7} sec, the dead time is approximately 0.2 μsec.

4. The response curve of counting rate versus applied voltage exhibits a very flat plateau (refer to Fig. 5-25A), the threshold of which depends on the x-ray wavelength and the amplifier gain. In the curves of Fig. Fig. 5-25A, the discrimination level is 5 V and the amplifier gain about 1000.

In three important respects the performance of a scintillation counter differs from that of a proportional counter:

1. The quantum-counting efficiency is between 90 and 100 per cent over a broad spectral range (Fig. 5-21), including the short-wavelength range encompassing $AgK\alpha$ and $MoK\alpha$ radiations, which are poorly detected by xenon-filled proportional counters.

2. The energy resolution as measured by the ratio $W/\bar{V}$ (see Section 5-3.2C) is considerably less sharp. Whereas for proportional counters $W/\bar{V}$ values of 15 or 20 per cent are practicable, for scintillation counters the resolution is generally about 50 per cent. This difference is illustrated by the solid-line curves in Fig. 5-26A and B.

3. Scintillation counters give much higher background counting rates, resulting from ohmic leakage, thermionic and field emission of electrons, and gaseous ionization. To achieve the optimum peak-to-background ratio, it is necessary therefore to employ relatively low operating voltages to suppress the noise pulses, and to operate at moderate amplifier gains to enhance the desired x-ray pulses as much as possible. With reference to Fig. 5-25A, the right-hand curve shows the noise in a typical scintillation counter in counts per second as a function of applied voltage. The noise counts become troublesome at longer wavelengths where the amplitudes of the noise pulses and low-energy x-rays are approximately equal. Thus it would not be possible to measure all the $TiK\alpha$ x-ray pulses (wavelength 2.75 Å) without encountering a noise background of about 9 cps, unacceptably large for the measurement of weakly diffracted x-rays. In conclusion, it may be emphasized that longer wavelengths can be much more satisfactorily measured with proportional counters because of their negligible instrumental noise levels.

Following is a suggestive procedure for putting a scintillation counter into operation [80]:

1. Set the analyzer circuit for the pulse-height discrimination mode.

2. Set the detector at a 2θ angle for the reception of a diffracted beam of moderate intensity.

3. With x-rays off and a fixed amplifier gain, prepare a noise curve by plotting counting rate against applied voltage. Above a particular voltage this curve rises rapidly to a level much higher than would be observed with a Geiger-Müller or proportional counter. For generally satisfactory diffraction measurements, the scintillator detector chosen should permit the desired x-ray wavelengths to be measured at an applied voltage low enough to keep the generation of noise pulses below a level of 2 or 3 cps.

4. Adjust the amplifier gain to a maximum value consistent with a minimum practicable value of the operating voltage. In this regard it

should be kept in mind that the photomultiplier tube is usually noisy compared with the amplifier. Most photomultiplier tubes can be permanently damaged by the application of voltages above about 1200.

5. With x-rays on and the diffracted beam entering the counter, start with the maximum gain and minimum voltage achieved in step 4 and plot counting rate against increasing voltage, the gain being held constant.

6. Select as the operating voltage a value on the plateau just low enough to hold the noise level below 2 to 3 cps.

E. Solid-State (Energy-Dispersive) Detectors. This topic actually embraces both the detectors and their utilization in x-ray diffraction measurements. Although the latter subject should logically appear in Section 5-4, in the interests of coherence we have chosen to present both aspects of the topic here.

The application of semiconductor detectors to x-ray diffraction is currently in its infancy but developing very rapidly; therefore we attempt only an introductory sketch of the state of the art. Giessen and Gordon [91] have suggested that x-ray diffraction measurements by energy-dispersive analysis be designated *spectrometric powder diffractometry*. Semiconductor detectors were originally developed for nuclear spectroscopy, but by the mid-1960s it was generally realized that such detectors also afforded special advantages in x-ray emission spectroscopy (fluorescence analysis), particularly as employed in an electron microprobe [92–95]. Only rather recently have serious efforts been made to utilize solid-state detectors in x-ray diffractometry [91, 93, 96].

For the detection of x-ray photons with energies in the diffraction range, lithium-drifted silicon [Si(Li)] detectors have generally superior characteristics and are most commonly used, although Ge(Li) detectors have advantages in some circumstances. The impinging x-ray photons create electron-hole pairs in the cooled (77°K) detector. The number of electron-hole pairs generated is proportional to the energy of the absorbed photon, and they are attracted to the detector electrode of opposite charge (bias), thus producing a pulse. The pulse amplitude in turn is proportional to the number of electron-hole pairs, hence also to the energy of the incident x-ray photon. Because of the proportionality of pulse height to photon amplitude, the detector is capable of differentiating x-ray energies by means of pulse-height analysis (see Section 5-3.4A) over a wide energy range.

As diagramed in Fig. 5-28, an unfiltered high-intensity polychromatic x-ray source, usually a target of molybdenum or tungsten (wolfram), irradiates a flat powder specimen at a convenient fixed angle θ. The

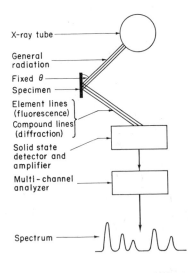

X-ray tube

General
radiation

Fixed θ

Specimen

Element lines
(fluorescence)

Compound lines
(diffraction)

Solid state
detector and
amplifier

Multi-channel
analyzer

Spectrum

Fig. 5-28. Schematic diagram of an energy-dispersive apparatus for x-ray powder diffractometry.

diffracted x-rays are recorded by the energy-dispersive detector, and its pulses appropriately amplified and fed into, and accumulated in, a multichannel energy analyzer which comprises 200 to 4000 channels, depending on the energy range to be scanned and the degree of resolution desired. The general-radiation spectrum of the source not only generates the x-ray diffraction pattern of the irradiated specimen, but also excites any x-ray fluorescence lines of the specimen's constituent atoms that fall within the energy range of the source (see Fig. 5-29B). Figure 5-29A shows the response of the detector to the undiffracted primary beam, while Fig. 5-29B shows the diffraction pattern of a platinum specimen together with its L-fluorescence lines. SWC is the short-wave (high-energy) cutoff of the primary beam corresponding to 45 keV. Figure 5-30 shows the diffraction pattern of α-quartz obtained by other investigators using a tungsten-target x-ray source and a Si(Li) detector. With multichannel energy-discrimination systems such as are employed with x-ray energy-dispersive detectors, the output may be displayed in the form of a histogram on a video screen or with an X-Y recorder, or it may be punched digitally on cards or paper tape for computer processing. Overlap of fluorescence spectral lines and diffraction lines from the specimen can be avoided by the proper choice of θ. A change in θ causes the diffraction pattern to shift, but leaves the positions of the fluorescence lines unchanged.

It is to be emphasized that energy-dispersive diffraction patterns such as those of Figs. 5-29 and 5-30 are obtained with the detector at a fixed angle and all lines accumulating simultaneously. With the use of very

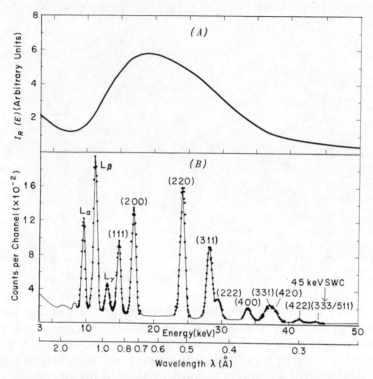

Fig. 5-29. (*A*) Response curve of Si(Li) semiconductor detector to undiffracted poly-chromatic beam from an iron target operated at 45 kV peak and 8 mA. (B) Diffraction pattern and *L*-fluorescence spectrum of platinum sheet obtained at $\sin \theta = 0.186$ with the same source and detector; *SWC*, short-wave cutoff of x-ray beam. Collection time was 1.5 hr, providing excellent statistical quality. (Courtesy of B. C. Giessen and G. E. Gordon and Philips Electronic Instruments [91].)

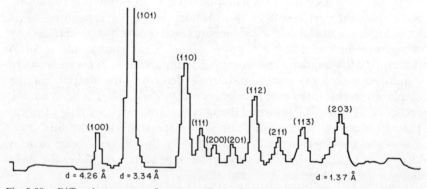

Fig. 5-30. Diffraction pattern of α-quartz obtained with a Si(Li) detector and a tungsten-target tube operated at 30 kV peak and 1 mA. Collection time 14 sec. Pattern displayed on a strip-chart recorder. (Courtesy of Nuclear Equipment Corporation.)

intense x-ray sources, such as rotating-anode tubes, and with the most sensitive detectors, it is possible to record a recognizable diffraction pattern within a few seconds. Thus the method is of unusual value in dynamical diffraction investigations wherein rapid changes in the spacings or intensities of lines occur. Another advantage of energy-dispersive over conventional diffractometry is that the entire pattern is recorded simultaneously and therefore a high degree of instrumental stability is not essential, as is true in the preparation of Debye-Scherrer photographic patterns. Also, pulse-height analysis is more efficacious with semiconductor detectors than with proportional counters because the attainable resolution, with a Si(Li) detector for example, is at present in the range of 160 to 200 eV, sharper by a factor of at least 5 [94, 97]. Figure 5-31 compares the response curves of typical scintillation, proportional, and Si(Li) detectors. The very superior resolving power of the Si(Li) detector is obvious.

A limitation on the usefulness of energy-dispersive powder diffractometry is imposed by the nonuniform response function $I_R(E)$ at different energies (wavelengths), as portrayed clearly in Fig. 5-29A. This response function is the product of two factors [91, 93]:

$$I_R(E) = I_{cont}(E) P_a(E), \qquad (5\text{-}26)$$

where $I_{cont}(E)$ is the intensity of the continuous x-radiation of energy E, and $P_a(E)$ is the probability of photopeak absorption of photons of energy E by the detector. Thus the $I_R(E)$ profile is the envelope of the

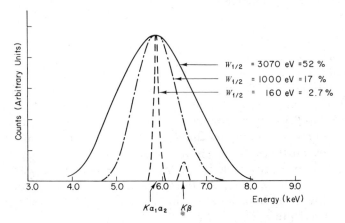

Fig. 5-31.   Comparison of the response of three energy-dispersive x-ray detectors to the manganese x-ray spectrum resulting from the decay of ^{55}Fe. NaI(Tl) scintillation counter, ———; proportional counter, ·—·—; small Si(Li) detector, ---. (Courtesy of R. S. Frankel and D. W. Aitken, *Appl. Spectrosc.*, **24**, 557 (1970).)

diffraction pattern, causing lines near the maximum at about 19 keV (Fig. 5-29A) to be relatively much more intense than those near the high-energy tail of the curve, for example, the 422 and combined 333/511 reflections of platinum. Another limitation of energy-dispersive diffractometry is the relatively small linear response range, which extends only to about 1000 cps at present; however, current efforts to improve this feature look promising. Thus at present the method is more adapted to the determination of interplanar (d) spacings than relative intensities. Nevertheless, energy-dispersive diffractometry holds promise for application to the qualitative identification of crystalline powders and other procedures requiring d spacings of only moderate accuracy.

For deriving d values from an energy-dispersed diffraction pattern, Bragg's law must be expressed in terms of the x-ray photon energy E[91, 93]:

$$E = \frac{hc}{\lambda} = \frac{12.398}{\lambda}. \tag{5-27}$$

For E expressed in kiloelectron volts and λ in ångström units, and since $\lambda = 2d \sin \theta$, we have

$$E = \frac{12.398}{2d \sin \theta} = \frac{6.199}{d \sin \theta},$$

or

$$d = \frac{6.199}{\sin \theta} \frac{1}{E}. \tag{5-28}$$

From equation 5-28 we note that the magnitude of the reciprocal-lattice vector $\mathbf{s}$ is proportional to E:

$$|\mathbf{s}| = \frac{1}{d} = \frac{\sin \theta}{6.199} E. \tag{5-29}$$

Finally, it is important to observe that solid-state detectors are capable of substantially improving the performance of conventional 2θ-dispersive diffractometers in certain respects[96]. Thus the resolution of a small Si(Li) detector is ample to discriminate completely against fluorescence radiation from the specimen and to render unnecessary the use of a β filter. Furthermore, such an apparatus can provide peak-to-background ratios comparable in quality to those obtainable with focusing crystal monochromators. At the same time, the peak intensities are stronger than those obtainable with a proportional counter in conjunction with a β filter.

For further information on these important new techniques, the reader must be referred to the literature (especially references [91–93,

96, 98, 99]) and to very informative and up-to-date brochures available from several manufacturers.*

5-3.3 Nonlinearity of Detector Response

In powder diffractometry this subject is important mainly when Geiger-Müller counters are employed, since their large resolving times of 100 μsec or more result in significant losses in counts even at rates as low as 100 cps. However, proportional, scintillation and solid-state detectors, with resolving times of the order of microseconds, suffer counting losses well under 1 per cent even at a counting rate of 5000 cps, which is only infrequently reached in powder diffractometry. Whatever the magnitude of the count losses, such failure of the detector system at higher counting rates to register all photons entering the counter is commonly referred to as *nonlinearity of detector response*.

Reasonably accurate corrections for resolving-time losses of Geiger-Müller counters can be calculated mathematically, provided the effective resolving time τ is known. Then, if N_0 is the observed counting rate and N the true counting rate, the following expression is reasonably accurate for corrections not exceeding about 10 per cent [83, 100]:

$$N = \frac{N_0}{1 - N_0\tau}. \tag{5-30}$$

It should be emphasized that equation 5-30 is not applicable to proportional, scintillation, or solid-state detectors, because the resolving time is more largely dependent upon the associated circuitry than the counter itself. Campbell [101] has shown that, for random arrival of x-ray photons, the standard deviation of the corrected counting rate N is approximately

$$\sigma_N \simeq \left[\frac{N(1 + N\tau)}{t}\right]^{1/2}, \tag{5-31}$$

t being the observation time. Equation 5-31 is asymptotically correct for $t \to \infty$ and therefore holds closely only for t large.

For Geiger-Müller counters nonlinearity losses can also be determined to a fair approximation by the method of multiple foils, first described by Lonsdale [102]. The theoretical basis for this method is as follows [88]. Before entering the counter tube, monochromatic x-rays of wavelength λ are reduced in intensity by a series of absorbing foils of uniform thickness x and linear absorption coefficient μ. If the response were linear,

---

*For example, we may cite Kevex Corporation, Analytical Instruments Division, 898 Mahler Road, Burlingame, California; Nuclear Equipment Corporation, 931 Terminal Way, San Carlos, California; Ortec, Inc., 100 Midland Road, Oak Ridge, Tennessee.

the relative counting rates for n and 0 foils would be equal to the relative intensities, which is to say

$$\frac{N_n}{N_0} = \exp(-n\mu x), \tag{5-32}$$

or, reducing to common logarithms,

$$\log_{10} N_n = \log_{10} N_0 - \frac{\mu x}{2.303} n. \tag{5-33}$$

Equation 5-33 shows that in the absence of response errors a plot of $\log_{10} N_n$ against n, the number of foils, should give a straight line with a slope of $-\mu x/2.303$.

Figure 5-32 illustrates how this method can be applied to the determination of approximate corrections for coincidence errors. The solid line is a plot of the experimentally observed counting rates for various

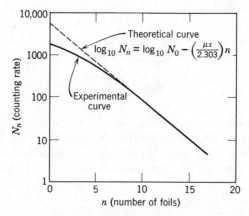

Fig. 5-32.   Plot of theoretical and experimental counting rates as a function of number of absorbing foils in the multiple-foil method for nonlinear response corrections. (Courtesy of L. E. Alexander, E. Kummer, and H. P. Klug, *J. Appl. Phys.*, **20**, 735 (1949).)

numbers of foils ranging from zero up to a number sufficient to reduce the counting rate to a very low level, for example, 5 cps. The beam intensity must be large enough to provide an intensity with zero foils that is at least as great as the largest anticipated in actual work. Sufficient intensity can be obtained by diffracting the x-ray beam from the $(10 \cdot 1)$ planes of a quartz plate with its surface ground parallel to $(10 \cdot 1)$, or from the (111) planes of a fluorite plate with its surface ground parallel to (111). After the experimental points have been plotted, the theoretical straight line is located as accurately as possible by extrapolating from the several experimental points at low counting rates (preferably not greater than about 50 cps).

The multiple-foil method as just described is not very accurate in practice because of two principal limitations: (1) The foil thickness is seldom known accurately and there is commonly a large variation from foil to foil. This is equivalent to saying that the absorption factor, $\exp(-\mu t)$, in principle invariant, actually varies from one foil to another. (2) Monochromatic incident radiation, reflected from a single crystal or polycrystalline specimen must be used, and second- and third-order harmonics are likely to be present. Short[103] has pointed out that objection (1) can in principle be overcome by plotting as the abscissa (Fig. 5-32) the accumulated absorber thickness rather than number of foils. Nevertheless, the accumulated thickness is also very difficult to determine accurately. Also, in practice objection (2) above can be met, when $CuK\alpha$ x-rays are employed, by using nickel foils and reducing the x-ray tube voltage below the level required to excite the $\lambda/3$ harmonic (approximately 25 kV peak). This is effective because it happens that the absorption factors of nickel for the λ and $\lambda/2$ components of copper radiation are equal[102].

Bragg[104], Short[103], and Burbank[105] have shown that it is practicable to determine nonlinearity corrections for any type of x-ray counter with the use of only a single absorbing foil. Their experimental procedures are similar and consist essentially of measuring the *apparent* absorption factor of a suitable absorber at various counting rates. The apparent absorption factor remains constant until the linearity range is exceeded, when it begins to increase rapidly (see Fig. 5-33). It is unnecessary to use strictly monochromatic radiation, provided the beam is attenuated by the reduction of the tube current and slit width[103]; nevertheless, if monochromatic x-rays diffracted by a powder or single crystal are employed, the counting rate can be conveniently increased by the removal of successive auxiliary absorbing foils. The calibration foil itself should be of sufficient thickness to absorb a fraction of from one-half to two-thirds of the incident-beam intensity. Employing a $MoK\alpha$ reflection from a lithium fluoride (LiF) crystal, Burbank[105] placed a 0.004-in. zirconium foil between the x-ray tube and the crystal.

We now outline the interpretative procedure suggested by Burbank [105]. Let I_0 and I_1 be the counting rates in counts per second for the incident and transmitted rays, respectively, and I_0^{loss} and I_1^{loss} the corresponding counting losses incurred as a result of nonlinear response. Then

$$I_1 + I_1^{loss} = (I_0 + I_0^{loss}) R_c, \qquad (5\text{-}34)$$

where R_c is the constant value of the ratio I_1/I_0 at counting rates low

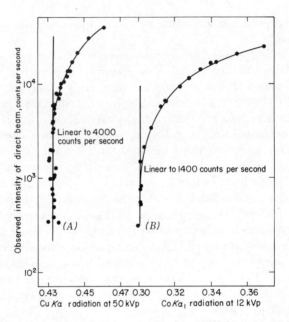

Fig. 5-33. Variation in the apparent absorption factor of 0.0007-in. nickel foil with increasing counting rate. (*A*) Full-wave rectified Cu$K\alpha$ radiation at 50 kV peak; (*B*) half-wave rectified Co$K\alpha_1$ radiation at 12 kV peak. (Courtesy of M. A. Short, *Rev. Sci. Instrum.*, **31** (6), 618 (1960).)

enough that there are no losses. Then the counting loss in I_0 is

$$I_0^{\text{loss}} = \frac{(I_1 + I_1^{\text{loss}})}{R_c} - I_0. \tag{5-35}$$

The solution of equation 5-35 can be obtained by an iterative process [103], but the following method is more direct [105]. Above the intensity region of linear response

$$I_1 = R_1 I_0, \quad I_2 = R_2 I_1, \quad \ldots, \quad I_n = R_n I_{n-1},$$

while within the linear region ($I \leqslant I_n$)

$$I_{n+1} = R_c I_n.$$

Within the linearity range I_0, I_1, I_0^{loss}, and I_1^{loss} of equation 5-35 may with equal validity be replaced, respectively, by I_1, I_2, I_1^{loss}, and I_2^{loss}, leading to the result:

$$I_0^{\text{loss}} = \frac{I_1 + I_1^{\text{loss}}}{R_c} - I_0 = \frac{I_1 + [(I_2 + I_2^{\text{loss}})/R_c] - I_1}{R_c} - I_0 = \frac{I_2 + I_2^{\text{loss}}}{R_c^2} - I_0. \tag{5-36}$$

It follows from equation 5-36 that

$$I_0^{\text{loss}} = \frac{I_1 + I_1^{\text{loss}}}{R_c} - I_0 = \frac{I_2 + I_2^{\text{loss}}}{R_c^2} - I_0 = \frac{I_n - I_n^{\text{loss}}}{R_c^n} - I_0 = \frac{I_n}{R_c^n} - I_0 \quad (5\text{-}37)$$

(since I_n is within the range of linear response).

Substitution in equation 5-37 of $R_n I_{n-1}$ for I_n, $R_{n-1} I_{n-2}$ for I_{n-1}, and so on, gives

$$I_0^{\text{loss}} = \frac{R_n I_{n-1}}{R_c^n} - I_0 = \frac{R_n R_{n-1} R_{n-2} \cdots R_2 R_1 I_0}{R_c^n} - I_0,$$

from which finally

$$\frac{I^{\text{loss}}}{I_0} = \frac{R_n R_{n-1} R_{n-2} \cdots R_2 R_1}{R_c^n} - 1 = K - 1. \quad (5\text{-}38)$$

The numerical value of the constant K can be obtained from a plot of the experimental values of I_0 against R. Then for a given value of I_0 one reads off R_1, for $R_1 I_0 = I_1$ one reads off R_2, and so on, until the linear range is entered at I_n. The numerical values of $R_1, R_2, \cdots, R_n$ are then substituted in equation 5-38 and it is solved for K. The accuracy of the correction given by equation 5-38 may be estimated with the aid of the expression [105],

$$d\left(\frac{I_0^{\text{loss}}}{I_0}\right) = \left[K \frac{dR_n}{R_n} + \cdots + \frac{dR_1}{R_1} - \frac{ndR_c}{R_c}\right]. \quad (5\text{-}39)$$

The principal source of error is in the experimental curve of I_0 versus R, because of uncertainty in R_c resulting from the need for long counting times to ensure good counting statistics and the attendant requirement of stringent current and voltage stabilization. Conservatively assuming the stabilities in current and voltage to be ± 0.1 per cent and dR_c/R_c to be accurate to ± 0.001, we may write equation 5-39 as

$$\text{Per cent error} = -Kn \frac{dR_c}{R_c} \times 100$$

$$= -Kn\,(\pm 0.001) \times 100$$

$$= \mp 0.1\,Kn. \quad (5\text{-}40)$$

The $\mp$ sign has the significance that, if R_c is too large the corrections will be too small, and the converse.

Using the foregoing procedure, Burbank [105] studied the linearity of a counting system consisting of a General Electric scintillation counter, a cathode follower, a preamplifier, and a Berkeley scaler. The incident intensity I_0 was supplied by reflecting $\text{Mo}K\alpha$ radiation from a

LiF crystal. A 0.004-in. zirconium foil was interposed between the x-ray tube and crystal. At 50 kV, R_c was found to be 0.339, and the response was linear to within ± 0.1 per cent up to 6300 cps. At 10,000 cps the correction was 0.3 per cent, and at 19,960 cps, 2.8 per cent.

Fukamachi[106] has developed a variation of the single-foil method for fast counters, which provides not only corrections for the observed counting rates but also a numerical value of the resolving time of the counting system itself for particular experimental conditions, including especially the voltage and wave form of the rectified x-ray source. The method furthermore provides a very accurate value of the absorption factor of the attenuating foil or any other specimen under study.

Returning to the application of equation 5-30 to the correction of the counting rates of Geiger-Müller counters, we must emphasize an important complication that arises for all x-ray generators other than those yielding true constant dc potentials. For half- or full-wave rectifying generators, photons are emitted from the targets in clusters or bursts, the frequency, duration and pulse-density configurations of which are functions of the wave form (form factor) of the energy through the x-ray tube. Figure 5-34 illustrates these effects by oscillographic pulse patterns from a radioactive source A, a full-wave rectified source of x-rays B, and a half-wave rectified source C. The pulse density is zero until the x-ray tube voltage reaches the critical excitation potential V_0 of the characteristic radiation employed, after which it rises from zero to a maximum value when V reaches the peak operating potential (V_m, or kV peak).

Such periodically varying x-ray sources have the effect of contributing a nearly constant factor k to the correction term $N_0\tau$ of equation 5-30[83, 88, 107–109]:

$$N = \frac{N_0}{1 - N_0 k\tau}. \tag{5-41}$$

The coefficient k, known as the *form factor* of the source, always assumes values larger than unity, which means that intensity variations of the source increase the effective resolving time of the apparatus ($\tau_{\text{eff}} = k\tau$). The form factor k is defined mathematically by

$$k^{1/2} = \frac{\text{root-mean-square intensity of source}}{\text{mean intensity}}. \tag{5-42}$$

It is evident from these results that it is necessary to know both the resolving time τ of the Geiger-Müller counter and the form factor k of the x-ray source before we can attempt to correct experimental counting rates from half- or full-wave rectified x-ray generators.

We now treat briefly the determination of the resolving time of a

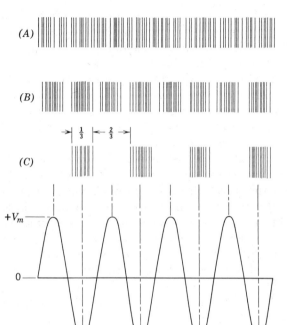

Fig. 5-34. Comparison of oscillographic pulse patterns from (A) a radioactive source, (B) a full-wave rectified source of x-rays, and (C) a half-wave rectified source of x-rays. (Courtesy of L. E. Alexander, E. Kummer, and H. P. Klug, *J. Appl. Phys.*, **20**, 735 (1949).)

Geiger-Müller counter to an accuracy of 10 per cent or better. Beers [110] has outlined a method in which two radioactive sources are counted separately and then together, giving counting rates N_1, N_2, and N_{12}. The resolving time is then calculated from the formula

$$\tau = \frac{N_1 + N_2 - N_{12}}{2N_1 N_2}.\qquad(5\text{-}43)$$

The application of this method entails several difficulties. Beers found that consistent results were realized only when pure β emitters were employed, a conclusion concurred in by some later investigators. Also, in dry atmospheres spurious counts may be obtained because of accumulation of charges on the counter window. This effect tends to be more pronounced when counts are recorded over long periods of time, such as are required for the present measurements because of the need to accumulate a very large number of counts in order to reduce statistical errors to a workable level. We obtained inconsistent results in making repeated tests of this sort with paired radium sources.

Barieau[111] modified the paired-source method of Beers so as to measure two beams of x-rays from the x-ray source itself by replacing the divergence slit with a plate containing two transmitting holes. This method directly determines the effective resolving time τ_{eff} of the Geiger-Müller counter coupled with the x-ray source under normal operating conditions. For a Norelco counter operated in conjunction with a full-wave rectified copper x-ray tube operating at 35 kV peak, Barieau obtained a value $\tau_{eff} = 215\ \mu\text{sec}$, which remained constant for counting rates between 180 and 750 cps.

In Section 5-3.2B we[88] described our use of the oscilloscopic technique of Stever[87] for the portrayal of Geiger-Müller pulse profiles, and also for the direct measurement of the inherent resolving time of the counter to an accuracy of ± 10 per cent or better. Pepinsky and co-workers[112] devised a refinement of Stever's technique in which the x-ray tube is pulsed into conduction by two sets of $\frac{1}{2}$-μsec, 1000-c triggers, the phases of which can be continuously varied with respect to each other. The two sets of x-ray pulses produce two standing pulse profiles on the oscilloscope screen and, by reducing the time separation until the second pulse just disappears, the investigator obtains a direct measurement of the dead time. Cochran[107] described a different method employing a γ-ray source to determine the resolving time; the results, though apparently reliable, were of somewhat less than 10 per cent accuracy.

The form factor k of a periodically varying x-ray source can be determined or estimated in several ways. It can be calculated from a knowledge of the wave form of the voltage across, and the current through, the x-ray tube, making use of empirical or assumed relations between voltage and intensity[83, 113]. It is also possible to measure k by varying τ; in this method a series of electronically defined paralysis times must be used, each being greater than the τ of the detector[83]. Some representative experimental values of k given by Arndt and Willis[83] are 3.2 and 3.4 for half-wave rectified sources, and 1.6 and 1.9 for full-wave rectified sources. We [88] have computed k for half- and full-wave rectified sources, on the assumption that the potential applied to the x-ray tube is sinusoidal and that the x-ray intensity varies as the square of the voltage in excess of the critical excitation potential[114]. For the representative case of the peak voltage equal to four times the critical excitation potential, this analysis of half-wave rectification leads to a limiting value of k of about 3.5 for vanishingly small counting rates, whereas for the largest counting rates likely to be met in practice k never assumes values smaller than about 3.3. The same assumptions as to wave form give 3.4 when k is computed by an analytical method from the defining expression 5-42. In full-wave rectified sources, the k values would be precisely one-half the above values, about 1.7. These numerical results agree well with the experimental values from Arndt and Willis cited above.

With a knowledge of τ and k for the Geiger-Müller counter and x-ray generator, respectively, observed counting rates can now be corrected by simply applying equation 5-41. For example, suppose that $\tau = 1.5 \times 10^{-4}$ sec and $k = 3.4$, both quantities being accurate to ± 10 per cent. At an observed counting rate of 400 cps, the fractional coincidence loss is

$$N_0 k\tau = 400 \times 3.4 \times 1.5 \times 10^{-4} = 0.204,$$

the accuracy being of the order of $\pm (10^2 + 10^2)^{1/2} = 14$ per cent. The true counting rate is then

$$N = \frac{400}{1 - 0.204} = 503,$$

which is accurate to about ± 3.5 per cent. This is approximately the largest correction for which expression 5-41 can be expected to yield even approximately correct results. The single-foil [103–105] or multiple-foil [88, 102] method of calibrating a Geiger-Müller counter for non-linear response is preferable when the coincidence loss exceeds 10 per cent.

Before concluding this section a brief discussion of multiple-chamber Geiger-Müller tubes is in order, even though such counters are now little used. As a first approximation the resolving time of a counter tube of n chambers is $\tau'/n^{1/2}$, τ' being the resolving time of one chamber. Because of the small dimensions of the chambers in such a tube, τ' is smaller than for a conventional single-chamber tube. Thus for $\tau' = 50 \, \mu$sec and $n = 10$, the net resolving time is about $50 \times 10^{-1/2} = 16 \, \mu$sec. For full-wave rectification this leads to coincidence losses of less than 1 per cent up to counting rates close to 400 cps. Hence, for a counter of these specifications, it is ordinarily unnecessary to correct counting rates smaller than about 500 cps. For reliable measurements at higher rates, τ and k should be accurately determined and corrections applied using equation 5-41, or alternatively the method of Barieau [111] can be applied.

Among other useful articles dealing with corrections for nonlinear response when periodically varying x-ray sources are employed, we may cite those by Trott [115] and Bernstein and Canon [116]. Eastabrook and Hughes [117] describe an ingenious circuit for eliminating resolving-time losses when a second Geiger-Müller counter is used to monitor the output of the x-ray tube while the first counter is measuring the diffraction pattern in the usual way. The circuitry permits neither counter to record a count while either counter is quenched, which results in the ratio of counts recorded being equal to the ratio of the x-ray photons received by the two counters. Furthermore, this ratio is independent of the quench time, the mean x-ray output, and the form factor of the source.

5-3.4 Monochromatizing Techniques

Although in principle polycrystalline x-ray diffraction techniques must be used in conjunction with strictly monochromatic x-radiation, in actuality such an ideal source of radiation is never attainable. Fortunately, several experimental devices and associated techniques can be employed that make it possible to approach true monochromatization to a degree depending on such factors as the rigorousness of the experimental conditions demanded, the pains the investigator is willing to exercise, and certain sacrifices he is prepared to accept, for example, monetary expense, beam intensity, speed of data collection, and so on. As a rough generalization, it may be said in regard to the diffraction patterns of well-crystallized substances that, the more nearly monochromatic the x-rays involved in the diffraction and detection processes, the higher (more favorable) the peak-to-background ratio are. This is true because the background originates very largely from the diffraction, or emission, by the specimen of wavelengths other than the primary monochromatic wavelength of the experiment. Furthermore, highly monochromatic techniques are absolutely essential in the quantitative evaluation of diffuse patterns produced by poorly or semicrystallized materials over wide angular ranges, and also by particulate and other inhomogeneous specimens at very small angles (small-angle scattering).

Bearing in mind, then, that *monochromatization* of x-rays is, strictly speaking, a misnomer (*wavelength discrimination* being possibly a more accurate term), and that actual experimental techniques always yield an imperfectly monochromatized beam, we shall describe the three principal techniques of current practical importance in diffractometry: (1) pulse-height discrimination (PHD) and pulse-height analysis (PHA); (2) filtration of the beam by elements of appropriate atomic number (in particular, Ross balanced filters); and (3) reflection of the direct or diffracted beam by crystals (crystal monochromators).

A. Pulse-Height Discrimination and Analysis. Considerable mention of this technique has already been made in Section 5-3.2 in conjunction with the discussion of proportional and scintillation counters and solid-state detectors, where it was pointed out that, when "monochromatic" x-rays are detected by one of these counters, the pulse amplitudes exhibit a Gaussian-like frequency distribution about the mean value, $\bar{V}$ (see Fig. 5-35 from the work of Miller[118], as well as Fig. 5-24). If the full-width at half-maximum height is designated W, and if decreasing values of $W/\bar{V}$ are taken as a measure of increasing resolution (see Section 5-3.2C), scintillation, proportional, and solid-state detectors possess increasing resolution in the order given (refer to Fig. 5-31).

Electronic discrimination between pulses of different amplitudes is effected by introducing into the counting circuitry following the linear amplifier (refer to Fig. 5-16) two auxiliary circuits, a *pulse-height discriminator*, which transmits only pulses above a certain base-line amplitude (V_B, Fig. 5-35) and a *pulse-height analyzer*, which blocks the transmission of all pulses with amplitudes larger than an upper limit V_C, thus establishing a transmitting energy window of width $V_C - V_B = \Delta V$. In the example shown in Fig. 5-35, if $V_B = 15$ and $\Delta V = 10$ V, only pulses with amplitudes between 15 and 25 V will be passed by the discriminator-

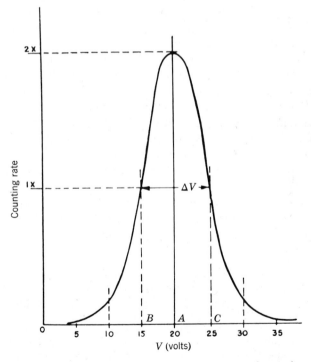

Fig. 5-35. Pulse-amplitude distribution generated when a monochromatic x-ray beam is measured by a proportional or scintillation counter. (Courtesy of D. C. Miller and Philips Electronic Instruments[118].)

analyzer window.* It is also seen from an inspection of Fig. 5-35 that even with a window width of 10 V an appreciable portion of the pulses produced by a monochromatic x-ray beam are lost. An increase in the window width increases the fraction of pulses counted, but only at the price of a further loss in resolution.

In practical operations it is the customary practice to adjust the window width so as to realize 85 to 90 per cent of the counting rate obtainable without PHD and PHA (with V_B set very low, but above the electronic noise level, and $\Delta V = \infty$). It should be stressed at this point that most relatively routine applications of x-ray diffractometry, such as compound identification and qualitative and quantitative analysis of polycrystalline mixtures, can be performed very adequately with the use of a β filter combined with PHD and PHA.†

From the foregoing discussion it can be appreciated that PHD and PHA are largely ineffectual in excluding pulses produced by wavelengths differing only slightly from the $K\alpha$ wavelength on which the window is centered, including $K\beta$ radiation. However, it is very effective in eliminating pulses from x-rays of much shorter or much longer wavelengths, for example, harmonics of $K\alpha$, large portions of the long- and short-wave continuous spectrum, and electronic noise pulses, the last-named being normally of much smaller amplitude than the pulses of $CuK\alpha$ and $MoK\alpha$ x-rays, for instance. Figure 5-36 shows the spectrum of a copper-target x-ray tube as detected using a β filter (A) without and (B) with PHD (actually PHD and PHA). The virtual deletion of the short-wavelength continuum contrasts sharply with the almost negligible attenuation of the residual $K\beta$ peak passed by the β filter. It is now pertinent to reiterate the observations made in Section 5-3.2 that, whereas the longer characteristic wavelengths of copper, iron, and so on, can be effectively measured with proportional counters and attendant PHD of intermediate resolution (about 20 per cent), "harder" x-rays such as the $K\alpha$ wavelengths of molybdenum and silver are much more efficiently detected with scintillation counters, although at the price of reducing the PHD resolution to about 50 per cent.

---

*Although not technically correct, it is common practice to use the terms pulse-height discrimination and pulse-height analysis interchangeably to refer *in toto* to the process of selecting only pulses falling within a given energy band ΔV, so that we may refer simply to the discriminator window or the analyzer window.

†E. L. Lippert (Owens-Illinois, Inc., Toledo, Ohio) has devised a graphic method which simplifies the selection of the base line and window setting of the pulse-height analyzer when used in conjunction with a scintillation counter. The procedure makes use of the fact that the pulse-amplitude distribution is Gaussian, which results in a plot of cumulative percentage counts versus base-line setting on normal probability graph paper being linear.

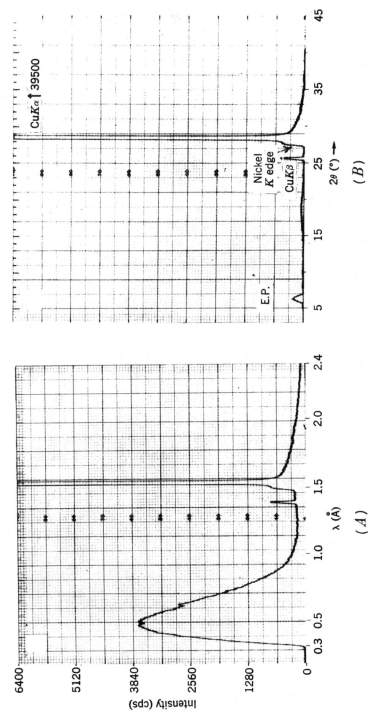

Fig. 5-36. Spectra from copper-target x-ray tube obtained with a scintillation counter without (A) and with (B) pulse-height discrimination. The small peak E.P. is due to the escape peak phenomenon. (After P. H. Dowling et al., *Philips Tech. Rev.*, **18**, 262 (1956–1957).)

The above treatment has been devoted to PHD with a single window, or channel, such as is normally employed in standard wavelength-dispersive diffractometry. However, with energy-dispersive solid-state detectors, because of their characteristically high PHD resolutions (2 to 3 per cent), it is practicable to make use of multichannel energy analysis, as has been explained in Section 5-3.2E.

B. Ross Balanced Filters. The specific usefulness of β filters combined with PHD/PHA in diffractometry has been pointed out in the preceding section, and a more general treatment of β filters, including the proper choice and thickness of materials, appeared in Section 2-3.6A (see especially Table 2-5). The subsequent treatment of balanced filters in the same section should be reviewed by the reader before he proceeds with the following material, which concentrates on various aspects of the application of balanced filters to x-ray diffractometry, particularly as related to PHD/PHA. The decision as to whether to place the filters in the primary or diffracted beam depends on several factors which are discussed at greater length below. In general, it may be said that convenience of installation and operation tend to favor their location in the direct beam, whereas the diffracted-beam location is preferable when it is essential to suppress fluorescence and incoherent x-radiation emitted by the specimen (refer also to Section 4-1.5).

In its simplest form diffractometric measurement with balanced filters consists of recording the counting rates while alternately interposing in the x-ray beam two filters whose absorption edges lie just above ($K\alpha$ filter) and just below ($K\beta$ filter) the wavelength of the desired $K\alpha$ radation. Table 2-6 lists appropriate filter materials and their proper thicknesses for the x-radiations commonly used. Figure 2-21 and the related text treat the particular case of balanced filters of nickel and cobalt for supplying relatively monochromatic ($K\alpha_1\alpha_2$) radiation from a copper-target tube. It must be reemphasized that Fig. 2-21 represents an idealized situation in which perfect balance is achieved on both sides of the pass band, whereas actual filters can be precisely balanced on only one side at a time[120]. This imperfection can be improved by utilizing PHD/PHA together with the balanced filter pair, which almost completely excludes all wavelengths considerably removed from the wavelength limits of the pass band. Nevertheless, it has no perceptible effect upon wavelengths near the pass band for reasons explained in Section 5-3.4A. Balancing of the filtration immediately below and above the pass band can be accomplished to a high degree by adding a third (neutral) filter which has no absorption edge near the pass band[120–122]. The following treatment is based on publications of Young[121]

and Arndt and Willis[122]. For filters and wavelengths of importance in
x-ray diffraction, the linear absorption coefficient varies approximately
as the third power of the wavelength, or

$$\mu = k\lambda^3, \tag{5-44}$$

in which k varies from one element to another and manifests a pronounced discontinuity at an absorption edge. Let subscripts 1 and 2
refer to the two primary filters of the Ross pair, and let unprimed and
primed quantities, respectively, refer to the wavelength regions below
and above the pass band. Then, from equation 5-44 we may write for the
first filter

$$\mu_1 t_1 = k_1 \lambda^3 t_1 \quad \text{and} \quad \mu_1' t_1 = k_1' \lambda^3 t_1, \tag{5-45}$$

and for the second filter,

$$\mu_2 t_2 = k_2 \lambda^3 t_2 \quad \text{and} \quad \mu' t_2 = k_2' \lambda^3 t_2. \tag{5-46}$$

If $\Delta(\mu t)$ and $\Delta(\mu' t)$ designate the differences in the absorption expo-
nents below and above the pass band, subtraction of equation 5-46 from
5-45 gives

$$\Delta(\mu t) = (k_1 t_1 - k_2 t_2)\lambda^3 \tag{5-47}$$

and

$$\Delta(\mu' t) = (k_1' t_1 - k_2' t_2)\lambda^3. \tag{5-47a}$$

To a reasonable approximation λ may be considered constant over the
pass-band range, from which $\Delta(\mu t) = \Delta(\mu' t)$, or

$$k_1 t_1 - k_2 t_2 = k_1' t_1 - k_2' t_2. \tag{5-48}$$

Solution of equation 5-48 for t_2 gives

$$t_2 = t_1 \left(\frac{k_1 - k_1'}{k_2 - k_2'} \right),$$

and substitution in equation 5-47 yields

$$\Delta(\mu t) = \left[k_1 t_1 - k_2 t_1 \left(\frac{k_1 - k_1'}{k_2 - k_2'} \right) \right] \lambda^3$$

$$= \left(\frac{k_1' k_2 - k_1 k_2'}{k_2 - k_2'} \right) \lambda^3 t_1$$

$$= \Delta(\mu' t).$$

For the third (neutral) filter we may write

$$\mu_3 t_3 = k^3 \lambda^3 t_3, \tag{5-50}$$

in which k_3 is the same on both sides of the pass band. The neutral filter is constructed such that

$$\mu_3 t_3 = \Delta(\mu t) = \Delta(\mu' t). \tag{5-51}$$

Then from equations 5-51 and 5-49 together with 5-50:

$$\Delta(\mu t) = k_3 \lambda^3 t_3$$

$$= \left(\frac{k_1' k_2 - k_1 k_2'}{k_2 - k_2'} \right) \lambda^3 t_1, \tag{5-52}$$

so that the required thickness of the neutral filter is given by

$$t_3 = \left(\frac{k_1' k_2 - k_1 k_2'}{k_2 - k_2'} \right) \frac{t_1}{k_3}. \tag{5-53}$$

An effective test of the balance achieved with three filters can be performed using a scintillation counter, a narrow PHA window of fixed width, and a source of continuous x-radiation, and then varying the PHD base line at a constant rate over the wavelength range encompassing the pass band [121, 122]. The wavelength discrimination of this arrangement, although not sharp, is ample for the purpose. It is important to test the balance at each stage of the adjustment of the primary filter-pair thicknesses. A valuable aid to achieving the optimal balance is to make the thicknesses of the three filters slightly smaller than the values actually required, and then rotating them by small amounts in suitable holders about an axis perpendicular to the x-ray beam until the final requisite absorptions have been precisely attained [121].

The overall process of balancing the filter triplet, then, consists of (1) varying the thicknesses of the two primary filters until $\Delta(\mu t) = \Delta(\mu' t)$ and so that the greater transmission occurs through the filter of lower atomic number, $Z - 1$ (α filter), and (2) adding the neutral filter and adjusting its thickness until the combined transmissions of the neutral and $Z - 1$ filters at a wavelength outside the pass band equal that of the $Z(\beta)$ filter. Gerrits and Bol [123] have given a method of testing the balance of a pair of Ross filters using PHD/PHA and x-radiation from a strongly fluorescing specimen. This technique was developed in connection with the study of diffraction patterns of liquids.

Figure 5-37, from the work of Young [121], is a plot of the intensities of continuous radiation transmitted by a well-balanced pair of zirconium and yttrium oxide (Y_2O_3) filters measured by a PHD base-line scan, aluminum being the neutral filter. The diagram shows perfect balance at all wavelengths outside the pass band. In this case the difference in transmission within the pass band results from air scattering, thermal diffuse scattering, disorder diffuse scattering, scattering from the colli-

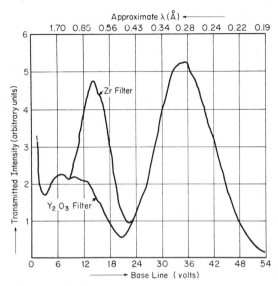

Fig. 5-37.   Wavelength scan for a Y_2O_3–Zr filter pair. The wavelength of the radiation detected by a scintillation counter was varied by changing the position of the narrow constant-width window of a pulse-height analyzer. Note the apparent broadening of the true pass band, 0.7277–$0.6888 = 0.0389$ Å, a consequence of the broad pulse-amplitude distributions (Fig. 5-35). (Courtesy of R. A. Young and the Akademische Verlagsgesellschaft, *Z. Kristallogr.*, **118**, 233 (1963).)

mator and any amorphous material on the crystal surface, and so on. At this point it is worth reiterating that, whereas the use of triple filters can provide excellent balance at wavelengths in the neighborhood of the pass band, it is nevertheless mandatory for best results that such a set of balanced filters be complemented with the simultaneous employment of PHD/PHA to exclude wavelengths arising from less satisfactory balancing at wavelengths considerably removed from the pass band.

Bol[124] investigated the factors bearing on the proper location of the filters (whether in the direct or diffracted beam), in particular the effects of fluorescence x-radiation emitted by the specimen and the filters themselves and of incoherent x-rays scattered by the specimen. The fraction ω of the incoherent x-radiation encompassed by the pass band diminishes from unity at small θ's to very small values at larger θ's. For commonly used characteristic radiations and their appropriate filter pairs, a correction for the factor ω is necessary only when the filters are placed in the diffracted beam. Table 5-7, reproduced from Bol[124], lists criteria for the optimal location of the filters in relation to the suppression of fluorescent x-rays as a function of the identity of the elements comprising the specimen. It can be seen that for any sample composition there exists a choice of characteristic wavelength and balanced filters that will eliminate fluorescence from the measured intensities. At the same time attention is directed to the fact that for certain combinations, designated by the numeral 4, fluorescence cannot be eliminated, in which case a change in x-ray target material is required.

Apparatus for the automatic operation of balanced filters in diffractometry has been designed by several investigators [125–128], and most of the major manufacturers and suppliers of diffractometric equipment offer such units, either as integral components of automatic diffractom-

Table 5-7.   Elimination of Fluorescence Radiation with Balanced Filters[a,b]

X-ray target	Fe	Co	Ni	Cu	Mo
Set of filters	Mn + Cr	Fe + Mn	Co + Fe	Ni + Co	Zr + Y

Element in sample

Atom No.	Symbol					
1–18	H . . . A					
19	K	2	2			
20–24	Ca . . . Cr	2	2	2	2	
25	Mn	1	2	2	2	
26	Fe	1	1	2	2	
27	Co	3	1	1	2	
28	Ni	3	3	1	1	
29	Cu	3	3	3	1	2
30–36	Zn . . . Kr	3	3	3	3	2
37	Rb		3	3	3	2
38–39	Sr, Y			3	3	2
40	Zr				3	1
41	Nb					3
42	Mo					1
43–45	Tc . . . Rh					3
46	Pb	2				3
47–48	Ag, Cd	2	2			3
49	In	2	2	2		3
50–57	Sn . . . La	2	2	2	2	3
58	Ce	4	2	2	2	3
59	Pr	4	2	2	2	
60	Nd	4	4	2	2	

Table 5-7 *continued*

X-ray target Set of filters		Fe Mn + Cr	Co Fe + Mn	Ni Co + Fe	Cu Ni + Co	Mo Zr + Y
Element in sample						
Atom No.	Symbol					
61	Pm	1	4	2	2	
62	Sm	1	4	4	2	
63	Eu	1	1	4	2	
64	Gd	1	1	4	4	
65	Tb	1	1	1	4	
66	Dy	1	3	1	2	
67	Ho	3	1	1	1	2
68	Er	1	1	3	1	2
69	Tm	1	3	1	1	2
70	Yb	3	1	1	1	2
71	Lu	3	1	1	3	2
72	Hf	3	1	3	1	2
73	Ta	3	3	1	1	2
74–75	W, Re	3	3	1	3	2
76–78	Os . . . Pt	3	3	3	1	2
79–83	Au . . . Bi	3	3	3	3	2

[a]Bol[124], p. 738.

[b]If fluorescence arises which cannot be eliminated with a pulse-height analyzer, this is indicated by a number: 1, filters to be placed in the primary beam; 2, filters to be placed in the scattered beam; 3, filters to be placed in one of the beams optionally; 4, elimination of fluorescence is not possible.

eters or as separate accessories.* One of the earliest instruments for automatically interchanging two balanced filters was designed by McKinstry and Short[125] and later made commercially available by Tem-Pres Research.* Tanaka et al.[126] describe an ingenious instrument which rotates a pair of Ross filters at a high speed synchronized with the frequency of an electronic gating circuit, which divides the

*For detailed information write to Tem-Pres Research Division, The Carborundum Company, 1401 So. Atherton St., State College, Pennsylvania, or to any of the major suppliers of x-ray diffraction equipment.

counts so that the signals transmitted by the two filters are separately registered by two scintillation counters, the *difference* in the two counting rates being automatically recorded. Hendricks, Arrington, and Mason [127] describe a mechanism and the necessary control circuit for driving systems of two or three balanced filters, provision being included for interfacing the equipment to fully automated diffractometers. Johnston and Romo's zirconium–yttrium filter changer [128], which is designed for use in the diffracted beam of a General Electric XRD-5 diffractometer, is relatively simple and inexpensive but still effectual; it has been applied to the study of diffraction by vitreous selenium specimens.

C. Crystal Monochromators. In the present state of the art of x-ray diffractometry, to achieve the purest possible monochromatization, it is necessary to use a crystal monochromator in the direct or diffracted beam together with PHD/PHA to remove harmonic wavelengths. The reader is referred to Section 2-3.6C for a general treatment of the factors determining the proper choice of crystal and the relative advantages of plane and curved crystals. A listing of various monochromator crystals and their characteristics is given in Table 2-7.

In contrast to photographic techniques, diffractometry presents the alternatives of placing the monochromatizing crystal in either the direct or diffracted beam, the latter position affording special advantages. Just as for balanced filters, these advantages include the elimination of fluorescence or radioactive background originating from the specimen and the suppression of incoherent radiation. Furthermore, (1) it is simpler to install the monochromator in the diffracted than in the incident beam, (2) crystals subject to deterioration under exposure to x-rays have longer life times in the diffracted beam, and (3) with refined geometrical adjustments it is possible to resolve the α_1 and α_2 components of the $K\alpha$ doublet, permitting measurement of only the α_1 diffraction.

Lang [129] has evaluated four geometrical arrangements of source, specimen, monochromator, and detector that may be employed in diffractometry with the monochromator crystal located in the diffracted beam. Of these the two of greatest importance in powder diffractometry are shown in Fig. 5-38A and B, the RR (reflection specimen–reflection monochromator) and TR (transmission specimen–reflection monochromator) geometries.

In the RR arrangement parafocused x-rays diffracted by the flat specimen AOB pass through the receiving slit S_1, as in the conventional Bragg-Brentano powder diffractometer; but instead of entering the counter directly, they first diverge and are then refocused by the

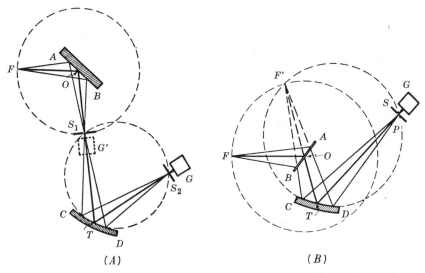

Fig. 5-38.   Two arrangements for x-ray diffractometry with a curved-crystal monochromator in the diffracted beam. (*A*) Reflection specimen–reflection monochromator (RR); (*B*)
transmission specimen–reflection monochromator (TR). (Courtesy of A. R. Lang, *Rev. Sci.
Instrum.*, **27** (1), 17 (1956).)

curved monochromator CTD to a focus at the counter slit S_2. The crystal
may be both bent and ground (Johansson[130]) or bent only (Johann
[131]), which gives a less perfect focus, but in either case the rays are
focused sufficiently to pass the narrow slit S_2 and enter the counter G.
In the RR arrangement there is no need to modify the basic geometry
of the conventional powder diffractometer, and the components of the
monochromator assembly, comprising the slits S_1 and S_2, the crystal,
and the detector, remain in a fixed relationship to one another and
revolve as a rigid unit about the diffractometer axis O. For a particular
wavelength the orientation of the singly curved crystal CTD is adjusted
so that its diffracting planes make the correct angle of incidence with
respect to the rays from the specimen diverging through the primary
receiving slit S_1, and the counter is set with its slit S_2 at the same angle
to the diffracting planes of the monochromator crystal.

From Fig. 5-38A it will be noted that the slits S_1 and S_2 and crystal surface all lie on a common circle of radius R_c, the relationship

$$D = R_c \sin \theta \qquad (5\text{-}54)$$

being satisfied. In this equation D is equal to the distances S_1T and S_2T,
and θ is the Bragg angle corresponding to the diffracting planes of the
monochromator crystal. By proper selection of the distances D and angle

θ, the monochromator assembly can be used with a number of characteristic wavelengths. As in conventional Bragg-Brentano parafocusing diffractometry, the RR arrangement is well suited to recording the diffraction patterns of moderately or highly absorbing specimens over a wide angular range in 2θ, extending from 10 or 15° to 165°. At lower angles and in fact at all angles for low-absorbing specimens such as carbonaceous or organic materials (e.g., bulk polymers), the TR arrangement (Fig. 5-38B) is to be preferred. For such specimens at angles even as high as 30°, the RR geometry is unfavorable because of pronounced broadening and distortion of the focus at S_1. For the TR arrangement a thin specimen ($t \leqslant 1$ mm) should be used to minimize geometrical broadening, and it must be rotated with the customary 1:2 relationship to the receiver assembly (in the TR geometry this means that the plane of the specimen maintains equal angles with the median lines FO and OT).

Lang[129] points out that another geometrical arrangement RT has the special merit of permitting measurements to be made at the highest back-reflection angles without interference between the monochromator assembly and the x-ray tube housing. Because of its geometrical compactness, this technique is characterized by a relatively short overall x-ray path length, thus minimizing background due to air scattering when long wavelengths such as those of Fe$K\alpha$ or Cr$K\alpha$ must be employed. For additional information on the RT arrangement the reader is directed to Lang's article[129].

For the RR technique Lang[129] employed as monochromator a quartz plate $35 \times 10 \times 0.3$ mm cut parallel to the ($10\bar{1}1$) planes and elastically bent to the proper radius with the aid of a crystal-bending and -supporting device based on the design described by Wooster, Ramachandran, and Lang[132]. Banerjee[133] constructed an RR monochromator for the General Electric goniometer, incorporating a ground and plastically bent rock salt crystal. Ogilvie[134] designed an RR-type monochromator which has been described in some detail by Koffman and Moll[135] and has since been distributed commercially.* Figure 5-39 shows the unit mounted on the standard detector-supporting bracket of the Norelco diffractometer. At first this unit was equipped with a singly bent LiF crystal, but more recently the option of a bent, highly oriented, pyrolitic graphite monochromator has been provided. Prior to their application in powder diffractometry, Canon [136] first showed that such pseudo-single crystals of graphite could

---

*Advanced Metals Research Corporation, 149 Middlesex Turnpike, Burlington, Massachusetts; and Philips Electronic Instruments, 750 So. Fulton Avenue, Mount Vernon, New York.

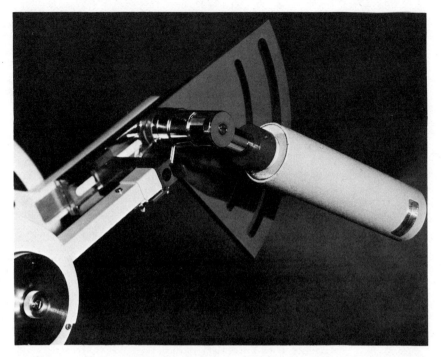

Fig. 5-39.   Diffracted-beam monochromator mounted on detector bracket of Norelco diffractometer. (Courtesy of Philips Electronic Instruments.)

serve very successfully as analyzing crystals in x-ray spectroscopy. Several advantages of graphite as a monochromator, including in particular its very high reflectivity, were cited in Section 2-3.6D. The relatively high intensities provided by graphite as compared with other monochromators can be seen from an inspection of Table 5-8[137, 138]. Also of special advantage, of course, is the relative stability of graphite under x-irradiation.

Figure 5-40, reproduced from an article by Proctor, Furnas, and Loranger[139], illustrates dramatically the reduction in fluorescence background that can be achieved by employing a crystal monochromator (graphite in this instance) in the diffracted beam when recording the diffraction pattern of ferric oxide (Fe_2O_3) with copper radiation. The intense iron K fluorescence evoked by $CuK\alpha$ radiation when simple filtration is used (Fig. 5-40B) yields a pattern with very unfavorable peak-to-background ratios. Diffracted-beam monochromators incorporating highly oriented pyrolytic graphite as well as other crystals, such as LiF, NaCl, α-quartz, pentaerythritol, may now be obtained from most

Table 5-8. Calculated Relative Diffracting Powers of Several X-Ray Monochromatizing Crystals for $\lambda = 1.54$ Å [137, 138]

Crystal	Diffracting Planes	Relative Diffracting Power
LiF	(200)	93
Graphite	(0002)	620
PET[a]	(002)	115
Diamond	(111)	120
Aluminum	(200)	24
Copper	(200)	71
Quartz	(10$\bar{1}$1)	43
NaCl	(200)	31
EDDT[b]	(020)	~62[c]

[a]Pentaerythritol.
[b]Ethylenediamine-D-tartrate.
[c]Approximated from L. S. Birks, *Electron Probe Microanalysis*, Interscience, New York, 1963, p. 76.

major suppliers of x-ray diffractometric apparatus, and inquiries for more detailed information may be directed to them. Readers interested in crystal monochromators for use in the direct beam are referred to the literature [140–142] as well as to commercial vendors.

5-4 CHOICE OF EXPERIMENTAL CONDITIONS AND PROCEDURES

The diffractionist is frequently faced with the decision as to whether to measure the peak (I_p) or the integrated (I_i) intensity of a reflection *hkl*. As a general rule, the latter choice is the correct one except in certain particular analytical applications or when only the peak line position is wanted. Without exception the accurate evaluation of a reflection intensity $I(hkl)$ and its related structure factor $F(hkl)$ must be based on measurement of the integrated intensity (see Section 3-3; also Section 10-6.2A and Fig. 10-17). Even in routine analytical applications peak intensities are acceptable only when they can be shown to be proportional to the corresponding integral intensities.

Differences in reflection breadth (W) result from any changes in the

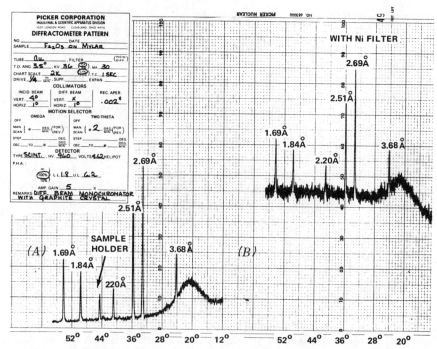

Fig. 5-40. Reduction of fluorescence background with diffracted-beam monochromator. Diffraction pattern of Fe_2O_3 on Mylar film prepared (A) with a graphite monochromator, and (B) with only a nickel β filter. (Courtesy of E. M. Proctor, T. C. Furnas, and W. F. Loranger, *Advances in X-Ray Analysis*, Vol. 14, Plenum Press, 1971, p. 38.)

various geometrical aberrations (see Fig. 5-11 and Table 5-3), as well as from variations in crystallite size and lattice perfection of the specimen under examination, and such variations in breadth affect the peak heights inversely. Therefore, as long as any uncertainty exists about the constancy of the factors just mentioned, the integrated intensity I_i must be taken as the only reliable measure of the reflection intensity $I(hkl)$.

In relation to the measurement of diffracted-beam intensities, we now emphasize a very significant point of superiority of the standard symmetrical-reflection parafocusing geometry customarily employed in powder diffractometry. Unlike the Debye-Scherrer and other diffraction techniques wherein a portion or all of the diffraction occurs by transmission through the specimen with a resulting dependence of the diffracted intensity on θ (refer to Section 3-3.4G), in the standard symmetrical-reflection technique the intensity is independent of θ, provided only that the minimum thickness of equation 5-11 is satisfied. Given this condition, it can easily be shown (see Section 6-8) that the intensity diffracted at any given angle 2θ is

$$I = \frac{I_0 A}{\mu}, \tag{5-55}$$

in which I_0 is the intensity that would be diffracted at the given angle 2θ by a unit volume of the specimen under hypothetical conditions of nonabsorption, A is the cross-sectional area of the incident beam at its point of intersection with the specimen, and μ is the linear absorption coefficient of the solid material composing the powder. In equation 5-55 the quantities I and I_0 may denote any desired measure of the reflection intensity, integrated, peak, and so on. If the thickness t of the specimen is insufficient to satisfy the criterion of equation 5-11, the following factor should be applied to convert the observed intensity I_t to the intensity corresponding to a specimen of effectively infinite thickness[143]:

$$\frac{I_\infty}{I_t} = [1 - \exp(-2\mu t \csc \theta)]^{-1}. \tag{5-56}$$

Inkinen[143a] has presented an assessment of the accuracy realizable in intensity measurements by powder diffractometry, and Jennings [143b] has discussed preliminary findings of the cooperative powder intensity project sponsored by the International Union of Crystallography (IUCr). The IUCr results tend to show that integrated intensities cannot be relied on to better than 5 per cent, and that techniques for measuring the incident-beam intensity appear to be more reliable than those for measuring relative intensities of reflections. For further comments on intensities measured by the symmetrical-reflection technique, the reader is referred to Section 6-8.

5-4.1. Statistical Accuracy of Counter Measurements

X-ray photons generated by the anode of an x-ray tube are emitted in a random manner both with respect to direction and time. Hence, if the intensity of a narrow beam is measured for very short time intervals, it will be found to fluctuate in a statistical manner above and below a certain mean level. When the measurements are made over longer time intervals, deviations from the mean intensity are found to be less, and at sufficiently long intervals become imperceptible.

These facts have an important bearing upon the measurement of diffraction intensities. When the recording is photographic, two important advantages are gained. One has been mentioned earlier, namely, that long-period variations in the x-ray output affect the intensities of all reflections equally, so that all may be directly compared. Another noteworthy advantage is that the total energy contributing to any particular reflection is the "true" average value, because of the large number of x-ray photons required for the production of a photographic line or spot. However, since the counter can record but one reflection at a time, time limitations generally prohibit the counting of sufficient photons to reduce statistical errors to the vanishing point. Hence it is important to bear in mind continually the magnitude of the statistical errors affecting any particular measurements and to appraise their influence upon the reliability of the results. The magnitude of the statistical error depends only upon the total number of quanta counted. For events occurring randomly in time sequence, the standard deviation of the number of events observed N from the true average number N_0 is given by

$$\sigma_N = N^{1/2}, \tag{5-57}$$

and the relative standard deviation is

$$\sigma_N(\text{rel}) = \frac{N^{1/2}}{N} = N^{-1/2}. \tag{5-58}$$

It follows that in order to measure an intensity with some desired degree of accuracy, a perfectly definite number of pulses must be counted. From equation 5-58 it is found that for various standard deviations the following numbers of counts are required.

Desired σ (per cent)	N required
0.2	250,000
0.4	62,500
0.6	27,790
0.8	15,625
1.0	10,000
1.5	4,444
2.0	2,500
3.0	1,111
4.0	625
5.0	400

When the background counting rate is appreciable, the uncertainty in the net peak height is greater than given above. Suppose that the total number of counts recorded at the top of a diffraction peak is 10,000, and that in an equal time interval 5000 counts are recorded at background. The *relative* standard deviation in the overall height is then

$$\frac{100}{10,000^{1/2}} = 1.00 \text{ per cent,}$$

and in the background

$$\frac{100}{5,000^{1/2}} = 1.41 \text{ per cent.}$$

The corresponding *absolute* standard deviations are:

In the overall height $\sigma_T = 10,000^{1/2} = 100.0$ counts.

In the background $\sigma_B = 5000^{1/2} = 70.7$ counts.

The absolute standard deviation in the *net* peak height σ_P is then

$$\begin{aligned}\sigma_P &= (\sigma_T{}^2 + \sigma_B{}^2)^{1/2} \\ &= (10,000 + 5000)^{1/2} = 122.5 \text{ counts.}\end{aligned}$$

This amounts to a percentage standard deviation of

$$\frac{122.5 \times 100}{10,000 - 5000} = 2.45 \text{ per cent.}$$

Note the great reduction in accuracy that has resulted from the high background. By recording 10,000 and 5000 counts, or a total of 15,000, a net accuracy of only 2.45 per cent has been achieved, whereas in the absence of an appreciable background this total count would have yielded an accuracy of about 0.8 per cent.

In general, if R is the ratio of total counting rate to background counting rate, and if N_T is the total number of counts recorded at the top of the peak (including background), the per cent standard deviation in the net peak height is given by

$$100 \, \sigma_P = \frac{100}{R-1} \left[\frac{R(R+1)}{N_T} \right]^{1/2}, \tag{5-59}$$

which can also be written in terms of N_T and N_B, the background count,

$$100 \, \sigma_P = \frac{100 \, (N_T + N_B)^{1/2}}{N_T - N_B}. \tag{5-60}$$

Figure 5-41 is a graphical exposition of equation 5-59. The curves may be used to determine rapidly the number of pulses that must be counted in order to attain a given net precision in the peak intensity for various ratios of total to background counting rate R. When the background

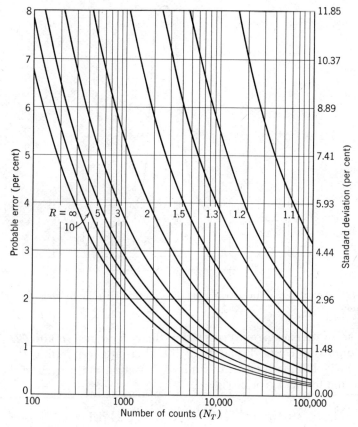

Fig. 5-41. Number of counts required to give a desired degree of accuracy for various ratios of total to background counting rate.

rate is small in comparison to the peak rate, the curve $R = \infty$ may be employed with little error.

These same considerations may be used to evaluate the statistical error in measuring the area under a diffraction maximum, that is, the integrated, rather than peak, intensity. Such measurements are valuable, for example, in the determination of crystallite size by the method of integral line breadths. Figure 5-42 illustrates how this can be accomplished. The peak BC and two regions of surrounding background AB and CD are scanned at some uniform and appropriate speed, and counts are totalized in the three regions. The sum of the angular ranges AB and CD must equal the angular range BC of the peak, and for the

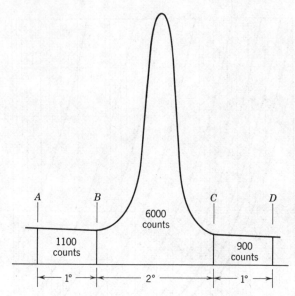

Fig. 5-42. Data required for determining the statistical error in the area of a diffraction maximum.

most accurate background correction the two ranges AB and CD should in addition be equal. Equation 5-59 is then applied, letting $R = 3$ and $N_T = 6000$. The net per cent standard deviation in the area of the peak (corrected for background) is

$$100\,\sigma_P = \frac{100}{2}\left[\frac{3(3+1)}{6000}\right]^{1/2} = 2.24 \text{ per cent.}$$

For a more detailed discussion of the statistical accuracy of counter measurements, the reader is referred to Mack and Spielberg[144].

When automatic strip-chart records are employed, the estimation of statistical errors is complicated by additional errors originating in the pulse-averaging circuit, which actuates both the rate meter and the recorder. These errors can be minimized by properly choosing the time constant RC of the pulse-averaging circuit. The evaluation of strip-chart records is discussed in greater detail in a later section.

5-4.2 The Specimen

This subject is of even more importance in diffractometric than in photographic powder techniques. The principal factors requiring attention in the preparation of the specimen are (1) crystallite, or grain, size; (2) sample thickness; (3) preferred orientation; (4) strain and cold-working (mainly in metals); and (5) surface planarity. Before reading what follows it would be well to review Section 4-1.3.

A. Preparation of Powders. It is essential for the most satisfactory results that (1) the number of crystallites contributing to each reflection be sufficiently large to generate signals of reproducible intensity, and that (2) preferred orientations of the crystallites be held to a minimum. Both conditions are more difficult to meet than when photographic methods are used. The first condition is equivalent to the Debye-Scherrer requirement that the powder be fine enough to avoid discontinuous or spotty lines. With Debye-Scherrer cameras, however, the condition is easily met with even a relatively small number of crystallites by rotating the cylindrical specimen about an axis normal to the incident beam. This enormously increases the effective number of orientations, so that even with particle dimensions as large as 40 or 50 μ the lines are smooth for most materials. Thus powders that have passed through a 325-mesh screen are nearly always fine enough for Debye-Scherrer photographs.

Powders meeting these specifications are not fine enough, however, for a flat diffractometer specimen. If the specimen is practically stationary during the measurement of a given diffraction peak, the number of < 325-mesh crystallites oriented so as to contribute diffracted rays to this peak is much too small to yield a reproducible and true average intensity. Often the variations extend over surprisingly large limits. Table 5-9[145] gives the results of intensity measurements on four quartz powder fractions of 15 to 50, 5 to 50, 5 to 15, and < 5 μ mean dimensions. Ten different samples of each fraction were measured under as nearly identical conditions as possible. The powder was in each case held in a specimen holder of the requisite dimensions to give a diffracted beam of maximum intensity (see equation 5-11). The intensity of the 3.34-Å (10·1) maximum was measured by scanning stepwise over the diffraction profile and plotting the counts, the area above background then being taken as proportional to the intensity. The effect of counting statistics upon the measured intensities was kept well below 1 per cent, by taking a sufficient number of counts at each point. The relationship between crystallite size and the reproducibility of the diffracted intensities is strikingly demonstrated. For the coarsest fraction of powder, the mean deviation in intensity from the average of the 10 values is 18.2 per cent, whereas for the finest fraction it is only 1.2 per cent. It will also be noticed that the average observed intensity is smaller for larger crystallite sizes, an effect that must be attributed to extinction.

The results given in Table 5-9 can be predicted with reasonable

Table 5-9. Intensity Measurements on Different Size Fractions of
            < 325-Mesh Quartz Powder[145][a]

Specimen No.	15 to 50 μ Fraction	5 to 50 μ Fraction	5 to 15 μ Fraction	<5 μ Fraction
1	7,612	8,688	10,841	11,055
2	8,373	9,040	11,336	11,040
3	8,255	10,232	11,046	11,386
4	9,333	9,333	11,597	11,212
5	4,823	8,530	11,541	11,460
6	11,123	8,617	11,336	11,260
7	11,051	11,598	11,686	11,241
8	5,773	7,818	11,288	11,428
9	8,527	8,021	11,126	11,406
10	10,255	10,190	10,878	11,444
Mean area	8,513	9,227	11,268	11,293
Mean deviation	1,545	929	236	132
Mean percentage deviation	18.2	10.1	2.1	1.2

[a]Tabulated values are areas in arbitrary units of the 3.34-Å maximum as counted
with a Geiger-counter diffractometer using Cu$K\alpha$ radiation.

accuracy by a theoretical treatment[145] based on the Laplacian probability equation, which leads to the following expression for the mean
relative deviation in diffracted intensity

$$U_m = 0.798 \left(\frac{q\mu \sum v_i f_i}{pA} \right)^{1/2}. \qquad (5\text{-}61)$$

In equation 5-61, $q = 1 - p$, where p is the fraction of crystallites oriented
(favorably) so as to contribute to a given reflection, q is the fraction
unfavorably oriented, μ is the linear absorption coefficient of the
powder, f_i is the volume fraction of the powder consisting of crystallites
of individual volumes v_i, and A is the cross-sectional area of the beam at
its point of intersection with the specimen. The quantities p and q cannot
be rigorously evaluated, but plausible values can be deduced in terms
of the geometrical conditions of diffraction and the rocking angle
$\Delta\theta_{1/2}$ of the crystallites.*

Equation 5-61 shows that the reproducibility of intensity measurements depends not only upon the crystallite size of the powder (the

*$\Delta\theta_{1/2}$ of the present notation is equivalent to $2w$ in Fig. 3-10.

dominant factor), but also upon the amount of powder irradiated and upon certain factors that determine p and q. These factors include, in particular, the rocking angle of the crystallites (a measure of the crystalline perfection as explained in Section 3-3.1) and the multiplicity factor of the reflecting planes (a function of crystal symmetry). The curves in Fig. 5-43 show the dependence of intensity variations upon the effective crystallite size of a powder for a set of typical experimental conditions and various values of the linear absorption coefficient μ. These curves are valid only if the sample is sufficiently thick to generate a diffracted ray of maximum intensity. An inspection of Fig. 5-43 and the data in Table 5-9 lead to the conclusion that with average powders there is high assurance of a mean intensity deviation smaller than 1 per cent if the effective crystallite dimension is less than 5 μ. When a reproducibility of 2 or 3 per cent is sufficient, the crystallite size may be as large as 10 μ.

These considerations show that powders passed through a 325-mesh screen (about 50 μ in lineal aperture) are still much too coarse for such precision. Prolonged grinding of 325-mesh powder in a power mortar or ball mill further reduces the crystallite size by considerable amounts, but in most cases a few coarse crystallites remain which continue to

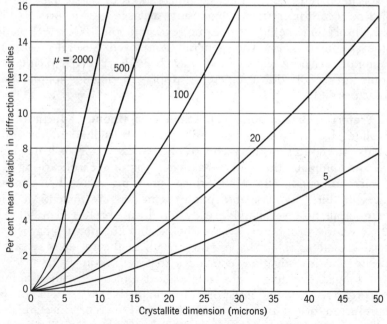

Fig. 5-43.   Dependence of deviation in diffraction intensities upon crystallite size for typical experimental conditions and various values of the linear absorption coefficient μ.

produce undesirable intensity fluctuations. Sedimentation techniques can be used to eliminate the coarser-sized fractions from a single-phase powder that exhibits a wide distribution of crystallite sizes (see Section 4-1.3). Rotation of the specimen about an axis normal to its face increases the effective number of crystallite orientations severalfold by bringing new crystals into position to reflect, but the improvement is far less conspicuous than that produced by rotating a cylindrical Debye-Scherrer specimen, which results in a vast increase in the effective number of orientations.

B. Rotation of the Specimen. A more extensive analysis of the problem of crystallite-size statistics by de Wolff[146] yielded an expression equivalent to equation 5-61 for stationary specimens, as well as theoretical predictions concerning the improvement in intensity deviations to be expected from rotating the flat specimen rapidly in a plane parallel to its surface. These expectations were subsequently substantiated by cooperative experimental measurements with Taylor and Parrish[147] on 30- to 50-μ silicon powder. The root-mean-square intensity deviations were observed to be reduced by factors of 7 or 8 for peak intensities, and 4 or 5 for integrated intensities. Thus rotation of the specimen reduces the intensity fluctuations to acceptable levels of 1 or 2 per cent for crystallite dimensions as large as 30 μ, provided the linear absorption coefficient does not exceed about 100 cm^{-1}. Rotation of the specimen is now very generally employed. Several devices for rotating or oscillating the specimen have been described[148], and most manufacturers of diffractometers supply specimen "spinners" of satisfactory design.

C. Preferred Orientation and the Specimen Mount. Reduction of the crystallite size by grinding or sedimentary fractionation has added benefits. It reduces primary and secondary extinction to a negligible level, and eliminates the need for taking into account errors arising from the microabsorption effect[149], which comes into play when two substances of different mass absorption coefficients are mixed. Most important of all, however, it aids greatly in reducing preferred orientation, a major source of intensity errors when examining powder particles of special crystallographic shapes. Birks[150], as the result of a careful experimental investigation, concludes that reduction of particle size by prolonged grinding is the single most effective means for minimizing preferred-orientation errors. He obtained a very great reduction in preferred orientation of both silver and mercuric acetate (both lathlike in crystal habit) by grinding for periods up to 30 min. The reader is cautioned, however, that prolonged grinding of soft crystalline substances, such as many organic compounds, causes lattice distortions and consequent line broadening (see also Section 4-1.3).

In spite of great size reduction due to grinding, crystals exhibiting extremely pronounced cleavage such as mica, kaolinite, and chrysotile, still orient preferentially, and no means has yet been devised for arriving at a truly random distribution of orientations with such materials. However, the reduction in size of coarser materials still has the beneficial effect of making the intensities of reflections from readily oriented materials more reproducible, provided a uniform mounting technique is adhered to.

The method of mounting the powder is of fundamental importance. Perhaps the most important consideration is the thickness of the powder layer. If maximum diffracted intensity is required, it is necessary that the thickness t be great enough to satisfy expression 5-11, namely:

$$t \geq \frac{3.2}{\mu} \frac{\rho}{\rho'} \sin \theta. \tag{5-11}$$

The reader should review the significance of the quantities ρ, ρ', and μ as given in Section 5-2.1D. When the available quantity of powder permits, it is usually preferable for several reasons to make the thickness large enough to conform to equation 5-11.

The "standard" specimen holders furnished by the manufacturers are rectangular aluminum plates about 2 mm thick containing a rectangular window (see Fig. 5-44D) in which the powder is placed. A thickness of 2 mm satisfies the requirement of equation 5-11 for all specimens save those consisting exclusively of light elements, such as organic compounds, most polymers, and carbonaceous materials. Specimen holders of this kind can be designed for loading of the powder from the front, edge, or back. Much accumulated experience tends to favor edge or back loading when preferred orientation needs to be minimized.

Figure 5-44A and B are sectional drawings of two biological drop slides with curved- and straight-walled specimen cavities, respectively. These are perhaps the most generally available specimen holders that are suitable for use in x-ray diffractometry in the proper circumstances. Only very fine and coherent powders can be successfully packed in a durable condition in a type-A holder; a more secure mounting can usually be achieved by mixing the powder with a few drops of a solution of some binder such as ethyl cellulose, collodion, or Duco cement. If preferred orientation is not of concern, the powder is packed into the slide cavity (of either type A or B, Fig. 5-44), and its surface made as plane as possible and flush with the surface of the slide by smoothing it

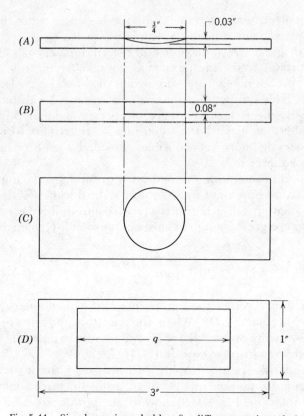

Fig. 5-44. Simple specimen holders for diffractometric analysis.

with a spatula or razor blade or by exerting a to-and-fro motion with an ordinary microscope slide.

If the standard reflection parafocusing technique is to be used at rather small Bragg angles, it is essential that a long, rectangular specimen cavity be used (Fig. 5-44D) if the theoretical maximum intensity is to be realized. From Fig. 5-45 it can be seen that if the horizontal divergence of the beam is limited to an angle γ, the beam will be entirely received by the powder surface only if θ is larger than $\sin^{-1}(\gamma l/2A)$. Thus if $2A = 4$ cm and $l = 16$ cm, for small Bragg angles, $\theta_{min} \cong \frac{16}{4} \times \gamma = 4\gamma$. If $\gamma = 1°$, $\theta_{min} = 4°$, and if $\gamma = 4°$, $\theta_{min} = 16°$, and so on. For ordinary work $2A$ is generally of the order of 2 cm, in which case abnormaly small intensities are obtained for $\sin \theta$ smaller than $\frac{16}{2} \times \gamma$. Then for $\gamma = 4°$,

$$\sin \theta_{min} = \frac{16}{2} \times \frac{4}{57.3} = 0.56;$$

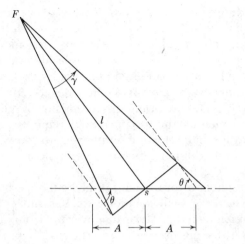

Fig. 5-45.   Length of sample illuminated by an x-ray beam of horizontal divergence γ.

therefore,

$$\theta_{min} = 34°.$$

And for $\gamma = 1°$,

$$\sin \theta_{min} = \frac{16}{2} \times \frac{1}{57.3} = 0.14;$$

therefore,

$$\theta_{min} = 8°.$$

These results illustrate why a narrow beam must be employed for accurate intensity measurements at small Bragg angles.

A special problem may be encountered at small Bragg angles when the specimen is relatively thick, namely, obstruction of a portion of the diffracted beam by the walls of the specimen cavity, with loss of intensity as well as generation of unwanted diffraction effects from the specimen holder. Careful attention to the geometrical conditions, especially depth of penetration of the specimen by the direct beam, is required to avoid this pitfall, as is explained in some detail by Vonk[40] and Langford and Wilson[41].

Okamoto, Kimura, and Nakajima[151] found that the use of a specimen mold producing a serrated surface for exposure to the x-ray beam is effective in reducing preferred orientation. For very orientation-prone substances such as yellow lead oxide, β aluminum, and mica it was sometimes found that dilution with a small proportion of flour was also helpful. The serration of the specimen surface broadened the reflections slightly but had no appreciable effect on their intensities or positions.

Byström-Asklund[152], among others, considers filling of the specimen cavity from the side (edgewise) to be more effective than other methods for the minimization of preferred orientation in mounting mineral powders, especially clays. During the filling process the open (front) face of the cavity in an aluminum holder is closed with a glass slide having a coarsely-ground surface in order to reduce the tendency for platelike crystallites to orient preferentially by gliding along its surface. At the same time the specimen holder and its supporting stand are subjected to mechanical vibration in order to compact the powder. For full details and additional literature references, the reader should consult reference[152].

There are a number of proponents of filling the specimen cavity from the back, of which we shall mention two. Vassamillet and King[153] suggest first stretching a thin Mylar film over the front of the window in the metal holder, having previously wet the surface of the plate with a very dilute solution of collodion in order to cause the Mylar to adhere to the metal, and then sifting the powder through a screen into the cavity from the back. It may be necessary to add a little very dilute adhesive to hold the powder in place. The sifting process causes the smallest particles to pass through first and produce a roughly plane surface for the larger particles to rest upon, thus reducing the degree of preferential orientation. The cavity is filled and the Mylar film stripped off, leaving a self-supporting layer of material with a plane front surface.

An extensive investigation of methods for mounting mineral powders led McCreery[154] to recommend the following procedure, which reduces preferred orientation by filling the cell from the back and produces a plane surface. The sample holder is a rectangular sheet of aluminum $1\frac{1}{2} \times 1 \times \frac{1}{8}$ in. in size with a $\frac{9}{16}$-in. hole drilled through the center. Other items of equipment used in setting up the powder include two microscope slides cut in half, a razor blade, a paper clip, small scissors, a stiff spatula, a roll of $\frac{3}{4}$-in. Scotch tape, and a special small sieve made from a DeVilbiss hose-coupling sleeve, which acts on the principle of a flour sifter and is used for filling the cell. The cell-filling procedure is then as follows (see Fig. 5-46):

1. A pencil notation on the face side of the aluminum cell serves for identification. The face is then covered with a clean glass slide and bound firmly at each end with tape. Doubling the tip of the tape on the back side of the cell will aid in removing it later.

2. Place the cell, face down, on a flat surface and sift an excess of powder into the cavity (Fig. 5-46 *A* and *B*).

3. Tamp the surplus gently but thoroughly with the edge of the

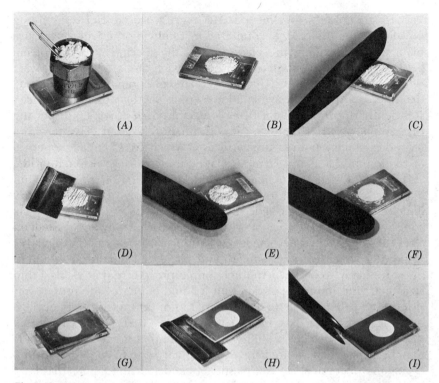

Fig. 5-46. McCreery's procedure for preparing specimens. (Courtesy of G. L. McCreery, *J. Am. Ceram. Soc.*, **32**, 141 (1949).)

spatula (*C*). This step is important because it causes the cavity to fill evenly with a minumum amount of flow at the glass face plate.

4. Slice off the surplus powder with the razor blade (*D*).

5. Sift on a loose layer of powder about $\frac{1}{16}$ in. thick (*E*). Press gently but firmly with the flat blade of the spatula to compress the powder (*F*). Slice off the surplus powder again and repeat with a fresh layer (*D, E, F*).

6. Loosen the tape at each end, wipe off the surplus powder, and cover with a clean glass slide (*G*).

7. Turn the assembly over, remove the tape from one end, and slip the razor blade under the glass face plate, thus holding the cell down while the glass is lifted off (*H*).

8. Two diagonally opposite corners of the cell are then bound with Scotch tape so that only two small patches of aluminum about $\frac{1}{8}$ in. square are covered. This keeps the tape out of the path of the x-ray beam. Excess tape is trimmed off with scissors (*I*).

Perhaps the greatest weakness in McCreery's procedure for mounting powders is the difficulty encountered in removing the front glass face plate without spoiling the plane face of the specimen. This problem becomes more severe when coarser powders are mounted.

Two other methods of reducing preferred orientation are: (1) diluting the powder with an adhesive solution so as to coat the individual particles sufficiently to prevent them from contacting each other, (2) dispersing the sample powder by mixing it with finely powdered glass or with a second crystalline powder the particles of which are essentially spherical or random in shape, for example, magnesium oxide or calcium fluoride.

If it is not essential to meet the thickness criterion for maximum intensity (equation 5-11), or if the available supply of powder is very limited, a glass or plastic slide may be coated with a thin film of some adhesive that dries rather slowly and a layer of powder dusted on after the adhesive has become tacky. This process yields a very thin layer of powder which is rather smooth and plane, but the diffracted intensities are likely to be low. Another simple procedure that yields a thin specimen is to mix the powder with a liquid binder of the proper consistency and then cause the thin slurry to flow as uniformly as possible over the surface of a glass microscope slide. It is necessary of course that the sample powder be insoluble in the mounting fluid. The glass slide may be replaced if desired by a rectangular slide of some other noncrystalline material such as a plastic. Some dispersing liquids that have been successfully used to mount inorganic materials are collodion, Duco cement diluted with amyl acetate, ethyl cellulose dissolved in toluene, and Parlodion (nitrocellulose) dissolved in amyl acetate. Organic substances insoluble in water may be dispersed in dilute aqueous solutions of tragacanth casein or hydroxyethyl cellulose. The supporting slide may be rocked as necessary to disperse the suspension as uniformly as possible, after which the sample is allowed to stand a few minutes in a horizontal position until the solvent evaporates. This mounting technique has the following objections: (1) it is highly conducive to preferred orientation, (2) it is likely to yield a diffracted ray of less than maximum intensity, and (3) it is difficult to make the specimen surface highly plane, with the result that the diffraction peaks may be broadened appreciably or displaced by small angular amounts from their correct positions. For some applications these disadvantages are offset by the simplicity of the method and the relatively small amount of powder required.

With regard to the overall problem of preferred orientation, it is probably fair to conclude that, no matter what procedure is followed or what special precautions are observed in the preparation of the diffractometric specimen, some degree of preferential orientation of particles possessing a pronounced anisotropic crystal habit will persist, or at least the possibility cannot be unequivocally ruled out. With this realization other investigators[155, 156] have suggested a different approach, namely, to test for the presence and extent of residual orientation by comparing the ratio of the integrated intensities of a number of pairs of peaks in reflection to their ratio when measured by transmission. Weiss[155] points out that if crystallite orientations are random, the ratio of the intensities of two peaks measured by the symmetrical-reflection geometry divided by the ratio of the corresponding intensities measured by the symmetrical-transmission geometry (Section 5-4.3) will be

$$\frac{1 - \exp\left(-\mu T / \sin \theta_1\right) \cos \theta_1 \exp\left(-\mu T / \cos \theta_1\right)}{1 - \exp\left(-\mu T / \sin \theta_2\right) \cos \theta_2 \exp\left(-\mu T / \cos \theta_2\right)}, \qquad (5\text{-}62)$$

θ_1 and θ_2 being the Bragg angles of the two reflections measured. If the ratios of *all pairs of peaks* agree well with equation 5-62, it may be concluded that the specimen is practically free of preferred orientation. However, if there is appreciable disagreement, some estimate of the residual degree of preferred orientation can be attempted.

Another method of dealing with preferred orientation is to quantitatively characterize it by appropriate intensity measurements, and then correct for it mathematically[157, 158]. To our knowledge, only Sturm and Lodding[157] have actually carried through such a numerical analysis, the specimen being pellets of kaolinite which displayed a simple cylindrically symmetrical mode of preferred orientation denoted *uniplanar* by Heffelfinger and Burton[159, 160]. Despite the evident success of this demonstration of their method, it seems that the exacting experimental requirements and computational work involved render it unsuited to routine analytical application. Another drawback is the inapplicability of their analysis to complex and unsymmetrical orientation modes. For a general treatment of the evaluation of preferred orientation and textures, the reader is referred to Chapter 10.

Massive polycrystalline specimens, such as fine-grained metals, rocks, and some polymers, make good diffraction specimens because the more-or-less random orientations of the grains comprising the structure simulate the distribution of orientations assumed by a large number of powder particles. Therefore specimens of this kind possessing one or more flat surfaces may be inserted directly in the diffractometer for analysis. Cold-worked metals, in particular filings, and cold-worked polymers tend to give broader maxima and higher background scatter as the result of lattice distortions. If compatible with the objectives of the diffraction studies involved, lattice distortions can be relieved by annealing or, if only surface layers of a metal are involved, by etching treatments. It must also be remembered that rolling, drawing, and other forms of directional cold-working introduce some degree of preferential orientation into the grain (crystallite) structure, and this causes variation in the diffracted intensities with the orientation of the specimen in the x-ray beam.

D. High-Temperature Techniques. Since many of the basic problems to be solved in the design of high-temperature photographic cameras and diffractometers are very similar, and because of stringent space limitations, we suggest the rereading of Section 4-5 during consideration of the following material. We shall also find it necessary to supplement the following condensed account of this subject by directing the reader's attention to selected literature references including several review articles [161–163].

Compared with photographic methods, diffractometry at high temperatures has the advantage of following solid-state phase transformations *dynamically*, permitting the dependence of intensity upon concentration to be determined quantitatively. Furthermore, the relatively large specimen dimensions improve the accuracy with which the temperature can be measured, and reduce inaccuracies resulting from crystallite growth, which often impair photographically derived results.

A high-temperature diffractometer furnance suited to a variety of investigations is undoubtedly of greatest general interest. Such a furnace must at least meet the following requirements [161]: (1) It must be possible to maintain furnace temperatures up to 1400°C; (2) the life of the heating element must be reasonably long; (3) the temperature of the sample must be ascertainable with a known accuracy and within acceptably narrow limits; (4) thermal gradients must be kept to a minimum; (5) the furnace should have a low heat capacity but at the same time maintain a constant sample temperature; (6) the furnace must be capable of operation in a vacuum or controlled atmosphere; (7) it should be possible to utilize either solid or powdered specimens; (8) the specimen must be easily accessible; (9) the furnace should be adaptable for use on one or more commercial wide-angle diffractometers; and (10) there should be provision for bringing the specimen surface into coincidence with the focusing circle *while at high temperatures*. The four methods that have been used for heating the specimen include [161]: (1) passage of an electric current through the specimen itself, (2) resistance heating, (3) high-frequency induction heating, and (4) focused high-intensity light (arc image or solar funace). Of these techniques 2 is the most widely applicable. The approximate maximum temperature at which a resistance furnace can be operated in an oxidizing atmosphere is 1600°C. At higher temperatures it becomes mandatory to employ a high vacuum or inert atmosphere (helium or, in appropriate circumstances, nitrogen).

Mauer and Bolz[164] have treated in some detail the problems to be dealt with in the temperature calibration of a diffractometer furnace by means of thermocouples. The chief sources of error are temperature gradients within the specimen and its holder, and unfortunately these gradients differ with the conductivity and emissivity of the materials concerned. Mauer and Bolz find that lattice constants of pure substances of known expansion coefficients are superior to melting points for the preparation of calibration curves, and they report on their experience with silver, platinum, and tungsten for calibrating a vacuum furnace with molybdenum windings up to a temperature of about 1400°C. The concensus of other investigators is that the errors in temperature measurements with optical pyrometers are likely to be greater than those determined with thermocouples, the uncertainty becoming relatively great at higher temperatures (1600 to 2500°C) [164–167], largely because of the difficulty in separating the blackbody-emitted light from the scattered light.

Many of the features of present-day, resistance-heated diffractometer furnaces actually represent developments of or improvements on the features of a furnace invented by Chiotti[168] some years ago. We here describe in some detail one modern resistance-heated furnace developed by McKinstry[163], which embodies several superior features. A commercial version of this furnace has been widely distributed over a period of several years.* Principal merits of this furnace include: (1) reduction of thermal gradients in the region of the specimen to a low level; (2) support of the sample independently of the heating chamber in such a way as to maintain alignment with the x-ray beam; (3) three sensitive alignment adjustments (controlled by the screws A_1, A_2, A_3, Fig. 5-47) which permit the surface of the flat specimen to be placed rather precisely in the plane of the direct beam at 0° 2θ and coincident with the goniometer axis; (4) ease of alignment of the specimen at any time during the operation of the furnace; and (5) adaptability for use with most commercial diffractometers.

With reference to Fig. 5-47, the specimen S is packed in a rectangular cavity in a nickel or platinum plate mounted on one end of a tubular support T. The other end of this tube is supported by the base, which is independent of the heating chamber and insulation. The platinum-wire heating element is internally wound on two hemispheres 5 cm in

*Tem-Pres Research Division, The Carborundum Company, 1401 So. Atherton St., State College, Pennsylvania.

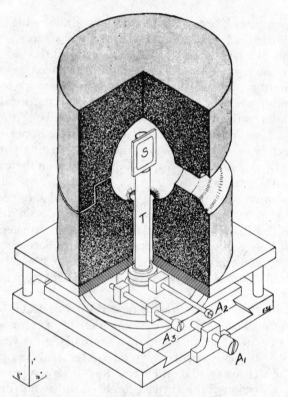

Fig. 5-47.   Cut-away isometric projection of high-temperature x-ray diffractometer furnace. (Courtesy of H. A. McKinstry, *J. Appl. Phys.*, **41**, 5074 (1970).)

diameter and supported by alumina cement. The two hemispheres are
placed in a formed insulating fire-clay brick. The sample-support tube
enters the heating chamber through a hole in the lower hemisphere. The
furnace can be operated continuously at any temperature up to 1400°C
(1850°C in one commercial model). Exploration of the specimen zone
with a chromel–alumel thermocouple showed that for furnace temperatures up to 1400°C the thermal gradients are less than 5°C/cm over a
distance of 1 cm from the sample in any direction.

Figure 5-48 is a diagram of the thyratron temperature control unit. A
mirror galvanometer is activated by an emf produced by a thermocouple
and a bucking emf. Light from the lamp illuminates the photocell, which
causes a phase shift in the grid circuit, which in turn yields a maximum
output. The two autotransformers make it possible for the 120 V, 5 A
thyratron to heat the 30 V, 10 A furnace. At 1000°C the temperature control was good to ±0.5° over a period of 24 hr. Furnaces of McKinstry's

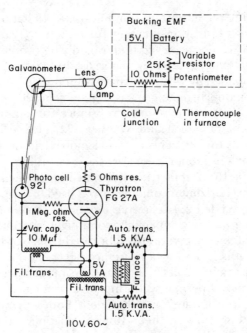

Fig. 5-48. Diagram of thyratron temperature controller. (Courtesy of H. A. McKinstry, *J. Appl. Phys*, **41**, 5074 (1970).)

design have been used in numerous investigations of high-temperature phase equilibria, thermal expansion, and crystal structure.

Heating by high-frequency induction has been utilized in designs described by Franklin and Lang[169] and by Debrenne, Laugier, and Chaudet[170]. The former furnace is adaptable to commercial diffractometers, can provide temperatures in excess of 1600°C, and can be evacuated to 5×10^{-6} torr* or operated with selected atmospheres. The latter furnace operates up to 2500°C and can be evacuated to 10^{-8} torr with a turbomolecular pump. Induction heating concentrates the power into a small volume, with a resulting rapid attainment of the desired temperature.

A simple focused-heat-source furnace was designed by Moss and Smith[171] for convenient use at moderate temperatures (to 425°C) with a Norelco (Philips) diffractometer. The more sophisticated design of Stecura[167] employs a carbon arc and light imaging to obtain temperatures as high as 3000°C. His furnace is incorporated in a diffractometer of Seemann-Bohlin parafocusing geometry (see Section 5-1.3),

*One torr equals the pressure of 1 mm of mercury at 0°C.

thereby providing the important simplification of a stationary specimen. Two advantages of such an imaging furnace are: (1) temperatures in excess of 2500°C can be achieved under oxidizing conditions, important for the investigation of refractory materials; and (2) the sample does not become contaminated by volatile furnace components.

For precise monochromatic measurements of the x-ray scattering curves of liquid metals, the unusual resistance furnace and diffractometer arrangement of North and Wagner[172] is especially suited. As shown in Fig. 5-49, the direct beam from a molybdenum target is focused by a singly bent LiF monochromator, the converging beam traversing the specimen in the symmetrical-transmission geometry (see Section 5-4.3). A Siemens horizontal goniometer was used to support the furnace and supply the needed 2:1 following motion of the counter and specimen. Depending on the compatability of a given specimen and its container, the specimen cell is variously fabricated from beryllium, graphite, or quartz, and it is heated by a resistance heater in the form of a thin-walled pyrolitic graphite tube. The transmission geometry and collimation are such as to permit scattering measurements over the range $2° < 2\theta < 90°$, appropriate for subsequent Fourier transformation to radial distribution curves when $MoK\alpha$ radiation, or another short wavelength, is employed.

Wilchinsky, Tsien, and Ver Strate[173] built a Nichrome-wound resistance furnace for the investigation of polymer properties up to at least 125°C. The unit was designed for a Norelco (Philips) diffractometer and has been used for studies of high-density polyethylene. Horne, Croft, and Smith[174] describe a thermoelectric thermostatic unit for controlling specimen temperatures between 0 and 70°C. A thin

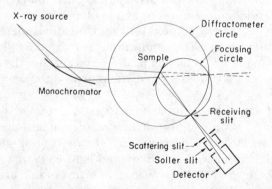

Fig. 5-49. Schematic diagram of transmission geometry, with crystal monochromator, used for the study of diffraction by liquids. (Courtesy of D. M. North and C. N. J. Wagner, *J. Appl. Crystallogr.*, **2**, 149 (1969).)

layer of sample is deposited on a lightweight (1.5-gram) copper plate. The n-type material is a doped bismuth telluride, and the p-type a doped antimony–bismuth telluride semiconductor.

Goldak's unusual apparatus[175] permits measurements throughout the range 77 to 600°K without remounting the specimen. The entire optical system, including x-ray tube, specimen, cryostat/heater, counter, and goniometer, is enclosed in a steel bell jar capable of evacuation to 1 μtorr. The apparatus utilizes a Rigaku-Denki horizontal diffractometer with standard symmetrical-reflection optics and automatic step-scanning with digital recording of the scaler output by a printer, tape, or IBM card punch or a magnetic tape writer. Goldak used an IBM 1620 computer and 807 plotter to plot the output counts as graphs resembling strip-chart records.

Another variable-temperature diffractometer worthy of mention is that built by Baun[176], based on a Philips goniometer, which permits elevated temperatures to 1600°C under vacuum to be achieved by indirect radiation heating, or temperatures down to − 196°C by replacing the heater with a large, hollow, specimen-supporting copper block containing channels for passage of cold or liquefied gases. The same block can be warmed to + 200°C when close temperature control is required.

For a partial listing of other high-temperature diffractometer furnaces, the reader's attention is directed to additional literature, one article describing a furnace with a novel diffractometer design[177], four describing units built for standard horizontal diffractometers[178], and seven describing units to fit Norelco/Philips vertical diffractometers[179].

E. Low-Temperature Techniques. In Section 5-4.2D three control systems were cited for maintaining specimen temperatures below, near, or above room temperature[174–176]. Like high-temperature diffractometry, low-temperature techniques permit solid-state phase transformations to be followed dynamically, an advantage not offered by film methods.

Nearly all the low-temperature diffractometric instrumentation described in the literature is designed for use with one or more commerical diffractometers. In general, low-temperature accessories divide themselves rather obviously into two categories, the *first* intended for measurements near 77°K (liquid nitrogen) or between 77°K and room temperature, and the *second* for work down to 4.2°K (liquid helium) or somewhat lower. Apparatus of the second class is characterized by a much higher degree of sophistication than the first and is more commonly designated by the term *cryostat*. At the same time accessories for work at 77°K or higher also differ greatly in complexity. We first cite very briefly several that are characterized by perhaps almost unexpected simplicity of construction, economy, and ease of operation[180–184], and then mention others of considerably greater sophistication[185–187] that are distinguished in particular by more accurate temperature control and measurement and by the capability of operation at any desired temperature between 77°K and room temperature.

The specimen-cooling chambers (cells) of Roessler and Bolling[180], Ghislain, Deléhouzée, and Deruyttere[181], and Brady and van Reuth [182] fit commercial diffractometers and are designed to operate only close to liquid-nitrogen temperature. The first two make liberal use of plastic foam in their construction and insulation and are very easy to build; in both units liquid nitrogen is contained in a plastic reservoir and cools the specimen by directly bathing the metal block in which it is mounted. In other low-temperature accessories of the first general category, the specimen-supporting block is cooled by circulation through it of dry nitrogen vapor which has first passed through a copper coil immersed in a bath of liquid nitrogen, Dry Ice and acetone, or ice and water. The temperature can be controlled to within reasonably narrow limits by regulating the flow of nitrogen gas. In the device built by Calhoun and Abrahams[183] for the Norelco/Philips goniometer, the back of the specimen holder is in contact with refrigerating fluid contained in a styrofoam-insulated cold cell. The rate of admission and egress of liquid nitrogen can be regulated so as to maintain a temperature of 77°K or higher. Frosting under humid conditions is prevented by passing dry nitrogen gas at room temperature between two cellophane windows located 10 and 15 mm above the specimen surface. The previously cited high-temperature accessory of Intrater and Hurwitt[179–4] provided for replacement of the heater with a cooling unit to yield specimen temperatures between 77°K and room temperature[184].

The considerably more complex apparatus of Jetter and co-workers [185], also designed for the Norelco/Philips diffractometer, encompasses the 2θ limits 25 to 140° and temperature control to ± 0.5°K over the range 77 to 475°K. The copper specimen holder is cooled by metallic conduction from the refrigerant reservoir; the cell is constructed of stainless steel, incorporates beryllium x-ray windows, and is evacuated to $0.1\,\mu$ (10^{-4} torr) or less to provide needed thermal insulation between the outer cell walls and the specimen. The temperature is controlled by heating the specimen block with a Nichrome resistance element while the reservoir is kept full of coolant. The temperature is measured with a copper–constantan thermocouple embedded in the specimen block or inserted in the powder itself. The cooling accessory described by

Bonfiglioli and Testard[186] has thermal capabilities similar to those of the apparatus just described, but the accessible 2θ range for any given measurement is very restricted. Another cooling accessory approximately equivalent to that of Jetter et al. [185], but with full $180°$ 2θ angular coverage and designed for the Siemens horizontal diffractometer, has been built by Baun and Renton[187]. They have used this apparatus to measure thermal expansion coefficients of binary alloys, and to determine the identifying powder patterns of numerous organic compounds that are liquid at room temperature.

We now turn to the relatively sophisticated cryostats of the second category. The first helium cryostat for diffractometry appears to be the apparatus constructed by Barrett[188, 189] at the University of Chicago about 1956 and used to measure the lattice constants of high-purity gallium and indium down to $2.4°K$ and to study recrystallization phenomena. The diffractometer cryostats built subsequently by other investigators[190–198] possess certain basic features of Barrett's apparatus, as well as additional developments and refinements, in particular greater flexibility and accuracy in the control and measurement of temperature and improvements in the handling of specimens.

As an example of one of these designs, we select the cryostat constructed for the General Electric horizontal diffractometer by Black et al. [190] at the National Bureau of Standards, and later improved in some important respects by Mauer and Bolz[191]. In the fabrication stainless steel was used wherever possible, and all but demountable joints were heli-arc welded and pretested for leaks. As shown in the sectional view of Fig. 5-50, the principal components of the cryostat are two concentric Dewar vessels, the inner (8) containing liquid helium and bearing at the bottom a hemispherical copper block (17), the vertical plane surface of which is coated with the specimen. The x-rays enter and leave the specimen chamber through beryllium windows (19) of sufficient length to permit measurements throughout the angular range -2 to $+168°$ 2θ. The helium Dewar is surrounded by a liquid-nitrogen vessel (9) of hollow cylindrical shape with appropriate copper radiation shields (7 and 15) attached. An x-ray window in the lower shield is covered with 0.00035-in. nickel foil (18) which shields the specimen from warm surfaces and also serves as a $K\beta$ filter for copper radiation. By this construction the helium vessel is almost entirely surrounded by surfaces at or near liquid-nitrogen temperature. A valve (10) at the back of the apparatus permits evacuation of the space around the specimen and around the nitrogen container, while the space around the helium vessel is evacuated through another valve (1). Because of the separating wall (14) this second space seldom requires pumping.

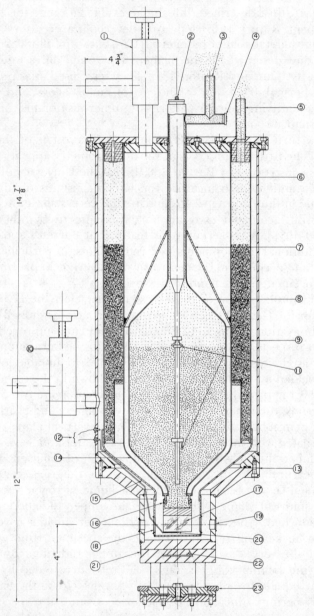

Fig. 5-50. Section through the x-ray-diffractometer cryostat of the National Bureau of Standards. (Courtesy of I. A. Black et al., *J. Res. Nat. Bur. Stand.*, **61**, 367 (1958).)

A thermocouple junction (20) (silver plus 0.37 atomic per cent gold versus gold plus 2.1 atomic per cent cobalt) is inserted in a well at the botton edge of the specimen surface, the leads being cemented to the copper block to keep them at the temperature of the junction. The liquid-helium level is monitored by measuring the resistance of two carbon resistors (11), which indicate when the $\frac{1}{4}$- and 2-liter levels are reached. The outputs from the thermocouple and liquid-level gage are recorded continuously on a chart synchronized with the diffractometer chart. The pressure in the specimen chamber is measured with a Philips cold-cathode ionization gage.

Translational alignment of the cryostat perpendicular to the specimen surface is permitted by a slide (22), and rotation around its vertical axis by a worm gear (23). The lower chamber can be detached at the flange (13) for cleaning and repairing of the specimen block, thermocouple, and lower radiation shield. A system of pulleys and a counterweight serve to balance the weight of the cryostat (68 lb), leaving it free to rotate with the goniometer table. After very thorough preliminary evacuation and flushing-out of the system and filling of the Dewar vessels, the specimen gas (or free radicals) is allowed to enter and deposit on the cold specimen-block surface until it loses its metallic appearance. About 15 or 20 min is required to deposit the 0.001- to 0.005-in. layer required for preparation of a diffraction pattern.

Several limitations of the original apparatus were corrected by improvements described by Mauer and Bolz[191] in 1961. These included (a) provision for isolation of the specimen block from the helium reservoir when desired, (b) facilities for heating the block electrically or cooling it to the refrigerant temperature, (c) addition of an electronic controlling circuit for the maintenance of a given temperature, and (d) separation of the specimen chamber from the insulating vacuum spaces to permit specimen vapor pressures as high as 1 atm to be tolerated. These modifications permitted the study of materials with vapor pressures of as much as 1 atm, and the maintenance of temperatures to an accuracy of $\pm 0.1°K$ in the range 4 to 20°K.

The application of the improved apparatus of Mauer and Bolz to the measurement of the lattice constant of neon in the temperature range 4.2 to 22°K is illustrated in Fig. 5-51. The temperature measurements are considered accurate to $\pm 0.1°K$, and the lattice constants to ± 0.001 Å. These results demonstrate the capability of the improved cryostat for maintaining a constant temperature with volatile specimens. For further details of construction and performance the reader is referred to the original articles[190, 191].

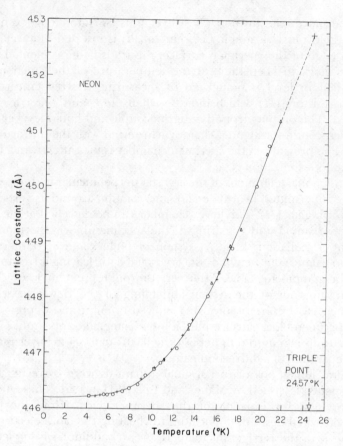

Fig. 5-51.　The lattice constant of neon as a function of temperature, showing the repro-ducibility among three experiments. The value at 4.2°K is the average of four measure-ments. That at the triple point is calculated from the density reported by Clusius. (Courtesy of F. A. Mauer and L. H. Bolz, *J. Res. Nat. Bur. Stand.*, **65C**, 225 (1961).)

Stochl and Ullman[192] have described a helium cryostat to fit a Norelco/Philips vertical goniometer, but the apparatus is apparently not capable of operation at temperatures intermediate between liquid-helium, liquid-nitrogen, and room temperatures. A cryostat resembling that of Black et al.[190, 191] has been built by Klug and Sears[193] in the authors' laboratory, and used to measure the lattice parameters, den-sities, and thermal expansion coefficients of polycrystalline xenon in the temperature range 5.5 to 75°K[194]. Gruber and associates[195] have constructed a helium cryostat of unusual design for the General Electric diffractometer. Their apparatus incorporates a removable helium-temperature specimen capsule which can be irradiated by fast neutrons at liquid-helium temperature and then transferred to the diffractometer cryostat for diffraction measurements at liquid-helium temperature without intermediate warm-up above 4.2°K. They have applied the equipment to the study of the effects of radiation damage on the lattice constants of materials. Another sophisticated liquid-helium cryostat, designed for diffractometric studies of metals and alloys at temperatures between 4.5 and 300°K, has been built by Zahkarov[196]. Between 4.5 and 10°K the temperature can be kept constant to ± 0.02°K.

Finally, a helium cryostat worthy of special note is that developed by King and Preece[77] for use with a Siemens horizontal diffractometer which permits the use of the double-scanning diffractometric technique [17, 76] on both sides of the direct beam up to $2\theta = 165°$, thus yielding high-precision lattice parameters at cryostatic temperatures. The alignment with respect to the x-ray source is not disturbed during insertion and removal of specimens. An improved version of the apparatus[197] provides for refinement of the alignment at cryogenic temperatures, and for lowering the specimen temperature to 1.5°K by pumping on the liquid helium contained in the hollow specimen block after its isolation from the main liquid-helium vessel. A further modification of this cryostat has been made by Abell and King[198] to permit metallic specimens to be strained in tension at temperatures down to 20°K during the registration of portions of their diffraction patterns.

For additional references concerned with low-temperature x-ray diffractometry, the reader is referred to two reviews having useful bibliographies[187-2, 189].

F. Other Special Specimen Techniques. Most of the apparatus for diffraction studies of substances under elevated pressures has been designed for photographic registration. Such techniques have already been treated in Section 4-7. Now we only comment on three high-pressure units designed specifically for counter measurements.

A small hydraulic press mounted as an integral part of the goniometer shaft of a Norelco/Philips diffractometer has been described by Steinfink and Gebhart[199]. The specimen powder is deposited on a lower beryllium platen, and pressure is exerted on it by an upper similar platen which fits inside the steel piston of the press. This apparatus was used to investigate the loss of interlayer water from a sodium vermiculite at 30,000 psi on the basis of changes in the intensities of sensitive peaks of the diffraction pattern. The tetrahedral anvil press of Barnett and Hall [200], already mentioned in Section 4-7, was used in conjunction with a General Electric CA-7 x-ray tube and scintillation counter to identify crystal structures and to determine volume compressibilities, lattice-parameter changes, and phase diagrams of materials at pressures up to 75 kbars and temperatures to 1000°C. For low-intensity patterns these investigators employed a slow scan speed with digital printout at 100-sec intervals, the count data being subsequently plotted. Christoe and Drickamer[201] have built a clamp cell for x-ray diffractometric or Mössbauer resonance measurements of specimens at low temperatures and at pressures up to 240 kbars.

A diffractometer specimen can be immersed in a *constant controlled atmosphere* in the specimen holder devised by Thompson and Mallett [202] for the Norelco/Philips goniometer. The holder is a precision Pyrex cell having an x-ray transmissive window and fitted with a positioning ring against which the sample is held with a vacuum putty. For regulation of humidity a saturated salt solution of appropriate composition is placed in the bottom chamber of the cell, while the atmosphere can be controlled by evacuating the system and adding a gas of known composition and pressure. For highly air-sensitive materials, the device designed by Norment and Henderson[203] for the Norelco/Philips goniometer permits the specimen to be loaded in the x-ray cell while in a dry box. The cell and its support are then transferred to the diffractometer while the specimen is kept immersed in a stream of flush gas carefully stripped of reactive contaminants, and the diffraction pattern is then recorded. Morrow[204] has built an admirably simple cell for maintaining a specimen under saturated atmospheres of water or organic compounds during diffraction measurements. Small modifications permit the cell to be used with either a vertical or horizontal diffractometer.

Rowland, Reinhardt, and Brooks[205] have constructed an *adsorption cell* for the Norelco/Philips diffractometer which permits the adsorption of condensable vapors by laminar materials, such as clays, to be followed by observing changes in the basal $(00l)$ d spacings. The cell, containing the mounted specimen, is attached to any standard adsorption apparatus suitable for condensable vapors, and the specimen is then saturated with the vapor at the desired partial pressure in an inert carrier gas. After the diffraction measurements the amount of vapor adsorbed is determined by desorption and elution of the clay with helium or nitrogen while simultaneously heating.

Kohler and Parrish[206] showed that the diffraction patterns of moderately *radioactive materials* can be measured satisfactorily with a proportional or scintillation counter used in conjunction with PHD/PHA (see Section 5-3.4A). By means of PHD/PHA the background from a Co^{60} sample was reduced from 785 to 26 cps, and the sensitivity of measurement of the $CuK\alpha$ radiation was increased by a factor of 2.5 to 3.0. For highly radioactive specimens, Bredig, Klein, and Borie[207] modified a horizontal diffractometer so as to cause the x-ray tube rather than the heavily lead-shielded detector to revolve about the specimen. The detector, a proportional or scintillation counter used with PHD/PHA, was employed either alone or in combination with a bent-crystal monochromator in the diffracted beam. A similar apparatus and technique were applied by Cummings, Kaulitz, and Sanderson[208] to the study of radioactive materials[208].

For measurement of the diffraction patterns of *liquids*, Kaplow and Averbach[209] have constructed a special diffractometer in which the specimen is held in a fixed horizontal position while the x-ray source and detector revolve about it in opposite directions and at equal velocity, thus preserving the Bragg-Brentano parafocusing condition. Their apparatus provides several advantageous features, including provision for a crystal monochromator in either the incident or diffracted beam, PHD/PHA for use with a scintillation counter, and a vacuum furnace for heating the specimen to 1100°C, including precise temperature regulation. The diffractometer was used to make precise measurements of the x-ray scattering curves of liquid mercury at room temperature and of a liquid 0.55 gold–0.45 tin alloy at 420°C.

5-4.3 Transmission Techniques[210]

Figure 5-52 shows the three principal geometrical arrangements of the specimen with respect to the x-ray source and counter that are useful in diffractometry under appropriate circumstances. Figure 5-52*A* is the widely used symmetrical-reflection technique of conventional powder diffractometry, which has been discussed almost exclusively to this point. In Section 5-4.2*C* it was explained that this geometrical arrangement suffers from two important limitations: (1) the excessive length of specimen required at small Bragg angles, and (2) the great thickness of low-absorbing specimens needed to satisfy equation 5-11, with resulting severe line broadening and other geometrical problems. These problems associated with low Bragg angles and specimens of low absorption for x-rays may be met by sacrificing the parafocusing intensity of arrangement *A* in Fig. 5-52 in favor of arrangement *B* or *C*, of which *C* is to be preferred as a general rule, for the zone of moderately small scattering angles ranging between 3 and 30° 2θ.

To preclude undue loss of resolution as a result of the absence of parafocusing geometry, the incident-beam divergence must be rather sharply restricted, generally being set from 0.2 to 1.0° 2θ depending on the

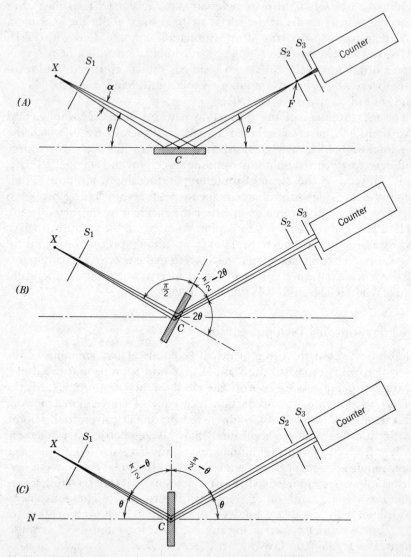

Fig. 5-52.   Three geometrical arrangements of specimen with respect to x-ray source and counter for diffractometry. (A) Symmetrical reflection; (B) normal-beam transmission; (C) symmetrical transmission. (Courtesy of L. E. Alexander and John Wiley and Sons, *X-Ray Diffraction Methods in Polymer Science*, 1969, p. 78.)

desired balance of intensity and resolution. Compared with the sym-metrical-reflection geometry, the symmetrical-transmission arrange-ment yields considerably lower intensities and demands correspondingly longer counting intervals to give acceptable statistical accuracy. Never-theless, it is the most generally acceptable technique for studying diffrac-tion effects from polymers, carbons, and most organic substances at the lower scattering angles.

For the symmetrical-transmission geometry it can be shown that the thickness required for maximum diffracted intensity is

$$t_m = \frac{1}{\mu \sec \theta},$$
(5-63)

and the absorption correction to be applied to intensities measured at some particular angle 2θ is:

$$\frac{I_{0°}}{I_{2\theta}} = \frac{\exp\left[-\mu t(1 - \sec \theta)\right]}{\sec \theta}.$$
(5-64)

For the derivation of these equations, the reader is referred to another source[210]. To avoid undue line broadening, it is good practice to employ a sample thickness of 1 to 2 mm, despite possible larger numerical values dictated by equation 5-63 for maximum intensity.

Lang[211] has described a modification of the symmetrical-transmission geometry (Fig. 5-52C) incorporating an unusual focusing feature made feasible by the use of a wide counter window. Ergun[212] corrects x-ray line profiles for instrumental broadening in transmission geometry by a procedure of successive convolutions (foldings) (see Section 9-1.2).

5-4.4 Continuous-Scan Techniques

The diffractometrist is always called upon to decide whether to record a diffraction pattern by continuous or step scanning. The output count data of a continuous scan are blended by a pulse-averaging circuit into a fluctuating current which is normally portrayed by a strip-chart recorder (Section 5-3.1). At fast scanning speeds this technique provides the entire diffraction pattern rather quickly, making it appropriate for preliminary examination of substances, particularly when the principal objective is their identification or qualitative evaluation. Very slow scans of selected portions of the pattern are suitable for finer differentiation of diffraction details, and for more quantitative measurements, for example, applications to quantitative analysis of mixed phases (Chapter 7), determination of precise interplanar spacings (Chapter 8), studies of crystallite size and lattice strains (Chapter 9), and small-angle scattering. These more quantitative applications, however, can be accomplished with highest accuracy by means of a step-scan mode with some form of digital registration of the output counts (see Section 5-3.1), because this technique makes possible a statistical evaluation of the precision of the results. The step-scan technique is treated in Section 5-4.5.

Regardless of the scan technique employed, the instrumental factors that decrease the breadth of the line profile (increase resolution) tend to be incompatible with those that promote intensity, as is evident from an inspection of Table 5-10. However, it will be noticed that two items, 3 and 4, though of basic importance in achieving high intensity, are of only secondary importance in the matter of resolution. It is right to deduce

Table 5-10. **Comparison of Conditions Promoting Resolution with Those Promoting Intensity of Diffracted Maxima**[a]

	Proper Adjustment of Factor to:	
Factor	Increase Resolution	Increase Intensity
1. Source		
Direction of view	Lateral (with Soller slits)*	Longitudinal*
Angle of view	3° or less	Not less than 6°
2. Receiving-slit width	Small*	Large*
3. Absorption of x-ray beam by sample	Thin layer of powder	Thick layer of powder*
4. Equatorial divergence, γ	Moderately small	Large*
5. Axial divergence, δ	Small	Large

[a]The more influential items are marked with an asterisk.

that as a general rule intensity can be gained with a minimum sacrifice of resolution by making use of a large equatorial divergence slit and a relatively thick sample. In adjusting the other factors, though, it is necessary to compromise between intensity and resolution.

Figure 5-53 illustrates the excellent resolution of reflections obtainable with a diffractometer adjusted for near-optimum performance. The specimen is fine quartz powder, and the instrumental conditions are as follows:

CuKα radiation
Source viewed laterally; Soller slits with aperture, $\Delta = 2.2°$
Take-off angle, $\alpha = 3°$; apparent angular width of source about 0.02°
Equatorial divergence, $\gamma = 1°$

Receiving slit width, $\nu = 0.025°\ 2\theta$
Scanning speed, $\omega = \frac{1}{4}°\ 2\theta/\text{min}$
Recorder chart speed $= 6$ in/hr.

The $\alpha_1\alpha_2$ doublet is observed to be partially resolved at $2\theta = 36°$ (the 11·0 reflection), and in fact resolution would commence below $30°$ were a peak present there. The best Debye-Scherrer cameras do not resolve the CuKα doublet at angles much below $2\theta = 90°$. The breadth of the 12·1 α_1 peak at half-maximum intensity is only about $0.10°$, whereas in Debye-Scherrer photographs the width of this line can hardly be reduced to less than $0.30°$. This degree of resolution permits the clear-cut separation of the well-known "five lines" of quartz, which appear within the space of $1°$ at a goniometer setting of about $68°$. Somewhat better resolution than this can be obtained by manually counting over the peaks and plotting the counting rates at small intervals, which is equivalent, however, to automatic step-scanning with digital registration of the output counts, a much more convenient technique (Section 5-4.5).

Compared with the intensities of Debye-Scherrer lines, the intensities of maxima generated by the parafocusing arrangement exhibit a falling-off with increasing diffraction angle. This difference stems from the dissimilar x-ray absorption characteristics inherent in the two techniques. The absorption of the x-ray beam is independent of angle in the parafocusing arrangement if the thickness of the specimen satisfies equation 5-11, but with ordinary Debye-Scherrer samples it diminishes with increasing angle. It is a common practice in preparing diffractometer strip-chart records to boost the intensity when entering the back-reflection region by considerably increasing the equatorial beam divergence γ, and in some cases by also widening the receiving slit. This results in an appreciable but not serious increase in the reflection breadth, so that only a slight loss of resolution results. Figure 5-54 shows another strip-chart record of quartz powder, in the angular region $0° < 2\theta < 90°$, obtained with a wider receiving slit in order to obtain greater intensity at a faster scanning speed. The loss of resolution under these conditions is more noticeable, as can be recognized by comparing the lowest angle at which the $\alpha_1\alpha_2$ doublet is resolved with that in Fig. 5-53.

The properties of the pulse-averaging circuit have a fundamental bearing upon the authenticity of strip-chart records. The reader should review the significance of the RC time constant as explained in Section 5-3.1. Small values of the time constant (e.g., less than 3 sec) result in a relatively small averaging-of-counts effect. This has two conspicuous effects upon the chart record. (1) The intensity profile appears jagged, or saw-toothed, because the signal emf fluctuates rapidly in a rather

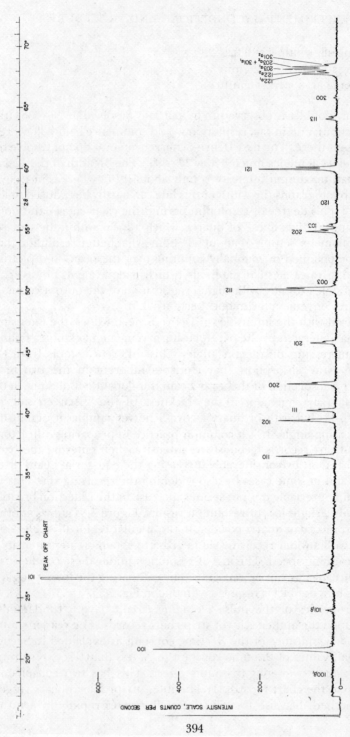

Fig. 5-53. Portion of the diffraction pattern of quartz powder prepared with a diffractometer adjusted for high resolution. Numbers above peaks are Miller indices. (Courtesy of Philips Electronic Instruments.)

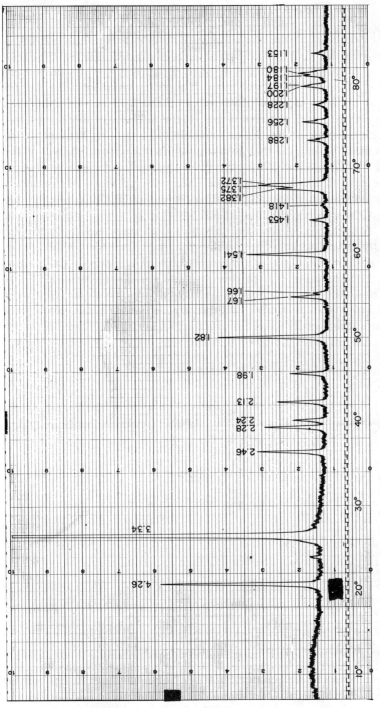

Fig. 5-54. Diffraction pattern of quartz powder in the region $2\theta = 0-90°$ prepared with a diffractometer adjusted for rapid scanning (2°/min). Numbers above peaks are d values in ångström units.

faithful portrayal of the statistical variations in the rate of arrival of counter pulses, and (2) the heights or areas of maxima represent the true totalized counts more accurately than if larger time constants were employed. Conversely, large values of the time constant cause the response of the strip-chart recorder to lag behind the actual time variation in counting rate, with the results that (1) the intensity profile is relatively smooth, (2) the peaks do not attain their full heights (phenomenon of "undershoot"), and (3) the positions of the peaks are shifted in the direction of scan. Undershoot can be made small or large according to the demands of the diffraction problem, but it can never be completely eliminated even if RC is set equal to zero, because the potentiometer circuit of the recorder also has a finite response time amounting to 1 or 2 sec for full-scale deflection, and this additional response lag comes into play if the time constant of the pulse-averaging circuit is made very small.

The phenomenon of undershoot is illustrated in Fig. 5-55, which shows the 11·2 maximum of quartz as recorded at a scanning speed of 2°/min, with small, medium, and large values of RC. These results are compared with a 3-min trace prepared with the receiver set at the top of the peak and using the medium value of the time constant. This trace evidently depicts the true height free from time-lag effects. By comparing the

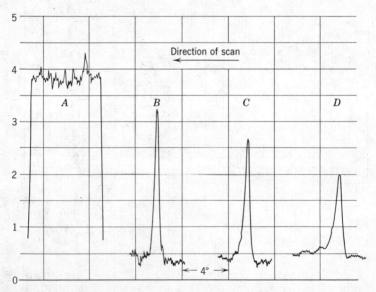

Fig. 5-55. Effect of time constant upon the profile of the 11·2 quartz maximum. B, C, and D were scanned at 2°/min with small, medium, and large time constants, respectively. For comparison A was prepared with the medium time constant and the receiver set at the top of the peak, the recorder being allowed to run for 3 min.

heights of the other peak traces with this equilibrium level, the progressive increase of the undershoot as the time constant increases is very
noticeable. In addition, these traces portray clearly the increasing background variations due to counting statistics, which appear as the time
constant decreases. Besides these effects, the response lag due to a large
time constant generates a special form of asymmetry in diffraction
maxima, which can be seen in Fig. 5-55 and is most noticeable in D. The
peak is rendered asymmetrical, the side scanned last in chronological
sequence being broader. Pronounced asymmetry of this kind throughout
a large angular range usually is an indication of a large time constant.
It is most noticeable when scanning at high speeds (1 or 2°/min) in conjunction with a time constant of 15 sec or more. The phenomena of
undershoot and, to a lesser extent, asymmetry also become more conspicuous with increasing scan speed.

These various time-lag effects demonstrate that diffraction patterns
are recorded most faithfully by keeping both the scanning speed and
the time constant small. In recorded traces the statistical fluctuations
may be allowed for by drawing the best smooth curve through the more-
or-less jagged line drawn by the pen.

Yet a third factor, the receiving-slit width, must be regulated with care
in order to provide faithful delineation of the diffraction pattern. In
this connection it is useful to denote the time required for the receiving
slit to traverse its own width v at a given angular scanning velocity
ω as the time width of the receiver at this velocity:

$$\text{Time width} = W_t = 60 \times \frac{v}{\omega} \qquad \text{seconds} \qquad (5\text{-}65)$$

In equation 5-65, ω should be expressed in degrees per minute and v
in degrees. With reference to Fig. 5-56, suppose that the true diffraction
profiles of two reflections separated by 0.10° are given by curve A, and
that they are measured at a scan speed of 0.50°/min with a receiving slit
of 0.05° angular width, so that $W_t = 6$ sec. Now if the time constant RC is
substantially less than 6 sec, the pulse-integrating circuit charges or dis-
charges rapidly enough so that the output emf, curve B, keeps fairly
close pace with the changing diffraction amplitudes shown in curve A.
This is possible not only because the time constant is sufficiently small,
but also because the slit width is small enough to reveal the finest diffraction detail. Both conditions must be fulfilled in faithfully registering the
details of a diffraction pattern by means of a strip-chart recorder. Curve
C in Fig. 5-56 shows the loss of detail suffered as the result of employing
an excessively large time constant, for instance, of the order of 16 sec
in the present example. Although the slit is narrow enough to give good

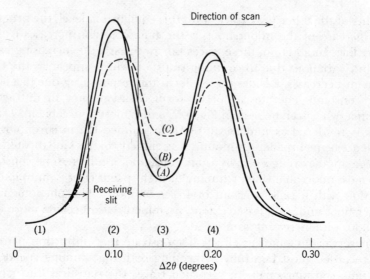

Fig. 5-56. Effect of scanning a pair of diffraction peaks of form (*A*), using a small value (*B*) and a large value (*C*) of the time constant.

resolution, the charging and discharging time is so long that the output emf fails to reach the maximum with the slit at (*2*) or the minimum at (*3*) by wide margins.

A good general rule to follow for the purpose of making the chart record a reasonably faithful picture of the time variation in the energy entering the receiving slit is to make the *RC* constant equal to one-half or less of the time width of the slit.* Table 5-11 suggests values of the time constant *RC* to be employed with various combinations of ν and ω values in order to satisfy this rule.

A typical present-day diffractometer[78] has the capability of utilizing the output of a high-precision linear rate meter (pulse-averaging circuit) to actuate the potentiometer recorder by means of either a linear or three-cycle logarithmic slide wire while the counter and specimen are either stationary or moving in the standard 2:1 relationship at constant velocity. Another optional mode causes the accumulated scalar counts to be printed out at the ends of constant, preselected angular intervals during the scan. A third mode provides for automatic oscillation through a chosen angular range, the scalar content being printed out when either end of the preset range is reached. With a linear slide wire the chart

---

*The value of conforming to a rule such as this was brought to our attention by H. W. Pickett of the General Electric Company.

Table 5-11.   Suggestive Choices of the Time Constant RC for Various Scanning
Speeds and Receiving-Slit Widths

Scanning Speed, ω (°/min)	Receiving-Slit ν (°)	Time Width of Slit, W_t (sec)	Maximum Recommended RC (sec)
2.0	0.20	6	3
	0.10	3	1.5
	0.05	1.5	< 1
	0.025		
1.0	0.20	12	6
	0.10	6	3
	0.05	3	1.5
	0.025	1.5	< 1
0.5	0.20	24	12
	0.10	12	6
	0.05	6	3
	0.025	3	1.5
0.2	0.20	60	30
	0.10	30	15
	0.05	15	7.5
	0.025	7.5	3.8

readings are true to within ± 0.25 per cent of full scale, which may be set
by a range selector switch to any one of a series of values ranging from
100 to 200,000 cps. This maximum value is far in excess of the strongest
powder reflection intensities likely to be encountered (seldom greater
than 10,000 cps). The logarithmic scale is good to only about ± 2 per cent
of the absolute reading, but it affords the advantage of encompassing all
intensities in one full-scale setting. For quantitative work the logarithmic
scale should be carefully calibrated.* Actually, logarithmic chart records
are little used for quantitative intensity comparisons.

A linear potentiometer output affords several practical advantages.
The relative intensities of two maxima may be directly compared by
measuring their relative heights or areas as they appear on the recorder
trace. Thus the integrated intensity of a reflection can be determined
simply and directly with the aid of a planimeter. A further advantage of
a linear-response circuit is the ease with which the width of a peak at
half-maximum intensity can be obtained by simply measuring the width

*A straight-forward calibration procedure is described in the first edition of this book
(1954), p. 312.

at a point midway between the apex of the peak and the background. The greatest disadvantage of this type of circuit is undoubtedly the limited range of intensities that can be registered with acceptable precision on the chart for any one amplitude adjustment. Although the recorder can detect a reflection with an intensity as small as 2 or 3 per cent of full-scale amplitude, the relative intensity of a weak signal has little quantitative value unless it amounts to at least 5 per cent of full-scale intensity. This means that the recorder is sensitive over a limiting intensity range of about 50-fold at any fixed amplitude adjustment and, moreover, that the useful range for quantitative comparisons is only about 20-fold. By comparing peaks of medium to strong intensity, for instance, 20 to 100 in arbitrary units, at a fixed amplitude adjustment corresponding to a full-scale reading of 100, and at the same time measuring weak to medium peaks, 1 to 20 units, for instance, at full scale = 20 or 30 units, it is possible to extend the useful quantitative range of the linear recorder to 50-fold or even more.

If the x-ray output and the heights of the x-ray and receiving slits are kept constant, the two basic factors governing the statistical accuracy of a strip-chart record are the time width of the receiving slit ($W_t = 60 \times v/\omega$ sec) and the equatorial divergence of the x-ray beam γ. The total number of counts received while scanning a small angular increment $\Delta 2\theta$ is directly proportional to γ and v, and inversely proportional to ω, which may be written

$$N = k \times \frac{60\gamma v}{\omega}.$$

The relative standard deviation in view of equation 5-58 is then

$$\sigma_N(\text{rel}) = \left(\frac{\omega}{k \times 60\gamma v}\right)^{1/2} = K\left(\frac{\omega}{\gamma v}\right)^{1/2} \tag{5-66}$$

$$= \frac{K'}{(\gamma W_t)^{1/2}}. \tag{5-67}$$

Expressions 5-66 and 5-67 convey the information that the accuracy of a recorded pattern, as it depends upon the instrumental conditions, can be improved equally by increasing either γ or v or by decreasing ω, all by proportional amounts.

The proper selection of the instrumental conditions ω, γ, and v depends upon the object of the diffraction measurements. Suppose one is confronted with the task of surveying a large portion of the diffraction pattern of a specimen in order to identify the crystalline constituents. The prime concern is, then, to obtain data rapidly because of the large angular range to be scanned, whereas the accuracy of line intensities and positions is of secondary importance. Accordingly, the scanning speed ω, must be relatively large, for instance, 2°/min. To compensate for the adverse effect of this high speed upon the accuracy of the diffraction details (see equation 5-66), both γ and ν may be increased. However, in view of the fact that increasing ν impairs the resolution more drastically than increasing γ, it is wise to increase γ to a value between 2 and 4°, at the same time increasing ν only moderately, to 0.10°, for instance. This makes the "accuracy index" of equation 5-66, $(\omega/\gamma\nu)^{1/2}$, equal to

$$\left(\frac{2}{2 \times 0.1}\right)^{1/2} = 10^{1/2} = 3.16.$$

This is evidently a measure of the magnitude of the statistical error under the given scanning conditions. The time width of the receiving slit is

$$W_t = \frac{60\,\nu}{\omega} = \frac{60 \times 0.1}{2} = 3 \text{ seconds},$$

which shows that the time constant should be kept small, 2 sec being satisfactory for this application.

The foregoing results are summarized in Table 5-12, along with suggested combinations of experimental conditions for other special purposes, including the accurate measurement of relative intensities and the optimum resolution of diffraction detail. The seventh column gives the size of the statistical-deviation function, which is proportional to the mean uncertainty in the recorded amplitudes. This deviation function may be allowed to assume fairly large values when preparing traces for simple qualitative analysis of crystalline mixtures. For high resolution it should be kept somewhat smaller, although it is still not of critical importance. It becomes of greatest concern in the accurate measurement of intensities and should then be kept at a minimum by employing slow scanning speeds and moderately large receiving-slit widths. In order to obtain high resolution, the great needs are for a slow scanning speed and a small receiving-slit width. The reader is reminded that at large Bragg angles spectral dispersion becomes a dominant factor in determining the experimental line breadth, so that larger values of γ, ν, and source width may be used without loss of resolution.

One factor concerned with chart-recording techniques has not been discussed up to this point, namely, chart speed. The most generally useful speed is about 30 in./hr. At higher scanning speeds (1 or 2°/min) this gives acceptable dispersion of the diffraction pattern for ordinary work. At low speeds ($\frac{1}{8}$ or $\frac{1}{4}$°/min) it results in the greater dispersion of diffraction detail needed for high resolution, accurate peak positions, or precise intensity measurements. Only infrequently is a reduced chart

Table 5-12. Suitable Experimental Conditions for Preparing Strip-Chart Records for Different Purposes

Problem	$\gamma(°)$	$\nu(°)$	$\omega(°/min)$	W_t (sec)	Maximum RC (sec)	Relative Standard Deviation $(\omega/\gamma\nu)^{1/2}$	Recorder Scale
1. To obtain a pattern over a large angular range for qualitative identification	$\{\begin{array}{c}2\\2\end{array}$	0.1 0.1	2 1	3 6	1.5–2.0 3	3.16 2.24	Linear or logarithmic
2. To measure accurately the relative integrated intensities of several sharp peaks	$\{\begin{array}{c}4\\4\\2\\2\end{array}$	0.05 0.10 0.05 0.10	$\frac{1}{8}$ $\frac{1}{4}$ $\frac{1}{8}$ $\frac{1}{4}$	24 24 24 24	12 12 12 12	0.79 0.79 1.12 1.12	Linear
3. Same as 2, but for broad peaks	$\{\begin{array}{c}4\\2\end{array}$	0.10 0.20	$\frac{1}{4}$ $\frac{1}{2}$	24 24	12 12	0.79 1.12	Linear
4. To obtain high resolution of diffraction detail	1 2 2	0.02 0.02 0.02	$\frac{1}{8}$ $\frac{1}{8}$ $\frac{1}{4}$	9.6 9.6 4.8	5 5 2–3	2.49 1.76 2.49	Linear
5. To determine precise lattice constants	1	≤ 0.035	$\frac{1}{8}$	≤ 17	8	≥ 1.88	Linear
6. To obtain a pattern over a large angular range for identification of minor constituents	4	0.1	2	3	1	2.24	Logarithmic

speed desirable. It is a decided asset, for example, when a large angular range is to be scanned at low speed. A fast chart speed would disperse the pattern out over an unwieldy length of chart paper without any advantage being gained. Figure 5-53 is a good example of a strip-chart record of a large angular range prepared at a low scanning speed ($\frac{1}{4}°/$ min) and low chart speed (6 in./hr).

For a detailed treatment of the choice of divergence, receiving, and antiscatter apertures in diffractometry, the reader is referred to an article by Parrish, Mack, and Taylor[213]. Vassamillet and King[214] have tabulated the available scanning speeds and receiving-slit widths and suggested time constants for several commercial diffractometers.

5-4.5 Step-Scan Techniques and Automation

Compared to continuous scanning, step-scan modes are more suited to the digital registration of scaler data, to computer processing including evaluation of statistical errors of counting, and to various degrees of automation. Formerly, counting by steps and recording the data were performed manually, a very tedious, time-consuming process, and one particularly vulnerable to operator errors. Present-day diffractometers provide facilities for automatically stepping by fixed angular increments over the desired portion of the diffraction pattern, each step being followed by printing or punching on paper tape the accumulated scaler count at that location. Many commercial diffractometers provide options for recording step-counted data on IBM cards or magnetic tapes. The principal applications of step-scan techniques are (1) measurement of reflection profiles with high precision for crystallite-size or lattice-strain calculations, (2) measurement of diffuse x-ray scattering over small angular ranges as in small-angle scattering studies, and (3) measurement of the broadened reflections and halos generated by semicrystalline or poorly crystallized materials such as polymers.

We point out again that all modern diffractometers provide for step counting in *fixed-time* or *fixed-count* modes. The former mode, although most commonly employed, suffers from the disadvantage that the relative statistical accuracy $N^{-1/2}$ varies with the intensity of each point measured. Thus, for the patterns of well-crystallized materials, background measurements are *relatively* much less accurate than peak measurements. Clearly, the *fixed-count* technique overcomes this failing by counting all points in the pattern to the same value of N. Attractive as this alternative may appear, it has the conspicuous weakness that very weak intensities, such as are usually encountered in the background, necessitate inordinately and usually impracticably long counting times.

Therefore the *fixed-count* mode finds its principal application in the measurement of relatively diffuse and continuous patterns, such as in many small-angle-scattering investigations and in the study of liquids and strained and poorly crystallized materials. We may also observe that the rapid development of computer control systems is so greatly enhancing the flexibility of the fixed-time counting mode as to make it adaptable to nearly all kinds of diffraction measurements.

The first stepping mechanisms to be developed provided for registration of the step pattern by a strip-chart recorder[148-2]. Mueller and associates[215] and McMillan[216] adapted Slo-Syn* stepping motors to drive General Electric and Norelco/Philips goniometers, respectively. Subsequent to these developments came more flexible and elaborate systems for driving diffractometers in step-scan modes, with printout of the scaler data[217–220]. Punched paper tape became an important medium for the read-in of preprogrammed control instructions and for the registration of the scaler output[217–220]. Frost and Whitehead's control unit[220], which fits a Norelco/Philips diffractometer with only minimal modifications, drives the goniometer rapidly to a predetermined angular region of interest and then causes it to step-scan over a selected portion of the diffraction pattern. The intensity data are punched out on paper tape and then processed by means of a computer program in Fortran IV language to yield peak positions, peak intensities, and integrated intensities for storage on magnetic tape.

Making use of a commercially available goniometer and counting components, Keller and Segmüller[221] designed an elaborate control and data processing system for the completely automatic step measurement of reflections and computer processing of the output, including Fourier analysis of the appropriately corrected profiles. In the block diagram of the apparatus (Fig. 5-57) the added control and output units are contained within the rectangle drawn with broken lines and include an electronic clock with an IBM 1624 paper tape punch and a step-scan programmer. The principal components of the electronic clock are commercially available items: a quartz-controlled sine-wave oscillator operated at a frequency of 1000 cps, a gating circuit, a binary counter with 20 stages, and a read-out unit. The clock can measure times up to 1048 sec, and its accuracy in measuring a 1-sec period is 0.1 per cent. All programmed steps are multiples of a basic step, which may be made as small as 960/degree or as large as 15/degree depending on the particular choice of gears and cam wheel. A given scan program may consist of either the basic steps, or multiple steps 2 to 10 times the basic

*The Superior Electric Company, Bristol, Connecticut.

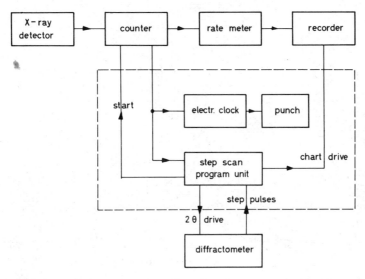

Fig. 5-57. Block diagram of Keller and Segmüller's automated measuring apparatus (*Rev. Sci. Instrum.*, **34** (6), 684 (1963).)

step, or combinations of both. In any case the stepping time is 2 sec, sufficient to permit punching out of the counts from the preceding step point. Figure 5-58 shows a typical line profile as monitored on the chart recorder. It illustrates the employment of single steps over the diffraction peaks, but multiple steps in the regions of lower and more uniform intensity.

We may also choose (somewhat arbitrarily for lack of space) as worthy of special mention the automated powder diffractometer system described by Rex[222]. It is built around a basic General Electric XRD-5 spectro-goniometer and is designed for rapid and almost uninterrupted digital recording of complete powder patterns of materials for qualitative analysis, in particular, reference phases. The output is registered on magnetic tape and, in addition to the *d* values and relative intensities, includes detailed information concerning the peak areas and shapes. Figure 5-59 is a flow diagram of the system, which is built around a Datex Corporation central control computer. Command instructions, as well as transfer of information from input to output, are accomplished by the reading of punched cards with a special IBM 026 card reader. Punched cards were found to be especially suitable for this purpose, inasmuch as each powder sample represents a discrete data block. A multiple automatic sample changer permits specimens to be removed and replaced during operation, making possible continuous measure-

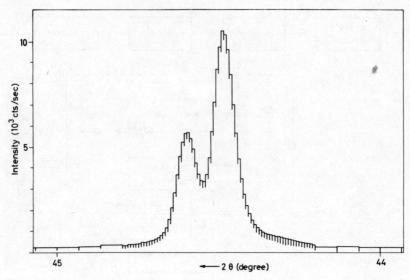

Fig. 5-58.   Typical step-scanned $K\alpha_1\alpha_2$ line profile as registered by a strip-chart recorder with two step lengths. (Courtesy of H. Keller and A. Segmüller, *Rev. Sci. Instrum.*, **34** (6), 684 (1963).)

ment and data registration over long periods of time. Rex reports one uninterrupted run lasting 18 days. The diffractometer of this system makes use of a crystal monochromator in the diffracted beam, as well as PHD/PHA (see Section 5-3.4A), to enhance the peak-to-background contrast. An outstanding feature of the data processing system is the use of a smoothing function and digital filtering to remove unwanted noise background, thus providing a "best" value of the background for correction of the overall reflection intensities. To guard against various types of errors in measurement and processing, the instrumental operation is monitored in several ways, one being the application of a tape edit/check program. Experience showed that repetitive measurement and time-averaging of the unpromising diffraction effects of poorly crystallized materials often gave useful patterns even though the conventional strip-chart records appeared largely amorphous. Such a sophisticated control and data processing system as this offers significant economic and technological advantages for a laboratory performing a large volume of qualitative diffraction analyses on a day-to-day basis.

The development of several excellent commercial automatic single-crystal diffractometers during the last decade lent added impetus to the automation of powder diffractometry, especially in the matter of on-line computer control. We cite two representative instances of such

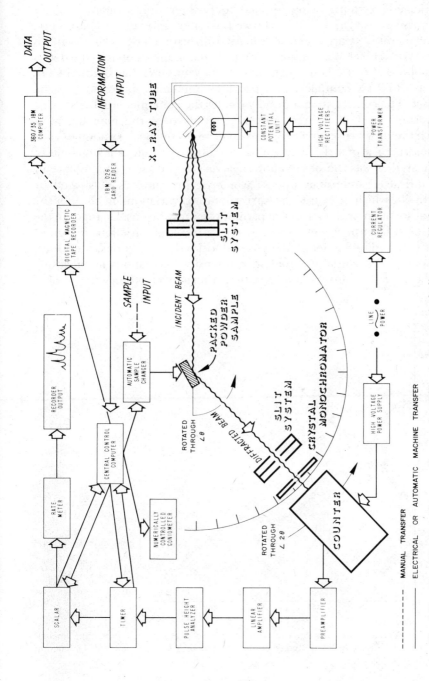

Fig. 5-59. Flow diagram of a complete powder diffraction system, including numerical control of the diffractometer and processing of the output data by computer. (Courtesy of R. W. Rex, *Advances in X-Ray Analysis*, Vol. 10, Plenum Press, 1967, p. 366.)

"packaged" systems being applied to powder diffractometry. Baker and associates[223], at the Harwell Atomic Energy Establishment, simultaneously control two automatic diffractometers on-line with an Elliott ARCH 9000B computer. The principal application is to the determination of precise lattice constants. This computer has both ALGOL and FORTRAN compilers and provides immediate processing of the output. The control system rapidly rotates the goniometer to the specific diffraction lines of interest and step-counts over the maxima, after which the data processing program determines the peak location of each by means of a quartic fit of the data points. A significant outcome of these studies was the observation of a very appreciable dependence of the diffracted intensities upon barometric pressure, as illustrated by Fig. 5-60, which is caused by corresponding variations in the density and absorption of the air in the path of the x-ray beams. Changes in the measured intensities were also noted as a result of variations in atmospheric humidity. Under changeable atmospheric conditions, therefore, it was found essential to avoid such intensity fluctuations by enclosing the beam path in an air-tight chamber. Alternatively, these effects could

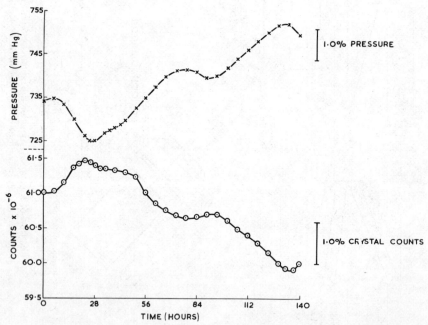

Fig. 5-60. Variation in x-ray intensity with barometric pressure for typical experimental conditions. (Courtesy of T. W. Baker et al., *Advances in X-Ray Analysis*, Vol. 11, Plenum Press, 1968, p. 359.)

have been eliminated by the use of time-averaging techniques of intensity measurement (see below).

Desper[224] gives a detailed description of his programming of a Picker four-circle automatic diffractometer controlled by a PDP-8S computer for studies of preferred orientation in polymers. This system makes use of teletype input/output and provides immediate processing of the output by utilizing the computer in a time-sharing operational mode. Some of the options provided by the programs include (1) 2θ step-scanning with variable step widths for randomly oriented specimens, (2) 2θ step-scanning together with azimuthal rotation of the specimen to average out intensity variations when preferred orientation is present, (3) determination of Legendre coefficients for oriented specimens, and (4) calculation of pole figures.

For full information on currently available diffractometer instrumentation, together with computer hardware and software for control and data processing, the reader is encouraged to contact any of the major manufacturers. We may add that as a general rule such package systems provide teletype input/output, but that for an additional cost manufacturers are usually able to supply fast paper or magnetic tape input/output accessories.

For the automatic collection of small-angle-scattering intensities over a limited angular range, Kratky and Kratky[225] constructed a photoelectrically actuated mechanical programming apparatus. Selective transmission of light through some of a large number of spirally arranged apertures on the surface of a motor-driven disk serves to switch the motor on and off, control the step size, actuate the scaler and printer, and so on, in a preprogrammed sequence. Later, in the same laboratory this optical-mechanical mechanism was replaced by a relatively sophisticated and more reliable electronic programmer described by Leopold[226]. In small-angle-scattering measurements of diffuse scattering with either apparatus, five step sizes were found to be adequate, the smallest being employed in the zone of highest intensity at the smallest angles, and progressively larger steps being utilized in direct relation to the decrease in intensity with scattering angle that is typical of such scattering.

In recent years an important advance has been made in the measurement of weak and/or diffuse x-ray scattering in the presence of appreciable electronic and other background noise. This technique is referred to as *continuous*, or *repetitive*, averaging of intensity measurements. It was first applied in nuclear magnetic resonance measurements[227] and then used in x-ray diffractometry by Peters and Milberg[228]. The success of the method rests on the principle that, as a selected portion of the pattern is scanned repeatedly, the true signal increases in proportion to the *number of scans*, whereas the noise increases only as the *square root of the number of scans*. In this way the effects of the well-known phenomenon of long-period drift, which so often characterizes protracted scattering measurements, can be avoided. Peters and Milberg describe a test experiment in which the amorphous scatter from glass was scanned 200 times continuously at a constant rate of 1.25°/sec. The successively measured counts at 21 points separated by a fixed angular increment of 1.25° were registered on paper tape and then separately summed and averaged at the 21 points by computer. Each 21-fold step scan required 52 sec. During the test experiment the x-ray tube current was artificially regulated so as to decrease at a rate of 10 per cent/hr from an initial value of 19 to a final value of 13 mA. As a control on the continuous-averaging measurements, the intensities were also measured under the same conditions of tube-current reduction by the standard single step-scan technique. The results, portrayed in Fig. 5-61*A* and *B*, are a dramatic illustration of the effectiveness of the continuous-averaging technique in combating long-period drift.

Weymouth and associates[229] improved the technique of Peters and Milberg by accumulating multiply scanned data with a 128-channel analyzer, and applied the method to the study of low-intensity thermal diffuse x-ray scattering. Their technique also differed in that the scanning was actually continuous, the counts collected in consecutive small uniform angular increments, rather than at fixed points, being accumulated by the multichannel analyzer. Berman and Ergun[230] used rather similar instrumentation in the measurement of diffuse scattering from carbons and coals. Their technique differed from that of Weymouth et al.[229] only in that the pattern was step-scanned and the counts at discrete angles totalized by the multichannel analyzer. Typical step-scan speeds were very high, about 150 steps/sec.

In conclusion, we may remark that until recently it was common practice to compensate for long-period drift in precise x-ray diffractometry by employing an incident-beam monitor. The monitor counter continuously recorded the x-ray intensity scattered at some fixed angle by a reference specimen placed in the direct beam, and when it had recorded a predetermined fixed count, a special switching circuit was actuated which shut off both the monitor counter and the specimen-scanning counter. The reference sample was sometimes placed in the same incident beam that irradiated the "unknown" specimen, and sometimes in the direct beam from the opposite x-ray tube port. Partly as a result of continuing improvement in the mechanical and electrical stability of diffractometer design, and even more as a consequence of the advent of

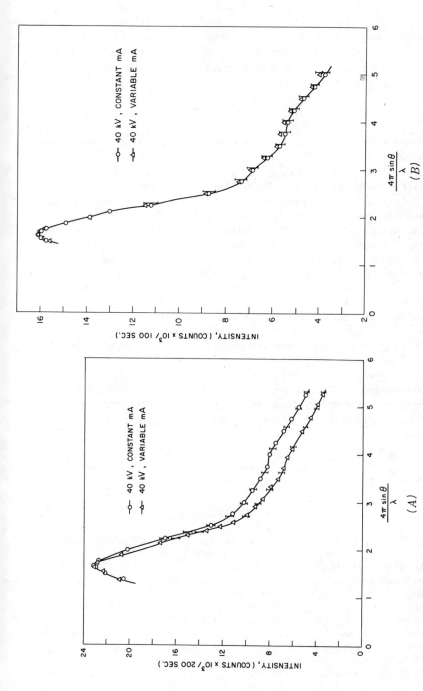

Fig. 5-61. Comparison of the x-ray scattering curves of glass as recorded (A) by the conventional slow step-scan technique, and (B) by the continuous-averaging scan technique. Each curve was recorded twice, first, with constant tube current and, second, with the tube current artificially reduced from 19 to 13 mA during the measurements. Error flags at data points represent ±3 standard deviations. (Courtesy of C. R. Peters and M. E. Milberg, *Rev. Sci. Instrum.*, **37** (9), 1186 (1966).)

411

continuous-averaging techniques (discussed above), beam monitors are gradually falling into disuse. Nevertheless, they may still serve as the only solution to the long-period drift problem under certain technical limitations or because of economic circumstances; hence some literature references are included [219, 231].

GENERAL REFERENCES

*1. W. Parrish (ed.), *Advances in X-Ray Diffractometry and X-Ray Spectrography*, Centrex Publishing Company, Eindhoven, The Netherlands, 1962.

2. L. F. Vassamillet and H. W. King, "Diffractometer Techniques," in *Handbook of X-Rays* (E. F. Kaelble, ed.), McGraw-Hill, New York, 1967, Chapter 9.

*3. A. J. C. Wilson, *Mathematical Theory of X-Ray Powder Diffractometry*. Philips Technical Library, Eindhoven, The Netherlands, 1963.

*4. U. W. Arndt and B. T. M. Willis, "Detectors," in *Single Crystal Diffractometry*, Cambridge University Press, London, 1966, Chapter 4.

5. V. E. Buhrke, "Detection and Measurement of X-Rays," in *Handbook of X-Rays* (E. F. Kaelble, ed.), McGraw-Hill, New York, 1967, Chapter 3.

6. D. W. Aitken, "Recent Advances in X-Ray Detection Technology," *IEEE Trans., Nucl. Sci. (U.S.A.)*, **15**(3), 10 (1968).

7. R. A. Young, "Balanced Filters for X-Ray Diffractometry", *Z. Kristallogr.*, **118**, 233 (1963).

8. A. R. Lang, "Diffracted-Beam Monochromatization Techniques in X-Ray Diffractometry," *Rev. Sci. Instrum.*, **27**, 17 (1956).

9. W. J. Campbell, S. Stecura, and C. Grain, "High-Temperature Furnaces for X-Ray Diffractometers," *Advances in X-Ray Analysis*, Vol. 5, Plenum Press, 1962, p. 169.

10. H. J. Goldschmidt, "Diffractometer Methods," in *High-Temperature X-Ray Diffraction Techniques*, Bibliography 1, International Union of Crystallography, Commission on Crystallographic Apparatus, 1964, Part II.

11. W. L. Baun and J. J. Renton, *Low-Temperature X-Ray Diffraction Techniques*, Wright-Patterson Air Force Base, Ohio, Technical Documentary Report No. ASD-TDR-63-278, April 1963.

SPECIFIC REFERENCES

[1]   W. H. Bragg and W. L. Bragg, *Proc. Roy. Soc.* (London), **88A**, 428 (1913).

[2]   W. H. Bragg, *Proc. Phys. Soc.* (London), **33**, 222 (1921).

[3]   H. Geiger and W. Müller, *Phys. Z.*, **29**, 839 (1928).

[4]   A. Huppertsberg, *Z. Phys.*, **75**, 231 (1932).

[5]   D. P. LeGalley, *Rev. Sci. Instrum.* **6**, 279 (1935).

[6]   H. Friedman, *Electronics*, April 1945, p. 132.

[7]   W. Parrish and S. G. Gordon, *Am. Mineral.*, **30**, 326 (1945).

[8]   For example, see W. Parrish (ed.), *Advances in X-Ray Diffractometry and X-Ray Spectrography*, Centrex Publishing Company, Eindhoven, The Netherlands, 1962.

[9]   W. Soller, *Phys. Rev.*, **24**, 158 (1924).

[10] W. Parrish, *Z. Kristallogr.* **127**, 200 (1968).

[11]   W. Parrish and M. Mack, *Acta Crystallogr.*, **23**, 687 (1967).

[12]   M. Mack and W. Parrish, *Acta Crystallogr.*, **23**, 693 (1967).

[13]   W. Parrish, M. Mack, and I. Vajda, *Norelco Rep.*, **14**, 56 (1967).

[14]   H. W. King, C. J. Gillham, and F. G. Huggins, *Advances in X-Ray Analysis*, Vol. 13, Plenum Press, New York, 1970, p. 550.

[14a] W. L. Baun and J. J. Renton, *J. Sci. Instrum.*, **40**, 498 (1963).

[15]   L. V. Azaroff, *Elements of X-Ray Crystallography*, McGraw-Hill, New York, 1968, pp. 369–373.

[16]   W. Parrish and K. Lowitzsch, *Am. Mineral.*, **44**, 765 (1959).

[17]   H. W. King and L. F. Vassamillet, *Advances in X-Ray Analysis*, Vol. 5, Plenum Press, New York, 1962, p. 78.

[18]   H. Neff, *Z. angew. Phys.*, **10**, 505 (1956).

[19]   C. O. Taylerson, *Machinist* (London), **71**, 1821 (1947).

[20]   C. E. Haven and A. G. Strang, *J. Res. Nat. Bur. Stand.*, **50**, 45 (1953).

[21]   L. F. Vassamillet and H. W. King, *Advances in X-Ray Analysis*, Vol. 6, Plenum Press, New York, 1963, p. 142.

[22]   W. Parrish, *X-Ray Reflection Angle Tables for Several Standards*, Philips Laboratories Technical Report No. 68, February 9, 1953.

[23]   L. F. Vassamillet and H. W. King in *Handbook of X-Rays* (E. F. Kaelble, ed.), McGraw-Hill, New York, 1967, Chapter 9.

[24]   B. W. Delf, *J. Sci. Instrum.*, **40**, 600 (1963).

[25]   S. Samson, *Rev. Sci. Instrum.*, **38**, 1273 (1967).

[26]   L. E. Alexander, *J. Appl. Phys.*, **19**, 1068 (1948).

[27]   L. E. Alexander, *J. Appl. Phys.*, **21**, 126 (1950).

[28]   L. E. Alexander, *J. Appl. Phys.*, **25**, 155 (1954).

[29]   H. Ekstein and S. Siegel, *Acta Crystallogr.*, **2**, 99 (1949).

[30]   R. C. Spencer, *Phys. Rev.*, **55**, 239 (1939); *J. Appl. Phys.*, **20**, 413 (1949).

[31]   A. J. C. Wilson, *Mathematical Theory of X-Ray Powder Diffractometry*, Philips Technical Library, Eindhoven, The Netherlands, 1963.

[32]   J. N. Eastabrook, *Brit. J. Appl. Phys.*, **3**, 349 (1952).

[33]   Reference 31, pp. 40–41.

[34]   E. R. Pike, *J. Sci. Instrum.*, **34**, 355 (1957); **36**, 52 (1959).

[35]   A. J. C. Wilson, *J. Sci. Instrum.*, **27**, 321 (1950).

[36]   L. E. Alexander and H. P. Klug, *Anal. Chem.*, **20**, 886 (1948).

[37]   D. T. Keating and B. E. Warren, *Rev. Sci. Instrum.*, **23**, 519 (1952).

[38]   M. E. Milberg, *J. Appl. Phys.*, **29**, 64 (1958).

[39]   R. H. Bragg and C. M. Packer, *Rev. Sci. Instrum.*, **34**, 1202 (1963).

[40]   C. G. Vonk, *Norelco Rep.*, **8**, 92 (1961).

[41]   J. I. Langford and A. J. C. Wilson, *J. Sci. Instrum.*, **39**, 581 (1962).

[42]   A. Hoyt, *Phys. Rev.*, **40**, 477 (1932).

[43]   A. J. C. Wilson, *Z. Kristallogr.*, **111**, 471 (1959).

[44]   J. Ladell, M. Mack, W. Parrish, and J. Taylor, *Acta Crystallogr.*, **12**, 567 (1959).

[45]   A. J. C. Wilson, *Proc. Phys. Soc.* (London), **78**, 249 (1961).

[46]   E. R. Pike, *Acta Crystallogr.*, **12**, 87 (1959).

[47]   E. R. Pike and A. J. C. Wilson, *Brit. J. Appl. Phys.*, **10**, 57 (1959).

[48]   B. W. Delf, *Brit. J. Appl. Phys.*, **12**, 421 (1961).

[49]   W. Parrish and A. J. C. Wilson in *International Tables for X-Ray Crystallography*, Vol. II, Kynoch Press, Birmingham, England, 1959, p. 216.

[50]   J. Ladell, W. Parrish, and J. Taylor, *Acta Crystallogr.*, **12**, 253 (1959).

[51]   J. I. Langford, *J. Sci. Instrum.*, **39**, 515 (1962).

[52] J. Ladell, W. Parrish, and J. Taylor, *Acta Crystallogr.*, **12**, 561 (1959).
[53] B. W. Delf, *Brit. J. Appl. Phys.*, **14**, 345 (1963).
[54] J. Taylor, M. Mack, and W. Parrish, *Acta Crystallogr.*, **16**, 1179 (1963).
[55] J. Taylor, M. Mack, and W. Parrish, *Acta Crystallogr.*, **17**, 1229 (1964).
[56] R. A. Young, R. J. Gerdes, and A. J. C. Wilson, *Acta Crystallogr.*, **22**, 155 (1967).
[57] A. J. C. Wilson, *Nature*, **193**, 568 (1962).
[58] A. J. C. Wilson, *Proc. Phys. Soc.* (London), **80**, 286 (1962).
[59] A. J. C. Wilson, *Proc. Phys. Soc.* (London), **81**, 41 (1963).
[60] A. J. C. Wilson, *Proc. Phys. Soc.* (London), **85**, 807 (1965).
[61] J. I. Langford, *J. Appl. Crystallogr.*, **1**, 131 (1968).
[62] M. Tournarie, *C. R. Acad. Sci., Paris*, **242**, 2016 (1956).
[63] H. Lipson and A. J. C. Wilson, *J. Sci. Instrum.*, **18**, 144 (1941).
[64] A. J. C. Wilson, *Proc. Camb. Phil. Soc.*, **36**, 485 (1940).
[65] A. J. C. Wilson, *Proc. Phys. Soc.* (London), **72**, 924 (1958).
[66] A. J. C. Wilson and B. W. Delf, *Proc. Phys. Soc.* (London), **78**, 1256 (1961).
[67] A. R. Lang, *J. Appl. Phys.*, **27**, 485 (1956).
[68] E. R. Pike and J. Ladell, *Acta Crystallogr.*, **14**, 53 (1961).
[69] W. Parrish, J. Taylor, and M. Mack, *Advances in X-Ray Analysis*, Vol. 7, Plenum Press, 1964, p. 66.
[70] M. Mack, W. Parrish, and J. Taylor, *J. Appl. Phys.*, **35**, 1118 (1964).
[71] Reference 31, p. 68.
[72] Reference 31, pp. 72–74.
[73] A. J. C. Wilson, *Brit. J. Appl. Phys.*, **16**, 665 (1965).
[74] A. J. C. Wilson, *Acta Crystallogr.*, **23**, 888 (1967).
[75] Reference 31, pp. 75–81.
[76] H. W. King and C. M. Russell, *Advances in X-Ray Analysis,* Vol. 8, Plenum Press, 1965, p. 1.
[77] H. W. King and C. M. Preece, *Advances in X-Ray Analysis*, Vol. 10, Plenum Press, 1967, p. 354.
[78] T. C. Furnas, Jr., and E. W. White, *Advances in X-Ray Analysis*, Vol. 4, Plenum Press, 1961, p. 521.
[79] E. W. Nuffield, *X-Ray Diffraction Methods*, Wiley, New York 1966, Part 2.
[80] V. E. Buhrke in *Handbook of X-Rays* (E. F. Kaelble, ed.), McGraw-Hill, New York, 1967, Chapter 3.
[81] S. C. Curran, *Sci. Prog.*, **42**, 32 (1954).
[82] W. Parrish and T. R. Kohler, *Rev. Sci. Instrum.*, **27**, 795 (1956).
[83] U. W. Arndt and B. T. M. Willis, *Single Crystal Diffractometry*, Cambridge University Press, 1966, Chapter 4.
[84] A. R. Lang, *J. Sci. Instrum.*, **33**, 96 (1956).
[85] Reference 79, pp. 177–195.
[86] J. Taylor and W. Parrish, *Rev. Sci. Instrum.*, **26**, 367 (1955).
[87] H. G. Stever, *Phys. Rev.*, **61**, 38 (1942).
[88] L. E. Alexander, E. Kummer, and H. P. Klug, *J. Appl. Phys.*, **20**, 735 (1949).
[89] W. Parrish, *Advances in X-Ray Analysis*, Vol. 8, Plenum Press, 1965, p. 118.
[90] S. Samson, *Rev. Sci. Instrum.*, **37**, 1255 (1966).
[91] B. C. Giessen and G. E. Gordon, *Norelco Rep.*, **17** (2), 17 (1970).
[92] H. R. Bowman, E. K. Hyde, S. G. Thompson, and R. C. Jared, *Science*, **151**, 562 (1966).
[93] B. C. Giessen and G. E. Gordon, *Science*, **159**, 973 (1968).
[94] R. S. Frankel and D. W. Aitken, *Appl. Spectrosc.*, **24**, 557 (1970).

[95]   R. H. Muller, *Anal. Chem.*, **38**, 155 (1966).
[96]   J. I. Drever and R. W. Fitzgerald, *Mater. Res. Bull.*, **5**, 101 (1970).
[97]   R. S. Frankel, *Am. Lab.*, May 1969.
[98]   D. W. Aitken, *IEEE Trans., Nucl. Sci., U.S.*, **15** (3), 10 (1968).
[99]   R. L. Heath, *Trans. Am. Nucl. Soc.*, **10**, 35 (1967).
[100] See, for example, S. A. Korff, *Electron and Nuclear Counters*, Van Nostrand Reinhold, New York, 1946, p. 140; L. I. Schiff, *Phys. Rev.*, **50**, 88 (1936); J. D. Kurbatov and H. B. Mann, *Phys. Rev.*, **68**, 40 (1945).
[101] L. L. Campbell, *Can. J. Phys.*, **34**, 929 (1956).
[102] K. Lonsdale, *Acta Crystallogr.*, **1**, 12 (1948).
[103] M. A. Short, *Rev. Sci. Instrum.*, **31**, 618 (1960).
[104] R. H. Bragg, *Rev. Sci. Instrum.*, **28**, 839 (1957).
[105] R. D. Burbank, *Rev. Sci. Instrum.*, **32**, 368 (1961).
[106] T. Fukamachi, *Jap. J. Appl. Phys.*, **8**, 851 (1969).
[107] W. Cochran, *Acta Crystallogr.*, **3**, 268 (1950).
[108] C. H. Westcott, *Proc. Roy. Soc.* (London), **194A**, 508 (1948).
[109] F. K. du Pré, *Philips Res. Rep.*, **8**, 411 (1953).
[110] Y. Beers, *Rev. Sci. Instrum.*, **13**, 72 (1942).
[111] R. E. Barieau, *Rev. Sci. Instrum.*, **26**, 729 (1955).
[112] R. Pepinsky, P. Jarmotz, H. M. Long, and D. Sayre, *Rev. Sci. Instrum.*, **19**, 51 (1948).
[113] U. W. Arndt, *J. Sci. Instrum.*, **26**, 45 (1949).
[114] C. S. Barrett, *Structure of Metals*, 2nd ed., McGraw-Hill, New York, 1952, pp. 52–53.
[115] B. B. Trott, *J. Sci. Instrum.*, **38**, 20 (1961).
[116] S. Bernstein and M. Canon, *Rev. Sci. Instrum.*, **33**, 112 (1962).
[117] J. N. Eastabrook and J. W. Hughes, *J. Sci. Instrum.*, **30**, 317 (1953).
[118] D. C. Miller, *Norelco Rep.*, **4** (2), 37 (1957).
[119] P. H. Dowling, C. F. Hendee, T. R. Kohler, and W. Parrish, *Philips Tech. Rev.*, **18**, 262 (1956–1957).
[120] P. Kirkpatrick, *Rev. Sci. Instrum.*, **10**, 186 (1939); **15**, 223 (1944).
[121] R. A. Young, *Z. Kristallogr.*, **118**, 233 (1963).
[122] Reference 83, Chapter 6.
[123] G. J. A. Gerrits and W. Bol, *J. Phys.*, **E2**, 175 (1969).
[124] W. Bol, *J. Sci. Instrum.*, **44**, 736 (1967).
[125] H. A. McKinstry and M. A. Short, *J. Sci. Instrum.*, **37**, 178 (1960).
[126] K. Tanaka, K. Katayama, J. Chikawa, and H. Suita, *Rev. Sci. Instrum.*, **30**, 430 (1959).
[127] R. W. Hendricks, J. S. Arrington, and W. J. Mason, *J. Appl. Crystallogr.*, **1**, 128 (1968).
[128] W. V. Johnston and P. C. Romo, *Rev. Sci. Instrum.*, **40**, 1285 (1969).
[129] A. R. Lang, *Rev. Sci. Instrum.*, **27**, 17 (1956).
[130] T. Johansson, *Naturwissenschaften*, **20**, 758 (1932); *Z. Phys.*, **82**, 507 (1933).
[131] H. T. Johann, *Z. Phys.*, **69**, 185 (1931).
[132] W. A. Wooster, G. N. Ramachandran, and A. R. Lang, *J. Sci. Instrum.*, **26**, 156 (1949).
[133] B. R. Banerjee, *Rev. Sci. Instrum.*, **26**, 564 (1955).
[134] R. E. Ogilvie, unpublished material.
[135] D. M. Koffman and S. H. Moll, *Norelco Rep.*, **11**, 95 (1964).
[136] M. Canon, *Advances in X-Ray Analysis*, Vol. 8, Plenum Press, 1965, p. 285.
[137] M. Renninger, *Z. Kristallogr.*, **107**, 464 (1956).
[138] R. W. Gould, S. R. Bates, and C. J. Sparks, *Appl. Spectrosc.*, **22**, 549 (1968).
[139] E. M. Proctor, T. C. Furnas, and W. F. Loranger, *Advances in X-Ray Analysis*, Vol. 14, Plenum Press, 1971, p. 38.

[140]   L. H. Schwartz, L. A. Morrison, and J. B. Cohen, *Advances in X-Ray Analysis*, Vol. 7, Plenum Press, 1964, p. 281.

[141]   D. M. Kheiker, L. A. Feigin, and I. I. Yakovlev, *Soviet Phys.—Crystallogr.*, **10**, 372 (1965).

[142]   S. Kavesh and J. M. Schultz, *Rev. Sci. Instrum.*, **40**, 98 (1969).

[143]   L. E. Alexander, *X-Ray Diffraction Methods in Polymer Science*, Wiley, New York, 1969, p. 81.

[143a]   O. Inkinen, *Acta Crystallogr.*, **A25**, 214 (1969).

[143b]   L. D. Jennings, *Acta Crystallogr.*, **A25**, 217 (1969).

[144]   M. Mack and N. Spielberg, *Spectrochim. Acta*, **12**, 169 (1958).

[145]   L. E. Alexander, H. P. Klug, and E. Kummer, *J. Appl. Phys.*, **19**, 742 (1948).

[146]   P. M. de Wolff, *Appl. Sci. Res.*, **B7**, 102 (1958).

[147]   P. M. de Wolff, J. M. Taylor, and W. Parrish, *J. Appl. Phys.*, **30**, 63 (1959).

[148]   E. F. Champaygne, *Rev. Sci. Instrum.*, **17**, 345 (1946); W. Parrish, *Philips Tech. Rev.*, **17**, 206 (1956); P. N. Bestelink and D. R. Holmes, *J. Sci. Instrum.*, **33**, 281 (1956); B. R. Banerjee, *Rev. Sci. Instrum.*, **29**, 438 (1958); R. A. Coyle, *J. Sci. Instrum.*, **39**, 90 (1962); E. Ho-Tun and A. J. Rossouw, *J. Phys.*, **E1**, 151 (1968).

[149]   G. W. Brindley, *Phil. Mag.*, **36**, 347 (1945); P. M. de Wolff, *Physica*, **13**, 62 (1947); Z. W. Wilchinsky, *Acta Crystallogr.*, **4**, 1 (1951).

[150]   L. S. Birks, Naval Research Laboratory, Washington, D. C., Report No. H-2517, April 20, 1945.

[151]   T. Okamoto, M. Kimura, and K. Nakajima, *J. Phys.*, **E3**, 414 (1970).

[152]   A. M. Byström-Asklund, *Am. Mineral.*, **51**, 1233 (1966).

[153]   Reference 23, pp., 9–33.

[154]   G. L. McCreery, *J. Am. Ceram. Soc.*, **32**, 141 (1949).

[155]   R. J. Weiss, *X-Ray Determination of Electron Distributions*, Wiley, 1966, p. 97.

[156]   W. Parrish, *X-Ray Analysis Papers*, Centrex Publishing Company, Eindhoven, The Netherlands, 1965, Chapter 7.

[157]   E. Sturm and W. Lodding, *Acta Crystallogr.*, **A24**, 650 (1968).

[158]   P. M. de Wolff and W. H. Sas, *Acta Crystallogr.*, **A25**, 206 (1969).

[159]   C. J. Heffelfinger and R. L. Burton, *J. Polymer Sci.*, **47**, 289 (1960).

[160]   Reference 143, p. 210.

[161]   W. J. Campbell, S. Stecura, and C. Grain, *Advances in X-Ray Analysis*, Vol. 5, Plenum Press, New York, 1962, p. 169.

[162]   H. J. Goldschmidt, *High-Temperature X-Ray Diffraction Techniques*, Bibliography 1, International Union of Crystallography, Commission on Crystallographic Apparatus, 1964, Part II, Diffractometer Methods; see also Chapter 4, reference 132.

[163]   H. A. McKinstry, *J. Appl. Phys.*, **41**, 5074 (1970).

[164]   F. A. Mauer and L. H. Bolz, *Advances in X-Ray Analysis*, Vol. 5, Plenum Press, New York, 1962, p. 229.

[165]   T. W. Baker, P. J. Baldock, and W. E. Spindler, *J. Sci. Instrum.*, **43**, 803 (1966).

[166]   E. A. Harper, *J. Phys.*, **E2**, 807 (1969).

[167]   S. Stecura, *Rev. Sci. Instrum.*, **39**, 760 (1968).

[168]   P. Chiotti, *Rev. Sci. Instrum.*, **25**, 683 (1954).

[169]   E. W. Franklin and S. M. Lang, *Advances in X-Ray Analysis*, Vol. 6, Plenum Press, New York, 1963, p. 250.

[170]   P. Debrenne, J. Laugier, and M. Chaudet, *J. Appl. Crystallogr.*, **3**, 493 (1970).

[171]   M. Moss and D. L. Smith, *Rev. Sci. Instrum.*, **36**, 1254 (1965).

[172]   D. M. North and C. N. J. Wagner, *J. Appl. Crystallogr.*, **2**, 149 (1969).

[173] Z. W. Wilchinsky, H. C. Tsien, and G. Ver Strate, *J. Phys.*, **E4**, 704 (1971).

[174] R. A. Horne, W. J. Croft, and L. B. Smith, *Rev. Sci. Instrum.*, **30**, 1132 (1959).

[175] J. A. Goldak, *J. Sci. Instrum.*, **41**, 722 (1964).

[176] W. L. Baun, *Advances in X-Ray Analysis*, Vol. 4, Plenum Press, New York, 1961, p. 201.

[177] R. H. Willens, *Rev. Sci. Instrum.*, **33**, 1069 (1962).

[178] J. Intrater, *Rev. Sci. Instrum.*, **32**, 982 (1961); R. E. Dreikorn, *Advances in X-Ray Analysis*, Vol. 5, Plenum Press, New York, 1962, p. 213; C. R. Houska and E. J. Keplin, *J. Sci. Instrum.*, **41**, 23 (1964); V. V. Zubenko, B. G. Grants, and M. M. Umanski, *Soviet Phys. — Crystallogr.*, **11**, 280 (1966).

[179] S. W. Kennedy and L. D. Calvert, *J. Sci. Instrum.*, **35**, 61 (1958); J. Spreadborough and J. W. Christian, *ibid.*, **36**, 116 (1959); J. N. Van Niekerk, *ibid.*, **37**, 172 (1960); J. Intrater and S. Hurwitt, *Rev. Sci. Instrum.*, **32**, 905 (1961); J. P. Holden, *J. Sci. Instrum.*, **41**, 706 (1964); H. Ebel and W. Novak, *Acta Phys. Austriaca*, **24**, 161 (1966); H. M. Fritz and W. M. Williams, *J. Phys.*, **E2**, 63 (1969).

[180] B. Roessler and G. F. Bolling, *Rev. Sci. Instrum.*, **35**, 230 (1964).

[181] R. Ghislain, L. Deléhouzée, and A. Deruyttere, *J. Sci. Instrum.*, **42**, 502 (1965).

[182] J. H. Brady and E. C. van Reuth, *J. Sci. Instrum.*, **43**, 833 (1966).

[183] B. A. Calhoun and S. C. Abrahams, *Rev. Sci. Instrum.*, **24**, 397 (1953).

[184] J. Intrater and A. Appel, *Rev. Sci. Instrum.*, **32**, 1065 (1961).

[185] L. K. Jetter, C. J. McHargue, R. O. Williams, and H. L. Yakel, Jr., *Rev. Sci. Instrum.*, **28**, 1087 (1957).

[186] A. F. Bonfiglioli and O. Testard, *Acta Crystallogr.*, **17**, 668 (1964).

[187] W. L. Baun and J. J. Renton, *Advances in X-Ray Analysis*, Vol. 7, Plenum Press, New York, 1964, p. 302; Wright-Patterson Air Force Base, Ohio, Technical Documentary Report No. ASD-TDR-63-278, April 1963.

[188] C. S. Barrett, *Acta Crystallogr.*, **9**, 671 (1956).

[189] C. S. Barrett, *Advances in X-Ray Analysis*, Vol. 5, Plenum Press, New York, 1962, p. 33.

[190] I. A. Black, L. H. Bolz, F. P. Brooks, F. A. Mauer, and H. S. Peiser, *J. Res. Nat. Bur. Stand.*, **61**, 367 (1958).

[191] F. A. Mauer and L. H. Bolz, *J. Res. Nat. Bur. Stand.*, **65C**, 225 (1961).

[192] C. A. Stochl and S. G. Ullman, *Rev. Sci. Instrum.*, **34**, 1134 (1963).

[193] H. P. Klug and D. R. Sears, Air Force Office of Scientific Research, Contract AF 49(638)-575, Final Report, April 30, 1962.

[194] D. R. Sears and H. P. Klug, *J. Chem. Phys.*, **37**, 3002 (1962).

[195] E. E. Gruber, T. H. Blewitt, J. A. Tesk, B. D. Sharma, and R. E. Black, *Rev. Sci. Instrum.*, **40**, 1429 (1969).

[196] A. I. Zahkarov, *Cryogenics*, February 1970, p. 65.

[197] H. W. King and C. M. Preece, *Siemens Rev.*, **35** (1968). Second Special Issue "X-Ray and Electron Microscopy News."

[198] J. S. Abell and H. W. King, *Cryogenics*, April 1970, p. 119.

[199] H. Steinfink and J. E. Gebhart, *Rev. Sci. Instrum.*, **33**, 542 (1962).

[200] J. D. Barnett and H. T. Hall, *Rev. Sci. Instrum.*, **35**, 175 (1964).

[201] C. W. Christoe and H. G. Drickamer, *Rev. Sci. Instrum.*, **40**, 169 (1969).

[202] M. A. Thompson and G. R. Mallett, *J. Sci. Instrum.*, **40**, 335 (1963).

[203] H. G. Norment and P. I. Henderson, *Rev. Sci. Instrum.*, **31**, 1153 (1960).

[204] M. L. Morrow, *Rev. Sci. Instrum.*, **41**, 781 (1970).

[205] R. A. Rowland, E. C. Reinhardt, and C. S. Brooks, *Rev. Sci. Instrum.*, **34**, 1047 (1963).

[206] T. R. Kohler and W. Parrish, *Rev. Sci. Instrum.*, **26**, 374 (1955).

[207] M. A. Bredig, G. E. Klein, and B. S. Borie, Jr., *Rev. Sci. Instrum.*, **26**, 610 (1955).

[208] W. V. Cummings, Jr., D. C. Kaulitz, and M. J. Sanderson, *Rev. Sci. Instrum.*, **26**, 5 (1955).

[209] R. Kaplow and B. L. Averbach, *Rev. Sci. Instrum.*, **34**, 579 (1963).

[210] Reference 143, pp. 68–72.

[211] A. R. Lang, *Rev. Sci. Instrum.*, **26**, 680 (1955).

[212] S. Ergun, *J. Appl. Phys.*, **40**, 293 (1969).

[213] W. Parrish, M. Mack, and J. Taylor, *J. Sci. Instrum.*, **43**, 623 (1966).

[214] Reference 23, pp. 9–26.

[215] M. H. Mueller, L. Heaton, and E. W. Johanson, *Rev. Sci. Instrum.*, **32**, 456 (1961).

[216] W. R. McMillan, *Rev. Sci. Instrum.*, **33**, 1294 (1962).

[217] W. A. Wooster, *Soviet Phys. — Crystallogr.*, **5**, 355 (1960).

[218] E. L. Lube and D. M. Kheiker, *Soviet Phys. — Crystallogr.*, **8**, 343 (1963).

[219] R. E. Ricketts, *J. Sci. Instrum.*, **39**, 532 (1962).

[220] M. T. Frost and D. G. Whitehead, *J. Phys.*, **E3**, 942 (1970).

[221] H. Keller and A. Segmüller, *Rev. Sci. Instrum.*, **34**, 684 (1963).

[222] R. W. Rex, *Advances in X-Ray Analysis*, Vol. 10, Plenum Press, New York, 1967, p. 366.

[223] T. W. Baker, J. D. George, B. A. Bellamy, and R. Causer, *Advances in X-Ray Analysis*, Vol. 11, Plenum Press, New York, 1968, p. 359.

[224] C. R. Desper, *Advances in X-Ray Analysis*, Vol. 12, Plenum Press, New York, 1969, p. 404.

[225] Ch. Kratky and O. Kratky, *Z. Instrumentenk.*, **72**, 302 (1964).

[226] H. Leopold, *Siemens Rev.*, **36** (1969). Third Special Issue "X-Ray and Electron Microscopy News," p. 51.

[227] M. P. Klein and G. W. Barton, Jr., *Rev. Sci. Instrum.*, **34**, 744 (1963).

[228] C. R. Peters and M. E. Milberg, *Rev. Sci. Instrum.*, **37**, 1186 (1966).

[229] J. W. Weymouth, J. Costello, S. Schuster, and P. Schulze, *Rev. Sci. Instrum.*, **39**, 476 (1968).

[230] M. Berman and S. Ergun, *Rev. Sci. Instrum.*, **40**, 1144 (1969).

[231] L. E. Alexander, S. Ohlberg, and G. R. Taylor, *J. Appl. Phys.*, **26**, 1068 (1955); T. F. Ford and T. G. Alexander, *J. Sci. Instrum.*, **33**, 204 (1956); E. R. Pike and J. W. Hughes, *J. Sci. Instrum.*, **36**, 212 (1959); A. E. Kiss and N. Patla, *J. Sci. Instrum.*, **38**, 78 (1961); B. W. Delf, *J. Sci. Instrum.*, **38**, 359 (1961); K. S. Chandrasekaran, *J. Sci. Instrum.*, **38**, 414 (1961).

CHAPTER 6

THE INTERPRETATION OF
POWDER DIFFRACTION DATA

The interpretation of a powder pattern can be a simple or a difficult operation, depending upon the number and structural complexity of the phases composing the specimen and upon the amount of information desired. Perhaps the simplest interpretation is to identify the pattern by comparing it with standard patterns of reference materials. In addition, it is commonly required that the interplanar spacings corresponding to the various lines be computed, and less frequently it is necessary to determine the Miller indices of the reflections in order that structural information may be obtained. Thus the unit-cell dimension and lattice type of a cubic substance can usually be arrived at in a straightforward manner from the diffraction pattern; similar information can often be deduced for tetragonal and hexagonal substances, and less frequently for orthorhombic materials. In favorable cases a knowledge of the line intensities and indices permits the complete unraveling of the structure of a cubic specimen, and more rarely the structures of tetragonal or hexagonal materials.

It is the purpose of this chapter to explain how the following basic kinds of information can be obtained from powder diagrams:

1. The interplanar (d) spacings of the reflections
2. The Miller indices of the reflections
3. The unit-cell dimensions and the lattice type
4. The intensities of the reflections.

The following more specialized interpretations are discussed in later chapters:

1. The qualitative identification of chemical compounds (Chapter 7)
2. The quantitative analysis of crystalline mixtures (Chapter 7)

419

3. The precise determination of lattice constants (Chapter 8)
4. The determination of crystallite size from line broadening (Chapter 9).

6-1 THE VIEWING AND PRECISION MEASUREMENT OF POWDER PHOTOGRAPHS

Suitable facilities must be provided for viewing the diffraction patterns. A viewing space at least 14 in. long should be available to accommodate the films prepared in 114.6-mm Debye-Scherrer cameras, the size most widely used in the United States. Figure 6-1 shows a commercial viewer suitable for examining from one to three such films or several 3×5 in. or 5×7 in. films. The viewing surface, which is 5×17 in. in size, consists of a single thickness of milky glass supported at an angle of 30° with the horizontal. A single 15-W daylight fluorescent lamp placed about 2 in. below the level of the film provides sufficiently uniform illumination of the entire viewing surface, and a rheostat is provided for varying the intensity. This illuminator can be improved by adding one or more pairs

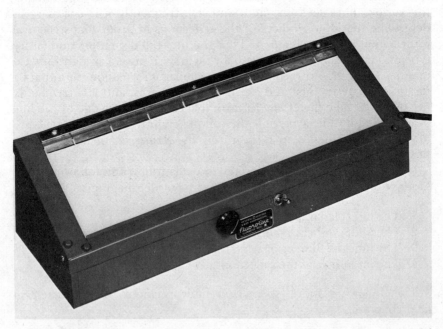

Fig. 6-1. Portable illuminator for viewing x-ray diffraction films. (Courtesy of General Electric Company.)

of spring clips to hold the films in position. This feature is especially necessary when a measuring scale is placed in contact with the film for reading 2θ or d values.

For simultaneously viewing a larger number of films, it is convenient to employ a commercial radiographic illuminator placed in a horizontal position. Thus a 14×17 in., milky-glass viewing surface will accommodate up to twelve 35-mm powder film strips or up to six 5×7 in. rectangular films. This type of illuminator is also improved by placing spring clips at intervals around the edges for holding films and measuring scales in a fixed position. If frequent use is to be made of such an illuminator, working conditions are rendered more comfortable by countersinking the box in a tabletop so that the viewing surface is approximately flush with the surface of the table.

Figure 6-2 shows a very useful built-in viewer for comparing large numbers of patterns. The viewer is shown as an integral part of a table with a 30×54 in. top, though it can be incorporated into a table of any desired dimensions. The large 20×30 in. viewing surface consists of a

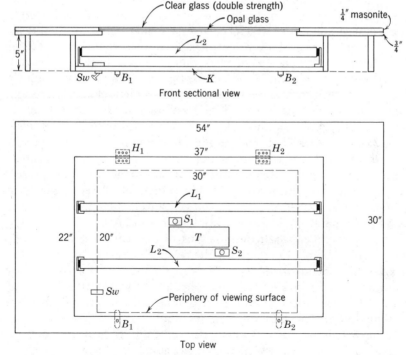

Fig. 6-2.   Large film viewer built in tabletop.

sheet of double-strength, clear glass resting on a sheet of milky glass. The two sheets of glass, with a total thickness of about $\frac{1}{4}$ in., rest on a ledge around the viewing box formed by stepping back the $\frac{1}{4}$-in. masonite layer of the tabletop about $\frac{1}{2}$ in. on all sides. The two 30-W white daylight fluorescent lamps L_1 and L_2, together with the transformer T, starters S_1 and S_2, and the rest of the electrical circuit are mounted on the 37 ×22 in. base board K, which is hinged at the back (H_1 and H_2). When in use, this base board is held up in place by the turn buttons B_1 and B_2, and it can be dropped downward when necessary for maintenance of the electrical components. Good reflectivity is achieved by painting the base board and all the electrical parts white. In order to avoid noticeable irregularity in the brightness of the viewing surface, the lamps should not be placed within 3 in of the glass.

Every x-ray diffraction laboratory should be supplied with a film measuring instrument capable of an accuracy of at least ±0.1 mm. This accuracy is adequate for indexing powder diagrams and for many measurements of a moderately precise nature. The instruments supplied by the manufacturers of x-ray diffraction equipment are rather similar in design and provide this degree of precision. Most of them accommodate Debye-Scherrer films up to 14 in. long, and the lines are located by means of a sliding finder which traverses a metal or glass scale engraved to 1.0 mm. By means of a vernier, the scale can be read with a precision of ±0.1 mm or, in some cases, ±0.05 mm. Though these specifications are satisfactory for many applications, they are not good enough for highly exact work such as the precision measurement of lattice constants (see Chapter 8). For these reasons it is recommended that the diffraction laboratory be supplied with a linear comparator or other type of measuring instrument which can be read to ±0.01 mm or less. The comparators procurable from scientific supply houses meet this specification, with precisions ranging from 0.01 to 0.001 mm. However, they present several serious disadvantages. (1) The range of travel is generally 100 mm or less, insufficient to measure a powder film from a 57.3- or 114.6-mm camera. (2) The screw translational mechanism is not adaptable to rapid movements from one part of the pattern to another. (3) The magnification is too great for the relatively grainy, diffuse lines of an x-ray diffraction pattern. (4) The price is too high for the average laboratory. Special comparators have been built which overcome some of these disadvantages[1], but they cannot be constructed in the average machine shop and the building costs are high.

Figure 6-3 shows an instrument of the required high precision for measuring 14-in. films, which can be economically constructed in the average machine shop[2]. Only standard available parts and simple machine work are involved. The support for the scale is a wooden case A

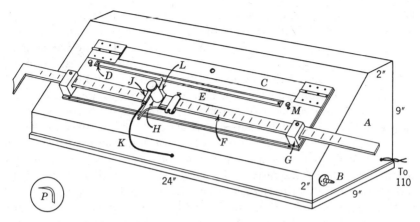

Fig. 6-3.   Device for the precision measurement of diffraction patterns. [Courtesy of Klug; reprinted by permission from *Ind. Eng. Chem. Anal. Ed.*, **12**, 753 (1940).]

with sloping front and a hinged back to provide access to the inside for maintenance. The film is supported on a piece of plate glass and illuminated by an 18-in. 15-W daylight fluorescent lamp through a piece of diffusing milky glass. If necessary, the accumulation of heat in the box can be relieved by blowing compressed air through a series of holes in a brass tube B, running the length of the case, the warm air escaping through holes bored through the back. When a fluorescent lamp is used there is ordinarily no need for cooling. C is a rectangular aluminum alloy plate, $18.5 \times 5 \times 0.188$ in., attached to the wooden case at three points. It has a 1×15 in. opening directly over the illuminated slot in the case. A piece of clear plate glass with its upper surface flush with the surface of C fills this opening and serves as a support for 14-in. powder films, which are temporarily held in place by the spring clips D at each end. E is a similar 18.5×5 in. aluminum alloy plate hinged to C at its upper edge and containing a 0.75×15 in. beveled slot centered over the opening in C. With E lifted up, the film is inserted and held in position by the slips D until the plate E is lowered, after which it is immovably clamped between C and E by the removable thumbscrews M, which are threaded into C. Plates C and E should be cut by sawing rather than by shearing to insure that their surfaces are flat and smooth.

The "heart" of the instrument is a 60-cm vernier caliper F,* graduated in 0.5-mm divisions and provided with a vernier reading to 0.02 mm, and by means of which measurements can be made to 0.01 mm by interpolation. This scale is clamped in position 0.75 in. above plate E by the slotted blocks G fastened to E. Attached to the movable jaw of the caliper is a

---

*Scale No. 122M, L. S. Starrett Company, Athol, Massachusetts.

1-in. 4 × magnifying lens in a tubular focusing mount *H*, for reading the vernier scale, which is illuminated by a low-voltage lamp in the side tube *J*. A bell push button or toggle switch at the left end of the box controls the light, which is energized by dry cells or a low-voltage transformer.

The movable jaw of the caliper is also drilled at *L* and provided with set screws for holding a reticle or pointer *P* for locating the diffraction lines. Any desired type of finder can be substituted for the pointer *P* shown in the figure. We have had success with both a pointer and a glass reticle engraved with two fine cross-lines oriented at about 60° to the film axis. If a pointer is used, it should be neither too sharp nor too broad. Experience suggests a point with a converging angle of 15° and a somewhat dull tip. In order to minimize parallax errors, the finder is set as close as possible to the film without making contact. The diffraction lines should be viewed with a low-power reading glass or other magnifier possessing a power appropriate to the sharpness and density of the line in question. As a general rule, it is unsatisfactory to mount a lens permanently in position over the finder, because the magnifying requirements vary so much with the line quality. This film viewer permits 14-in. or larger Debye-Scherrer patterns to be measured with sufficient accuracy for the precision determination of lattice constants according to any of the techniques described in Chapter 8. For 7- or 8-in. films, however, the discrimination is not sufficient for such work, and a linear comparator must be employed instead.

Other precision measuring devices for powder diffraction films have been reported in the literature [3–6]. Perdok and Boom [5] describe an improved eyepiece graticule for such instruments. In the design of Armstrong and Davis [3] the eyepiece crosshairs are replaced by a projected light spot which is easier to set on faint lines, eliminates parallax, and allows the use of a binocular eyepiece. Several equipment suppliers list precision measuring devices, but their range in some instances is too small for 114.6-mm films. The precision measuring feature is frequently incorporated in a comparator- or densitometer-type instrument which is designed also to measure intensities accurately (Section 6-7.4).

6-2 DETERMINATION OF INTERPLANAR (*d*) SPACINGS

6-2.1 Debye-Scherrer Patterns

Interplanar spacings are derived from the observed reflection angles *θ*, with the aid of the Bragg law,

$$d = \frac{\lambda}{2} \frac{1}{\sin \theta}. \tag{6-1}$$

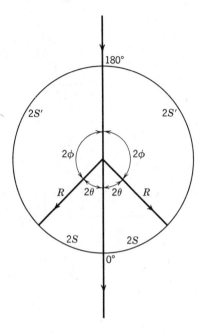

Fig. 6-4.

The process consists of three steps: (1) a linear measurement on the film, (2) conversion of the linear measurement to the equivalent Bragg angle θ, and (3) calculation of d from equation 6-1. Let $2S$ be the linear equivalent on the film of the angle of deviation 2θ, as shown in Fig. 6-4, and let $2S'$ be equivalent to 2ϕ. In a particular experimental situation the nature of the linear measurement and its conversion to the angle θ depend upon the position of the film in the camera. Figure 6-5 shows the four commonly employed film positions with the corresponding appearance of the diffraction pattern obtained with each. Two symmetrically arranged film strips may be employed in cameras of unusually large diameter, because of the unwieldy dimensions assumed by a single film. The 19-cm cameras developed in W. L. Bragg's laboratories[7, 8a] utilize this arrangement. The great majority of cameras used in the United States, are of 57.3- and 114.6-mm diameter and employ a single film placed in one of the positions $A, B, C,$ or D in Fig. 6-5.

When using position A, measurement of the distance between corresponding lines gives $4S$ directly, from which θ can be obtained by making use of the effective film radius R:

$$\theta = \frac{4S}{4R} \text{ radians} = \frac{57.296}{4R} \times (4S) \text{ degrees.} \tag{6-2}$$

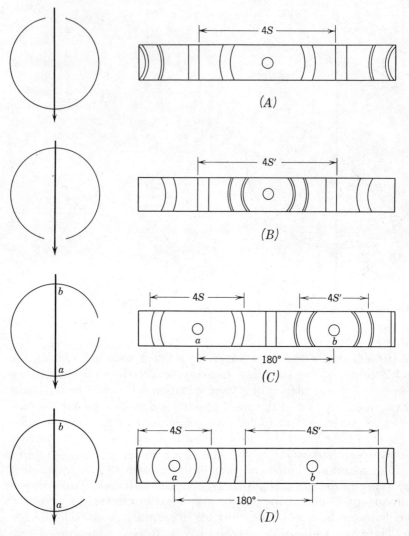

Fig. 6-5. The common Debye–Scherrer film positions: (A) regular; (B) precision; (C) asymmetric (Straumanis); (D) asymmetric (Wilson).

In the precision position B, the distance between corresponding lines is 4S′, from which

$$\phi = \frac{57.296}{4R} \times (4S') \text{ degrees.} \tag{6-3}$$

and

$$\theta = 90 - \phi \text{ degrees.} \tag{6-4}$$

The asymmetric position C offers the great advantage that an independent knowledge of R is not necessary. For pairs of lines centered about a, equation 6-2 applies, and for pairs about b, equations 6-3 and 6-4 apply. The points a and b can be accurately determined by averaging the midpoints of the various pairs of lines, and the linear distance ab then equals $180°$. The effective radius of the film is

$$R = \frac{ab}{\pi},\qquad(6\text{-}5)$$

and θ is calculated from equation 6-2, or 6-3 and 6-4, depending upon the position of the line pair concerned. Because of this self-calibrating property, the asymmetric film position is recommended as being of the most general utility for the preparation of Debye-Scherrer patterns. Since each film is self-calibrated, allowance is automatically made for differences in film shrinkage resulting from variations in the film-processing conditions. Wilson[8b] modified the Straumanis asymmetric arrangement, as portrayed in D, to enable the film radius to be determined when no reflections are present at 2θ values larger than $90°$.

With the film in positions A or B, the effective film radius must be determined by some calibrating procedure. Perhaps the most satisfactory is the marking of the film with fiducial spots separated by an accurately known angle. Bradley and Jay[9] and Buerger[10] have described methods for accomplishing this. For example, at two points on the circumference of the camera separated by an angle Δ, two tiny holes are provided which can be opened to the light, thereby producing fiducial spots on the film (see Fig. 6-6). When the slightly tapered friction plug P is withdrawn a few seconds' exposure to a weak light produces a round spot on the film where it comes in contact with the no. 80 drill hole. Figure 6-6A shows appropriate positions for pairs of calibrating marks with the film in the regular (r and r') and precision (p and p') positions. To illustrate for a film in the regular position, the diffraction pattern together with the distance L_r is measured, after which the effective film radius R is calculated from the relationship:

$$\frac{R}{L_r} = \frac{1}{\Delta_r},\qquad(6\text{-}6)$$

Δ_r being given in radians. Similarly, for the precision positions with openings at p and p':

$$\frac{R}{L_p} = \frac{1}{\Delta_p}.\qquad(6\text{-}7)$$

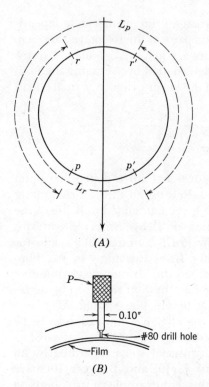

(A)

(B)

Fig. 6-6.   Arrangement for printing fiducial marks on a film.

For highest accuracy Δ_r and Δ_p should be made large enough to place the fiducial spots rather near the ends of the film. An angle of 310° has proved satisfactory in practice.

The angles Δ_r and Δ_p can be determined conveniently and accurately by exposing a strip of Eastman printing-out proof paper placed in the camera in the film position concerned. Special care must be taken that the paper makes good contact with the camera wall at all points. This paper is well adapted to this calibration since it requires no processing, and several minutes' exposure to bright light is required to produce a well-defined spot. The distance between the spots is measured immediately after exposure, which eliminates the possibility of changes in the dimensions of the paper, and thereby furnishes accurate values of the distances L_r and L_p (see Fig. 6-6A). The camera diameter D can be measured with high accuracy, using a good inside caliper, after which the fiducial angles Δ_r and Δ_p can be calculated with the aid of the relationships

$$\frac{\Delta_r}{360°} = \frac{L_r}{\pi D} \tag{6-8}$$

and

$$\frac{\Delta_p}{360°} = \frac{L_p}{\pi D}.$$ (6-9)

Two alternative procedures may be used to calibrate the angle subtended by a pair of fiducial spots. The first is based upon the preparation of the diffraction pattern of a standard substance, together with the fiducial marks on a single piece of film. With a knowledge of the precise angles of the diffraction lines in the neighborhood of the marks, the angles of the marks can be obtained by simple linear interpolation. Quartz and sodium chloride are commonly used calibrating materials, the former being the more satisfactory for films in the precision position because of the larger number of diffraction lines at small angles. The use of a calibrating substance has two principal objections, the first being the residual uncertainty in the lattice constant resulting from the presence in the structure of foreign ions, a subject that is discussed in Section 8-3.1. Actually, this objection is not serious when using diffraction lines in the low-angle region where the sensitivity of angle to changes in lattice spacing is least. However, in the back-reflection region appreciable errors may be introduced. A second objection is the necessity for overcoming angular errors resulting from displacement of the lines, especially at small angles, due to absorption of the x-ray beam by the sample. This is best done by preparing the specimen according to the specifications of Straumanis for eliminating angular errors (see Section 8-3.2).

Another procedure is to measure the angles Δ directly with the aid of an accurately graduated spectrometer or goniometer turntable[8a]. The camera is mounted on the table, and its position adjusted until the camera cylinder axis coincides with the rotation axis. The telescope is next focused on first one calibration hole and then the other, the angle Δ then being the difference between the two angular readings. The same procedure can be used to measure the angle between the calibrating knife edges of a 19-cm camera[7, 8a]. In a camera of this kind, pairs of knife edges at low angles and at high angles cast sharp shadows on the film, which mark the extreme ends of the recorded pattern. The shadow edges are measured along with each diffraction pattern, so that the effective radius of each film can be calculated from a knowledge of the precise angles subtended by these two pairs of knife edges. The use of two film strips introduces complications which cannot be treated in detail here. Suffice it to say that the approximation can be accepted that the linear distance between the low-angle knife edges is a constant quantity[8a]. This simplification does not affect the accuracy of the

Table 6-1.   Measurement of RbCl Film and Calculation of d Spacings[a,b]

Line	Scale Reading (cm) Left	Middle	Right	4S (cm)	2S (cm)	θ (°)	$\sin\theta$	d(Å)
1	8.19	12.86		4.67	2.335	11.710	0.20306	3.7965
2	7.82	13.22		5.40	2.700	13.541	0.23413	3.2927
3	6.65	14.38		7.73	3.865	19.383	0.33188	2.3229
4	5.96	15.10		9.14	4.570	22.919	0.38943	1.9796
5	5.73	15.30		9.57	4.785	23.997	0.40669	1.8956
6	4.95	16.10		11.15	5.575	27.959	0.46884	1.6443
7		16.66			6.138	30.782	0.51178	1.5064
8		16.83			6.308	31.635	0.52451	1.4698
9		17.50			6.978	34.995	0.57351	1.3442
10		17.99			7.438	37.302	0.60602	1.2721
11		18.79			8.268	41.465	0.66216	1.1643
12		19.27			8.748	43.872	0.69305	1.1124
13		19.42			8.898	44.624	0.70245	1.0975
14		20.05			9.528	47.783	0.74061	1.0409
15		20.51			9.988	50.090	0.76706	1.0050
$16\alpha_1$		20.67			10.148	50.893	0.77597	0.9927
$17\alpha_1$		21.32			10.788	54.102	0.81007	0.9509
$18\alpha_1$		21.82			11.298	56.660	0.83542	0.9220

Line	Scale Reading (cm) Left	Middle	Right	4S' (cm)	ϕ (°)	θ (°)	$\sin\theta$	d (Å)
$19\alpha_2$		21.85	35.07	13.22	33.149	56.851	0.83725	0.9223
$20\alpha_1$		21.98	34.94	12.96	32.497	57.503	0.84342	0.9133
$21\alpha_2$		22.03	34.91	12.88	32.297	57.703	0.84529	0.9135
$22\alpha_1$		22.71	34.23	11.52	28.886	61.113	0.87557	0.8797
$23\alpha_2$		22.75	34.18	11.43	28.661	61.339	0.87747	0.8800
$24\alpha_1$		23.27	33.67	10.40	26.078	63.922	0.89820	0.8576
$25\alpha_2$		23.34	33.61	10.27	25.752	64.248	0.90068	0.8574
$26\alpha_1$		24.35	32.60	8.25	20.687	69.313	0.93552	0.8234
$27\alpha_2$		24.41	32.52	8.11	20.336	69.664	0.93767	0.8235
$28\alpha_1$		25.09	31.85	6.76	16.951	73.049	0.95656	0.8053
$29\alpha_1$		25.40	31.54	6.14	15.396	74.604	0.96411	0.7990
$30\alpha_2$		25.51	31.43	5.92	14.844	75.156	0.96663	0.7989
$31\alpha_1$		27.03	29.90	2.87	7.197	82.804	0.99212	0.7764
$32\alpha_2$		27.28	29.65	2.37	5.943	84.057	0.99463	0.7764

[a]Cu$K\alpha$ radiation, film in asymmetric position, camera diameter 114.6 mm.
[b]Scale reading for 0° from lines 1–6: 10.522 cm. Scale reading for 180° from lines 19–32: 28.468 cm. $R = ab/\pi = (28.468 - 10.522)/3.1416 = 5.7124$ cm; $\theta = (57.296/4R) \times (4S) = 2.5075 \times (4S)$ degrees; $\phi = (57.296/4R) \times (4S') = 2.5075 \times (4S')$ degrees; $\theta = (57.296/2R) \times (2S) = 5.0150 \times (2S)$ degrees. Cu$K\alpha = 1.54184$ Å; Cu$K\alpha_1 = 1.54056$ Å; Cu$K\alpha_2 = 1.54439$ Å.

measurements to a significant degree, and it permits the data from a 19-cm camera to be evaluated in the same manner as data from smaller cameras with a single film in the regular or precision position.

Table 6-1 illustrates the calculation of d spacings from measurements of a Debye-Scherrer pattern of a specimen of <.325-mesh rubidium chloride (RbCl) powder. The film was placed in the asymmetric position and the effective radius determined from pairs of lines about the beam entrance and beam exit. A camera of approximately 57.3-mm radius was used. The data involved in the calculations are tabulated in a convenient manner. Except for the method of determining the effective radius, data from films in the other positions in Fig. 6-5 are calculated in an entirely analogous fashion.

The calculation of d spacings can be shortened by making use of currently available tables of interplanar spacings as a function of θ or 2θ [11]. For ordinary work of a nonprecision nature, a valuable simplification can be achieved by employing cameras of approximately 114.6- or 57.3-mm diameter, since these dimensions make 1 mm on the film equal 2 or 1° of arc, respectively. The reader can obtain these results by substituting 57.296 or 114.592 for $2R$ in equations 6-2 and 6-3. Figure 6-7A shows a special millimeter scale, which can be constructed of cardboard or transparent plastic, for directly reading the diffraction angles 2θ from Debye-Scherrer films of 114.6-mm diameter. Before readings are taken the

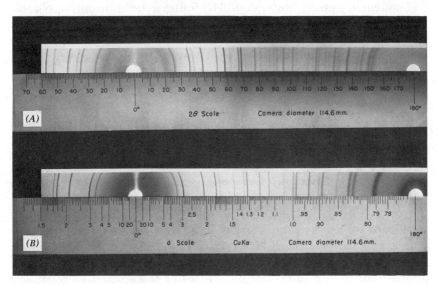

Fig. 6-7.   Scales for measuring Debye–Scherrer patterns. (A) 2θ scale. (B) d scale for Cu$K\alpha$ radiation.

scale is first moved until corresponding lines on each side of the point $2\theta = 0°$ give the same reading. In the figure the scale is shown in position for measuring a film prepared in the asymmetric position, but it can be applied equally well to films prepared in the regular and precision positions, except that for the precision position the angle measured directly is 2ϕ rather than 2θ. By constructing a scale to read in half millimeters instead of millimeters, films of 57.3-mm diameter can be measured directly.

Direct measurements of the kind just described are completely satisfactory only if the effective film diameter is equal to 57.3 or 114.6 mm within narrow limits. These dimensions can be realized by making the camera diameter somewhat larger, in order to allow for the thickness of the film and film shrinkage. The amount of shrinkage suffered by a film depends principally upon the washing time[12], and it generally ranges between 0.2 and 0.3 per cent if the washing time is not less than 30 or more than 60 min. The thickness of double-coated x-ray film is about 0.25 mm. Hence, in order to achieve a mean effective film diameter of 114.6 mm, the camera diameter should be

$$(114.6 \times 1.0025) + 0.25 = 115.1 \text{ mm},$$

and for a film diameter of 57.3 mm, it should be 57.6 mm. If cameras of these diameters are employed and if all films are processed according to the same time schedule (especially as regards washing), a millimeter or half-millimeter measuring scale will be found to be uniformly applicable to all films to within 0.1 or 0.2 per cent. This accuracy is ample for the common applications of the Debye-Scherrer technique to qualitative chemical identification. By using such a scale, the process of deducing d spacings from a diffraction pattern consists of only two steps, (1) measuring the angle 2θ with the scale, and (2) looking up the equivalent d spacing in a table[11].

The process of measuring d spacings can be reduced to one step by constructing scales reading in d spacings directly (see Fig. 6-7B). Such scales have the same limiting accuracy as angle-measuring scales but have the disadvantage that a different scale is required for each radiation. Nevertheless, in laboratories where large numbers of powder diagrams must be examined, it is economical of time to construct such scales for the commonly used radiations. We have found scales made of white cardboard to be very satisfactory in most respects. The lines and numerals can be marked clearly with a sharp, hard lead pencil, and even with daily use such a scale remains legible for a year or more. When a scale becomes soiled, it is discarded and a new one constructed, an operation requiring only about 2 hr, once the calculation of the line positions has

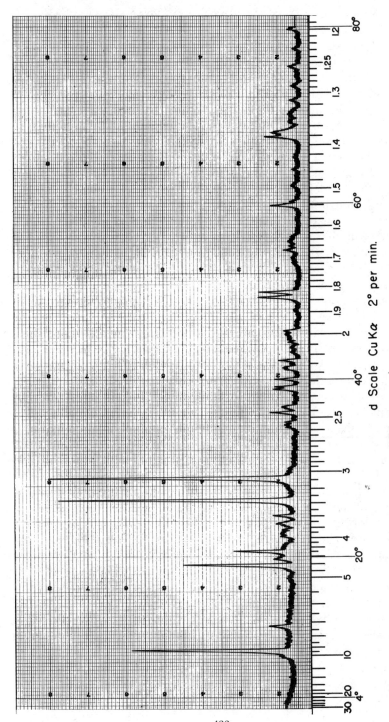

d Scale CuKα 2° per min.

Fig. 6-8.   Scale for measuring d spacings from an x-ray diffractometer chart.

been performed. We found transparent sheet Lucite to be an unsatisfactory material for film-measuring scales because of its tendency to shrink with the passage of time. Sets of transparent plastic d scales for the common radiations and film diameters close to 114.6 mm have also appeared on the market.* The user selects the particular scale of the set that has a length best suited to the film being measured, a procedure that allows for differences in film shrinkage and small differences in camera diameter. Transparent scales are difficult to read when superposed upon films of dense background, but this disadvantage is possibly offset by the full-length view of the lines that is permitted, a help in discerning faint lines. Numerical data for the construction of d scales for different radiations and for any desired camera diameter can be calculated with the aid of the Bragg law.

Diffractometer strip-chart records are particularly easy to interpret, since the chart paper is printed with a net from which 2θ values can be read directly. By using the data for the construction of d scales for measuring photographic patterns, it is very easy to construct cardboard scales for the direct measurement of d spacings from strip-chart records. The scale in most demand is ordinarily that for $CuK\alpha$ radiation and for a scanning speed of 1 or 2°/min and some relatively fast chart speed, say, 30 in/hr. Figure 6-8 shows such a $CuK\alpha$ scale in position for the measurement of a counter diffractometer chart record.

Because of absorption of the primary and diffracted x-ray beams by the sample, Debye-Scherrer lines are generally displaced to angles higher than those calculated theoretically, the effect being most pronounced at small angles. The amount of the displacement depends not only upon the Bragg angle but also upon the diameter and mass absorption coefficient of the specimen. Empirical curves for correcting for this effect can be constructed by preparing the diffraction pattern of a reference substance of accurately known unit-cell dimensions, such as quartz or sodium chloride, and then comparing the observed and theoretical line positions over the range of angles in question. The differences are plotted against 2θ, $\sin \theta$, or some other angular function, after which the difference curve can be used to correct the diffraction patterns of other specimens. The success of this method depends upon the degree of similarity between the diameters and linear absorption coefficients of the sample in question and the reference specimen. Therefore it is best to allow for expected variations in these properties by preparing several correction curves.

*Obtainable from Nelson P. Nies, 969 Skyline Drive, Laguna Beach, California.

6-2.2 Monochromatic-Pinhole (Flat-Film) Patterns

The forward- and back-reflection techniques are described in Section 4-3, and the reader should compare Fig. 6-9A and B, with Figs. 4-34 and 4-36, respectively. Considering first the forward-reflection technique, a powder halo of radius r is produced on a flat film at a distance D from the sample X by x-rays diffracted at the angle 2θ. Evidently,

$$2\theta = \tan^{-1}\frac{r}{D}, \qquad (6\text{-}10)$$

from which d can be obtained by reference to a table or by calculation with the aid of the Bragg law. For accurate work it is necessary that D be precisely known. In most cases this distance is best determined by mixing the powder under examination with a reference substance of known lattice spacing. From the measured radius r_s and theoretical diffraction angle $2\theta_s$ of a reflection from the reference substance, the sample-to-film distance can be computed from the relation

$$D = \frac{r_s}{\tan 2\theta_s}. \qquad (6\text{-}11)$$

Though the admixture of a reference substance is not recommended for the precise calibration of the radius of a Debye-Scherrer pattern, it is generally satisfactory for the forward-reflection monochromatic-pinhole technique because only low-angle reflections are recorded, for which small variations in the lattice constants exert but negligible effects upon diffraction angles. Hence D can be ascertained with an accuracy limited only by the accuracy with which r_s can be measured.

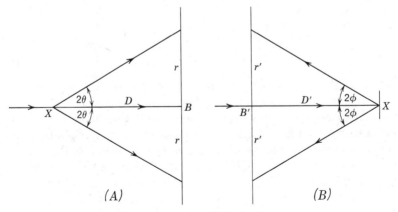

Fig. 6-9.   (A) Forward-reflection geometry. (B) Back-reflection geometry.

By far the most numerous applications of the back-reflection method are to metals, and two methods for determining the sample-to-film distance for such samples have been described in Section 4-3.2. When the method of calibrating powders or paints is used, the distance D' can be determined from the observed radius r'_s and theoretical angle $2\phi_s$ of a reflection from the reference substance by means of the relation

$$D' = \frac{r'_s}{\tan 2\phi_s}$$

$$= \frac{r'_s}{\tan(180° - 2\theta_s)}. \qquad (6\text{-}12)$$

Increasing use is now being made of special gages or templates for directly fixing the distance D'. This has come about because the common applications to metals require relatively high precision in the comparison of d spacings of back-reflections, whereas the absolute values may be of secondary importance. A precision distance gage serves this purpose admirably by permitting a given sample-to-film distance to be reproduced repeatedly within a close tolerance. When it occasionally becomes necessary to determine the absolute value of an interplanar spacing with optimum precision, a calibrating substance should be used. In this case the precise lattice spacing of this substance must first be obtained by one of the precision Debye-Scherrer techniques described in Chapter 8.

If for any reason it is necessary to examine a powder by the back-reflection monochromatic-pinhole technique, distance gages are of no value and a reference powder should be admixed for the accurate determination of D'. In any event, once D' is known, the angle 2ϕ of a reflection of the "unknown" is given by

$$2\phi = \tan^{-1}\frac{r'}{D'}$$

$$= 180° - 2\theta, \qquad (6\text{-}13)$$

from which d can be obtained in the usual way. Although designed especially for the measurement of Debye-Scherrer films, the precision measuring instrument described in Section 6-1 can be used equally well for measuring flat-film diagrams. Most of the commercial instruments with an accuracy of 0.05 to 0.10 mm are also adaptable to the measurement of such patterns, although in a few cases the precision may not be high enough.

6-3 INDEXING CUBIC POWDER PATTERNS

6-3.1 Reciprocal-Lattice Picture of Diffraction by a Cubic Powder

At the beginning of Chapter 4 the concept of the reciprocal lattice was used to demonstrate the origin of powder diffraction diagrams. Straumanis[13] has shown how the concept can be of additional help in understanding and analyzing these patterns. Figure 6-10 illustrates in greater

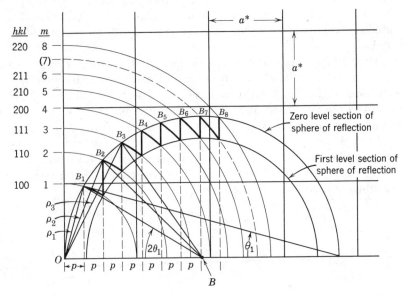

Fig. 6-10.   Reciprocal lattice representation of diffraction by a simple cubic lattice.

detail the graphical construction of a portion of a cubic diffraction pattern. The zero-level reciprocal lattice net of interval a^* is shown. The points 100, 010, and 001 generate a sphere of radius $\rho_1 = 1/a = a^*$, a circular section of which is shown. This circle cuts the sphere of reflection at B_1, generating a reflection at the angle $2\theta_1$ (angle OBB_1). The next circle of radius $\rho_2 = (1/a)\sqrt{2} = a^*\sqrt{2}$, generated by planes of the form $\{110\}$, cuts the sphere of reflection at B_2. The third circle is generated by planes $\{111\}$ and has the radius $\rho_3 = (1/a)\sqrt{3} = a^*\sqrt{3}$. It is noted that this circle does not intersect any point on the diagram, which is to be expected since the generating points $\{111\}$ do not lie in the zero-level reciprocal lattice net. By continuing in this way, successive circles are found to possess the radii: $\rho_4 = a^*\sqrt{4}$, $\rho_5 = a^*\sqrt{5}$, $\rho_6 = a^*\sqrt{6}$, $\rho_8 = a^*\sqrt{8}$, and so on. In general,

$$\rho_m = a^*\sqrt{m}, \tag{6-14}$$

where $m = h^2 + k^2 + l^2$, which shows why there is no ρ_7 (because 7 cannot be represented as the sum of the squares of three integers). If the process is continued to its logical conclusion, the sphere of reflection will be cut by a series of circles of increasing radii, the largest permissible radius being 2, the diameter of the sphere of reflection. Planes do not produce reflections when $(h^2 + k^2 + l^2) > 4/(a^*)^2$. From the preceding discussion it is evident that the intersection points $B_1, B_2, B_3, \ldots, B_{max}$ locate all the possible reflections from a simple cubic lattice with the reciprocal constant a^*.

Figure 6-10 has other interesting properties. If a circle is drawn with center at B and a radius equal to the radius of the circle formed by the intersection of the first reciprocal lattice level with the sphere of reflection, all the possible reflections from a simple cubic lattice of reciprocal constant a^* are arrived at by following the indicated zigzag course (heavy black line) between the two concentric circles with centers at B. If perpendiculars are dropped to the horizontal axis from the points $B_1, B_2, B_3, \ldots, B_{max}$, it will be found that all the intervals between successive perpendiculars are equal. Thus

$$\sin \theta_1 = \frac{\rho_1}{2} = \frac{a^*}{2} = \frac{p}{a^*},$$

and

$$p = \frac{(a^*)^2}{2} = \frac{1}{2a^2}. \tag{6-15}$$

Likewise, Fig. 6-10 shows that

$$\frac{1 - mp}{1} = \cos 2\theta_m,$$

from which

$$pm = 1 - \cos 2\theta_m. \tag{6-16}$$

Without any knowledge of the unit-cell dimension or lattice type, it is possible to use relation 6-16 to index a cubic pattern. From measurements of the lines on the film, values of 2θ are calculated, and the pm's of each line obtained. The smallest value of Δpm between consecutive lines nearly always corresponds to the true value of p. With p known, and expression 6-16 cast in the form

$$m = \frac{1 - \cos 2\theta_m}{p} = h^2 + k^2 + l^2, \tag{6-17}$$

the m's of the various lines are readily identified, and in turn their indices hkl. If the smallest Δpm actually happens to correspond to $2p$ or $3p$, it may take possibly two or three trials to index the film completely. The procedure has been carried through in detail for a powder pattern of RbCl by Klug and Alexander[14].

It is important to note here that pm is just one of several quantities related to the diffraction angle θ, which are useful in indexing powder patterns. For instance, for the cubic case

$$2pm = ma^{*2} = Q_m = \frac{1}{d_m{}^2} = \frac{4\sin^2\theta_m}{\lambda^2} = \rho_m{}^2. \tag{6-18}$$

The quantities Q and $\sin^2\theta$ are in common use in indexing procedures. The reader is reminded that the complete numerical equality of the above quantities depends on the value of the constant k^2 in the expression 1-34 for the reciprocal lattice vector distance ρ. For many calculations it is conveneint to set constant k^2 equal to unity, while at other times $k^2 = \lambda$ is useful, as in expressions 3-27 and 3-29.

6-3.2 Indexing a Cubic Pattern by $\sin^2\theta$ Ratios

Equation 6-18 indicates an integral relationship between the $\sin^2\theta$ values values of the various lines, and the situation is simply demonstrated in the geometry of the crystal lattice. According to the Bragg equation and the expression for an interplanar spacing of a cubic crystal (see Table 1-3), two reflections $(h_1k_1l_1)$ and $(h_2k_2l_2)$ satisfy the equations

$$\sin^2\theta_1 = \frac{\lambda^2}{4a^2}(h_1{}^2 + k_1{}^2 + l_1{}^2),$$

$$\sin^2\theta_2 = \frac{\lambda^2}{4a^2}(h_2{}^2 + k_2{}^2 + l_2{}^2), \tag{6-19}$$

in which the order of reflection n has been combined with the Miller indices. Division of the first equation by the second gives

$$\frac{\sin^2\theta_1}{\sin^2\theta_2} = \frac{m_1}{m_2}, \tag{6-20}$$

wherein m_1 and m_2 are the integral sums of the squares of the respective Miller indices. This indexing procedure consists essentially of determining the simplest integral relationship existing between the experimental $\sin^2\theta$'s that does not yield forbidden integers (7, 15, 23, 28, etc.). It is of course necessary that all the data be referred to a common wavelength, a condition that was assumed in deriving equation 6-20 from equation 6-19. Thus, for some constant spacing d, a value of $\sin^2 \theta$ for one wavelength is related to $\sin^2 \theta$ for a second wavelength as follows:

$$\frac{\sin^2 \theta_1}{\sin^2 \theta_2} = \frac{\lambda_1^2}{\lambda_2^2}. \tag{6-21}$$

This relationship follows directly from the square of the Bragg law written for two wavelengths.

Experimental data from a tantalum powder pattern are used in Table 6-2 to illustrate this indexing procedure. The observed $\sin^2 \theta$ ratios differ more or less from ideal values (designated m in the table) largely as the result of line displacements due to absorption of the x-ray beam by the

Table 6-2.   Indexing of a Tantalum Powder Pattern by the Method of $\sin^2 \theta$ Ratios[a]

Line	Radiation	$\sin \theta$	$\sin^2 \theta$	$\sin^2 \theta \rightarrow \alpha_1$	Calculated Ratios		m	(hkl)
					I	II		
1	α	0.33563	0.11265	0.11246	1.03	2.05	2	110
2	α	0.47157	0.22238	0.22201	2.03	4.06	4	200
3	α	0.57580	0.33155	0.33100	3.02	6.05	6	211
4	α	0.66346	0.44018	0.43945	4.02	8.03	8	220
5	α	0.74044	0.54825	0.54734	(5.00)	(10.00)	10	310
6	α	0.81024	0.65649	0.65540	5.99	11.97	12	222
7	α	0.87357	0.76312	0.76185	6.96	13.92	14	321
8	α_1	0.93303	0.87054	0.87054	7.95	15.91	16	400
9	α_2	0.93575	0.87563	0.87130	7.96	15.92	16	400
10	α_1	0.98907	0.97826	0.97826	8.94	17.87	18	411
11	α_2	0.99164	0.98335	0.97848	8.94	17.88	18	411

[a]Wavelengths employed: Cu$K\alpha$, 1.54184 Å; Cu$K\alpha_1$, 1.54056 Å; Cu$K\alpha_2$, 1.54439 Å. To convert α to α_1 multiply $\sin^2 \theta$ by 0.99834. To convert α_2 to α_1 multiply $\sin^2 \theta$ by 0.99505. The lattice type is identified as body-centered.

sample. These discrepancies can be kept smaller by selecting an intermediate $\sin^2 \theta$ value as the point of reference, as has been done in the present instance, line 5 at $2\theta = 95°$ being used. When the line of reference is such an intermediate one, it results in the experimental $\sin^2 \theta$ ratios becoming progressively larger than the theoretical integers at smaller angles, and smaller at larger angles. The simplest set of approximately integral ratios is given in column I, but this choice is shown to be incorrect by the presence of the integer 7 (calculated 6.96), which is a number not expressible as the sum of three squares. The simplest set of ratios that does not include a forbidden number is therefore the set listed in column II, which are the doubled values from I. The Miller indices (hkl) are deduced directly from the true integers m by inspection. From the occurrence of only even values of $h + k + l$, the lattice type is identified as body-centered (see Table 3-2 and Appendix VI).

A related scheme has been proposed by Bloss[33]. The quotients d_1^2/d_1^2, $d_1^2/d_2^2, \ldots$, d_1^2/d_n^2 are calculated, where d_1 is the largest interplanar spacing observed on the pattern and $d_2, \ldots, d_n$ are successively smaller ones. The relationship to the $\sin^2 \theta$ method is seen from the following expression,

$$\frac{d_1^2}{d_2^2} = \frac{m_2}{m_1} = \frac{\sin^2 \theta_2}{\sin^2 \theta_1},$$

which follows from equations 6-1 and 6-20. Indexing is carried out by comparing the set of d_1^2/d_n^2 quotients with a table of values of m_n/m_1. In general it is not known whether the $h_1k_1l_1$ reflection is 100, 110, 111, 200, 210, 211, or a still more complicated reflection. Hence it is necessary to prepare a table of m_n/m_1 values in which each of these (and higher index planes) is $h_1k_1l_1$. Thus the column in which $h_1k_1l_1$ is 100 contains only integers. If $h_1k_1l_1$ is 110, the quotients will be integers plus half intervals; if $h_1k_1l_1$ is 111, the quotients will be integers plus one-third intervals; if $h_1k_1l_1$ is 200, the quotients will be integers plus one-quarter intervals; and in general, if $(h_1^2 + k_1^2 + l_1^2) = N_1$, the quotients will be integers plus $1/N_1$th intervals.

One proceeds to compare the series of d_1^2/d_n^2 quotients with the successive columns of m_n/m_1 values in the table, starting with the column of smallest m_1 value, $m_1 = 1$. The first column that accounts for all the observed quotients provides the true indices of the largest interplanar spacing $h_1k_1l_1$, and the indices of the remaining reflections are thereby identified. An advantage of the Bloss method is that reflections for which 2θ and $h^2 + k^2 + l^2$ are large are indexed as easily as those for which 2θ and $h^2 + k^2 + l^2$ are small. Further details can be found in reference[33].

6-3.3 Determination of the Unit-Cell Dimension a

From the observed $\sin \theta$ value of any line in an indexed cubic pattern, the unit-cell dimension can be determined by employing the Bragg equation in the form

$$a = \frac{\lambda}{2} \frac{\sqrt{h^2 + k^2 + l^2}}{\sin \theta}, \tag{6-22}$$

or from the d spacing of any line by making use of the cubic formula for the interplanar spacing,

$$a = d\sqrt{h^2 + k^2 + l^2}$$
$$= d\sqrt{m}. \qquad (6\text{-}23)$$

The computations can be facilitated by employing the numerical data of Appendix VI. In Table 6-3 experimental tantalum spacings, calculated with the relation $d = \lambda/(2 \sin \theta)$, have been converted to the equivalent a values by means of equation 6-23. Because of certain systematic sources

Table 6-3. Calculation of the Lattice Constant a of Tantalum from Experimental d Values

Line	$\theta(°)$	$d(\text{Å})$	$m = h^2 + k^2 + l^2$	$\sqrt{m}$	$a(\text{Å})^b$
1	19.611	2.2969	2	1.4142	3.2483
2	28.136	1.6348	4	2.0000	3.2696
3	35.156	1.3389	6	2.4495	3.2796
4	41.564	1.1620	8	2.8284	3.2866
5	47.769	1.0412	10	3.1623	3.2926
6	54.119	0.9515	12	3.4641	3.2961
7	60.876	0.8825	14	3.7417	3.3021
8	68.912	0.8256	16	4.0000	3.3024
9	69.349	0.8252	16	4.0000	3.3008
10	81.520	0.7788	18	4.2426	3.3041
11	82.588	0.7787	18	4.2426	3.3037

[a]Refer to Table 6-2.
[b]The "best" value of a lies in the neighborhood of 3.304 Å. This is larger than the accepted lattice constant of pure tantalum because of the presence of hydrogen and possibly other impurities in solid solution.

of error in the Debye-Scherrer technique, the a's display a drift toward higher values as θ increases. In Chapter 8 these effects are explained in detail, together with procedures for correcting the apparent a values and arriving at a best value. For the present suffice it to say that the systematic errors tend to vanish at $\theta = 90°$, with the result that the most accurate values of a are obtainable from back-reflection lines. On the basis of the last pair of lines ($\theta = 81.5$ and $82.6°$) the true value of a lies in the neighborhood of 3.304 Å. For more precise methods of determining lattice constants from powder photographs, the reader is referred to Chapter 8.

6-3.4 Indexing a Cubic Pattern When a Is Known

When a is known, the indexing can be performed with certainty and ease by comparing d spacings calculated from relationship 6-23 with the observed spacings. Or alternatively values of $\sin^2 \theta$ calculated with the aid of equation 6-22 in the form

$$\sin^2 \theta = \frac{\lambda^2(h^2 + k^2 + l^2)}{4a^2} \tag{6-24}$$

can be compared with the experimental values. In either method the observed quantities exhibit a drift with respect to the calculated values as a function of θ for the same reasons cited above for the drift of observed $h^2 + k^2 + l^2$ integers and a values.

Sometimes the size of a *possible* unit cell is known, but it is uncertain whether the true a is actually equal to this value or rather to some integral

Table 6-4. Proper Choice of Unit-Cell Dimension for the Case of a Cubic Face-Centered Lattice and $d_{100}/n = 2.815 \, \mathring{A}$[a]

Observed $d(\mathring{A})$	$n = 1$			$n = 2$		
	m	hkl	Calculated $d(\mathring{A})$	m	hkl	Calculated $d(\mathring{A})$
3.203				3	111	3.251
2.781				4	200	2.815
1.974				8	220	1.991
1.689				11	311	1.697
1.616	3	111	1.625	12	222	1.625
1.403	4	200	1.408	16	400	1.408
				19	331	1.291
1.255				20	420	1.259
1.147				24	422	1.149
				27	511, 333	1.082
0.994	8	220	0.995	32	440	0.995
				35	531	0.952
0.938				36	600, 442	0.938
0.890				40	620	0.890
				43	533	0.859
0.849	11	311	0.849	44	622	0.849
				48	444	0.812
				51	711, 551	0.789
0.781				52	640	0.781

[a]Conclusion: $n = 2$; $a = 5.630$ Å.

fraction or multiple of it. For example, suppose that the lattice is known to be face-centered, and that reflections from the (100) face of a single crystal have shown that the unit cell has one of the dimensions 2.815, 5.630, 8.445, 11.260 Å, and so on, depending upon whether the reflection of 2.815 Å spacing is the first, second, third, fourth, and so on, order of (100). This situation may be expressed as

$$\frac{d_{100}}{n} = 2.815 \text{ Å},$$

where n is an unknown small integer. The first step, as shown in Table 6-4, is to calculate the interplanar spacings for a face-centered lattice, assuming $n = 1$, and compare them with the observed spacings. It is seen that when this is done only four of the 13 reflections are explained. The next step is to assume $n = 2$, recalculate the spacings, and again attempt to match the observed and calculated spacings. It is seen from the table that this leads to a fit for every reflection. Accordingly, the proper choice of n is seen to be 2, which is equivalent to $a = d_{100} = 5.630$ Å, because this is the smallest unit cell that satisfies all the observed reflections.

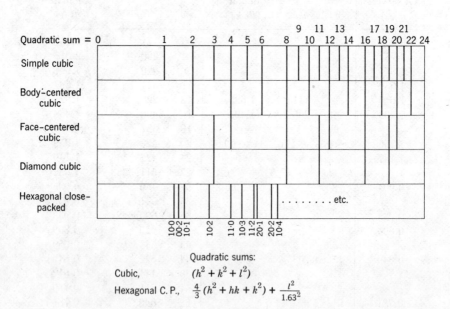

Quadratic sums:

Cubic, $(h^2 + k^2 + l^2)$

Hexagonal C. P., $\frac{4}{3}(h^2 + hk + k^2) + \frac{l^2}{1.63^2}$

Fig. 6-11. Powder patterns of the several cubic lattices and the hexagonal close-packed structure.

6-4 DETERMINATION OF LATTICE TYPE

It is helpful to learn to recognize the lattice type of a cubic substance from the appearance of the powder diagram. This is especially true of structures composed of but one kind of atom, since in this case the relative intensities as well as the relative interplanar spacings vary little from one substance to another. Figure 6-11 shows the characteristic arrange-

Table 6-5. **Characteristic Powder Patterns of the Cubic Lattice Types and the Hexagonal Close-Packed Structure**

		Cubic Lattices				Hexagonal Close-Packed Structure	
m	(hkl)	Ratios of Interplanar Spacings, $d/a = 1/\sqrt{h^2+k^2+l^2}$				$(hk \cdot l)$	Ratios of Interplanar Spacings
		Simple	Face-centered	Body-centered	Diamond		
1	100	1.000				10·0	1.000
2	110	0.707		0.707		00·2	0.942
3	111	0.577	0.577		0.577	10·1	0.882
4	200	0.500	0.500	0.500		10·2	0.686
5	210	0.447				11·0	0.577
6	211	0.408		0.408		10·3	0.532
(7)						11·2	0.492
8	220	0.354	0.354	0.354	0.354	20·1	0.483
9	300, 221	0.333				00·4	0.471
10	310	0.316		0.316		20·2	0.442
11	311	0.302	0.302		0.302	10·4	0.427
12	222	0.289	0.289	0.289			
13	320	0.277					
14	321	0.267		0.267			
(15)							
16	400	0.250	0.250	0.250	0.250		
17	410, 322	0.243					
18	411, 330	0.236		0.236			
19	331	0.229	0.229		0.229		
20	420	0.224	0.224	0.224			
21	421	0.218					
22	332	0.213		0.213			
(23)							
24	422	0.204	0.204	0.204	0.204		

ments of lines in the powder patterns of simple cubic, body-centered cubic, face-centered cubic, and diamond cubic lattices, and of the hexagonal close-packed structure. The hexagonal close-packed structure, although strictly speaking not pertinent to the present discussion, has been included because of its great importance as one of the three most common elemental structures. It must be remembered that for all except elemental structures the relative intensities of the lines vary greatly from one pattern to another, making it more difficult to recognize the several lattice patterns from visual inspection alone. Under these circumstances it is helpful to calculate the ratios of the d spacings observed, and to compare them with the theoretical values for each of the lattice types. These theoretical ratios have been compiled in Table 6-5.

6-5 INDEXING NONCUBIC POWDER PATTERNS

6-5.1 Indexing Noncubic Patterns When the Unit-Cell Dimensions Are Known

Under these circumstances any powder pattern can in principle be surely and completely indexed by comparing the observed d spacings with those computed, using the appropriate formula (see Table 1-3) or by comparing the observed and theoretical values of $\sin^2 \theta$. In practice, however, this procedure is not necessarily completely successful, because of the likelihood of line superpositions. The method is likely to be more successful when applied to hexagonal and tetragonal crystals than to orthorhombic, monoclinic, or triclinic crystals, because of the greater complexity of the patterns of the latter. In Table 6-6 the $\sin^2 \theta$ method is used to index a portion of the powder pattern of (tetragonal) potassium dihydrogen phosphate (KH_2PO_4). It is noted that an unequivocal assignment of indices cannot be made for lines 7 and 13. Thus it is uncertain whether line 7 should be assigned the indices 130, 301, or both, although the magnitude of the drift term $\Delta \sin^2 \theta$ favors the latter choice. An even greater degree of uncertainty exists about line 13. These uncertainties are rather typical of the difficulties encountered when applying this method to noncubic patterns. Little difficulty is experienced in indexing the low-angle lines, but the number of ambiguous cases increases with angle because of the greater frequency of superpositions. It is helpful of course to eliminate as many index triplets as possible on the basis of space-group extinctions whenever the symmetry is known. Line superpositions can also be reduced by employing a large camera radius and other experimental conditions which promote resolution.

Table 6-6. Indexing of the Powder Diagram of Tetragonal $KH_2PO_4{}^{a,b}$

Line	$(\sin^2 \theta)_{obs}$	$(\sin^2 \theta)_{calc}$	$\Delta \sin^2 \theta (\times 10^4)$	(hkl)
1	0.0236	0.0229	+7	101
2	0.0437	0.0428	+9	200
3	0.0668	0.0657	+11	121
4	0.0712	0.0700	+12	112
5	0.0867	0.0856	+11	220
6	0.0927	0.0915	+12	202
7	0.1093	$\begin{cases} 0.1070 \\ 0.1086 \end{cases}$	$\begin{cases} +23 \\ +7 \end{cases}$	$\begin{cases} 130 \\ 301 \end{cases}$
8	0.1212	0.1202	+10	103
9	0.1512	0.1513	−1	321
10	0.1559	0.1566	−7	132
11	0.1633	0.1630	+3	123
12	0.1708	0.1712	−4	400
13	0.1945	$\begin{cases} 0.1941 \\ 0.1946 \end{cases}$	$\begin{cases} +4 \\ -1 \end{cases}$	$\begin{cases} 141 \\ 004 \end{cases}$
14	0.2064	0.2057	+7	303
15	0.2139	0.2141	−2	240

[a]Lattice constants known: $a = 7.43$, $c = 6.97$ Å. $(\sin^2 \theta)_{calc} = (1/a^2) \ (\lambda/2)^2 \ [(h^2+k^2) + (a/c)^2 l^2] = 0.010704 \ [(h^2+k^2) + 1.1364 l^2]$.

[b]Space-group absences: (hkl) present only with $h+k+l = 2n$; $(0kl)$ present only with $k+l = 2n$; $(h00)$ present only with $h = 2n$; $(hh0)$ present only with $h = 2n$.

6-5.2 Graphical Methods of Indexing

Most graphical methods are based upon the logarithmic forms of the Bragg equation appropriate to the several crystal systems:

Tetragonal,
$$\log d = \log a - \frac{1}{2} \log \left[(h^2+k^2) + \frac{l^2 a^2}{c^2} \right]. \tag{6-25}$$

Hexagonal,
$$\log d = \log a - \frac{1}{2} \log \left[\frac{4}{3} (h^2 + hk + k^2) + \frac{l^2 a^2}{c^2} \right]. \tag{6-26}$$

Orthorhombic,
$$\log d = \log b - \frac{1}{2} \log \left[\frac{h^2 b^2}{a^2} + k^2 + \frac{l^2 b^2}{c^2} \right]. \tag{6-27}$$

The quadratic forms of the Bragg equation have also found application in graphical and mechanical schemes for indexing noncubic patterns, and they are the basis for most analytical methods of indexing:

Tetragonal, $\qquad\qquad \sin^2\theta = A(h^2+k^2)+Cl^2,$ $\qquad\qquad$ (6-28)

where

$$A = \left(\frac{\lambda}{2a}\right)^2 \quad \text{and} \quad C = \left(\frac{\lambda}{2c}\right)^2.$$

Hexagonal, $\qquad\qquad \sin^2\theta = A(h^2+hk+k^2)+Cl^2,$ $\qquad\qquad$ (6-29)

where

$$A = \frac{1}{3}\left(\frac{\lambda}{a}\right)^2 \quad \text{and} \quad C = \left(\frac{\lambda}{2c}\right)^2.$$

Orthorhombic, $\qquad\qquad \sin^2\theta = Ah^2+Bk^2+Cl^2,$ $\qquad\qquad$ (6-30)

where

$$A = \left(\frac{\lambda}{2a}\right)^2, \quad B = \left(\frac{\lambda}{2b}\right)^2, \quad \text{and} \quad C = \left(\frac{\lambda}{2c}\right)^2.$$

Graphical procedures are nearly always successful when applied to two-parameter (tetragonal and hexagonal) problems, except when the cell dimensions are very large or the extinguished reflections very numerous. The solution of three-parameter (orthorhombic) problems is much more cumbersome, as well as less certain.

The Hull-Davey charts [15, 16], based upon expressions 6-25 to 6-27, have been rather widely used. For tetragonal or hexagonal crystals, the charts are prepared by plotting the axial ratio c/a against $\log d$ for all the possible index triplets hkl constituting a primitive lattice pattern. The "constant" term $\log a$ of course varies from one tetragonal substance to another and may be arbitrarily set equal to any value, such as zero, in preparing the charts. An example of such a two-parameter chart is shown in Fig. 6-12. To index a pattern believed to be tetragonal, the first step is to plot the observed d spacings along the edge of a strip of paper to the same logarithmic scale as that of the chart (see Fig. 6-12). While being kept in a horizontal position, the strip is then moved over the chart until a fit is found between all the plotted d spacings and the lines on the graph. It can be seen that this procedure is the mechanical equivalent of varying the parameters a and c/a until the experimental data agree with the theoretical. Once the fit has been obtained, the indices of the lines and the approximate axial ratio can be read directly from the chart. These results, however, should never be considered final but need to be confirmed by numerical calculations of a more precise nature. Figure 6-12 shows the position of the strip on the tetragonal Hull-Davey

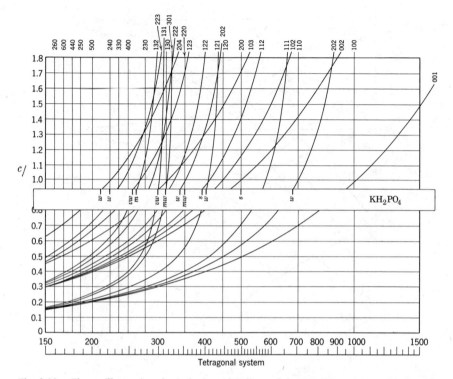

Fig. 6-12.   Figure illustrating the indexing of 12 lines of the KH_2PO_4 pattern with the aid of a Hull–Davey chart.

chart when a fit was found for the d spacings of KH_2PO_4. The axial ratio is found to be about 0.94. It should be noted that at ratio $c/a = 1$ the tetragonal chart can be used to index cubic crystals.

Orthorhombic patterns require charts capable of portraying the effects of three parameters, b, a/b, and c/b (see equation 6-27). This can be accomplished, but only in an approximate way, by constructing a series of charts, each for a different fixed value of c/b and with abscissa $\log d$ and ordinate a/b[17]. It is evident that a perfect fit can seldom be hoped for, since each chart must serve over an appreciable range of c/b values. Nevertheless, it is often possible to arrive at an approximate fit, which must then be verified and the parameters refined by numerical calculations. The principal weakness of this method of indexing orthorhombic patterns is that it is very time-consuming. In addition, the choice of the right fit is frequently less certain than for two-parameter problems.

Bjurström[18] has made use of the quadratic equations 6-28, 6-29, and 6-30 to design simple straight-line charts for indexing two- and three-

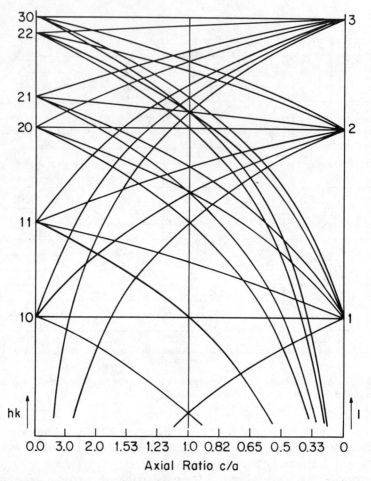

Fig. 6-13.   Bunn-Bjurström logarithmic chart for indexing tetragonal patterns. (Courtesy of C. Bunn; *Chemical Crystallography*, 1961, Clarendon Press, Oxford.)

parameter patterns. Although easy to prepare, the Bjurström charts often proved confusing to use, and Bunn[19] improved them by making the c/a, hk, and l directions logarithmic (Fig. 6-13). Many consider the Bunn-Bjurström charts more satisfactory than the Hull-Davey charts, but most workers find graphical methods generally of limited utility, unless they have many very simple powder patterns to analyze. Graphical methods of indexing noncubic patterns not only have the disadvantage of being time-consuming, but they also hold the definite risk of a wrong solution, especially when a large number of reflections are extinguished. Additional references [20–27] on graphical procedures are listed for those interested.

6-5.3 Analytical Methods of Indexing: Tetragonal, Hexagonal, and Orthorhombic Patterns

Various investigators have experimented with numerical methods of indexing two- and three-parameter patterns. A common principle employed has been comparing differences in the experimental $\sin^2 \theta$'s and inferring the proper indices by noting the existence of significant integral ratios among them [28]. More systematic mathematical procedures, actually extensions of this same principle, have been devised [29–31]. These methods are rather straightforward when applied to tetragonal and hexagonal patterns, but more cumbersome and somewhat less systematic when applied to orthorhombic patterns. Nevertheless, they have repeatedly demonstrated their efficacy in solving orthorhombic problems, provided a reasonable amount of perseverance is exercised. The most difficult cases of course involve large unit-cell dimensions, or large numbers of extinguished reflections, or both. As an example of a workable analytical method, Lipson's technique [30], which is basically similar to Hesse's [29], is now described, together with an actual application to a three-parameter powder pattern.

For an orthorhombic specimen, equation 6-30 is applicable:

$$\sin^2 \theta_{hkl} = q_{hkl} = Ah^2 + Bk^2 + Cl^2.$$

For simplicity of notation $\sin^2 \theta$ is designated q in the ensuing treatment. The problem is to find values of the coefficients A, B, and C that account for all the observed q's when h, k, and l assume various integral values. It is essential to the success of this and similar methods that the values of q be known with an accuracy of at least ± 0.0010, and preferably ± 0.0005.

At the outset an estimate of the magnitude of the constants A, B, and C should be made. Suppose that there are M lines on the powder photograph with q less than q_m. The number of possible lattice points in reciprocal space with q less than q_m is approximately equal to the volume in reciprocal space of a sphere with radius ρ_m (equivalent to q_m) divided by the volume of the reciprocal unit cell, V^*. From Fig. 6-14 it can be seen that the volume of the sphere is

$$V = \tfrac{4}{3} \pi \rho m^3$$

$$= \tfrac{4}{3} \pi (2qm^{1/2})^3,$$

since $q_m = \sin^2 \theta_m = (\rho_m/2)^2$. Hence the number of reciprocal points

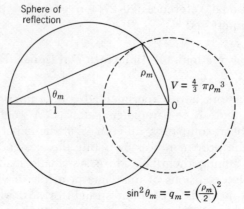

Fig. 6-14.

sought is given by

$$\frac{V}{V^*} = \frac{4}{3}\pi(2q_m^{1/2})^3 \frac{1}{V^*}. \tag{6-31}$$

The actual number of observed lines, however, is less than this because of the effect of the multiplicity factor for powder reflections and because of space-group and accidental extinctions. These factors vary from one structure to another, but their combined effect for orthorhombic crystals can be allowed for with sufficient accuracy for the present purpose by applying a factor of $\frac{1}{16}$. This follows because the number of reflections is reduced by a factor of roughly 8, the multiplicity factor for a general reflection in the orthorhombic system and another factor of about 2 as a result of systematic and random extinctions. The approximate number of lines to be expected with q less than q_m is then

$$M \cong \frac{1}{16}\frac{4}{3}\pi(2q_m^{1/2})^3\frac{1}{V^*}$$

$$\cong \frac{2\pi}{3}\frac{q_m^{3/2}}{V^*}. \tag{6-32}$$

Since $V^* = \lambda^3/abc = 8\sqrt{ABC}$, and assuming A, B, and C to be of the same order of magnitude,

$$M \cong \frac{2\pi}{3} \cdot \frac{q_m^{3/2}}{8A^{3/2}},$$

from which

$$A \cong \frac{0.4q_m}{M^{2/3}}. \tag{6-33}$$

Lipson's method of arriving at satisfactory values of A, B, and C consists of systematically calculating the differences between the observed q's and examining these differences for frequently occurring values which are related integrally to one another, such as

$$
\begin{aligned}
q_{1kl} - q_{0kl} &= A, \\
q_{2kl} - q_{0kl} &= 4A, \\
q_{3kl} - q_{0kl} &= 9A, \\
q_{4kl} - q_{0kl} &= 16A,
\end{aligned}
\tag{6-34}
$$

and similar relationships involving B and C. These equations are seen to follow directly from equation 6-30 upon substitution of the proper indices. Once relationships of these types have been discovered and possible values of A, B, and C determined, it is usually possible to identify one or more of the pinacoidal reflections, $h00$, $0k0$, or $00l$, by noting among the experimental q values themselves some of the same multiples of the proposed values of A, B, and C. For example, if a line is present with $q = A$, its indices are almost certainly 100; if $4A$, 200; and if $9A$, 300; and so on. The principal point of inefficiency in this procedure is that, in place of selecting the desired relationships of the 6-34 type, it is quite possible to be led astray by the following types, which possess the same integral relationships:

$$
\begin{aligned}
q_{hkl} - q_{00l} &= Ah^2 + Bk^2, \\
q_{2h,2k,l} - q_{00l} &= 4(Ah^2 + Bk^2), \\
q_{3h,3k,l} - q_{00l} &= 9(Ah^2 + Bk^2), \\
q_{4h,4k,l} - q_{00l} &= 16(Ah^2 + Bk^2).
\end{aligned}
\tag{6-35}
$$

Thus the reflections $00l$, $11l$, $22l$, and $33l$, or the set $00l$, $12l$, $24l$, and $36l$, satisfy these requirements, and, furthermore, the quantity $Ah^2 + Bk^2$ or its small integral multiples are likely to be found among the observed q's (reflections 110, 220, 330, 440, and so on, or 120, 240, 360, 480, and so on). Fortunately, the trial-and-error procedure of differentiating between the desired relationships of type 6-34 and those of the unwanted 6-35 type need not be unduly time-consuming. More false moves will be made if the number of 6-34 relationships is meager as the result of extinctions, particularly when pinacoidal reflections $h00$, $0k0$, and $00l$ are absent in the lower orders.

Lipson's technique of indexing can be understood best by solving an actual problem. For this purpose the published data of Jacob and Warren[32] for uranium are used. Table 6-7 lists the experimental $\sin^2 \theta$ (q) values for the first 30 lines of the powder pattern. Next the differences between the q's should be tabulated, the scheme given in

Table 6-7.   Values of $\sin^2 \theta$ for the First 30 Lines of the
Uranium Pattern[a]

Line	$\sin^2 \theta$	Line	$\sin^2 \theta$	Line	$\sin^2 \theta$
1	0.0684	11	0.2882	21	0.4790
2	0.0904	12	0.2936	22	0.4909
3	0.0935	13	0.3025	23	0.5022
4	0.0972	14	0.3090	24	0.5687
5	0.1149	15	0.3263	25	0.5800
6	0.1667	16	0.3613	26	0.5931
7	0.1880	17	0.3744	27	0.6038
8	0.2282	18	0.3878	28	0.6228
9	0.2532	19	0.4469	29	0.6472
10	0.2763	20	0.4573	30	0.6656

[a]Computed from published data of Jacob and Warren [32]

Table 6-8 being suggestive. For lack of space the table includes only the differences derived from the first 12 lines of the pattern, but in actual practice it is helpful to extend it to include all 30 lines. The Δq's should also be graphed according to the scheme shown in Fig. 6-15. Using a scale of not less than 1 cm to 0.01 in q, mark the observed q's on the edge of a long strip of paper as shown in A. The q's up to at least one-half the maximum value are now transferred to the lowest horizontal line on a sheet of graph paper (B in the figure). The strip is then shifted up to the next horizontal line and over until q_1 coincides with the zero ordinate, after which the q's are marked on the graph paper again. Next q_2 is placed at the zero ordinate with the strip on the third horizontal line, and the q's are plotted again, and so on. The process is continued until each value of q has successively been placed at the zero ordinate. To facilitate the interpretation of this Δq chart, each point may be plotted as a short horizontal bar, the length of which represents the experimental error range, for instance, ± 0.0005. Frequently occurring Δq's can now be easily detected by noting at what values of Δq (horizontal axis) a vertical line intersects a number of q bars. Table 6-9 lists the frequently occurring Δq's revealed by the complete diagram, which is shown only in part in Fig. 6-15. In Table 6-10 are given the possibly significant values of Δq, which are related by integral factors of 4, 9, and 16. The asterisked Δq's deserve special attention because they have counterparts among the q's, thus raising the definite possibility that these counterparts are first-, second-, or fourth-order pinacoidal reflections (reflections of the type 100, 200, or 400).

Table 6-8.   Partial Table of Differences between $\sin^2 \theta$ Values of the Uranium Pattern

Line No.	1	2	3	4	5	6	7	8	9	10	11
1											
2	0.0220										
3	0.0251	0.0031									
4	0.0288	0.0068	0.0037								
5	0.0465	0.0245	0.0214	0.0177							
6	0.0983	0.0763	0.0732	0.0695	0.0518						
7	0.1196	0.0976	0.0945	0.0908	0.0731	0.0213					
8	0.1598	0.1378	0.1347	0.1310	0.1133	0.0615	0.0402				
9	0.1848	0.1628	0.1597	0.1560	0.1383	0.0865	0.0652	0.0250			
10	0.2079	0.1859	0.1828	0.1791	0.1614	0.1096	0.0883	0.0481	0.0231		
11	0.2198	0.1978	0.1947	0.1910	0.1733	0.1215	0.1002	0.0600	0.0350	0.0119	
12	0.2252	0.2032	0.2001	0.1964	0.1787	0.1269	0.1056	0.0654	0.0404	0.0173	0.0054

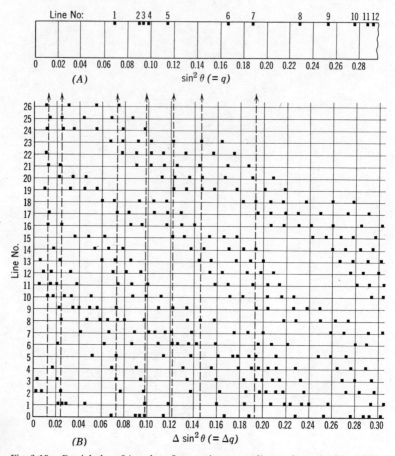

Fig. 6-15.   Partial plot of Δq values for uranium according to the procedure of Lipson.

Before proceeding further, an estimate of the magnitude of A, B, and C should be made with the aid of equation 6-33. The result for the first 30 lines of the uranium pattern is

$$A \cong \frac{0.4 \times 0.6656}{30^{2/3}} = 0.0276.$$

.26624

This result may now be profitably compared with the values of A, B, or C, which can be deduced from the asterisked Δq's in Table 6-10 on the tentative assumption that they correspond to pinacoidal reflections. These possible A, B, and C values are given in Table 6-11. Except for 0.0245, all are seen to be considerably larger than the mean expected value, 0.0276. Such a larger value is not impossible but, if found to be

Table 6-9. Frequently Occurring Δq Values in Uranium Pattern

Δq	No. of Times	Δq	No. of Times	Δq	No. of Times
0.0119	5	0.0980	5	0.2038	5
0.0218	4	0.1216	7	0.2086	4
0.0242	9	0.1382	4	0.2184	5
0.0440	4	0.1458	4	0.2601	5
0.0656	4	0.1599	5	0.2775	4
0.0693	4	0.1761	4	0.2942	4
0.0736	5	0.1820	4	0.3638	5
0.0781	4	0.1860	4	0.3883	4
0.0862	4	0.1941	5		
0.0901	4	0.2002	4		

Table 6-10. Frequently Occurring Δq Values Related by Factors of 4, 9, or 16[a]

1×	4×	9×	16×
0.0218	0.0862	0.1941	
0.0242	0.0980*	0.2184	0.3883*
0.0440	0.1761		
0.0656	0.2601		
0.0693*	0.2775*		
0.0736	0.2942*		
0.0901*	0.3638*		

[a]Values of Δq marked with an asterisk have counterparts among the q's.

Table 6-11. Values of the Constants A, B, or C Corresponding to Possible Pinacoidal Reflections

First Order	Second Order	Fourth Order
	$0.0980/4 = 0.0245$	$0.3883/16 = 0.0243$
0.0693	$0.2775/4 = 0.0694$	
	$0.2942/4 = 0.0736$	
0.0901	$0.3638/4 = 0.0910$	

correct, it should be counterbalanced by one or two constants with values considerably less than 0.0276. It is evident that only in the rather unusual case of an orthorhombic cell possessing nearly equal dimensions are A, B, and C nearly equal and, furthermore, only then might it be expected that A, B, and C will all have the same approximate magnitude as that calculated with equation 6-33, namely, 0.0276.

On the basis of the preliminary evidence presented so far, it seems extremely likely that 0.0245 is one of the constants being sought. Employing this assumption, line 4 can be assigned the indices, 200, and the pairs 3 and 1, 5 and 2, and 9 and 8 (see Table 6-8) are probably of the types $1kl$ and $0kl$. Furthermore, line 18, with $q = 0.3878 \cong 4 \times 0.980$, is probably 400. The lack of a line with $q = 0.0245$ is not surprising as this merely signifies the absence of the 100 reflection, a very common extinction. Similarly, the lack of a q value of 0.2184 denotes the extinction of the 300 reflection. However, in Table 6-8 the Δq of 0.2198 $\cong 9 \times 0.0245$ for lines 11 and 1 makes it probable that they are, respectively, reflections of the types $3kl$ and $0kl$. Of the first 12 lines, presumably all of simple indices, 5 lines have thus been tentatively eliminated from the need of consideration as potential $0k0$ or $00l$ reflections (namely, lines 3, 4, 5, 9, and 11).

On the basis of size, the most likely value of one of the remaining constants is 0.0694, as it is the smallest of the remaining constants proposed in Table 6-11. If B is assigned this value, line 10, with $q = 4 \times 0.0694$, has the indices 020. However, this makes 010 the indices of line 1 ($q = 0.0684$), which is a very likely extinction, especially in view of the fact that 100 is absent on the basis of the postulated value of A. Evidently, lines 1 and 10 can also be indexed as 020 and 040, respectively, in which case $B = 0.0694/4 = 0.01735$. This figure is in better agreement with the expected magnitude of A, B, and C, 0.0276. If B is set equal to 0.01735, the absence of this value among the frequently occurring Δq's denotes a scarcity of reflection pairs of the type $h0l$ and $h1l$. However, the presence of four reflection pairs of the type $h0l$ and $h2l$ is indicated by four Δq's of 0.0693 (Table 6-9). Proceeding for the time being on the assumption that $B = 0.01735$, lines 1 and 10 have the respective indices 020 and 040, as pointed out above. Furthermore, the indices of line 6 are now also determined as 220, since line 4 is 200 and lines 6 and 4 are of the respective types $h2l$ and $h0l$, judging from the fact that their Δq is 0.0695 (Table 6-8). It has also been shown previously that lines 1 and 11 are probably of the types $0kl$ and $3kl$. Combination of this information with the newly postulated indices of line 1, namely, 020, fixes the indices of line 11 as 320. Though the choice between $B = 0.0694$ or 0.01735 cannot be made with certainty at this stage, the preceding analysis tends to favor the second value. It has just been shown that this leads to conclusions that partially or completely fix the indices of lines 1, 6, 10, and 11. Of the first 12 lines of the pattern therefore, only lines 2, 7, 8, and 12 remain to be considered in the search for $00l$ reflections.

Since the proposed values of A and B, respectively 0.0245 and 0.01735, are both less than 0.0276, it is likely that C is larger. Reference to Table 6-10 discloses two promising integral ratios of 4 remaining between pairs of Δq's, 0.0736 and 0.2942, and 0.0901 and 0.3638. Consider the first pair. If 0.0736 derives from $hk0$ and $hk1$ reflections, and 0.2942 from $hk0$ and $hk2$ reflections, then $C = 0.2942/4 = 0.0736$, which is larger than 0.0276 as anticipated, but possibly by too large a margin. However, if 0.0736 and 0.2942 are attributed to the respective pairs $hk0$ and $hk2$, and $hk0$ and $hk4$, then $C = 0.2942/16 = 0.018$, which is *smaller* than 0.0276. It is difficult to choose between these alternatives. A sound procedure is to begin by testing the value of C corresponding to the smaller unit-cell dimension c. If part but not all of the reflections can be indexed on this basis, this dimension should then be doubled and the indexing repeated. Accordingly, let it be proposed that $C = 0.0736$, which corresponds to the smaller c dimension (see equation 6-30). On this basis line 12, with $q = 0.2936$, is 002. From Table 6-8 line pairs 6 and 3, and 7 and 5, appear to be of the types $hk1$ and $hk0$. However, since $3A = 0.0735$, it appears that they may equally well be of the types $2kl$ and $1kl$. This ambiguity can be resolved on the basis of the indices previously assigned to lines 6 and 5, respectively, 220 and $1kl$. These indices agree with the second alternative and, furthermore, demonstrate that line 3 is 120 and that line 7 is of the type $2kl$.

Table 6-12 summarizes the index information available as the result of the proposals concerning A, B, and C made to this point. It is seen that

Table 6-12. Tentative Postulates Regarding the Indices of Lines of the Uranium Pattern

Line	q	Proposed (hkl)	Line	q	Proposed (hkl)
1	0.0684	020	8	0.2282	$0k_2l_2$
2	0.0904	$0k_1l_1$	9	0.2532	$1k_2l_2$
3	0.0935	120	10	0.2763	040
4	0.0972	200	11	0.2882	320
5	0.1149	$1k_1l_1$	12	0.2936	002
6	0.1667	220	18	0.3878	400
7	0.1880	$2k_1l_1$			

specific numerical indices have been assigned to 7 of the first 12 lines, and certain relationships among the other 5 have been deduced. The self-consistency of these results vouches strongly for their correctness. It remains to prove their authenticity by comparing the observed q's with those calculated using equation 6-30. This procedure consists in trying to match the observed q's with the sums of three terms of the forms Ah^2, Bk^2, and Cl^2, employing various simple values of the indices. The matching process is accelerated by first preparing a working table of the possible values of these three terms as a function of the index h, k, or l, as shown in Table 6-13. The results of these calculations are given in Table 6-14. The matching of experimental and calculated q's is evidently good enough to verify the rightness of the indexing, but the first values of A, B, and C can be refined to give much better agreement. The next to the last column gives the improved values of q obtained when the constants $A = 0.0243$, $B = 0.0172$, and $C = 0.0734$ are used. The remaining lines in the pattern can be indexed in the same way. The indices agree with those assigned by Jacob and Warren, except for the choice of axes.

In the foregoing example the reader has been led to the correct solution of a three-parameter problem without any wasted effort pursuing erroneous bypaths. The consequent absence of trial-and-error experiences is admittedly somewhat artificial, but this course has been chosen to save the reader unnecessary confusion in familiarizing himself with a topic that is necessarily complex at best. Even so, it is evident from the example described above that the solution of a three-parameter problem is far from systematic. True, the several preliminary steps can be executed in a straightforward manner. They include: (1) the calculation of the mean anticipated magnitude of A, B, and C; (2) the search for frequently occurring Δq values; and (3) the proposing of likely values of A, B, and C by noting sets of Δq's that stand in the ratios $1:4:9$, and so on. However, the process of discriminating between the various possibilities and selecting the proper constants is one that not

Table 6-13.   Possible values of Ah^2, Bk^2, and Cl^2

h, k, or l	Ah^2	Bk^2	Cl^2
1	0.0245	0.01735	0.0736
2	0.0980	0.0694	0.2944
3	0.2205	0.1562	0.6624
4	0.3920	0.2776	

Table 6-14. Calculation of q's to Match Experimental Values, and Assignment of Indices[a]

Line	Observed q	Ah^2	Bk^2	Cl^2	Calculated Values of q on Basis of:		Proposed Indices
					First Constants	Revised Constants	
1	0.0684		0.0694		0.0694	0.0688	020
2	0.0904		0.0174	0.0736	0.0910	0.0906	011
3	0.0935	0.0245	0.0694		0.0939	0.0931	120
4	0.0972	0.0980			0.0980	0.0972	200
5	0.1149	0.0245	0.0174	0.0736	0.1155	0.1149	111
6	0.1667	0.0980	0.0694		0.1674	0.1660	220
7	0.1880	0.0980	0.0174	0.0736	0.1890	0.1878	211
8	0.2282		0.1562	0.0736	0.2298	0.2282	031
9	0.2532	0.0245	0.1562	0.0736	0.2543	0.2525	131
10	0.2763		0.2776		0.2776	0.2752	040
11	0.2882	0.2205	0.0694		0.2899	0.2875	320
12	0.2936			0.2944	0.2944	0.2936	002
18	0.3878	0.3920			0.3920	0.3888	400

[a]First set of constants: $A = 0.0245$, $B = 0.01735$, $C = 0.0736$. Refined set of constants: $A = 0.0243$, $B = 0.0172$, $C = 0.0734$. Values of Ah^2, Bk^2, and Cl^2 in body of table are derived from the first set of constants.

only requires considerable individual initiative but also entails, in most cases, some trial and error. The application of Lipson's method to tetragonal and hexagonal structures is much simpler and demands less ingenuity. As an aid in the use of the Lipson-Hesse method, Zsoldos [34] considered the most probable differences likely to occur. Studies by Barabash and Davydov[35] apply Neskuchaev's method[36] to indexing powder patterns of intermediate symmetry.

6-5.4 Analytical Methods of Indexing: Monoclinic and Triclinic Patterns

The orthorhombic indexing problem has proved to be one of considerable complexity. Evidently, the indexing of monoclinic patterns with four parameters, and of triclinic patterns involving six parameters, is very difficult. A number of investigators, nevertheless, have tackled the problem[37–49]. Vand[38, 39] has described several methods for indexing powder patterns of long-spacing compounds. The long spacing permits the reliable determination of one edge of the unit cell. With one edge of the cell known, the problem of finding the other two edges is then attacked. Zsoldos[42] started from Vand's "third graphical method"[39] and developed a general method for indexing powder patterns of low-symmetry crystals. This method is probably most advantageous in the monoclinic case.

Ito[40, 41] attacked the problem without making any assumptions as to the symmetry and dimensions of the unit cell. Each line of a powder photograph corresponds to a vector in reciprocal space. Three non-coplanar vectors define the edges of a possible unit cell, and three additional vectors fix their interaxial angles. Ito's method seeks to select suitably six lines of the powder pattern, and obtain therefrom the corresponding unit cell. Once a unit cell has been obtained, it should be possible to index all lines of the pattern. Based on the approach of Ito[40, 41] and Runge[37], de Wolff[43, 44] suggested a more elegant method for constructing the reciprocal lattice of a crystal, using concepts of zone relations. Viswanathan[47], in an elaborate study, has sought to systematize the de Wolff method, and Visser[49] has tabulated many of the constants and quantities involved in indexing calculations. Vand and Johnson[48] applied Diophantine equations to the triclinic indexing problem in a method somewhat allied to that of de Wolff.

Since space limitations do not permit even brief details of the above methods, it is most useful to outline Ito's method as a basis for understanding the general low-symmetry indexing problem. Also, the Ito method is often the basis of automatic computer indexing procedures for powder patterns (Section 6-6). If the symmetry of the crystal is not known, Ito's method provides a practical procedure for indexing the powder pattern. Tests can be applied to determine whether the crystal symmetry is cubic, tetragonal, or hexagonal, in which cases simpler procedures (previously described) can be used. When the suggested symmetry is monoclinic or triclinic, Ito's method is called for.

The powder photograph is treated as having triclinic symmetry. In equation 6-18 the quantity Q_{hkl} was defined as

$$Q_{hkl} = \frac{1}{d^2_{hkl}} = \frac{4 \sin^2 \theta_{hkl}}{\lambda^2} = \rho^2_{hkl}. \tag{6-36}$$

For a triclinic lattice

$$\begin{aligned} Q_{hkl} = \; & h^2 a^{*2} + k^2 b^{*2} + l^2 c^{*2} \\ & + 2\,hka^*b^* \cos \gamma^* \\ & + 2\,klb^*c^* \cos \alpha^* \\ & + 2\,lhc^*a^* \cos \beta^*, \end{aligned} \tag{6-37}$$

where $a^*, b^*, c^*, \alpha^*, \beta^*, \gamma^*$ are the reciprocal axial lengths and angles to be determined. The first step is to prepare a list of all the Q's, listing them in order of increasing magnitude. Next, three of the smallest Q's are chosen and assumed to have indices 100, 010, and 001. These three Q's permit a^*, b^*, and c^* to be determined, since from equation 6-37

$$Q_{100} = a^{*2} \qquad Q_{010} = b^{*2} \qquad Q_{001} = c^{*2}. \qquad (6\text{-}38)$$

Evaluation of the axial angles proceeds as follows. Suppose the angle α^* between b^* and c^* is to be determined first. Angle $\alpha^* \neq 90°$ can be determined from the Q values for planes 011 and 01$\bar{1}$, if they can be identified. From 6-37

$$Q_{011} = b^{*2} + c^{*2} + 2\,b^*c^* \cos \alpha^*$$
$$= Q_{010} + Q_{001} + 2\,b^*c^* \cos \alpha^* \qquad (6\text{-}39)$$

and

$$Q_{01\bar{1}} = Q_{010} + Q_{001} - 2\,b^*c^* \cos \alpha^*. \qquad (6\text{-}40)$$

On subtracting equation 6-40 from equation 6-39

$$Q_{011} - Q_{01\bar{1}} = 4\,b^*c^* \cos \alpha^*, \qquad (6\text{-}41)$$

and

$$\cos \alpha^* = \frac{Q_{011} - Q_{01\bar{1}}}{4\,b^*c^*}. \qquad (6\text{-}42)$$

A word as to how to locate the values Q_{011} and $Q_{01\bar{1}}$: It is seen from equation 6-37 that if angle α^* were 90° the reflections 011 and 01$\bar{1}$ would coincide and

$$Q'_{011} = Q'_{01\bar{1}} = b^{*2} + c^{*2}. \qquad (6\text{-}43)$$

Hence the required Q's are symmetrical with respect to the value $b^{*2} + c^{*2}$, and should be searched for around this value in the table of Q's. If symmetrically placed values for Q_{011} and $Q_{01\bar{1}}$ cannot be found, it may mean that one member of the pair, for instance 011, is missing. The next step is to search for quantities symmetrically placed about $4b^{*2} + c^{*2}$ or $b^{*2} + 4c^{*2}$. If successful, this search would locate planes 021, 02$\bar{1}$, or 012, 01$\bar{2}$. The other two angles β^* and γ^* are derived in similar fashion from Q's symmetrically related to quantities $h^2a^{*2} + l^2c^{*2}$ and $h^2a^{*2} + k^2b^{*2}$, respectively.

Lack of success in determining the axial angles may mean that the initial three reflections chosen are not all first-order reflections. In this instance one might assign other indices $h00$, $0k0$, and $00l$ to them and repeat the search for related pairs of simple prism reflections. If the search is still unsuccessful, it may be necessary to assign the indices $h00$, $0k0$, and $00l$ to a new triplet of lines.

When a^*, b^*, and c^* have been chosen as described above, and the axial angles α^*, β^*, and γ^* have been determined from them, a reciprocal cell and lattice have been defined for the pattern. To check whether the defined cell is the correct cell for the pattern, a complete set of Q values is calculated and compared with the observed Q's. If all lines of the pattern can be indexed, and if satisfactory agreement (possibly after suitable refinement of the cell parameters) of the observed and calculated Q's is achieved, then a preliminary unit cell has been obtained. The preliminary cell often does not exhibit the symmetry of the lattice; it is merely one of the infinite number of possible cells for the specific lattice of the crystal being considered. It may be, however, a cell simple enough for the investigator's needs; if not, it must be reduced to a more suitable cell.

It should be pointed out that if a nonprimitive cell is used to index the lines of a powder pattern many lines will not be indexed. Indeed, the indexed lines will comprise only $1/N$th of the total number of observed lines, where N is the multiplicity of the nonprimitive cell. Thus failure to index an appreciable number of lines in a pattern suggests that a nonprimitive cell was chosen, and a new triplet of lines for 100, 010, and 001 must be chosen and the analysis repeated.

This outline of Ito's method may make it appear that it is a simple procedure to carry out manually. However, this is far from the actual case. In addition to the problem of the choice of the 100, 010, and 001 triplet, difficulties arise from weak or extinguished reflections not recorded, and from inaccuracies in the observed Q's. Indeed, we feel that workers in the future will almost universally want to avail themselves of automated computer methods for solving monoclinic and triclinic patterns, this being the only practical, economic, and efficient approach to the problem. Those readers who would like to study the arithmetical details of an application of Ito's method are directed to reference [50].

Earlier it was mentioned that quick tests should be made to determine whether a powder pattern is cubic, tetragonal, or hexagonal before proceeding to the Ito method. If such higher symmetry is indicated, a great amount of labor can be saved. If graphical methods are available in the laboratory, it is a simple method to plot the data in suitable form for examination on cubic, tetragonal, or hexagonal charts (Section 6-5.2). If charts are not available, it is still a simple matter to test the data analytically for cubic, tetragonal, and hexagonal symmetry. Testing for orthorhombic symmetry requires the use of the Lipson method, a more time-consuming procedure. Further details on testing for high symmetry are presented in reference [50].

The preliminary unit cell obtained from the Ito method and other general indexing procedures is often not the most desirable cell for describing the crystal concerned. It is merely a possible cell and does not necessarily exhibit the true symmetry of the lattice. In other words, the arbitrarily chosen triplet of Q's may lead to a cell that appears to be triclinic in symmetry, when in fact the crystal has higher symmetry. If the preliminary cell is unsuitable for describing and characterizing the crystal under study, it should be transformed into a more suitable one. A standard cell, known as the *reduced cell*, is available for this purpose. The reduced cell has an elegant mathematical crystallographic basis, and has been widely discussed in the crystallographic literature [51–56]. The *International Tables* [55] define the reduced cell as a primitive cell based on the three shortest lattice vectors, and for which α, β, and γ are all $\geqq 90°$ and the direction cosines of [111] are all positive. The reduced cell bears a definite geometrical relationship to the crystal symmetry.

Delaunay[57] first described a method for reducing an arbitrary primitive cell of a lattice to its reduced cell. Others[58–61] have discussed Delaunay's method, introduced new procedures, and, on occasion modified the definition of the reduced cell. The determination of reduced cells is complicated and often of limited interest, and the objectives of this book do not warrant the space required for its detailed discussion. Fortunately, Azároff and Buerger[59] have presented an excellent introduction to the whole subject.

6-6 AUTOMATED COMPUTING PROCEDURES FOR INDEXING POWDER PATTERNS

With the advent of electronic computers, it was inevitable that they would be applied to automated indexing. Modern computers are easy to program for the simple numerical operations to be performed in indexing powder patterns. At first programs were written for individual steps in the indexing procedure. The operator made appropriate decisions as required, and participated in and intervened in the process of computation as it proceeded. Later programs were coded to make many of the decisions by examining the next possible assignment. Many of the working programs have not been published. Periodically, however, the International Union of Crystallography publishes a world list of crystallographic computer programs[62]. A dozen or more powder indexing programs were listed in 1966, and copies of most of these programs were available. It is appropriate here to indicate the nature of a few indexing programs and the success to be expected in their use. The reader will appreciate that computer programs for use in x-ray diffraction are always in a state of flux, constantly being improved or changed to fit new models of computers. Accordingly, the worker wishing to automate his indexing of powder patterns must (1) survey the computer facilities available to him, (2) learn some of the rudiments of computer programming, and (3) familiarize himself with the nature of the indexing programs available. Usually, it is possible to obtain a working program and adapt it to one's situation with only minor programming problems.

6-6.1 Programs for Patterns of Orthorhombic and Higher Symmetry

A computer indexing program, INDEX, by Goebel and Wilson[65, 66] for patterns of orthorhombic and higher symmetry has proved satisfactory for various investigators. The program approaches the indexing on the basis of the highest lattice symmetry, cubic, and proceeds through tetragonal, hexagonal, and orthorhombic as required. It is written in FORTRAN and is automatic and continuous. A given set of data generally produce many indexings, so that the operator is faced with a final problem of choosing a valid indexing. The data, 2θ values, should be precision data good to about 0.0005 in $\sin^2 \theta$. Suspected extraneous reflections, such as those from impurities, should not be included in the 2θ list as they may prevent a valid indexing. The operator has many program command options, such as to index on a single system instead of proceeding successively from the cubic to the orthorhombic case. An error term E, the maximum deviation between observed and calculated $\sin^2 \theta$ values to be tolerated, may be chosen.

A brief outline of INDEX for the orthorhombic case is of interest. Starting with a list of $\sin^2 \theta$ values in ascending order, the first three reflections are assigned the indices 100, 010, and 001 respectively. These three values yield calculated provisional values of the lattice constants. These lattice constants are used to calculate indices for the fourth reflection. The chosen indices for the fourth reflection are those whose calculated $\sin^2 \theta$ value differs from the observed $\sin^2 \theta$ value by less than the chosen error term E, the accepted set of indices being that which gives the best $\sin^2 \theta$ agreement. Next a new set of provisional lattice constants is calculated by a least-squares fit of the four indexed reflections. Then the fifth reflection is indexed on the basis of the new provisional lattice constants, and a least-squares fit of the five indexed reflections provides another provisional set of lattice constants. By iteration of this procedure, all reflections are indexed. A least-squares fit of all the data is carried out, and the final lattice constants are used to reindex the entire list. This is necessary because some initially accepted indices may no longer be acceptable because of subsequent changes in the lattice constants. Reindexing is continued until lattice constant changes are less than 0.001 Å, or until five reindexings have been completed. Usually, no more than two reindexings are required. After an indexing is completed, the value of E is replaced by $E/2$ and the indexing procedure is repeated until E reaches its minimum value.

If the chosen triplet of reflections, 100, 010, and 001, is correct, indexing proceeds smoothly. If this first choice is incorrect, another triplet of indices must be chosen, and the indexing repeated. The INDEX program in the orthorhombic section provides a subroutine (named FETCH) which furnishes a list of reflection triplets of which no index is greater than two. Each triplet is presented as a 3×3 matrix. No permutations of indices that represent an interchange of the crystallographic axes are included. The list, however, contains matrices that represent dependent equations. Such matrices serve as a means of checking the dependency of the first three reflections. If dependency is found, other combinations of reflections 1, 2, 4; 1, 3, 4; and 2, 3, 4 are used in attempting to index the pattern. If necessary, the fifth line may be called into use, with all its combinations with the first four lines. Also, the program may double the original E in an attempt to obtain a successful indexing. If no indexing is achieved after all these attempts, it means that no independent set of three equations exists among the first five reflections. At this point the indexing attempt ceases. Typical indexing problems are reported to have required about 3 min on an IBM 7090 computer. Holland and Gawthorp[67] have modified the Goebel-Wilson program for use on the IBM 360 series of computers.

Tannenbaum, Lemke, and Kramer[63] briefly described a program for indexing tetragonal and hexagonal patterns. The program was based on the numerical method of Hesse[29] and was written for the IBM 709 computer. If the precision of the data is high, error in $\sin^2 \theta$ about ±0.0005, the program works well. Running time for tetragonal samples is reported to be about 75 sec, and for hexagonal samples about 2 min. Other programs written primarily for tetragonal, hexagonal, and orthorhombic patterns are those of Lefker[70], Hardcastle and Stock [71], and Hoff, Wallace, and Kitchingham[72].

6-6.2   Programs for Patterns of Low Symmetry

In 1957 and 1958, de Wolff[43, 44] reported a method for calculation of zone relations among powder diffraction lines and used a computer in the study. The general approach was that of Ito[40, 41], and the method enabled the Bravais lattice type to be determined rather automatically by arranging the Q's in a schematized reciprocal lattice. The presence of symmetry causes the Q's to form sets of equal values for equivalent reciprocal-lattice points. Details were further discussed in a survey[68], and Visser[69] finally described a fully automatic program based on de Wolff's method. The program consists of the following steps: (1) Zones are found and reduced. (2) Tests determine whether any or both of the base vectors should be halved. Parameters are refined by the least-squares method. The probability that the zone is found by pure chance is calculated. (3) Next, pairs of zones with a common reciprocal-lattice row line are found, and the angle between these zones determined. (4) The lattices found are reduced, and transformed if necessary. (5) Finally, an attempt is made to index the first 20 lines of the pattern. After a least-squares refinement of the pattern, the indexing is repeated. The number of lines actually indexed is noted, and the figure of merit calculated.

The figure of merit is a simplified criterion developed by de Wolff [75] for the reliability of a powder pattern indexing:

$$M_{20} = \frac{Q_{20}}{2\bar{\epsilon} N_{20}}. \tag{6-44}$$

Here M_{20} is the figure of merit based on $20\,Q$ values, Q_{20} being the Q value of the twentieth observed and indexed line. N_{20} is the number of different calculated Q values up to Q_{20}, and $\bar{\epsilon}$ is the average discrepancy in Q for these 20 lines. From confirmed indexings values of $M_{20} = 10$ to 60 appear to guarantee the essential correctness of the indexing, provided there are not more than two spurious lines (unindexed lines) below Q_{20}. Values of M_{20} below 6 must be regarded with suspicion, and values below 3 as of little significance. As many as five spurious or impurity lines apparently can be present on occasion without preventing a successful indexing and a corresponding moderately high M_{20} value.

Visser's program (written in ALGOL 60), when used on a Telefunken TR4 computer, is reported usually to require less than 3 min of machine time for an indexing.

The program of Haendler and Cooney[64], while requiring operator participation, has been the basis of several successful automatic indexing programs. It was prepared as a series of programs for the IBM 1620 computer on the basis of Ito's method[40, 41]. Calculations are based on the reciprocal cell expression 6-37 for Q_{hkl} for a triclinic lattice. Two lines are designated, $Q_{100} = a^{*2}$ and $Q_{010} = b^{*2}$, as corresponding to two vectors of the unit cell. The computer searches for two other lines evenly spaced about the sum $Q_{100} + Q_{010}$, subject to certain assigned arbitrary limits defining "evenly spaced." If such a pair of lines is found, the larger is assigned indices 110 and the smaller $1\bar{1}0$, and the reciprocal-cell angle γ^* can be determined by

$$\cos \gamma^* = \frac{Q_{110} - Q_{1\bar{1}0}}{4a^*b^*}.$$

Another Q is now designated Q_{001}, and paired first with Q_{100} and then with Q_{010} to find pairs of lines evenly spaced about the sums $Q_{100} + Q_{001}$ and $Q_{010} + Q_{001}$. Success in these searches leads to calculation of reciprocal angles β^* and α^*, respectively. The set of nine Q values thus obtained determines a reciprocal unit cell. A large number of such sets can usually be formed from the Q's of a powder pattern. To determine which set defines a correct cell, the operator forms all possible internally consistent sets of nine indexed Q values leading to a reciprocal cell, and judiciously selects a few for testing. Finally, those reciprocal cells not eliminated by testing procedure are, one by one, entered into the triclinic indexing program. Haendler and Cooney include programs for obtaining the reduced cell. Indexing time might require 1 day with IBM-1620 machine time of 3 hr for a triclinic cell.

A program, INDX, by Roof[73] has been found to handle readily the triclinic patterns and patterns of higher symmetry as well. INDX, a modification of the Haendler and Cooney[64] program, is composed of seven sequential programs so arranged that a special path is automatically taken through them in a particular indexing application. At the start the program reads and numerically identifies M observed Q values. From every possible sum of pairs of N different Q values, the program subtracts each of the M observed Q's. The differences resulting from the subtractions from a given Q sum are searched for pairs of values of the same magnitude (within a preset toleration limit) but opposite sign. Values of the two Q's in the sum and the two Q's forming the evenly spaced pair about the sum are stored. The program next searches these stored sets of four numbers and forms all possible internally consistent sets of nine numbers. "Internally consistent" here refers to the fact that the two Q's forming a sum may be possible lattice axes, and that the two Q's forming an evenly spaced pair about the sum may determine the angle between the possible lattice axes. Nine suitable Q values define a unit cell. Such sets of nine Q values are stored for future examination.

The program continues by computing the acute reciprocal cell represented by each set of nine Q values. After storing the result the cell angles of each acute cell are changed from all acute to all obtuse. Any case in which the sum of the three cell angles exceeds 357° is discarded; otherwise the obtuse cell data are stored. Thus, at this point, a list of possible or trial cells is stored in the computer. These trial cells are then tested by attempting to calculate Q values that match the observed Q's (within plus or minus a prescribed error term). No output occurs unless a certain desired percentage (e.g., 90 per cent) of the observed and calculated Q's match. If some lesser percentage of match (e.g., 80 per cent) is achieved, the program will refine the reciprocal cell constants by least-squares procedures and recalculate the Q's. If the desired 90 per cent match is not obtained after a preset number of refinement cycles, the next unit cell is processed. In the above calculation the Q values are not processed by the classical expression, equation 6-37, but by the equivalent expression involving the sum of the squares of the components of a reciprocal vector in orthogonal space

$$Q = (\Delta V1)^2 + (\Delta V2)^2 + (\Delta V3)^2,$$

where

$$\Delta V1 = ha^* + kb^* \cos \gamma^* + lc^* \cos \beta^*$$
$$\Delta V2 = kb^* \sin \gamma^* - lc^* \sin \beta^* \cos \alpha$$
$$\Delta V3 = lc^* \sin \alpha \sin \beta^*. \tag{6-45}$$

Another portion of the program transforms the trial reciprocal unit cells into the corresponding real space unit cells. The last portion of the program finds the reduced cell of a given lattice. This part also forms the Niggli matrix[52, 56] from the reduced cell constants. Tables of Niggli matrices[56, 74] may then be consulted to ascertain whether the cell whose Niggli matrix has been generated may be transformed to a cell of higher symmetry.

Roof[73] provides a simple example of the use of INDX in indexing the powder pattern of compound κ ($PuZr_{\sim 3}$) in the plutonium–zirconium alloy system. No assumptions are made as to symmetry. The 17 smallest Q values are chosen, and 5 Q values are used in sums to generate evenly spaced pairs of Q's. Thus $M = 17$ and $N = 5$ in the program. These data lead to 30 evenly spaced pairs of Q values, which in turn generate 257 sets of 9 internally consistent Q's and 429 possible cells to be tested. Of these, 54 cells satisfactorily index 90 per cent or more of the 17 listed Q values. However, only 15 of these cells are unique. Next, the 15 unique cells are tested against all 27 observed Q values in the pattern. Only 2 of these cells prove able to index all 27 Q's. One cell is hexagonal, with $a = 5.05$, $c = 3.12$ Å, and the other is orthorhombic, with $a = 4.37$, $b = 2.52$, and $c = 3.12$ Å. Either is a satisfactory solution, since an orthorhombic cell is always an alternative representation of a hexagonal cell.

INDX is written in FORTRAN IV and programmed for the CDC 6600 computer. As an idea of the computing time required, note that the

CDC 6600 computer generates and tests cells at the rate of approximately 40 cells/sec. The input cards allow the operator considerable latitude in describing the conditions for an indexing run. In another example described by Roof[73], 8282 possible cells were tested in ~196 sec, but no promising unit cell was found among them, although several triclinic cells passed the testing. In such a situation there is still the possibility that the smallest Q's are not first-order reflections. To investigate this possibility they are assumed to be second-order lines, and the implied first-order lines are calculated. In the example mentioned five such first-order lines arise from the assumption. When these lines are added to the observed Q's and the new list is passed through the program, 91 cells are obtained that index all 29 of the 29 Q values used. Moreover, the smallest cell of highest symmetry is an orthorhombic cell. The 63 different cells in the list of 91 are then tested against the 45 Q's making up the entire pattern. The small orthorhombic cell is still the smallest and most symmetric cell that indexes all 45 lines. Its dimensions are $a = 5.43$, $b = 14.84$, $c = 10.86$ Å, and its space group was determined to be *Cmca*.

These two examples illustrate how readily Roof's program handles an indexing problem without any assumptions as to the pattern's symmetry. The symmetry is revealed in the course of passing the data through the program. If the true cell is triclinic (or monoclinic), this should be revealed by the only reduced cell satisfying the data being triclinic (or monoclinic) and not transformable into a cell with higher symmetry.

In automated practice the indexing of monoclinic patterns is seemingly the least reliable, apparently because of the systematic extinctions so prevalent in the monoclinic case[105, 106]. The strategy of Smith and Kahara[105, 106] in attacking this problem is to locate the 020 reflection, which occurs far more frequently in monoclinic patterns than the 010 reflection. The relations

$$2Q_{020} + Q_{h10} = Q_{h30} \qquad (6\text{-}46)$$

and

$$3Q_{020} + Q_{h20} = Q_{h40} \qquad (6\text{-}47)$$

serve as 020 detectors. These relations are equally true if l is the running index instead of h. Their program rapidly tries low 2θ reflections as the 020 reflection in combination with all lines as $h10$ or $h20$ (or $01l$ and $02l$) reflections, and looks for coincidences between the generated values, $2Qj + Qi$ and $3Qj + Qi$, and the experimental Q's. A coincidence in the table of $2Qj + Qi$ data means that potentially 020, $h10$, and $h30$ have been found (or their corresponding reflections in the $0kl$ zone). This coincidence may be solved to yield a value of b^* and a^*, in which case a possible $hk0$ zone has been located. All coincidences are investigated, and all possible zones $hk0$ and $0kl$ are formed.

The next step is to try to determine the monoclinic angle β^* for a given combination of $hk0$ and $0kl$. The program tries various simple $h0l$ and hkl values for those reflections not already indexed as belonging to the orthogonal zones. Some zone combinations do not yield a β^* value and are rejected. Combinations that yield a β^* value are further used to calculate a unit-cell volume. The various unit-cell volumes are ordered according to magnitude, and indexing of the entire pattern is attempted starting with the smallest cell. The program finally lists as output the least-squares refined cell constants and indices of (up to) the 10 smallest cells.

Smith and Kahara have tested the program on more than 70 monoclinic patterns, and very quickly and automatically achieved success in approximately 60 per cent of the patterns. Hopefully, with improvements in the program, the percentage of successful indexing will ultimately go up to 80 to 90 per cent. The program is very fast, requiring less than 1 min of CDC-6600 time per pattern. It works best with highly accurate d data, such as from Guinier and other focusing cameras. The program tolerates a small number of impurity lines.

The reader can now appreciate that the field of automated indexing is in a state of development which appears very promising and hopeful. It is not possible at this stage to label particular programs as *best*. We have simply chosen to give these brief descriptions as illustrative of several serious efforts in computer indexing. Readers are invited also to consider the following additional references: Werner[107], Taupin[108], Cole and Villiger[109], and McLachlan and Chen[110].

6-7 THE MEASUREMENT OF INTENSITIES FROM PHOTOGRAPHIC BLACKENING

6-7.1 Introduction

Before studying the following, the reader should review Section 2-3.7. The density of photographic blackening depends not only upon the exposure, $E = It$, but also upon several other factors such as the quality of the radiation employed, the nature of the photographic emulsion, and the conditions under which the film is processed. Therefore, in order to compare intensities accurately, it is necessary first of all that all these other factors, as well as the time t, be kept constant. In most powder diffraction experiments t is the same for all the lines or spots on any given film, so that relative intensities can be determined once the relationship between exposure and photographic density has been found. The usual diffraction experiments do not require a knowledge of the absolute intensity, which is to say, the ratio between the energy in the diffracted ray and that in the primary x-ray beam.

All visual and photometric methods of translating film densities into intensities are based upon the use of a reference pattern consisting of a graded series of spots corresponding to known exposures. Because, as explained in Chapter 2, the reciprocity law holds with considerable accuracy in the x-ray region, such a reference scale can be prepared by employing a beam of constant intensity and recording the spots for a known series of exposure times. Assuming the reciprocity law to hold, the exposures of the various spots are the same as if the exposure time had been kept constant and the beam intensity had been varied by the same increments.

6-7.2 Preparation of a Graded Intensity Scale

A simple way to prepare such an intensity scale is shown in Fig. 6-16. The film is wrapped in a light-tight paper or aluminum-foil envelope E which slides freely in the metal jacket M. This jacket protects all the film from the x-ray beam save that portion exposed by the aperture A which can be made circular, oval, or rectangular in shape according to the shape of the spots desired. It is important that the entire area of the aperture be irradiated with x-rays of uniform intensity, accomplished by employing a rather divergent beam and placing the film holder as far from the x-ray source as necessary. Thus in the figure the beam is

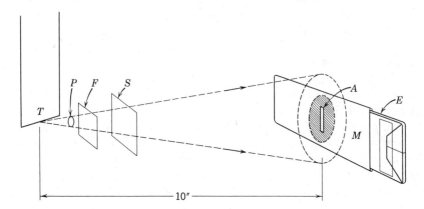

Fig. 6-16.   Simple scheme for preparing a graded intensity scale.

represented as being limited only by P, the port of a commercial x-ray tube housing. Under such working conditions precautions must be exercised to protect both the operator and the projecting end of the film packet E from direct and scattered x-rays by means of suitable absorbing shields. The central shaded area of the beam in the figure represents the zone of uniform intensity in which the aperture A is to be centered. The indicated source-to-film distance of 10 in. is only suggestive and may be varied to suit the particular circumstances. F is a metal filter for reducing the intensity of the x-ray beam sufficiently so that the exposures may be of practicable lengths. With a copper anode operating at 35 kV peak and 20 mA, a suitable thickness for a nickel filter is of the order of 0.025 in., the precise value varying with the other experimental conditions. S represents a simple, manually operated lead shutter for regulating the x-ray exposures. Some preliminary experiments must be made in order to regulate the conditions so that a barely discernible spot is produced with the shortest convenient exposure, 2 sec, for instance. The adjustment of the experimental conditions is less critical and can be performed in less time, if a packet of several films is employed. For example, if four films are used, each with a transmission factor of $\frac{1}{3}$, the intensity incident upon the last film is $(\frac{1}{3})^3 = \frac{1}{27}$ of that incident upon the first. This allows considerable latitude in the experimental conditions within which a satisfactory range of densities is likely to be obtained on one or another of the films in the packet.

Using a copper-target tube operating at 35 kV peak and 20 mA, a source-to-film distance of 10 in., and a nickel filter 0.023 in. thick, we have obtained satisfactory intensity scales with the following exposures: 2, 4, 8, 12, 18, 24, 32, 44, 60, 90, 120, 160, 220, and 300 sec. The x-ray voltage and current should be kept as constant as possible during the course of the exposures. The film used should be fresh and of the same brand and type as that used in the experimental work to which the scale is to be applied. Furthermore, a standard procedure should be used in processing the intensity-scale film and all the films to which this intensity scale is to be applied. In particular, it should be remembered that the developing time has a pronounced effect upon both the density of the pattern and the gamma (γ) of the response curve (see Fig. 2-24). Increasing the developing time increases both the density and the gamma within certain limits, with the result that a change takes place in the ratio of intensities corresponding to a given difference in density. Consequently, the use of a scale processed with one developing time will lead to errors if applied to a film developed for a different period of time.

6-7.3 Visual Estimation of Intensities

In this method of estimating intensities, the spot scale is superposed upon the pattern, and the density of a given reflection is visually matched with that of a spot on the graded scale. Care must be taken that the spot concerned is placed close to the reflection concerned and where the background is apparently equal to that beneath the reflection itself. When applied to patterns with low backgrounds and reflections of moderate density, this method is capable of an accuracy of 15 to 20 per cent. It becomes inaccurate when applied to reflections on a dense background or to excessively dense reflections on any kind of background. In general, good precision is not possible when the absolute photographic density is much greater than unity.

There is considerable advantage in making the size and shape of the spots on the scale about the same as those of the reflections to be measured. Thus narrow rectangular spots or lines are best for powder patterns, whereas circular or elliptical spots are more suitable for oscillation or Weissenberg photographs. A refinement of this principle consists in using an actual reflection from a powder or single crystal as the basis for an intensity scale. The scale should be prepared in the same camera and with the same type of specimen as those used for the diffraction patterns themselves, resulting in a close resemblance between the spots of the scale and the reflections of the pattern, which makes the matching process easier and more accurate. In preparing such an intensity scale, it is necessary to screen the film properly so as to eliminate all but the desired reflection and avoid the building up of undesirable background. Means must also be provided for moving the film between the various graded exposures.

A serious failing of photographic methods of intensity measurement is the limited range of intensities obtainable from any one film. Whereas intensities differing by a ratio of 100 to 1 and more are frequently encountered in powder diffraction work, accurate photographic measurements cannot be made on a single film of intensities differing by a ratio of more than about 20 to 1. In single-crystal methods, intensity differences of 1000 to 1 are not uncommon. The effective range of photographic measurements can be greatly extended by means of the multiple-film technique[76, 77]. Radiations of longer wavelengths (copper, iron, cobalt, and chromium) are very appreciably absorbed by a single thickness of x-ray film, so that a series of patterns of graded intensity can be prepared by simultaneously exposing two to six films. More penetrating radiations such as $MoK\alpha$ and $AgK\alpha$ can be absorbed sufficiently by interleaving the layers of film with absorbing foils. For example, 0.0025 in. of copper reduces the intensity of Mo$K\alpha$ radiation by a factor of 3.

The absorption factor per layer of film must be determined experimentally for x-rays of a given wavelength by comparing the apparent intensities of 12 to 15 reflections as registered on adjacent films of the multiple-film pack. Only reflections of intermediate intensity should be included in the determination of the film factor (the ratio of incident intensity to transmitted intensity), because very faint and very intense reflections cannot be measured with sufficient accuracy. If a microphotometer or densitometer is to be employed in the contemplated intensity measurements, it should likewise be used in determining the film factor. For present-day double-coated Kodak No Screen x-ray film, the transmission factor (reciprocal of the film factor) for crystal-reflected Cu$K\alpha$ radiation is approximately 0.26. If the films of the multiple pack are separated by paper spacers, the transmission factor is decreased further. In applying the multiple-film technique, we must recognize an important source of error which comes into play when the diffracted x-rays impinge upon the film at angles differing much from 90°. Because of the increased absorbing path length, the film factor is increased for such rays, and this should be allowed for by an appropriate geometrical correction. Debye-Scherrer patterns are not subject to errors of this kind. Variations of the multiple film technique have been reported by Iball[78] and by Davis and Armstrong[79]. By using films of different speeds in the multiple pack it is possible to obtain greater attenuation with fewer films. For instance, Iball used an Ilford G film backed by an an Ilford B film and achieved a density ratio of 8:1 with Cu$K\alpha$ radiation, whereas many commonly used x-ray films require a pack of three or four films to give the same ratio. In photographing elements that fluoresce strongly[79], patterns of superior quality may be obtained by a proper choice and order of films of different speeds used in a pack.

In conclusion, it must be emphasized that visual or photometric matching of the densities of reflections with the densities of an intensity scale tends to give a measure of the peak rather than the integrated intensity. When integrated intensities are sought, this procedure gives fairly reliable results only when the reflections are all of the same size and shape. When the area of a spot or the width of a line varies from one reflection to another for either geometrical or structural reasons, it becomes very difficult to estimate integrated intensities by this method.

6-7.4 Photometer Techniques

The peak and integrated intensities of powder reflections can be measured more accurately from photographs by employing a micro-photometer. Such an instrument measures the fraction T of the incident light transmitted through a very limited region of an absorbing medium. In the application to x-ray films, the absorbing medium is the silver deposited in the emulsion of the photographic film. The optical system of a very simple microphotometer is shown in Fig. 6-17. Light from the single-wire filament W of the illuminating lamp is collimated by the lens L_1 and brought to a focus on the film F by means of the condensing lens L_2. The rays transmitted by the film diverge and are recollimated by the lens L_3 and focused by L_4 in the plane of the slit S. This slit excludes extraneous light and transmits the main beam, which diverges and is received on the active surface of a photocell P. The photoelectric current is measured by means of a sensitive galvanometer, the throw of the gal-vanometer being taken as proportional to the transmitted intensity.

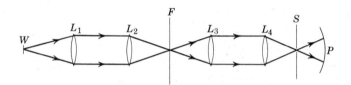

Fig. 6-17.   Optical system of a simple microphotometer.

Mechanical means must be provided for translating the film so that any desired portion of the pattern is brought into the path of the light beam. This is preferably accomplished by means of a worm gear which pro-vides a known distance of translation per revolution. Jay has described a simple photometer of this type suitable for examining x-ray films [80].

Photometers are subject to several sources of error. Indeed, Wooster [81], after a theoretical examination of the limitations of microdensito-metry, concluded that an overall accuracy of better than 5 per cent is difficult to attain in such measurements. There may be fluctuations or gradual changes in the intensity of the light source resulting from variations in the line voltage, blackening of the lamp, and so on. Modern photometers incorporate features designed to overcome these diffi-culties. In one type of instrument the light transmitted through the film is balanced by means of a bridge circuit against a known light intensity, which is regulated by means of a calibrated optical wedge. The intensity of the x-ray reflection is then read directly from the position of the calibrated wedge. Errors due to variations in the intensity of the light source can be reduced by stabilizing the line voltage, or they can be eliminated by using the same lamp to illuminate both the sample and the optical wedge. Thus the light from one lamp can be divided into two beams for this purpose with the aid of a beam splitter. In one design differences in the sensitivity of the two photocells, another source of error, are avoided by employing only one photocell and alternately sending the reference and "unknown" beams into it by means of a movable mirror. Present-day commercial microphotometers* incorporate additional features promoting ease of operation and interpretation of the results. The most satisfactory instruments for measuring x-ray films incorporate a means for scanning the pattern at a constant speed, while at the same time the photoelectric current is amplified and registered with a strip-chart recorder. The literature contains descriptions of several microphotometers suitable for scanning powder patterns [80, 82–91].

Where the volume of powder diffraction patterns to be measured warrants the elaborate equipment and expense, it is possible to automate the operation rather completely. Frevel[92] has described a system that converts digitized film data from a precision microphotometer, with the aid of a computer, into a printout of interplanar spacings and corresponding peak intensities. Inspection of the trace of the automatic scan allows the diffractionist to detect overlapping peaks and to reject any spurious peaks due to scratches or blemishes on the film. Overlapping reflections may be resolved, and a manually punched card inserted in the computer input deck. Frevel's previous technique was to measure the film twice on a precision comparator and then estimate intensities with a comparison scale. This combined procedure took three to four times as long as the automated procedure. Reproducibility and precision are reported to have improved with the automation. Segmüller and Cole[93] have described an elaborate group of procedures to run an automated microdensitometer on a shared-time computer system.

Powder lines are more easily photometered than single-crystal re-

*Among the commercial microphotometers available are: STRIPSCAN, a single-scan digital microdensitometer, Optronics International, Inc., 18 Adams Street, Burlington, Massachusetts. High-precision recording microphotometer, Model 24-450, Jarrell-Ash Division, Fischer Scientific Company, 590 Lincoln Street, Waltham, Massachusetts. Recording photometer with KOMPENSOGRAPH, Siemens Aktiengesellschaft, Schöneberger Str. 2-4, Berlin, Germany. In the United States, from Siemens Corporation, 186 Wood Avenue South, Iselin, New Jersey. Recording microphotometer, Joyce, Loebl and Co., Ltd., Princesway, Team Valley, Gateshead, County Durham, England. In the United States, from Joyce Loebl and Company, Inc., 111 Terrace Hall Avenue, Burlington, Massachusetts.

flections, but for optimum accuracy certain conditions must be fulfilled. First, the comparative graininess of x-ray films requires that the light beam illuminate an area of emulsion large enough to curtail variations in output resulting from the absorption of the individual grains[94]. For the lines of an average powder diagram, a recatangular illuminated area 0.2×2.0 mm in size is satisfactory. The long axis of course is aligned with the lines of the pattern. If the lines are very sharp, the width may need to be reduced to 0.1 mm in order not to sacrifice resolution; however, it may be possible to compensate for the lost width by increasing the length of the illuminated zone to 4 or 5 mm, provided the lines to be measured are very nearly straight. For very diffuse patterns, such as those produced by amorphous and liquid substances, the width may be made as large as 0.5 mm without impairing the resolution.

Microphotometers designed for spectrographic measurements in which the optical system projects a sharp image of the illuminating filament on the film can still be satisfactorily applied to powder patterns by throwing the filament image somewhat out of focus in order to illuminate more grains. Spectrographic densitometers in which the illuminated area is defined by a slit can usually be adapted to the measurement of diffraction patterns by increasing the slit dimensions. If this boosts the transmitted light energy beyond the capacity of the receiving components, it can be reduced by means of a neutral density filter of the proper density. Powder diffraction patterns are generally more dense than spectrographic patterns, and therefore most commercial microphotometers and densitometers are designed for relatively high transparencies. This factor must be taken into account in building an instrument for measuring x-ray films, or in modifying most commercial units. During measurement x-ray films must be held accurately in the focal plane of a focusing-type microphotometer by clamping them sandwich style between two plane glass plates. Any large deviations from planarity can lead to significant errors in the measured intensities.

Because of the dependence of photographic density upon the film-processing conditions, it is necessary for highly accurate work that the intensity scale of graded spots be printed directly on the film on which the pattern is recorded. This is most conveniently done with the aid of a rotating stepped sector[95–97], as shown in Fig. 6-18A. Such a sector can be cut from a $\frac{1}{4}$-in. brass disk in such a way that the successive steps produce exposures of the film that bear a known relationship to each other. For example, the steps of the sector may provide exposures differing by factors of 2 (1, 2, 4, 8, 16, 32, etc.) or $\sqrt{2}$ (1.00, 1.41, 2.00, 2.83, 4.00, 5.66, 8.00, etc.). When employing the stepped-sector technique, the diffraction pattern is first prepared in the usual way, except

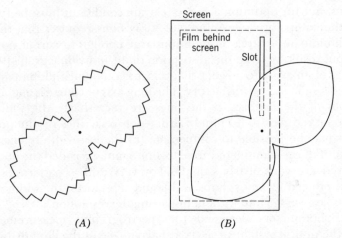

Fig. 6-18. Arrangements for printing intensity calibration strips on x-ray photographs. (*A*) Sector wheel to give a stepped wedge. (*B*) Cam to give a continuous wedge. (Courtesy of C. Bunn, *Chemical Crystallography*, 2nd ed., and Clarendon Press, Oxford.)

that a small region of the film is protected from radiation with a lead strip. After the diffraction exposure the film is protected from light with black paper and placed in position for the stepped-sector exposure. All the film except the reserved strip is shielded from the x-rays by means of a slotted brass screen, and the stepped intensity wedge is then produced on this reserved zone by exposing it to a uniform x-ray beam while the sector is rotated in front of the slot. A uniform x-ray beam of low intensity can be obtained by placing the sector and film at a considerable distance from the source and employing a filter of the proper thickness, as explained in Section 6-7.2 for the preparation of a graded-intensity spot scale. After the film has been processed, both the diffraction pattern and the stepped wedge are microphotometered, and from the recorder trace or plot of the latter a curve is prepared by means of which the intensities of the diffraction pattern can be read off.

It should be pointed out that the characteristic curve of photographic film for x-rays is largely independent of wavelength[98], so that it is permissible to prepare an intensity scale with an x-ray beam of different quality from that used in the experimental work. Nevertheless, any uncertainty in this regard can be removed by employing the same source of x-rays for both purposes, and it is recommended that this be done whenever feasible. In any event it is not necessary to use strictly monochromatic x-rays for preparing intensity scales[95, 98]. The reader is referred to Fig. 12-6 for microphotometer records of an intensity scale and two powder patterns of a carbon-black specimen. Figure 6-19*A*

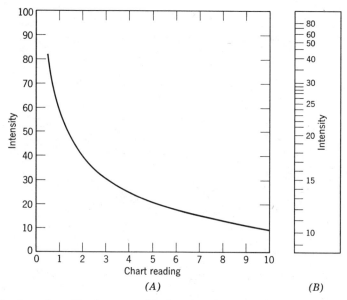

Fig. 6-19.   Intensity calibration curve for the microphotometer record of a carbon black pattern (refer to $CuK\alpha$ curve of Fig. 12-6). (B) Scale for reading intensities from the microphotometer chart.

shows the corresponding intensity calibration curve for $CuK\alpha$ radiation, from which the scale of Fig. 6-19B was constructed for directly reading the intensities from the microphotometer pattern.

It is important to allow for film background correctly. The proper method is first to convert all the photographic densities into relative intensities with the aid of the intensity scale, and then subtract the background intensity from the combined line-plus-background intensity. In the early days of quantitative spectrographic analysis, the background was sometimes allowed for by subtracting the background *density* from the line-plus-background *density*. In spectrography most measurements are made in the region of linearity between D and $\log E$ (see left curve in Fig. 2-24), from which it is seen that subtracting the background density D_B from the line-plus-background density D_{L+B} is equivalent to dividing E_{L+B} by E_B, an obviously incorrect procedure for deducting the background intensity from the overall intensity[99]. In contrast to these results, most x-ray diffraction measurements are made in the relatively low density region, and Van Horn[98] has confirmed the findings of earlier investigators by demonstrating that for a large number of coatings of various x-ray film types a linear relationship exists between D and E. By proper choice of the processing conditions, it is

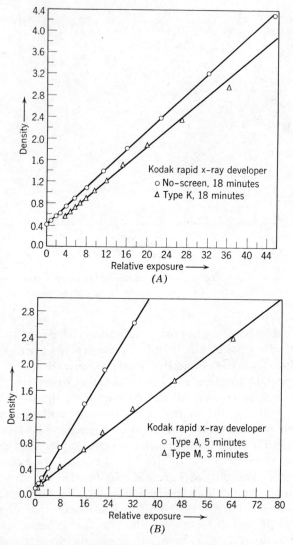

Fig. 6-20. Plots illustrating the extended linear ranges of four film types. In (A) the lower curve is displaced four units to the right for clarity. [Courtesy of M. Van Horn, *Rev. Sci. Instrum.*, **22**, 809 (1951).]

possible to preserve this linearity to within 3 per cent up to densities at least as high as 2.5 for any of the standard x-ray films (see Fig. 6-20). This range encompasses all densities small enough to be useful in the accurate measurement of diffraction intensities. The dependence of D upon E in this range may be written

$$D = KE + D_0. \qquad (6-48)$$

where D_0 is the density produced by development of an unexposed portion of the film. D_0 is the ordinate intercept of the D versus E (exposure) plots of Fig. 6-20 as well as the intercept on the density axis of the toe portion of the D versus log E curve of Fig. 2-24 marked *fog*. This residual film density is due to all causes of photographic blackening other than the x-ray exposure. The transmission of this unexposed portion of the film may likewise be designated T_0 to distinguish it from a true optical transmission of unity, or 100 per cent.

In most photometers the deflection of the galvanometer or the pen of the potentiometer recorder is a linear function of the film transmission. Figure 6-21A represents a potentiometer chart with the full-scale reading of 100 set equal to an optical transmission of 1.0, or an optical density of 0. The points T_0, T_B, and T_{L+B} represent, respectively, the optical transmissions of an unexposed portion of the film, background on the exposed portion, and a line. For sake of illustration these have been assigned representative numerical values of 0.70, 0.40, and 0.10, respectively.

Figure 6-21B shows the chart with full-scale reading set equal to T_0 (microphotometer "zeroed" in the unexposed portion of the film). This adjustment is recommended for most work, because no errors result if the film response deviates from linearity between D and E. The carbon-black patterns of Fig. 12-6 were recorded with the chart zeroed close to

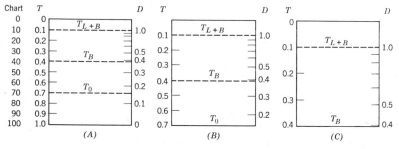

Fig. 6-21.   Three methods of adjusting the full-scale reading of the potentiometer recorder in microphotometering x-ray films; 100 chart units set equal to transmissions of (*A*) unity, (*B*) unexposed portion of film, (*C*) film background.

the T_0 transmission level. Comparison of A and B of Fig. 6-21 shows that this adjustment of the chart response has the advantage of providing greater contrast between the transmission of the background and that of the line plus background. With the full-scale reading of the chart adjusted to a transmission of T_0, equation 2-23 gives for the *apparent* density when the microphotometer is set on the line,

$$D_{L+B} = \log_{10} \frac{T_0}{T_{L+B}},$$

and, when it is set on the background,

$$D_B = \log_{10} \frac{T_0}{T_B}$$

Assuming the existence of linearity between D and E, as is at least approximately true in most x-ray patterns, and referring to equation 6-48, we find the net line intensity to be given by

$$
\begin{aligned}
I_L &= I_{L+B} - I_B \\
&= k(E_{L+B} - E_B) \\
&= \frac{k}{K}(D_{L+B} - D_0 - D_B + D_0) \\
&= \frac{k}{K}\left(\log_{10}\frac{T_0}{T_{L+B}} - \log_{10}\frac{T_0}{T_B}\right) \\
&= \frac{k}{K}\log_{10}\frac{T_B}{T_{L+B}}.
\end{aligned}
\tag{6-49}
$$

Continuing to grant the existence of linearity between D and E, we can achieve the same result more simply by zeroing the microphotometer in the background (Fig. 6-21C). In this case the *apparent* density when the densitometer is set on the line is

$$D_{L+B} = \log_{10} \frac{T_B}{T_{L+B}}. \tag{6-50}$$

Then, as in the previous paragraph,

$$I_L = \frac{k}{K}(D_{L+B} - D_0 - D_B + D_0). \tag{6-51}$$

But the *apparent* density of the background is

$$D_B = \log_{10}\frac{T_B}{T_B} = \log_{10} 1 = 0,$$

so that substitution of equation 6-50 in equation 6-51 results in

$$I = \frac{k}{K} \log_{10} \frac{T_B}{T_{L+B}},$$

which is identical with equation 6-49.

It must be strongly emphasized that the simplification in procedure of zeroing the microphotometer in the background must not be employed when there is any doubt that the film response between D and E is linear. It can easily be shown that in the region of linearity between D and $\log E$ the practice of zeroing in the background is equivalent to subtracting densities rather than intensities, thus giving rise to significant errors[99]. The procedure to be generally recommended, therefore, is that of zeroing to the density level of an unexposed portion of the film.

With all possible care applied to the measurement of line intensities, the intensities so obtained will be the correct intensities only if proper attention has been given to the powder sample and to preparation of its diffraction pattern. The diffracted intensity is dependent upon particle diameter, linear absorption coefficient, degree of compaction of the powder, crystal size, primary extinction coefficient, and grain size for polyphased particles. Wilchinsky[100] has treated the effects of crystal, grain, and particle size on intensity. In general, the integrated intensities are smaller for coarser powders than for fine powders or a solid block, an effect qualitatively explained by the entrapment of the x-rays in the crevices between the particles. Järvinen and Merisalo[101], in attempting to avoid the problems of sample porosity and inhomogeneity, compressed the samples heavily, which introduced excessive preferred orientation. Corrections had to be worked out for the orientation effects to deduce the true intensities. The use of intensifying screens can introduce errors in the intensities and may even lead to faulty line breadths[102]. A Fluorazure screen has been observed to reduce the breadth of lines in an iron pattern to less than half the true value, a calcium tungstate screen, however, reduces breadths by only about 3 per cent. The use of black paper to protect films during diffraction exposures is a possible source of error[103]. Some papers are not completely light-proof, but contain a large number of minute pinholes. Also, the mass per unit area of the protective paper is not constant, and its absorption of the scattered and diffracted radiation varies over its surface. The effects of these irregularities in the paper may be superposed on the diffraction pattern or the background at the point of measurement. Other abnormalities in diffraction patterns are discussed in Section 6-10.

6-8 THE MEASUREMENT OF INTENSITIES WITH THE X-RAY DIFFRACTOMETER

This subject has been treated rather fully in Chapter 5, where it was pointed out that direct measurements of x-ray intensity with the aid of a counter diffractometer are inherently more accurate than estimates of intensity from photographic blackening. An additional advantage of diffractometer techniques, which was not discussed in Chapter 5, is concerned with the absorption correction.

In Chapter 3 a description was presented of the various factors that affect the intensities of diffracted x-rays for the several methods of crystal analysis. For the Debye-Scherrer method these include (1) the Lorentz factor, (2) the polarization factor, (3) the temperature factor, (4) the structure factor (including the atomic scattering factor), (5) the multiplicity factor, and (6) the absorption factor. Factors 2, 3, and 4 are independent of the diffraction technique employed, and the forms of factors 1, 2, and 4 are rather accurately known. The numerical value of 5 for the powder method is a unique and known function of the crystal symmetry and the indices of the reflection. The absorption factor $A(\theta)$ for the Debye-Scherrer technique is, of all the factors, the most difficult to appraise accurately. In general, it is a rather complicated function of the diffraction angle θ and of the dimensions and absorption co-efficient of the specimen (see Section 3-3.4G). In this connection the peculiar advantage of the diffractometer technique when applied to powders is not only that the absorption factor can be expressed simply, but also that it is independent of θ.

This is readily shown as follows. Consider a flat powder specimen the thickness of which is great enough to give maximum diffracted intensity. A criterion for this condition is that

$$\mu t \gtrsim 3.2 \frac{\rho}{\rho'} \sin \theta$$

(see Section 5-2.1D). In this equation t is the sample thickness in centi-meters, μ and ρ are, respectively, the linear absorption coefficient and the density of the solid material composing the powder, and ρ' is the density of the powder including interstices. The linear absorption coefficient of the powder including interstices is

$$\mu' = \frac{\rho'}{\rho} \mu.$$

As shown in Fig. 6-22, if A is the cross-sectional area of the beam at its point of intersection with the sample, the diffracting volume of a layer

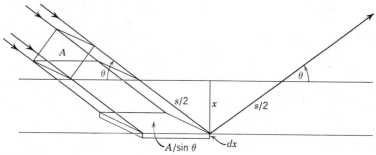

Fig. 6-22.

of powder of thickness dx at depth x is

$$dV = \frac{A}{\sin \theta} \, dx,$$

or, since $x = \frac{1}{2}s \sin \theta$,

$$dV = \tfrac{1}{2}A \, ds.$$

If we let I'_0 be the intensity diffracted at the angle θ by unit volume of powder under hypothetical conditions of nonabsorption, the intensity diffracted by the element dV is $\tfrac{1}{2}I'_0 A \, ds$ and, when the absorbing path length s is taken into account, it is

$$dI = \tfrac{1}{2}I'_0 A e^{-\mu's} \, ds. \tag{6-52}$$

The total intensity diffracted at the angle θ by the entire sample is then

$$I = \int_{s=0}^{s=\infty} dI = \frac{I'_0 A}{2\mu'}.$$

In terms of the solid material composing the powder, both I'_0 and μ' are decreased proportionally with respect to I_0 and μ, so that

$$I = \frac{I_0 A}{2\mu}. \tag{6-53}$$

In equation 6-53, I_0 has the same meaning as I'_0 except that I_0 refers to the solid material only. It is then seen that, with respect to absorption, I is inversely proportional to μ and independent of angle, so that $A(\theta)$ of equation 3-77 takes the form

$$A(\theta) = \frac{K}{\mu}. \tag{6-54}$$

I_0 of expression 6-53 is a complicated function involving the energy in the incident beam and the several factors other than absorption that affect the intensity of the diffracted beam.

Braybrook[104] has warned of the possible effect of the filter absorption edge in integrated intensity measurements. The absorption edge of the filter results in a sudden diminution of the background intensity on the low-angle side of a line. If its position is sufficiently close to the line, the background count on this side of the line may be considerably less than the true background. Care is urged in selecting the position of the background count when a $K\beta$ filter is used.

6-9 PUTTING INTENSITIES ON AN ABSOLUTE SCALE

Equation 3-77 can be used to compute relative numerical values of the structure factors F_{hkl} from the observed intensities of powder reflections. For many structural studies and other investigations, this is all that is required, but for more precise studies it is necessary to relate the relative F's to the amplitude scattered by a single electron (see Section 3-3.3). Fine powders are ideal subjects for accurate intensity measurements because of the virtual absence of both primary and secondary extinction in most cases (see Section 3-3.2). If a diffractometer is employed, the additional advantage can be gained of practically eliminating uncertainties in the absorption factor $A(\theta)$.

Reliable relative intensities of this sort can be placed on an absolute scale by comparing them with the intensity of a reflection from a standard powder, the F of which is accurately known[111]. A suitable standard substance is finely powdered sodium chloride[112]. Let the subscripts x and s denote, respectively, the "unknown" and the standard substances. Then equation 3-77 can be applied to any particular reflections of both materials as follows:

$$F_x{}^2 = \frac{2}{j_x} \cdot \left(\frac{2 \sin^2 \theta \cos \theta}{1+\cos^2 2\theta}\right)_x \cdot \frac{1}{A(\theta)_x} \cdot I_x. \tag{6-55}$$

$$F_s{}^2 = \frac{2}{j_s} \cdot \left(\frac{2 \sin^2 \theta \cos \theta}{1+\cos^2 2\theta}\right)_s \cdot \frac{1}{A(\theta)_s} \cdot I_s. \tag{6-56}$$

Division of equation 6-55 by equation 6-56 gives

$$F_x{}^2 = F_s{}^2 \cdot \frac{j_s}{j_x} \cdot \frac{\left(\dfrac{\sin^2 \theta \cos \theta}{1+\cos^2 2\theta}\right)_x}{\left(\dfrac{\sin^2 \theta \cos \theta}{1+\cos^2 2\theta}\right)_s} \cdot \frac{A(\theta)_s}{A(\theta)_x} \cdot \frac{I_x}{I_s}. \tag{6-57}$$

Errors due to uncertainty in $A(\theta)_s$ and $A(\theta)_x$ can be minimized when the photographic method is used, by mixing the two powders and using two reflections at very nearly the same angle. If the powders are photographed separately, their absorption factors must be very precisely known. When the x-ray powder diffractometer is used to measure the two powders separately,

$$\frac{A(\theta)_s}{A(\theta)_x} = \frac{\mu_x}{\mu_s},$$

provided the sample thickness is great enough. If μ_x and μ_s are not known with the desired precision, the method of mixtures may be used, in which event μ is the same for reflections of both substances and $\mu_x/\mu_s = 1$. Brindley and Spiers have critically examined the sources of error in these two techniques for placing the intensities of powder reflections on an absolute basis[113]. The principal difficulty in the method of mixtures, which is preferred over the method of substitution, is the need for an extremely small mean particle size ($< 0.1\ \mu$ for most materials). From accurate experimental measurements of I_x and I_s and a theoretical knowledge of F_s and the other quantities contained in the right-hand side of equation 6-57, F_x can be calculated in electron scattering units. It was noted in Chapter 3 that Robertson and other workers have made use of a similar method for putting the intensities of single-crystal reflections on an absolute basis (see Section 3-3.3).

6-10 SPECIAL SCATTERING AND DIFFRACTION EFFECTS

6-10.1 Background Effects

Several factors contribute to scattering and diffraction at angles other than those of the discrete Bragg reflections produced by the characteristic radiation employed. The more important effects include (1) lattice imperfections of various kinds, (2) diffraction of the general radiation by the crystal planes, together with discontinuities introduced into this pattern by the K absorption edges of the elements composing the specimen and the silver and bromine present in the photographic emulsion, (3) x-ray scatter by the air in the camera, (4) secondary fluorescence radiation from the specimen, and (5) Compton (incoherent) scatter. Effects, 2, 4, and 5 have already been discussed in a fundamental way in Sections 2-3.3 and 2-3.4. The reader would do well to review this material before proceeding with the following brief treatment of some of the practical aspects of background scatter.

A. Background Due to Lattice Imperfections. In its broadest sense this topic includes all departures of the actual scattering centers in a structure from the points associated with the ideal lattice. Hence it includes the diffuse scattering due to thermal motions of the atoms, as discussed in Sections 2-3.4 and 3-3.4C. This effect tends to become more pronounced as the melting point of a crystalline solid is approached. Randomness of stacking of the layers in a layer structure or imperfections in the mutual orientations of the layers lead both to broadening of certain reflections and to increased background scatter, especially at low angles. Substances exhibiting various degress of such randomness are montmorillonite and other layer silicates, graphite, and cobalt (see Section 6-10.2D)[114, 115]. Randomness in the population of atomic sites in three-dimensional structures also leads to distortion and broadening of certain powder reflections, as well as to increased background[114].

This topic if broadly defined also encompasses the increase in background and decrease in Bragg-reflection intensity arising from the presence of uncrystallized or poorly crystallized material in an otherwise crystalline specimen. The x-ray scattering by such disordered material resembles the scattering by a liquid or a gas, so that the angular intensity distribution is similar to that of the squared atomic scattering factor as a first approximation. Figure 6-23 shows two patterns of carbon, A being ungraphitized and containing a large proportion of disordered material, and B being partially graphitized and containing relatively little disordered material. Interlayer randomness of orientation may also contribute to the higher background of A at low angles.

Cold working of metals introduces strains into the lattice, displacing the atoms from their ideal sites. This kind of lattice imperfection also

Fig. 6-23.   Debye-Scherrer patterns of two carbon specimens. (A) contains a larger proportion of disordered carbon than (B), which produces less background scatter at low angles.

produces line broadening, and the scattering level between the peaks may be raised appreciably. Under these circumstances it becomes difficult to determine the true background level and resolve from it that portion of the scattering associated with the Bragg reflections[116].

B. Background Due to General Radiation. In powder photographs the various wavelengths of the general radiation are diffracted by the different crystallographic planes over a continuous angular range, with the region of maximum intensity at a smaller angle than the angles of the stronger characteristic Bragg reflections. This is one of the principal reasons for the strong general blackening observed in most powder patterns about the beam-exit position. This blackening is stronger when unfiltered radiation is used, because β filters absorb a considerable fraction of the general radiation. This background can be almost completely eliminated if desired by making use of a crystal monochromator. In other than strictly monochromatic diffraction work, the maximum intensity of the characteristic radiation with respect to the general radiation can be achieved by adjusting the operating potential to a point between 3.5 and 5.0 times the critical excitation potential of the target material. For further details and recommended operating potentials, see Section 2-3.1B.

C. Absorption Discontinuities. Because of the overlapping of the continuous-radiation spectra diffracted by the various sets of crystallographic planes, these discontinuities are ordinarily inconspicuous in powder photographs even though they produce irregularities in the otherwise smooth contour of the general-radiation background[117]. They are most noticeable in the diffraction patterns of the elements and when unifiltered radiation is used. Metals and alloys are most likely to reveal these effects.

Discontinuities due to absorption by elements in the sample or to the silver and bromine in the photographic emulsion can be recognized by their appearance, coupled with their angular position. K or L absorption discontinuities due to the sample resemble two-dimensional lattice reflections in shape, terminating sharply on the low-angle side but falling off gradually in intensity on the high-angle side. K absorption edges due to the silver and bromine in the film emulsion are of similar shape but are reversed in position, the sharp edge being on the high-angle side. Both kinds of absorption edges are observed at angles smaller than those of the stronger monochromatic reflections in the pattern. The angular positions of such discontinuities can be calculated from a knowledge of the d spacings of the stronger reflections produced by the characteristic radiation. These same sets of planes also diffract the general radiation with greatest intensity. For example, suppose that the first strong characteristic line corresponds to $d_1 = 2.425$ Å. With the silver and bromine K absorption edges occurring at 0.486 and 0.920 Å, respectively, the Bragg law can be used to predict the smallest angles at which these edges should be observed on the film:

$$(\sin \theta)_{Ag} = \frac{\lambda_{Ag}}{2d_1} = \frac{0.486}{2 \times 2.425} = 0.100.$$

$$(\sin \theta)_{Br} = \frac{\lambda_{Br}}{2d_1} = \frac{0.920}{2 \times 2.425} = 0.190.$$

Should there be one or more additional strong characteristic lines at d spacings smaller than 2.425 Å and well separated from it, these sets of planes may also generate visible discontinuities on the film at larger angles. The positions of discontinuities due to absorption by elements in the sample can be predicted in the same way. In any case of course it is necessary for the observation of these effects that the general radiation emitted by the x-ray tube contain the wavelength in question in sufficient strength.

Figure 6-24A shows a number of absorption edges in the pattern of molybdenum produced by the general radiation of a copper-target tube. The intense K edge of molybdenum is diffracted visibly by at least three

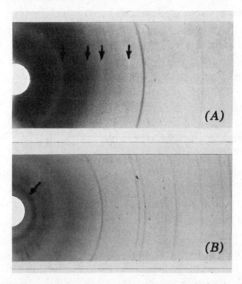

Fig. 6-24. (A) Absorption discontinuities observed with a molybdenum specimen and a copper-anode tube operated at 35 kV. (B) Combination of silver and antimony absorption edges to simulate a diffraction line. Antimony specimen, Cu$K\alpha$ radiation.

sets of planes, (110), (211), and (310). Figure 6-24B shows a striking effect resembling a low-angle diffraction line which was actually produced by two overlapping absorption edges, the K edge of silver in the emulsion at 0.486 Å and that of antimony (the sample) at 0.407 Å.

D. Air Scatter. A portion of the energy in the direct beam is dissipated by coherent and incoherent scattering from the gas molecules com-composing the air. The extent of this air scatter increases with increasing wavelength, so that, for example, it is much more noticeable for chromium than for molybdenum radiation. Because of the almost complete absence of order in a gaseous scattering medium, the angular distribution of this scattered energy is continuous and resembles in a general way the scattering factor (f_0) curve for independent atoms, being largest at small angles (see Fig. 3-15). Background of this kind can be kept to a minimum in ordinary Debye-Scherrer cameras by designing the collimator and exit tubes as described in Chapter 4 so as to permit the smallest possible portion of the air-scattered rays from reaching the film. When exceptionally low background is mandatory at small Bragg angles, as, for example, when faint reflections of large d spacing are to be observed, it is often worthwhile to evacuate the camera or fill it with a gas of low scattering power, such as hydrogen or helium[118].

E. Secondary Fluorescence Radiation. Background due to this cause is uniformly distributed over the entire angular range. As explained in Section 2-3.4, it can be avoided or kept small in most cases by selecting a target material the atomic number of which is *not* 2, 3, or 4 larger than that of important elements in the specimen. The effect is most severe when $\Delta Z = Z(\text{target}) - Z(\text{element in specimen}) = 2$ or 3. It is generally not very troublesome when $\Delta Z = 4$ or 5, and it is scarcely detectable for larger differences. Because the atomic numbers of some of the more useful target materials, chromium, iron, cobalt, and copper are the same as or close to the atomic numbers of many of the most prevalent metallic elements encountered in x-ray diffraction samples, it follows that the possibility of fluorescence difficulties when employing these radiations must be taken into account continually. On the contrary, short wavelengths such as Ag$K\alpha$ or Mo$K\alpha$ provoke little trouble of this sort, although an interesting exception is the excitation of the lead L_l spectrum (absorption edge at 0.78 Å) by Mo$K\alpha$ radiation (0.71 Å). The most commonly encountered fluorescence difficulty, that of FeK fluorescence when Cu$K\alpha$ radiation ($\Delta Z = 3$) is used, poses a special problem in the widespread applications of diffraction in iron and steel analysis. Co$K\alpha$ ($\Delta Z = 1$) and Fe$K\alpha(\Delta Z = 0)$ are most generally used in this field. For alloy steels containing considerable amounts of chromium, manganese, and nickel it is becoming a common practice to employ $CrK\alpha$ radiation for most studies.

The yield coefficient of K fluorescence increases with increasing atomic number[119] so that considerable background may be generated in certain cases of this kind wherein the wavelength of the exciting radiation is considerably smaller than that of the absorption edge concerned. Examples of this kind are the heavy backgrounds observed on patterns of compounds containing bromine and rubidium, respective absorption edges at 0.920 and 0.815 Å, no matter what the radiation employed. The exciting radiation in these cases is the general radiation itself. The peak intensity falls between 0.45 and 0.55 Å for applied voltages of 30 to 50 kV, and there is considerable energy at wavelengths up to 0.80 Å in this voltage range.

A generally helpful device for reducing fluorescence background is to interpose aluminum foil or, even better, a foil of the β-filter element, between the specimen and the film. The improvement is especially good when $\Delta Z = 5$ or more and the fluorescence is of moderate intensity. The fluorescence background is reduced then because the aluminum or other filter always absorbs more strongly the wavelengths longer than the wavelength of the characteristic radiation employed, which of course includes the fluorescence radiation.

6-10.2 Reflections of Unusual Character

A. Spotty Lines. As explained in Section 4-1.3, discontinuous or spotty powder lines are an indication of large crystallite size. Patterns B and C in Fig. 4-18 illustrate this effect. Rotated samples begin to display spottiness when the mean crystallite dimension exceeds about $45\,\mu$, and stationary samples show this effect for dimensions as low as $5\,\mu$, in some cases. When < 325-mesh powder produces a very spotty pattern, it is an almost certain sign that the specimen did not rotate during the exposure. The degree of spottiness in a stationary powder pattern can be used as the basis for a scheme of classifying powders according to size in the subsieve and lower-siéve range[120].

B. Arclike Lines. If some or all of the lines are not of uniform intensity but resemble short arcs rather than uniform segments of circles, it is an indication of preferred orientation. This phenomenon is most frequently observed for crystallites of a platelike or flaky habit, and when the powder is mounted on the outside of a cylindrical supporting fiber, causing the flakes to adhere with their flat faces parallel to the surface of the fiber.

C. Broadened Lines. Broadened lines may arise from very small crystallite size, or lattice distortions, or both. If the lines are very broad and diffuse, the first explanation is certainly the major one. In the absence of lattice strains, well-crystallized material should give reasonably sharp lines at all angles if the crystallite dimension is larger than 0.1 μ. As the dimension falls below this level, the sharpness of the back-reflection lines is first affected, and the $\alpha_1\alpha_2$ doublet eventually ceases to be resolved. At sizes much less than 0.01 μ, the back-reflection lines disappear entirely, and the low-angle lines become very wide and more diffuse. Theoretical calculations of the diffraction effects to be expected from extremely minute crystallites have been calculated for the face-centered cubic case by Germer and White[121] and for the body-centered cubic case by Morozumi and Ritter[122]. For instance, crystallites less than about 20 Å in diameter (containing less than about 60 unit cells) give very broadened lines and lack some of the lines that characterize the normal powder pattern of the material. Indeed, not until the crystallites are 50 Å in diameter or greater (> 1000 unit cells) is the pattern characteristic in appearance for the material, although the lines are still strongly broadened. Line broadening of this kind can be used as a rather accurate criterion of crystallite size in the colloidal range. For a detailed treatment of this subject the reader is referred to Chapter 9. Figure 6-25A shows a pattern displaying pronounced line broadening of this kind.

If some lines are noticeably wider than others at approximately the same Bragg angle, it may indicate either that the crystallites have a characteristic shape involving some long and some short dimensions, or that certain kinds of lattice imperfections are present. Figure 6-25B illustrates the effect of lattice imperfections. The specimen is cobalt, and the pattern, instead of displaying only lines of the α (hexagonal) form or those of the β (cubic) form, shows all the lines of both structures. However, the lines common to both structures are sharp, whereas the noncommon lines are diffuse and, in some cases, abnormally weak. These effects can be explained by a structure in which layers stacked in hexagonal close packing alternate in a random manner with layers stacked in cubic lose packing. Lele and co-workers[123] have presented discussions of the types of stacking faults and the theoretical interpretation of their diffraction effects.

D. Two-Dimensional Lattice Lines. Lines of this kind are visible in the pattern of anodic nickel hydroxide shown in Fig. 6-25C. They are characterized by terminating sharply on the low-angle sides but falling off gradually in intensity on the high-angle sides. Such reflections are pro-

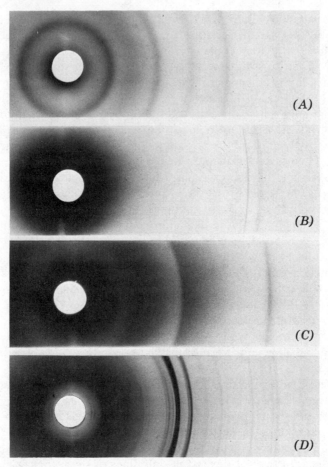

Fig. 6-25. Patterns illustrating various special diffraction effects. (A) Lines broadened due to small crystallite size (α-Al$_2$O$_3$·H$_2$O). (B) Sharp and broad lines due to lattice imperfection (cobalt). (C) Pattern containing two-dimensional lattice lines (anodic nickel hydroxide). (D) Pattern showing lines doubled due to absorption by the specimen support (stearic acid, Cr$K\alpha$ radiation).

duced by two-dimensional lattices (random layer lattices). Other materials that give reflections of this kind are carbon blacks and certain clays, including in particular montmorillonite. The crystallites of these substances develop rather perfectly in two dimensions, but growth in the third dimension is either absent or extremely small. These sheetlike crystals of negligible thickness tend to accumulate in parallel groups in which the adjacent sheets have no fixed orientation with respect to reach other except that they are mutually parallel. They thus resemble a stack of cards thrown haphazardly on a table. Warren[124] and Wilson[125] have theoretically investigated the problem of diffraction by a random layer lattice. Ergun[126] has discussed the abnormal profiles of $hk0$ reflections in powder patterns of layered structures, and Mitra[127] has demonstrated that the photographic powder technique can be applied to the study of diffraction line profiles, although diffractometric techniques are usually preferred (Section 5-2).

E. Splitting of Lines. This occurs as shown in Fig. 6-26 when the sample powder or its supporting fiber is highly absorbing to x-rays. The incident rays are completely absorbed by the thick central portion of the sample, whereas smaller and larger amounts of radiation, respectively, are transmitted by the edge regions a and b where the absorbing paths are shorter. The result is a line split into a weaker low-angle component a' and a stronger high-angle component b'. Such lines are seldom observed for penetrating radiations such as $AgK\alpha$ and $MoK\alpha$, but are commonly observed for soft radiations with specimens of moderate to high absorbing power. Figure 6-25D shows doubled lines of this kind obtained with $CrK\alpha$ radiation and crystalline organic powder mounted on the outside of a Pyrex capillary, which strongly absorbs the soft radiation. It is evidently important that this kind of geometrical doubling of diffraction lines be positively recognized and distinguished from

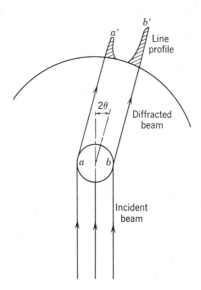

Fig. 6-26.   Doubling of a Debye-Scherrer line due to total absorption by the center of the specimen.

true doubling due to structural features of the sample material. Spurious doublets and line displacement can arise from the Eberhard effect[128]. This phenomenon, when present, results from diffusion problems in the thick emulsion of the x-ray film during the development process.

6-10.3 Spurious Lines

A. Lines Due to Misalignment of Camera Elements. The most common cause of trouble of this type is sufficient misalignment of the collimator and beam-exit tubes to allow the direct beam to strike some portion of the tip of the exit tube. The result is the generation of a somewhat diffuse diffraction pattern of the metal composing the exit tube, usually brass. Since the brass "sample" is not at the center of the camera, the lines are observed to be shifted from the theoretical positions by varying amounts.

B. Diffraction Effects from the Sample Mount. Any material bathed by the x-ray beam other than the pure specimen itself contributes scattering of some kind to the diffraction pattern. Glass or silica fibers and thin-walled tubes used to support the sample powder generate a broad halo centered at about 4.2 Å, in addition to weaker diffuse scatter at other angles. Collodion and Parlodion tubes produce a principal broad halo at 4.35 Å, in addition to fainter diffuse effects. Cellulose acetate produces similar amorphous halos. Ethyl cellulose and polyethylene are unsatisfactory materials for supporting powder samples because they are crystalline enough to produce rather sharp diffraction patterns. Regardless of the type of sample tube employed, a good practice is to keep the wall thickness as small as possible in order to minimize the intensity of the x-ray scatter, and the same rule applies to the thickness of supporting fibers. Similarly, amorphous materials like Canada balsam and gum tragacanth, used as binders for molded specimens, should be used in minimum proportions. All amorphous substances used in sample-mounting techniques are objectionable on at least one score, namely, their contribution to the blackening of the film at low Bragg angles. As noted above, this region is already likely to be darkened excessively as a result of general radiation diffraction and air scatter, making it difficult to detect weak lines.

A final word of warning should be given in regard to lubricating materials such as graphite or sodium stearate used to facilitate the removal of plastic sample tubes from the metal wires on which they are formed. A portion of the lubricant usually adheres to the inside wall of the tube and may contribute to the diffraction pattern if present in considerable amount. Graphite powder is a prime offender in this respect, its strongest reflection at 3.35 Å being the first to appear.

C. Diffraction from Radiation Contaminants. Ideally only pure mono-chromatic radiation should be employed in powder diffraction experiments. The most commonly encountered additional radiation is the $K\beta$. Its removal, when desired, by means of filters or monochromatizing crystals has already been thoroughly discussed. Sometimes a target surface prepared by electrodeposition of the desired element upon a copper or iron anode will be sufficiently thin or porous to emit weakly the characteristic radiation of the anode substrate.

Table 6-15.   Tungsten L Spectrum Lines to Be Considered in Con-taminated X-ray Sources[a]

Tungsten L Radiation	Intensity	Wavelength (Å)	$\lambda(\mathrm{Cu}K\alpha)/\lambda(\mathrm{W})$
α_2	w	1.48742	1.0366
α_1	vs	1.47635	1.0443
β_1	s	1.28176	1.2029
β_2	m	1.24458	1.2388
β_3	w	1.26247	1.2213
γ_1	w	1.09851	1.4035

[a] $d_{\mathrm{W}\rightarrow\mathrm{Cu}} = [1.5418/\lambda(\mathrm{W})] \times d.$

After several hundred hours of use, a hot-cathode tube may begin to give evidences of contamination by tungsten. As explained in Section 2-2.2B, volatilization of tungsten from the filament eventually leads to some contamination of the target unless the tube is especially designed to prevent this from occurring. In general, serious radiation contamina-tion does not develop until a tube has seen at least 1000 hr of service. At the usual operating potentials of 30 to 50 kV, the WK spectrum is not aroused, but instead the interfering radiations are those of the L spec-trum. Table 6-15 lists the tungsten L lines of sufficient intensity to be detectable in a considerably contaminated source, together with the proper factors to give the apparent d spacings when all lines are assumed to be due to Cu$K\alpha$ radiation. From the relative intensities it is seen that the α_1 and β_1 wavelengths are the first to produce detectable lines when the radiation is unfiltered. When a nickel filter is used, as with a copper target tube, the nickel K absorption edge at 1.488 Å greatly weakens the tungsten L lines. The β filters for the other longer wavelength radiations (iron, cobalt, and chromium) tend to reduce the intensity of all the tungsten L lines more or less.

GENERAL REFERENCES

*1. L. V. Azároff and M. J. Buerger, *The Powder Method in X-ray Crystallography*, McGraw-Hill, New York, 1958, Chapters 7–12, 16.

2. R. W. M. D'Eye and E. Wait, *X-ray Powder Photography in Inorganic Chemistry*, Academic Press, New York; Butterworths, London, 1960, Chapter 4.

*3. H. Lipson and H. Steeple, *Interpretation of X-ray Powder Diffraction Patterns*, Macmillan, London, 1970.

4. L. I. Mirkin, *Handbook of X-ray Analysis of Polycrystalline Materials*, Fizmatgiz, Moscow, 1961; English translation by J. E. S. Bradley, Consultants Bureau, New York, 1964.

5. C. S. Barrett and T. B. Massalski, *Structure of Metals*, McGraw-Hill, New York, 1966, Chapter 7.

6. J. Brentano, "The Determination of the Atomic Scattering Power for X-rays from Powders of Gold, Silver, and Aluminium for CuKα Radiation," *Phil. Mag.*, (7) **6**, 178 (1928). An early but excellent paper on powder line intensity measurements.

7. H. Lipson and A. J. C. Wilson, "The Derivation of Lattice Spacings From Debye-Scherrer Photographs," *J. Sci. Instrum.*, **18**, 144 (1941).

8. J. M. Robertson, "Technique of Intensity Measurements in X-Ray Crystal Analysis by Photographic Methods," *J. Sci. Instrum.*, **20**, 175 (1943).

9. M. Straumanis, "The Construction of Reciprocal Lattice Nets from Debye-Scherrer Films," *Z. Kristallogr.*, **104**, 167 (1942).

SPECIFIC REFERENCES

[1] M. Straumanis and A. Ievinš, *Die Präzisionsbestimmung von Gitterkonstanten nach der asymmetrischen Methode*, Springer, Berlin, 1940; reprinted by Edwards Brothers, Ann Arbor, Michigan, 1948, pp. 27–30; J. E. Sears and A. Turner, *J. Sci. Instrum.*, **18**, 17 (1941); M. E. Straumanis, *J. Appl. Phys.*, **20**, 726 (1949).

[2] H. P. Klug, *Ind. Eng. Chem., Anal. Ed.*, **12**, 753 (1940).

[3] W. E. Armstrong and R. J. Davis, *J. Sci. Instrum.*, **35**, 59 (1958).

[4] M. J. Columbe, *Rev. Sci. Instrum.*, **30**, 181 (1959).

[5] W. G. Perdok and G. Boom, *J. Sci. Instrum.*, **37**, 134 (1960).

[6] E. Aruja, *J. Sci. Instrum.*, **40**, 263 (1963).

[7] A. J. Bradley, H. Lipson, and N. J. Petch, *J. Sci. Instrum.*, **18**, 216 (1941).

[8a] H. Lipson and A. J. C. Wilson, *J. Sci. Instrum.*, **18**, 144 (1941).

[8b] A. J. C. Wilson, *Rev. Sci. Instrum.*, **20**, 831 (1949).

[9] A. J. Bradley and A. H. Jay, *Proc. Phys. Soc.* (London), **44**, 563 (1932).

[10] M. J. Buerger, *Am. Mineral.*, **21**, 11 (1936).

[11] *International Tables for X-ray Crystallography*, Vol. II, Kynoch Press, Birmingham, England, 1959, p. 202; or *Internationale Tabellen zur Bestimmung von Kristallstrukturen*, Vol. II, Gebrüder Borntrager, Berlin, 1935, pp. 588–608; *Tables for Conversion of X-ray Diffraction Angles to Interplanar Spacing*, National Bureau of Standards Applied Mathematics Series, No. 10, September 20, 1950, U.S. Government Printing Office, Washington, D.C.

[12] H. H. Claassen and K. E. Beu, *Rev. Sci. Instrum.*, **17**, 307 (1946).

[13] M. Straumanis, *Z. Kristallogr.*, **104**, 167 (1942); *Am. Mineral.*, **37**, 48 (1952).

[14] H. P. Klug and L. E. Alexander, *X-ray Diffraction Procedures*, 1st ed., Wiley, New York, 1954, pp. 340, 341.

[15] A. W. Hull and W. P. Davey, *Phys. Rev.*, **17**, 549 (1921); W. P. Davey, *Gen. Elec. Rev.*, **25**, 564 (1922).

[16] W. P. Davey, *A Study of Crystal Structure and Its Applications*, McGraw-Hill, New York, 1934, pp. 596–613.

[17] J. O. Wilhelm, *Trans. Roy. Soc. Can.*, **21**, 1 (1927).

[18] T. Bjurström, *Z. Phys.*, **69**, 346 (1931).

[19] C. W. Bunn, *Chemical Crystallography*, 2nd ed., Oxford University Press, London 1961, pp. 142–144, Appendix 3.

[20] E. A. Owen and G. D. Preston, *Proc. Phys. Soc.* (London), **35**, 101 (1923).

[21] L. V. Azároff and M. J. Buerger, *The Powder Method in X-ray Crystallography*, McGraw-Hill, New York, 1958, Chapter 7.

[22] F. Ebert, *Z. Kristallogr.*, **111**, 15 (1958).

[23] P. P. Williams, *Acta Crystallogr.*, **12**, 250 (1959).

[24] M. Cernohorsky, *Acta Crystallogr.*, **14**, 1081 (1961).

[25] R. Ferro, *Acta Crystallogr.*, **14**, 1093 (1961).

[26] N. C. Schieltz, *Advances in X-ray Analysis*, Vol. 7, Plenum Press, New York, 1963, p. 94.

[27] R. P. Williams, *Z. Kristallogr.*, **123**, 342 (1966).

[28] A. J. Bradley and A. Taylor, *Phil. Mag.*, (7)**23**, 1049 (1937).

[29] R. Hesse, *Acta Crystallogr.*, **1**, 200 (1948).

[30] H. Lipson, *Acta Crystallogr.*, **2**, 43 (1949).

[31] A. J. Stosick, *Acta Crystallogr.*, **2**, 271 (1949).

[32] C. W. Jacob and B. E. Warren, *J. Am. Chem. Soc.*, **59**, 2588 (1937).

[33] F. D. Bloss, *Am. Mineral.*, **54**, 924 (1969).

[34] L. Zsoldos, *Acta Crystallogr.*, **16**, 572 (1963).

[35] I. A. Barabash and G. V. Davydov, *Acta Crystallogr.*, **23**, 6 (1967); **A24**, 608 (1968).

[36] V. Neskuchaev, *Zh. tekh. Fiz.*, **1**, 105 (1931).

[37] C. Runge, *Phys. Z.*, **18**, 509 (1917).

[38] V. Vand, *Acta Crystallogr.*, **1**, 109 (1948).

[39] V. Vand, *Acta Crystallogr.*, **1**, 290 (1948).

[40] T. Ito, *Nature*, **164**, 755 (1949).

[41] T. Ito, *X-ray Studies on Polymorphism*, Maruzen, Tokyo, 1950.

[42] L. Zsoldos, *Acta Crystallogr.*, **11**, 835 (1958).

[43] P. M. de Wolff, *Acta Crystallogr.*, **10**, 590 (1957).

[44] P. M. de Wolff, *Acta Crystallogr.*, **11**, 664 (1958).

[45] I. P. Vyrodov, *Soviet Phys. — Crystallogr.*, **7**, 386 (1962).

[46] I. P. Vyrodov, *Soviet Phys. — Crystallogr.*, **9**, 316 (1964).

[47] K. Viswanathan, *Am. Mineral.*, **53**, 2047 (1968).

[48] V. Vand and G. G. Johnon, Jr., *Acta Crystallogr.*, **A24**, 543 (1968).

[49] J. W. Visser, *J. Appl. Crystallogr.*, **2**, 142 (1969).

[50] See reference 21, pp. 111–123.

[51] A. Johnsen, *Kristallstruktur, Fortschr. Mineral.*, **5**, 36–40 (1916).

[52] P. Niggli, *Kristallographische und strukturtheoretische Grundbegriffe*, Handbuch der Experimentalphysik, Vol. 7, Teil 1, Akademische Verlagsgesellschaft, Leipzig, 1928, pp. 108–176.

[53] V. Bolashov, *Acta Crystallogr.*, **9**, 319 (1956).

[54] M. J. Buerger, *Z. Kristallogr.*, **109**, 42 (1957).

[55] *International Tables for X-ray Crystallography*, Vol. I, Kynoch Press, Birmingham, England, 1952, p. 530.

[56] R. B. Roof, Jr., *A Theoretical Extension of the Reduced-Cell Concept in Crystallography*, Los Alamos Scientific Laboratory Report, LA-4038 (1969). Available from Clearinghouse for Federal Scientific and Technical Information, National Bureau of Standards, U.S. Department of Commerce, Springfield, Virginia. Printed Copy $3.00; microfiche $0.65.

[57] B. Delaunay, Z. *Kristallogr.*, **84**, 109 (1933).

[58] A. L. Patterson and W. E. Love, *Acta Crystallogr.*, **10**, 111 (1957).

[59] See reference 21, Chapters 11 and 12.

[60] Y. Takéuchi, Z. *Kristallogr.*, **127**, 276 (1968).

[61] A. Santoro and A. D. Mighell, *Acta Crystallogr.*, **A26**, 124 (1970).

[62] D. P. Shoemaker (ed.), *World List of Crystallographic Computer Programs*, 2nd ed., International Union of Crystallography, 1966. Available from A. Oosthoek's Uitgevers Mij. N.V., Doomstraat 11–13, Utrecht, The Netherlands, or from Polycrystal Book Service, P.O. Box 11567, Pittsburgh, Pennsylvania. Price 10 Netherlands Guilders, U.S. $3.00, U.K. £1.)

[63] I. R. Tannenbaum, B. J. Lemke, and D. Kramer, *Acta Crystallogr.*, **14**, 1287 (1961).

[64] H. M. Haendler and William A. Cooney, *Acta Crystallogr.*, **16**, 1243 (1963).

[65] J. B. Goebel and A. S. Wilson, *INDEX*, U.S. Atomic Energy Commission, Research and Development Report, BNWL-22, 1965.

[66] J. B. Goebel and A. S. Wilson, *J. Met.*, p. 10, May (1967).

[67] H. J. Holland and J. A. Gawthrop, *J. Appl. Crystallogr.*, **2**, 81 (1969).

[68] P. M. de Wolff, *Advances in X-ray Analysis*, Vol. 6, Plenum Press, New York, 1963, p.1.

[69] J. W. Visser, *J. Appl. Crystallogr.*, **2**, 89 (1969).

[70] R. Lefker, *Anal. Chem.*, **36**, 332 (1964).

[71] K. I. Hardcastle and A. D. Stock, *J. Phys.*, **E1**, 951 (1968).

[72] W. D. Hoff, W. Wallace, and W. J. Kitchingham, *J. Sci. Instrum.*, **42**, 171 (1965); W. D. Hoff and W. J. Kitchingham, *J. Sci. Instrum.*, **43**, 952 (1966).

[73] R. B. Roof, Jr., *INDX*, Los Alamos Scientific Laboratory Report, LA-3920, 1968. Available from Clearinghouse for Federal Scientific and Technical Information, National Bureau of Standards, U.S. Department of Commerce, Springfield, Virginia. Printed Copy $3.00; microfiche $0.65.

[74] See reference 21, Chapter 11.

[75] P. M. de Wolff, *J. Appl. Crystallogr.*, **1**, 108 (1968).

[76] J. J. de Lange, J. M. Robertson, and I. Woodward, *Proc. Roy. Soc.* (London), **171A**, 398 (1939).

[77] J. M. Robertson, *J. Sci. Instrum.*, **43**, 175 (1943).

[78] J. Iball, *J. Sci. Instrum.*, **31**, 71 (1954).

[79] R. J. Davis and W. E. Armstrong, *J. Sci. Instrum.*, **31**, 305 (1954).

[80] A. H. Jay, *J. Sci. Instrum.*, **18**, 128 (1941).

[81] W. A. Wooster, *Acta Crystallogr.*, **17**, 878 (1964).

[82] G. M. B. Dobson, *Proc. Roy. Soc.* (London), **104A**, 248 (1923).

[83] H. R. Ronnebeck, *J. Sci. Instrum.*, **20**, 154 (1943).

[84] J. W. Ballard, H. I. Oshry, and H. H. Schrenk, *U.S. Bur. Mines, Rep. Invest.* 3638 (April 1942).

[85] A. Taylor, *An Introduction to X-ray Metallography*, Chapman and Hall, London, 1945, pp. 93–96.

[86] P. Rosenblum and A. deBretteville, Jr., *Rev. Sci. Instrum.*, **20**, 321 (1949).

[87]   A. Taylor, *J. Sci. Instrum.*, **28**, 200 (1951).

[88]   E. Alexander and B. S. Fraenkel, *Rev. Sci. Instrum.*, **26**, 895 (1955).

[89]   J. R. Brown, H. K. Moneypenny, and R. J. Wakelin, *J. Sci. Instrum.*, **32**, 55 (1955).

[90]   W. A. Wooster, *J. Sci. Instrum.*, **32**, 457 (1955).

[91]   A. Franks, D. C. Barnes, and W. Wilson, *J. Sci. Instrum.*, **41**, 685 (1964).

[92]   L. K. Frevel, *Anal. Chem.*, **38**, 1914 (1966).

[93]   A. Segmüller and H. Cole, *Advances in X-ray Analysis*, Vol. 14, Plenum Press, New York, 1971, p. 338.

[94]   K. E. Beu, *Rev. Sci. Instrum.*, **24**, 103 (1953).

[95]   A. Bouwers, Z. *Physik*, **14**, 374 (1923); *Physica*, **3**, 313 (1923); **5**, 8 (1925).

[96]   C. W. Bunn, *Chemical Crystallography*, 2nd ed., Oxford University Press, 1961, p. 207 and Fig. 115.

[97]   H. Morimoto and R. Uyeda, *Acta Crystallogr.*, **16**, 1107 (1963).

[98]   M. H. Van Horn, *Rev. Sci. Instrum.*, **22**, 809 (1951).

[99]   N. H. Nachtrieb, *Principles and Practice of Spectrochemical Analysis*, McGraw-Hill, New York, 1950, pp. 137–139.

[100]   Z. W. Wilchinsky, *Acta Crystallogr.*, **4**, 1 (1951).

[101]   M. Järvinen and M. Merisalo, *Acta Crystallogr.*, **25A**, S18 (1969).

[102]   R. I. Garrod and J. H. Auld, *Brit. J. Appl. Phys.*, **5**, 454 (1954).

[103]   H. Ruck, M. Kouris, and R. St. J. Manley, *J. Sci. Instrum.*, **37**, 223 (1960).

[104]   R. F. Braybrook, *J. Sci. Instrum.*, **32**, 365 (1955).

[105]   G. S. Smith and E. Kahara, American Crystallographic Association, Program and Abstracts, Summer Meeting, Ames, Iowa, August 15–20, 1971, Abstract E10.

[106]   G. S. Smith and E. Kahara, *A Brief Account of Automated Computer Indexing of Powder Patterns: The Monoclinic Case*, Chemistry Department Technical Note No. 71-77, Lawrence Livermore Laboratory, Livermore, California, December 3, 1971.

[107]   P-E. Werner, *Z. Kristallogr.*, **120**, 375 (1964).

[108]   D. Taupin, *J. Appl. Crystallogr.*, **1**, 178 (1968).

[109]   J. F. Cole and H. Villiger, *Mineral. Mag.*, **37**, 300 (1969).

[110]   D. McLachlan and S. Chen, *Advances in X-ray Analysis*, Vol. 14, Plenum Press, New York, 1971, p. 11.

[111]   R. W. G. Wyckoff, *Structure of Crystals*, 2nd ed., Chemical Catalog Company, New York, 1931, pp. 176–179.

[112]   R. W. James and E. M. Firth, *Proc. Roy. Soc.* (London), **117A**, 62 (1927).

[113]   G. W. Brindley and F. W. Spiers, *Proc. Phys. Soc.* (London), **50**, 17 (1938).

[114]   A. J. C. Wilson, *X-ray Optics*, Methuen, London, 1949, Chapters 5 and 6.

[115]   G. W. Brindley, *X-ray Identification and Crystal Structures of Clay Minerals*, Mineralogical Society, London, 1951, Chapters 11 and 12.

[116]   See, for example, B. L. Averbach and B. E. Warren, *J. Appl. Phys.*, **20**, 1066 (1949).

[117]   A. Baxter and J. C. M. Brentano, *Phil. Mag.*, (7)**24**, 473 (1937).

[118]   D. M. C. MacEwan, *Chem. Ind.* (London), **65**, 298 (1946).

[119]   I. Backhurst, *Phil. Mag.*, (7)**22**, 737 (1936); R. J. Stephenson, *Phys. Rev.*, **51**, 637 (1937).

[120]   H. P. Rooksby, *Roy. Soc. Arts*, **90**, 673 (1942); A. Taylor, *X-ray Metallography*, Wiley, New York, 1961, pp. 663–674.

[121]   L. H. Germer and A. H. White, *Phys. Rev.*, **60**, 447 (1941).

[122]   C. Morozumi and H. L. Ritter, *Acta Crystallogr.*, **6**, 588 (1953).

[123]   S. Lele and P. Rama Rao, *Scripta Met.*, **4**, 43 (1970); S. Lele, *Acta Crystallogr.*, **A26**, 344 (1970); S. Lele, B. Prasad, and T. R. Anantharaman, *Acta Crystallogr.*, **A25**, 471 (1969).

[124]   B. E. Warren, *Phys. Rev.*, **59**, 693 (1941).
[125]   A. J. C. Wilson, *Acta Crystallogr.*, **2**, 245 (1949).
[126]   S. Ergun, *J. Appl. Crystallogr.*, **3**, 153 (1970).
[127]   G. B. Mitra, *Brit. J. Appl. Phys.*, **14**, 529 (1963).
[128]   T. W. Baker, *Brit. J. Appl. Phys.*, **7**, 150 (1956).

CHAPTER 7

QUALITATIVE AND QUANTITATIVE ANALYSIS
OF CRYSTALLINE POWDERS

7-1 ROUTINE QUALITATIVE IDENTIFICATION OF
CRYSTALLINE POWDERS

The routine application of powder diffraction techniques for the identification of polycrystalline materials dates from 1938, when the pioneering work of Hanawalt, Rinn, and Frevel[1] of the Dow Chemical Company, Midland, Michigan, was published. Indeed, it is surprising that such routine procedures were not devised much earlier. The simplicity and advantages of the powder diffraction method for chemical analysis were pointed out in 1919 by Hull[2], who emphasized that: (1) the powder diffraction pattern is characteristic of the substance, (2) each substance in a mixture produces its pattern independently of the others, (3) the diffraction pattern indicates the state of chemical combination of the elements in the material, (4) only a minute amount of sample is required, and (5) the method is capable of development as a quantitative analysis. Many, during the 20 intervening years, did use the method in a limited fashion for identification and analysis, but its extensive use had to await the accumulation of an adequate file of standard reference patterns. Moreover, to be really useful, such a library of patterns should number in the thousands, so that the assembling of such a quantity of data would be an unthinkable task for any one person. The outstanding contribution of Hanawalt, Rinn, and Frevel (HRF) was a compilation of tested data on 1000 chemical compounds to form the basis of a reference-pattern library. More important still, they devised a simple scheme for classifying data from thousands of patterns, so that the data are easy to use in identifying an unknown, even when the unknown is a mixture. At this time Harcourt[3] was actively preparing tables for the identification of ore minerals by x-ray powder patterns.

The great utility of the HRF scheme was immediately obvious, and it provided a tremendous stimulus to the analytical use of diffraction data. This served to sharpen the need for a larger library of patterns, since an unknown could not be identified unless its pattern was present in the library. At this point a joint committee of the American Society for Testing Materials (ASTM) and the American Society for X-ray and Electron Diffraction* undertook the task of standardizing the HRF procedure and of publishing a card file of the pattern data. Every effort was made also to increase the number of compounds covered, and when the first edition of the ASTM diffraction data cards appeared in 1942, it listed about 1300 compounds. With the further collaboration of the British Institute of Physics, a supplement to this file was published in 1945, and in 1950 a second supplement, together with a revision of the first two sets, appeared covering in all approximately 4000 compounds.

Over the years this diffraction file has been regularly enlarged and revised. It is now known as the Powder Diffraction File (PDF) and presently (1971) consists of 21 sets of data containing over 21,500 numeric patterns of crystalline materials. Several years ago it seemed best to place the collecting, editing, publishing, and distribution of the powder diffraction data on an international basis. So the Joint Committee on Powder Diffraction Standards (JCPDS) was established as an international organization and ultimately constituted as a Pennsylvania non-profit corporation in 1969 with headquarters in Swarthmore, Pennsylvania.† A group at the U.S. National Bureau of Standards (NBS) has long been charged with checking and correcting the data in the file, and with the preparation of high-quality standard powder patterns for the file[4, 5]. For the purposes of this book, the PDF is described in its latest form, after which precise details for its use are given.

7-1.1 The JCPDS Powder Diffraction File (PDF)

The original scheme of HRF[1] classified the data for each pattern on the basis of the three most intense diffraction lines. The file also contained for each pattern the d spacing for every line and the relative intensities of the lines based on the strongest line as 100. An unknown material was tentatively identified by finding in the file a compound whose three strongest lines matched in both d spacing and relative intensity the three strongest lines of the unknown pattern. Identification was

---

*Predecessor of the American Crystallographic Association.

†Information regarding the purchase of the PDF may be had by addressing the Joint Committee on Powder Diffraction Standards, 1601 Park Lane, Swarthmore, Pennsylvania. The PDF is supplied on 3×5 in. plain cards, on 4×6 in. edge-codeable cards, as a microfiche file, in book form, or on magnetic tape.

unequivocally established by a systematic comparison of all lines of the unknown and standard patterns, which should agree throughout as to spacing and intensity. Use of HRF's data in this fashion required the preparation of an index book of about 462 pages as an aid in locating the matching standard pattern. The great labor involved in the preparation of this index book by each user of the data was in part responsible for the publication of the data as a card file in 1942.

Samples of the present (1971) format of the cards in the plain card form of PDF are illustrated in Fig. 7-1. The various areas of the 3×5 in. card (Fig. 7-1A) are reserved for specific kinds of data:

Spaces 1a, 1b, 1c: Interplanar d spacings of the three strongest lines
Space 1d: Largest d spacing for the material
Spaces 2a, 2b, 2c, 2d: Intensities of the above lines relative to the strongest line as 100
Space 3: Various data on experimental conditions used
Space 4: Crystallographic data on the material
Space 5: Optical information
Space 6: Chemical analysis; chemical treatment of sample, source, and so on
Space 7: Chemical formula and name.
Space 8: "Dot" formula and name if mineral; organic structural formula. A star in the upper right corner indicates the data are of high accuracy, and the symbol (o) indicates data of less reliability
Space 9: Tabulation of observed d spacings, relative intensities, and Miller indices
Space 10: Serial number of the card within the set

Not all the foregoing data are on every card, but such as are available have been entered. Figure 7-1B reproduces a typical card from the file.

The various sets of cards may be purchased separately, with the cards arranged in the order of increasing serial number within each set. For greatest convenience of course the user will want to own all available sets. The cards are used with an index, which is supplied in revised form as each new set of data appears. The 1971 index[6], containing the data of the first 21 card sets, is printed in two volumes, an inorganic index and an organic index. Each volume of the index contains a numerical index of the diffraction data and an alphabetical index with the names and formulas of the compounds. The file number and fiche number of each entry is also included (Fig. 7-2).

The numerical index lists the d spacings and relative intensities of the eight most intense lines of the pattern (Fig. 7-2A). Relative intensities are listed as subscripts, intensities being rounded off to the nearest

10

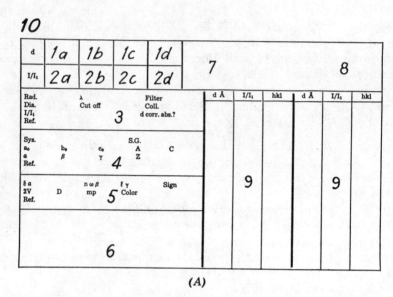

d	1a	1b	1c	1d		
I/I₁	2a	2b	2c	2d	**7**	**8**

	d Å	I/I₁	hkl	d Å	I/I₁	hkl
Rad. λ Filter Dia. Cut off Coll. I/I₁ **3** d corr. abs.? Ref.						

Sys.   S.G.
a₀   b₀   c₀   A   C
a   β   γ   **4**   Z
Ref.

δa   nωβ   ℓγ   Sign
2V   D   mp   **5** Color
Ref.

9   **9**

6

(A)

4-0726 MINOR CORRECTION

d 4-0726	2.67	1.89	3.09	3.087	KF	✦
I/I₁ 4-0726	100	63	29	29	POTASSIUM FLUORIDE	

Rad. CuKα₁   λ 1.5405   Filter Ni
Dia.   Cut off   Coll.
I/I₁ G. C. DIFFRACTOMETER   d corr. abs.?
Ref. SWANSON AND TATGE, JC FEL. REPORTS, NBS 1950

Sys. CUBIC   S.G. O$_h^5$ – FM3M
a₀ 5.347   b₀   c₀   A   C
a   β   γ   Z 4
Ref. IBID.

δa   nωβ 1.352 ℓγ   Sign
2V   D 2.524 mp   Color
Ref. IBID.

NBS ANALYSIS SHOWS ABOUT 0.1% NA IN SAMPLE.
AT 26°C
TO REPLACE 1-1056, 1-1069, 2-0966

d Å	I/I₁	hkl	d Å	I/I₁	hkl
3.087	29	111			
2.671	100	200			
1.890	63	220			
1.612	10	311			
1.542	17	222			
1.337	8	400			
1.227	2	331			
1.1946	14	420			
1.0912	8	422			
1.0297	3	511			
0.9452	3	440			
.9037	4	531			
.8915	5	600			
.8455	5	620			
.8060	4	622			

(B)

Fig. 7-1. (*A*) Format of the JCPDS PDF data card. (*B*) Typical data card: potassium fluoride (KF). (Courtesy of Joint Committee on Powder Diffraction Standards.)

integer after taking the maximum intensity to be 10 (10 is represented by x). To minimize intensity variations arising from differences in techniques, there are three entries of each reference pattern in the numerical index. The three most intense lines of the pattern (the first three d values on the left in the index) are arranged in the following sequences

$$d(A), d(B), d(C)$$
$$d(B), d(C), d(A)$$
$$d(C), d(A), d(B)$$

where A, B, C denote the descending series of intensities. The entire file of patterns is then arranged according to the "Hanawalt" grouping scheme, which is essentially, but not identically, that of HRF[1]. The first d spacing determines the Hanawalt group to which the pattern belongs. The pattern's position in each group is determined by the second d spacing arranged in decreasing sequence. If a small group of patterns have identical second d spacings, the entries are arranged in order of decreasing first d spacings. In sorting the file into Hanawalt groups, an overlap has been permitted which corresponds to about twice the experimental error in a d spacing from a pattern prepared by careful but routine procedures. Thus, in Hanawalt group 3.19 to 3.15 Å (Fig. 7-2A), d values as high as 3.21 Å and as low as 3.13 Å have been included. The symbols ⋆ and o preceding an entry have the significance of the same symbols in the upper right corner of the data cards. Entries pre-

Hanawalt Grouping Scheme

d (Å)	Hanawalt Groups
999.99–10.00	One group
9.99–8.00	One group
7.99–6.00	Grouped in steps of 1.00 Å
5.99–5.00	Grouped in steps of 0.50 Å
4.99–4.60	One group
4.59–4.30	One group
4.29–3.90	Grouped in steps of 0.20 Å
3.89–3.60	Grouped in steps of 0.15 Å
3.59–3.40	Grouped in steps of 0.10 Å
3.39–3.32	One group
3.31–3.25	One group
3.24–1.80	Grouped in steps of 0.05 Å
1.79–1.40	Grouped in steps of 0.10 Å
1.39–1.00	Grouped in steps of 0.20 Å

									Formula	File No.	Fiche No.
	$3.20x$	10.50_5	2.58_1	4.90_1	1.58_1	3.55_1	2.93_1	2.80_1	$(BiO)_2CrO_4$	1-738	I- 4-E 8
o	3.18_8	$10.40x$	4.04_8	3.68_6	5.18_6	8.04_4	6.14_4	3.41_4	$(Cu,Co,Ni)_5As_4O_{15}\cdot 9H_2O$	11-165	I- 39-C 9
	3.17_3	$10.40x$	4.46_3	2.54_2	6.33_2	4.13_2	3.22_2	2.58_2	$(Mg,Al)(Si,Al)O(OH)\cdot 8H_2O$	21-957	I-139-B 2
	3.19_8	$10.30x$	3.71_8	1.338	2.328	5.286	4.006	1.966	$AsHP_2O_8\cdot 2H_2O$	21- 55	I-131-D 6
	3.20_5	$10.20x$	5.02_9	6.62_3	2.04_4	3.37_3	3.12_3	4.02_2	$Ca(UO_2)_2(VO_4)_2\cdot 8H_2O$	6- 17	I- 19-E10
*	3.21_9	10.10_5	$4.54x$	3.06_5	3.65_2	7.17_2	2.96_2		$C_4H_{20}IN_9S_4$	20-1481	I-130-E10
	$3.15x$	10.10_7	$3.34x$	6.32_7	5.02_7	3.12_7	3.08_7	2.77_7	$K_3HP_2O_7$	16- 265	I- 73-F 1
*	3.20_6	$9.97x$	5.00_7	3.12_6	6.94_4	3.33_4	7.20_4	3.08_3	$C_9H_{32}N_{16}O_{38}Tl_2$	20-1493	I-130-F 4
	$3.20x$	9.60_7	1.92_3	9.59_7	2.52_2	4.81_2	1.48_2	4.41_1	$Na_2Al_2(Al_2Si_2)O_{10}(OH)_2$	19-1181	I-113-E 1
	$3.19x$	$9.60x$	$4.79x$	$2.88x$	2.98_6	2.82_6	2.52_6	2.21_6	$Hg_3O_2(NO_3)_2\cdot H_2O$	13- 470	I- 55-B 6

(A)

	Name	Formula	d_1	d_2	d_3	File No.	Fiche No.
*	Chloride Bromide : Rubidium Iridium	$Rb_2(IrCl_3Br_3)$	5.66x	2.51x	1.79x	18-1118	I- 98-D 9
*	Chloride Bromide : Rubidium Iridium	$Rb_2(IrClBr_5)$	5.79x	2.55x	1.81x	18-1109	I- 98-D 5
*	Chloride Bromide : Silver	$Ag(Cl,Br)$	2.81x	1.996	1.26_4	14- 255	I- 58-D 2
*	Chloride : Cadmium	$CdCl_2$	5.85x	2.65_9	3.27_7	9- 401	I- 33-D 5
o	Chloride : Cadmium Aluminum	$Cd(AlCl)_2$	5.74x	2.86_8	1.88_7	14- 421	I- 59-F 3
o	Chloride : Cadmium Arsenic	$Cd_2(AlCl_4)_2$	3.24x	8.80_8	4.59_6	14- 305	I- 58-F 3
*	Chloride : Cadmium Arsenic	Cd_3AsCl_3	3.38x	2.02_8	2.01_8	19- 183	I-104-E 2
*	Chloride : Cadmium Arsenic	Cd_2AsCl_2	3.85x	3.05_9	3.40_8	19- 182	I-104-E 2
*	Chloride : Cadmium Arsenic	$Cd_4As_2Cl_3$	3.57x	3.20_6	1.87_5	19- 184	I-104-E 3
	Chloride : Cadmium Bismuth Oxide	$Cd_{1.25}Bi_{1.5}O_2Cl_3$	3.52x	2.11x	2.86_7	12- 119	I- 45-B 6

(B)

Fig. 7-2. Format of the 1971 PDF inorganic index. (A) A few lines from Hanawalt group 3·19 to 3·15 Å, p. 271, of the numerical index. (B) A few entries from p. 853 of the alphabetical index. (Courtesy of Joint Committee on Powder Diffraction Standards.)

511

ceded by the symbol c are calculated data. The alphabetic index also lists the d spacings and intensities of the three strongest lines of the pattern, in addition to the name and formula of the material (Fig. 7-2B). For further descriptive information on the file and indexes, the reader is referred to the descriptive cards accompanying the file and to the descriptive sections of the indexes.

7-1.2 Experimental Technique of the PDF Method

A. Preparation of the Diffraction Pattern. The experimental preparation of powder diffraction patterns has already been presented in considerable detail: photographic patterns, Sections 4-1.3 to 4-1.6; diffractometer patterns, Sections 5-4 and 5-2.4. The reader is urged to study these sections carefully, as they will form the basis for a successful technique in powder diffraction studies. It is desirable in identification applications to follow a standardized procedure such as that described by HRF[1]. The Hanawalt method was originally devised for the photographic powder pattern technique, but it is readily adapted to the diffractometric technique. The diffractometer, however, is usually found not to be as sensitive as photographic methods in detecting components present in a mixture in small amounts.

Modern diffractometers provide higher intensities as a result of their multiple "Soller" slits, and the geometry of the instrument simplifies the absorption correction so that it becomes independent of θ (Section 6-8). Diffractometer intensity data are superior to those obtained from Debye-Scherrer patterns, because of the diffractometer's more accurate direct recording of x-ray intensities, and its simpler absorption relations which lead directly to the correct relative intensities for the lines. Since the Debye-Scherrer intensities of the PDF have not usually been corrected, they will be at variance with those obtained from the same sample on the diffractometer. This is generally not serious in the interpretation of the diffractometer patterns, but must be kept in mind, and may at times necessitate comparison of the unknown pattern with diffractometer patterns from authenticated samples.

Most of the data in HRF's[1] original file were obtained with MoKα radiation, because it represented a satisfactory compromise between penetrating ability and range of diffraction angles for the simple inorganic substances that made up most of the file. As the file increased in size and more complex compounds were added, it was often desirable to use longer x-ray wavelengths, such as copper, iron, cobalt, and chromium radiations, in order to spread out the pattern as much as possible and to avoid losing important large d spacings. The relative line intensities obtained with one of these radiations, however, are not directly comparable with those obtained with another radiation. Originally, a table for converting $CuK\alpha$ relative intensities to a $MoK\alpha$ basis was supplied with the file, but now $CuK\alpha$ is the predominant radiation used in preparing the patterns in the PDF. The wavelength used in obtaining a pattern is usually listed on the card, and if serious intensity discrepancies arise, they can be resolved by repeating the photograph or diffractometer trace with the same wavelength used in the preparation of the data card. In photographic work sufficient exposure should be given to provide a film of good intensity. If, however, an incorrect exposure is made, it is best to repeat the film rather than to try to correct it by reduction or intensification methods.

B. Measurement of Lines on Films and Diffractometer Charts. The PDF numerical index discusses the measurement of films and the quality of d-value data required. Careful, routine laboratory work usually provides d values with uncertainties Δd comparable to those listed in Table 7-1, which are completely acceptable for PDF studies. We have discussed in detail the precision measurement of photographic powder patterns and the determination of d spacings therefrom in Sections 6-1 and 6-2.1. Measurements with a good film-measuring device (Fig. 6-3) or with a recording microphotometer yield d values with uncertainties usually much smaller than those in Table 7-1. The d data from direct reading d scales, however, are often entirely satisfactory for routine identification analysis, although the approximate Δd error with these scales may be as much as twice the Debye-Scherrer values of Table 7-1.

For the measurement of line intensities on powder films, the Hanawalt method[1] recommends visual estimation. The human eye is so sensitive in matching line blackness against a standard scale over a very wide range that it is often perferred to the microphotometer in identification

Table 7-1. Δd Values Attainable in Routine Work[a]

d Spacing (Å)	114-mm Diameter Debye Camera, $\pm \Delta d\,(\text{Å})$	114-mm Diameter Guinier Camera $\pm \Delta d\,(\text{Å})$
1	0.002	0.001
2	0.004	0.002
3	0.010	0.003
4	0.018	0.005
5	0.025	0.006

[a]Suggested by the PDF numerical index.

analysis. Details of intensity measurements on films are presented in Section 6-7. Intensities of lines on diffractometer traces are measured as heights of the lines above background if the material is above the colloidal range (10^{-5} to 10^{-7} cm) in crystallite size. When the crystallites of a component are in the colloidal size range, its lines are broadened and peak areas rather than heights are then a more accurate measure of intensity. It must be kept in mind also that crystallites above about 20 μ cause diffractometer intensities to fluctuate sharply as a function of the orientation of the larger crystallites, whereas it affects Debye-Scherrer lines seriously (by making them spotty) only above 50 to 60 μ. Details of diffractometer intensity measurements are presented in Sections 5-4 and 5-2.4.

C. Identification Interpretation of the Data. With the d values tabulated in decreasing order, and the relative intensities recorded on a scale of 100 for the strongest line, the identification of the diffracting phases in the sample can proceed. Suggested procedures are outlined in the numerical index, and the individual user will naturally evolve certain methods of his own. The interpretation, nevertheless, is carried through in much the same fashion whether the pattern is that of a single substance or of a mixture, although mixtures of compounds are usually more difficult to analyze and identify than single compounds. Four routine steps generally identify a single compound, or a component of a mixture. Designating the strongest, second-strongest, third-strongest, and so on, d values of the pattern as d_1, d_2, d_3, and so on, the steps are as follows: (1) the proper Hanawalt group for d_1 is located in the numerical index; (2) the values of d_2 and d_3 are next sought within the group; (3) when a match of d values is found for d_1, d_2, and d_3, relative intensities are compared; (4) agreement of the intensities as well as the d values suggests the identity of the phase, and confirmation is obtained by reference to the compound's data card in the file. There must be agreement within experimental error for all lines of the compound's pattern both as to d spacing and relative intensity. Because of the errors of measurement in the d values, it is always necessary to consider a small range of values on each side of the observed value of d. By way of illustration, a few typical sets of powder diffraction data are carried through the identification procedure.

I. *Single component.* Tabulated data for the unknown pattern are listed in Table 7-2. The three strongest lines of the pattern are $d_1 = 2.685$, $d_2 = 1.901$, and $d_3 = 3.100$. Since the value 2.685 may be in error by as much as ± 0.02, it is necessary to scan the approximate range 2.70 to 2.66 in Hanawalt group 2.69 to 2.65. In the inorganic numerical index under

Table 7-2.   Unknown Pattern I

d (Å)	I/I_1	d (Å)	I/I_1
3.100	25	1.092	10
2.685	100	1.029	4
1.901	60	0.945	3
1.618	10	0.907	5
1.540	18	0.895	6
1.333	8	0.840	7
1.225	4	0.801	6
1.195	16		

group 2.69 to 2.65, the reader locates d_2 subgroup 1.91 to 1.89 (pp. 469, 470) and finds the following list of compounds with d_3 in the range 3.12 to 3.08. It is observed that of these 12 compounds only potassium fluoride (KF) has a sequence of relative intensities in agreement with those of the unknown pattern.

Substance	Pattern No.	d (Å)	Relative Intensities
Unknown	I	2.685, 1.901, 3.100	100-60-25
$(Sc_2Se_3)S$	18-1160	2.70, 1.91, 3.100	100-100-80
Rb_2TeBr_6	9-378	2.69, 1.90, 3.110	90-50-100
UO_2	13-225	2.69, 1.90, 3.09	50-50-100
$Rb_3Sb_2Br_{11}$	11-669	2.67, 1.90, 3.12	100-80-100
$KZnPO_4$	20-1447	2.66, 1.90, 3.10	60-40-100
$Pb_3Nb_2O_8$	16-104	2.66, 1.90, 3.09	100-100-100
Cs_2MoBr_6	21-216	2.69, 1.89, 3.12	80-60-100
Cs_2SeBr_6	9-375	2.68, 1.89, 3.10	80-50-100
KF	4-726	2.67, 1.89, 3.09	100-60-30
$Bi_2Sn_2O_7$	17-457	2.67, 1.89, 3.08	100-100-100
Rb_2SbBr_6	2-583	2.66, 1.89, 3.09	100-100-100
$Pr_2Zr_2O_7$	20-1362	2.66, 1.89, 3.08	80-80-100

The agreement is good enough to warrant comparison of all the data with that of pattern 4-726 (Fig. 7-1B). When this is done agreement is satisfactory (within experimental error) for all lines, and no lines of the unknown pattern are unaccounted for. This confirms the identity of the unknown as KF.

II. *Mixture of two components (no superposed pattern lines).* Table 7-3 reproduces data for such an unknown mixture. In the d_1 range 2.74 to

Table 7-3.   Unknown Pattern II

$d\,(\text{Å})$	I/I_1	$d\,(\text{Å})$	I/I_1
3.355	2	1.363	4
3.015	5	1.291	8
2.731	100	1.237	3
2.459	50	1.181	1
2.362	38	1.082	3
2.140	8	1.066	1
1.740	1	1.055	2
1.670	20	0.963	1
1.506	15	0.911	1
1.422	15		

2.71 of Hanawalt group 2.74 to 2.70, a search of the d_2 range 2.47 to 2.45 reveals no compound with a d_3 line matching any intensity sequence of the stronger lines of the unknown pattern. This suggests that line 2.459 is not the d_2 line of a component, but probably the d_1 line of a component. Continuing with the same Hanawalt group and investigating as d_2 the range 2.37 to 2.35, several compounds with d_3 lines near 1.66 to 1.68 appear to be possible components. Their three strongest lines compare with lines in the unknown pattern as follows:

Substance	Pattern No.	$d\,(\text{Å})$	Relative Intensities
Unknown	II	2.73, 2.36, 1.67	100-38-20
$NaErO_2$	21-1112	2.74, 2.37, 1.68	80-100-90
Ag_2O	12-793	2.73, 2.37, 1.67	100-40-20
$NaYbO_2$	19-1260	2.72, 2.35, 1.67	100-100-100
$HfH_{1.70}$	5-641	2.71, 2.35, 1.67	100-80-80
$NaTlO_2$	17-828	2.71, 2.35, 1.67	90-90-100
CdO	5-640	2.71, 2.35, 1.66	100-90-40
$(ZrC)8F$	19-1487	2.71, 2.35, 1.66	100-80-50
Ca_2Si	3-798	2.71, 2.35, 1.66	100-80-40

It is immediately evident that all except Ag_2O are in unsatisfactory agreement with the unknown intensities, and that the Ag_2O pattern deserves further investigation. Comparison of Table 7-3 with pattern 12-793 for Ag_2O, Table 7-4, shows good agreement for each Ag_2O line with a line of the unknown. Ag_2O is thus identified as a component of the mixture.

Table 7-4. Ag_2O, Pattern 12-793

$d(\text{Å})$	I/I_1	$d(\text{Å})$	I/I_1
3.348	4	1.184	1
2.734	100	1.086	4
2.367	35	1.059	2
1.674	18	0.9667	2
1.427	12	0.9115	2
1.367	6		

The remaining unknown lines are now retabulated (Table 7-5), and their intensities put on the basis of 100 for the strongest line. The three strongest lines are $d_1 = 2.46$, $d_2 = 2.14$, and $d_3 = 1.51$. Reference to Hanawalt group 2.49 to 2.45 and subgroup 2.15 to 2.13 shows that only cuprous oxide (Cu_2O), pattern 5-0667, has a line d_3 at 1.51 and the proper sequence of intensities for the three strongest lines. Comparison

Table 7-5.   Remaining Unidentified Lines of
Pattern II

$d(\text{Å})$	I/I_1	$d(\text{Å})$	I/I_1
3.015	$5 \times 2 = 10$	1.506	$15 \times 2 = 30$
2.459	$50 \times 2 = 100$	1.291	$8 \times 2 = 16$
2.140	$18 \times 2 = 36$	1.237	$3 \times 2 = 6$
1.740	$1 \times 2 = 2$	1.066	$1 \times 2 = 2$

of the Cu_2O pattern with the remaining unknown lines satisfactorily accounts for all of them, and establishes the second component as Cu_2O. Obviously, a qualitative emission spectrographic analysis indicating the major, minor, and trace elements in the unknown would be a useful confirmation of the diffraction results.

III. *Mixture of two components (superposed lines).* Assume that the data listed in Table 7-6 have been obtained from an unknown pattern. Because two lines in the pattern, 2.708 and 1.670, each have relative intensities of 100, it is desirable to investigate first Hanawalt group range 2.72 to 2.70 in its d_2 subgroup range 1.68 to 1.66, and then Hanawalt group range 1.68 to 1.66 in its d_2 range 2.72 to 2.70. Several patterns with d_1 and d_2 values matching the unknown are readily eliminated by the presence of strong lines not observed in the unknown pattern. After such a study of the index, the following seven substances with a third line matching a strong line in the unknown have been listed:

Substance	Pattern Number	$d(\text{Å})$	Relative Intensities
Unknown	III	2.71 1.67, 2.35	100-100-90
$Sc_{0.3-0.5}C$	20-1028	2.71, 1.66, 2.34	100-60-70
Na_2PbO_3	8-245	2.70, 1.66, 2.34	80-100-80
CdO	5-640	2.71, 1.66, 2.35	100-40-90
$PrSn_3$	2-1416	2.72, 1.67, 1.42	80-80-100
$HfH_{1.70}$	5-641	2.71, 1.67, 2.35	80-100-80
Ca_2Si	3-798	2.71, 1.66, 2.35	100-40-80
(ZrC)8F	19-1487	2.71, 1.66, 2.35	100-50-80

The fact that none of the intensity triplets above agrees with the corresponding intensities of the unknown triplet is evidence of probable superposition of lines, and it is necessary to compare the pattern of each of the substances above with the unknown to see whether there is a match with the rest of the pattern. It is observed that Na_2PbO_3 has four lines (1.41, 1.05, 0.95, 0.90) as strong as line 2.61, and is thus eliminated by these large discrepancies in relative intensities. Probably, $PrSn_3$ can safely be thrown out because its line 0.91 has a relative intensity of 100, against the value 8 in the unknown. A comparison of the d spacings of the remaining five substances (Table 7-7) with those of the unknown pattern illustrates one of the serious difficulties often encountered in analyzing powder mixtures. Without consideration of relative intensities, it appears that any of the five substances might be components of

Table 7-6.   Unknown Pattern III

$d(\text{Å})$	I/I_1	$d(\text{Å})$	I/I_1
3.05	29	1.322	14
2.708	100	1.251	13
2.621	70	1.174	5
2.352	90	1.136	54
2.176	58	1.081	35
1.941	55	1.051	12
1.780	27	1.020	13
1.670	100	1.004	29
1.536	70	0.957	10
1.419	40	0.904	8
1.370	16	0.832	5
1.351	12		

Table 7-7.   Diffraction Data for Five Possible Components Compared
with Data for Unknown III[a]

Unknown III $d(\text{Å})$	$Sc_{0.3-0.5}C$ $d(\text{Å})$	CdO $d(\text{Å})$	$HfH_{1.70}$ $d(\text{Å})$	Ca_2Si $d(\text{Å})$	$(ZrC)8F$ $d(\text{Å})$
3.05					
2.708_{10}	2.71_{10}	2.71_{10}	2.71_{10}	2.71_{10}	2.71_{10}
2.621					
2.352_9	2.344_7	2.35_9	2.35_8	2.35_8	2.345_8
2.176					
1.941					
1.780					
1.670_{10}	1.658_6	1.66_4	1.665_8	1.66_4	1.66_5
1.536					
1.419_4	1.415_6	1.416_3	1.419_8	1.42_3	1.415_5
1.370					
1.351_1	1.352_2	1.355_1	1.360_5	1.36_1	1.355_2
1.322					
1.251					
1.174_1	1.175_1	1.174_1	1.177_3	1.18_1	1.174_1
1.136					
1.081_4	1.075_2	1.077_1	1.081_5	1.08_1	1.077_2
1.051_1	1.049_3	1.05_1	1.053_5	1.05_1	1.050_2
1.020					
1.004					
0.957_1	0.958_3	0.958_1	0.962_5	0.959_1	0.959_2
0.904_1	0.902_3	0.904_1	0.906_5	0.905_1	0.904_2
0.832_1	0.829_2	0.830_1	0.832_3		0.830_1

[a]Subscripts provide the relative intensities of the lines, on the basis of 10 representing the intensity of the strongest line in the pattern.

the mixture. Moreover, consideration of the line intensities (subscripts in Table 7-7) can hardly distinguish between CdO and Ca_2Si. In this instance it is known that only one of the five can be a component, since the unknown is a two-component mixture and the remaining lines of the pattern must be those of the second component.

All this points to the great desirability, and in this case the absolute necessity, of having some preliminary knowledge of the composition of the unknown, such as simple qualitative chemical or spectrographic analysis can supply. It is assumed therefore that a brief analysis has revealed the presence of cadmium, manganese, and titanium in unknown III, but no more than trace amounts of any other metallic element. With such information available it is obvious that cadmium oxide (CdO) is one component of the unknown, and that lines 1.670, 1.419, and 1.081 appear to be superpositions. The identification is precisely confirmed by comparison with the data on CdO file card 5-0640.

After the CdO lines are removed from the unknown-III pattern and the CdO intensity content of the superposed lines is subtracted, the remaining unknown lines are tabulated (Table 7-8) and their intensities placed on the basis of 100 for the strongest line. Consultation of the Hanawalt group range 2.63 to 2.61 and d_2-subgroup range 1.56 to 1.52 uniquely identifies the second component (the lines in Table 7-8) as manganese titanate (Mn_2TiO_4) (card 2-0983). For Mn_2TiO_4, $d_1 = 2.61$, $d_2 = 1.540$, $d_3 = 2.18$, relative intensities, 100-100-80, and the agreement with the other lines of the Mn_2TiO_4 pattern is equally satisfactory.

Table 7-8.   Remaining Unidentified Lines of Pattern III

$d(Å)$	I/I_1	$d(Å)$	I/I_1
3.05	$29 \times 1.43 = 42$	1.370	$16 \times 1.43 = 23$
2.621	$70 \times 1.43 = 100$	1.322	$14 \times 1.43 = 20$
2.176	$58 \times 1.43 = 83$	1.251	$13 \times 1.43 = 19$
1.941	$55 \times 1.43 = 79$	1.136	$54 \times 1.43 = 77$
1.780	$27 \times 1.43 = 39$	1.081	$(35-9) \times 1.43 = 37$
1.670	$(100-43) \times 1.43 = 82$	1.020	$13 \times 1.43 = 19$
1.536	$70 \times 1.43 = 100$	1.004	$29 \times 1.43 = 42$
1.419	$(40-28) \times 1.43 = 17$		

More complicated mixtures are often encountered than the examples just discussed, but the procedure follows the same course in every instance. It has been found practicable to analyze mixtures for as many as four or five components, but naturally the difficulty of analysis increases with the number of components.

7-1.3   Computer Applications in the PDF Method

As the number of standard reference patterns in the PDF increased, users began to look for more rapid and efficient techniques for searching the file. Likewise, the difficulties in sorting out and identifying the phases increase rapidly as the number of phases present in an unknown mixture increases. The solution to both problems logically appears to be computer automation. Frevel[7] therefore devised a computer program to carry out the entire searching-and-matching process of the phase identification. The input information consisted of the d and I values for the lines in the pattern and elemental information about the sample. The output of the program listed in report form the experimental diffraction data in juxtaposition to the 10 most intense lines for each matched standard.

Frevel[8] next demonstrated that the process of measuring and photometering the diffraction pattern could be automated and coupled with computerized processing of the digitized data. Rex[9] developed a numerical control x-ray powder diffractometry system which stores the diffractometer data output on magnetic tape ready for processing. Following these successes, Frevel and Adams[10] presented a program written in ALGOL 60 for the Burroughs B 5500 computer which carries out completely objective quantitative comparisons of powder patterns with standard patterns. Criteria are established for assigning a merit rating $\Delta\theta_r$ for the d-spacing data, and a similar merit rating $\Delta\tau_r$ for the I values of the lines. For powder data of the quality of the NBS data [4, 5] the $\Delta\theta_r$ value is about 0.0002, but for most of the PDF patterns $\Delta\theta_r$ varies between 0.0008 and 0.0015. Similarly, data of NBS quality rate a $\Delta\tau_r$ value of about 0.020, whereas most PDF standards rate a $\Delta\tau_r$ value between 0.060 and 0.250. Fortuitous superpositions of lines from two or more phases are resolved by an iterative correction. The output of the program provides a printout of the individual matches with the original diffraction data and a second printout of the decomposed pattern corrected for any superposed lines. The entire d,I-match program on the Burroughs 5500 computer requires about 11.5 sec processing time and 18 sec input/output time per 100 reflections.

An example of the results of Frevel and Adam's program is the partial printout of the decomposed match of a sample designated B-Z (Table 7-9). The data in the first three columns of the table were obtained by automated processing of the powder pattern of B-Z. The sample was readily identified as a mixture of zinc oxide (ZnO) and barium sulfate ($BaSO_4$). The standard pattern for $BaSO_4$ lacked the 101 reflection ($d = 5.57$ Å) and gave unresolved peaks for lines 16, 28, 29. Frevel and Adams calculated the d spacings and I_c (italics, Table 7-9) for these double peaks. The last column in Table 7-9 provides a measure of the mismatch of the intensities, $I_v - \sum_p (I_c)_p$, the results being considered satisfactory. Usually, for an unresolved reflection the observed peak intensity is less than the sum of the calculated peak intensities of its components.

Table 7-9. Partial Printout of Decomposed Match of Sample B-Z[10]

	Sample		ZnO		BaSO$_4$		
ν	d_ν (Å)	I_ν	d_s(Å)	I_c	d_s (Å)	I_c	$I_\nu - \sum\limits_p (I_c)_p$
2	5.572	1.6			5.567	3.2	−1.6
3	4.426	15.5			4.435	18.0	−2.5
4	4.329	41.4			4.336	39.2	2.2
6	3.893	64.4			3.896	56.2	8.2
7	3.770	9.2			3.769	12.7	−3.5
8	3.573	33.5			3.575	28.6	4.9
9	3.439	125.3			3.441	106.1	19.2
10	3.313	83.0			3.316	70.0	13.0
12	3.098	121.9			3.101	103.9	18.0
13	2.831	56.7			2.833	49.9	6.8
14	2.813	29.6	2.816	24.7			4.9
16	2.724	60.1			2.727	46.7 + 9.7[a]	(13.4)3.7
17	2.602	21.4	2.602	19.5			1.9
19	2.475	56.4	2.476	34.8	2.479	13.8	7.8
20	2.448	1.3			2.443	1.1	0.2
21	2.419	0.4					0.4
23	2.321	14.2			2.323	13.8	0.4
24	2.301	4.3			2.302	4.2	0.1
25	2.277	4.7			2.279	6.4	−1.7
26	2.207	26.9			2.209	23.3	3.6
27	2.168	1.6			2.168	3.2	−1.6
28	2.119	59.4			2.121	40.3 + 30.7[a]	(19.1) − 11.6
29	2.104	57.6			2.106	41.4 + 34.0[a]	(16.2) − 17.8
30	2.053	13.0			2.054	17.0	−4.0
33	1.929	5.8			1.930	6.4	−0.6
34	1.910	9.3	1.911	10.1			−0.8
36	1.888	0.9					0.9
37	1.854	10.7			1.856	14.8	−4.1

[a]Italics represent I_c values of superposed lines, which had to be calculated separately since the program matches only one calculated d value with a single d_ν.

Another computerized identification system for multicomponent powder patterns has been provided by Johnson and Vand[11]. This program, already adopted by many laboratories, is written in FORTRAN IV for the IBM 360/67 computer. It has successfully identified up to six components in a powder mixture in less than 2 min running time. It operates with ordinary diffraction laboratory techniques of sample preparation, and without corrections for absorption, particle size, and so on. Data may be collected with Guinier or Debye-Scherrer cameras, or by counter diffractometer. With poorer-quality data several incorrect patterns are usually retrieved, but the correct results are properly retrieved. In exceptional cases the user may have to remove incorrectly retrieved patterns manually on the basis of chemical knowledge from spectroscopic or other analysis. The remaining group of possibly four or five patterns is then examined in greater detail with a diffractometer. This program has been available through the JCPDS. Additional references on computer automation applications can be found in the General References section of this chapter.

7-1.4 Complications and Limitations of the PDF Method

The complications and limitations of chemical analysis by powder diffraction are largely of two types: (1) difficulties in the registration of the diffraction patterns, and (2) problems in interpretation. Since both these aspects have been considered in detail by Frevel[12], the present treatment aims merely to acquaint the analyst with them. Some of the effects have already had general discussion elsewhere in this book (Section 6-10), but in other instances the serious analyst will also want to consult Frevel's article.

The filtered radiation used in most work can lead to several effects because of characteristic x-ray absorption. In particular, these effects can result from the absorption edges of the silver bromide emulsion ($AgK_{abs} = 0.486$ Å, $BrK_{abs} = 0.920$ Å), of the elements in the filter ($ZrK_{abs} = 0.689$ Å, $NiK_{abs} = 1.488$ Å), and of the elements in the sample. Closely similar is the effect of dirty radiation as, for example, when the target is contaminated with tungsten (sputtered from the filament), thereby producing a small amount of tungsten L-series radiation. Insufficient purity of the target metal itself, for instance, a cobalt target containing small amounts of iron, can give the same sort of difficulty. All the above tend to produce additional weak spectra associated with the most intense reflections, and care must be exercised to distinguish them from weak $K\alpha$ or $K\beta$ reflections of the primary radiation.

Fluorescence existed in the sample can contribute a troublesome general background which tends to obscure the pattern. This problem and its avoidance have been discussed in Sections 2-3.4 and 6-10.1E. It is to be noted that such fluorescence can arise from either filtered radiation or crystal-monochromatized radiation, and that the position of the filter is important in avoiding it. Thus, in registering patterns of molybdenum compounds with $MoK\alpha$, the strong short-wavelength continuous radiation in the beam may excite considerable K radiation from the sample, which may increase the background severalfold if the zirconium filter is placed directly in front of the film rather than in front of the slit system. Usually the availability of two radiations, for example, $MoK\alpha$ and $CuK\alpha$, is adequate to avoid serious cases of fluorescence.

Registration of the pattern with sufficient resolution that its lines are readily measured is very important. The geometrical factors of camera design that promote line resolution have been considered in Section 4-1.2A, and of these the most obvious is increase in the camera or film radius. With the same camera the pattern may be further spread out by using radiation of a longer wavelength. This becomes especially important with the complex patterns of materials having large unit cells. So many organic compounds fall into this class that the subject is given special treatment in the next section of this chapter.

In sample preparation attention must be given to instability of the sample. The sample must not be allowed to change its composition or nature during the preparation, mounting, and photographing. Deliquescent materials must be handled in a dry-atmosphere box and sealed in the specimen tube. Efflorescent materials can be loaded wet admixed with a little starch to prevent crystal growth. Substances subject to oxidation can be handled and loaded in an inert atmosphere. Volatile samples are best preserved by completely sealing in glass capillaries. Contamination during crushing and grinding must be avoided by proper choice of equipment. In one instance a sample of quartz ground several hours in a tungsten carbide mortar picked up nearly 1 per cent of tungsten and about 3.5 per cent of cobalt (from the cobalt binder used). Crushing generally introduces less contamination than mortar grinding. Metallic iron splinters introduced in using a Plattner mortar may be removed by means of a magnet. Metallic filings usually require a vacuum-anneal at a suitable temperature to remove the strains of cold work. Other less frequently met sample preparation problems are mentioned by Frevel[12].

Finally, one of the limitations of pattern preparation is the nature of the material as a diffractor of x-rays. Many crystalline substances give such sharp powder patterns that they are detectable when present to the extent of 1 to 2 per cent, or less, in a mixture. Other materials give such poor patterns that, although they can be readily identified when alone, they may not be detected when present in a mixture even to the extent of 50 per cent. Little work has been done to determine the amounts of crystalline substances determinable by x-rays in the presence of other materials. An interesting study by Gorbunov and Tsyurupa[13], however, sought to determine the minimum detectable amounts of several crystalline minerals when admixed with amorphous matter. A few of their results are:

Minimum Per Cent Detectable

Amorphous material	Quartz	Kaolinite	Montmorillonite
Humic acid	2	2	5
Hydrated iron oxide	6	20	25

Also of interest in this connection is a study by Parrish and Taylor[13a] of the factors in detecting low concentrations of phases in powder diffractometry.

The analyst is frequently aware that the pattern intensity and the components identified are not sufficient to account for the entire composition of the sample. In these instances chemical or emission spectrographic analysis can often confirm these suspicions. Indeed, standard emission spectrographic analysis is recommended as a routine supplement to all powder diffraction identification work.

Obviously, the reliability of a chemical analysis by the PDF method depends upon the validity of the standard patterns. Each standard, to be acceptable, should have had a precise chemical analysis to establish its chemical identity and formula. This was especially emphasized in the HRF compilation, but the present file contains many data not so examined. Common compositional errors are: (1) incorrect degree of hydration; (2) original substance altered by reaction with water, oxygen, carbon dioxide, and so on; (3) admixture with contaminating substances; and (4) solid solution effects. The analyst must use with caution data not substantiated by good analytical data. Occasionally, there are errors in the diffraction data for a compound. They have arisen frequently from failure to record the innermost reflections of a pattern, or from pseudo doublets due to absorption. The PDF, however, is being constantly improved by the research associateship at the NBS, which reviews and corrects conflicting data and makes new standard patterns [4, 5].

Other inherent limitations in the use of the standard file arise from solid solution, isomorphism, and structural similarity. Qualitative analytical data from chemical, spectroscopic, and spot tests are important in differentiating such cases. Common examples of solid solution ambiguities are lead and the 85 lead-15 sodium alloy; iron and iron–chromium alloys; and so on. Precise back-reflection techniques will measure the d shifts between standard and solid solution. Frevel has listed numerous isomorphism ambiguities, among them silver and gold, cobalt and nickel, iron and chromium, ZnF_2 and CoF_2, $CoCl_2 \cdot 6H_2O$ and $NiCl_2 \cdot 6H_2O$. Structural similarity frequently leads to patterns with closely similar d values. In these instances, however, there are usually radical intensity differences for important corresponding lines. Some examples cited by Frevel are Si, β-ZnS, and CuCl; CoO, Cu_2O, and FeO; CdO and Ag_2O; Al and LiF. Additional ambiguities arise between the numerous spinel-type compounds, and in the pairs Fe_3O_4 and γ-Fe_2O_3, and α-iron and ferrite.

A most serious limitation to the method is the problem of the substance whose pattern is not in the file. There is then of course always the negative information that the unknown is not one of the 21,500 patterns in the file. Positive identification, however, is likely to be difficult (or impossible) unless analytical data suggest likely choices of authentic samples for comparison with the unknown. In some instances the method of identification by isomorphism (see next section) can be applied.

7-1.5 Special Identification Techniques

Several specialized identification techniques warrant only brief treatment here, but those with special interests in them may consult the attached references for further details.

A. Compound Identification by Isomorphism. Frevel developed procedures for indexing and comparing[14] diffraction patterns of isomorphous substances, which have become the basis of an identification scheme[15, 16, 16a] frequently applicable where the patterns sought are not in the catalogue file. The scheme systematically compares the unidentified pattern with representative patterns of the various known crystal structures in an attempt to establish isomorphism between the unknown phase and one of the standard file patterns. For such structure-type identification, diagrams of representative known structure types are prepared with the diffraction lines plotted on a log d scale (abscissa). At the same time the relative line intensities (arithmetically averaged values for representative members of the isomorphous group) are depicted on the vertical scale. Frevel and co-workers have published such diagrams for 33 cubic[15] and 40 tetragonal[16] structure types and have even extended the method to orthorhombic, monoclinic, and triclinic phases[16a].

The general procedure for identifying a noncatalogued pattern is as follows. (1) The investigator plots the log d values and their relative intensities on a narrow strip of paper. (2) The pattern is verified as cubic or tetragonal by comparison with an index scale on the plots of the standard patterns. (3) Using the appropriate plot of standard diffraction types, the investigator searches for an isomorphous proto-type of the unknown phase. In this comparison all lines and their relative intensities must be accounted for. (4) Once the isomorphous structure type is known, a relatively few typical general formulas may be written expressing the chemical stoichiometry of the material. For instance, if the pattern is isomorphous with the rutile type **C4**, it is probably a difluoride MF_2, or an oxide of general formula MO_2 or $MM'O_4$. If, then, qualitative chemical or spectrographic data are available, it may be possible to deduce the formula of the unknown compound. (5) Final confirmation of the unknown phase involves synthesis of the proposed compound and comparison of its diffraction pattern with that of the unknown compound.

B. Procedures for Organic Compounds. At first, routine identifica-tion by powder diffraction methods was mostly confined to inorganic compounds and minerals. This was partly due to the fact that so few pattern data were available on organic compounds. The original HRF[1] compilation of 1000 patterns listed only 49 organic compounds. Al-though the present (1971) PDF contains more than 6000 organic pat-terns, it is obviously a wholly inadequate library of organic patterns, considering that there are well over a half million known organic compounds. A further serious deterrent to routine diffraction analysis in the organic field has been the complex nature of the powder patterns, resulting from the low symmetry and large unit cells of the crystals. Indeed, it was not certain for several years that the data, once obtained, would suffice to distinguish between compounds of similar composition, and particularly between consecutive members of an homologous series and between structural (molecular) isomers. Many studies, how-ever, indicate that organic powder patterns, except in rare cases, are exact enough to identify unambiguously a pure compound, and some work may even be done with binary mixtures. McKinley, Nickels, and Sidhu[17] have studied the identification of a series of phenols; Clark, Kaye, and Parks[18] have investigated aldehydes and ketones; and Matthews and Michell[19] have worked on the anilides of saturated fatty acids. With a modified technique it has even been possible to identify surface-active agents and detergents[20]. Hofer and Peebles have made important applications to the identification of aromatic hydrocarbons[21]. Smith and Heady[22] showed that the components of frozen samples of liquid shale-oil distillate fractions could be identi-fied from their diffractometer patterns. Williams[23] has supplied 20 patterns of substituted barbituric acids, and Baun[24] has identified 22 crystalline ferrocenes. Klesper, Corwin, and Iber[25] extended these procedures to porphyrins, and Parsons and co-workers[26] in a series of 10 articles have provided a library of data on 532 steroids. These data of course have been added to the PDF as they appeared.

The study by Matthews and Michell[19] is particularly significant, since it stresses the importance of using longer-wavelength radiation for greater dispersion of the patterns (Fig. 7-3). Molybdenum radiation is seen to be virtually useless, copper radiation is fair, and chromium radiation is preferred for such patterns. Similar pattern dispersion could be accomplished by the use of larger camera diameters, but this requires greater refinement of the collimating system to give suitable definition of the pattern, and the greater target-to-specimen and specimen-to-film distances increase the exposure time. If we use a chromium target tube with a beryllium window and a vanadium pentoxide filter (to remove the $CrK\beta$ radiation), the exposure in a camera of 5.73-cm radius is 2 to 3 hr.

The prevalence of polymorphism among organic crystalline compounds must always be kept in mind. Once polymorphism is recognized

Fig. 7-3. Effect of change in x-ray wavelength on the dispersion and resolution of a pattern of an etioporphyrin. Top, $MoK\alpha$; middle, $CuK\alpha$; bottom, $CrK\alpha$; camera diameter, 11.46 cm.

in a compound, it should lead to no further trouble, since the pattern for each polymorphic form of a substance is specific. Williams[23] found that the chief difficulties of polymorphism could be avoided by standardizing crystallization methods.

Obviously, until the powder library of organic patterns is many times larger than at present, workers in organic-compound identification may find it necessary to prepare their own library of patterns in the special field of their interest. When the importance of the investigation warrants the additional work involved, powder diffraction methods should prove a most important analytical tool, as evidenced by references [17–26].

C. Identification of Clay Minerals. Identification of clay minerals is one of the most specialized and complex of all powder diffraction identification problems, so that only the briefest outline is possible here. Those who wish to do serious work in the field must first consult the excellent reviews and summaries of the subject[27, 28] as an introduction to its voluminous literature.

The minerals making up clays and soils belong to the same structural types as the micas and the chlorites, being layer structures with pronounced basal cleavages and lamellar or platy habits. Several of the commoner clay minerals with their layer thicknesses are listed in Table 7-10. These long spacings, found close to the undeviated beam, are the characteristic spacings which must be observed for identification. In addition, there may be imperfections arising from random isomorphous substitution, from irregular superposition of the layers, and from stacking of layers of different composition. These imperfections complicate the pattern and its interpretation[28].

Table 7-10. Some Common Clay and Soil Minerals

Name	Formula	Approximate Layer Spacing (Å)	
		Original	After Heating to 500°C
Kaolinite	$Al_2(Si_2O_5)(OH)_4$	7.2	Destroyed
Micas	$KAl_2(AlSi_3O_{10})(OH)_2$	10	10
Chlorites	$Mg_3(AlSi_3O_{10})(OH)_2 \cdot Mg_2Al(OH)_6$	14	14
Vermiculite	$(Mg,Fe)_3(Al,Si)_4O_{10} \cdot (OH)_2 \cdot 4H_2O$	14	10
Montmorillonite	$(Mg,Ca)Al_2Si(Si_4O_{10})(OH)_8$	Variable[a]	9.4–10

[a]Dry hydrogen-saturated clay, 9.4 Å; wet clay, 20 Å.

The diffraction pattern may be recorded either photographically in Debye-Scherrer cameras, or by means of a counter diffractometer with a strip chart. Cameras to permit registration of the long spacings must not have too large a blind space at the beam-exit position. For a 5.73-cm radius camera and $CuK\alpha$ radiation, this space must be less than 8.8 mm to record a 20-Å line. Actually, it is desirable to be able to record lines up to 50 Å, which under these conditions requires a blind space of < 3.2 mm. In addition, the beam must be well collimated, and it is good practice to evacuate the camera to prevent air scatter. Most commercially available cameras require modification to meet these specifications.

Special preparation or treatment of the sample is often desirable. Oriented aggregates of the platelike crystals enhance the basal reflections considerably, so that sometimes they are the only ones visible. Such oriented specimens are prepared by allowing a suspension of clay to settle onto a flat glass plate, or by permitting a dispersion to dry out gradually. Oxides and hydroxides of iron and aluminum are frequently leached out of the sample by various treatments. Another valuable treatment is the selective thermal destruction of clay components. Table 7-10 lists the approximate basal spacings after heating to 500°C. It is noted that kaolinite is completely destroyed, leaving an amorphous residue, whereas the other minerals lose adsorbed water but not, in general, structural OH groups. Montmorillonite forms a stable complex with glycerol in which the 14-Å spacing is increased to 17.7 Å, whereas treatment of vermiculite with NH_4^+ ion displaces its 14-Å line to about 11 Å. All these methods are important in the following standard procedure based on suggestions of MacEwan[27] and Walker[29].

1. A powder photograph of the whole untreated clay is made. If much organic matter is present, it should be removed by preliminary treatment with 30-vol (9 per cent) hydrogen peroxide. Nonclay minerals can be identified on the pattern and a start made on the clay minerals.

2. Dissolve out the sesquioxides (with ammonium oxalate-oxalic acid solution of pH 3.3 to avoid attacking chlorites), boil gently for a few minutes with a solution of an ammonium salt, treat the washed and dried residue with glycerol, and photograph. Lines at 11, 14, and 17.7 Å indicate vermiculite, chlorite, and montmorillonite, respectively. If chlorite is indicated by this test, it is always advisable to check the identification by heat treatment.

3. A photograph is prepared of an oriented aggregate that has been heated to 500°C for several hours and immediately sealed off or protected from moist air. The 14-Å line of chlorite persists through this treatment.

This program necessitates the preparation of three photographs or spectrometer patterns, and MacEwan feels the number of patterns can seldom safely be reduced. The investigator will find the tabulated data of Brown[28], as well as the PDF, most useful in these identifications. Another useful source of data is the American Petroleum Institute Research Project 49 reports[30].

7-2 QUANTITATIVE ANALYSIS OF POWDER MIXTURES

Quantitative analytical chemistry readily gives the elemental composition of a material, but usually has great difficulty in distinguishing the chemical identity of the various phases in a mixture, and in determining the precise amounts of each phase present. Powder x-ray diffraction analysis, however, is seemingly the perfect technique for crystalline-mixture analysis, since each component of the mixture produces its characteristic pattern independently of the others, making it possible to identify the various components by unscrambling their superposed patterns as already described. Moreover, the intensity of each component's pattern is proportional to the amount present (except for an absorption correction), so that a quantitative analysis for the various components may be developed. Thus analytical determinations such as quartz in the presence of mineral silicates, mixed alloy phases of different proportions of the same elements, and the relative amounts of polymorphs in mixtures are handled routinely by diffraction, but are difficult or impossible by chemical methods.

Although Hull[2] in 1919 pointed out these unique features of the powder pattern, no significant work in quantitative analysis seems to have been described until 1936 when Clark and Reynolds published their scheme for mine-dust analysis[31]. This scheme and others that followed shortly were based on the use of an internal standard to correct for absorption, and on intensity data obtained by microphotometering selected diffraction lines on the film. The work of Clark and Reynolds gave considerable impetus to quantitative x-ray analysis because it awakened workers to the possibilities of diffraction methods. The intensity data from photographic-microphotometric methods, however, always lacked something in desired precision, so it was not until the advent of the Geiger-counter diffractometer[32] in 1945 that x-ray quantitative analysis really came into its own. Now, with the diffractometer intensity measurements can be made just about as precise as the worker's patience and the job demand. After a discussion of the special aspects of absorption in analysis, the most useful diffractometric techniques are presented.

7-2.1 Basic Aspects of Absorption in Quantitative Analysis

Even a slight experience with powder diffraction patterns of mixtures leads to the observation that absorption effects are present which usually prevent us from directly comparing line intensities of a component in a mixture with the pattern of the component prepared under identical conditions. Specifically, when the mixture contains both a weak and a strong absorber, lines of the weakly absorbing component appear weaker, and those of the strongly absorbing component stronger, than expected from a linear relationship between composition and pattern intensity for each component. In one noteworthy instance a photograph of stannic oxide (SnO_2) in dried lung tissue indicated 20 to 25 per cent SnO_2 by direct comparison with a pure SnO_2 pattern, but correction for absorption effects revealed its true amount to be less than 1 per cent.

In spite of such pitfalls in the application of the technique, the simple but important mathematical relationships between pattern intensity and the absorptive properties of the sample (as applied to analytical procedure) were little investigated until Alexander and Klug[33] presented considerations for the case of diffraction from a flat cake of powder, the arrangement employed in diffractometric practice. Prior to this, Brentano[34] had discussed theoretically the measurement of the absolute intensities of x-rays diffracted by the components of a binary powder mixture. In a similar manner, Glocker[35] and Schäfer[36] showed that the fundamental intensity formulas of Laue could be applied in the quantitative diffraction analysis of binary powder mixtures and alloys. These earlier investigators, however, did not extend their treatment to the point of developing a systematic practical scheme of analysis. The most useful theoretical findings of Alexander and Klug are presented here in brief.

The following nomenclature is used:

$i,j,k, \cdots$ = lowercase subscripts designating lines in the diffraction pattern

$J,K,L, \cdots$ = uppercase subscripts designating components of a powder mixture

M = subscript referring to the "matrix", that is, the sum of all the components other than J

S = subscript referring to a component employed as a reference standard

iJ = subscript referring to line i of component J

ρ_J = density of (solid) component J

f_J = volume fraction of component J in a mixture

x_J = weight fraction of component J in a mixture

μ_J = linear absorption coefficient of (solid) component J for the x-ray wavelength used

$\mu_J^* = \mu_J/\rho_J$ = mass absorption coefficient of component J for the x-ray wavelength used

$\bar{\mu}$ = linear absorption coefficient of a sample consisting of several components

$\bar{\mu}^*$ = mass absorption coefficient of a sample consisting of several components

$$\bar{\mu}^* = \sum_{J=1}^{N} x_J\mu_J^* \tag{7-1}$$

I_{iJ} = intensity of line i of component J of a mixture

$(I_{iJ})_0$ = intensity of line i of pure component J.

The sample is assumed to be a uniform mixture of n components with a particle size small enough that extinction and so-called microabsorption effects are negligible, and of such thickness as to give maximum diffracted intensities (see Section 5-4.2A). With such a powder cake the total intensity of x-rays diffracted by the Jth component of the mixture by some selected plane $(hkl) = i$ is given by

$$I_{iJ} = \frac{K_{iJ}f_J}{\bar{\mu}} \tag{7-2}$$

where K_{iJ} depends upon the nature of component J and the geometry of the apparatus. If x_J is the weight fraction and ρ_J the density of the Jth component, it may be demonstrated that

$$f_J = \frac{x_J/\rho_J}{\sum_1^N (x_J/\rho_J)}. \tag{7-3}$$

Likewise, $\bar{\mu}$ is

$$\bar{\mu} = \frac{\sum x_J(\mu_J/\rho_J)}{\sum (x_J/\rho_J)} = \frac{\sum x_J(\mu_J^*)}{\sum (x_J/\rho_J)}, \tag{7-4}$$

with $\bar{\mu}$, μ_J, and μ_J^* having the significance given in the nomenclature list above. Substituting relations 7-3 and 7-4 in 7-2 gives

$$I_{iJ} = K_{iJ}\frac{x_J/\rho_J}{\sum x_J(\mu_J^*)}. \tag{7-5}$$

A very elegant way of handling a mixture of N components by equation 7-5 is to regard it as composed of just two components, the component to be analyzed for, component J, and the sum of the other components, which may be designated the matrix and referred to by the subscript M. Then the weight fraction of the Jth component in the matrix is

$$(x_J)_M = \frac{w_J}{w_M} = \frac{Wx_J}{W(1-x_J)} = \frac{x_J}{1-x_J}, \qquad (7\text{-}6)$$

where W and w_J are the weights of the sample and of the Jth component, respectively. Also, the mass absorption coefficient of the matrix is

$$\mu_M^* = \mu_2^*(x_2)_M + \mu_3^*(x_3)_M + \mu_4^*(x_4)_M + \cdots = \frac{\sum\limits_{2}^{N} \mu_J^* x_J}{1-x_1}, \qquad (7\text{-}7)$$

and the relationship between $\bar{\mu}^*$ and μ_M^* is seen to be (by equation 7-1)

$$\bar{\mu}^* = \sum_{J=1}^{N} x_J \mu_J^* = x_J \mu_J^* + (1-x_J)\mu_M^*. \qquad (7\text{-}8)$$

Thus, in terms of the component to be analyzed for, component J, equation 7-5 becomes

$$I_{iJ} = \frac{K_{iJ}x_J}{\rho_J[x_J(\mu_J^* - \mu_M^*) + \mu_M^*]}, \qquad (7\text{-}9)$$

or

$$I_{iJ} = \frac{K_{iJ}x_J}{\rho_J \mu^*} \qquad (7\text{-}10)$$

Expressions 7-9 and 7-10 are two equivalent forms of the basic relationship underlying all quantitative analysis with an x-ray diffractometer. They relate the intensity of the ith line of a selected unknown component J to its concentration and density, and to its mass absorption coefficient and the mass absorption coefficient of the matrix, or to the mass absorption coefficient of the unknown mixture.

Consideration of equations 7-9 and 7-10 leads to three important analytical cases, depending on the number of components and the equality or nonequality of μ_J^* and μ_M^* or $\bar{\mu}^*$. Each of these cases in turn permits or requires a particular procedure.

A. Mixtures of N Components; $\mu_J^* = \bar{\mu}^*$. In such mixtures the absorbing power of the unknown component is the same as that of the entire mixture. These are rather infrequent cases, such as mixtures of the various polymorphic or allotropic forms of a substance. By starting with equation 7-10, it is easy to show that the ratio of intensities of the ith line of J in a mixture to the same line in the pure component J is

$$\frac{I_{iJ}}{(I_{iJ})_0} = \frac{\mu_J^*}{\bar{\mu}^*} x_J. \qquad (7\text{-}11)$$

Since in this case $\mu_J^* = \bar{\mu}^*$, it is evident that the weight fraction x_J of component J in such a mixture is directly equal to the intensity ratio $I_{iJ}/(I_{iJ})_0$, and that direct linear analysis is possible.

The authors[33] strikingly demonstrated the validity of equation 7-11 by diffraction data from mixtures of quartz and cristobalite, polymorphs of silica, for which $\mu^* = 34.4$ for CuKα radiation. Three synthetic mixtures were prepared containing 25, 50, and 75 per cent quartz powder admixed with α-cristobalite.* These mixtures were analyzed in triplicate for quartz following the general Geiger-counter diffractometric procedures of the day. All intensity measurements were made on the 3.34-Å quartz line by manually scanning. Sufficient counts were taken at each point to keep the Geiger-counter statistical errors small, and the recorded counts were corrected for the resolving time of the counter. Figure 7-4 reveals that the experimental points lie very close to the predicted straight line.

*Johns Manville Celite, Type I, a finely divided commercial thermal insulating powder, was used. Diffraction analysis showed it to consist largely (90 to 95 per cent) of finely divided α-cristobalite. It also contains a very small amount of clay and possibly a little amorphous silica, but the absorbing powers of these impurities differ but little from the absorbing power of cristobalite.

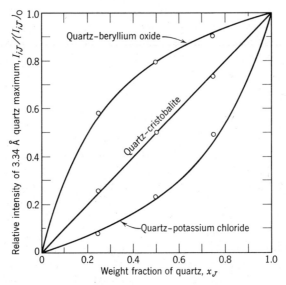

Fig. 7-4.   Comparison of theoretical intensity–concentration curves (solid lines) with experimental measurements (open circles) for several binary mixtures. [From L. Alexander and H. Klug, *Anal. Chem.*, **20**, 886 (1948); by permission of the American Chemical Society.]

B. Mixtures of Two Components; $\mu_1^* \neq \mu_2^*$. In these instances the identity of both components is known. The intensity–concentration curve, however, is no longer linear because of the inequality in the absorbing powers of the unknown and the diluent. Despite this fact, direct analysis is possible by reference to a calibration curve prepared from synthetic mixtures of the two components. It is possible also to calculate the theoretical intensity–concentration curve. For the pure component 1

$$(I_{i1})_0 = \frac{K_{i1}}{\rho_1 \mu_1^*} \tag{7-12}$$

and for a binary mixture with a weight fraction x_1 of this component

$$I_{i1} = \frac{K_{i1} x_1}{\rho_1 [x_1 (\mu_1^* - \mu_2^*) + \mu_2^*]}. \tag{7-13}$$

Division of equation 7-13 by 7-12 gives the expression for the theoretical analysis curve:

$$\frac{I_{i1}}{(I_{i1})_0} = \frac{x_1 \mu_1^*}{x_1 (\mu_1^* - \mu_2^*) + \mu_2^*}. \tag{7-14}$$

A comparison of the theoretical intensity–concentration curve as calculated by equation 7-14 with experimentally determined points is presented in Fig. 7-4 for the binary mixtures SiO_2–BeO ($\mu_1^* > \mu_2^*$) and SiO_2–KCl ($\mu_1^* < \mu_2^*$). The mass absorption coefficients of these compounds for $CuK\alpha$ radiation are: $BeO = 7.9$, $SiO_2 = 34.4$, and $KCl = 125$. The intensity measurements were made in the manner described under case A. It is to be noted from Fig. 7-4, as well as from equation 7-14, that the intensity ratio is greater than x_1 when $\mu_1^* > \mu_2^*$, and less than x_1 when $\mu_1^* < \mu_2^*$.

C. Mixtures of N Components ($N > 2$); $\mu_1^* \neq \mu_M^*$. This is the general case. The absorption coefficient of the unknown component is different from that of the matrix, and the latter is unknown. These circumstances call for indirect analysis by means of an internal standard. The internal-standard technique has been applied in emission spectrographic analysis for some years[37], and its use in x-ray diffraction seems perfectly valid on more-or-less intuitive grounds. Alexander and Klug, however, in a brief theoretical treatment[33] demonstrated that it is easily justified.

Suppose an internal standard, component S, is added to the sample in known amount, and that the volume fractions of the unknown and internal-standard components after such addition are f_J' and f_S, the volume fraction of component J in the original sample being f_J. From equation 7-2 it is seen that

$$I_{iJ} = \frac{K_{iJ}f'_J}{\bar{\mu}} \quad \text{and} \quad I_{kS} = \frac{K_{kS}f_S}{\mu}. \tag{7-15}$$

Dividing I_{iJ} by I_{kS} and substituting f'_J and f_S from equation 7-3 gives

$$\frac{I_{iJ}}{I_{kS}} = \frac{K_{iJ}x'_J\rho_S}{K_{kS}\rho_J x_S}, \tag{7-16}$$

which, on solving for x'_J, yields

$$x'_J = \frac{K_{kS}\rho_J x_S}{K_{iJ}\rho_S} \times \frac{I_{iJ}}{I_{kS}} = k' \times \frac{I_{iJ}}{I_{kS}}, \tag{7-17}$$

provided x_S is held constant. It is the weight fraction x_J in the original sample, however, that is sought, and it is related to x'_J as follows:

$$x_J = \frac{x'_J}{1 - x_S}. \tag{7-18}$$

When equation 7-18 is combined with equation 7-17,

$$x_J = \frac{k'}{1 - x_S} \times \frac{I_{iJ}}{I_{kS}} = k \times \frac{I_{iJ}}{I_{kS}}. \tag{7-19}$$

Thus, when the internal standard is added in a constant proportion x_S, the concentration of component J is a linear function of the intensity ratio I_{iJ}/I_{kS}. The correctness of this relation is demonstrated in Fig. 7-5 by a calibration curve for quartz analysis, using fluorite (CaF_2) as an internal standard. Synthetic mixtures of quartz and calcium

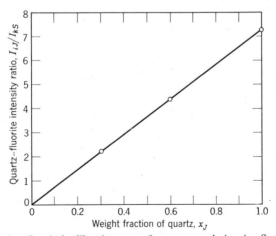

Fig. 7-5.   Linearity of typical calibration curve for quartz analysis using fluorite as internal standard. [From L. Alexander and H. Klug, *Anal. Chem.*, **20**, 886 (1948); by permission of the American Chemical Society.]

carbonate, containing 30, 60, and 100 per cent quartz, were each admixed with fluorite in the constant proportion $x_S = 0.20$. The intensity ratio of the 3.34-Å quartz line and the 3.16-Å fluorite line was determined diffractometrically as previously described. Each plotted point is the average of 10 determinations. The predicted linearity of the internal-standard calibration curve is thus completely substantiated.

Although this same analysis of absorption effects was not carried through for the Debye-Scherrer cylindrical sample, there appear to be no valid reasons why these three analytical cases cannot be handled in the same general fashion when photographic-microphotometric methods are employed.

7-2.2 Photographic-Microphotometric Technique

All the earlier work in quantitative analysis by x-ray diffraction was carried out by photographic-microphotometric technique. However, because of the unquestioned superiority of counter diffractometers to photographic cameras for the measurement of powder diffraction intensities, diffractometry has almost entirely replaced photographic techniques in quantitative analysis. It seems appropriate here, then, to refer the reader to the detailed presentation of photographic-microphotometric quantitative analysis procedures by Klug and Alexander in the first edition of this text[38], and to mention some of the main contributors in this area.

Most of the early quantitative procedures were based on the work of Clark and Reynolds[31] and of Ballard and co-workers[39–40]. Their wide applicability resulted from the use of the internal-standard method, the wedge-type specimen holder, and the standard Debye-Scherrer camera. Gross and Martin[41] developed an ingenious variation of the internal-standard method, necessitating the preparation of only a single synthetic specimen of known composition, but unfortunately it requires each time a previous comparison with the pure standard under identical tube operation, a condition not easily fulfilled. Various workers, particularly Taylor[42] and Brindley[43], have investigated certain theoretical aspects dealing with grain and crystal size and their influence on absorption (microabsorption) as applied to mixed powders and alloys. These factors, largely taken care of in mixed powders by the internal-standard technique, may become important in alloy analysis where the use of an internal standard may be difficult or impossible. de Wolff[44], in addition to consideration of theoretical aspects, has presented various practical details in connection with the use of the Guinier focusing camera (Section 4-2). More recently, Jahanbagloo and Zoltai[45] chose film methods to demonstrate a quantitative analysis technique based on the use of calculated x-ray powder patterns.

7-2.3 Counter Diffractometric Technique

At the time of the introduction of the Geiger-counter diffractometer in 1945, Klug and Alexander were strongly interested in the determination of quartz in industrial dust. They were thus among the first to investigate the whole problem of precise quantitative analysis with the diffractometer [33, 46–50]. Many others have since made important contributions, and quantitative analysis is now a mature area of powder diffraction. A comprehensive treatment of the theory and operation of the diffractometer has been provided in Chapter 5, hence the discussion here is limited to details of diffractometric technique as related to quantitative analysis.

A. Instrumental Requirements. A modern x-ray diffractometer with a narrow "line" focus and limited horizontal and vertical angular divergence is presupposed. It is important that the x-ray generator and counter circuits be highly stabilized. Since Geiger tubes suffer from large resolving-time losses, most instruments now have proportional or scintillation counters with resolving times of 1 or 2 μsec and a linear response extending up to 6000 counts/sec or higher. For quantitative intensity measurements such high linearity of response is of paramount importance.

When the size of the specimen permits, a relatively large divergence of the primary beam should be provided (2 to 4°) in order to irradiate a larger volume of the powder and thus accomplish a better statistical sampling of the components present. Large primary-beam divergence together with rotation of the specimen in its own plane [51] is also effective in reducing intensity fluctuations arising from undesirably large crystallite sizes [48, 52, 53]. In general, the crystallite size of the component to be measured should not exceed 5 μ for a stationary specimen with $\bar{\mu}^*$ equal to about 40 cm²/gram, while for larger values of $\bar{\mu}^*$ the crystallite size must be even smaller. When the specimen is rotated in a moderately divergent beam, the maximum crystallite size for $\bar{\mu}^* = 40$ cm²/gram can be increased to 10 or 15 μ. In this regard we must distinguish the actual crystallite, or grain, size of a specimen from the particle size. Thus, in many fine chemical precipitates and clay minerals, the ultimate crystallite dimension is much smaller than necessary to satisfy the above requirement, even though the particle size is much greater, say, 40 or 50 μ. Each particle, then, consists of a large number of crystallites.

B. General Recommendations on Procedure. When quantitative analyses of a nonroutine nature are to be made, it is recommended that a preliminary photographic diffraction pattern be prepared in order to furnish information on the chemical composition and physical state of the sample. Photographic information of particular value includes (1) the approximate weight fraction of the component to be analyzed for, (2) the possible presence of interfering lines due to matrix components, and (3) the approximate crystallite sizes of the constituents as revealed by the spottiness of their respective lines. The last-mentioned information shows whether or not the crystallite dimension of the "unknown" component satisfies the criterion cited above for statistically valid intensities and assists the analyst in determining what degree of size reduction is required prior to analysis. Figure 4-18A to C shows the dependence of line spottiness upon the particle (crystallite) dimensions for quartz specimens.

Quantitative analyses should be based on integrated rather than peak line intensities, which is to say, areas rather than heights. In either case the background is determined from measurements on both sides of a given line and subtracted. Peak measurements are unreliable because the height decreases and the width increases as the crystallite size falls below approximately 1000 Å (0.1 μ). The integrated intensity may be measured from a strip-chart record of the line profile, or by recording the total counts obtained while the counter scans across the line position.

Analyses should always be made in duplicate or triplicate, with the sample newly mounted for each measurement. Due attention must be given to keep the statistical errors of counting small. When N counts are recorded, the absolute standard deviation σ is $N^{1/2}$, and the relative σ is $N^{-1/2}$. When the background count N_B is not negligible with respect to the total count N_T, the relative standard deviation becomes greater:

$$\sigma_{\text{rel}} = \frac{(N_T + N_B)^{1/2}}{N_T - N_B}. \tag{7-20}$$

When the nature of the specimen is conducive to a direct method of analysis, such a procedure tends to be more accurate than indirect analysis with addition of a standard substance, because this entails additional errors in weighing, in mixing, and in measurement of the intensity ratio I_{iJ}/I_{kS}. Because of differences in particle size, shape, and density between the standard substance and the other components of the sample, it is extremely difficult to achieve homogeneous mixing. When dry mixing is not entirely successful, a liquid dispersing agent with nonsolvating characteristics can be helpful during the process of mixing in a mortar or ball mill. We have found the Wigglebug "spectroscopic" mixer to be effective in many cases for dry mixing of powders for x-ray analysis.

The size of the crystallites or particles composing the powder has been shown to be of great influence. Table 5-9 demonstrates strikingly the effect of variations in the crystallite dimensions of quartz powder upon the reproducibility of the intensity of the 3.34-Å reflection. For the 15 to 50-μ powder, the mean deviation in intensity from the average of 10 measured values as 18.2 per cent, whereas for the < 5-μ powder it was only 1.2 per cent. A significant factor affecting the *average* experimental intensity is also illustrated by Table 5-9, namely, *primary extinction* (see Section 3-3.2). The intensity of reflection from powders of substances that crystallize with a high degree of perfection (quartz and calcite are good examples) decreases when the crystallite size exceeds 10 or 15 μ. Like the errors arising from crystallite size statistics, errors caused by primary extinction are negligibly small if the crystallite size is not allowed to exceed 5 or 10 μ.

The particle sizes of the components of a mixture affect the accuracy of intensity measurements through another phenomenon which cannot be ignored, the *microabsorption*, or *particle-absorption*, effect. Brindley's treatment[43] is generally recognized as constituting a valid, and at least reasonably quantitative, exposition of this complicated subject, and de Wolff[54] and Wilchinsky[55] have contributed further refinements of the theory. Following Brindley, and considering for simplicity only a two-component system (which leads to the correct general conclusions), the ideal intensity ratio I_{iJ}/I_{jK} is diminished or enhanced by a micro-absorption factor

$$K = \frac{\tau_J}{\tau_K} = \frac{V_K \int_0^{V_J} e^{-(\mu_J - \bar{\mu})x} \, dv}{V_J \int_0^{V_K} e^{-(\mu_K - \bar{\mu})x} \, dv}, \tag{7-21}$$

wherein V_J and V_K are the volumes of the individual particles and x is the path of the x-ray beam within one particle. This equation is capable of solution only by analytical means for certain symmetrical particulate shapes.

On the basis of a radius R_K equal to a fixed value of 2 μ and with $\bar{\mu} = \frac{1}{2}(\mu_J + \mu_K)$ for representative calculations, the curves of Fig. 7-6 are obtained. We see that the factor K affecting I_{iJ}/I_{jK} depends both upon the size of the J particles and the contrast in absorbing power of components J and K, namely, $\mu_J - \mu_K$. If $\mu_J > \mu_K$, the intensity ratio is diminished, and if $\mu_J < \mu_K$, it is enhanced. From Fig. 7-6 it is evident that the change in intensity can be minimized either by keeping R_J very small, or by applying x-ray analysis only to systems in which μ_J and μ_K do not differ greatly, or both. Suppose, for example, that we consider values of τ_J/τ_K between 0.95 and 1.05 to be acceptable for a particular analytical prob-

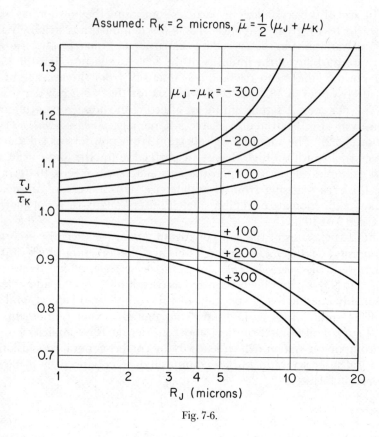

Fig. 7-6.

lem. Then for $|\mu_J - \mu_K| \leqq 100$, R_J's as large as $5\,\mu$ are tolerable, whereas when $|\mu_J - \mu_K| = 200$, R_J should not exceed $2\,\mu$. In a study closely related to Brindley's, Shimazu[55a] has taken account of several additional aspects neglected by Brindley.

It can also be seen from a consideration of Fig. 7-6 that larger particle sizes and contrast in absorption are tolerable so long as the systems being analyzed do not vary greatly in either parameter, for the reason that the microabsorption factor, although relatively large, tends to be constant. Brindley[56] proposes as a general criterion for maximum acceptable particle size in quantitative x-ray analysis

$$t_{max} = \frac{1}{100\,\bar{\mu}},\qquad(7\text{-}22)$$

$\bar{\mu}$ being the average linear absorption coefficient of the solid material composing the powder. This rather severe limitation can sometimes be ameliorated by changing to a more penetrating x-radiation, which reduces $\bar{\mu}$ (e.g., Mo$K\alpha$ instead of Cu$K\alpha$). We may state as a general conclusion that x-ray analysis by the internal-standard method will be least affected by microabsorption errors if the μ's of the components, including the internal-standard substance, do not differ greatly.

Preferred orientation is another source of error which must be dealt with. Particles possessing special and characteristic shapes, especially platelike and needlelike particles, tend to assume a preferred mode of orientation when mounted for diffractometric analysis in the usual way, which combines pressing the powder into the specimen cavity from the front side (side exposed to x-rays) and smoothing the powder surface by some kind of rubbing motion. Any degree of preferred orientation results in deviations of the reflection intensities from their ideal values based on randomized orientations; in general, some lines of the pattern are intensified and others weakened. When the degree of orientation is not great, an averaging of the $x_{i,j}$'s deduced from several lines of the unknown may provide a sufficiently accurate analysis. Sometimes it suffices to fill the specimen cavity from the back side or end[57], sifting the powder in through a 140-mesh screen. In more severe cases of orientation, good results can sometimes be obtained by mixing the powder in a ratio of perhaps 2:1 with a thermoplastic cement having a low flow-point temperature and a small absorption coefficient for x-rays.* However, Sturm and Lodding[57a], considering the platelike particles of silicates, contend it is advantageous to work with oriented specimens whose orientation can be characterized quantitatively, rather than with so-called randomly oriented samples which are rarely ideally random. Arnell[58] and Durnin and Ridal[59] have treated the problem of relatively low degrees of preferred orientation in quantitative metallic-phase analysis, and Dickson[60] has considered very high degrees of preferred orientation, such as arise during heavy cold-rolling of metals and alloys.

The attainment of an optimal crystallite size range of about 5 to 20 μ is a prerequisite which is possibly peculiar to successful quantitative analysis of quartz in mixed powders. As explained earlier, primary extinction in coarser quartz powders is likely to result in pronounced intensity losses. However, an intensity loss of similar magnitude has been found to occur when the average crystallite size is much less than 5 μ[61,62]. This diminution in intensity is believed to be due to the presence of an amorphous surface layer on each particle. For this reason caution must be exercised in prolonged grinding or ball-milling of quartz that the

*For example, Lakeside Brand thermoplastic cement 70C or 30L, available from Hugh Courtright and Company, 7652 So. Vincennes Ave., Chicago, Illinois.

average crystallite size is not reduced so much as to evoke appreciable intensity losses.

For the more routine aspects of diffractometric powder sample preparation, the reader is referred to Section 5-4.2.

7-2.4 Outline of Important Analytical Procedures

A. Direct Analysis when $\mu_J^* = \bar{\mu}^*$. Equation 7-11 then becomes $I_{iJ}/(I_{iJ})_0 = x_J$, and its most useful application is to polymorphic forms of the same compound. In Section 7-2.1A and Fig. 7-4, mixtures of quartz and cristobalite, polymorphic forms of silica, have been treated by this relation. Another rather common example of such determinations is the analysis of mixtures of anatase and rutile, polymorphs of titanium dioxide. In actual practice the analysis of anatase–rutile mixtures is not as straightforward as just presented because of difficulty in obtaining the pure polymorphs[63].

B. Direct Analysis of Two-Component Systems, $\mu_1^* = \mu_2^*$. The applicable equation is 7-14,

$$\frac{I_{i1}}{(I_{i1})_0} = \frac{\mu_1^* x_1}{x_1(\mu_1^* - \mu_2^*) + \mu_2^*}, \tag{7-23}$$

which was used in Section 7-2.1B and Fig. 7-4 to compare theoretical intensity–concentration curves (solid lines) with experimental measurements (open circles) for the binary mixtures SiO_2–BeO and SiO_2–KCl. An especially favorable application of this procedure is to the measurement of a component 1 of high absorption coefficient present in rather small concentrations with a second component of low absorption. Under these conditions $I_{i1}/(I_{i1})_0$ is much larger than x_1, and even very small concentrations may be measured with considerable accuracy. Figure 7-7 shows the character of the 311 reflection of metallic platinum (component 1) as recorded with a counter diffractometer when present in low concentrations in an alumina matrix (component 2). The mass absorption coefficients are $\mu_1^* = 200$ and $\mu_2^* = 31$ cm²/gram. The effect of a decrease in crystallite size of platinum from 150 to 80 Å in decreasing the height and increasing the breadth of the reflection is clearly evident in the illustration. Figure 7-8 shows this effect more clearly for specimens containing 0.25 per cent platinum of four different crystallite sizes, 50, 80, 100, and 150 Å. These calibrating line profiles were prepared by Van Nordstrand, Lincoln, and Carnevale for the analysis of metallic platinum supported on activated alumina catalysts[64].

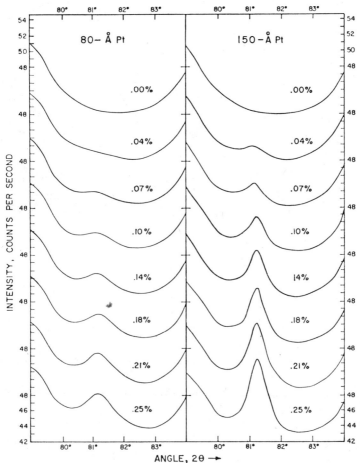

Fig. 7-7. Reference patterns for 80 and 150 Å platinum on low-temperature (480°C) alumina. [Courtesy of Van Nordstrand, Lincoln, and Carnevale, *Anal. Chem.*, **36**, 819 (1964); by permission of the American Chemical Society.]

C. Direct Analysis by Absorption-Diffraction, Multicomponent Systems. In referring to equation 7-11,

$$x_J = \frac{\bar{\mu}^*}{\mu_J^*} \frac{I_{iJ}}{(I_{iJ})_0},$$

it is seen that, since μ_J^* is a known quantity, x_J in a multicomponent mixture can be determined directly from the intensity ratio $I_{iJ}/(I_{iJ})_0$, provided $\bar{\mu}^*$, the mass absorption coefficient of the mixture, can be

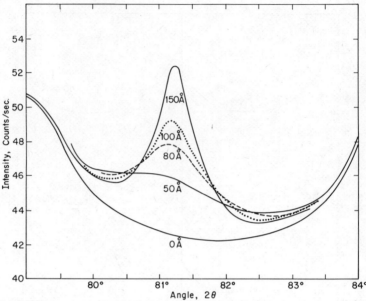

Fig. 7-8.   Reference patterns for 0.25 per cent platinum metal of various crystallite sizes on 480°C alumina. [Courtesy of Van Nordstrand, Lincoln, and Carnevale, *Anal. Chem.*, **36**, 819 (1964); by permission of the American Chemical Society.]

accurately measured or calculated. Such direct analysis, called absorption-diffraction analysis, has been developed and applied extensively by Leroux, Lennox, and colleages [65–67], but it entails some special experimental difficulties and sources of error. They discovered that plots of $\log(\bar{\mu}^*/\mu_j^*)$ versus $\log[I_{iJ}/(I_{iJ})_0]$ departed from the theoretical slope of -1 required by equation 7-11 and necessitated the introduction of an empirical constant which varied with the radiation employed. These departures from ideality were later explained, at least qualitatively, by a more accurate analysis of the diffraction geometry employed [67], but other factors undoubtedly contributed. A general limitation of the method of Leroux, Lennox, and others is its applicability only to specimens possessing rather low absorption coefficients. Ergun and Tiensuu [68] have shown that large errors can be made in measuring the absorption coefficients of materials possessing inhomogeneities of colloidal dimensions, because of the resulting small-angle scattering of x-rays. On account of the difficulties just cited, direct analysis of multicomponent mixtures by a diffraction-absorption technique cannot be generally recommended, even though it can give satisfactory results under favorable and reproducible circumstances (references [65,66], and see also the work of Niskanen [57] and of Williams [69]).

Williams[69] introduced a simplification in the absorption-diffraction technique which reduces errors by permitting the absorption coefficient of the powder mixture to be measured with the same specimen mount and geometrical arrangement used for the measurement of the analytical intensities. For these purposes a layer of powder considerably thinner than specified by equation 5-11 is mounted on the surface of a fine-grained metal, permitting measurements of diffracted rays from the powder and from the metal after attenuation by traversing the powder. Under these conditions, for a thickness of powder t, bulk density $\bar{\rho}$, and mass absorption coefficient (including interstices) $\bar{\mu}^*$, the ratio of the intensities of a metal reflection with and without the specimen in place is given by

$$\frac{T_S}{(T_S)_0} = \exp\left(-\frac{2t\mu^*\rho}{\sin\theta_S}\right). \tag{7-24}$$

If A is the cross-sectional area of the direct beam at the sample, and if B and W are the area and weight, respectively, of the powder specimen illuminated, it can be shown that the weight fraction x_J of an "unknown" is related to the intensity ratio $I_{iJ}/(I_{iJ})_0$ of its ith line by the equation

$$x_J = -\frac{I_{iJ}}{(I_{iJ})_0} \times \frac{B\sin\theta_S}{AW} \times \frac{\log[T_S/(T_S)_0]}{1-[T_S/(T_S)_0]^r}, \tag{7-25}$$

in which $r = (\sin\theta_S)/(\sin\theta_{iJ})$.

If in equation 7-25 we define P_{iJ} by

$$P_{iJ} = (I_{iJ})_0 \frac{A}{B}, \tag{7-26}$$

we have

$$x_J P_{iJ} = -\frac{I_{iJ}\sin\theta_S}{W} \times \frac{\log[T_S/(T_S)_0]}{1-[T_S/(T_S)_0]^r}. \tag{7-27}$$

In equation 7-27 P_{iJ} is a constant for line i of a particular analyte J, provided the incident radiation and A and B remain invariant.

Williams applied this method successfully to the measurement of quartz, and sometimes other components, in ceramics. The powder was held in a flat-bottomed circular recess 2.22 cm in diameter and typically 0.025 cm deep in a soft copper holder. Nickel-filtered CuK radiation was used, and the specimen was rotated to improve crystallite-size statistics. The usefulness of this analytical technique depends upon the reproducibility of P. Table 7-11 illustrates the results of 10 determinations of P for the 38.86° 2θ line of sodium fluoride (NaF) in various mixtures of NaF and quartz (SiO_2), variations being made in the choice of line $\sin\theta_S$, the degree of dilution with gum tragacanth, and the speed of scanning. The degree of constancy in P is seen to be very good. Table 7-12 shows

Table 7-11. P for NaF Analysis[a] [69]

No.	x	P
1	0.6	335.9
2	0.6	341.6
3	0.5	340.9
4	0.5	346.4
5	0.5	335.7
6	0.3	337.3
7	0.3	342.8
8[b]	0.3	347.0
9[c]	0.3	688.5
10	0.1	340.1

[a] Glancing angle, θ_S, 21.6°, scanning rate, $\frac{1}{4}°2\theta$/min.
[b] Glancing angle, 25.2°.
[c] Scanning rate, $\frac{1}{8}°\,2\theta$/min.

Table 7-12.   Experimental Data for Calculating Compositions of Standard Sodium Fluoride-Quartz Mixtures[a] [69]

Composition							W	Calculated Composition	
x_{NaF}	x_{SiO_2}	T_s/T_0	F_{NaF}^b	$F_{SiO_2}^b$	I_{NaF}	I_{SiO_2}	(mg)	x_{NaF}	x_{SiO_2}
0.1	0.4	0.0976	0.4031	0.3811	6357	29901	79.4	0.095	0.386
0.1	0.4	0.9032	0.4084	0.3882	6387	30241	79.7	0.096	0.396
0.2	0.3	0.1147	0.3813	0.3570	13950	24461	77.4	0.201	0.303
0.2	0.3	0.1060	0.3916	0.3693	13701	25084	81.9	0.192	0.304
0.2	0.3	0.1093	0.3880	0.3646	13983	24597	80.5	0.197	0.300
0.3	0.2	0.1328	0.3620	0.3364	21946	17639	80.7	0.288	0.198
0.3	0.2	0.1060	0.3916	0.3693	22698	18124	82.2	0.317	0.219
0.3	0.2	0.1241	0.3708	0.3461	21805	18182	80.1	0.296	0.211
0.4	0.1	0.1227	0.3721	0.3746	32257	10102	87.3	0.403	0.108
0.4	0.1	0.1340	0.3591	0.3353	31600	10409	85.4	0.389	0.110
0.4	0.1	0.1259	0.3692	0.3440	30668	10004	87.8	0.378	0.106

[a] Integrated intensities were obrained from a single scan over the peak.
[b] The significance of F is explained in Williams' article [69].

the results of a further test of the method using a series of standard mixtures of NaF and quartz containing 20, 40, 60, and 80 per cent NaF and in each case diluted with an equal weight of gum tragacanth. An examination of the analytical results calculated in the last two columns shows that they are accurate to within about ±4 per cent of the weights of the various mixtures.

D. Internal-Standard Analysis for One Component of a Multi-component System, No Interfering Lines. Internal-standard analysis is the most reliable procedure when (1) the number of samples to be analyzed for a given component J is large, and (2) the compositions of the samples vary greatly or, at least, cannot be known in advance. For this procedure, considering the ith line of a component J and the kth line of a standard component S, equation 7-10 leads to

$$\frac{I_{jj}}{I_{kS}} = K' \frac{x_j}{x_S}. \tag{7-28}$$

We illustrate the procedure with analyses of industrial and community dusts for quartz, a long-established analytical program at Mellon Institute [46, 47, 50]. Because of the wide variety of industrial processes and raw materials responsible for the dusts, the internal-standard method is mandatory in most instances.

Figure 5-54 shows part of the x-ray diffraction pattern of α-quartz (Q) as prepared with a diffractometer using CuKα radiation. In order of decreasing importance for analysis, the lines with d spacings of 3.34, 1.82, and 4.26 Å are employed. The internal standard used is reagent grade calcium fluoride (CaF_2) (F) which has such a small crystallite size as to produce a slight degree of line broadening. The 3.16- and 1.93-Å lines of CaF_2 were used with the 3.34-, 1.82-, and 4.26-Å lines of quartz to prepare several calibration curves of I_{iQ}/I_{kF} versus x_Q. The usual practice has been to add 0.25 gram CaF_2 per gram of sample to be analyzed, so that $x_F = 0.20$.

Figure 7-5 shows a typical calibration curve of $I_{3.34\,Q}/I_{3.16\,F}$ versus x_Q prepared with the aid of standard mixtures. Figure 7-9 presents duplicate strip-chart records of the 3.34 Q and 3.16 F lines of an unknown sample prepared at a scanning speed of $\frac{1}{4}$° min. It has been found sufficiently accurate to take the peak area as proportional to the product of the width at half-maximum intensity and height above background. From the intensity ratios 1.03 and 1.09 for the two determinations (and a suitable calibration curve), the quartz concentration for the sample concerned was calculated as 19.5 and 20.8 per cent. Table 7-13 compares the actual and analyzed percentages of quartz in 10 known mixtures

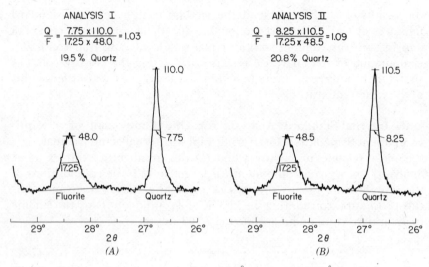

Fig. 7-9. Duplicate strip-chart scans of the 3·34 Å quartz and 3·16 Å fluorite lines of an unknown sample.

Table 7-13. Analyses of Test Mixtures for Quartz by Manual Counting Technique[50]

Mixture No.	Diluent	Quartz, Actual	Per Cent Analyzed	Mixture No.	Diluent	Quartz, Actual	Per Cent Analyzed
1	Al_2O_3	2	2.5	6	MgO	40	40.1
2	MgO	5	5.2	7	Bentonite	50	50.1
3	MgO	10	9.7	8	$CaCO_3$	60	59.0
4	$CaCO_3$	15	15.5	9	$CaCO_3$	80	77.0
5	Bentonite	30	31.4	10	Al_2O_3	90	93.0

composed of quartz and various diluents. This analytical procedure has been proved accurate to about ±5 per cent of the amount of quartz present for quartz concentrations greater than 20 per cent.

E. Internal-Standard Analysis for One Component of a Multi-component System, Interfering Lines of Unknown and Standard. Copeland and Bragg[70] have shown that unresolved lines of components J and S are useful in analysis, since the intensities of the several lines in the pattern of either J or S bear fixed ratios to each other. Thus suppose that lines iJ and kS are more or less superposed but that lines $1J$ and $1S$ are well resolved. Then the following fixed relationships can be defined,

$$I_{iJ} = \alpha_i I_{1J} \quad \text{and} \quad I_{kS} = \beta_k I_{1S},$$

and the sum of the intensities of lines iJ and kS is

$$I_{iJ} + I_{kS} = \alpha_i I_{1J} + \beta_k I_{1S}. \tag{7-29}$$

Division of equation (7-29) by I_{1S} and substitution of

$$\frac{I_{1J}}{I_{1S}} = \frac{K_{1J}\rho_S}{K_{1S}\rho_J}\frac{x_J}{x_S}$$

gives

$$\frac{I_{1J} + I_{kS}}{I_{1S}} = \left(\alpha_i \frac{K_{1J}\rho_S}{K_{1S}\rho_J}\right)\frac{x_J}{x_S} + \beta_k. \tag{7-30}$$

We illustrate the use of superposed lines in analysis from the work of Copeland and Bragg on Portland cement compositions[70]. Consider specifically the analysis for calcium hydroxide $[Ca(OH)_2]$ using magnesium hydroxide $[Mg(OH)_2]$ as internal standard. Let I_{11} and I_{21} be the intensities of the 4.90- and 2.63-Å lines, respectively, of $Ca(OH)_2$, and I_{1S} and I_{2S} the intensities of the 4.76- and 2.38-Å lines of $Mg(OH)_2$. These lines are the strongest observed in their respective patterns, but I_{11} and I_{1S} are partially superposed so that only their sum can be measured. Although both I_{11} and I_{1S} arise from cleavage planes and therefore are very sensitive to preferred orientation, they were useful in the analysis because they were most likely to be free from interferences by lines of the anhydrous silicate compounds.

Three calibration curves, represented by the following equations, were prepared from known mixtures of $Ca(OH)_2$ and $Mg(OH)_2$, designated components 1 and S respectively:

$$R_1 = \frac{I_{21}}{I_{2S}} = Kx \tag{7-31}$$

$$R_2 = \frac{I_{11} + I_{1S}}{I_{2S}} = \alpha Kx + \beta \tag{7-32}$$

$$R_3 = \frac{I_{11} + I_{1S}}{I_{21}} = \alpha + \frac{\beta}{Kx} \tag{7-33}$$

In these equations $x =$ grams $Ca(OH)_2$ per gram $Mg(OH)_2$. Equation (7-31) was the one normally employed, while the other two were used to detect errors caused by preferred orientation. When R_1 could not be determined because of interference with lines of other components of the specimens, R_2 or R_3 could usually be used. The calibration and test mixtures were prepared by weighing and mixing the purified components in an atmosphere free from carbon dioxide and water vapor, and the diffractometer specimens were covered with thin sheets of polyethylene to protect them from the atmosphere during the x-ray measurements. Table 7-14 gives the results obtained by analyzing six test mixtures composed of $Ca(OH)_2$ and various matrix components

Table 7-14.   Analysis of Test Mixtures for $Ca(OH)_2$ [70]

| | | Ca(OH)$_2$ Found (per cent) | | | |
| | | Manual Recording | | Automatic Recording | |
Diluent	Ca(OH)$_2$ Taken, per cent	Using R_1, or R_2 and R_3 (corrected)	Using R_2	Using R_2 and R_3 (uncorrected)	Using R_2
β-Ca$_2$SiO$_4$	17.9	22.2	19.6	33	25
Ca$_2$Si$_2$O$_7$·3H$_2$O	39.4	37.8	37.4	44	42
Portland cement	9.9	10.2	9.9	(Negative)	9
Ca$_2$SiO$_5$	29.8	36.8	29.4	127	24
CaSiO$_3$·H$_2$O	39.8	40.9	39.9	45	42
MgO	64.5	68.0	63.0	64	62

likely to be found in Portland cements. Only two analyses are presented for each mixture because the three experimental ratios can lead to only two independent determinations. The ratios R_1 and R_3 involve the 2.63-Å line of $Ca(OH)_2$, which is overlapped by lines of some of the components of the test mixtures. The data in the fourth column are uncorrected for such instances of superposition, whereas the data in the second column are corrected by applying equation 7-30. The greater accuracy (2 to 3 per cent) afforded in this analysis by manual recording is strikingly evident.

F. Simultaneous Analysis for Several Components with Allowance for Line Superpositions[70]. Suppose that N components are to be determined, and that line i is a composite line with contributions from each of the components. Then

$$I_i = I_{i1} + I_{i2} + \cdots + I_{iJ} + \cdots + I_{iN} \qquad (7\text{-}34)$$

and I_{iJ} is given by equation 7-10:

$$I_{iJ} = \frac{K_{iJ}x_J}{\rho_J \bar{\mu}^*} = C_{iJ}x_J. \qquad (7\text{-}35)$$

In order to determine N components, at least n lines must be measured, so that the following set of equations may be solved:

$$
\begin{aligned}
I_1 &= C_{11}x_1 + C_{12}x_2 + \cdots + C_{1N}x_N \\
I_2 &= C_{21}x_1 + C_{22}x_2 + \cdots + C_{2N}x_N \\
&\;\cdots\cdots\cdots\cdots\cdots\cdots\cdots\cdots\cdots \\
I_n &= C_{n1}x_1 + C_{n2}x_2 + \cdots + C_{nN}x_N
\end{aligned}
\qquad (7\text{-}36)
$$

In equation 7-36 the constant C_{iJ} corresponding to a line i of component J is given by

$$C_{iJ} = \frac{K_{iJ}}{\rho_J \mu^*}. \qquad (7\text{-}37)$$

Thus equations (7-36) represent a scheme of direct analysis, such as an absorption-diffraction method. Given experimental values of $\bar{\mu}^*$ for the sample, and ρ_J and $(I_{iJ})_0$ for pure component J ($x_J = 1$), K_{iJ} can be calculated using equation 7-35, leading to a value of the constant C_{iJ} defined by 7-37. Once all the C_{iJ}'s have been determined from the pure components, equations 7-36 can be used for multicomponent analyses.

Copeland and Bragg[70] show how this scheme can be extended to permit internal-standard analysis by the addition of an internal-standard component $N+1$ furnishing an *unsuperposed* line of intensity I_{n+1}. The intensities on the left-hand side of equation 7-36 are then replaced by ratios I_i/I_{n+1} and the coefficients C_{iJ} by ratios

$$\frac{C_{iJ}}{C_{n+1,\,N+1}}.$$

G. Analysis by Dilution of Sample with y_J Grams of Unknown per Gram of Sample[66]. When the composition of a multicomponent system varies greatly and only a few samples are to be analyzed for component J, one can avoid the labor of setting up a calibration curve for an internal-standard substance by adding a known weight of the unknown itself. The method is particularly useful when the original weight fraction x_J of the unknown is relatively small. Considering only the unknown J and a second component L, suppose that y_J grams of J are added per gram of sample. Then the new weight fractions of J and L are:

$$x'_J = \frac{x_J + y_J}{1 + y_J} \quad \text{and} \quad x'_L = \frac{x_L}{1 + y_J}. \qquad (7\text{-}38)$$

The new intensity ratio for lines iJ and kL is then:

$$\frac{I_{iJ}}{I_{kL}} = \frac{K_{iJ}\rho_L}{K_{kL}\rho_J}\frac{x_J+y_J}{x_L}$$

$$= \text{constant} \times (x_J+y_J) \qquad (7\text{-}39)$$

Figure 7-10 shows that a plot of I_{iJ}/I_{kL} versus y_J is a straight line whose abscissa intercept is $-x_J$. Thus the magnitude of the intercept is the concentration of J in the original sample.

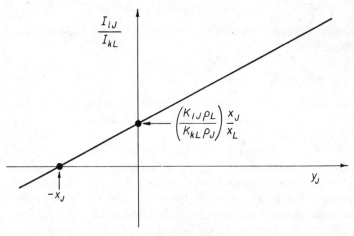

Fig. 7-10.

7-2.5   Selected Examples and Applications

A. Dust Analysis. Because of the seriousness of the silicosis hazard to individuals who must work in the presence of siliceous dusts, x-ray diffraction analysis for silica has had a very large amount of study, and a rich literature has appeared[31, 39, 46].* These analytical schemes in general are based on the internal-standard technique with either photographic or spectrometric diffraction methods.

Of the many hundreds of dust analyses made in our laboratory, that of Industrial Hygiene Foundation Sample 450, whose spectrometer trace is depicted in Fig. 7-11, is a simple example. This sample was received as <5-μ powder.† As a first step such samples are routinely

---

*See also the General References at the end of the chapter.
†Most interest centers on the so-called respirable fraction of the dust, the fraction that is small enough in particle size to have a fair chance of reaching the lungs on inhalation. For this fraction the upper limit is generally set at 5μ. Hence the collected dust samples are usually sized by sedimentation procedures to provide the <5-μ fraction for quartz analysis. It is desirable to have at least 25 mg for a diffractometric analysis.

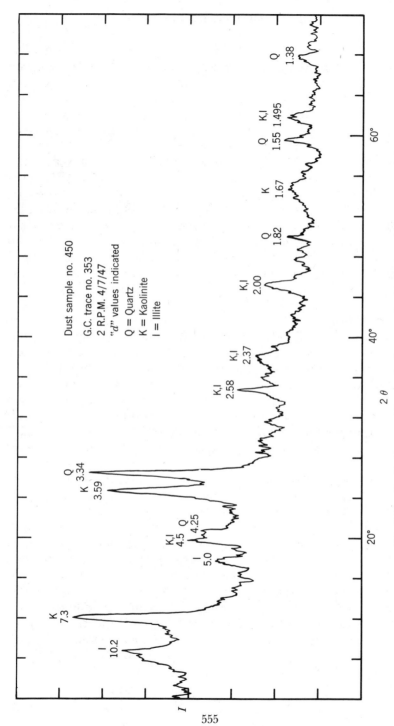

Dust sample no. 450

G.C. trace no. 353
2 R.P.M. 4/7/47
"*d*" values indicated

Q = Quartz
K = Kaolinite
I = Illite

Fig. 7-11.   Preliminary rapid scan of a dust sample, the first step in its analysis.

555

scanned at $2°/\text{min}$ over the 2θ range from $72°$ down to $4°$. This trace (Fig. 7-11) allows the investigator to establish immediately whether appreciable amounts of quartz are present, and to identify other major components of the dust (kaolinite and illite in this example). A second use of this preliminary scan is to make a rough estimate of the amount of quartz present. This is done by direct comparison (disregarding absorption effects) with a pure quartz pattern scanned the same day under the same conditions. For this sample the rough estimate was 21 per cent quartz. The trace also indicates whether any line superposition with the internal standard is to be expected. Sample 450 is seen to be free of lines at the position of the 3.16-Å fluorite line (at $2\theta = 28.25°$), thus making fluorite a suitable standard. After these preliminary studies the required weight of CaF_2 is weighed out, mixed with the sample, and the intensity ratio I_Q/I_F determined by manual counting. This led to an analyzed value of 19 per cent quartz, the value reported for the sample.*

The other forms of silica, such as amorphous silica, cristobalite, and tridymite, are also a similar hazard when present in dusts. Only cristobalite and tridymite, however, can be detected and determined by x-ray diffraction. Both of these are considerably rarer than quartz in dust samples, and tridymite is much rarer than cristobalite. Cristobalite occurs in many samples of fired clay products. One of the few observations of tridymite industrially is its occurrence in the central portion of silica bricks which have served as the lining of a glass-melting furnace for a period of time. Thus it may be found in dust collected during dismantling of the worn-out lining. It is also occasionally found in foundry dusts, being transformed from quartz during molding operations. From our experience a fair estimate of the relative frequency of appearance of the crystalline forms of silica in industrial dusts generally would be: quartz: cristobalite: tridymite = 100:10:1. One may prepare standard curves for use in cristobalite and tridymite analysis similar to those for quartz if the volume of work warrants it. In many instances cristobalite occurs in a matrix of essentially the same absorption coefficient as silica itself, and thus permits direct analysis against the pure compound. In one study of 20 samples of this nature[46], which gave good agreement for quartz content by both this direct method and the internal-standard method, cristobalite was found in 6 samples, and its determination here by the direct method may be considered completely reliable. The reader is referred to this same communication[46] for a discussion of line superposition in quartz, cristobalite, and tridymite

---

*Actually, a partial superposition of the quartz line by the third-order 10-Å illite line causes this value to be slightly high.

determinations. Both Talvitie and Brewer[71] and Bradley[72] have discussed the analysis of very small samples for quartz and cristobalite.

B. Retained Austenite in Steel. An analytical problem of considerable complexity is the determination of retained austenite in steel. Austenite is an interstitial solid solution of carbon in γ-iron (f.c.c.). Below 723°C it normally decomposes into ferrite, a solid solution of carbon in α-iron (b.c.c.), and cementite (Fe_3C). During quenching or extremely rapid cooling, these two products may not have time to form, with the result that there is an unstable body-centered tetragonal product, martensite, usually with residual untransformed austenite (Fig. 7-12). Because of the significant effects of this retained austenite on the properties of the steel, it is important to be able to measure quantitatively the amount present. Although dilatometric[73] and magnetic[74] measurements have been used for quantitative determination of austenite, they are volume properties and do not permit explorations of variation from point to point in a specimen. Quantitative x-ray diffraction, however, seems to be the perfect technique for this determination, except for the problem of calibrating the method. It is virtually impossible to obtain at

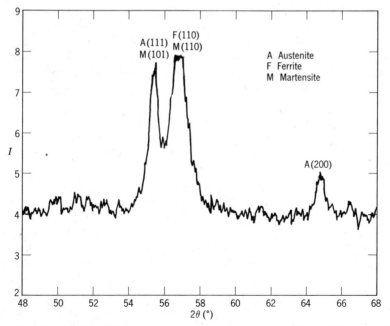

Fig. 7-12.   Portion of a diffractometer trace of a steel sample containing residual austenite. Fe$K\alpha$ radiation.

room temperature a 100 per cent austenite in an ordinary carbon steel. To alleviate this difficulty the Institute for Materials Research, U.S. National Bureau of Standards, has undertaken to produce a series of reference standards containing known amounts of austenite. The first of these reference samples, SRM-485, nominally 4 per cent austenite, has been described by Hicho et al. [75].

An early treatment of the x-ray method was given by Gardner, Cohen, and Antia[76], who used point counting and reflectance measurements for calibration. The point-counting and lineal methods of analysis have been critically studied by Howard and Cohen[77]. These methods are very laborious, and they become less reliable below 10 to 15 per cent austenite. Averbach and Cohen[78] have attempted an x-ray determination of retained austenite by integrated intensities, using the coexisting martensite as an internal standard.

Among the most critical studies of the x-ray diffraction determination of retained austenite are those of Durnin and Ridal[59] and of Arnell[58]. Durnin and Ridal used a diffractometer whose output stability was ±0.03 per cent. Like Averbach and Cohen, they compare integrated intensities of selected hkl reflections of the austenite phase and the martensite phase. They use care with respect to microabsorption, extinction, the Debye-Waller temperature factor, and correction of the atomic scattering factor in the vicinity of an absorption edge. Experimentally, they take into account the effect of alloying elements, the presence of carbides, preferred orientation, specimen surface condition, and the choice of radiation. Finally, they have used lineal analysis and magnetic analysis for comparison with their x-ray results. The interested reader should refer to the original article[59] for details. Table 7-15 presents a comparison of some of Durnin and Ridal's results. It is seen that point counting loses its accuracy below about 10 per cent and was not used on the NCMV sample. To measure retained austenite with the degree of accuracy reported in Table 7-15 requires 2 to 5 hr per determination. If such accuracy is not needed, the time can be reduced to 1 hr or less. Arnell[58] has presented in detail an effective way to correct for preferred orientation, and Dickson[60] has considered the corrections for extremely high degrees of orientation. Kelly and Short[89] have described a paper-tape-controlled diffractometer setup for the measurement of retained austenite.

C. Organic Mixtures. One of the first examples of organic quantitative analysis by diffraction methods was the determination of sodium penicillin G described by Christ, Barnes, and Williams[79]. At least five types of penicillin may be produced in the natural fermentation process

Table 7-15. Durnin and Ridal's Retained Austenite Results [59]

Specimen	Retained Austenite (Per cent)		
	X-ray	Point Counting	Magnetic Susceptibility
16.8 per cent Ni–Fe			
1. Side 1	37.3 ± 0.5	35.3 ± 2.1	37.2 ± 0.3
Side 2	32.7 ± 0.8	30.3 ± 1.9	
2. Side 1	26.8 ± 0.4	28.1 ± 1.9	25.0 ± 0.4
Side 2	26.8 ± 0.4	25.0 ± 1.8	
3. Side 1	16.5 ± 0.3	19.7 ± 1.6	13.7 ± 0.4
Side 2	16.0 ± 0.4	19.1 ± 1.5	
4. Side 1	10.7 ± 0.4	11.0 ± 1.2	8.4 ± 0.5
Side 2	10.5 ± 0.5	9.5 ± 1.1	
5. Side 1	8.2 ± 0.4	8.3 ± 1.0	6.3 ± 0.5
Side 2	8.0 ± 0.4	8.0 ± 1.0	
6. Side 1	6.7 ± 0.4	7.5 ± 1.0	3.9 ± 0.4
Side 2	6.6 ± 0.5	7.3 ± 0.9	
NCMV			
1. Side 1	4.1 ± 0.3		4.4 ± 0.5
Side 2	3.6 ± 0.4		
2. Side 1	$2.2 + 0.3$		5.2 ± 0.5
Side 2	2.2 ± 0.3		
3. Side 1	1.6 ± 0.2		5.6 ± 0.5
Side 2	1.4 ± 0.3		

under certain conditions, and it was hoped the diffraction technique could be used to determine quantitatively the relative amounts of each. Studies revealed, however, that the powder patterns of the forms G,* dihydro F, K, and F are all characterized by an exceptionally strong line at $2\theta = 5.70°$, but are otherwise weak and poor for analytical purposes. The remaining form X has a decidedly different pattern from the first four. It was immediately clear that the method could best be applied to determine the total amount of crystalline penicillin, other than X, in an unknown. Since this type of determination was of considerable practical importance, the method was developed for use with the recording Geiger-counter diffractometer.

To attain reproducible intensity measurements at the rather small

*These various forms of penicillin are further identified as follows: G = benzyl, dihydro F = n-amyl, F = 2-pentenyl, K = n-heptyl, and X = p-hydroxybenzyl.

angle ($2\theta = 5.70°$) required careful adjustment of the sample–x-ray-beam geometry and a special specimen holder. An addition of 1 or 2 per cent of carbon black was found to be an aid in grinding the penicillin to the required fineness. Other details are presented in the reference, which reports the estimated accuracy as approximately ± 2.5 per cent in the range of 40 to 100 per cent crystalline content.

Lutz, Hunter, and Eddy[80] have developed a direct method for the determination of free urea in urea adducts of ethyl stearate. A standard curve was prepared from pressed disks of known mixtures of urea and urea-free ethyl stearate adduct. The ratio of the height of the 4.0-Å peak of urea to that of the 4.1-Å peak of the adduct was obtained from slow scans for six different orientations of the sample disk. A similar study by Chambers[81] sought to determine dialkyldimethylammonium-urea adduct in urea. Again, automatic diffractometric scanning was used, and the technique was based on an internal standard, reagent grade CaF_2. For additional details of these studies, the reader should consult the references cited.

The paper industry long needed a correlation between the crystallinity and texture of various wood pulps and the properties of paper sheets made from these pulps. In a study by Clark and Terford[82], the percentage of amorphous cellulose phase in pulps has been determined experimentally by x-ray diffraction. The technique involves quantitative calibration and scattering correction standards based upon the ratio of the 002 peak intensity of crystalline cellulose to that of the halo at 19° for amorphous cellulose. The details are far too lengthy for presentation here, but certain aspects of the technique can be mentioned. Air-dried pulp samples were further dried at 110°C and then molded into cylindrical pellets of approximately 0.165-cm diameter using a 1 per cent gum arabic solution as a binder. Diffraction patterns of the cellulose pellet and a nickel foil reference standard were recorded on a single film simultaneously. The background of the cellulose intensity was corrected for air, thermal, and Compton scattering through the use of a single sugar crystal of appropriate thickness. The per cent of crystalline material in the pulp is ultimately expressed by the equation

$$\text{Per cent crystallinity} = \frac{I'_c}{I'_c + I_a} \times 100 \qquad (7\text{-}40)$$

where I'_c and I'_a are, respectively, the corrected crystalline and amorphous intensities. When samples are run in triplicate the reproducibility is evident from the following results:

Sample No.	Crystallinity (per cent)	σ (per cent)
5A	63.6, 63.9, and 65.1	0.8
10	62.7, 62.9, and 64.3	0.9
14	59.3, 56.9, and 57.2	1.3

The per cent of amorphous phase present is of course obtained by difference. For a comprehensive treatment of the determination of the amounts of crystalline and amorphous phases present in cellulosic and polymeric materials, the reader is referred to Alexander [83].

D. Miscellaneous Inorganic Analyses. In the evaluation of a new bauxite deposit, thousands of drill-hole samples must be examined. To analyze them chemically is tedious and expensive. Thus Black [84] chose x-ray diffractometric analysis as apparently very promising because it gives direct information on the compounds present, rather than only the elements. His investigation led to a remarkable "assembly-line" technique for this routine analysis. The minerals determined are: gibbsite ($Al_2O_3 \cdot 3H_2O$); boehmite ($Al_2O_3 \cdot H_2O$); kaolinite ($Al_2O_3 \cdot 2SiO_2 \cdot 2H_2O$); hematite ($Fe_2O_3$) plus goethite ($Fe_2O_3 \cdot H_2O$); and quartz ($SiO_2$). Anatase ($TiO_2$) (not determined) is present also but usually in nearly constant amount. Bauxite proved to be a very favorable material for diffractometric analysis, and Black reported that a team of two technicians can analyze about 80 samples per day by his scheme.

In a study by Herbstein, Smuts, and Van Niekerk [85], iron-based Fischer-Tropsch catalysts are diffractometrically analyzed for crystalline α-iron; magnetite (Fe_3O_4); pseudocementite (Fe_3C); Hägg carbide (Fe_2C); and Eckström-Adcock carbide (FeC). Pure reference samples are not required. When an elementary analysis of the catalyst is available, the total time for a determination is about 4 hr. When the chemical composition is not known, the analysis requires about 6 hr. The estimated standard deviations are 3 to 10 per cent of component concentrations.

Nearly any type of siliceous mineral or material is susceptible to diffractometric analysis, provided it is crystalline. For instance, Niskanen [57] successfully made a quantitative mineral analysis of kaolin-bearing rocks using the direct-analysis method of Leroux, Lennox, and co-workers [65–67]. The average time for the complete analysis of one rock sample was approximately $2\frac{1}{2}$ man-hours. Copeland and co-workers [86] developed a method for the quantitative determination of the four major phases of Portland cement by combined x-ray and chemical analysis. The x-ray procedures used were those of Copeland and Bragg [70], already briefly described. It is interesting to note that they found the material formerly called glass or glassy phase by cement chemists to be microcrystalline to x-rays. The phases determined are tricalcium silicate or alite, β-dicalcium silicate or belite, tricalcium aluminate, and ferrite.

E. Analysis of Solid-Solution Phases. Many pairs of compounds (and of metals) form continuous series of solid solutions, often over the entire range from one pure end member to the other. Important metallic phases and commercial materials, such as phosphors, photoconductors, and so on, frequently are solid solutions, so it is necessary to be able to determine precisely and promptly their composition. Usually, the analysis is based on a precise determination of the lattice constant of the solid solution. The lattice constant of the solid solution series varies continuously (linearly if Vegard's law holds) from one pure end member to the other. Once a standard curve (lattice constant versus mole per cent of the end members) has been prepared for the series, the composition of a given solution in the series is available when its lattice constant has been determined. Details of the precise determination of lattice constants are given in Chapter 8. Harrison and Curtis[87] and Cherin, Davis, and Bielan[88] have discussed the rapid determination of solid-solution compositions, illustrating their procedures with solid solutions of the ZnS–CdS series.

GENERAL REFERENCES

1. J. V. Smith, "The Powder Diffraction File—Phase Identification from Diffraction Data," *Norelco Rep.*, **15**, 40 (1968).

*2. L. V. Azároff and M. J. Buerger, *The Powder Method in X-ray Crystallography,* McGraw-Hill, New York, 1958, Chapter 13.

3. B. D. Cullity, *Elements of X-ray Diffraction,* Addison-Wesley, Reading, Massachusetts, 1956, Chapter 14.

4. L. G. Berry and R. M. Thompson, *X-ray Powder Data for Ore Minerals: The Peacock Atlas,* Geological Society of America, New York, 1962.

5. H. Lipson and H. Steeple, *Interpretation of X-ray Powder Diffraction Patterns,* Macmillan, London, 1970, Chapter 10.

6. G. J. C. Frohnsdorf and P. H. Harris, "Use of Digital Techniques to Aid in the Phase Analysis of Multicomponent Mixtures by X-ray Diffraction," in *Developments in Applied Spectroscopy,* Vol. 3, Plenum Press, New York, 1964, pp. 58–68.

7. S. B. McCaleb, "X-ray Diffraction Automation and Its Use in Clay Mineralogy," in *Clays and Clay Minerals,* Vol. 25, Pergamon Press, London, 1966, pp. 123–130.

8. D. K. Smith, "Computer Simulation of X-ray Diffractometer Traces," *Norelco Rep..,* **15**, 57 (1968).

9. D. M. C. MacEwan, "The Identification and Estimation of the Montmorillonite Group of Minerals, with Special Reference to Clay Soils," *Chem. Ind.* (London), **68**, 298 (1946).

10. C. E. Imhoff and L. A. Burkardt, "Crystalline Compounds Observed in Water Treatment." *Ind. Eng. Chem.*, **35**, 873 (1943).

11. H. J. Goldschmidt and G. T. Harris, "An Examination of Mechanical Wear Products," *J. Sci. Instrum.*, **18**, 94 (1941).

12. R. F. Karlak and D. S. Burnett, "Quantitative Phase Analysis by X-ray Diffraction," *Anal. Chem.*, **38**, 1741 (1966).

SPECIFIC REFERENCES

[1] J. D. Hanawalt, H. Rinn, and L. K. Frevel, *Ind. Eng. Chem., Anal. Ed.*, **10**, 457 (1938).

[2] A. W. Hull, *J. Am. Chem. Soc.*, **41**, 1168 (1919).

[3] G. A. Harcourt, *Am. Mineral.*, **27**, 63 (1942).

[4] H. E. Swanson et al., *Nat. Bur. Stand. Circ. 539*, Vols. 1–10 (1953–1960) U.S. Government Printing Office, Washington, D.C.

[5] H. E. Swanson et al., *Nat. Bur. Stand. Monogr.* 25, Sections 1–9, (1962–1971), U.S. Government Printing Office, Washington, D.C.

[6] L. G. Berry (ed.), *Index (Inorganic or Organic) to the Powder Diffraction File 1971*, Publication PDlS-2li, 2 Vols., Joint Committee on Powder Diffraction Standards, 1601 Park Lane, Swarthmore, Pennsylvania.

[7] L. K. Frevel, *Anal. Chem.*, **37**, 471 (1965).

[8] L. K. Frevel, *Anal. Chem.*, **38**, 1914 (1966).

[9] R. W. Rex, *Advances in X-ray Analysis*, Vol. 10, Plenum Press, New York, 1967, p. 366.

[10] L. K. Frevel and C. E. Adams, *Anal. Chem.*, **40**, 1335 (1968).

[11] G. G. Johnson, Jr., and V. Vand, *Advances in X-ray Analysis*, Vol. 11, Plenum Press, New York, 1968, p. 376.

[12] L. K. Frevel, *Ind. Eng. Chem. Anal. Ed.*, **16**, 209 (1944).

[13] N. I. Gorbunov and I. G. Tsyurupa, *Dokl. Akad. Nauk SSSR*, **65**, 81 (1949).

[13a] W. Parrish and J. Taylor in *X-ray Microscopy and X-ray Microanalysis* (A. Engström, V. Cosslett, and H. Pattee, eds.), Elsevier, Amsterdam, 1960, p. 458; also in *X-ray Analysis Papers*, Centrex Publishing Company, Eindhoven, The Netherlands, 1965, Chapter 9.

[14] L. K. Frevel, *J. Appl. Phys.*, **13**, 109 (1942).

[15] L. K. Frevel, *Ind. Eng. Chem., Anal. Ed.*, **14**, 687 (1942).

[16] L. K. Frevel, H. W. Rinn, and H. C. Anderson, *Ind. Eng. Chem., Anal. Ed.*, **18**, 83 (1946).

[16a] L. K. Frevel, *Anal. Chem.*, **44**, 1850 (1972).

[17] J. B. McKinley, J. E. Nickels, and S. S. Sidhu, *Ind. Eng. Chem., Anal. Ed.*, **16**, 304 (1944).

[18] G. L. Clark, W. I. Kaye, and T. D. Parks, *Ind. Eng. Chem., Anal. Ed.*, **18**, 310 (1946).

[19] F. W. Matthews and J. H. Michell, *Ind. Eng. Chem., Anal. Ed.*, **18**, 662 (1946).

[20] T. F. Boyd, J. M. MacQueen, and I. Stacy, *Anal. Chem.*, **21**, 731 (1949).

[21] L. J. E. Hofer and W. C. Peebles, *Anal. Chem.*, **23**, 690 (1951); **24**, 822 (1952).

[22] H. N. Smith and H. H. Heady, *Anal. Chem.*, **27**, 883 (1955).

[23] P. P. Williams, *Anal. Chem.*, **31**, 140 (1959).

[24] W. L. Baun, *Anal. Chem.*, **31**, 1308 (1959).

[25] E. Klesper, A. H. Corwin, and P. K. Iber, *Anal. Chem.*, **33**, 1091 (1961).

[26] J. Parsons, M. L. Polk, and W. T. Beher, *Henry Ford Hosp. Med. J.*, **16**, 215 (1968). This is the 10th in a series published in this Journal by Parsons et al.

[27] D. M. C. MacEwan, *Research*, **2**, 459 (1949). An excellent but very brief review.

[28] G. Brown (ed.) *X-ray Identification and Crystal Structures of Clay Minerals*, 2nd ed., Mineralogical Society, London, 1961, 544 pp. A most complete monograph (with

bibliography) compiled by leading workers in the field. Indispensable for serious work in clay mineral identification.

[29] G. F. Walker, *Nature*, **164**, 577 (1949).

[30] Reference Clay Minerals, American Petroleum Institute Research Project 49, Preliminary Reports. Particularly No. 7, P. F. Kerr et al., *Analytical Data on Reference Clay Minerals*, Columbia University, New York, 1950.

[31] G. L. Clark and D. H. Reynolds, *Ind. Eng. Chem., Anal. Ed.*, **8**, 36 (1936).

[32] H. Friedman, *Electronics*, April 1945, p. 132.

[33] L. Alexander and H. P. Klug, *Anal. Chem.*, **20**, 886 (1948).

[34] J. C. M. Brentano, *Phil. Mag.*, (7) **6**, 178 (1928); *Proc. Phys. Soc.* (London), **47**, 932 (1935).

[35] R. Glocker, *Metallwirtschaften*, **12**, 599 (1933).

[36] K. Schäfer, *Z. Kristallogr.*, **99**, 142 (1938).

[37] G. Scheibe, *Chemische Spektralanalyse, physikalische Methoden der analytischen Chemie*, Vol. I, Akademische Verlagsgesellschaft, Leipzig, 1933, p. 108.

[38] H. P. Klug and L. E. Alexander, *X-ray Diffraction Procedures*, 1st ed., Wiley, New York, 1954, pp. 416–422.

[39] J. W. Ballard, H. I. Oshry, and H. H. Schrenk, *U. S. Bur. Mines Rep. Invest.*, 3520 June (1940); 3638 April (1942); *J. Opt. Soc. Am.*, **33**, 667 (1943).

[40] J. W. Ballard and H. H. Schrenk, *U. S. Bur. Mines Rep. Invest.*, 3888 June (1946).

[41] S. T. Gross and D. E. Martin, *Ind. Eng. Chem., Anal. Ed.*, **16**, 95 (1944).

[42] A. Taylor, *Phil. Mag.*, (7) **35**, 215, 632 (1944).

[43] G. W. Brindley, *Phil. Mag.*, (7) **36**, 347 (1945).

[44] P. M. de Wolff, *Contributions to the Theory and Practice of Quantitative Determinations by the X-ray Powder Diffraction Method*, Doctoral Thesis, Technical University of Delft, 1951. 70 pp.

[45] I. C. Jahanbagloo and T. Zoltai, *Anal. Chem.*, **40**, 1739 (1968).

[46] H. P. Klug, L. Alexander, and E. Kummer, *J. Ind. Hyg. Toxicol.*, **30**, 166 (1948).

[47] H. P. Klug, L. Alexander, and E. Kummer, *Anal. Chem.*, **20**, 607 (1948).

[48] L. Alexander, H. P. Klug, and E. Kummer, *J. Appl. Phys.*, **19**, 742 (1948).

[49] L. Alexander, E. Kummer, and H. P. Klug, *J. Appl. Phys.*, **20**, 735 (1949).

[50] H. P. Klug, *Anal. Chem.*, **25**, 704 (1953).

[51] *International Tables for X-ray Crystallography*, Vol. III, Kynoch Press, Birmingham, England, 1962, pp. 162–165, 175–192.

[52] W. Parrish, *Philips Tech. Rev.*, **17**, 206 (1956).

[53] P. M. de Wolff, *Appl. Sci. Res.*, **B7**, 102 (1958).

[54] P. M. de Wolff, *Physica*, **12**, 62 (1947).

[55] Z. W. Wilchinsky, *Acta Crystallogr.*, **4**, 1 (1951).

[55a] M. Shimazu, *Mineral. J.*, **5**, 180 (1967); M. Shimazu and S. Hosoya, *Mineral. J.*, **5**, 239 (1968).

[56] G. W. Brindley, in reference 28, p. 492.

[57] E. Niskanen, *Anal. Chem.*, **36**, 1268 (1964).

[57a] E. Sturm and W. Lodding, *Acta Crystallogr.*, **A24**, 650 (1968).

[58] R. D. Arnell, *J. Iron Steel Inst.*, **206**, 1035 (1968).

[59] J. Durnin and K. A. Ridal, *J. Iron Steel Inst.*, **206**, 60 (1968).

[60] M. J. Dickson, *J. Appl. Crystallogr.*, **2**, 176 (1969).

[61] G. Nagelschmidt, R. L. Gordon, and O. G. Griffin, *Nature*, **169**, 539 (1952).

[62] R. L. Gordon and G. W. Harris, *Nature*, **175**, 1135 (1955); Safety in Mines Research Establishment, *Res. Rep.*, 138, 1–67 (1956).

[63] R. A. Spurr and H. Myers, *Anal. Chem.*, **29**, 760 (1957).

[64] R. A. Van Nordstrand, A. J. Lincoln, and A. Carnevale, *Anal. Chem.*, **36**, 819 (1964)

[65] J. Leroux, D. H. Lennox, and K. Kay, *Anal. Chem.*, **25**, 740 (1953).

[66] D. H. Lennox, *Anal. Chem.*, **29**, 767 (1957).

[67] J. Leroux and M. Mahmud, *Appl. Spectrosc.*, **14**, 131 (1960).

[68] S. Ergun and V. H. Tiensuu, *J. Appl. Phys.*, **29**, 946 (1958).

[69] P. P. Williams, *Anal. Chem.*, **31**, 1842 (1959).

[70] L. E. Copeland and R. H. Bragg, *Anal. Chem.*, **30**, 196 (1958).

[71] N. A. Talvitie and L. W. Brewer, *Amer. Ind. Hyg. Assoc. J.*, **23**, 58 (1962).

[72] A. A. Bradley, *J. Sci. Instrum.*, **44**, 287 (1967).

[73] P. Gordon, M. Cohen, and R. S. Rose, *Trans. ASM*, **31**, 161 (1943).

[74] E. Maurer and K. Schroeter, *Stahl Eisen*, **26**, 929 (1929); K. Tamaru and S. Sekito, *Sci. Rep. Tôhoku Imp. Univ.*, **20**, 377 (1931); H. Lange, *Arch. Eisenhüttenw.*, **15**, 263 (1933); O. Zmeskal and M. Cohen, *Rev. Sci. Instrum.*, **13**, 346 (1942).

[75] G. E. Hicho, Y. Yakowitz, S. D. Rasberry, and R. E. Michaelis, *Advances in X-ray Analysis*, Vol. 14, Plenum Press, New York, 1971, p. 78.

[76] F. S. Gardner, M. Cohen, and D. P. Antia, *Trans. AIME*, **154**, 306 (1943).

[77] R. T. Howard and M. Cohen, *Met. Technol.*, **14** (August 1947), T. P. No. 2215.

[78] B. L. Averbach and M. Cohen, *Met. Technol.*, **15** (February 1948), T. P. No. 2342.

[79] C. L. Christ, R. B. Barnes, and E. F. Williams, *Anal. Chem.*, **20**, 789 (1948).

[80] D. A. Lutz, J. J. Hunter, and C. R. Eddy, *Anal. Chem.*, **37**, 274 (1965).

[81] L. M. Chambers, *Anal. Chem.*, **39**, 1650 (1967).

[82] G. L. Clark and H. C. Terford, *Anal. Chem.*, **27**, 888 (1955).

[83] L. E. Alexander, *X-ray Diffraction Methods in Polymer Science*, Wiley, New York, 1969, Chapter 3.

[84] R. H. Black, *Anal. Chem.*, **25**, 743 (1953).

[85] F. H. Herbstein, J. Smuts, and J. N. Van Niekerk, *Anal. Chem.*, **32**, 20 (1960).

[86] L. E. Copeland, S. Brunauer, D. L. Kantro, E. G. Schulz, and C. H. Weise, *Anal. Chem.*, **31**, 1521 (1959).

[87] F. W. Harrison and B. J. Curtis, *Brit. J. Appl. Phys.*, **13**, 247, 376 (1962).

[88] P. Cherin, E. A. Davis, and C. Bielan, *Anal. Chem.*, **40**, 611 (1968).

[89] C. J. Kelly and M. A. Short, *Advances in X-ray Analysis*, Vol. 15, Plenum Press, New York, 1972, p. 102.

CHAPTER 8

THE PRECISION DETERMINATION OF
LATTICE CONSTANTS

The accurate determination of lattice constants has assumed increased importance over the years. A very precise knowledge of the unit-cell dimensions of the elements and many simpler chemical compounds is of great theoretical importance in investigations of the solid state. For example, such data have been essential to the development of more satisfactory concepts of bonding energies in crystalline solids. Precision measurements are also invaluable in the study of interstitial and sub-stitutional solid solutions, and the true densities and thermal expansion coefficients of many materials can be ascertained by x-ray means when conventional methods are inapplicable because the physical state of the substances is unsuitable for one reason or another (powders, porous solids, etc.). Numerous other applications could be cited.

So important is precision lattice-parameter determination that an international project was conducted by the IUCr Commission on Crystallographic Apparatus between 1956 and 1960. Under the organi-zation of William Parrish, 16 laboratories from 9 countries participated in the project. The aim of the project was to determine what precision various laboratories could attain in measuring the lattice constants of simple crystalline powder samples. Each laboratory was given uniform powder samples of diamond, silicon, and tungsten, and used the same values of x-ray wavelengths, coefficients of thermal expansion, and refraction corrections. Parrish[1], in reporting on the results, listed the agreement among the laboratories as only about 1 part in 10,000, a value much lower than the precision generally reported by the individual laboratories. For instance, the mean value of a for silicon at 25°C from 25 determined values was 5.43054 Å with the standard deviation $\sigma = \pm 0.00017$.

It seems that advances in techniques since this study now make it possible to measure lattice constants with somewhat higher precision *and* absolute accuracy. Measurement to 1 part in 20,000 is not very high precision. Good practice should yield results to 1 part in 50,000 or 100,000, and possibly to 1 part in 200,000. Many claims for very high precision in the past, however, must be examined with care before being accepted.

8-1 GENERAL CONSIDERATIONS

High accuracy in lattice-parameter determination involves two separable problems, wavelength accuracy and accuracy of Bragg angle measurement. The wavelength problem is mainly the responsibility of the x-ray spectroscopist. The diffractionist's chief concern with wavelength is to establish a one-to-one correspondence between wavelength distribution and the diffraction profile distribution. In other words, the same profile feature used to measure the x-ray wavelength must be used to measure the diffraction line position. X-ray spectral lines have been measured by peak positions (determined in several different ways), and by the center of gravity or *centroid* of the line. The centroid method (Section 5-2.2A and C) is theoretically the most precise, but it is so laborious to carry out that x-ray spectroscopists doubt that any appreciable group of x-ray wavelengths will ever be determined by centroids. Present x-ray wavelengths were determined by the "centerline" peak* technique. Thus Bearden claims a precision *and accuracy* for the CuKα wavelength value of about 1 part in 200,000. All wavelengths in Table 2-3 are centerline peak data, and the centerline peak technique should be used in all diffraction profile measurements when these measurements are used in precision lattice-constant determinations.

Preliminary to the precise calculation of unit-cell dimensions from the observed positions of lines on a powder photograph it is, of course, necessary that the Miller indices of each reflection be known. For methods of indexing powder photographs, the reader is referred to Chapter 6. Taking crystals of the cubic system as the simplest example, we can calculate a value of the lattice constant a for every line on the film by employing equation 6-22:

$$a = \frac{\lambda}{2} \frac{\sqrt{h^2 + k^2 + l^2}}{\sin \theta}. \tag{8-1}$$

The a values computed in this way are found to vary from one reflection to another, the variations being of two kinds, *random* and *systematic*. Random errors decrease in magnitude as θ increases, the effect being very

---

*The peak position is determined by extrapolation of profile chord midpoints to the experimental peak profile[2, 3].

striking as θ approaches $90°$. Systematic errors manifest a definite dependence upon θ, and they also tend to a minimum as θ approaches $90°$.

The increase in precision of interplanar-spacing measurements with increasing angle θ is readily demonstrated by differentiating the Bragg equation with respect to θ:

$$n\lambda = 2d \sin \theta,$$

$$0 = 2 \frac{\Delta d}{\Delta \theta} \sin \theta + 2d \cos \theta,$$

$$\frac{\Delta d}{d} = -\frac{\cos \theta}{\sin \theta} \Delta\theta = - \cot \theta \, \Delta\theta. \tag{8-2}$$

Since $\log d = \log a - \frac{1}{2} \log (h^2 + k^2 + l^2)$ for the cubic system, it is evident

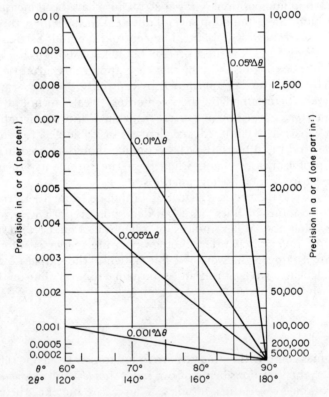

Fig. 8-1. Precision in a as a function of θ or 2θ for various precisions in θ according to equation 8-3. (From K. Beu, *Encyclopedia of X-rays and Gamma Rays*, by G. Clark,© 1963 by Litton Educational Publishing, Inc. Reprinted by permission of Van Nostrand Reinhold Company.)

that $\Delta d/d = \Delta a/a$, hence

$$\frac{\Delta a}{a} = - \cot \theta \, \Delta\theta. \tag{8-3}$$

Equation 8-3 shows that, for a given precision of measurement of the diffraction angle $\Delta\theta$, the precision in a is proportional to $\cot\theta$. Figure 8-1 shows a plot of this function, based on several precisions of measurement. It is clear that a much more precise value of the lattice constant a can be obtained by employing a line in the neighborhood of 80° than one at, for instance, even 60°.

This discussion of random and systematic errors serves to emphasize two basic considerations in precision lattice-constant measurements. First, the nature of the systematic errors must be ascertained and properly allowed for in some manner. Second, greatest weight must be accorded the measurements at high angles where the sensitivity of the lattice constant to the angular position of the line is greatest. In general, if precisions of about $\pm 0.05°$ θ are satisfactory, the systematic errors will largely "take care of themselves" when one employs modern diffraction equipment, moderately careful experimental technique, and any of several calculating procedures. At the 0.01° θ level, however, it is important to evaluate the magnitudes of the systematic errors for the experimental conditions used. At the 0.001° θ level, *all* systematic errors must be closely evaluated, and valid statistical tests made to ascertain that such errors have been removed from the data.

8-2 SOURCES OF SYSTEMATIC ERRORS IN THE DEBYE-SCHERRER METHOD

The various sources of systematic errors are now considered. In the main, the treatment of Bradley and Jay[5] is followed. The significant sources of error are:

1. Radius errors
2. Film shrinkage
3. Eccentricity of the specimen
4. Absorption of the x-ray beam by the specimen
5. Radial (horizontal) divergence of the beam
6. Axial (vertical) divergence of the beam.

8-2.1 Radius Errors and Film Shrinkage

Errors 1 and 2 are of the same nature and may be discussed together. Consider first the case of forward reflections, the film being in the so-called regular position (Fig. 8-2A). If shrinkage of the film is disregarded

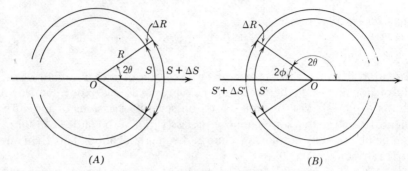

Fig. 8-2.   Effects of radius errors and film shrinkage.

for the moment, the supposed film radius is $R + \Delta R$ and the actual radius R. For a pair of lines separated by an interval S on the film, the supposed angle θ is given by $S/4(R + \Delta R)$, whereas the true angle is $S/4R$. The error in measuring θ is then

$$\Delta\theta_R = \theta(\text{apparent}) - \theta(\text{true})$$

$$= \frac{S}{4(R + \Delta R)} - \frac{S}{4R}$$

$$= -\frac{S}{4R}\left(\frac{\Delta R}{R + \Delta R}\right).$$

With sufficient exactness this may be written

$$\Delta\theta_R = -\theta\frac{\Delta R}{R}. \tag{8-4}$$

Similarly, as a result of *uniform* film shrinkage, the apparent value of S will be $S + \Delta S$, ΔS being negative. Then

$$\Delta\theta_S = \theta(\text{apparent}) - \theta(\text{true})$$

$$= \frac{S + \Delta S}{4R} - \frac{S}{4R}$$

$$= \frac{\Delta S}{4R}.$$

But, since $\theta = S/4R$,

$$\Delta\theta_S = +\theta\frac{\Delta S}{S}, \tag{8-5}$$

where ΔS is negative. The combined error in θ, as a result of a radius error ΔR and a shrinkage error ΔS, is then

$$\Delta\theta = \Delta\theta_R + \Delta\theta_S$$

$$= \theta\left(\frac{\Delta S}{S} - \frac{\Delta R}{R}\right). \tag{8-6}$$

Substitution of equation 8-6 in equation 8-3 gives for the corresponding relative error in a

$$\frac{\Delta a}{a} = \left(\frac{\Delta R}{R} - \frac{\Delta S}{S}\right) \theta \cot \theta. \tag{8-7}$$

For back-reflection lines and for the film in the precision position (Fig. 8-2B), it can be shown in similar fashion that

$$\frac{\Delta a}{a} = \left(\frac{\Delta S'}{S'} - \frac{\Delta R}{R}\right)\left(\frac{\pi}{2} - \theta\right) \cot \theta. \tag{8-8}$$

It should be noted that functions 8-7 and 8-8, although of different form, both approach zero as θ approaches 90°, demonstrating that radius and uniform film-shrinkage errors vanish at $\theta = 90°$.

Precision-built cameras and precise film measurements can minimize the radius errors, and Borisov and Golovachev[6] have presented a refined determination of the effective camera radius. Both Hägg[7] and Scott and Beu[8] have printed scales on the film as an aid in correcting for linear film shrinkage. Jellinek[9], Scott and Beu[8], and others have pointed out that nonlinear shrinkage can also occur, particularly in the vicinity of holes punched in the film. Accordingly, Straumanis[10] suggests that holes in precision films should not be punched but carefully drilled with sharp "cork borer"-type drill bits.

8-2.2 Specimen Eccentricity

Eccentricity of the specimen means that owing to faulty construction the rotation axis of the specimen is displaced from the axis of the film-supporting surface. With reference to Fig. 8-3A, suppose the sample axis is displaced from the film axis O to the point Z, the polar coordinates of

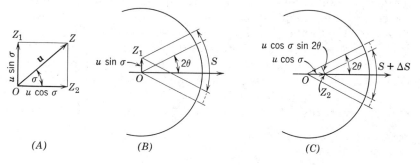

(A)   (B)   (C)

Fig. 8-3.   Effect of sample eccentricity upon forward reflections.

which are $\mathbf{u}$, σ.* The vector $\mathbf{u}$ may be resolved into components $OZ_1 = u \sin \sigma$ and $OZ_2 = u \cos \sigma$, respectively, perpendicular and parallel to the incident beam. From Fig. 8-3B it can be seen that the effect of the perpendicular component is to shift both lines of a pair by almost identical amounts (for reasonably small values of OZ_1), so that the measured distance S is unchanged. Hence this component of the eccentricity introduces no appreciable error into the measured angle.

However, the parallel component OZ_2 shifts both lines of a pair either toward or away from each other (Fig. 8-3C). The magnitude of the error in S is

$$\Delta S = -2OZ_2 \sin 2\theta$$

$$= -2u \cos \sigma \sin 2\theta,$$

and the equivalent error in θ is

$$\Delta \theta = -\frac{2u \cos \sigma \sin 2\theta}{4R}$$

$$= -\frac{u}{R} \cos \sigma \sin \theta \cos \theta. \tag{8-9}$$

The relative error in a is then found to be

$$\frac{\Delta a}{a} = -\cot \theta \, \Delta \theta$$

$$= \frac{u \cos \sigma}{R} \cos^2 \theta. \tag{8-10}$$

For back-reflection lines and for the film in the precision position, entirely similar reasoning shows the relative error in a is

$$\frac{\Delta a}{a} = -\frac{u \cos \sigma}{R} \cos^2 \theta. \tag{8-11}$$

This expression differs from equation 8-10 only in sign. It is noted that the error due to eccentricity of the specimen also vanishes at $\theta = 90°$.

Because the Bradley and Jay[5] eccentricity correction function above is only approximate, being based on the assumption of negligible eccentricity perpendicular to the primary x-ray beam and a primary beam consisting of only parallel rays, Beu and Scott[11] presented an exact analytical method for correcting cylindrical film measurements for sample eccentricity. The method requires accurate measurements of the

---

*This angle σ is not to be confused with the standard deviation σ commonly used to express the reliability of results.

camera radius in three directions. These data permit the analytical calculation of the eccentricity vector and angle (u, σ), the true camera radius R, and the corrections $\Delta\phi$ or $\Delta\theta$. The mathematical details are too lengthy to warrant presentation here. In summary, they find that the Bradley and Jay correction is satisfactory with the average camera, if the magnitude of the eccentricity vector u is not greater than 0.01 mm, and if the eccentricity correction need not be more accurate than $\Delta\phi = \Delta[(\pi/2) - \theta] = 0.01°$. If one requires corrections on the order of $\Delta\phi = 0.001°$ and $u \geq 0.01$ mm, then the exact method must be used to calculate accurate corrections. Beu and Scott[12] have demonstrated that precision cameras can be built in which $u \leq 0.00005$ in. $= 0.0013$ mm, and for which eccentricity corrections are not greater than $+0.0015°\phi$.

8-2.3 Sample Absorption and Radial Divergence of the Beam

The effect of absorption of the x-ray beam by a cylindrical sample is shown in Fig. 8-4. In the absence of absorption the incident beam AA' strikes the sample C and gives rise to a diffracted beam which impinges upon the film over the interval $E'D'$, the geometric center being at C'. A highly absorbing powder, however, causes the diffraction to take place almost exclusively from the surface of the sample, with the result that the observed line extends only over the zone $B'D'$. In short, absorption shifts the center of gravity of the line to an angle somewhat *larger* than the theoretical angle 2θ. For samples of moderate or low absorbing power the line displacement is less than that shown in Fig. 8-4. From the figure it can be deduced that the line displacement decreases as 2θ increases.

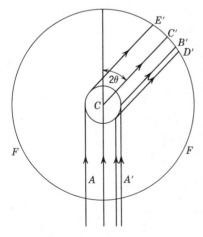

Fig. 8-4. Line displacement due to absorption.

A. Bradley and Jay's Approximate Treatment. Space does not permit a rigorous analysis of the absorption error, but the somewhat approximate method of Bradley and Jay for a highly absorbing sample will serve rather satisfactorily to show the angular dependence of the errors resulting from the simultaneous action of absorption and beam divergence. Referring to Fig. 8-5, consider a cylindrical powder specimen immersed in a beam diverging from a point source X. If the specimen were only a point at A, it would give rise to a diffraction line at B, the theoretical position. As a result of beam divergence and the finite diameter of the sample, the reflection extends from F to G, with the greatest intensity rather near F for a highly absorbing sample. Designating this point of maximum intensity by P (not shown in Fig. 8-5), we see that it is necessary to determine the extent of the line displacement BP.

If for the moment the incident beam is assumed to be parallel, the diffracted rays will produce a line extending from K to L. Since the angle

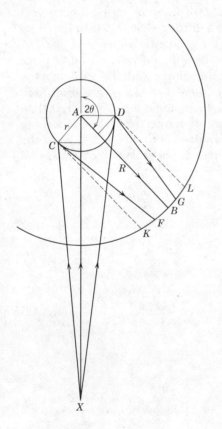

Fig. 8-5.

$CAD = 2\theta$, it can be seen from the figure that the displacements of the line edges are $BK = r$ and $BL = -r \cos 2\theta$, r being the radius of the specimen. The added effect of beam divergence is to move the line boundaries K and L to F and G, respectively. The extent of these displacements, KF and LG, can be evaluated as follows. An examination of the geometry of Fig. 8-5 shows that

$$\angle GDL = \angle AXD \cong \frac{r}{AX}$$

and

$$\angle FCK = \angle AXC \cong -\frac{r}{AX} \cos 2\theta,$$

from which

$$LG = \frac{r}{AX} R$$

and

$$KF = -\frac{r}{AX} R \cos 2\theta,$$

R being the camera radius. The net displacement of the low-angle edge of the line G from its theoretical position is then

$$BG = LG - BL$$

$$= r\left(\frac{R}{AX} + \cos 2\theta\right). \tag{8-12}$$

In this analysis the convention is adhered to of calling a displacement toward a larger angle 2θ positive, and toward a lower angle negative. Similarly, the displacement of the high-angle edge of the line F is

$$BF = BK - KF$$

$$= r\left(1 + \frac{R}{AX} \cos 2\theta\right). \tag{8-13}$$

The displacement of the geometrical center of the line is given by the mean of equations 8-12 and 8-13. By trigonometric manipulation this is found to be

$$\tfrac{1}{2}(BF + BG) = \frac{r}{2}(1 + \cos 2\theta)\left(1 + \frac{R}{AX}\right)$$

$$= r \cos^2 \theta \left(1 + \frac{R}{AX}\right). \tag{8-14}$$

However, the net displacement of the line is not best defined as the mean of BF and BG, but rather as the distance from B to the center of gravity of the line or to the point of maximum intensity P. This latter point Bradley and Jay place, without proof, at a distance from B of

$$BP = \frac{r \sin 2\theta}{2\theta}\left(1 + \frac{R}{AX}\right). \tag{8-15}$$

Figure 8-6 shows the relationship between Bradley and Jay's peak intensity function, the geometrical center, and the low- and high-angle boundaries of the diffraction line as a function of θ for a typical experimental case ($R = 50$, $AX = 100$, and $r = 0.25$ mm, and very high absorption). In order to give the curves more general significance, the displacements are plotted as fractions of the sample radius $\Delta r/r$. It should be noted that Bradley and Jay's approximate peak-intensity function does indeed fall close to the high-angle side of the reflection except at very large Bragg angles, as expected for reflections from a highly absorbing sample.

If this result is assumed valid, the equivalent angular error is found to be

$$\Delta\theta = \frac{BP}{2R}$$

$$= \frac{r}{2R}\left(1 + \frac{R}{AX}\right)\frac{\sin\theta\cos\theta}{\theta}. \tag{8-16}$$

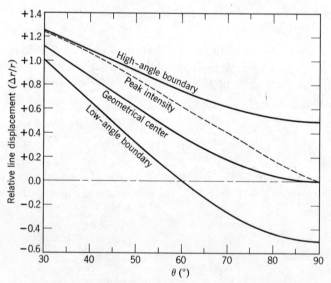

Fig. 8-6.   Bradley and Jay's analysis of line displacement due to sample absorption.

According to equation 8-3, the relative error in a is then

$$\frac{\Delta a}{a} = -\cot\theta\,\Delta\theta$$

$$= -\frac{r}{2R}\left(1+\frac{R}{AX}\right)\frac{\cos^2\theta}{\theta}. \tag{8-17}$$

These equations show that absorption errors, like the others discussed previously, vanish at $\theta = 90°$.

B. More Rigorous Investigations of the Absorption Error. Warren [13] and Taylor and Sinclair[14] have examined more rigorously the effect of sample absorption upon line position. Warren considered in detail the two cases of parallel radiation and radiation diverging from a point source, and he appraised the line shift in terms of the position of the *center of gravity* of the diffracted beam. Taylor and Sinclair's treatment differs from Warren's by, first, considering the effect of an x-ray source of finite extent and, second, defining line shift in terms of the displacement of the *point of maximum intensity*. Buerger[15] has also investigated the absorption error, but his definition of the line shift as the displacement of the *geometric center* of the bundle of diffracted rays seems less justifiable than either of the other definitions and leads to a different result.

The results of the various investigations are compared in Fig. 8-7, W referring to Warren, TS to Taylor and Sinclair, B to Buerger, and BJ to Bradley and Jay. As in Fig. 8-6 the displacements are plotted as fractions

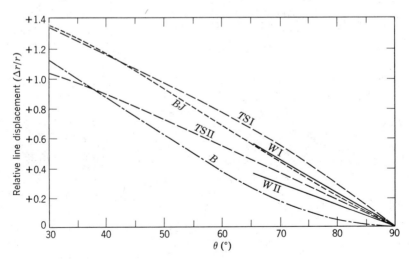

Fig. 8-7. Various analyses of line displacement compared.

of the sample radius, $\Delta r/r$. The characteristics of the six curves are presented in Table 8-1. In all but curve TS II, a highly absorbing sample is assumed, and even in this one case $\mu r = 2$ represents considerable absorption, so that in a general way at least this curve is still comparable with the others. Curves TS I and W I differ principally because of the different conceptions of the peak position, respectively, *point of maximum intensity* and *center of gravity*. The conditions prevailing in most Debye-Scherrer diffraction work are probably portrayed most faithfully by curve TS II. The very close agreement of W I and TS II with Bradley and Jay's curve BJ is a further illustration of the essential correctness of their approximate treatment of absorption errors, at least for $\theta > 60°$. Curve W II is significant mainly in connection with the parallel radiation that can be produced by crystal monochromatization. Curve B is Buerger's displacement curve, which gives the location of the midpoint of the bundle of diffracted rays and follows directly from Bradley and Jay's derivation without any approximations.

Taylor and Sinclair's curve II is very nearly a linear function of ϕ, or $(\pi/2) - \theta$, for $\theta > 60°$, and Warren's curves I and II are both rather accurately linear in $\tan \phi$, hence approximately linear in ϕ for the same angular range. For moderate absorptions Warren found an even more exact linear dependence of $\Delta r/r$ upon ϕ, and only for low absorption (the

Table 8-1.   Key to the Curves in Fig. 8-7

Curve	R/AX	Sample Absorption (μr)	X-ray Source	Radiation	Equation of $\Delta r/r$ Curve if Definable
TS I	0.45	∞	Point	Divergent	
W I	0.50	∞	Point	Divergent	$\dfrac{\pi}{4}\left(1+\dfrac{R}{AX}\right)\tan\dfrac{\phi}{2}$
BJ		∞	Point	Divergent	$\left(1+\dfrac{R}{AX}\right)\dfrac{\sin 2\theta}{2\theta}$
TS II	0.63	2	Exponential	Divergent	$k\left(\cos\theta+\dfrac{\sin 2\theta}{2\theta}\right)$
W II		∞		Parallel	$\dfrac{\pi}{4}\tan\dfrac{\phi}{2}$
B		∞	Point	Divergent	$\left(1+\dfrac{R}{AX}\right)\cos^2\theta$

exceptional experimental case) did he discover a different functionality, namely, a proportionality to sin ϕ. Buerger's result alone is not in keeping with the general tenor of these findings. By way of summarizing these important investigations, it can be said that, when the true line position is defined as either the *point of maximum intensity* or the *center of gravity*, mathematical analysis demonstrates a very close linear dependence of the line displacement upon ϕ, or $(\pi/2) - \theta$, for Bragg angles larger than about 60 or 65°. Furthermore, it is again noteworthy that these analyses indicate the vanishing of absorption errors when $\theta = 90°$.

Straumanis[16, 17] attempts to eliminate almost completely the absorption error experimentally by using a very small low-absorbing sample (Section 8-3.2A). Others rely upon various extrapolation techniques to correct for absorption, and Beu[4] calculates profiles due to absorption and radial divergence using a modified Taylor-Sinclair[14] line contour matrix method. In Beu's method the experimentally determined absorption coefficient of the sample powder and the capillary or fiber on which the powder is mounted, the sample diameter, and the camera dimensions are taken into account.

8-2.4 Axial Divergence of the Beam

In comparison with the factors discussed above, axial (vertical) divergence is a negligible source of error except when slits of considerable height are employed. Bradley and Jay[5] have shown that the center of blackening of a line is shifted by an angular amount

$$\Delta\theta = -\frac{1+x}{96}\left(\frac{h}{R}\right)^2 \cot 2\theta \tag{8-18}$$

due to axial divergence h of the reflected beam in a camera of radius R. The quantity x is a fraction between 0 and 1, the precise value of which depends upon whether the beam is assumed to be uniform or to fade out at the edges. The equivalent relative error in the lattice constant beomes

$$\frac{\Delta a}{a} = -\cot\theta\,\Delta\theta$$

$$= \frac{1+x}{192}\left(\frac{h}{R}\right)^2 (\cot^2\theta - 1). \tag{8-19}$$

Evidently, at large angles this factor decreases the measured value of a, since $\cot^2\theta - 1 = -1$ at 90°. It is noteworthy that this is the first error that does not vanish at $\theta = 90°$. Instead it is zero at 45° and is positive at lower angles and negative at higher angles. The best way of combating this source of error is virtually to eliminate it by using only pinhole collimators with small apertures. Under such circumstances h/R is so small that the error in a is trivial in comparison with those arising from sample eccentricity and absorption.

Table 8-2 summarizes the mathematical expressions for the angular errors and the relative error in the lattice constant arising from the several sources that have been discussed in the preceding pages.

Beu et al.[18] have attacked the problem of axial divergence and developed an exact graphical method for calculating axial divergence profiles. For precision-type powder cameras, the profiles can easily be plotted with an accuracy greater than the best precision of measurement of the diffraction lines (better than $0.002° \theta$). The method is based on the assumptions that (1) the x-ray source is a line of uniform brightness, and (2) the sample is a line parallel to the source. In the case of a spot focus, the profile becomes very narrow. The reader is referred to reference [18] for details of the equations and the graphical procedure. Straumanis and Ievinš[16] have claimed that they eliminated the systematic errors by precise experimental technique. Hence Beu et al.[18] investigated the lattice constant of chromium reported by Straumanis and Weng[19], based on the 321 diffraction line at about $175° 2\theta$ for $CuK\alpha_1$ radiation.

Table 8-2.   Mathematical Expressions for the Various Errors in θ and a Encountered in the Debye-Scherrer Method[a]

Source of Error	$\Delta\theta$	$\Delta a/a$
Radius errors	$\left(\dfrac{\pi}{2}-\theta\right)\dfrac{\Delta R}{R}$	$-\dfrac{\Delta R}{R}\left(\dfrac{\pi}{2}-\theta\right)\cot\theta$
Uniform film shrinkage	$-\left(\dfrac{\pi}{2}-\theta\right)\dfrac{\Delta S'}{S'}$	$\dfrac{\Delta S}{S}\left(\dfrac{\pi}{2}-\theta\right)\cot\theta$
Sample eccentricity	$\dfrac{u}{R}\cos\sigma\sin\theta\cos\theta$	$-\dfrac{u\cos\sigma}{R}\cos^2\theta$
Absorption and radial divergence of the the beam	$\dfrac{r}{2R}\left(1+\dfrac{R}{AX}\right)\dfrac{\sin\theta\cos\theta}{\theta}$	$-\dfrac{r}{2R}\left(1+\dfrac{R}{AX}\right)\dfrac{\cos^2\theta}{\theta}$
Axial divergence of the beam	$-\dfrac{1+x}{96}\left(\dfrac{h}{R}\right)^2\cot 2\theta$	$\dfrac{1+x}{192}\left(\dfrac{h}{R}\right)^2(\cot^2\theta-1)$

[a]Principally after Bradley and Jay[5].

They found that Straumanis and Weng were justified in neglecting axial divergence even at the high anle of 175° 2θ. Indeed, it can be generally stated that the axial divergence corrections may usually be neglected in the 2θ range of 120° to 175° in high-precision work, if a spot focus is used with the collimator geometry of Straumanis and Weng[19], and if the measurements are made in the equatorial plane. In another study Langford, Pike, and Beu[20] have presented an analytic method for deriving axial-divergence line profiles, centroid shifts, and variances for cylindrical powder cameras.

8-3 METHODS OF CORRECTING FOR ERRORS IN THE DEBYE-SCHERRER METHOD

Three principal approaches have been made to the problem of correcting for the various errors that affect the determination of lattice constants by the Debye-Scherrer technique. They include (1) the use of calibrating substances, (2) refinements in experimental technique, and (3) mathematical procedures for minimizing the errors.

8-3.1 Use of Calibrating Substances

This method is now mainly of historical interest as far as highly precise measurements are concerned, although it still is occasionally employed for special purposes. It consists in utilizing a substance possessing a known lattice constant to calibrate the film. The substance, commonly sodium chloride (NaCl) or quartz, is preferably mixed with the material to be investigated to yield a composite powder pattern. Formerly, a septum technique was often employed by means of which the two substances simultaneously generated their respective diffraction patterns side by side on a single strip of film[21], but this method has fallen into disuse in favor of the simpler method of mixtures. Lines of the standard substance are used to construct a smooth curve of the true Bragg angle or some related quantity, such as d, $\sin \theta$, or $\sin^2 \theta$, as a function of linear position on the film. The true angles characterizing the lines of the unknown substance can then be determined, in principle at least, by direct reference to this calibration curve. The method of mixtures was first employed by Gerlach and Pauli[22], and subsequently it was used by Ott [23], Havighurst, Mack, and Blake[24], and others to determine the lattice constants of the alkali halides with a degree of precision not previously attained.

The fundamental objection to the use of a calibrating substance is that such a method cannot give results of greater accuracy than the accuracy with which the lattice constant of the calibrating substance is known. Furthermore, the purity, hence the lattice spacing, of any reference substance is likely to be in doubt by a smaller or larger amount. For example, the alkali halides readily form solid solutions with each other, and even slight contamination of this sort noticeably affects the lattice constants. Also rock salt and chemically prepared NaCl do not possess identical spacings. As regards quartz, which has been much used as a reference material, it has been recently found that the lattice dimensions of clear, colorless specimens vary perceptibly with the presence or absence of minute amounts of foreign ions[25].

8-3.2 The Straumanis Method of Refined Experimental Technique

By exceptionally precise camera construction together with the utilization of several refinements of experimental technique, it is possible to reduce lattice-spacing errors to the vanishing point and obtain highly accurate spacings directly from the observed positions of lines in the back-reflection region. The foremost exponent of this mode of attack on the problem of errors is Straumanis, who began developing the method about 1934 at the University of Latvia (Riga). Previous to this, Küstner[26] and Blake[27] had used cylindrical specimens of small diameter (as little as 0.2 mm) to reduce the displacement of lines due to absorption in the sample. Blake in addition eliminated sample eccentricity mechanically and allowed for film shrinkage and radius errors by a special method of calibration. In 1927, van Arkel[28] introduced the van Arkel, or precision, film position (see Fig. 6-5B) by means of which an accuracy approaching 0.01 per cent could be achieved without the use of a calibrating substance. There followed an increasing use of the lines at high values of θ, with the realization that most of the errors in line position vanish at $\theta = 90°$. This added refinement permitted an accuracy as high as 0.005 per cent to be realized in favorable cases.

A. Essential Features of the Straumanis Method[4, 16, 17]. (1) Errors due to uniform film shrinkage and inexact knowledge of the camera and film radii are eliminated by placing the film in the asymmetric, or Straumanis, position (see Fig. 6-5C). This position of the film is the most satisfactory one to employ, regardless of which method of precision measurement is adopted (see the remainder of this chapter and especially the summary of methods in Table 8-11). (2) Sample eccentricity is reduced to the vanishing point by highly precise centering of the sample spindle in the camera. (3) Line shifts due to absorption in the sample are eliminated by utilizing lines in the back-reflection region, decreasing the sample diameter, and making the sample more transparent to x-rays.

The sample, 0.2 mm or less in diameter, consists of a thin layer of finely sieved powder adhering to the outside of a practically nonabsorbing beryllium–lithium–boron glass hair having a diameter of about 0.08 mm [29]. Figure 8-8 is a magnified view of a completed powder mount 0.12 mm in diameter. (4) For cameras 114.6 mm in diameter and larger, a method of measuring line positions with a precision between 0.01 and 0.02 mm must be employed. Although the accuracy required depends in part upon the camera diameter, for any practicable camera dimensions this necessitates either the use of a comparator of high precision or the construction of a film-measuring instrument of high accuracy. (5) In addition to the foregoing features of technique, it is absolutely necessary to insure that the sharpness of the lines is not impaired as a result of thermal expansion and contraction of the lattice during the course of the exposure. Furthermore, in order that lattice-constant measurements may be of any general significance, they must be directly comparable with the results of other investigations. This requires not only constancy of temperature during an exposure, but also a knowledge of what the temperature is. Variations of several degrees affect the fourth decimal place of the lattice constant in typical cases, and variations of even a few tenths of a degree affect the fifth. Hence, on the grounds of both optimum accuracy and precision, effective control of the specimen temperature to $\pm 0.1°C$, or better, is mandatory.

Straumanis and Ievinš[16, 17] have described several thermostatic devices for controlling camera temperature, including both air and water as the circulating media. The usual thermostatic control techniques are so widely known that details are not presented here. Klug and Alexander[30], in the first edition of this book, present brief details of Straumanis' techniques. Beu, Musil, and Scott[31] placed a thermistor (mounted in a hypodermic needle) in the camera about 1 mm from the

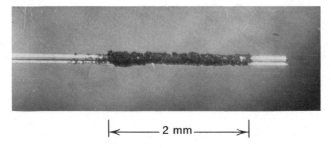

|←——— 2 mm ———→|

Fig. 8-8. Completed mount of WO_2 powder on a Lindemann glass fiber (20×). [Courtesy of M. Straumanis, *J. Appl. Phys.*, **20**, 726 (1949).]

sample and just outside the x-ray beam. Sample temperatures were then read to 0.05°C from a chart record during the exposure.

As a result of a careful study, Straumanis came to the conclusion that a large camera diameter is not necessary for optimum accuracy of measurement[17]. Using aluminum of 99.998 per cent purity as the subject, he found that cameras 57.7 mm and more in diameter all gave equally accurate results, provided only that a comparator with a discrimination of ±0.01 mm or better was employed. If such an instrument is not available nearly equal accuracy can be realized by employing a 114.6-mm camera in conjunction with a measuring instrument possessing a discrimination of ±0.02 mm. The essential features of such an instrument are described in Section 6-1.

B. Illustrative Film Measurements and Calculations. Figure 8-9 and Table 8-3, taken together, illustrate the measurements and calculations involved in a precision determination of the lattice constant of high-purity aluminum by the Straumanis method[32]. Figure 8-9 represents the appearance of the diffraction pattern. The insert shows how the incident beam entered the cylindrically disposed film at B and departed at A. Unfiltered CuK radiation was employed in order to provide a favorable arrangement of lines in the back-reflection region. The camera

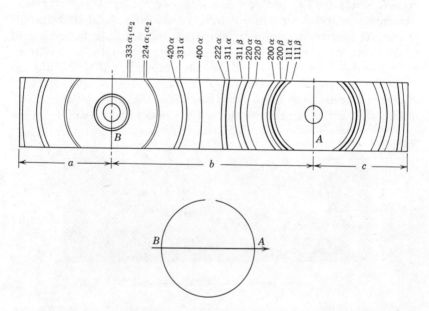

Fig. 8-9.   Debye-Scherrer pattern of aluminum. Film in asymmetric position.

diameter was 57.7 mm, the sample diameter 0.18 mm, the sample temperature 23.10°C, and the exposure time 8 hr. The course of the determination involved first measuring the five lines 111 β, 200 β, 200 α, 220 β, and 220 α on each side of line A (Fig. 8-9) to obtain the mean position of A. Likewise the lines 224 α_1, 224 α_2, 333 α_1, and 333 α_2 about B (Fig. 8-9) were measured to provide the mean value of line B. In a fashion analogous to Table 6-1, these values yield the distance BA of Fig. 8-9, one-half the film circumference, as 90.205 mm. In turn the degrees per millimeter on the film are: 180/90.205 mm = 1.99545. Table 8-3 gives the steps in the calculation of the Bragg angle θ for each of the lines in the far back-reflection region, followed by the computation of the corresponding lattice constants corrected for refraction (see Section 2-3.5). From equation 2-18 the value of δ for aluminum is found to be 8.4×10^{-6}, hence the observed values of a must be increased by an amount $4.04 \times 8.4 \times 10^{-6} = 0.00003$ Å.

Because of the much greater sensitivity of the lattice constant to line position at very high angles, Straumanis has found that the ultimate precision can be attained only for lines in the far back-reflection region ($\theta > 80°$). Hence in Table 8-3 only the two values of a deduced from the

Table 8-3. Precision Determination of the Lattice Constant of Aluminum, Using Back-Reflection Lines (Method of Straumanis[a][16])

| | Reflections about the point B | | | |
Item	224 α_1	224 α_2	333 α_1	333 α_2
Position of line in region b	120.986	120.63	108.44	107.45
Position of line in region a	78.334	78.72	90.91	91.89
4ϕ in mm	42.652	41.91	17.53	15.56
ϕ in degrees ($4\phi \times 0.498863$)	21.2775	20.9073	8.7451	7.7623
θ in degrees	68.7225	69.0927	81.2549	82.2377
sin θ	0.931834	0.934159	0.988374	0.990837
a (kX)	4.04132	4.04132	4.04125	4.04127
a (Å)	4.04948	4.04948	4.04941	4.04943
a corrected for refraction (Å)	4.04951	4.04951	4.04944	4.04946
Best value of a (using only the 333 lines)			4.04945 Å	

[a]See Fig. 8-9.

333 α_1 and 333 α_2 lines would normally be averaged together to give a best value of the lattice constant from this particular film. These illustrative calculations also serve to emphasize another principal feature of the Straumanis technique, namely, the importance of choosing the correct radiation to yield good-quality reflections in the far back-reflection region.

8-3.3 The Convolution-Film Method with the Likelihood Ratio Method

Beu and his associates[4,33] developed a film method for lattice-parameter determination at the $0.001° \theta$ level based on a modified Straumanis technique[16,17]. The Straumanis experimental refinements are used. Only diffraction lines with $\theta > 60°$ are measured, and these are corrected for residual systematic errors. The corrected measurements are then used to calculate the maximum likelihood estimate of the lattice constant $\hat{a}$ by means of the likelihood ratio method.

A. The Convolution-Film Method (CFM). The convolution-film method (CFM) uses centerline peak measurements. Next, error profiles are determined which are convoluted to obtain a measure of the total displacement of the measured peak position from its true 2θ position. Profiles resulting from absorption and radial divergence are calculated using a modified Taylor-Sinclair line contour matrix technique[14]. The axial divergence profiles are calculated for the sample and camera geometry as described by Beu et al.[18,20]. The source profile, as viewed by the sample, is obtained from pinhole images of the x-ray tube target by microphotometering these images. The convolution profile syntheses are carried out as described in Section 5-2.1.

Since the Lorentz and polarization factors do not have profiles, they are applied multiplicatively point by point to the profile in the vicinity of the peak. The dispersion correction can be ignored at the $0.001° \theta$ level. The eccentricity correction of Beu and Scott[11] is then added, after the above corrections by convolution and multiplicative procedures. The refraction correction is applied to the final corrected a value. No corrections are required for camera radius, zero position, and uniform film shrinkage when the Straumanis technique is used. Lines on the film are measured to about $0.002° \theta$, using a manual densitometer (capable of an accuracy of $0.001° \theta$) and printed calibration lines on the film[8]. Non-uniform film shrinkage in the vicinity of a diffraction line is also corrected by this measurement procedure. Finally, the diffraction lines, measured with a precision of about $0.002° \theta$ and corrected for systematic errors with an accuracy of about $0.001° \theta$, are ready for calculating a by the likelihood ratio method.

B. The Likelihood Ratio Method (LRM). The relatively poor agreement in the results of the IUCr Project on Lattice Parameters[1] convinced Beu and his associates[34] that systematic errors in the measurements of the various laboratories had not been eliminated. They point out that all previous lattice-parameter techniques have lacked a suitable statistical means for testing the accuracy of the final calculated lattice-parameter values. Even a least-squares extrapolation, with or without weighting, despite its statistical implications, does not provide data for assessing the parameter accuracy. Thus it has been difficult or impossible to determine whether or not the systematic errors have actually been removed within the precision of measurement.

Accordingly, Beu, Musil, and Whitney[34] presented the results of a study seeking to evaluate critically the maximum practical precision and accuracy attainable in lattice-parameter determinations. Their approach uses a modified Straumanis technique (or the convolution-film method) and involves measuring back-reflection line positions on films to a precision of about $\pm 0.001°\ \phi$. Each measurement is corrected for geometrical systematic errors, and the lattice parameter is calculated from the corrected data by means of a rigorous statistical method called the likelihood ratio method (LRM). The LRM in turn provides an accurate estimate of the lattice parameter $\hat{a}$ and its variance S_a^2. It also provides a statistical test (the likelihood ratio test, LRT) as to whether or not the systematic errors have been eliminated from the measurements.

The LRM is based on the principle that the lattice parameter of a given crystalline substance should be the same regardless of the hkl reflection from which it is calculated, provided the angular position values are free from all systematic errors. Indeed, the agreement among themselves of lattice constants calculated from several different reflections is an indication that the systematic errors have truly been removed from the corrected angles within the precision of measurement.

Details of the LRM and LRT can be presented here only in great brevity. Users of the technique will want to study the various original articles[34–36] and their appendixes. In practice the Bragg angle ψ for each chosen $h_i k_i l_i$ reflection is measured n_i times, and its variance S_i^2 calculated:

$$S_i^2 = \frac{1}{n_i} \sum_\alpha (\psi_{i\alpha} - \psi_i)^2. \qquad (8\text{-}20)$$

Here $\psi_{i\alpha}$ is the αth measurement of the ith diffraction angle, corrected for known systematic errors, and ψ_i is the average of n_i measurements of $\psi_{i\alpha}$. Next the maximum likelihood estimates of the parameters $\hat{a}$, $\hat{\sigma}_i$, $\hat{e}_i$, and $\hat{\theta}_i$ are calculated from the relations:

$$\hat{a} \sin \hat{\theta}_i = K_i = \frac{n\lambda \sqrt{(h^2 + k^2 + l^2)}}{2} \qquad \text{(cubic case)} \qquad (8\text{-}21)$$

$$\hat{\sigma}_i^2 = S_i^2 \qquad (8\text{-}22)$$

$$\hat{e}_i = \psi_i - \hat{\theta}_i \qquad (8\text{-}23)$$

$$\sum_i \hat{e}_i = 0. \qquad (8\text{-}24)$$

In the above θ_i is the true but unknown value of the Bragg angle, e_i is the unknown systematic error in θ_i, and the method assumes that $\sum_i e_i = 0$. The first step is to determine $\hat{a}$ by estimating a, calculating the θ_i, summing the e_i values, and plotting $\sum_i e_i$ versus a for several estimated values of a (Fig. 8-10).

The hypothesis H is then made that there are no remaining systematic errors in the corrected $\psi_{i\alpha}$ values:

$$H: \qquad e_1 = e_2 = e_3 = \cdots = e_m = 0. \qquad (8\text{-}25)$$

Under hypothesis H the maximum likelihood estimates $\hat{a}$ and $\hat{\sigma}_i^2$ are determined using a function $W(a)$, where

$$W(a) = \sum_i n_i \ln \left[1 + \frac{(\psi_i - \theta_i)^2}{S_i^2} \right]. \qquad (8\text{-}26)$$

A graph of $W(a)$ versus a is examined for a minimum value of $W(a)$ in the vicinity of $\hat{a}$. The value of a at this minimum is $\hat{\hat{a}}$, and the minimum value of $W(a)$ is designated $W(\hat{\hat{a}})$ or W_m (Fig. 8-11). It is now possible to calculate $\hat{\hat{\theta}}_i$ from

$$\hat{\hat{a}} \sin \hat{\hat{\theta}}_i = K_i = \frac{n\lambda \sqrt{h^2 + k^2 + l^2}}{2} \qquad \text{(cubic case)}, \qquad (8\text{-}27)$$

and from the values of $\hat{\hat{\theta}}_i$

$$\hat{\hat{\sigma}}_i^2 = S_i^2 + (\psi_i - \hat{\hat{\theta}}_i)^2 \qquad (8\text{-}28)$$

and

$$S_a^2 = \frac{\hat{\hat{a}}^2}{\sum_i (n_i / \hat{\hat{\sigma}}_i^2) \tan^2 \hat{\hat{\theta}}_i}. \qquad (8\text{-}29)$$

The final test that $\hat{\hat{a}}$ and S_a have significance is accomplished by the LRT (Mood[37]). Hypothesis H is tested by means of W_m, which can be shown to have a distribution approximately like chi-square (χ^2) with $m - 1$ degrees of freedom, where m is the number of diffraction lines measured. The value of W_m is compared with a critical value w_ϵ, where ϵ

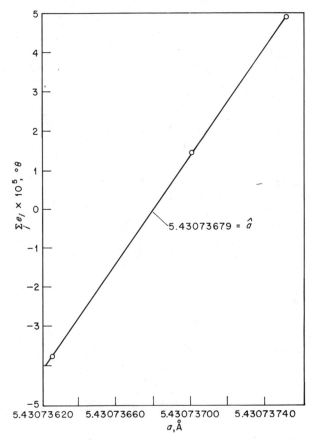

Fig. 8-10. $\sum_i e_i$ versus a for silicon lattice-constant data corrected for refraction only (expanded scale). [Courtesy of K. Beu, F. Musil, and D. Whitney *Acta Crystallogr.*, **15**, 1292 (1962).]

is the significance level chosen for the test by the investigator (commonly $\epsilon = 0.05$). Hodgman[38] has provided a table from which values of w_ϵ may be obtained.

If $W_m \geq w_\epsilon$, there is $100(1 - \epsilon)$ per cent chance that systematic errors remain in the corrected measurements and calculations. However, if $W_m < w_\epsilon$, one may conclude with the same per cent confidence that the values of $\hat{a}$ and S_a have no remaining systematic errors. When the LRT shows systematic errors still remain, it is necessary to reevaluate and improve the techniques for removing such errors.

C. Application of the Convolution-Film Method to IUCr Silicon. Beu had participated in the IUCr lattice parameter project[1]. It was to be expected therefore that Beu and Scott[33] would apply their highly

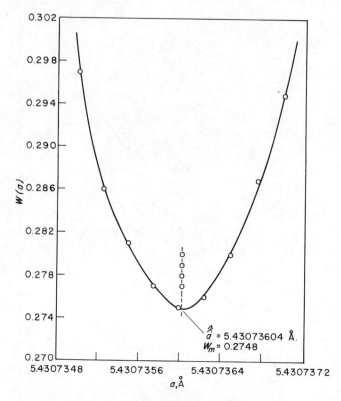

Fig. 8-11. $W(a)$ versus a for silicon lattice-constant data corrected for refraction only (expanded scale). [Courtesy of K. Beu, F. Musil, and D. Whitney, *Acta Crystallogr.*, **15**, 1292 (1962).]

precise CFM to a redetermination of the lattice parameter of IUCr silicon. A refined experimental technique for taking powder photographs, described by Scott and Beu[8], was used, and reflection angles were measured by the centerline peak technique to about $0.002°\ \theta$. Beu and Scott listed 18 factors that may contribute to systematic errors in lattice-parameter determination by the CFM. However, with the refined experimental procedures used, this number is reduced to five, of which only three contribute directly to systematic error in the Bragg angle. Of these three, only one, the absorption profile, requires extensive calculations. The course of the determination and the results are best presented by reference to Table 8-4.

It is seen in Table 8-4 that when no correction is made, or when only absorption is corrected for, the function $W_m > w_\epsilon$, indicating that

Table 8-4. Systematic Error Corrections and LRM Analysis of IUCr Silicon Data [33] [a,b]

Type of Correction	Reflection	Peak Correction, $^\circ\phi$	Corrected Angle, $^\circ\phi$	Standard Deviation	$\hat{e}_i, ^\circ\phi$	W_m
None	(620)		26.2090	0.0035	-0.0084	
	(533)		21.5364	0.0031	-0.0039	
	(444)		10.6696	0.0017	$+0.0123$	
						20.0
Absorption	(620)	$+0.0263$	26.2353		$+0.0072$	
	(533)	$+0.0210$	21.5574		$+0.0038$	
	(444)	$+0.0045$	10.6741		-0.0110	
						17.5
						$5.99 = w_\epsilon$
Absorption	(620)	$+0.0126$	26.2216		-0.0012	
and x-ray	(533)	$+0.0108$	21.5472		0.0000	
source (all	(444)	$+0.0031$	10.6727		$+0.0012$	
corrections)						0.86

[a]Data gathered at 32°C.
[b]$\hat{a}_1 = 5.430356$ Å (corresponding to $W_m = 0.86$, corrected to 25°C, corrected for refraction, and using $\lambda = 1.540510$ Å as proposed for the IUCr lattice-constant study). $\hat{a}_2 = 5.430540$ Å (corrected as above, but using Bearden's $\lambda = 1.540562$ Å). S_a, standard deviation estimate of $\hat{a} = 0.000014$ Å. 95 per cent confidence limits of $\hat{a} = \pm0.000028$ Å.

systematic errors still remain in the data. However, when absorption and all x-ray source errors are corrected, $W_m < w_\epsilon$, and it may be concluded that $\hat{a}$ and S_a are free of systematic errors (within 95 per cent confidence limits). The lattice constant $\hat{a}_1$ was calculated using the value $\lambda = 1.540510$ Å proposed by Parrish[1] for the international precision lattice-constant study. The preferred lattice constant is $\hat{a}_2$, based on Bearden's[3] highly precise and accurate centerline peak value $\lambda = 1.540562$ Å for CuKα_1. This silicon lattice constant, $\hat{a}_2 = 5.430540$ Å, is believed to be accurate on an absolute basis to 1 part in 200,000. The coefficient of thermal expansion for silicon used in the calculations is $\alpha = 2.7 \times 10^{-6}$, as determined by Taylor, Mack, and Parrish[39].

8-3.4 Use of Extrapolation Methods

A. Bradley and Jay's Extrapolation against $\cos^2 \theta$ [5]. Bradley and Jay's analysis of the systematic errors in the Debye-Scherrer method led to the defining of each error as a fairly definite function of the Bragg angle θ (see Table 8-2). Their procedure was, first, to minimize radius errors and errors due to uniform film shrinkage by marking the film with calibration spots at known angles and, second, to allow for eccentricity, absorption, and beam divergence by plotting values of a for lines above $\theta = 60°$ against $\cos^2 \theta$ and extrapolating linearly to $\cos^2 \theta = 0$. By reference to Table 8-2 it is seen that Bradley and Jay's combined error equation can be written

$$\frac{\Delta a}{a} = -\left(\frac{u \cos \sigma}{R} + \frac{r}{2\theta R} + \frac{r}{2\theta A X}\right) \cos^2 \theta, \qquad (8\text{-}30)$$

if it be assumed that radius and film-shrinkage errors are removed experimentally and that axial divergence is negligibly small. This function is not truly linear in $\cos^2 \theta$, but for $\theta > 60°$ it is sufficiently linear to justify a straight-line extrapolation procedure, at least when one or more lines at very high angles are present. A large amount of experimental work since the time of Bradley and Jay's original investigation has been performed with this extrapolation procedure, and the general success that has attended its use attests to its inherent soundness.

The more exact investigations of Warren[13] and of Taylor and Sinclair[14] relative to the absorption error, discussed earlier, revealed a nearly linear dependence of peak displacement due to this cause upon ϕ, or $(\pi/2) - \theta$, for $\theta > 60°$. This may be expressed as

$$\frac{\Delta r}{r} = k\phi,$$

which reflects as an angular error of the same form:

$$\Delta \theta = k' \frac{\Delta r}{r} = K' \phi.$$

If the expression 8-3 is used, the relative error in the lattice constant due to absorption becomes

$$\frac{\Delta a}{a} = -\cot \theta \Delta \theta$$
$$= -K' \phi \cot \theta$$
$$= -K' \phi \tan \phi. \qquad (8\text{-}31)$$

Also, from Table 8-2, any residual errors due to radius errors and film shrinkage take the form

$$\frac{\Delta a}{a} = \left(\frac{\Delta S'}{S'} - \frac{\Delta R}{R}\right)\left(\frac{\pi}{2} - \theta\right) \cot \theta$$
$$= \left(\frac{\Delta S'}{S'} - \frac{\Delta R}{R}\right) \phi \tan \phi. \qquad (8\text{-}32)$$

Furthermore, sample-eccentricity errors are given by

$$\frac{\Delta a}{a} = -\frac{u \cos \sigma}{R} \cos^2 \theta$$

$$= -\frac{u \cos \sigma}{R} \sin^2 \phi,$$

and for moderately small values of ϕ this can be approximated by

$$\frac{\Delta a}{a} = -\frac{u \cos \sigma}{R} \phi \tan \phi, \tag{8-33}$$

Hence, for small ϕ's, all the systematic errors can be lumped together into one expression

$$\frac{\Delta a}{a} = K\phi \tan \phi,$$

which is essentially the same as

$$\frac{\Delta a}{a} = K \sin^2 \phi$$
$$= K \cos^2 \theta. \tag{8-34}$$

The preceding analysis substantiates the validity of Bradley and Jay's procedure of plotting the calculated a's against $\cos^2 \theta$ and extrapolating linearly to $\cos^2 \theta = 0$. Actually, there is a slight preference for plotting a against $\phi \tan \phi$, particularly when sample eccentricity is known to be small, but in practice a plot against $\phi \tan \phi$ is not visibly more linear than one against $\cos^2 \theta$ (or $\sin^2 \phi$) for moderately small values of ϕ. Figure 8-12 shows a Bradley-and-Jay-type extrapolation against $\cos^2 \theta$ for a sample of rubidium bromide. The diffraction pattern was obtained in a Debye-Scherrer camera of 114.6-mm diameter, using CuKα radiation and a sample diameter of 0.3 to 0.4 mm.

It should be emphasized that such extrapolations against $\cos^2 \theta$ are

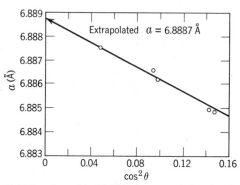

Fig. 8-12. Rubidium bromide (25.9°C) data extrapolated against $\cos^2 \theta$.

most successful when (1) there are numerous, well-distributed lines in the region $\theta = 60$ to $90°$ and (2) there is at least one reliable line with $\theta > 80°$. Under these conditions, and with angle measurements to the $0.01°$ θ level, the position of the extrapolation line is fixed with certainty, and an accuracy of about 1 part in 20,000 is usually realized. Such accuracy is satisfactory for much lattice-parameter work. If the number of lines present is insufficient or they are improperly distributed, it may yet be possible to obtain a good result by (1) using not only $K\alpha_1$ and $K\alpha_2$ lines but also $K\beta$ lines, (2) changing to a different characteristic radiation, or (3) employing an alloy target if one is available. Carapella[40] has published charts to facilitate the correct choice of radiation for precisely determining the lattice constants of cubic materials of any given unit-cell dimension and lattice type.

B. Extrapolation against $(\cos^2 \theta)/\sin \theta + (\cos^2 \theta)/\theta$. A valuable contribution to extrapolation techniques was made independently by Taylor and Sinclair[41], and by Nelson and Riley[42]. Attacking the problem from different angles, these investigators showed that when absorption is the principal source of error in a Debye-Scherrer photograph, and provided the x-ray source has an exponential intensity profile, $\Delta a/a$ is a highly linear function of $(\cos^2 \theta)/\sin \theta + (\cos^2 \theta)/\theta$ down to very low values of θ. For all practical purposes lines at all angles may be employed in making the extrapolation. This discovery has made possible more accurate determinations of the lattice constants of substances possessing such relatively high crystal symmetry that the number of reflections in the region $\theta = 60$ to $90°$ is inadequate for a reliable $\cos^2 \theta$ extrapolation. Furthermore, this method for the first time permits direct precision determinations of the lattice constants of many noncubic crystals from plots of a, b and c values, respectively, from $h00$, $0k0$, and $00l$ reflections spaced over large angular ranges. Although it ignores eccentricity and certain other systematic errors, in favorable cases an extrapolation against the function $(\cos^2 \theta)/\sin \theta + (\cos^2 \theta)/\theta$ can fix the lattice constant with a precision somewhat better than that of the $\cos^2 \theta$ procedure, possibly as high as 1 part in 50,000.

The excellent linearity obtainable with this function is illustrated in Fig. 8-13 by means of four extrapolations for annealed Cu_9Al_{14} filings from the work of Nelson and Riley[42]. In order to test the power and reproducibility of the method, the filings were used to prepare four cylindrical samples of different diameters and dilutions, so that line displacement due to sample absorption differed from one diffraction pattern to another. These variations resulted in pronounced differences in the slopes of the plots, but the figure shows in a striking manner how all

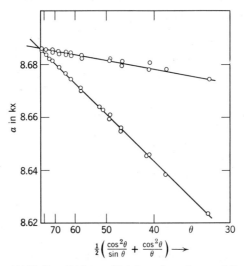

(A) Medium thickness diluted specimen (upper plot).
Thick diluted specimen (lower plot).

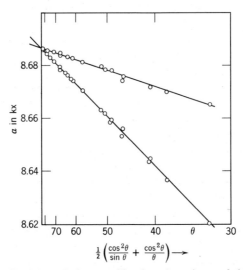

(B) Medium thickness undiluted specimen (upper plot).
Thick undiluted specimen (lower plot).

Fig. 8-13. Large-scale plots of a against $\frac{1}{2}\{[(\cos^2\theta)/(\sin\theta)]+[(\cos^2\theta)/\theta]\}$ for lines with $\theta > 30°$. Annealed Cu_9Al_{14} filings. [Courtesy of J. Nelson and D. Riley, *Proc. Phys. Soc.* (London), **57**, 160 (1945).]

Table 8-5. Values of a for Cu_9Al_{14} Obtained from Extra-polations against $(\cos^2 \theta)/\sin \theta + (\cos^2 \theta)/\theta$[a,b]

Diameter of Specimen (mm)	Dilution with Gum Tragacanth	Temperature (°C)	a (kX)	a (Å)
0.59	yes	15.8	8.6863	8.7038
1.46	yes	15.4	8.6861	8.7036
0.45	no	16.2	8.6866	8.7041
1.35	no	16.4	8.6864[c]	8.7039[c]

[a]Data from Nelson and Riley [42].
[b]Camera diameter 19 cm.
[c]Mean value of a, 8.6864 kX (8.7039 Å).

the points of intersection with the axis of ordinates are practically identical. The extrapolated a values are given in Table 8-5, together with the characteristics of the different samples and the temperatures during the exposures. It is evident that the agreement between the a values is excellent in spite of great variations in the samples and appreciable differences in the temperatures of the exposures. The maximum deviation from the mean a value, 8.6864 kX (8.7039 Å), is only about 1 part in 30,000.

In conclusion, several suggestions of a general nature may be given for attaining optimum results with the $(\cos^2 \theta)/\sin \theta + (\cos^2 \theta)/\theta$ extrapolation technique. First, whenever practicable, use a relatively thin and/or low-absorbing specimen in order to minimize the slope of the plot; the steeper the slope, the less accurately the extrapolated a value can be read from the graph. Second, choose the radiation so as to yield at least one good line in the far back-reflection region ($\theta > 80°$) in order to fix with highest accuracy the point of intersection of the plot with the axis of ordinates. Third, employ only lines between $\theta = 30$ and $90°$ and extrapolate linearly to zero value of the function $(\cos^2 \theta)/\sin \theta + (\cos^2 \theta)/\theta$. Fourth, be reasonably certain that the x-ray tube focus is exponential in character. With sealed-off tubes the focus is most likely to be exponential, but a gas tube may show a more nearly uniform focus. If there is doubt as to whether the x-ray tube focus is more nearly uniform or exponential in character, first prepare a powder photograph of some cubic crystal and draw extrapolation curves of a versus the two functions $(\pi/2 - \theta) \cot \theta$ and $(\cos^2 \theta)/\sin \theta + (\cos^2 \theta)/\theta$ to ascertain which gives the more linear plot over a large angular range. An exponential focus tends to give a more linear result with the second function. If the first function gives the straightest plot, then it should always be used instead of the second function whenever the same x-ray tube is employed.

C. Cohen's Least-Squares Extrapolation[43]. If the angular dependence of the combined systematic errors is expressible as a simple mathematical function, the effect of these errors can be eliminated by a least-squares analytical treatment of the data. The procedure is in principle the analytical equivalent of a graphical extrapolation, but it is superior to such methods in that it is applicable to noncubic crystals. Nevertheless, because of the added complexity of such calculations, the method is most useful in the determination of cubic lattice constants.

Cohen's method can be most readily understood by deriving the mathematical expressions for the cubic case. Suppose that the combined systematic errors take the form

$$\frac{\Delta d}{d} \propto \cos^2 \theta, \tag{8-35}$$

which has been shown earlier to be a justifiable assumption for $\theta > 60°$ when the principal error arises from absorption in the sample. By squaring the Bragg equation and taking logarithms, the following equation is obtained:

$$2 \log d = -\log \sin^2 \theta + 2 \log \frac{n\lambda}{2}. \tag{8-36}$$

Differentiation gives

$$\frac{2\Delta d}{d} = -\frac{\Delta \sin^2 \theta}{\sin^2 \theta} + \frac{2\Delta\lambda}{\lambda},$$

or, since $\frac{\Delta\lambda}{\lambda}$ can be assumed equal to zero,

$$\frac{2\Delta d}{d} = -\frac{\Delta \sin^2 \theta}{\sin^2 \theta}. \tag{8-37}$$

Equating of formulas 8-35 and 8-37 gives

$$\frac{\Delta \sin^2 \theta}{\sin^2 \theta} \propto \cos^2 \theta$$

or

$$\Delta \sin^2 \theta \propto \sin^2 \theta \cos^2 \theta,$$

from which

$$\Delta \sin^2 \theta = D \sin^2 2\theta. \tag{8-38}$$

The meaning of equation 8-38 is that the observed $\sin^2 \theta$ value for any given line above $\theta = 60°$ will be in error by the amount $D \sin^2 2\theta$ as a result of the combined action of the systematic errors. This state of affairs can be expressed mathematically by adding an error term $D \sin^2 2\theta$ to the Bragg equation in its quadratic form. For cubic crystals this gives

$$\frac{\lambda^2}{4a^2} (h^2 + k^2 + l^2)_i + D \sin^2 2\theta_i = \sin^2 \theta_i \qquad (8\text{-}39)$$

or

$$A_0 \alpha_i + D \delta_i = \sin^2 \theta_i, \qquad (8\text{-}40)$$

if $A_0 = \lambda^2/4a^2$, $\alpha_i = (h^2 + k^2 + l^2)_i$, and $\delta_i = \sin^2 2\theta_i$. The subscript i indicates that there is an equation for each line of the diffraction pattern, each with its own values of α, δ, and $\sin^2 \theta$. D is called the "drift" constant, and is a fixed quantity for any one film but differs from one film to another. The variables A_0 and D can be determined by solving simultaneously a pair of equations from any two lines of the pattern; but by solving all possible pairs of equations for A_0 and D and averaging the results, more accurate values can be arrived at. The corresponding value of the lattice constant is then obtained from the relation $a^2 = \lambda^2/4A_0$.

A better procedure for evaluating a, however, consists in applying the least-squares principle in order to minimize the effect of random observational errors. As the result of such errors, relationship 8-40 will not hold exactly for any particular reflection i, but instead the quantity $A_0 \alpha_i + D \delta_i - \sin^2 \theta_i$ will differ from zero by a small amount, v_i:

$$A_0 \alpha_i + D \delta_i - \sin^2 \theta_i = v_i. \qquad (8\text{-}41)$$

According to the theory of least squares, the best values of the coefficients A_0 and D are those for which the sum of the squares of the random observational errors is a minimum:

$$\sum v_i^2 = \text{a minimum} = \sum (A_0 \alpha_i + D \delta_i - \sin^2 \theta_i)^2. \qquad (8\text{-}42)$$

The condition for this to be true is that the first derivatives of $\sum v_i^2$ with respect to the variables A_0 and D shall be equal to zero. Hence

$$\frac{\partial}{\partial A_0} \left(\sum v_i^2 \right) = \sum \alpha_i (A_0 \alpha_i + D \delta_i - \sin^2 \theta_i)$$

and

$$\frac{\partial}{\partial D} \left(\sum v_i^2 \right) = \sum \delta_i (A_0 \alpha_i + D \delta_i - \sin^2 \theta_i).$$

Equating of these expressions to zero gives the pair of normal equations

$$A_0 \sum \alpha_i^2 + D \sum \alpha_i \delta_i = \sum \alpha_i \sin^2 \theta_i,$$

$$A_0 \sum \alpha_i \delta_i + D \sum \delta_i^2 = \sum \delta_i \sin^2 \theta_i. \qquad (8\text{-}43)$$

To compute the best value of a for a number of lines in the region $\theta > 60°$, it is necessary, then, to evaluate the coefficients $\sum \alpha_i^2$, $\sum \alpha_i \delta_i$, $\sum \alpha_i \sin^2 \theta_i$, $\sum \delta_i^2$, and $\sum \delta_i \sin^2 \theta_i$, substitute them in the two normal equations 8-43, and solve simultaneously for A_0. The value of a is then obtained with the aid of the relation $a^2 = \lambda^2/4A_0$.

The application of Cohen's method to a cubic substance is illustrated in Table 8-6, with data from a Debye-Scherrer pattern of spectrographically pure lead. Three pairs of $\alpha_1 \alpha_2$ doublets of satisfactory quality appear in the region $\theta > 60°$. In order that all the coefficients of the normal equations may for convenience be of the same general magnitude, δ is arbitrarily defined as $10 \sin^2 2\theta$ instead of $\sin^2 2\theta$. Reference to equation 8-40 shows that this has the effect of making the calculated value of the drift constant D one-tenth its true magnitude, but this has no effect on the calculated values of A_0 and a. Before the coefficients involving $\sin^2 \theta$ can be computed, it is necessary that all the $\sin^2 \theta$ values be normalized to a common wavelength. In the present instance the $\sin^2 \theta$'s for the α_2 reflections have been converted to the equivalent $\sin^2 \theta$'s for

Table 8-6. Cohen Least-Squares Calculation of the Lattice Constant of Spectrographic Lead[a,b]

Line	hkl	α	λ	$\theta(°)$	$\sin^2 \theta$	$\sin^2 \theta \to K\alpha_1$	δ
1	531	35	$K\alpha_1$	67.080	0.84833	0.84833	5.1
2	531	35	$K\alpha_2$	67.421	0.85258	0.84835	5.0
3	600	36	$K\alpha_1$	69.061	0.87230	0.87230	4.5
4	600	36	$K\alpha_2$	69.467	0.87698	0.87263	4.3
5	620	40	$K\alpha_1$	79.794	0.96861	0.96861	1.2
6	620	40	$K\alpha_2$	80.601	0.97332	0.96849	1.0

[a]Film no. 398; camera diameter $= 114.6$ mm; 1 mm on film $= 1.00307°$, $CuK\alpha$ radiation; 25°C.
[b]$\sum \alpha^2 = 8242$; $\sum \alpha\delta = 758.3$; $\sum \delta^2 = 92.2$; $\sum \alpha \sin^2 \theta = 199.6853$; $\sum \delta \sin^2 \theta = 18.3767$. Normal equations: $8242.000\,A_0 + 758.3\,D = 199.6853$; $758.3\,A_0 + 92.2\,D = 18.3767$; $A_0 = 0.0242082$; $D = 0.000213$; $a = 4.9505_2$ Å; $a = 4.9506_6$ Å (corrected for refraction).

the α_1 wavelength by multiplying them by $(\lambda\alpha_1/\lambda\alpha_2)^2$. For CuK$\alpha$ radiation this factor is 0.99503. It is important to remember that the value of δ for any given line depends upon the *observed* angle of that line, hence δ must be computed from the observed value of $\sin^2\theta$ rather than from the value of $\sin^2\theta$ normalized to a different wavelength. It is also important to note in equation 8-40 that the term $D\delta_i$ is of much smaller magnitude than the term $A_0\alpha_i$, with the result that values of δ_i need only be calculated to two significant figures, whereas $\sin^2\theta$ is ascertained to five or six significant figures. The lattice constant calculated by Cohen's method, 4.9505_2 Å, compares well with the value 4.9506_0 ($\pm0.0001_0$) Å deduced by a $\cos^2\theta$ graphical extrapolation.

The effect of a radius error of considerable size is practically eliminated by applying either a graphical extrapolation or Cohen's analytical method to the diffraction data. This can be demonstrated by imposing an artificial radius error on a set of Debye-Scherrer data and then repeating the calculation of a. In Table 8-7 the least-squares calculations of Table 8-6 for lead are repeated after first imposing an arbitrary radius error of 0.2 per cent, which reduces the degrees per millimeter on the film from 1.00307 to 1.00107. It is seen that the resultant slight decrease in the calculated value of a (about 2 parts in 500,000) is far less than the experimental error. However, when the lattice constant is calculated from but one line at $\theta = 80°$, a radius error of 0.2 per cent leads to a final error of 6 parts in 100,000. These results illustrate that the employment of sound graphical or analytical extrapolation techniques eliminates the need for determining the film radius with great precision.

The least-squares analytical method is utilized in a different manner in Table 8-8. The subject is a cubic solid solution with the composition 41.33

Table 8-7.   Recalculation of Table 8-6 after Initially Imposing a Radius Error of 0.2 Per Cent[a]

$\Sigma\,\alpha^2 = 8242$	$\Sigma\,\alpha\sin^2\theta = 199.6012$
$\Sigma\,\alpha\delta = 761.8$	$\Sigma\,\delta\sin^2\theta = 18.4512$
$\Sigma\,\delta^2 = 93.2$	

Normal equations:

$8242.000\,A_0 + 761.8\,D = 199.6012$

$\quad\ 761.8\,A_0 + 93.2\,D = 18.4512$

$A_0 = 0.0242085$

$D = 0.000099$

$a = 4.9505_0$ Å (0.2 per cent radius error)

$a = 4.9505_2$ Å (no radius error)

[a] 1 mm on film = 1.00107°.

per cent KBr and 58.67 per cent RbBr. The powder was mounted on the outside of a Pyrex capillary about 0.3 mm in diameter and a diffraction diagram prepared with CuKα radiation in a Debye-Scherrer camera of 114.6-mm diameter. The line quality was fair but not nearly as good as that of the spectrographic-lead pattern just considered, and there was a scarcity of even fair-quality lines in the region $\theta > 60°$. Therefore it was decided to employ the extrapolation function $(\cos^2 \theta)/\sin \theta + (\cos^2 \theta)/\theta$ and utilize as many lines as possible in the larger angular region $\theta = 30°$ to 90°. Figure 8-14 shows the results of this extrapolation when the nine lines of acceptable quality are used. Because of the reduced line sharpness, the scatter of a values is greater than that observed for RbBr (Fig. 8-12) or for Cu_9Al_{14} (Fig. 8-13), but nevertheless a fairly well-defined value of $a = 6.743_6 (\pm 0.000_7)$ Å is indicated.

For the least-squares analysis, equation 8-35 is replaced by

$$\frac{\Delta d}{d} \propto \frac{\cos^2 \theta}{\sin \theta} + \frac{\cos^2 \theta}{\theta}, \tag{8-44}$$

from which it is found that

$$\Delta \sin^2 \theta = D \sin^2 2\theta \left(\frac{1}{\sin \theta} + \frac{1}{\theta} \right) \tag{8-45}$$

Table 8-8. Cohen Least-Squares Calculation of the Lattice Constant of a KBr-RbBr Solid Solution, Using the Error Term $\Delta \sin^2 \theta = D \sin^2 2\theta$ $(1/\sin \theta + 1/\theta)^a$

Line	hkl	α	λ	$\theta(°)$	$\left(\dfrac{\cos^2 \theta}{\sin \theta} + \dfrac{\cos^2 \theta}{\theta} \right)$ $\to K\alpha$	$\sin^2 \theta$ $\to K\alpha$	δ
1	420	20	Kα	30.763	2.86	0.26162	7.5
2	422	24	Kα	34.067	2.38	0.31378	7.5
3	440	32	Kα	40.304	1.73	0.41840	7.2
4	600	36	Kα	43.308	1.47	0.47049	6.9
5	620	40	Kα	46.312	1.25	0.52289	6.5
6	622	44	Kα	49.315	1.06	0.57503	6.1
7	820,644	68	$K\alpha_1$	70.371	0.209	0.88866	1.9
8	820,644	68	$K\alpha_2$	70.812	0.208	0.88906	1.8
9	822,660	72	$K\alpha_1$	75.692	0.106	0.94051	1.1

^{a}Camera diameter = 114.6 mm; 1 mm on film = 1.00122°; CuKα radiation; 26.2°C.

$^b\Sigma \alpha^2 = 21264$; $\Sigma \alpha\delta = 1668.0$; $\Sigma \delta^2 = 299.5$; $\Sigma \alpha \sin^2 \theta = 277.9082$; $\Sigma \delta \sin^2 \theta = 21.8042$. Normal equations: $21264.00 A_0 + 1668.0 D = 277.9082$; $1668.0 A_0 + 299.5 D = 21.8042$; $A_0 = 0.0130674$; $D = 0.0000260$; $a = 6.743_7$ Å.

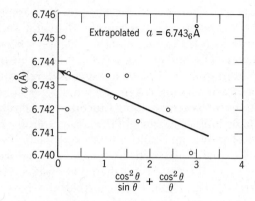

Fig. 8-14. Lattice constant of a KBr–RbBr solid solution at 26.2°C as deduced from a wide-angle extrapolation.

by the same method used in deriving equation 8-38. The least-squares calculations can now be performed with the aid of the normal equations 8-43 as before, the only difference being that δ is now set equal to $\sin^2 2\theta[(1/\sin\theta) + (1/\theta)]$. The course of the computations can easily be followed in Table 8-8. It is noted that, since most of the lines were obtained with unresolved $K\alpha$ radiation, the labor-saving convention has been adopted of relating all values of $\sin^2 2\theta[(1/\sin\theta) + (1/\theta)]$ to $K\alpha$ rather than to $K\alpha_1$, as is more satisfactory when only high-angle lines are employed. For convenience δ is defined as $10\sin^2 2\theta[(1/\sin\theta) + (1/\theta)]$, making the calculated value of D one-tenth its true value. The result of the Cohen analysis, 6.743_7 Å, is in excellent agreement with the graphical finding, $6.743_6 (\pm 0.000_7)$ Å.

One evident disadvantage of the method is that it assigns equal weight to all reflections regardless of whether they can be measured with good or poor accuracy. On the contrary, extrapolation methods instantly reveal any significant errors in indexing, measurement, or calculation. A good policy is to eliminate from the least-squares analysis data from any diffraction lines of decidedly inferior quality. Hess[44], however, not content with such approximate measures, critically examined the problem of weighting the measurements in the Cohen method. Hess pointed out that Cohen's scheme may lead to lattice constants that differ somewhat from the most probable value, and frequently also to an underestimation of their precision. Cohen's normal equations 8-43 imply that only the $\sin^2\theta$ quantities are subject to error, whereas the δ_i terms also are subject to errors in the angular measurements. Hess' analysis, however, shows that ignoring the uncertainties in the δ's is equivalent to an alternate approximation that can be fully justified. Hess also felt that Cohen's minimization of the sum of the equally weighted squares of the residuals violated basic least-squares principles. Hence he sought new formulas which satisfied the principle of minimizing the sum of the *weighted* squares of the residuals of the angular measurements.

Utilizing Deming's[45] Gaussian solution of the nonlinear least-squares problem, Hess obtained the analogous new set of normal equations for the Cohen procedure applied to the cubic case and back-reflection Debye-Scherrer lines:

$$\Delta A \sum \alpha_i^2 w_\gamma + \Delta D \sum \alpha_i \delta_i w_\gamma = \sum \alpha_i F_0 w_\gamma,$$

$$\Delta A \sum \alpha_i \delta_i w_\gamma + \Delta D \sum \delta_i^2 w_\gamma = \sum \delta_i F_0 w_\gamma, \qquad (8\text{-}46)$$

where ΔA and ΔD are changes obtained in assumed values of the parameters A_0 and D. Thus

$$\Delta A = A_\alpha - A_0 \qquad \text{and} \qquad \Delta D = O - D. \qquad (8\text{-}47)$$

Also

$$F_0 = A_\alpha \alpha - \gamma, \qquad (8\text{-}48)$$

where $\gamma = \sin^2 \theta = \cos^2 \phi$. In Hess' procedure the Cohen weights w_γ (i.e., unity) are replaced in the Debye-Scherrer case with the weight w_s cosec$^2 \phi$, where w_s is usually unity. The same weight is used with a symmetrical back-reflection camera, and the weight for a flat-film back-reflection camera becomes w_s cosec$^2 \phi$ sec$^4 \phi$. Complete mathematical details are presented in Hess' article[44]. The slightly greater computational effort required for the method has been completely justified by detectable improvement in the precision of the computed lattice constants.* Weyerer[46] has also discussed lattice-constant determination by nonlinear graphic and least-squares extrapolations.

Computer techniques can be introduced to advantage in the least-square lattice-constant calculations. For instance, the Cohen procedure [43] with the Taylor and Sinclair error correction[41] has been programmed in Mercury Autocode for the Manchester University Atlas computer by Hoff, Stratton, and Kitchingman[47]. A FORTRAN code for the IBM 704 was developed by Vogel and Kempter[48] for the Hess method. Mueller, Heaton, and Miller[49] have prepared elaborate

---

*Note added in galley proof. In a reassessment of the geometrical and statistical aspects of the accuracy of Debye-Scherrer powder data Langford and Wilson[78] conclude that Cohen's use of unit weights is justified in general, and Langford[79] further reports a procedure for obtaining the standard deviations of lattice constants derived by Cohen's method.

programs for the IBM 704. These permit calculation of lattice parameters and standard errors for crystals of orthorhombic and higher symmetry by least-squares procedures and use the Hess weighting scheme. The programs also provide for the use of as many as three systematic-error correction terms at one time. Calculation time for a single solution is a matter of a few seconds making it possible to obtain several solutions based on different combinations of correction terms and weighting factors. The "best" lattice parameter may be chosen as the one whose combination of correction terms leads to the smallest v_i's in the relationship

$$v_i = \sin^2 \theta_i (\text{computed}) - \sin^2 \theta_i (\text{observed}) \tag{8-49}$$

when applied to several diffraction lines. A similar program has been developed by Elliott[50] for the IBM 7090 computer.

8-4 PRECISE LATTICE CONSTANTS BY OTHER FILM TECHNIQUES

Other film techniques, especially the symmetrical parafocusing back-reflection method (Section 4-2), have found use in precision lattice-constant determination. The systematic errors of this method are similar in many respects to those of the Debye-Scherrer method. Angle measurements are conveniently expressed as $\phi - \tfrac{1}{2}\pi - \theta$. Film shrinkage corrections may be made by means of calibrated knife edges. The specimen is sometimes mixed with a standard substance for calibration. Careful

Table 8-9. Systematic Errors in the Symmetrical Back-Reflection Focusing Method[a]

Source of error	Effect on ϕ[b]	Variations of d error with ϕ
Film shrinkage	—	$\phi \tan \phi$
Low-angle knife-edge calibration	$+$ or $-$	$\phi \tan \phi$
Specimen transparency	$+$	$\tan^2 \phi$
Specimen displacement		$\tan 2\phi \tan \phi$
Outside true circumference	$+$	
Inside true circumference	$-$	
Beam divergence in plane $\perp$ to camera diameter	$+$	$\tan 2\phi \tan \phi$

[a]Courtesy of Parrish and Wilson[57], and Kynoch Press.
[b]$+$, toward larger ϕ; $-$, toward smaller ϕ. ($\phi = \tfrac{1}{2}\pi - \theta$.)

preparation of the specimen is necessary to assure that it has the correct curvature and that its front surface lies accurately on the same circle as the film's front surface. Single-coated film should be used, or the displaced back image of double-coated film should be stripped or its development prevented[51]. Axial divergence should be kept small. Obviously, many of these errors can be minimized by good camera construction. Table 8-9 summarizes briefly the systematic errors in the symmetrical back-reflection focusing method. Limited space precludes the mathematical development of these error expressions, but the interested reader can readily follow the derivations in articles by Saini[52], Cohen [43], and Jette and Foote[53]. In the first edition of this book[54], we discussed these errors and presented their angular dependence functions in the form $\Delta \sin^2 \theta$ for immediate use in Cohen's least-squares method. Franks[55] has described a very precisely constructed symmetrical back-reflection focusing camera which eliminates or greatly reduces instrumental and shrinkage errors.

Some investigators have used the flat-plate back-reflection method (Section 4-3.2), but the small number of lines ordinarily recorded makes extrapolation procedures impractical. For accurate lattice-parameter work, accurate instrument construction and careful calibration are necessary. The principal corrections for flat-plate back-reflection camera measurements are listed in Table 8-10. The calibration may be accomplished by coating the specimen surface with a thin film (~ 0.002 cm

Table 8-10. Corrections for Measurements with Flat-Plate Back-Reflection Camera[a]

Source of error	Correction[b]
Film shrinkage	Use low-angle knife edges or other fiducial marks.
Specimen-to-film distance	Use accurate mechanical gauge or calibrating substance.
Specimen transparency	Decrease D by $2 \tan 2\phi/\mu(1 + \sec 2\phi)$, or use thin specimen or longer wavelength.
Inclination of incident beam to specimen normal	Construct camera so that inclination of central ray is less than $0.1°$.
Beam divergence	Limit to small values or reduce D by $\frac{3}{2}D \tan^2 \alpha$.

[a]Courtesy of Parrish and Wilson[57], and Kynoch Press.
[b]D, measured ring diameter; $\phi = \frac{1}{2}\pi - \theta$; μ, linear absorption coefficient; α, semiangle of divergence.

thick) of a material such as powdered aluminum or silver. The investigator then measures one diffraction ring from the standard and one from the specimen. The angle ϕ is obtained from the relation

$$\tan 2\phi = \frac{D}{D_c} \tan 2\phi_c, \qquad (8\text{-}50)$$

where the D's are the ring diameters, and the subscript c refers to the calibrating substance. Raeuchle[56] has discussed the accurate measurement of these rings on flat films. Again we have displaced images because of the double coating of standard x-ray film, and the back image must be eliminated[51]. Of course, temperature control is necessary for any precision work. Rovinskii and Kostyukova[58], in using this method for large samples of irregular shape, have developed special techniques for precision lattice-constant studies.

8-5   PRECISE LATTICE CONSTANTS FROM DIFFRACTOMETRIC MEASUREMENTS

Although Wilson[59] pointed out in 1950 that there appeared to be no reason why the Geiger-counter "spectrometer" (diffractometer) should not permit the same accuracy as powder cameras in lattice-spacing measurements, there was little use of diffractometric measurements for precise lattice-constant determination until the late 1950s and early 1960s. During this time, however, the aberrations of diffractometer profiles were being studied with great care. These aberrations and sources of error in counter diffractometry are discussed at length in Chapter 5. The major error sources we list as follows: (1) instrument misalignment, (2) missetting of the 2:1 drive, (3) error in the $0°\ 2\theta$ position, (4) flat-specimen error, (5) specimen transparency, (6) axial divergence, (7) specimen displacement, (8) rate-meter recording, and (9) dispersion and the Lorentz factor. The investigator is referred to Sections 5-1 and 5-2 for details on the diffraction profile aberrations, and to Section 5-4 for operating techniques in diffractometry. The accurate determination of interplanar d spacings is discussed in Section 5-2.3.

The problem of what feature of the diffraction profile to use in Bragg angle measurements in diffractometry is still with us. Although the centroid of the diffraction line (discussed in Section 5-2.2) has certain advantages for measurement of the angular position of the line, it is much less easily measured than the centerline peak position (Section 8-1). Moreover, our most accurate x-ray wavelength data (Table 2-3) have been determined by the centerline peak method. Thus, by using the centerline peak method in diffractometry, a one-to-one relationship is maintained between wavelengths and Bragg angles, once the line profiles have been corrected for angular aberrations. It is recommended that all precision lattice-constant investigators applying diffractometric methods use the centerline peak method. However, for those whose accuracy requirements seem to demand the time-consuming centroid measurements, two x-ray wavelengths, Cu$K\alpha$ and Fe$K\alpha$, have been provided by Mack, Parrish, and Taylor[60] from centroid studies of Bearden's two-crystal spectrometer profiles. Additional spectral data are needed before reference centroid wavelengths can be established.

An early attempt at precision determination of lattice constants with a Geiger-counter x-ray diffractometer was that of Smakula and Kalnajs [61]. They aligned the diffractometer with special care, determined its zero point by two methods, maintained the temperature constant within 0.05°C, and obtained line positions (2θ) with a reported accuracy of $\pm 0.001°$. Lattice constants were obtained by the $\cos^2 \theta$ extrapolation according to Wilson's equation[59]. The vertical (axial) divergence error, as computed by Eastabrook[62], could be neglected. They concluded that a precision in lattice-constant measurement of ± 0.00002 Å could be obtained with the Geiger-counter diffractometer, but that the absolute value was limited to 0.004 per cent by the uncertainty of the x-ray wavelengths then available. Similarly, Geiss[63] sought to establish a procedure for simply and precisely determining lattice parameters with a commercial diffractometer. Using a General Electric XRD-5 diffractometer, the most important step in the analysis was found to be the alignment of the instrument. After a precise initial alignment, the daily alignment check was a routine matter. To assure specimen flatness, Geiss took a solid piece of aluminum and inscribed it with shallow, narrow, saw-blade cuts in a cross-hatched fashion. When these grooves are filled with powder, a solid surface is approximated. Moreover, the surface is very flat. Geiss estimated the precision of his technique for lattice parameters of cubic materials to be ± 0.000025 Å or less. In the light of later precise diffractometric lattice-constant studies, the precision claimed for these two studies seems to be a bit on the optimistic side.

Otte[64] and King and Vassamillet[65] introduced the technique of double-scanning diffractometry. This technique requires a diffractometer that can scan both sides of the direct beam, making it possible to locate precisely the $0° 2\theta$ position by comparing as few as three pairs of low-angle reflection peaks measured on each side of the beam. Using an R. C. A. Siemens Crystalloflex IV diffractometer, whose scanning range extended from -100 to $+165° 2\theta$, King and Vassamillet explored the feasibility of this method for precise lattice-constant determination by a study of high-purity silver (99.999 per cent). The sample temperature was controlled to ±0.3°C, and peak positions were measured to ±0.005° 2θ. Each peak was double-scanned (i.e., in both the clockwise and counterclockwise directions), and reflections on both sides of the direct beam were scanned. Wilson's [59] analysis for eliminating systematic errors due to absorption and the use of a flat specimen was applied, and the data were extrapolated against the function $\cos^2\theta$. For values of θ greater than about 45°, the data conformed closely to a straight line. Their final value of the lattice parameter of pure silver at 25°C was 4.0862 ± 0.0002 Å. It was concluded, too, that a precision considerably better than the 1 part in 20,000 observed for silver might be achieved through improved procedures. For instance, the temperature might be controlled to ±0.1°C, the peak positions might be established to ±0.001° 2θ, and the validity of the extrapolation functions for eliminating systematic errors might be reexamined.

Accordingly, Vassamillet and King [66] carried through a critical evaluation of the following sources of error which enter into the diffractometric determination of precise lattice spacings:

1. Errors due to physical effects.
2. Errors arising from geometry.
3. Errors inherent in the instrument.
4. Errors associated with alignment.
5. Direct errors in the measurement of 2θ.
6. Errors involved in extrapolation procedures.

The reader must be referred to their article [66] for details, but this study alerted investigators particularly to the errors inherent in the diffractometer instrument (see also Section 5-1.4C). For instance, imperfections in the gears may nearly double the error in the final extrapolated lattice-parameter value. Uneven wear of the teeth on the main gear, and even variations in the oil film on the gear, can lead to random errors. Systematic errors can rise if the gear circle possesses some slight elipticity, or if its center is not exactly coincident with the axis of the diffractometer, or if the 2:1 following mechanism is inaccurate. Testing for these errors is not a simple procedure, but Vassamillet and King's study of several diffractometers in the Pittsburgh area revealed the necessity for a thorough calibration of the instrument before undertaking really precise work.

Following the error studies of Vassamillet and King [66], King and Russell [67] developed a double-scanning technique yielding lattice parameters to 1 part in 100,000 on a semiroutine basis. A Siemens diffractometer and its accompanying x-ray tube arrangement were modified to permit a scanning range over Bragg angles from -163 to $+163°\ 2\theta$. A modified specimen holder was designed which kept the specimen displacement error insignificantly small. The scanning procedure adopted was a counterclockwise scan, that is, toward lower angles on the negative side, and toward higher angles on the positive side. Both the dynamic (instrumental) errors and the static (alignment) errors are removed by this scanning technique. Conditions of the scan were: $\frac{1}{8}°/$ min; time constant, 8 sec; chart speed, 1 cm/min (agreed to within $0.005°$ 2θ with those measured at $\frac{1}{4}°$/min, 4 sec, 2 cm/min). Peak positions of high-angle reflections were measured. When the horizontal divergence was $< 2°$, and peak positions were measured to $\pm0.005°\ 2\theta$, the remaining systematic errors could be eliminated by using any of the extrapolation functions[67], and when the peaks are double-scanned on both sides of the beam, the simple $\cos^2\theta$ extrapolation is entirely adequate.

King and Russell[67] applied the above technique to a study of the lattice parameters of the IUCr specimens of silicon and tungsten and achieved a reproducibility of better than 1:150,000. In a later study, King

	Silicon at 25°C	Tungsten at 25°C
King and Russell	$a = 5.43059 \pm 0.00002$ Å	$a = 3.16524 \pm 0.00002$ Å
IUCr mean value	$a = 5.43054 \pm 0.00017$ Å	$a = 3.16522 \pm 0.00009$ Å

and Preece[68] applied the double-scanning diffractometric method to lattice-parameter measurements at liquid-helium temperatures. The lowest temperature achieved with the cryostat was 18°K, but temperatures could be maintained constant to $\pm0.5°$K during the measurements. A filling of liquid helium ($2\frac{1}{2}$ liters) maintains the cryostat at helium temperatures for 6 hr. Other refrigerants may be used for steady temperatures of 77°, 195°K, and so on. It is believed that the double-scanning technique as described by King et al.[65, 67] is potentially capable of yielding the most precise diffractometric lattice constants.

8-6 THE PRECISION DETERMINATION OF LATTICE CONSTANTS OF NONCUBIC MATERIALS

For crystals of lower than cubic symmetry, it is necessary to determine more than one lattice constant, and in the monoclinic and triclinic systems there are in addition axial angles to be measured. As a general rule, powder diffraction data are not suitable for the precision measurement of crystals belonging to the orthorhombic, monoclinic, and triclinic systems, although in some cases orthorhombic substances can be successfully handled. This is largely due to the great complexity of the diffraction patterns, which precludes the unambiguous indexing of the essential back-reflection lines. Such complex cases can be more satisfactorily solved with the aid of rotating-crystal diagrams or single-crystal diffractometric measurements, provided of course that single crystals are obtainable. Since most workers will probably choose single-crystal procedures for crystals of lower symmetry, we feel that a detailed treatment of the noncubic cases is not warranted here. The topic was treated at length, with illustrative examples, in the first edition of this book[69]. It suffices here to present a few details concerning the handling of tetragonal and hexagonal powder patterns, and to give additional references for study by those interested.

A situation amenable to a simple approach is that wherein reflections of the types $hk0$ and $00l$ are present at angles in the neighborhood of $\theta = 80°$. Sometimes a proper choice of radiation serves to supply the desired reflections on a single photograph; in other cases two exposures with different radiations may be required. In either event both a and c can be precisely determined by applying the Straumanis technique for cubic materials.

If several $hk0$ and several $00l$ reflections are observed in the range $\theta = 30$ to $90°$, it is practicable to deduce accurate values of the lattice constants by performing two extrapolations (of a and c) against the wide-angle function $(\cos^2 \theta)/\sin \theta + (\cos^2 \theta)/\theta$ in the manner described earlier for cubic materials. Under such circumstances it is unnecessary to eliminate rigidly absorption and other errors by experimental means, as is required of the Straumanis method, wherein only a very few high-angle reflections are utilized. Alternatively, a Cohen's least-squares treatment may be employed by making use of the wide-angle error term

$$\Delta \sin^2 \theta = D \sin^2 2\theta \left(\frac{1}{\sin \theta} + \frac{1}{\theta}\right). \qquad (8\text{-}51)$$

If two good-quality reflections of different indices $h_1 h_1 l_1$ and $h_2 k_2 l_2$ can be obtained at angles in the neighborhood of $\theta = 80°$, the precision technique of Straumanis[4, 16, 17] can be used to supply very accurate values of the corresponding Bragg angles θ_1 and θ_2. From these data alone a can be calculated (tetragonal case) from the equation

$$a = \frac{\lambda}{2} \sqrt{\frac{l_1^2(h_2^2 + k_2^2) - l_2^2(h_1^2 + k_1^2)}{l_1^2 \sin^2 \theta_2 - l_2^2 \sin^2 \theta_1}}. \qquad (8\text{-}52)$$

Using the values of h, k, l, and θ appropriate for either reflection, we find that c is then given by

$$c = \frac{\lambda a l}{\sqrt{4a^2 \sin^2 \theta - \lambda^2 (h^2 + k^2)}}. \qquad (8\text{-}53)$$

Sometimes two suitable reflections produced by a single wavelength are not present, but two or more reflections can be obtained if we utilize two wavelengths (for example, β and α_2 lines as well as α_1 lines). The desired reflections can also be obtained by employing an alloy target which emits the characteristic spectra of two or three elements. Another procedure consists in preparing two photographs with different radiations but at the same controlled sample temperature. In situations of this sort in which two reflections $h_1k_1l_1$ and $h_2k_2l_2$, corresponding to wavelengths λ_1 and λ_2, are utilized, it is easy to show from the two Bragg equations,

$$\sin^2\theta_1 = \left(\frac{\lambda_1^2}{4}\right)\left(\frac{h_1^2+k_1^2}{a^2}+\frac{l_1^2}{c^2}\right),$$

$$\sin^2\theta_2 = \left(\frac{\lambda_2^2}{4}\right)\left(\frac{h_2^2+k_2^2}{a^2}+\frac{l_2^2}{c^2}\right),$$

(8-54)

that

$$a = \frac{\lambda_1\lambda_2}{2}\sqrt{\frac{l_1^2(h_2^2+k_2^2)-l_2^2(h_1^2+k_1^2)}{\lambda_1^2 l_1^2 \sin^2\theta_2 - \lambda_2^2 l_2^2 \sin^2\theta_1}}.$$

(8-55)

After a has been calculated, the value of c can again be determined with the aid of equation 8-53.

Straumanis and Ieviņš[70] applied this method to the two-parameter problem of tetragonal β-tin. Two $CuK\alpha$ resolved doublets, reflections 503 and 271, were measured. These experimental results provided four pairs of data, from which four values of a and four values of c could be determined, leading to parameters (and σ's) $a = 5.83140 \pm 0.00005$ and $c = 3.18116 \pm 0.00029$ Å corrected to 25°C. The hexagonal case can be treated in an entirely analogous manner, and the corresponding expressions for a and c are only slightly more complicated.

If very few $hk0$ and $00l$ reflections are available, so that neither direct-extrapolation nor simple Straumanis techniques are applicable, general reflections hkl can usually be employed to obtain a and c by a method of successive extrapolations proposed by Lipson and Wilson[71]. The method was illustrated in the first edition of the book with data from hexagonal germanium dioxide (GeO_2)[72]. In many cases hexagonal and tetragonal crystals produce several satisfactory $hk0$ reflections but only one $00l$ reflection in the region $\theta = 30$ to $90°$. Under these circumstances Taylor and Floyd[73] suggest making an accurate extrapolation of a with the aid of the function $(\cos^2\theta)/\sin\theta + (\cos^2\theta)/\theta$, after which c is obtained by drawing a straight line through the c value of the single $00l$ reflection, the slope of this line being calculated from the slope of the a line. The relationship between the slopes of the a and c extrapolations can be deduced from equation 8-3. An example applied to lattice-constant data on tetragonal potassium dihydrogen phosphate is given in references[73, 74]. Bacon[75] has shown how the slope of the extrapolation can be ascertained in favorable cases by adding a cubic substance as an internal standard.

Cohen's analytical least-squares method[43] can be applied directly to noncubic crystals, provided only that the patterns can be indexed and that a sufficient number of lines is present. Details, illustrated by data from hexagonal GeO_2, are provided in reference[76]. Computer techniques based on the Cohen method are often programmed to handle noncubic data as well as cubic data (see Hoff, Stratton, and Kitchingman [47], Vogel and Kempter[48], and Mueller, Heaton, and Miller[49]).

8-7   SUMMARY

The foregoing discussion of the several techniques of precision lattice-constant determination may well leave the reader confused as to just what method or methods should be applied under any particular set of experimental circumstances. It is hoped that Table 8-11 will serve as a guide to the reader in making a good choice. For clarity and simplicity only the Debye-Scherrer and diffractometric techniques are considered in this summary. As the experimentalist gains facility with these methods, the extension to parafocusing cameras will become apparent.

In using Table 8-11 the diffractionist must keep in mind the following important facts:

I. In many experimental cases two or more procedures are of equal efficacy. For example, the choice between the Straumanis technique and the CFM is largely up to the discretion or personal preference of the investigator. For the most part, only those workers who are called upon with some degree of frequency to measure lattice constants precisely will find it feasible to equip themselves with all the refinements of technique and instrumentation that are essential to the successful application of these methods. The majority of diffractionists, who either make precise measurements only infrequently or who require in any case only moderately precise lattice constants, can obtain the desired information with minimum difficulty by employing an extrapolation or least-squares analytical procedure.

II. The extrapolation or analytical methods will yield results of highest accuracy only when certain basic refinements of experimental technique are observed. For the Debye-Scherrer method these include: (a) the use of a precision-constructed camera so that residual radius

Table 8-11. Guide to the Choice of the Proper Method of Precision Lattice-Constant Determination as Dictated by the Observed Line Distribution

Code No.	Debye-Scherrer Method	Precision Attainable	Section Described
1	Straumanis technique	Up to 1:200,000	8-3.2
2	Convolution-film method with likelihood ratio method	Up to 1:200,000	8-3.3
3	Debye-Scherrer technique (DST), plus $\cos^2 \theta$ extrapolation	1:20,000	8-3.4A
4	DST, plus least-squares analysis, $\Delta \sin^2 \theta = D \sin^2 2\theta$ (also with weighting)	Up to 1:50,000	8-3.4C
5	DST, plus $(\cos^2 \theta/\sin \theta) + (\cos^2 \theta/\theta)$ extrapolation	Possibly up to 1:50,000	8-3.4B
6	DST, plus least-squares analysis, $\Delta \sin^2 \theta = D \sin^2 2\theta$ $[(1/\sin \theta) + (1/\theta)]$ (also with weighting)	Possibly up to 1:100,000	8-3.4C

Code No.	Diffractometric Methods	Precision Attainable	Section Described
7	Double-scan technique with extrapolation	1:100,000 to 1:150,000	8-5
8	Simple, careful, single-scan technique with extrapolation	Up to 1:50,000	8-5

Crystal System	Observed Line Distribution	Code Numbers of Recommended Methods	Possible Alternatives
Cubic	Two or more well-spaced $\alpha_1\alpha_2$ doublets in region $\theta = 60$–$90°$	2, 3, 4, 5, 6, 7, 8	1, if at least one line is at $\theta > 75°$
Cubic	One $\alpha_1\alpha_2$ doublet in region $\theta = 60$–$90°$ or two very close together	2, 5, 6, 7, 8	1, if at least one line is at $\theta > 75°$
Tetragonal or hexagonal	One ($hk0$) doublet and one ($00l$) doublet in region $\theta = 60$–$90°$	1	
Tetragonal or hexagonal	Several ($hk0$) and several ($00l$) lines distributed in region $\theta = 30$–$90°$	5, 6, 7, 8	
Tetragonal or hexagonal	Two or more (hkl) and/or ($h0l$) doublets in region $\theta > 75°$	1	See Section 8-6

and eccentricity errors are very small, (b) the employment of the asymmetric film position, (c) the use of the smallest practicable sample diameter to minimize absorption errors, (d) the use of small collimator apertures, and (e) the maintenance of constant sample temperature within limits prescribed by the desired accuracy of the lattice constants being measured. For the diffractometric technique these include: (a) accurate alignment of the diffractometer, (b) careful calibration of the instrument, (c) accurate determination of the $0°$ 2θ position, (d) a good flat specimen accurately parallel to the diffractometer axis, (e) a good choice of scanning conditions, (f) small ($<2°$) radial divergence of the beam, and (g) suitably constant sample temperature.

III. The investigator must to the best of his ability select the radiation that will provide the most favorable distribution of lines. In many cases only one radiation is feasible at all, which may serve to fix the best procedure. In other cases the particular method preferred by the worker may be the deciding factor in the choice of radiation.

IV. Only part of the special cases of tetragonal and hexagonal crystals outlined in Section 8-6 are specifically summarized in Table 8-11. Hence the investigator faced with a tetragonal or hexagonal lattice-constant measuring problem should consult the examples of Section 8-6 as well as Table 8-11.

In concluding this chapter some remarks should be made concerning the degree of precision that is theoretically attainable in the determination of lattice constants. From a pure consideration of the possible reflnements in experimental techniques or improvements in extrapolation procedures, it appears that no sharply defined limit exists, and indeed optimistic estimates have been made of limiting precisions as high as $\Delta a/a = 0.5 \times 10^{-5}$ (1 part in 200,000) and better. However, such estimates presuppose no restrictions on the sharpness of back-reflection lines save those prescribed by camera geometry, the crystallite size or lattice perfection of the sample powder, and the graininess of the photographic film. In actuality a further and inexorable limit is imposed by the spectral width of the so-called monochromatic x-radiation employed in preparing the diffraction photographs. Ekstein and Siegel[77] have shown that this leads to an ultimate precision in $\Delta a/a$ amounting to 2 or 3 parts in 100,000 for any one back-reflection line. Evidently, then, a greater precision in the lattice constant can be arrived at only by averaging the results from several reflections at very high Bragg angles ($\theta > 80°$) or from several photographs. Therefore, unless the spectral purity can be improved in some way, the ultimate practicable limit in the precision of lattice-constant determinations appears to lie in the neighborhood of 1×10^{-5}.

GENERAL REFERENCES

*1. L. V. Azároff and M. J. Buerger, *The Powder Method in X-ray Crystallography*, McGraw-Hill, New York, 1958, Chapters 14 and 15.

2. B. D. Cullity, *Elements of X-ray Diffraction*, Addison-Wesley, Reading, Massachusetts, 1956, Chapter 11.

3. H. Lipson and H. Steeple, *Interpretation of X-ray Powder Diffraction Patterns*, Macmillan, London, 1970, Chapter 6.

4. M. E. Straumanis, "Parameters of Crystal Lattices. II. High Precision Measurements by the Asymmetric Diffraction Method," and K. E. Beu, "Parameters of Crystal Lattices. III. "X-ray Diffraction Methods with Estimates of Accuracy and Precision," in *Encyclopedia of X-rays and Gamma Rays* (G. L. Clark, ed.) Van Nostrand Reinhold, New York, 1963, pp. 700–727.

5. "The Precision Determination of Lattice Parameters," Nine papers from a conference of the International Union of Crystallography held in Stockholm, June 9–12, 1959; *Acta Crystallogr.*, **13**, 818–850 (1960).

6. W. Parrish and A. J. C. Wilson, "Precision Measurement of Lattice Parameters of Polycrystalline Specimens," in *International Tables for X-ray Crystallography*, Vol. II, Kynoch Press, Birmingham, England, 1959, Section 4.7, pp. 216–234.

7. W. Parrish, J. Taylor, and M. Mack, 'Dependence of Lattice Parameters on Various Angular Measures of Diffractometer Line Profiles," in *Advances in X-ray Analysis*, Vol. 7, Plenum Press, New York, 1964, pp. 66–84.

8. L. F. Vassamillet and H. W. King, "Precision X-ray Diffractometry Using Powder Specimens," in *Advances in X-ray Analysis*, Vol. 6, Plenum Press, New York, 1963, pp. 142–157.

SPECIFIC REFERENCES

[1] W. Parrish, *Acta Crystallogr.*, **13**, 838 (1960).

[2] J. A. Bearden, *Phys. Rev.*, **43**, 94 (1933).

[3] J. A. Bearden, *Rev. Mod. Phys.*, **39**, 78 (1967); USAEC Report NYO-10586, 1964, Clearinghouse for Federal Scientific Information, National Bureau of Standards, Springfield, Virginia. Price $8.45.

[4] K. E. Beu, in *Encyclopedia of X-rays and Gamma Rays* (G. L. Clark, ed.), Van Nostrand Reinhold, New York, 1963, p. 709.

[5] A. J. Bradley and A. H. Jay, *Proc. Phys. Soc.* (London), **44**, 563 (1932).

[6] S. V. Borisov and V. P. Golovachev, *Soviet Phys. – Crystallogr.*, **3**, 386 (1958).

[7] G. Hägg, *Rev. Sci. Instrum.*, **18**, 371 (1947).

[8] D. L. Scott and K. E. Beu, *Experimental Techniques for the Precise and Accurate Determination of Lattice Parameters by Film Powder Methods*, USAEC Report GAT-T-1305/Oral, 1965, Goodyear Atomic Corporation, P.O. Box 628, Piketon, Ohio.

[9] M. H. Jellinek, *Rev. Sci. Instrum.*, **20**, 368 (1949).

[10] M. E. Straumanis, *Acta Crystallogr.*, **13**, 818 (1960).

[11] K. E. Beu and D. L. Scott, *Acta Crystallogr.*, **15**, 1301 (1962).

[12] K. E. Beu and D. L. Scott, *The Precise and Accurate Determination of Lattice Parameters Using Film Powder Diffraction Methods: The Eccentricity Correction*, USAEC Report GAT-392, Part IIA, 1962, Goodyear Atomic Corporation, P.O. Box 628, Piketon, Ohio.

[13] B. E. Warren, *J. Appl. Phys.*, **16**, 614 (1945).

[14] A. Taylor and H. Sinclair, *Proc. Phys. Soc.* (London), **57**, 108 (1945).

[15] M. J. Buerger, *X-ray Crystallography*, Wiley, New York, 1942, pp. 402–407.

[16] M. Straumanis and A. Ieviņš, *Die Präzisionsbestimmung von Gitterkonstanten nach der asymmetrischen Methode*, Springer, Berlin, 1940; reprinted by Edwards Brothers, Ann Arbor, Michigan, 1948.

[17] M. E. Straumanis, *J. Appl. Phys.*, **20**, 726 (1949).

[18] K. E. Beu, D. K. Landstrom, D. R. Whitney, and E. R. Pike, *Acta Crystallogr.*, **17**, 639 (1964).

[19] M. E. Straumanis and C. C. Weng, *Acta Crystallogr.*, **8**, 367 (1955).

[20] J. I. Langford, E. R. Pike, and K. E. Beu, *Acta Crystallogr.*, **17**, 645 (1964).

[21] W. P. Davey, *A Study of Crystal Structure and Its Applications*, McGraw-Hill, New York, 1934, pp. 113–114 and 158–163.

[22] W. Gerlach and O. Pauli, *Z. Phys.*, **7**, 116 (1921).

[23] H. Ott, *Ber. bayer. Akad. Wiss.*, **31** (1924).

[24] R. J. Havighurst, E. Mack and F. C. Blake, *J. Am. Chem. Soc.*, **46**, 2368 (1924).

[25] H. D. Keith, *Proc. Phys. Soc.* (London), **63B**, 208 (1950).

[26] H. Küstner, *Phys. Z.*, **23**, 257 (1922).

[27] F. C. Blake, *Phys. Rev.*, **26**, 60 (1925).

[28] A. E. van Arkel, *Z. Kristallogr.*, **27**, 25 (1927).

[29] For the preparation of Lindemann glass and suitable sample-mounting fibers, see A. Schleede and M. Wellmann, *Z. Kristallogr.*, **83**, 148 (1932), as well as reference 16, p. 33.

[30] H. P. Klug and L. E. Alexander, *X-ray Diffraction Procedures*, 1st ed., Wiley, New York, 1954, pp. 455–458.

[31] K. E. Beu, F. J. Musil, and D. L. Scott, *Modifications of a Commercial Powder Diffraction Camera for Precise and Accurate Lattice Parameter Measurements*, USAEC Report GAT-T-973, Goodyear Atomic Corporation, P.O. Box 628, Piketon, Ohio.

[32] Reference 16, pp. 30–32.

[33] K. E. Beu and D. L. Scott, *The Convolution-Film Method for the Precise and Accurate Determination of Lattice Parameters and Its Application to IUCr Silicon*. USAEC Report GAT-T-1306/Oral, 1965, Goodyear Atomic Corporation, P.O. Box 628, Piketon, Ohio.

[34] K. E. Beu, F. J. Musil, and D. R. Whitney, *Acta Crystallogr.*, **15**, 1292 (1962).

[35] K. E. Beu, F. J. Musil, and D. R. Whitney, *Acta Crystallogr.*, **16**, 1241 (1963).

[36] K. E. Beu and D. R. Whitney, *Acta Crystallogr.*, **22**, 932 (1967).

[37] A. M. Mood, *Introduction to the Theory of Statistics*, McGraw-Hill, New York, 1950, pp. 257–259.

[38] C. D. Hodgman, *Handbook of Chemistry and Physics*, 40th ed., Chemical Rubber Company, Cleveland, 1959, p. 218.

[39] J. Taylor, M. Mack, and W. Parrish, *Acta Crystallogr.*, **17**, 1229 (1964).

[40] L. A. Carapella, *J. Appl. Phys.*, **11**, 510 (1940).

[41] A. Taylor and H. Sinclair, *Proc. Phys. Soc.* (London), **57**, 126 (1945).

[42] J. B. Nelson and D. P. Riley, *Proc. Phys. Soc.* (London), **57**, 160 (1945).

[43] M. U. Cohen, *Rev. Sci. Instrum.*, **6**, 68 (1935); **7**, 155 (1936).

[44] J. B. Hess, *Acta Crystallogr.*, **4**, 209 (1951).

[45] W. E. Deming, *Statistical Adjustment of Data*, Wiley, New York, 1943.

[46] H. Weyerer, *Z. Kristallogr.*, **109**, 338, 354 (1957).

[47] W. D. Hoff, R. P. Stratton, and W. J. Kitchingman, *J. Sci. Instrum.*, **41**, 695 (1964).

[48] R. E. Vogel and C. P. Kempter, *Acta Crystallogr.*, **14**, 1130 (1961).

[49] M. H. Mueller, L. Heaton, and K. T. Miller, *Acta Crystallogr.*, **13**, 828 (1960).

[50] R. P. Elliott, *Advances in X-ray Analysis*, Vol. 8, Plenum Press, New York, 1965, p. 134.

[51] M. L. Polk, *Rev. Sci. Instrum.*, **40**, 1645 (1969); W. Parrish, *Norelco Rep.*, **2**, 67 (1955).

[52] H. Saini, *Helv. Phys. Acta*, **6**, 597 (1933).

[53] E. R. Jette and F. Foote, *J. Chem. Phys.*, **3**, 605 (1935).

[54] Reference 30, pp. 474–477.

[55] A. Franks, *Advances in X-ray Analysis*, Vol. 3, Plenum Press, New York, 1960, p. 69.

[56] R. F. Raeuchle, *Rev. Sci. Instrum.*, **24**, 875 (1953).

[57] W. Parrish and A. J. C. Wilson, *International Tables for X-ray Crystallography*, Vol. II, Kynoch Press, Birmingham, England, 1959, p. 216.

[58] B. M. Rovinskii and E. P. Kostyukova, *Soviet Phys.—Crystallogr.*, **3**, 383, (1958); **8**, 200 (1963).

[59] A. J. C. Wilson, *J. Sci. Instrum.*, **27**, 321 (1950).

[60] M. Mack, W. Parrish, and J. Taylor, *J. Appl. Phys.*, **35**, 1118 (1964).

[61] A. Smakula and J. Kalnajs, *Phys. Rev.*, **99**, 1737 (1955).

[62] J. N. Eastabrook, *Brit. J. Appl. Phys.*, **3**, 349 (1952).

[63] R. H. Geiss, *Advances in X-ray Analysis*, Vol. 5, Plenum Press, New York, 1962, p. 71.

[64] H. M. Otte, *J. Appl. Phys.*, **32**, 1536 (1961).

[65] H. W. King and L. F. Vassamillet, *Advances in X-ray Analysis*, Vol. 5, Plenum Press, New York, 1962, p. 78.

[66] L. F. Vassamillet and H. W. King, *Advances in X-ray Analysis*, Vol. 6, Plenum Press, New York, 1963, p. 142.

[67] H. W. King and C. M. Russell, *Advances in X-ray Analysis*, Vol. 8, Plenum Press, New York, 1965, p. 1.

[68] H. W. King and C. M. Preece, *Advances in X-ray Analysis*, Vol. 10, Plenum Press, New York, 1967, p. 354.

[69] Reference 30, pp. 477–487.

[70] Reference 16, pp. 55–56; or reference 30, pp. 478–481.

[71] H. Lipson and A. J. C. Wilson, *J. Sci. Instrum.*, **18**, 144 (1941).

[72] Reference 30, pp. 481–483.

[73] A. Taylor and R. W. Floyd, *Acta Crystallogr.*, **3**, 285 (1950).

[74] Reference 30, pp. 483–485.

[75] G. E. Bacon, *Acta Crystallogr.*, **1**, 337 (1948).

[76] Reference 30, pp. 485–487.

[77] H. Ekstein and S. Siegel, *Acta Crystallogr.*, **2**, 99 (1949).

[78] J. I. Langford and A. J. C. Wilson, *J. Appl. Crystallogr.*, **6**, 197 (1973).

[79] J. I. Langford, *J. Appl. Crystallogr.*, **6**, 190 (1973).

CHAPTER 9

CRYSTALLITE SIZE AND LATTICE
STRAINS FROM LINE BROADENING

For a polycrystalline specimen consisting of sufficiently large and strain-free crystallites, diffraction theory predicts that the lines of the powder pattern will be exceedingly sharp (breadths measured in seconds of arc). In actual experiments lines of such sharpness are never observed because of the combined effects of a number of instrumental and physical factors [net profile $g(\epsilon)$] that broaden the "pure" diffraction line profile $f(\epsilon)$ (see Sections 5-2.1 and 5-2.2). Of more fundamental importance to the present discussion is the fact that the shape and breadth of the pure profile $f(\epsilon)$ are determined both by the mean crystallite size, or distribution of sizes, characterizing the specimen, and by the particular imperfections prevailing in the crystalline lattice. Hence, in principle at least, an appropriate analysis of the line profile $f(\epsilon)$ should yield such information as the mean crystallite dimension or size of the coherent crystalline domains, distribution of crystallite sizes, and the nature and extent of lattice imperfections.

Clearly, we are faced with the twofold task of (I) extracting the pure diffraction profile $f(\epsilon)$ from the experimental profile $h(\epsilon)$, and (II) extracting size and imperfection information from the profiles $f(\epsilon)$ of suitable lines of the pattern. We now deal with these two topics in the order given. Operation I is feasible because of the convolution relationship among $h(\epsilon)$, $g(\epsilon)$, and $f(\epsilon)$ (see Section 5-2.1):

$$h(\epsilon) = \int_{-\infty}^{+\infty} g(\eta)f(\epsilon - \eta)\, d\eta. \tag{9-1}$$

Thus $f(\epsilon)$ can be obtained from $h(\epsilon)$ by the process of deconvolution (unfolding). Operation II resolves itself naturally into two cases, (A) only size or imperfection broadening present, and (B) both causes of broadening present simultaneously, requiring resolution of the two effects. Further, in regard to Operation II, lattice imperfections are

618

of several kinds, each demanding specialized analytical treatment. Because of space limitations we treat in this chapter only two, which are believed to be of most interest to the general diffractionist: simple lattice strains (microstrains) and turbostratic stacking disorder, a form of faulting common in carbons and clay minerals. For a most valuable and authoritative introduction to diffraction analysis of the various modes of faulting, which finds its principal applicability in the field of metals, the reader is referred to Warren[1].

9-1 DETERMINATION OF THE PURE LINE PROFILE

That this operation is an indispensable step preparatory to the calculation of crystallite-size and lattice-strain parameters has been emphasized above. Within certain mathematical limitations imposed principally by (1) the presence of random experimental errors (so-called "noise") in the curves $h(\epsilon)$ and $g(\epsilon)$, and (2) the practical necessity of evaluating the integral of equation 9-1 over a finite rather than infinite angular range, this equation can be solved for $f(\epsilon)$ more or less rigorously by several methods of accomplishing the unfolding [2]. The two methods most commonly applied to x-ray data are Fourier transformation[3, 4], commonly referred to as the method of Stokes[4], and the iterative method of successive foldings of $h(\epsilon)$ with $g(\epsilon)$, first described by Burger and van Cittert[5] for spectroscopic corrections, and recently applied to x-ray diffraction profiles by Ergun[6].

These relatively powerful methods were formerly of little practical value because of the extensive and tedious computations involved, but today, because of the general accessibility of digital computers, they comprise essential tools for deducing numerical crystallite-size and lattice-strain parameters with optimum absolute significance. However, when only relative numerical results are useful, as in many analytical and routine applications, these sophisticated procedures may justifiably be circumvented or approximated by substituting simple equations or graphs based on line profiles of assumed analytical forms. Therefore we first discuss the Fourier-transform and iterative-folding methods and then give a brief treatment of simpler analytical procedures for deducing the breadth of the pure diffraction profile.

9-1.1 The Fourier-Transform Method[3, 4, 7–9]

With reference to equation 9-1, let the Fourier transforms of $h(\epsilon)$, $g(\epsilon)$, and $f(\epsilon)$ be $H(\zeta)$, $G(\zeta)$, and $F(\zeta)$, respectively. By application of

Fourier's integral theorem, it can be shown that [10]

$$F(\zeta) = \frac{H(\zeta)}{G(\zeta)}, \tag{9-2}$$

and furthermore [11]

$$f(\epsilon) = \frac{1}{\sqrt{2\pi}} \int_{-\infty}^{+\infty} F(\zeta) \exp(-2\pi i \epsilon \zeta) \, d\zeta \tag{9-3}$$

$$= \frac{1}{\sqrt{2\pi}} \int_{-\infty}^{+\infty} \frac{H(\zeta)}{G(\zeta)} \exp(-2\pi i \epsilon \zeta) \, d\zeta. \tag{9-4}$$

By employing the exponential form of the Fourier integral, generality is preserved, permitting the integrals to be applied to asymmetrical as well as symmetrical functions.

According to Stokes' method, the integrals are replaced by the corresponding summations, and the limits of ϵ are changed from $\pm\infty$ to $\pm\frac{1}{2}\epsilon_m$, the minimum values of ϵ beyond which the intensity can be considered to have fallen to its background value. Equation 9-4 may then be written as

$$f(\epsilon) = \frac{1}{\sqrt{2\pi}} \sum_{\zeta} \frac{H(\zeta)}{\epsilon_m G(\zeta)} \exp\left(\frac{-2\pi i \epsilon \zeta}{\epsilon_m}\right) \Delta\zeta. \tag{9-5}$$

In practice the constant factors $1/\sqrt{2\pi}$, $1/\epsilon_m$, and $\Delta\zeta$ can be eliminated from the computations, since they affect only the height of a line profile and not its shape. This gives

$$f(\epsilon) = \sum_{\zeta} \frac{H(\zeta)}{G(\zeta)} \exp\left(\frac{-2\pi i \epsilon \zeta}{\epsilon_m}\right), \tag{9-6}$$

in which

$$H(\zeta) = \sum_{\epsilon} h(\epsilon) \exp\frac{2\pi i \epsilon \zeta}{\epsilon_m} \tag{9-7}$$

and

$$G(\zeta) = \sum_{\epsilon} g(\epsilon) \exp\frac{2\pi i \epsilon \zeta}{\epsilon_m}. \tag{9-8}$$

Now, since

$$e^{iz} = \cos z + i \sin z, \tag{9-9}$$

it is evident that each of the series above consists of real and imaginary parts; for example,

$$G(\zeta) = \sum_{\epsilon} g(\epsilon) \cos\frac{2\pi\epsilon\zeta}{\epsilon_m} + i \sum_{\epsilon} g(\epsilon) \sin\frac{2\pi\epsilon\zeta}{\epsilon_m}, \tag{9-10}$$

which may be represented by

$$G(\zeta) = G_r(\zeta) + iG_i(\zeta). \tag{9-11}$$

Similarly, $H(\zeta)$ may be written

$$H(\zeta) = H_r(\zeta) + iH_i(\zeta). \tag{9-12}$$

From equations 9-11 and 9-12

$$F(\zeta) = \frac{H(\zeta)}{G(\zeta)} = \frac{H_r + iH_i}{G_r + iG_i} = \frac{(H_r + iH_i)(G_r - iG_i)}{G_r^2 + G_i^2}.$$

Evidently, then F has the real and imaginary components

$$F_r = \frac{H_r G_r + H_i G_i}{G_r^2 + G_i^2} \tag{9-13}$$

and

$$F_i = \frac{H_i G_r - H_r G_i}{G_r^2 + G_i^2}. \tag{9-14}$$

These equations provide a basis for the numerical calculation of F_r and F_i from the Fourier series G_r, G_i, H_r, and H_i, which can be evaluated from experimental data.

To return to equation 9-6, the desired pure diffraction profile $f(\epsilon)$ can be computed as follows:

$$f(\epsilon) = \sum_{\zeta} F(\zeta) \exp \frac{-2\pi i \epsilon \zeta}{\epsilon_m}$$

$$= \sum_{\zeta} [F_r(\zeta) + iF_i(\zeta)] \left(\cos \frac{2\pi \epsilon \zeta}{\epsilon_m} - i \sin \frac{2\pi \epsilon \zeta}{\epsilon_m} \right)$$

$$= \underbrace{\sum_{\zeta} F_r(\zeta) \cos \frac{2\pi \epsilon \zeta}{\epsilon_m}}_{(A)} + \underbrace{\sum_{\zeta} F_i(\zeta) \sin \frac{2\pi \epsilon \zeta}{\epsilon_m}}_{(B)}$$

$$\underbrace{-i \sum_{\zeta} F_r(\zeta) \sin \frac{2\pi \epsilon \zeta}{\epsilon_m}}_{(C)} + \underbrace{i \sum_{\zeta} F_i(\zeta) \cos \frac{2\pi \epsilon \zeta}{\epsilon_m}}_{(D)}. \tag{9-15}$$

But $F(-\zeta)$ is the complex conjugate of $F(\zeta)$ (see equation 9-2), and therefore $F_r(-\zeta) = F_r(\zeta)$ and $F_i(-\zeta) = -F_i(\zeta)$. Since the summations extend over all positive and negative values of ζ, it follows that terms (C) and (D) in equation 9-15 reduce to zero, and $f(\epsilon)$ can be calculated from the equation

$$f(\epsilon) = \sum_{\zeta} F_r(\zeta) \cos \frac{2\pi\epsilon\zeta}{\epsilon_m} + \sum_{\zeta} F_i(\zeta) \sin \frac{2\pi\epsilon\zeta}{\epsilon_m}. \qquad (9\text{-}16)$$

It is necessary to measure the diffracted intensity over the experimental profiles at small intervals. It is convenient to divide the angular range ϵ_m within which a line has measurable intensity into 60 parts and to select an artificial unit of angle so small that $\epsilon_m = 60$ units and so that intensity measurements are made at intervals $\Delta\epsilon$ equal to unity. These experimental intensity data then become the Fourier coefficients for the series H_r, H_i, G_r, G_i, which assume the forms

$$H_r(\zeta) = \frac{1}{60} \sum_{-30}^{+30} h(\epsilon) \cos \frac{2\pi\epsilon\zeta}{60},$$

$$H_i(\zeta) = \frac{1}{60} \sum_{-30}^{+30} h(\epsilon) \sin \frac{2\pi\epsilon\zeta}{60},$$

$$G_r(\zeta) = \frac{1}{60} \sum_{-30}^{+30} g(\epsilon) \cos \frac{2\pi\epsilon\zeta}{60},$$

$$G_i(\zeta) = \frac{1}{60} \sum_{-30}^{+30} g(\epsilon) \sin \frac{2\pi\epsilon\zeta}{60}. \qquad (9\text{-}17)$$

It should be kept in mind that $h(\epsilon)$ refers to an experimental intensity profile broadened as a result of small crystallite size, or lattice distortions, or both, and $g(\epsilon)$ refers to the profile of a reference specimen that is free of size and distortion broadening. The values of $H_r, H_i, G_r,$ and G_i computed with equation 9-17 over the required range of ζ values are next used to compute the corresponding values of $F_r(\zeta)$ and $F_i(\zeta)$ with equations 9-13 and 9-14. These F_r and F_i values then become the Fourier coefficients for evaluating the pure diffraction profile $f(\epsilon)$, which can be expressed as a Fourier series in the following "working" form of equation 9-16:

$$f(\epsilon) = \frac{1}{60} \sum_{\zeta} F_r(\zeta) \cos \frac{2\pi\epsilon\zeta}{60} + \frac{1}{60} \sum_{\zeta} F_i(\zeta) \sin \frac{2\pi\epsilon\zeta}{60}. \qquad (9\text{-}18)$$

Figure 9-1 shows illustrative h, g, and f profiles from such a Fourier analysis of a diffraction line of cold-worked copper as reported by Stokes [4]. The broadened profile h of cold-worked copper filings is "unfolded" to give the pure profile f by reference to the corresponding line profile g of well-annealed copper. These results illustrate a valuable characteristic

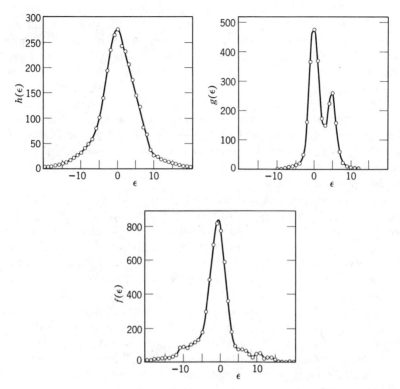

Fig. 9-1.   Line profiles h, g, and f involved in a Fourier analysis of line shapes for a cold-worked copper specimen. (Courtesy of A. R. Stokes, *Proc. Phys. Soc.* (London), **A61**, 382 (1948).)

of the Fourier method, namely, that any degree of doubling or broadening of a line due to the $K\alpha_1 K\alpha_2$ doublet is automatically allowed for; the deduced profile f will be of the nature expected for the normal wavelength distribution prevailing in either the $K\alpha_1$ or $K\alpha_2$ radiation.

One condition for the success of the Fourier-transform method is that the measurement range $\pm\frac{1}{2}\epsilon_m$ be divided into a sufficiently large number of intervals; otherwise the summations (9-17) will not be accurate representations of the components of the Fourier coefficients F_r and F_i required in the evaluation of $f(\epsilon)$. This condition is fulfilled if $H(\zeta)$ and $G(\zeta)$ fall effectively to zero before the ζ values exceed the range limits $\pm\frac{1}{2}\epsilon_m$. If this is not the case, it becomes necessary to divide the angular range into a larger number of increments, for example, 120 rather than 60. The number of increments must exceed twice the value of ζ for the Fourier component of lowest frequency that is effectively equal to zero.

We cannot overemphasize the adverse effects of random experimental errors in the $h(\epsilon)$ profile, which tend to produce high-frequency oscillations in the Fourier transform $H(\zeta)$. The overall result of this phenomenon, including inaccurate establishing of the background level, is that the uncertainty in the unfolded profile $f(\epsilon)$ is always greatest near the limits of the $\pm\frac{1}{2}\epsilon_m$ range, manifesting as diffraction ripples [see tails of $f(\epsilon)$ curve, Fig. 9-1]. These consequences of noise in the experimental data can be reduced by careful smoothing of the $h(\epsilon)$ and $g(\epsilon)$ curves through the data points at the larger values of ϵ[2]. In fact, satisfactory unfolding can seldom be accomplished without smoothing of the experimental curves, including special attention to the placement of the background.

A general property of any method for deriving the $f(\epsilon)$ curve from experimental measurements of $h(\epsilon)$ and $g(\epsilon)$, including the Fourier-transform method, is that the accuracy of the results decreases with increasing width of the instrumental profile $g(\epsilon)$ relative to the width of the experimentally observable specimen profile $h(\epsilon)$. This is equivalent to the familiar commentary on crystallite-size measurements from line broadening that the accuracy diminishes as the size increases. For this reason it is of utmost importance in line-profile analyses of crystallite size and strain that the experimental techniques be selected with a view to holding the instrumental broadening to a minimum. Undoubtedly, the most generally satisfactory technique fulfilling this requirement is parafocusing x-ray diffractometry with step-scanning (Section 5-4.5). Where photographic registration of the pattern is mandatory, a monochromatic focusing camera such as that of Guinier-de Wolff may be recommended (Section 4-2).

Returning to the Fourier analysis of Stokes, we note that in general the Fourier coefficients,

$$F(\zeta) = \frac{H(\zeta)}{G(\zeta)},$$

of the desired curve $f(\epsilon)$ are complex numbers, hence require the calculation of both cosine and sine terms from the experimental curves $h(\epsilon)$ and $g(\epsilon)$. These cosine and sine coefficients must then be combined as indicated by equations 9-13 and 9-14 to obtain the coefficients $F_r(\zeta)$ and $F_i(\zeta)$ from which the profile $f(\epsilon)$ is synthesized (equations 9-16 and 9-18). It follows that the computations would be greatly simplified (by elimination of the sine terms) if the functions f, h, and g were symmetrical. Now experience has shown that in actual fact many experimental line profiles would be symmetrical to a good approximation were it not for the asymmetry introduced by the α_1- and α_2-component profiles of the $K\alpha$ doublet. Rachinger[12], Papoulis[13], and Keating[14] have developed methods of rather general applicability for eliminating the α_2 contribution from the combined $K\alpha_1\alpha_2$ functions $h(\epsilon)$ and $g(\epsilon)$. Mitchell and de Wolff[15] outline an unfolding method for eliminating any generalized spectral dispersion effect from single-crystal or powder reflection profiles or from wide-range powder diffraction patterns. The application of their method is contingent upon the prior correction of the diffraction effects for instrumental broadening. Three methods of correcting line profiles for the entire $K\alpha$ satellite group of wavelengths have been described by Edwards and Toman[16] for use with the variance method of profile analysis (refer to Section 5-2.2B and C). The analytical procedures of Keating[14] and of Mitchell and de Wolff[15] can be used to correct the entire diffraction pattern or any portion thereof, such as a single line profile; thus they are useful in radial distribution studies involving the measurement of relatively diffuse diffraction patterns in their entirety. However, for the present purpose of handling single line profiles, the method of Rachinger is effective and more easily applied and therefore is now described.

The Rachinger Correction[12, 7]. The following discussion is based mainly on Warren's treatment[7]. It is assumed that the $K\alpha_1$ and $K\alpha_2$ line profiles are identical in shape and not necessarily symmetrical, and that the α_2 profile is of one-half the intensity of the α_1 profile and shifted from it toward larger angles by

$$\Delta2\theta = 2 \tan \theta \left(\frac{\Delta\lambda}{\lambda}\right), \tag{9-19}$$

$\Delta\lambda$ being the dispersion separation $\lambda(\alpha_2) - \lambda(\alpha_1)$ in ångströms. In the example in Fig. 9-2, the left-hand curve represents the combined contributions of the α_1 and α_2 profiles, which are shown separately at the right. The separation $\Delta2\theta$ corresponds to the wavelength difference expressed on the 2θ scale (equation 9-19). We now lay off intervals of width $\Delta2\theta/m$, beginning at the left end of the combined (experimentally observed) curve at a point where the intensity may be assumed to have reached zero, and then number $0, 1, 2, 3 \cdots, i, \cdots, n$ to the right-hand extremity of the profile. The symbol m represents a small integer the size of which is determined by the magnitude of $\Delta2\theta$ for the reflection concerned. At very small angles it may be assigned a value of unity, whereas at intermediate angles a value of 2 or 3 is satisfactory. Only at rather large 2θ's does m need to exceed 3, and then the angular separation of the α_1 and α_2 profiles is likely to be so large, and their overlap so small, as to permit resolution of the α_1 profile by direct inspection. In any case m must be chosen so as to be compatible with a total number of intervals n that is large enough to delineate clearly the complete line

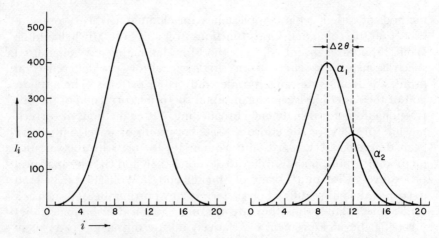

Fig. 9-2. Profiles illustrating the Rachinger correction. On the right the α_1 and α_2 contributions are shown separately; on the left they are combined as in an experimental measurement. Curves plotted from data of Table 9-1.

profile, typically in the range 20 to 30. In the example in Fig. 9-2 and Table 9-1, $n = 21$ and $m = 3$.

Now let I_i be the intensity of the experimental profile at point i in the foregoing series of n intervals, and let $I_i(\alpha_1)$ be that part of I_i due to α_1 alone. Then

$$I_i(\alpha_1) = I_i - \tfrac{1}{2} I_{i-m}(\alpha_1). \qquad (9\text{-}20)$$

Thus, for the case $m = 3$ and $n = 21$, the values of $I(\alpha_1)$ for successive points beginning at the left-hand end of the experimental profile are:

$$I_0(\alpha_1) = 0$$
$$I_1(\alpha_1) = I_1$$
$$I_2(\alpha_1) = I_2$$
$$I_3(\alpha_1) = I_3$$
$$I_4(\alpha_1) = I_4 - \tfrac{1}{2} I_1(\alpha_1)$$
$$I_5(\alpha_1) = I_5 - \tfrac{1}{2} I_2(\alpha_1)$$

. . . .

$$I_i(\alpha_1) = I_i - \tfrac{1}{2} I_{i-3}(\alpha_1)$$

. . . .

$$I_{21}(\alpha_1) = I_{21} - \tfrac{1}{2} I_{18}(\alpha_1) = 0 - 0 = 0$$

Table 9-1.   Illustrative Calculation of the Rachinger
Correction for $K\alpha_1\alpha_2$ Broadening[a]

Interval number (i)	$\delta2\theta(°)$	I_i	$\frac{1}{2}I_{i-m}(\alpha_1)$	$I_i(\alpha_1)$
0	0	0	0	0
1	0.1	2	0	0
2	0.2	8	0	8
3	0.3	22	0	22
4	0.4	55	1	54
5	0.5	115	4	111
6	0.6	205	11	194
7	0.7	317	27	290
8	0.8	424	56	368
9	0.9	496	97	399
10	1.0	513	145	368
11	1.1	474	184	290
12	1.2	394	200	194
13	1.3	295	184	111
14	1.4	199	145	54
15	1.5	119	97	22
16	1.6	64	56	8
17	1.7	29	27	2
18	1.8	11	11	0
19	1.9	4	4	0
20	2.0	1	1	0
21	2.1	0	0	0

[a]Conditions: Gaussian profiles, $\Delta2\theta = 0.3°$, $n = 21$, $m = 3$; $I_i(\alpha_1)$ calculated with equation 9-20.

Table 9-1 illustrates the simple calculations involved in the Rachinger analysis of an idealized line profile of 2.1° overall width synthesized from two Gaussian profiles 1.8° wide and displaced relative to each other by $\Delta2\theta = 0.3°$. There are 21 intervals of width

$$\frac{\Delta2\theta}{m} = \frac{0.3°}{3} = 0.1°.$$

The numerical data on Table 9.1 are portrayed in Fig. 9-2.

When the foregoing procedure is applied to actual experimental data, it is capable of yielding $I_i(\alpha_1)$ values of good accuracy from the low-angle limit ($i = 0$) up to a point about midway down the high-angle side of the profile, after which the differences between the I_i and

$\frac{1}{2}I_{i-m}(\alpha_1)$ curves become so small (see Table 9-1 for idealized situation) that random errors of measurement are likely to produce meaningless fluctuations in the $I_i(\alpha_1)$ curve, which therefore should be artificially smoothed out. As pointed out earlier, if the $I(\alpha_1)$ profile turns out to be symmetrical, only cosine terms need to be included in evaluating the Fourier coefficients for the Stokes correction, which will then yield a symmetrical $f(\epsilon)$ profile. In the Stokes analysis the origins of the $g(\epsilon')$ and $h(\epsilon'')$ profiles should be chosen at their midpoints, even though they may fall at different points on the 2θ scale.* Warren[7] points out that the $f(\epsilon)$ profile will then be displaced from $\epsilon = 0$ by the angular separation of the origins of the $g(\epsilon')$ and $h(\epsilon'')$ functions. Computer programs for performing the Fourier transform computations of Stokes[4] may obe obtained from several sources.†

9-1.2 The Method of Iterative Folding[5, 6]

This method seems to have been employed first by Burger and van Cittert[5], who applied it to the elimination of instrumental broadening from spectroscopic patterns. Because of the nonavailability of digital computers at the time, they invented an optical-analogue apparatus for performing the integrations. Only rather recently, the iterative method was first applied to x-ray diffraction patterns by Ergun[6], who made very effective use of high-speed computers. The acknowledged success and ease of implementation of his computerized procedure amply justify its description here in some detail. In fact, the present outlook is that the simplicity and other advantages of the iterative folding method are such as to recommend it in preference to Stokes' Fourier–transform method. At the outset we may take special note of the simplification and drastic reduction in the overall computational labor effected by choosing the iterative-folding in preference to the transform method.

The only special initial conditions to be met are that $g(\epsilon)$ have bounded support and that it be normalized:

$$\int g(\epsilon)\, d\epsilon = 1.$$

Then a first approximation of the effect of folding $g(\epsilon)$ and $f(\epsilon)$ to give $h(\epsilon)$ can be obtained by subtracting the fold of g and h, $\widehat{gh}$, from h. This

---

*The variables ϵ, ϵ', and ϵ'' all have the dimensions of $\Delta 2\theta$, but are here distinguished from one another to indicate possible differences in their origin points, $2\theta_0$, $2\theta_0'$ and $2\theta_0''$.

†W. D. Hoff and W. J. Kitchingman, Metallurgy Department, University of Manchester Institute of Science and Technology, Manchester, England[9] (Atlas Autocode, for Atlas computer); H. McKinstry, Materials Research Laboratory, The Pennsylvania State University, University Park, Pennsylvania (Fortran).

difference added to h gives the first approximation to f and may be designated the first unfold,

$$f_1 = f_0 + (h - \widehat{gf_0}) = h + (h - \widehat{gh}). \tag{9-21}$$

The second approximation of f is obtained by subtracting gf_1 from h and adding this second residual to f_1. Thus, if we define the first residual by $u_0 = h - \widehat{gf_0} = h - \widehat{gh}$, the next by $u_1 = h - \widehat{gf_1}$, and the nth by $u_n = h - \widehat{gf_n}$, successive iterations may be expressed:

$$\begin{aligned}
f_1 &= f_0 + u_0 = h + (h - \widehat{gh}) \\
f_2 &= f_1 + u_1 = f_1 + (h - \widehat{gf_1}) \\
f_3 &= f_2 + u_2 = f_2 + (h - \widehat{gf_2})
\end{aligned} \tag{9-22}$$

. . . .

$$f_{n+1} = f_n + u_n = f_n + (h - \widehat{gf_n})$$

Iterations are continued until some convergence criterion is satisfied, for example, until further iterations produce no change in $f(\epsilon)$, or until the difference between $h(\epsilon)$ and $\widehat{g(\epsilon)f_n}(\epsilon)$ reaches values of the same magnitude as the statistical accuracy of $h(\epsilon)$ and $g(\epsilon)$. It can be concluded that these two functions are compatible if convergence is indeed achieved after a reasonable number of cycles, usually from two to four, for typical x-ray profiles h and g. On the contrary, if additional cycles continue to produce changes in f, the functions h and g are thereby shown to be incompatible and the broadening function g invalid, thus rendering the non- or quasi-convergent values of f unreliable.

Some further comments are in order in regard to the practical evaluation of equations 9-22. First, the folding integrations $\widehat{gf_n}$ may be performed numerically using the trapezoidal or Simpson's rule[17]. The requirement that $g(\epsilon)$ have bounded support* as well as being normalized may be expressed

$$\int_{-a}^{+b} g(\epsilon) \, d\epsilon = 1. \tag{9-23}$$

A special problem also arises in the successive foldings, namely, that with each fold the range of $f_n(\epsilon)$ is reduced by $a + b$ relative to the range of $f_{n-1}(\epsilon)$. For example, if the range of $h(\epsilon)$ extends from c to d, the domain

---

*The concept of bounded support may be explained by saying that, both below and above the limits $-a$ and $+b$ of the integration range, the function $g(\epsilon)$ must assume a uniform and minimum value extending for at least one domain $a + b$. In integrating over a diffraction profile, this minimum value must be zero, inasmuch as $g(\epsilon)$ is measured above the prevailing background.

of $\widehat{g(\epsilon)h}(\epsilon)$ will be from $c+b$ to $d-a$. To eliminate this reduction in range with each fold, $f(\epsilon)$ may be set equal to $h(\epsilon)$ in the intervals $\epsilon = c$ to $c+b$ and $\epsilon = d-a$ to d. This simplification is acceptable provided these intervals contain no profile details such as maxima or minima.

In order that the $f(\epsilon)$ solution arrived at by iterative folding be meaningful, it is necessary that the residual u_n converge to zero uniformly. Because of the lack of a mathematical analysis of this factor, Ergun[6] tested the convergence numerically using folds of various known functions and thereby demonstrated that convergence was readily obtained provided these functions were everywhere differentiable. Since this is a characteristic property of x-ray line profiles, Ergun concluded that application of the iterative-folding method to such diffraction data should be convergent and easy to accomplish.

We now illustrate the iterative-folding method with Ergun's analysis [6] of the combined 10/004 reflections of a carbon black, which are largely unresolved as experimentally observed (Fig. 9-3). Because of the low absorption of carbon for the x-rays used, β-filtered Mo$K\alpha$, the measurements were made diffractometrically employing transmission geometry (refer to Section 5-4.3), a beam divergence of $1°$, and specimens in the form of slabs 2 mm thick. The instrumental broadening function $g(s)$ was obtained from measurements of the 111 reflection of 5- to 10-μ diamond powder (Fig. 9-3) carried out under the same experimental conditions including sample dimensions and density. The near identity of the angular locations of the diamond and carbon black reflections is of course also essential to ensure the validity of the instrumental broadening correction. From Fig. 9-3 we note that the 2θ scale of experimental measurements has been replaced by the familiar spherical reciprocal-space variable, $s = 2(\sin \theta)/\lambda$, which is more satisfactory for the analysis of broad and diffuse reflections that encompass a considerable angular range, such as in radial-distribution studies (see Chapter 12). Thus the three profiles of the iterative analysis may be expressed functionally as $h(s)$, $g(s)$, and $f(s)$, or, alternatively, as functions of t, which has the same dimensionality as s but relates to an origin defined as the theoretical location of the 111 reference reflection of diamond,

$$t(0) = s_{111} = \left(\frac{2 \sin \theta}{\lambda}\right)_{111} = \frac{1}{d_{111}}$$

(refer to the upper scale of Fig. 9-3). If it turns out that $t(0)$ does not coincide with the observed peak position of the 111 line profile because of misalignment or other geometrical aberrations, this fault will be rectified in the unfolding process.

Figure 9-4 shows the original experimental carbon-black profile $h(s)$,

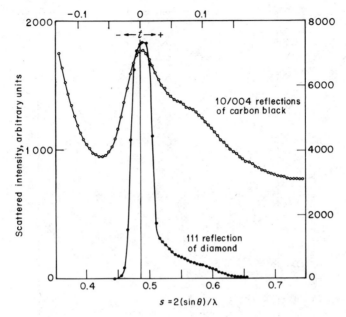

Fig. 9-3.   The 10/004 reflections of carbon black and 111 reflection of powdered diamond
in transmission geometry. (Courtesy of S. Ergun [6]; reprinted by permission of the Bureau
of Mines, U.S. Department of the Interior.)

the first fold $\widehat{g(s)h(s)}$, and the first unfold

$$f_1(s) = 2h(s) - \widehat{g(s)h(s)}.$$

Figure 9-5 compares the original curve $h(s)$, the second unfold

$$f_2(s) = f_1(s) + [h(s) - \widehat{g(s)f_1(s)}],$$

and the fourth (and final) unfold

$$f_4(s) = f_3(s) + [h(s) - \widehat{g(s)f_3(s)}].$$

The maximum residual $u_4(s)$ in the fourth unfold was less than 0.5 per
cent, well within the precision with which the experimental points were
determined, and therefore a satisfactory criterion of a convergent result.
This was confirmed by the performance of a fifth and a sixth unfold, the
latter exhibiting no discernible change from the fourth.

An inspection of Fig. 9-5 reveals that the final unfold [pure diffraction
profile $f_4(s)$] is much sharper than the observed line profile $h(s)$, and that
it displays a greater degree of resolution of the 10 and 004 component
reflections. The $f_4(s)$ line profile is now suitable for the analysis of such

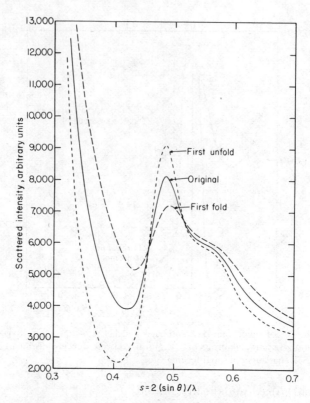

Fig. 9-4. The 10/004 reflections of carbon black. Original profile, $h(s)$; first fold, $\widehat{g(s)h(s)}$; first unfold, $f_1(s) = 2h(s) - \widehat{g(s)h(s)}$. (Courtesy of S. Ergun[6]; reprinted by permission of the Bureau of Mines, U.S. Department of the Interior.)

broadening factors as mean crystallite size (size of the coherently scattering domains), size distribution, and lattice imperfections.

To establish further the reliability of the results of iterative folding, Ergun[6] performed such an analysis of the synthesized fold of the 111 diamond reflection with the 10/004 profile calculated for a representative turbostratically stacked carbon-black model (see Section 9-2.5B). This model consisted of disk-shaped graphite layers 20 Å in diameter and turbostratically aggregated into stacks of different heights consisting of from 2 to 20 layers. On the assumption of an arbitrary distribution of stack heights, the profiles of the resulting 10 and 004 reflections were calculated with the aid of equations derived by Warren and Bodenstein [18, 19]. These theoretical profiles were then combined and corrected for the appropriate experimental factors to normalize them to the conditions prevailing during the measurement of the 111 diamond

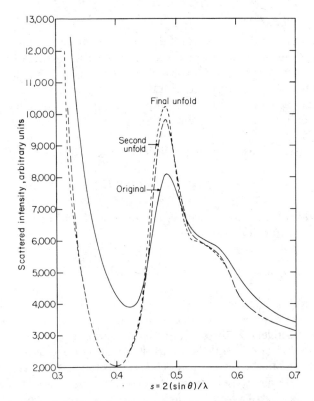

Fig. 9-5.   The original profile and second and fourth unfolds of the 10/004 reflections of carbon black. (Courtesy of S. Ergun[6]; reprinted by permission of the Bureau of Mines, U.S. Department of the Interior.)

reflection, and then folded with the diamond 111 profile to yield a synthetic 10/004 "experimental" profile $h(s)$. The iterative folding procedure was then applied to the synthesized profile $h(s)$, with the diamond 111 line profile filling the role of $g(s)$, as before. The resulting $f_m(s)$ line profile was convergent, and showed very good agreement with the original 10/004 carbon-black model profile. We may emphasize again that the iterative-folding method of performing a deconvolution may be applied just as well to entire diffraction patterns as to isolated line profiles. For further details of the method as applied to x-ray diffraction patterns, the reader is referred to Ergun[6].*

*Copies of a computer program for deconvolution by iterative folding, in Fortran or Algol language, may be obtained on request to Dr. S. Ergun, Pittsburgh Energy Research Center, Bureau of Mines, U.S. Department of the Interior, Pittsburgh, Pennsylvania.

9-1.3 Simplified Methods

Simplified methods circumvent the determination of the entire line profile $f(\epsilon)$ and depend instead on direct measurements of the profile width, of which three types were defined in Section 5-2.2B, the half-maximum width $W_{1/2}$, the integral breadth W_i, and the variance W. Of these the most convenient to use, although the least amenable to mathematical analysis, is $W_{1/2}$ (see Fig. 5-13). Thus $W_{1/2}$ is most generally employed for routine measurements of crystallite size and strain, particularly when relative values suffice. The integral breadth,

$$W_i(2\theta) = \frac{1}{I_p} \int I(2\theta) \, d(2\theta), \qquad (9\text{-}24)$$

is the width of a rectangle having the same area and peak height as the actual line profile. The variance,

$$W_{2\theta} = \frac{\int (2\theta - <2\theta>)^2 I(2\theta) \, d(2\theta)}{\int I(2\theta) \, d(2\theta)}, \qquad (9\text{-}25)$$

has important mathematical advantages (see Section 5-2.2B), but its evaluation involves lengthy calculations (for example, see Table 5-2) and depends sensitively on the choice of background level with an attendant truncation operation. Both $W_{1/2}$ and W_i are also dependent on the background level, but the former is much less sensitive to this factor.

It should be pointed out that the above three measures of line breadth may be expressed in terms of any convenient angular variable such as 2θ, θ, $s = 2(\sin\theta)/\lambda$, or $S = 4\pi(\sin\theta)/\lambda$. However, one must take care to make allowance for the particular variable employed when deriving numerical values of the pure line breadth β from the experimental breadth, and in subsequently extracting size and strain parameters. We now proceed to describe some simplified procedures for deducing approximate values of $\beta_{1/2}$ or β_i from the corresponding observed breadths $W_{1/2}(2\theta)$ or $W_i(2\theta)$, respectively. A similar derivation of β from W, the variance, does not appear to be feasible at present.

These procedures are based on specific assumptions concerning the shapes of the line profiles h, g, and f, with the consequence that the derived $\beta_{1/2}$ or β_i values have correspondingly restricted absolute significance. The chief application of these "streamlined" methods therefore is to analytical measurements wherein the paramount need is to acquire numerical results speedily and with good *relative accuracy*. In the following discussion the breadths of the h, g, and f line profiles are for the sake of simplicity of notation designated B, b, and β, respectively, with subscripts 1/2 or i as required.

A. Gaussian or Cauchy Profiles. The two most commonly assumed line shapes are the Gaussian,

$$I(\epsilon) = I_p(\epsilon) \exp\left(- k^2\epsilon^2\right), \qquad (9\text{-}26)$$

and the Cauchy,

$$I\epsilon = I_p(\epsilon)/(1 + k^2\epsilon^2). \qquad (9\text{-}27)$$

Application of the convolution integral, equation 9-1, shows that for h, g, and f all Gaussian

$$B^2 = b^2 + \beta^2, \qquad (9\text{-}28)$$

and for all Cauchy profiles

$$B = b + \beta, \qquad (9\text{-}29)$$

the relationships holding for both half-maximum and integral breadths.

Since the actual profiles are never pure Gaussian or pure Cauchy, these simple formulas have only limited and rather uncertain practical value. Nevertheless, the Gaussian approximation 9-28 has over the years been very widely invoked in more routine analytical and order-of-magnitude calculations. This practice has some justification in the Debye-Scherrer technique, wherein equation 9-26 constitutes a more realistic approximation of the instrumental profile than does 9-27, particularly for low-absorbing specimens and/or back-reflection angles. In general, though, it must be realized that the g profile for Debye-Scherrer measurements is profoundly modified by the dimensions and absorption coefficient of the specimen, the profile of the x-ray source, the collimation geometry, and so on[20]. Likewise, the geometrical features of diffraction in a high-resolution parafocusing diffractometer cause the g profile to depart widely from Gaussian and become highly unsymmetrical (see Section 5-2.1 and especially Figs. 5-11 and 5-12). Existing evidence tends to show that the $f(\epsilon)$ profile generated by a distribution of crystallite sizes more nearly approaches the Cauchy than the Gaussian form[21, 22]. At this point, therefore, we must conclude that equations 9-28 and 9-29 are to be used, if at all, only with a full recognition of their very considerable limitations.

B. Jones' Correction Curves for Debye-Scherrer Lines[21]. The Debye-Scherrer technique is markedly inferior to high-resolution diffractometry for measuring line profiles because of its relatively large instrumental function g. Nevertheless, Jones' method of calculating and depicting the instrumental broadening correction deserves consideration here because of its attractiveness and general applicability to other diffraction techniques. Beginning with the convolution equation 9-1, 9-1, Jones proved that the integral breadths B_i, b_i, and β_i of the h, g, and f

profiles, respectively, are related as follows:

$$\frac{\beta_i}{B_i} = \frac{\int f(\epsilon)g(\epsilon)\,d\epsilon}{\int g(\epsilon)\,d\epsilon},\qquad(9\text{-}30)$$

$$\frac{b_i}{B_i} = \frac{\int f(\epsilon)g(\epsilon)\,d\epsilon}{\int f(\epsilon)\,d\epsilon}.\qquad(9\text{-}31)$$

The function $g(\epsilon)$ is invariant for a given set of experimental conditions and is determined by measuring the intensity distribution in the line profile of a strain-free material of sufficiently large crystallite size that pure diffraction broadening is effectively absent (about $1000\,\text{Å}$ for Debye-Scherrer and $3000\,\text{Å}$ for high-resolution diffractometric techniques). Using a Debye-Scherrer camera, Jones microphotometered the line of such a reference substance at $\theta = 80°$, where the $K\alpha_1\alpha_2$ doublet was well resolved, and used the $K\alpha_1$ profile as the invariant reference function $g(\epsilon)$. Then he assumed $f(\epsilon)$ to be either Gaussian or Cauchy in form and evaluated the integrals of equations 9-30 and 9-31 for a range of crystallite sizes by appropriately varying the integral breadth of the function $f(\epsilon)$ (by changes in the coefficient k of equations 9-26 and 9-27). In Fig. 9-6 the resulting curves of β_i/B_i versus b_i/B_i are plotted for

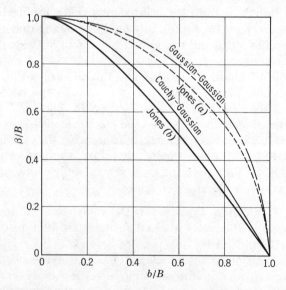

Fig. 9-6. Curves for correcting integral breadths of Debye-Scherrer lines for instrumental broadening. Jones (a): $f(\epsilon)$ Gaussian. Jones (b): $f(\epsilon)$ Cauchy. Gaussian-Gaussian: $g(\epsilon)$ and $f(\epsilon)$ both Gaussian. Cauchy-Gaussian: $f(\epsilon)$ Cauchy and $g(\epsilon)$ Gaussian.

Gaussian (*a*) and Cauchy (*b*) pure diffraction profiles. For comparison the curves for *f* Cauchy and *g* Gaussian, and for *g* and *f* both Gaussian (equation 9-28), are also shown. The latter curve is obtained by plotting equation 9-28 in the form

$$\frac{\beta_i}{B_i} = \left(1 - \frac{b_i^2}{B_i^2}\right)^{1/2}. \tag{9-32}$$

As an illustration of the use of these curves, suppose that the crystallite size of a sample of magnesium oxide (MgO) is to be determined. Finely pulverized quartz powder is selected as the reference material, and a Debye-Scherrer pattern is prepared of the mixed powders, using $CuK\alpha$ radiation. The pure diffraction width of the 220 MgO line at $2\theta = 62.2°$ is to be deduced by comparison with the 12·1 quartz line at 60.0°, nearby lines being selected so as to make the geometrical conditions as similar as possible. From a microphotometer trace of these lines, it is found that $B = 0.80°$ (MgO) and $b = 0.48°$ (quartz). If for the present we neglect broadening of the lines due to the $K\alpha$ doublet, by reference to Fig. 9-6 the following results are obtained:

	b/B	β/B	$\beta(°)$
Jones' b	0.60	0.51	0.41
Jones' a	0.60	0.75	0.60
Gaussian	0.60	0.80	0.64

These results may be compared with those obtained for a much broader 220 MgO peak. Suppose that $B = 1.60°$, whereas b of course is unchanged. Then:

	b/B	β/B	$\beta(°)$
Jones' b	0.30	0.83	1.33
Jones' a	0.30	0.93	1.49
Gaussian	0.30	0.96	1.53

It will be noted that the difference between the results obtained from curves *a* and *b* is relatively much smaller for the broader MgO line. This illustrates how the accuracy of a crystallite-size determination increases as the size becomes smaller. When B is very large, b/B becomes very small, and all three curves yield practically identical results. However, for large crystallite sizes b/B approaches unity, and the β values derived from the three curves differ widely.

C. The $K\alpha_1\alpha_2$ Doublet Correction for the Debye-Scherrer Technique.

The values of β determined in the manner just described are subject to error due to separation of the $K\alpha$ doublet. This factor is usually inconsequential at low Bragg angles, where the angular separation of α_1 and α_2 is small, but it becomes progressively more important as θ increases (see Fig. 9-7). At angles sufficiently large it is often possible to circumvent this problem by measuring the separate profile of either an α_1 or α_2 line, but this recourse fails if the pure diffraction broadening is too great. In some cases a satisfactory $K\beta$ line may be obtained by employing unfiltered radiation. However, a correction for $K\alpha$-doublet broadening needs to be applied in a large proportion of line-profile measurements. For good reliability without excessive computational labor, the method of Rachinger[12, 7] (see Section 9-1.1A) is to be recommended. Corrections of less accuracy but still good to a fair approximation may be much more conveniently applied by reference to correction curves based on profiles of certain standard shapes, for example, those of Fig. 9-8.

In Fig. 9-8, d is the angular separation of the $K\alpha_1\alpha_2$ doublet, b_0 and B_0 are the breadths of the g and h profiles as experimentally observed, and

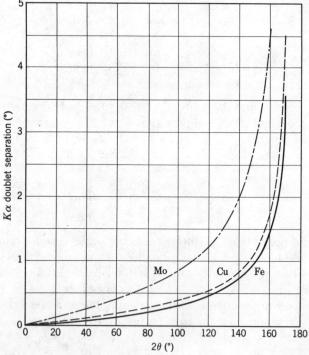

Fig. 9-7. Angular separation of the $K\alpha$ doublet as a function of 2θ for three common radiations.

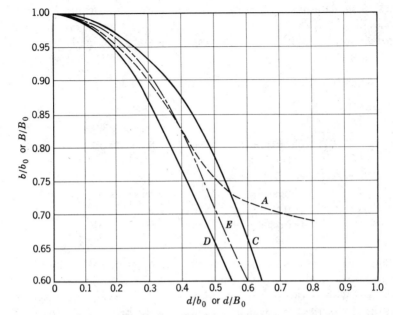

Fig. 9-8.   Curves for correcting line breadths for $K\alpha$-doublet broadening. Angular separation of the $K\alpha$ doublet d; breadths of the g and h profiles as observed, b_o and B_o; as corrected, b and B. See text for explanation of curves A, E, C, and D.

b and B are the corresponding corrected breadths. Curve A is that of Jones[21] for the integral breadths of back-reflection, Debye-Scherrer f and g profiles of a particular arbitrary shape. Curves C and D have been calculated for f and g both Gaussian and both Cauchy, respectively, while E refers to both profiles intermediate in shape between Gaussian and Cauchy. Curves E, C, and D refer to breadths at half-maximum intensity, unlike A. Of these curves A and E are probably most generally useful, E being applicable not only to Debye-Scherrer patterns but also in low-resolution diffractometry (refer to Section 5-1.2 and Fig. 5-4A).

The magnitude of this correction will be made plain by applying it to the MgO and quartz data discussed above. From standard tables[23] the separations of the $CuK\alpha$ doublet at 60 and 62° are $d = 0.17$ and $0.18°$ 2θ. Then, from curve A of Fig. 9-8, the breadths $b_o = 0.48°$, $B_o = 0.80°$, and $B_o = 1.60°$ lead to the results:

$$
\begin{array}{lll}
b_o = 0.48° & B_o = 0.80° & B_o = 1.60° \\
d/b_o = 0.35 & d/B_o = 0.23 & d/B_o = 0.112 \\
b/b_o = 0.86 & B/B_o = 0.94 & B/B_o = 0.98 \\
b = 0.41° & B = 0.75° & B = 1.57°
\end{array}
$$

These calculations show how relatively more important $K\alpha$-doublet broadening is for a narrow line than for a broad one. Continuing with the determination of β, the following results are obtained by reference to Fig. 9-6:

	b/B	β/B	β(°)
B = 0.75°			
Jones' b	0.55	0.57	0.43
Jones' a	0.55	0.78	0.58
Gaussian-Gaussian	0.55	0.83	0.62
B = 1.57°			
Jones' b	0.26	0.86	1.35
Jones' a	0.26	0.95	1.49
Gaussian-Gaussian	0.26	0.96	1.51

The β values above are presumably more accurate than the ones calculated previously, because the effect of the $K\alpha$ doublet has been reckoned with. However, the change in the results is not conspicuous, and the differences are in fact negligible for the broader line ($B = 1.57°$).

In regard to the comparative validity of Jones' curves a and b and the Gaussian-Gaussian curve for the determination of the pure diffraction breadth from Debye-Scherrer line profiles, Jones' curve b is to be preferred for a fairly wide distribution of crystallite sizes (which doubtless describes the constitution of most actual specimens). In the comparatively rare event of a very narrow range of sizes, or of a single discrete size, Jones' curve a or the Gaussian-Gaussian correction is to be preferred. It should also be emphasized that the Jones' corrections are based upon the observed profile of a line in the back-reflection region, where the $K\alpha$ doublet is resolved and where the effects of absorption of of the x-ray beam by the specimen exert a minimum effect upon the ideally symmetrical profile. If the measurements are made in the low-angle region where highly asymmetrical line profiles are produced by moderately or highly absorbing specimens [20], it is necessary to redetermine the calibration curve of β/B versus b/B on the basis of an observed instrumental profile $g(\epsilon)$ at the particular angle concerned. The instrumental line profile is relatively constant and more or less independent of the particular details of the Debye-Scherrer technique only in the far back-reflection region.

D. Correction Curves for Diffractometer Line Profiles. Because of its superiority over other techniques for the measurement of line profiles, we consider almost exclusively high-resolution geometry ("line" source,

Soller slits—see Fig. 5-4B). Relative to low-resolution geometry we may mention that the correction curve designated Cauchy-Gaussian, lying somewhat above Jones' curve *b* in Fig. 9-6, is appropriate.*

Convolution calculations under high-resolution conditions lead to two limiting curves for low and high Bragg angles:

1. Low angles, spectral breadth insignificant.
2. High angles, spectral breadth important.

For a discussion of convolution synthesis of line profiles under these conditions, the reader is referred to Section 5-2.1 and to the literature [26]. At low or moderate Bragg angles, the instrumental profile $g(\epsilon)$ is produced by the joint action of several factors of approximately equal weight. Since some of these factors are asymmetric, the resulting low-angle profile is also more or less asymmetric, this effect being greatest at the smaller angles where the asymmetric factors have the most weight. The convolutions of such profiles with pure diffraction profiles $f(\epsilon)$ of various breadths lead to the correction curve designated "low-angle reflections" in Fig. 9-9. These curves are valid to the extent that the $f(\epsilon)$

*This curve gives integral-breadth corrections for Cauchy $f(\epsilon)$ and Gaussian $g(\epsilon)$ profiles. It was derived analytically by Ruland[24], who has also published correction curves for a large number of other profile couplets of possible practical interest[25]. For a further discussion see also the first edition of this book (1954), pp. 507–509.

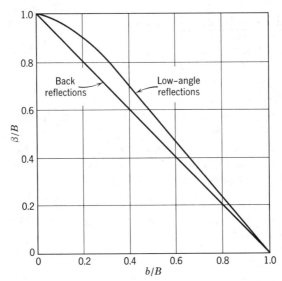

Fig. 9-9.   Curves for correcting x-ray diffractometer line breadths for instrumental broadening under conditions of high resolution.

profiles assume the Cauchy form (equation 9-27), and there is a small difference between the curves for half-maximum and integral breadths.

At high Bragg angles broadening due to the various instrumental factors becomes small in relation to broadening resulting from the natural spectral distribution in the radiation. To a fair approximation a pure spectral line such as $K\alpha_1$ or $K\alpha_2$ may be assigned a Cauchy profile, as explained in the last paragraph of Section 5-2.1F. A similar conclusion was reached by Compton and Allison[27] from double-crystal spectrometer measurements. This means that to a first approximation at large Bragg angles the observed profile is simply the convolution of two $1/(1+k^2\epsilon^2)$ profiles, one due to the crystallite-size distribution and the other to the spectral distribution in the incident radiation. Computation of a series of such convolutions yields the result labeled "back-reflections" in Fig. 9-9, a straight line from the upper left-hand to the lower right-hard corner of the chart. This correction applies to both widths at half-maximum intensity and integral breadths. It is equivalent to

$$\frac{\beta}{B}+\frac{b}{B}=1,$$

which reduces to the simple additivity rule of equation 9-29.

It must be emphasized again that this additivity formula has no general applicability in diffraction work but is strictly valid only when both the functions $f(\epsilon)$ and $g(\epsilon)$ are of the form $1/(1+k^2\epsilon^2)$. This condition is approached rather closely in diffractometer measurements at back-reflection angles for the case of a narrow source profile. It is probable that in any actual crystallite-size determination at high angles the optimum correction curve lies somewhere between the two curves of Fig. 9-9 but rather closer to the lower. Although the exact site of this curve varies somewhat with small deviations of the functions $f(\epsilon)$ and $g(\epsilon)$ from the form $1/(1+k^2\epsilon^2)$, it is best from the practical standpoint to adopt the straight-line function for all measurements in the angular range between 145 and 180°. The correction curve marked "back-reflections" is of special importance because it applies to the angular region in which $\cos\theta$ has minimum values, thereby permitting measurements of relatively large crystallite dimensions (see Section 9-3.4).

9-2 DETERMINATION OF CRYSTALLITE SIZE AND LATTICE IMPERFECTIONS SIMULTANEOUSLY

9-2.1 Introduction

If a ductile material, in particular a metal, is internally stressed by cold work or other treatment, the original rather sharp lines of the polycrystalline ("powder") diffraction pattern become broadened, indicating that the original degree of perfection of the crystallites, or metallic grains, has been lost. There have been several suggestions advanced as to the nature of the imperfections:

i. The material is broken up into crystallites, or coherently diffracting domains in the case of metallic grains, so small (say, 100 to 1000 Å) that diffraction broadening occurs.

ii. The material is broken up into crystallites about $1\,\mu$ (10,000 Å) in size, with lattice spacings differing from one to another.

iii. The crystallites composing the material remain fairly large (about $1\,\mu$ in size), but they (a) are individually elastically distorted, (b) suffer deformation faulting, or (c) undergo both effects.

Unquestionably, the weight of experimental evidence to date favors the joint action of operations i and iii and tends to rule out ii as a participating effect. A reasonable general interpretation[28] is that cold work produces dislocation arrays, which have the effect of subdividing the original crystallites (grains) into much smaller coherently diffracting domains. These domains are visualized as being sufficiently disoriented from one to the next that they scatter incoherently (without a consistent phase relationship) with respect to one another. Thus the domains behave like very small crystals, so that in the following discussion we use the terms *crystallites* and *domains* interchangeably and in the same sense. Finally, the dislocations that cause the subdivision of the original larger crystallites into domains also produce tensile and compressive strains within the domains.

The formation of small crystallites (domains) and the elastic strains introduced by dislocations both result in broadening of the line profile. It is the *size* and *strain* factors that are given principal attention in the remainder of this chapter. Deformation faulting (iiib above) and its effects on the line profiles are much more complex phenomena which are largely outside the province of this book. For an excellent introductory treatment with appropriate references to other literature, the reader is referred to Chapter 12 of Warren[1].

9-2.2. The Fourier Method of Warren and Averbach

A vast amount of experience accumulated over the past 2 decades supports the preeminence of the Fourier method of Warren and Averbach[29] in this field. The theory was originally developed for the investigation of cold-work distortion in metals, but has since found application to other strained materials, both with and without the simultaneous presence of small crystallite size. Furthermore, in the absence of strains, it is useful for the measurement of the number-average crystallite size and, in favorable circumstances, for an estimate of the size distribution [28–33]. Because of its importance we now give a reasonably complete account of the Warren-Averbach theory, following rather closely the presentation in Warren [28]. Only simple lattice strains within crystallites (or coherent domains in a metal) and crystallite size are encompassed in Section 9-2.2. The diffraction effects of faulting are treated very briefly in Section 9-2.5.

A. Derivation of the Fourier Series Expression. It can be shown that a general treatment of the present problem is equivalent to a treatment of any crystal as if it were orthorhombic with reflection from the $(00l)$ planes. For a given crystallite the vector location of the origin of a particular unit cell $m_1 m_2 m_3$ is defined as

$$\mathbf{R}_m = m_1 \mathbf{a}_1 + m_2 \mathbf{a}_2 + m_3 \mathbf{a}_3 + \boldsymbol{\delta}(m_1 m_2 m_3), \tag{9-33}$$

where $\boldsymbol{\delta}$ allows for the possibility of differing distortion displacements for the various unit cells, and m_1, m_2, and m_3 are the numbers of unit cells along the $\mathbf{a}_1$, $\mathbf{a}_2$, and $\mathbf{a}_3$ directions, respectively (see Fig. 9-10). If the distortions *within* the unit cells are considered to be negligible, the intensity in electron units is given by

$$I_{eu} = F^2 \sum_m \sum_{m'} \exp\left[\frac{2\pi i}{\lambda}(\mathbf{s} - \mathbf{s}_0)(\mathbf{R}_{m'} - \mathbf{R}_m)\right]. \tag{9-34}$$

In order to allow for broadening of the reciprocal-lattice nodes (line broadening in powder diffraction), it is necessary to represent the diffraction vector $(\mathbf{s} - \mathbf{s}_0)/\lambda$ in terms of continuous variables h_1, h_2, and h_3, rather than integers $h_1 k_1 l_1$. Then, if $\mathbf{a}_1^*$, $\mathbf{a}_2^*$, and $\mathbf{a}_3^*$ are the corre-

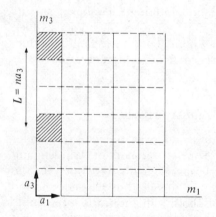

Fig. 9-10. Representation of the crystal as an assemblage of columns of cells along the a_3 direction. (Courtesy of B. E. Warren and Addison–Wesley [28].)

sponding reciprocal-lattice translations,

$$(\mathbf{s} - \mathbf{s}_0)/\lambda = h_1\mathbf{a}_1^* + h_2\mathbf{a}_2^* + h_3\mathbf{a}_3^*$$

and equation 9-34 becomes

$$I_{eu} = F^2 \sum_m \sum_{m'} \exp\{2\pi i[(m_1' - m_1)h_1 + (m_2' - m_2)h_2 + (m_3' - m_3)h_3]$$

$$+ \frac{\mathbf{s} - \mathbf{s}_0}{\lambda}(\boldsymbol{\delta}_{m'} - \boldsymbol{\delta}_m)\}. \tag{9-35}$$

Let the distortion displacement vector be expressed by

$$\boldsymbol{\delta}_m = X_m\mathbf{a}_1 + Y_m\mathbf{a}_2 + Z_m\mathbf{a}_3.$$

Then if in equation 9-35 we approximate $(\mathbf{s} - \mathbf{s}_0)/\lambda$ by its average value $l\mathbf{a}_3^*$ (refer to Fig. 9-11), the scalar product may be written

$$\frac{\mathbf{s} - \mathbf{s}_0}{\lambda}\boldsymbol{\delta}_m = l\mathbf{a}_3^*(X_m\mathbf{a}_1 + Y_m\mathbf{a}_2 + Z_m\mathbf{a}_3) = lZ_m,$$

and 9-35 becomes

$$I_{eu}(h_1h_2h_3) = F^2 \sum_m \sum_{m'} \exp\{2\pi i[(m_1' - m_1)h_1 + (m_2' - m_2)h_2 + (m_3' - m_3)h_3$$

$$+ l(Z_{m'} - Z_m)]\}. \tag{9-36}$$

We note that only the component $Z_{m'} - Z_m$ of the distortion displacement perpendicular to the reflecting planes appears in the intensity expression.

Making use of Warren's powder pattern power theorem (Appendix X), we can express the total power in a powder reflection as an integral

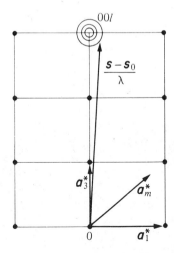

Fig. 9-11. Representation of a 00l reflection in a two-dimensional section of an orthorhombic reciprocal lattice. (Courtesy of B. E. Warren and Addison-Wesley [28].)

of the power distribution $P(2\theta)$:

$$\int P(2\theta)\, d(2\theta) = \frac{I_e M R^2 \lambda^3 p(hkl)}{4v_a} \int\int\int \frac{I_{eu}(h_1 h_2 h_3)}{\sin\theta}\, dh_1\, dh_2\, dh_3. \qquad (9\text{-}37)$$

The multiplicity factor $p(hkl)$ confines the integration to one equivalent region of reciprocal space. The intensity of a $00l$ reflection is vanishingly small except for small h_1 and h_2, and the length of the diffraction vector is given to a good approximation by its vertical ($\mathbf{a}_3^*$) component,

$$s = \frac{|\mathbf{s} - \mathbf{s}_0|}{\lambda} = \frac{2\sin\theta}{\lambda} = |h_1 \mathbf{a}_1^* + h_2 \mathbf{a}_2^* + h_3 \mathbf{a}_3^*| \simeq h_3 |\mathbf{a}_3^*|, \qquad (9\text{-}38)$$

as can be seen from Fig. 9-11. Letting

$$\frac{2\sin\theta}{\lambda} = h_3 |\mathbf{a}_3^*|,$$

we obtain

$$dh_3 = \frac{\cos\theta\, d(2\theta)}{\lambda |\mathbf{a}_3^*|},$$

which when substituted in equation 9-37 gives

$$P(2\theta) = \frac{I_e M R^2 \lambda^2 p(hkl) \cos\theta}{4v_a |\mathbf{a}_3^*| \sin\theta} \int\int I_{eu}(h_1 h_2 h_3)\, dh_1\, dh_2. \qquad (9\text{-}39)$$

But the measured power, $P'(2\theta)$, is actually proportional to the power per unit length of diffraction circle at a distance R from the sample, or

$$P'(2\theta) = \frac{P(2\theta)}{2\pi R \sin 2\theta}.$$

With this expression, a constant K defined by

$$K = \frac{I_e R \lambda^2 p(hkl)}{|6\pi v_a |\mathbf{a}_3^*|},$$

and the interference function as expressed by equation 9-36, the experimentally observable power $P'(2\theta)$ becomes:

$$P'(2\theta) = \frac{KMF^2}{\sin^2\theta} \int\int \sum_m \sum_{m'} \exp\{2\pi i[(m_1' - m_1)h_1 + (m_2' - m_2)h_2$$

$$+ (m_3' - m_3)h_3 + l(Z_{m'} - Z_m)]\}\, dh_1\, dh_2. \qquad (9\text{-}40)$$

In order that all the diffracted power belonging to the reflection $00l$ be included, it is necessary to integrate equation 9-40 over h_1 and h_2 from

$-\frac{1}{2}$ to $+\frac{1}{2}$, which yields

$$P'(2\theta) = \frac{KMF^2}{\sin^2 \theta} \sum_m \sum_{m'} \frac{\sin \pi (m_1' - m_1)}{\pi (m_1' - m_1)} \frac{\sin \pi (m_2' - m_2)}{\pi (m_2' - m_2)}$$

$$\times \exp \{2\pi i[(m_3' - m_3)h_3 + l(Z_{m'} - Z_m)]\}.$$

Since the sine quotients are equal to unity for $m_1' = m_1$ and $m_2' = m_2$ and zero otherwise, $P'(2\theta)$ simplifies to

$$P'(2\theta) = \frac{KMF^2}{\sin^2 \theta} \sum_{m_1} \sum_{m_2} \sum_{m_3} \sum_{m_3'} [\exp \{2\pi i l[Z(m_3') - Z(m_3)]\}$$

$$\times \exp \{2\pi i(m_3' - m_3)h_3\}]_{m_1 m_2}. \tag{9-41}$$

At this point it is useful, as first suggested by Bertaut[34] to visualize the crystallite as an assemblage of columns of unit cells parallel to the $\mathbf{a}_3$ (and $\mathbf{a}_3^*$) direction sufficient in number to traverse the crystallite from one surface to the other. This concept is pictured schematically in Fig. 9-10 by an $\mathbf{a}_1\mathbf{a}_3$ section of a rectangular assemblage of unit cells. The summation over m_3 and m_3' gives the contribution of all pairs of cells in a given column $m_1 m_2$, so that the summations over m_1 and m_2 then give the contributions of all columns composing the crystallite. If we now let

$$n = m_3' - m_3 \quad \text{and} \quad Z_n = Z(m_3') - Z(m_3),$$

and denote by N_n (sample) the number of cells *in the whole sample* with an nth neighbor in the same column, so that $\langle \exp 2\pi i l Z_n \rangle$ may represent an average over all pairs of nth neighbors in the same column throughout the entire sample, equation 9-41 may be further simplified to:

$$P'(2\theta) = \frac{KF^2}{\sin^2 \theta} \sum_{-\infty \atop n}^{+\infty} N_n (\text{sample}) \langle \exp 2\pi i l Z_n \rangle \exp 2\pi i n h_3. \tag{9-42}$$

In equation 9-42 the fourfold summation of equation 9-41 and factor M, which denotes the number of crystallites in the sample, have been replaced by a single summation over all values of n.

It is now pertinent to let $N =$ number of cells in the sample, $N_3 =$ average number of cells per column, and $N_n =$ average number of n pairs per column. Introduction of these quantities into equation 9-42 gives the result:

$$P'(2\theta) = \frac{KNF^2}{\sin^2 \theta} \sum_{-\infty \atop n}^{+\infty} \frac{N_n}{N_3} \langle \exp 2\pi i l Z_n \rangle \exp 2\pi i n h_3.$$

Expansion of the exponentials into their real and imaginary parts gives

$$P'(2\theta) = \frac{KNF^2}{\sin^2 \theta} \sum_{-\infty}^{+\infty} \frac{N_n}{N_3} [\langle \cos 2\pi l Z_n \rangle \cos 2\pi n h_3 - \langle \sin 2\pi l Z_n \rangle \sin 2\pi n h_3$$

$$+ i(\langle \cos 2\pi l Z_n \rangle \sin 2\pi n h_3 + \langle \sin 2\pi l Z_n \rangle \cos 2\pi n h_3)]. \quad (9\text{-}43)$$

Now every pair of cells m_3' and m_3 appears in the summation twice, as $n = m_3' - m_3$ and as $-n = m_3 - m_3'$. Also, $Z_n = Z(m_3') - Z(m_3)$ and $Z_{-n} = Z(m_3) - Z(m_3')$, from which it follows that $Z_{-n} = -Z_n$, hence the imaginary terms of equation 9-43 cancel out. At this point we may introduce the Fourier coefficients

$$A_n = \frac{N_n}{N_3} \langle \cos 2\pi l Z_n \rangle \quad \text{and} \quad B_n = -\frac{N_n}{N_3} \langle \sin 2\pi l Z_n \rangle \quad (9\text{-}44)$$

and express $P'(2\theta)$ as a Fourier series:

$$P'(2\theta) = \frac{KNF^2}{\sin^2 \theta} \sum_{-\infty}^{+\infty} (A_n \cos 2\pi n h_3 + B_n \sin 2\pi n h_3). \quad (9\text{-}45)$$

It should be noted that the Fourier series is expressed in terms of the variable (see again equation 9-38)

$$h_3 = \frac{2 \sin \theta}{\lambda |\mathbf{a}_3^*|} = \frac{2|\mathbf{a}_3| \sin \theta}{\lambda},$$

and n is the harmonic number corresponding to the separation $L = na_3$ between a pair of cells in a column parallel to $\mathbf{a}_3$ and $\mathbf{a}_3^*$ (perpendicular to the 00l planes).

By examining equation 9-45 we see that a set of uncorrected Fourier coefficients can be obtained by plotting the experimentally measured quantity $P'(2\theta)(\sin^2 \theta)/F^2$ against values of $h_3 = 2a_3(\sin \theta)/\lambda$ between $l - \frac{1}{2}$ and $l + \frac{1}{2}$ [i.e., across the powder line profile $h(\epsilon)$ over a sufficiently wide range centered about the theoretical peak location $2\theta_0$ of the 00l reflection concerned]. Then the correction for instrumental broadening $g(\epsilon)$ is applied by one of the two unfolding procedures described in Section 9-1. If the Fourier-transform (Stokes) method is employed, the Fourier coefficients A_n and B_n of the pure diffraction profile $f(\epsilon)$ are obtained directly, ready for the determination of the crystallite size and strain parameter by the methods described below. The averages $\langle \cos 2\pi l Z_n \rangle$ and $\langle \sin 2\pi l Z_n \rangle$ of equations 9-44 are averages over pairs of nth neighbors in *all columns in the sample*. If for a given value of n, positive and negative values of Z_n are equally likely, the sine coefficients B_n are zero. If this is not the case, the sine coefficients will cause a displacement and/or asymmetry of the line profile. In cold-worked metals, if we exclude the possibility of stacking faults, positive and negative strains Z_n will occur with approximately equal probability and the sine coefficients may be assumed negligible in relation to the cosine coefficients. Accordingly, we proceed with the analysis of the cosine coefficients A_n.

B. Separation of Size and Distortion Components. The cosine coefficient is the product of N_n/N_3, which depends only on the column length and is therefore a size coefficient, and $\langle \cos 2\pi l Z_n \rangle$, which depends only on the distortion in the domain (crystallite) and is thus a distortion coefficient. Denoting size and distortion by superscripts S and D, respectively, we have then

$$A_n{}^S = \frac{N_n}{N_3} \quad \text{and} \quad A_n{}^D = \langle \cos 2\pi l Z_n \rangle, \tag{9-46}$$

and the cosine coefficient that is determined experimentally may be denoted as the product:

$$A_n = A_n{}^S A_n{}^D. \tag{9-47}$$

We suppose with Warren and Averbach[29] that several orders of $00l$, say 001, 002, and 003, have been measured and their cosine coefficients A_n derived therefrom. It is possible to separate the coefficients $A_n{}^S$ and $A_n{}^D$ because the size coefficient is independent of order l, whereas the distortion coefficient $\langle \cos 2\pi l Z_n \rangle$ is a function of l and approaches unity as l goes to zero. On the assumption that l and n are sufficiently small that the product $l Z_n$ is also small,

$$\langle \cos 2\pi l Z_n \rangle \rightarrow 1 - 2\pi^2 l^2 \langle Z_n{}^2 \rangle,$$

and in logarithmic form,

$$\ln \langle \cos 2\pi l Z_n \rangle = \ln(1 - 2\pi^2 l^2 \langle Z_n{}^2 \rangle) = -2\pi^2 l^2 \langle Z_n{}^2 \rangle.$$

Then for small values of l and n the logarithm of the measured Fourier coefficient (equation 9-47) may be written:

$$\ln A_n(l) = \ln A_n{}^S - 2\pi^2 l^2 \langle Z_n{}^2 \rangle. \tag{9-48}$$

Now for each fixed value of n the values of $\ln A_n(l)$ are plotted against l^2 (see Fig. 9-12), whereupon the l^2 zero intercepts give the size coefficients $A_n{}^S$ and the slopes give values of $-2\pi^2 l^2 \langle Z_n{}^2 \rangle$. By way of interpretation of the foregoing results, $L = na_3$, represented schematically in Fig. 9-10, is the undistorted distance between the pair of cells m_3' and m_3, but distortion changes this distance by $\Delta L = a_3 Z_n$. Then $\epsilon_L = \Delta L/L$ is the component of strain in the a_3 direction averaged over the distance L. It

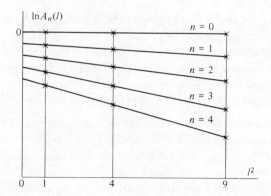

Fig. 9-12. Logarithmic plot for separating crystallite size and distortion effects when multiple $00l$ orders are available. (Courtesy of B. E. Warren and Addison–Wesley [28].)

follows also that since $a_3 Z_n$ is the change in length of a column of length $a_3 n$, $\epsilon_L = Z_n/n$, and $\langle Z_n{}^2 \rangle$ may be replaced by $n^2 \langle \epsilon_L{}^2 \rangle$. Thus the *initial slopes* of the $\ln A_n(l)$ versus l^2 plots give values of $-2\pi^2 l^2 n^2 \langle \epsilon_L{}^2 \rangle$, from which mean-square values of the component of strain $\langle \epsilon_L{}^2 \rangle$ may be calculated.

It is very important to note that equation 9-48 is valid only for small l and n, so that the plot of $\ln A_n(l)$ versus l^2 is necessarily straight only in the low l region. However, this is precisely the region where linearity is required in order to effect a reliable extrapolation to the $l^2 = 0$ intercept. Of special interest is a hypothetical Gaussian strain distribution, which may be formulated

$$p(Z_n) = \frac{a}{\pi^{1/2}} \exp\left(- a^2 Z_n{}^2\right).$$

Then

$$\langle Z_n{}^2 \rangle = \frac{a}{\pi^{1/2}} \int_{-\infty}^{+\infty} Z_n{}^2 [\exp\left(- a^2 Z_n{}^2\right)]\, dZ_n = \frac{1}{2a^2},$$

and

$$\langle \cos 2\pi l Z_n \rangle = \frac{a}{\pi^{1/2}} \int_{-\infty}^{+\infty} \cos 2\pi l Z_n [\exp\left(- a^2 Z_n{}^2\right)]\, dZ_n,$$

so that the distortion coefficient assumes the value

$$A_n{}^D(l) = \exp\frac{- \pi^2 l^2}{a^2} = \exp\left(- 2\pi^2 l^2 \langle Z_n{}^2 \rangle\right). \tag{9-49}$$

Evidently, then, for a Gaussian strain distribution equation 9-48 is exact for all values of l and n, and the plots of $\ln A_n(l)$ against l^2 will be linear out to indefinitely large values of l^2.

We consider now the size coefficient A_n^S of the cosine Fourier coefficient A_n (equation 9-47). With Warren[28, 31] we follow a treatment first given by Bertaut[30] in terms of the concept of the crystal as being comprised of columns of unit cells parallel to $\mathbf{a}_3$. If there are n_i columns composed of i cells and if $p(i)\, di$ denotes the fraction of columns with a length between i and $i + di$, A_n^S may be expressed as

$$A_n^S = \frac{1}{N_3} \int_{i=|n|}^{\infty} (i - |n|) p(i)\, di. \tag{9-50}$$

It is observed that the A_n^S coefficients, which were derived from the plots of $\ln A_n(l)$ versus l^2, vary smoothly with n as illustrated by Fig. 9-13, so that it is possible to view A_n^S as a continuous quantity. In order to obtain the average column length $N_3 a_3$, we need to determine the derivative of A_n^S with respect to n (regarded as a continuous variable) We rewrite equation 9-50 as

$$A_n^S = \frac{1}{N_3} \left[\int_{i=|n|}^{\infty} i p(i)\, di - |n| \int_{i=|n|}^{\infty} p(i)\, di \right], \tag{9-51}$$

in which the variable i appears in the lower limit of the integral. Therefore we may employ the theorem

$$y = \int_{\epsilon}^{\infty} f(x)\, dx, \qquad \frac{dy}{d\epsilon} = -f(\epsilon)$$

and obtain from equation 9-51:

$$\frac{dA_n^S}{dn} = \frac{1}{N_3} \left[-|n| p(|n|) - \int_{i=|n|}^{\infty} p(i)\, di + |n| p(|n|) \right]$$

$$= -\frac{1}{N_3} \int_{i=|n|}^{\infty} p(i)\, di. \tag{9-52}$$

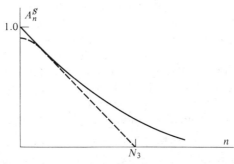

Fig. 9-13. Plot of the Fourier size coefficient A_n^S versus n. The abscissa intercept of the initial slope is the mean column length number N_3. (Courtesy of B. E. Warren and Addison–Wesley [28].)

The derivative as n approaches zero is then

$$\left(\frac{dA_n{}^S}{dn}\right)_{n\to 0} = -\frac{1}{N_3}\int_{i=0}^{\infty} p(i)\,di = -\frac{1}{N_3}. \tag{9-53}$$

Thus, with reference to Fig. 9-13, the initial slope of the $A_n{}^S$ versus n curve is $-1/N_3$, and furthermore N_3 is the $A_n{}^S = 0$ intercept of the initial slope. Thus the initial slope of the curve of $A_n{}^S$ plotted against n yields $N_3 a_3$, the average crystallite (domain) dimension perpendicular to the $00l$ planes (in the $\mathbf{a}_3$ direction of the hypothetical orthorhombic crystal). This result is independent of the crystallite size and shape distributions.

Differentiation of equation (9-52) gives

$$\frac{d^2 A_n{}^S}{dn^2} = \frac{p(|n|)}{N_3}, \tag{9-54}$$

which is a column length, or crystallite size, distribution. As pointed out by Warren[28], when size and strain broadening are simultaneously present, this method of deriving the size distribution is of only very limited utility. However, Bienenstock[32, 33] has shown that *in the absence of strains* the Fourier coefficients of a broadened line can be handled in such a way as to yield a size distribution that is meaningful and also includes the smaller sizes. Implicit in Warren and Averbach's treatment is the substitution of $(\pi h_3)^2$ for $\sin^2 \pi h_3$, which is not a good approximation for very broad lines (very small crystallite sizes). A brief statement of Bienenstock's method is given in Section 9-3.3.

With respect to equation 9-54 and Fig. 9-13, it is very important to note that since $p(|n|)$ cannot be negative, the second derivative likewise can never be negative, from which it follows that the plot of $A_n{}^S$ against n can be concave upward but never downward. Such a so-called "hook effect," or downward curve near $n = 0$ (dotted line in Fig. 9-13), usually indicates that the A_0 coefficient of the data is too small relative to the other A_n coefficients. Since A_0 is proportional to the area under the line profile, this is most likely to be a consequence of the background being placed too high, which in turn is a probable effect of the overlapping of the long tails of the line profiles. Warren suggests determining an upper permissible limit of the background level by measuring the pattern of an annealed sample under identical experimental conditions, including the use of crystal-monochromatized radiation.

C. Generalization of the Warren-Averbach Theory[28]. The foregoing theoretical results may be expressed in a generalized manner so as to apply to any powder-pattern reflection. Parallel to the notation used to this point, we first introduce fictitious quantities a_3', $a_3^{*\prime}$, l', h_3' and

n'. The new angular variable is then

$$h_3' = 2a_3' \frac{\sin \theta}{\lambda},$$

and the experimental line profile, corrected for instrumental broadening, may then be represented as shown in Fig. 9-14. At the peak position, $\theta = \theta_0$, we let $h_3' = l'$. Next a large enough interval in h_3', extending from $l' - \frac{1}{2}$ to $l' + \frac{1}{2}$, is selected to encompass the entire observable line profile.

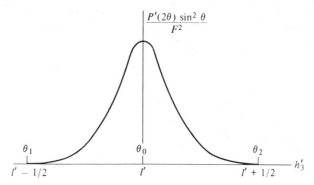

Fig. 9-14. The corrected experimental line profile represented in terms of the fictitious quantities h_3' and l'. (Courtesy of B. E. Warren and Addison–Wesley [28].)

With the angles θ_1 and θ_2 corresponding, respectively, to the limits $l' - \frac{1}{2}$ and $l' + \frac{1}{2}$, we have

$$l' - \frac{1}{2} = \frac{2a_3'}{\lambda} \sin \theta_1,$$

$$l' = \frac{2a_3'}{\lambda} \sin \theta_0,$$

$$l' + \frac{1}{2} = \frac{2a_3'}{\lambda} \sin \theta_2,$$

from which

$$\frac{1}{2} = \frac{2a_3'}{\lambda} (\sin \theta_2 - \sin \theta_0). \tag{9-55}$$

This equation defines a_3'.

The Fourier coefficients $A_{n'}$ are evaluated over the interval $l' - \frac{1}{2}$ to $l' + \frac{1}{2}$, and since $L = n'a_3'$ is a real distance in the $\mathbf{a}_3$ direction, we may multiply all n' values by the value of a_3' obtained from equation 9-55 and designate the coefficients A_L. Furthermore, a_3'/l' is the real interplanar spacing d, and Z'_n/n' is the component of strain ϵ_L, so that $\langle\cos 2\pi l' Z'_n\rangle$ can be replaced by $\langle\cos(2\pi\epsilon_L L/d)\rangle$, and equation 9-48 can be written

$$\ln A_L \frac{1}{d} = \ln A_L{}^S - \frac{2\pi^2\langle\epsilon_L{}^2\rangle L^2}{d^2}. \qquad (9\text{-}56)$$

Employing several reflection orders as before, we can now perform the extrapolation typified by Fig. 9-12 by plotting $\ln A_L(1/d)$ against $(1/d)^2$. Treating the size coefficients obtained from the ordinate intercepts in a manner parallel to that described above (see Fig. 9-13 and related text), we plot the values of $A_L{}^S$ versus L as shown in Fig. 9-15, obtaining directly from the abscissa intercept of the *initial slope* the column length $\langle D\rangle$ perpendicular to the reflecting planes.

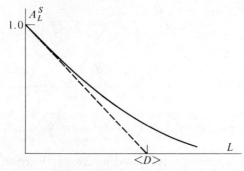

Fig. 9-15.   Plot of the Fourier size coefficient $A_L{}^S$ against L, where $L = n'a'_3$ is a real distance along the columns of cells. (Courtesy of B. E. Warren and Addison–Wesley [28].)

For cubic crystals, which are of special importance in the study of metals, equation 9-56 may be transformed to the convenient form

$$\ln A_L(h_0) = \ln A_L{}^S - \frac{2\pi^2\langle\epsilon_L{}^2\rangle L^2 h_0{}^2}{a^2}, \qquad (9\text{-}57)$$

where $h_0{}^2$ is the quadratic index sum $h^2 + k^2 + l^2$ and $d = a/h_0$. From a set of multiple orders, for example, 100, 200, 300 for $h00$, or 110, 220, 330 for $hh0$, and so on, we plot $\ln A_L(h_0)$ against $h_0{}^2$ and obtain $A_L{}^S$ from the ordinate intercept. Although, in general, an extrapolation from three or more orders is to be preferred, in the case of cubic substances one must usually be satisfied with two orders because the third order of any simple reflection is always blended with another reflection at the same Bragg angle (thus 100, 200, and 300/221, or 110, 220, and 330/411).

Occasionally, cubic substances are subject to the right mode of preferred orientation to resolve the third order adequately.

9-2.3 Use of Variance of the Line Profile

At the present time increasing attention is being paid to the variance W as a measure of line-profile dispersion, or breadth[35, 36]. A preliminary discussion of this topic has been given in Section 5-2.2B and C, including a numerical example of the calculations involved for a typical line profile (Table 5-2). The variance, or reduced second moment, of the line profile on the 2θ scale may be expressed

$$W_{2\theta} = \langle (2\theta - \langle 2\theta \rangle)^2 \rangle$$
$$= \frac{\int (2\theta - \langle 2\theta \rangle)^2 I(2\theta)\, d(2\theta)}{\int I(2\theta)\, d(2\theta)}, \qquad (9\text{-}58)$$

in which $\langle 2\theta \rangle$ is the centroid. The variance of a line profile has the great advantage over the half-maximum breadth $B_{1/2}$ and integral breadth B_i that it can be corrected for instrumental broadening by simple subtraction of the instrumental-profile variance rather than by unfolding. Likewise, when small crystallite size (superscript S) and lattice distortions (superscript D) simultaneously act to broaden the pure diffraction profile, its net variance W^{SD} is simply the sum of the variances W^S and W^D.*

The theoretical advantage gained as a result of the additivity of variances is possibly more than offset in practical applications by the sensitive dependence of the line-profile variance upon the choice of background level[40–43]. This is the consequence of the rapid change in the range of integration $-\sigma_1$ to $+\sigma_2$ corresponding to a given change in background level. It is pertinent to note here that the sensitivity of the integral breadth to the choice of background level is considerably less than that of the variance. For a more thorough treatment of the errors affecting the experimental determination of the line profile, especially in relation to the optimization of the background level, the reader is referred to additional articles in the literature[40, 43a].

If the tails of the profile approach the background according to an

---

*More recent investigations of Wilson[37, 38] and Edwards and Toman[39] have revealed second-order nonadditivity properties of the line-profile variance. Such errors can be reduced by extending the range of measurement, but the value of this strategem has an ultimate limit imposed by the necessity of truncating experimental profiles. It is doubtful that the accuracy of typical experimental data is sufficient to justify efforts to correct for nonadditivity in ordinary analytical work.

inverse-square law, the variance is a linear function of the range, $\sigma_1 + \sigma_2$. If, as is commonly true, $\sigma_1 = \sigma_2$, we may let the range of measurement $\sigma_1 + \sigma_2 = \sigma$ and express the linearity of the variance-range function in the form [41]*:

$$W = W_0 + k\sigma. \tag{9-59}$$

In equation 9-59 the numerical values of W_0 and k depend upon the instrumental aberrations and physical condition of the specimen. The validity of this equation depends strongly on the establishment of the background level so as to ensure the asymptotic behavior of W as a function of σ. It should also be emphasized that for the variance to be finite it is necessary that the line profile be truncated (see Section 5-2.2B and references cited there). For a valuable treatment of practical aspects of the determination of the variance, the reader is referred to Langford and Wilson [41].

Despite the above exposition of the rather stringent limitations affecting the practical use of the variance as a measure of the line breadth, the weight of experimental evidence increasingly supports its validity, when proper account is taken of its limitations, as an independent check on the Warren-Averbach Fourier method for the simultaneous determination of crystallite size and lattice distortions from line broadening.

We shall summarize the treatment of Wilson [36] for line broadening caused only by small crystallite size, and then give its extension to broadening arising from lattice distortions [44]. Before proceeding with Wilson's findings, we find it pertinent to refer to the historically important result of Scherrer [45] for broadening resulting from small crystallite size alone,

$$L = \frac{K\lambda}{\beta \cos \theta}, \tag{9-60}$$

wherein θ and λ have their usual meanings, L is the mean dimension of the crystallites composing the powder, β is the breadth (commonly $\beta_{1/2}$ or β_i) of the pure diffraction profile on the 2θ scale in radians, and K a constant approximately equal to unity and related both to the crystallite shape and to the way in which β and L are defined. It is sometimes advantageous to modify equation 9-60 so as to express the reflection breadth in δs units [$s = 2(\sin \theta)/\lambda$] as follows:

$$L = \frac{K}{[(2 \cos \theta)/\lambda] \, \delta\theta} = \frac{K}{\delta s}. \tag{9-61}$$

---

*Langford and Wilson [41] suggest that in practice the range be initially set equal to 10 times the separation of the $K\alpha$ doublet, after which the upper limit should be adjusted until the centroid of the reflection is within $\pm 0.005° \, 2\theta$ of the midpoint of the range.

A simplified derivation of the Scherrer equation is given in Section 9-3.1.

A. Contribution of Crystallite Size to the Variance. Returning now to the variance as a measure of line breadth, we first employ a device of Stokes and Wilson[46, 47] by defining a volume $V(t)$ common to a crystallite of volume V and its "ghost" displaced a distance t in a direction perpendicular to the reflecting planes, as shown by the shaded region in Fig. 9-16. Then, for a cube of edge length p, the "true" crystallite size is clearly

$$p = [V(0)]^{1/3}. \tag{9-62}$$

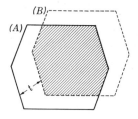

Fig. 9-16. Crystal (A) of volume V and its "ghost" (B) displaced a distance t. (Courtesy of A. J. C. Wilson and Methuen[47].)

Furthermore, it can be readily shown that[36]

$$\frac{-V'(0)}{V(0)} = \frac{1}{p}, \tag{9-63}$$

$V'(0)$ being the derivative of $V(t)$ with respect to t for $t = 0$. On the basis of this result for a cube, it is natural for *other reflections* and *other crystal shapes* to define the "apparent" crystallite size as

$$E = \frac{-V(0)}{V'(0)}, \tag{9-64}$$

from which it follows that the Scherrer constant is equivalent to

$$K = \frac{-pV'(0)}{V(0)}$$

$$= \frac{-V'(0)}{[V(0)]^{2/3}}. \tag{9-65}$$

From this point of view, then, the Scherrer constant is the ratio by which the apparent crystallite size must be multiplied to give the true size.

Regardless of the precise meanings assigned to β and L of the Scherrer equation 9-60, we see that for a given value of the dimension L the line breadth on the 2θ scale varies as $1/\cos\theta$, so that the broadening caused by small crystallite size is most conspicuous at large Bragg angles. In common practice K of equation 9-60 or 9-61 is set equal to unity, which confers on L the specific significance of the volume-average crystallite dimension L_{hkl} perpendicular to the reflecting planes.

Wilson[36] has shown that when small crystallite size is the only source of line broadening, there is no displacement of the line and the expression for the variance of the profile becomes:

$$W = -\frac{1}{4\pi^2}\left[2(\sigma_1+\sigma_2)\frac{V'(0)}{V(0)}+\frac{V''(0)}{V(0)}\right], \qquad (9\text{-}66)$$

which may be written

$$W = W_0+\tfrac{1}{2}(\sigma_1+\sigma_2)W_1 \qquad (9\text{-}67)$$

if we let

$$W_1 = -\frac{V'(0)}{\pi^2 V(0)} \quad\text{and}\quad W_0 = -\frac{V''(0)}{4\pi^2 V(0)}. \qquad (9\text{-}68)$$

The reader will note the equivalence of equations 9-67 and 9-59. By comparison with equations 9-64 and 9-65

$$W_1 = \frac{1}{E\pi^2} = \frac{K}{\pi^2 p}. \qquad (9\text{-}69)$$

With reference to the second term of equation 9-66, $V''(0)/V(0)$ is a constant specifying the limit, as $t \to 0$, of the fraction of the volume of the crystals for which the thickness measured normal to the reflecting planes is between t and $t+dt$. If we let

$$T_a = p^2\frac{V''(0)}{V(0)}, \qquad (9\text{-}70)$$

W_0 (equations 9-67 and 9-68) assumes the form

$$W_0 = -\frac{T_a}{4\pi^2 p^2}, \qquad (9\text{-}71)$$

and substitution of equations 9-69 and 9-71 in 9-67 gives for the variance of the line profile (in δs units) due to small crystallite size:

$$W_s{}^S = \frac{\tfrac{1}{2}(\sigma_1+\sigma_2)K}{\pi^2 p} - \frac{T_a}{4\pi^2 p^2}. \qquad (9\text{-}72)$$

The quantity T_a is termed the *taper parameter* by Wilson[36], because it is a measure of the initial rate at which the cross section of the crystals decreases with t (the reader should refer to Fig. 9-16 and the related text for the precise significance of t). In practically all experimental situations, the magnitude of the second term of equation 9-72 is very small relative to the first. For small crystallites the ratio of the second term to the first ranges from zero to possibly 10 per cent for different shapes. Therefore it has become almost standard practice in crystallite-size analysis to neglect the second term.

The crystallite-size variance in $\delta 2\theta$ (expressed in radians) may be obtained from equation 9-72 by employing the relationship

$$W_{2\theta}^S = \frac{\lambda^2}{\cos^2\theta} W_s{}^S,$$

which gives

$$W_{2\theta}^S = \frac{K\lambda\Delta 2\theta}{2\pi^2 p \cos\theta} - \frac{T_a\lambda^2}{4\pi^2 p^2 \cos^2\theta}, \tag{9-73}$$

$\Delta 2\theta$ being the range of integration. Although the taper parameter T_a varies with the shape of the crystallite, its sensitivity is too small to permit useful distinctions to be made using actual experimental data. On the basis of *integral breadths* numerical calculations of the Scherrer constant K, sometimes referred to as the shape factor, have been calculated by Stokes and Wilson[46] for seven reflections from cubic crystals having the shapes of cubes, tetrahedra, and octahedra. Tournarie[40] has determined numerical values of the *variance* Scherrer constant for the same crystal shapes and reflections, as well as a number of additional reflections. Their results are tabulated in Table 9-2 together with reference values for spheres calculated for $\beta_{1/2}$, β_i, and W. The reader is reminded that these numerical results refer to the pure diffraction profile corrected for instrumental broadening.

Table 9-2. Values of the Scherrer Constant K for Cubic Crystals of Common Regular Shapes[a,b]

hkl	Cube β_i	Cube W^S	Tetrahedron β_i	Tetrahedron W^S	Octahedron β_i	Octahedron W^S
100	1.000	1.000	1.387	2.080	1.101	1.651
110	1.061	1.414	0.981	1.471	1.038	1.167
111	1.155	1.732	1.201	1.802	1.144	1.430
210	1.073	1.342	1.240	1.860	1.108	1.477
211	1.153	1.633	1.132	1.698	1.106	1.348
221	1.143	1.667	1.156	1.733	1.119	1.376
310	1.067	1.265	1.316	1.973	1.114	1.566

[a]Calculated for several reflections and for the integral breadth β_i[46] and variance W^S[36, 40]. The crystal dimension L (or p) is defined as $V^{1/3}$, the so-called "true" size.
[b]For spheres: $K = 0.893$ ($\beta_{1/2}$), 1.075 (β_i), $1.209(W)$.

B. Contribution of Lattice Distortions to the Variance. Wilson[44] has shown that for local strains, $e = \Delta d/d$, that are not too large the reflection is shifted (to a good approximation) by

$$\langle s \rangle = \frac{\langle e \rangle}{d},$$

d being the interplanar spacing. At the same time the reflection is broadened (to a good approximation) by

$$W_s^D = \frac{\langle e^2 \rangle - \langle e \rangle^2}{d^2} \tag{9-74}$$

in $(\delta s)^2$ units. In these equations $\langle e^2 \rangle$ is the variance of the lattice-strain distribution. In a polycrystalline specimen with the appropriate numerical value assigned to d, the mean value of the local strains $\langle e \rangle$ is zero, so that equation (9-74) becomes effectively

$$W_s^D = \frac{\langle e^2 \rangle}{d^2}$$

$$= \frac{4 \sin^2 \theta}{\lambda^2} \langle e^2 \rangle. \tag{9-75}$$

Furthermore, since

$$\delta s = \frac{2}{\lambda} \cos \theta \delta \theta,$$

the variance in δ^2 (2θ) units (expressed in radians squared) is related to W_S^D by

$$W_{2\theta}^D = \frac{\lambda^2}{\cos^2 \theta} W_s^D, \tag{9-76}$$

which when combined with equation (9-75) yields

$$W_{2\theta}^D = 4 \tan^2 \theta \langle e^2 \rangle. \tag{9-77}$$

Equations 9-76 and 9-77 hold as long as $\cos \theta$ does not vary appreciably over the range of the line profile. If it does vary considerably, the observed line profile must be replotted as a function of s before calculating the variance[44].

In view of the additivity of the size and distortion variances, equations 9-73 and 9-77 may be combined to give for the overall line-profile variance resulting from both small crystallite size and strains:

$$W_{2\theta} = \frac{K\lambda \Delta 2\theta}{2\pi^2 p \cos \theta} - \frac{T_a \lambda^2}{4\pi^2 p^2 \cos^2 \theta} + 4 \tan^2 \theta \langle e^2 \rangle. \tag{9-78a}$$

As explained previously, the second term, which contains the taper parameter T_a, is commonly neglected in practical applications. It should also be remarked that in actual practice the variance method appears to be more reliable for the derivation of strain than of size parameters. This is a consequence of the greater sensitivity of the size parameter to the angular range of measurement.

The numerical solution of the size and strain parameters is conveniently carried out by neglecting T_a and arranging equation 9-78a in the form

$$\frac{W_{2\theta}}{\Delta 2\theta} \frac{\cos \theta}{\lambda} = \frac{1}{2\pi^2 p} + \frac{4 \sin \theta \tan \theta}{\lambda (\Delta 2\theta)} \langle e^2 \rangle, \qquad (9\text{-}78\text{b})$$

after which a plot of $W_{2\theta} (\cos \theta)/\lambda(\Delta 2\theta)$ against $4(\sin \theta \tan \theta)/\lambda (\Delta 2\theta)$ is made, p being obtained from the ordinate intercept and $\langle e^2 \rangle$ from the slope.

9-2.4 Method of Integral Breadths

The separation of size and strain broadening by the method of integral breadths depends upon explicit assumptions as to the shape of the broadening profile due to each effect. Therefore, in general, the numerical solutions likewise exhibit a dependence upon the assumptions and tend to differ in a systematic manner both from each other as well as from the solutions obtained by Fourier analysis or variance.

Wilson[48] has shown that the integral profile breadth in s units generated by distortions (strains) alone may be expressed:

$$(\delta s)_i{}^D = 4e \frac{\sin \theta}{\lambda} = 2es = \frac{2e}{d_{hkl}}. \qquad (9\text{-}79)$$

In equation 9-79, s, θ, λ, and d_{hkl} have their usual meanings and $e \simeq (\Delta d/\bar{d})_{hkl}$ is an approximate upper limit of the lattice distortions. Parallel to the relationship of equation 9-77 to 9-75, equation 9-79 may be expressed on the 2θ scale in radians as

$$(\delta 2\theta)_i{}^D = 4e \tan \theta. \qquad (9\text{-}80)$$

Stokes and Wilson[49] have defined an "apparent strain" η by

$$\eta = (\delta 2\theta)_i{}^D \cot \theta, \qquad (9\text{-}81)$$

which may be written in the manner of equation 9-80 as

$$(\delta 2\theta)_i{}^D = \eta \tan \theta. \qquad (9\text{-}82)$$

Evidently, then

$$\eta = 4e. \qquad (9\text{-}83)$$

Stokes and Wilson also show that in terms of the root-mean-square strain $\langle \epsilon^2 \rangle^{1/2}$

$$\eta = 2(2\pi)^{1/2}\langle \epsilon^2 \rangle^{1/2}$$
$$= 5.0\langle \epsilon^2 \rangle^{1/2}. \tag{9-84}$$

From these relations between η and e and between η and $\langle \epsilon^2 \rangle^{1/2}$, it follows that

$$e = 1.25\langle \epsilon^2 \rangle^{1/2}, \tag{9-85}$$

as Buchanan, McCullough, and Miller[50] have stated. According to the Scherrer relationship (equation 9-60), the pure integral line breadth resulting from small crystallite size is

$$(\delta 2\theta)_i^S = \frac{K\lambda}{L \cos \theta} \tag{9-60}$$

on the 2θ scale, or

$$(\delta s)_i^S = \frac{K}{L} \tag{9-61}$$

on the s scale. If at least two orders of reflection are available and if only one source of broadening is operative (strain or size, but not both), it is possible in principle to distinguish which factor is present by ascertaining whether the breadth of the pure diffraction profile increases with θ as $\tan \theta$ or $1/\cos \theta$.

As pointed out above, when both size and strain broadening are present, it becomes necessary to make an assumption as to the shapes of the two contributing line profiles. Commonly, they are presumed to be either both Cauchy or both Gaussian, whereupon we have for their overall broadening effect $(\delta s)_0$, respectively:

$$(\delta s)_0 = (\delta s)^S + (\delta s)^D \quad \text{(Cauchy/Cauchy)}, \tag{9-86}$$
$$(\delta s)_0^2 = [(\delta s)^S]^2 + [(\delta s)^D]^2 \quad \text{(Gaussian/Gaussian)} \tag{9-87}$$

In these equations we have employed the variable s (with units in Å^{-1}), and for simplicity we have dropped the subscript i, indicative of integral breadths.

Theoretical considerations and experimental results both tend to show that strain broadening may usually be approximated rather well by a Gaussian function, whereas the effects of small crystallite size and size distributions more closely resemble a Cauchy broadening profile. Ruland[24] and Schoening[51] have independently demonstrated that the integral breadth of the convolution of a Gaussian function of breadth B_G and a Cauchy function of breadth B_C is given by

$$B = B_G \frac{\exp\left[-(B_C/B_G)^2/\pi\right]}{1 - \text{erf}\,(2/\pi)^{1/2}\,(B_C/B_G)}, \tag{9-88}$$

in which erf (x) is the error function, defined by

$$\text{erf } (x) = \frac{1}{(2\pi)^{1/2}} \int_0^x \exp \left(-t^2/2\right) \, dt.$$

The numerical evaluation of equation 9-88 is awkward and laborious; however, Halder and Wagner[52] have shown that to a satisfactory approximation it may be simplified to the parabolic relationship:

$$\frac{B_C}{B} = 1 - \left(\frac{B_G}{B}\right)^2. \tag{9-89}$$

For Cauchy size broadening and Gaussian strain broadening, then,

$$\frac{B^S}{B} = 1 - \left(\frac{B^D}{B}\right)^2. \tag{9-89a}$$

Numerical evaluation proves that the values of B^S and B^D derived from equation (9-89a) differ from the values derived from equation (9-88) by at most 10 per cent. Trömel and Hinkel[53] have also proposed the use of the approximation equation 9-89a for the separation of size and strain broadening.

Figure 9-17 compares the curves of B^S/B versus B^D/B corresponding to

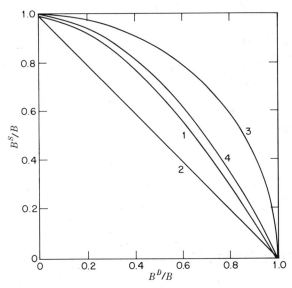

Fig. 9-17.   Plots of B^S/B versus B^D/B. (*1*) Cauchy-Gaussian (equation 9-88); (*2*) Cauchy-Cauchy (equation 9-86); (*3*) Gaussian-Gaussian (equation 9-87); (*4*) Cauchy-Gaussian parabolic approximation (equation 9-89a). (Courtesy of N. C. Halder and C. N. J. Wagner, *Acta Crystallogr.*, **20**, 312 (1966).)

equations 9-86, 9-87, 9-88, and 9-89a. The close agreement between the numerical values from equations 9-88 and 9-89a is evident. In this connection the reader is reminded that in Section 9-1.3 an identical curve (marked Cauchy-Gaussian in Fig. 9-6) was presented as one of several options for the correction of experimental line profiles for instrumental broadening. The use of a parabolic equation of the form of equation 9-89a for this purpose has been shown by Wagner and Aqua [54] to yield integral breadths agreeing closely with those obtained by the Stokes Fourier unfolding procedure. In regard to equations 9-88 and 9-89a, we emphasize that B, B^S, and B^D refer to integral breadths characterizing a given pure diffraction profile (line profile corrected for instrumental broadening) and, further, that these quantities may be expressed in either δs or $\delta(2\theta)$ units.

If the size and distortion line profiles are both presumed to be Cauchy, substitution of equations 9-61 and 9-79 in 9-86 yields

$$(\delta s)_0 = \frac{1}{L} + 2es \quad \text{(Cauchy/Cauchy),} \tag{9-90}$$

wherein K of equation 9-61 has been set equal to unity. On the assumption of both profiles being Gaussian, substitution of equations 9-61 and 9-79 in 9-87 gives

$$(\delta s)_0{}^2 = \left(\frac{1}{L}\right)^2 + (2es)^2 \quad \text{(Gaussian/Gaussian).} \tag{9-91}$$

Thus on the basis of the Cauchy/Cauchy hypothesis, we can plot $(\delta s)_0$ against s and obtain from the ordinate intercept the mean crystallite dimension L and from the slope the "maximum" lattice distortion e. On the other hand, for the Gaussian/Gaussian assumption, we utilize a plot of $(\delta s)_0{}^2$ against s^2 in the same manner. For a Cauchy size broadening function $(\delta s)^D$ we substitute equations 9-61 and 9-79 in 9-89 and obtain

$$\frac{K}{(\delta s)_0 L} = 1 - \left[\frac{2es}{(\delta s)_0}\right]^2$$

$$= 1 - \frac{4e^2}{d_{hkl}^2 (\delta s)_0{}^2} \quad \text{(Cauchy/Gaussian).} \tag{9-92}$$

Once more employing the relation

$$\delta s = \frac{2}{\lambda} \cos \theta \delta\theta,$$

we can convert equation 9-92 to the following form with the breadth in $\delta 2\theta$ (radians) rather than δs units,

$$\frac{K\lambda}{(\delta 2\theta)L\cos\theta_0} = 1 - \frac{16e^2}{(\delta 2\theta)^2\cot^2\theta_0}. \tag{9-92a}$$

In this expression θ_0 is the position of the peak maximum. Substitution of pairs of $(\delta s)_0$ and s values from two or more reflection orders in equation 9-92, or of triplets of $\delta 2\theta$, $\cos\theta_0$, and $\cot^2\theta_0$ values from two or more orders of reflection in equation 9-92a, then leads to numerical solutions for L and e. Alternatively, equation 9-92a may be arranged in the form

$$\frac{(\delta 2\theta)^2}{\tan^2\theta_0} = \frac{K\lambda}{L}\left(\frac{\delta 2\theta}{\tan\theta_0\sin\theta_0}\right) + 16e^2, \tag{9-92b}$$

from which any available orders of a given reflection may be used to construct a linear plot of $(\delta 2\theta)^2/\tan^2\theta_0$ against $(\delta 2\theta)/\tan\theta_0\sin\theta_0$. From the slope $K\lambda/L$ and ordinate intercept $16e^2$, the size and distortion parameters in a direction normal to the diffraction planes may be determined.

In concluding this section we emphasize that in equations 9-90 through 9-92b the "maximum strain" e may be replaced by $\frac{1}{4}\eta$, $1.25\langle\epsilon^2\rangle^{1/2}$, or any other equivalent specified measure of the lattice strain. Also, it is common practice to set K equal to unity.

9-2.5 Determination of Faulting in Layered Structures

In a material consisting of well-defined layers or one that may be regarded structurally as composed of layers, faulting (crystallographic misplacement of successive layers) may occur as a result of cold work (*deformation faulting*) or growth (*twin faulting*). The effects of the various types of faulting on the diffraction pattern is a subject of special importance in the field of metals. With reference to the [111] direction in a face-centered cubic metal as a typical example, Fig. 9-18 shows in an oversimplified and schematic manner the arrangements of successive ABCABCABC... layers (*a*) in the correct (unfaulted) crystallographic sequence, (*b*) in a deformation fault and (*c*) in a growth, or twin, fault. Similar effects are observed in body-centered cubic and hexagonal close-packed structures in particular crystallographic directions.

The most extreme and dramatic type of layer faulting is exemplified by carbon blacks and activated carbons, in which successive layers are displaced and rotated parallel but otherwise more or less randomly with respect to each other. These are termed *random-layer* or *turbostratic structures*. This kind of faulting, although usually lesser in degree, is also observed in other lamellar substances, notably in clays and micas. Turbostratic faulting is discussed at somewhat greater length below.

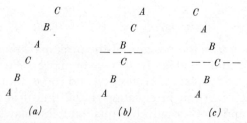

Fig. 9-18. Schematic diagram of arrangements of successive 111 layers A, B, C. (*a*) The correct face-centered cubic sequence $ABCABC$; (*b*) a deformation fault; (*c*) a growth, or twin, fault. (Courtesy of B. E. Warren and Addison–Wesley [28].)

A. Deformation and Twin Faulting. Because of space limitations we now give only a brief introduction to faulting in metals, taking cubic structures as the representative example. An extensive theoretical treatment of faulting in face-centered cubic, body-centered cubic, and hexagonal close-packed metals has been given by Warren [55]. If α and β are the respective probabilities of encountering a deformation fault and a twin fault between two successive layers, a *compound fault probability*, $1.5\alpha + \beta$, can usually be determined in a straightforward manner from high-quality experimental data. The first step is to determine the Fourier coefficients A_n and separate the size and distortion coefficients A_n^S and A_n^D in the manner described in Section 9-2.2B. However, when both size and faulting effects are present, the size coefficient assumes the form [54, 55]

$$A_L^S = 1 - L\left(\frac{1}{D} + \frac{1.5\alpha + \beta}{a} V_{hkl}\right), \qquad (9\text{-}93)$$

in which $L = |n|a_3$ as before (a length perpendicular to the reflecting planes), a is the cubic cell edge, and the coefficient V_{hkl} has a numerical value depending on the lattice type, face-centered cubic or body-centered cubic, and the indices hkl. The values of A_L^S are then plotted against L, a step equivalent to that portrayed in Fig. 9-13 in the absence of faulting, to determine the mean column length $N_3 a_3$ (designated D in equation 9-93). In the present situation, however, the initial slope and abscissa intercept of the A_L^S versus L plot provide an *effective* crystallite (domain) size $\bar{D}_{\text{eff}}$, a composite of the true mean dimension D and the faulting effect, which simulates a crystallite-size term. These observations are made clear by differentiating equation 9-93 with respect to L:

$$-\frac{dA_L^S}{dL} = \frac{1}{D_{\text{eff}}} = \frac{1}{\bar{D}} + \frac{1.5\alpha + \beta}{a} V_{hkl}. \qquad (9\text{-}94)$$

Since D_{eff} is a function of hkl, the two quantities $\bar{D}$ and $1.5\alpha + \beta$ can be evaluated by making use of two or more reflections, for example, 111 and 200.

Besides causing broadening of reflections, deformation faulting also shifts the peak positions by small amounts in positive or negative directions on the 2θ scale. From the change in angular separation of two judiciously selected reflections, one displaced positively and the other negatively, a numerical value of α can be derived from the theoretical expressions for the displacements. Thus it is possible to arrive at a corresponding value of β by difference, once the compound-fault probability has been found by application of equation 9-94. For the theoretical background and full details, the reader is referred to Warren's treatment[55].

B. Random-Layer (Turbostratic) Structures. Mention of random-layer structures has been made in Section 6-10.2. In the first comprehensive treatment of diffraction by such structures, Warren showed that two kinds of reflections are obtained, ordinary crystalline-type reflections and diffuse two-dimensional reflections, the latter terminating abruptly on the low-angle sides but falling off gradually in intensity on the high-angle sides (see Fig. 6-25C). If axis c is perpendicular to, and axes a and b lie on the plane of, the layer, the crystalline reflections will be of the type $00l$ and the two-dimensional reflections will be of the index type hk. No general hkl reflections will be observed. The size of a two-dimensional layer *in the plane of the layer* can be calculated from the observed breadth of an hk reflection, using the Scherrer equation but with an appropriate value of the shape factor K.

With the aid of certain approximations, Warren[56] first derived an expression for the diffraction profile of an hk reflection from a two-dimensional lattice of limited extent. He showed that the peak of a "two-dimensional" reflection hk is displaced from the theoretical position θ_0 of the equivalent three-dimensional lattice reflection $hk0$ by the amount

$$\Delta (\sin \theta) = \frac{1.16\lambda}{L_{hk}}$$

and that the breadth of the reflection on the 2θ scale at half-maximum intensity is

$$\beta_{1/2} = \frac{1.84\lambda}{L_{hk} \cos \theta}, \tag{9-95}$$

which is seen to be the Scherrer equation 9-60, with $K = 1.84$. Here L_{hk} is the mean dimension of the two-dimensional layers in a direction perpendicular to the diffracting lattice row lines hk. Subsequently,

Wilson[57], Diamond[58], Warren and Bodenstein[18, 19], Ruland[59], Perret and Ruland[60], and Ergun[61] all made contributions to the refinement of Warren's early results, including special attention to the analytical expression of the hk profile for small layers, to the calculation of line profiles, and to applications of the theory to carbons.

Of most interest in the present discussion are the new values deduced for the half-maximum-intensity Scherrer constant for two-dimensional structures: 1.91 for disks[18], 2.16 for a Cauchy intensity distribution normal to an hk reciprocal-lattice rod[59], and an average value of 1.98 (± 0.06) from recent theoretical work[60] on a number of carbon reflections. Thus it can be seen that a generally acceptable value of K is 2, just twice that suggested as a general working value for three-dimensional structures. In the event of truly random orientations of the two-dimensional layers, no $00l$ reflections would be produced, but in the great majority of cases the layers tend to stack together into parallel-layer groups as might be expected. These turbostratic pseudocrystallites possess no interlayer order other than the parallelism and degree of separation of the layers. Such parallel-layer groups give rise to more-or-less diffuse $00l$ reflections because of the limited number of layers and also, in some cases, because of a lack of constancy of the interlayer spacing. The mean dimension of a parallel-layer group perpendicular to the layers can be calculated from the breadth of a $00l$ reflection in the usual way by employing equation 9-60 or 9-61.

9-2.6 Very Defective Lattices

Ergun[62] has presented evidence that certain substances that do not yield sharp diffraction effects are better regarded as highly defective than as consisting of small crystallites diffracting incoherently with respect to one another. Especially when other physical evidence is lacking for fragmentation or for the existence of a reasonable distribution of crystallite sizes, the concept of large but defective domains is more realistic. Under these conditions the dimension that is to be sought is not the crystallite size but the defect-free distance L', which Ergun shows can be defined by

$$g(l) = \exp\frac{-l}{L'}, \tag{9-96}$$

in which $g(l)$ is the probability that a distance l can be traversed without encountering a defect.

If P is the volume fraction occupied by defects (concentration of defects), m the number of defects per unit volume, and r the defect radius,

$$P = \tfrac{4}{3}\pi r^3 m.$$

But for a cylinder of unit volume and unit cross-sectional area $L' = 1/\pi m r^2$, whence

$$P = \frac{4r}{3L'}. \tag{9-97}$$

Thus, for example, if $r = 1$ Å and $L' = 50$ Å, the defect concentration is somewhat less than 3 per cent.

Anticipating a result from Section 12-1.1, the Debye equation for the intensity scattered by a rigid structure consisting of one kind of atom and assuming all orientations with equal probability may be written

$$j(s) = 1 + \sum_q n(l_q) \frac{\sin 2\pi s l_q}{2\pi s l_q}, \tag{9-98}$$

whereas for a highly defective structure the terms of the summation in equation 9-98 are modified by the coefficients $g(l_q)$:

$$j(s) = 1 + \sum_q g(l_q) n(l_q) \frac{\sin 2\pi s l_q}{2\pi s l_q}. \tag{9-99}$$

As an illustration of the application of equations 9-98 and 9-99, Fig. 9-19 compares the experimental profile of the 002 reflection of a heat-treated carbon black with theoretical profiles calculated with these equations. The much better agreement of experiment with equation 9-99 than with 9-98 is obvious, thus supporting a carbon-black model consisting of extensive, but faulty, stacks rather than small, regular stacks of layers of uniform height. Although better agreement with the experimental line profile would be obtained from a distribution of small stack heights (pseudocrystallite dimensions L_c), electron-microscope observations and heat-treatment studies both lend additional evidence in favor of extensive but highly faulted stacks. For additional details of the theory, including extraction of L' from the diffraction data and for applications to other materials, the reader is referred to Ergun [62].

Omitted from this volume is an important alternative approach to the analysis of lattice distortions, the theory of paracrystalline diffraction. Regretfully, this omission has been necessary for several reasons, including in particular spatial limitations and also the general realization that the theory has still not matured to the point of providing a truly firm and otherwise satisfactory basis for analytical applications to three-dimensional structures. Nevertheless, there can be no doubt of its validity in the realm of what are essentially one-dimensional structures (natural and synthetic fibers) and, with some reservations, of its applic-

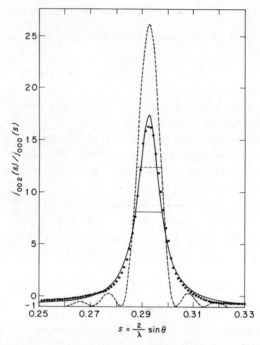

Fig. 9-19. Comparative profiles of the 002 reflection of a heat-treated carbon black. Experimental data (•); calculated from equation 9-99, defect broadening (———); calculated from equation 9-98, crystallite-size broadening (----). (Courtesy of S. Ergun [62]; reprinted by permission of the Bureau of Mines, U.S. Department of the Interior.)

ability to two-dimensional structures. The early foundations for the concept of the paracrystal and its diffraction effects were laid by Zernike and Prins [63], Kratky [64], Hermans [65], Hosemann [66], and Porod [67, 68]. Since 1950 the principal impetus to the further evolution of the theory of paracrystalline diffraction has been supplied by Hosemann and his associates, in particular Bonart, Bagchi, and Wilke [69–74]. The interested reader is referred to the cited literature, and especially to Hosemann and Bagchi [72]. Like the Warren-Averbach theory and the variance theory, the paracrystalline diffraction theory provides a means of separation of size and distortion parameters. Experience has tended to show that paracrystalline size parameters agree well with values deduced by the Warren-Averbach Fourier method, while the paracrystalline distortion parameters not only deviate considerably from the Warren-Averbach results but seem to be relatively insensitive to the experimental data at low-distortion levels. A brief description of the paracrystalline theory, including its application to certain polymers, has been given by Alexander [75].

9-2.7 Illustrative Analyses

A. Cold-Worked Copper–Silicon Single Crystal, Fourier Method (Warren and Averbach [76]). This early study by Warren and Averbach, although applied to a deformed single crystal rather than a polycrystalline material, constitutes an ideal illustration of the use of the Fourier method because five resolved orders of the 111 reflection are available for the plotting of $\ln A_L(h_0)$ versus h_0^2 (see equation 9-57 and associated text). The crystal, having the composition 98% copper–2% silicon, was rolled to 50 per cent reduction in the $[\bar{1}12]$ direction and parallel to the (110) planes. For the five reflection profiles analyzed, 111, 222, 333, 444, and 555, h_0^2 has the respective values 3, 12, 27, 48, and 75.

Figure 9-20 shows the logarithmic plots of $A_L(h_0)$ against h_0^2 for four values of $L = nd_{111}$. From the limiting slopes of the curves as $h_0^2 \to 0$ (see broken lines), we arrive at the four estimates of the root-mean-square (rms) lattice strain $\langle \epsilon_L^2 \rangle^{1/2}$, in the direction normal to the (111) planes, as outlined in Table 9-3. The four values suggest a "best" rms value of about 0.0015.

From the ordinate intercepts of Fig. 9-20, the Fourier size coefficients A_L^S corresponding to $L = 25, 50, 75,$ and 100 Å are found to be, respectively, 0.99, 0.95, 0.94, and 0.89. In Fig. 9-21 these values are plotted against L; from the slope of the best straight line through these points $\langle L \rangle \simeq 1000$ Å, the mean crystallite (grain) dimension perpendicular to

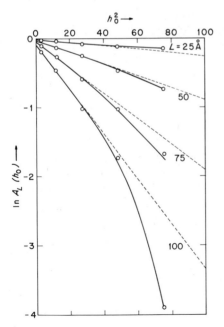

Fig. 9-20. Plot of $\ln A_L(h_0)$ versus h_0^2 for the first five orders of 111 and four values of $L = nd_{111}$. Single crystal, 98% copper–2% silicon rolled to 50 per cent reduction. (Courtesy of B. E. Warren and B. L. Averbach, *J. Appl. Phys.*, **23**, 497 (1952).)

Table 9-3. Calculation of rms Lattice Strain Normal to the (111) Planes of a Rolled Cu–Si Single Crystal[a,b]

$L(\text{Å})$	$L^2(\text{Å}^2)$	Slope for $h_0^2 \to 0$	$2\pi^2(L/a)^2$	$\langle \epsilon_L^2 \rangle \times 10^6$	$\langle \epsilon_L^2 \rangle^{1/2} \times 10^3$
25	625	− 0.0024	948	2.53	1.59
50	2,500	− 0.0085	3,790	2.24	1.50
75	5,625	− 0.0184	8,520	2.16	1.47
100	10,000	− 0.323	15,150	2.11	1.45

[a]Data from Figs. 1 and 2 of reference [76].
[b]From equation 9-57: $\langle \epsilon_L^2 \rangle = -(\text{slope}) \times (a^2/2\pi^2L^2)$. Lattice constant a assumed to be 3.608 Å.

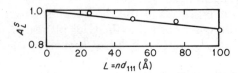

Fig. 9-21. Plot of A_L^S versus $L = nd_{111}$. Single crystal, 98% copper–2% silicon, rolled to 50 per cent reduction. (Courtesy of B. E. Warren and B. L. Averbach, *J. Appl. Phys.*, **23**, 497 (1952).)

the (111) planes. Thus the broadening of the *hhh* reflection profiles as a result of rolling is shown to be primarily the result of distortion and only secondarily the result of small crystallite size.

B. Deformed Thoriated Tungsten, Variance Method (Halder and Mitra[77]). Tungsten filaments containing 1.11 per cent thoria were slowly hammered and then drawn into uniform wires 0.3 mm in diameter at about 750°C. For a line-broadening reference standard a cold-worked specimen was annealed in a vacuum furnace at 1140°C for 12 hr and then slowly cooled. The diffraction patterns were recorded photographically in an evacuated Unicam 9-cm camera with nickel-filtered CuKα radiation. The line profiles were determined by careful microphotometering of the patterns, special attention being given to the tails of the reflections. The variances of the line profiles of the annealed specimens were subtracted from the variances of the corresponding lines of the cold-worked samples (see equation 9-58 and explanatory text). In order to facilitate the calculations, Halder and Mitra[77] expressed equation 9-78a in the form

$$W_{2\theta}^* = \frac{W_{2\theta}}{d^2} = \frac{K\lambda(\delta 2\theta)^*}{2\pi^2 p \cos \theta} - \frac{T_a^* \lambda^2}{4\pi^2 p^2 \cos^2 \theta} + \frac{4\tan^2 \theta}{d^2} \langle e^2 \rangle, \quad (9\text{-}78c)$$

in which

$$(\delta 2\theta)^* = \frac{\delta 2\theta}{d^2} \quad \text{and} \quad T_a^* = \frac{T_a}{d^2}.$$

Table 9-4.   Calculation of Corrected Line-Profile Variances[a]

hkl	$\theta(°)$	$(\delta 2\theta)^*$ (rad Å⁻²)	$W_{2\theta}^*$ (10^{-5} rad² Å⁻²)		
			Observed	Geometrical	Corrected
110	20.20	0.00908	0.6926	0.2124	0.4802
200	29.12	0.01278	1.5890	0.5213	1.0677
211	36.55	0.01430	2.6532	0.6517	2.0015
220	40.47	0.01508	3.8351	1.1336	2.7015
310	50.32	0.01570	8.7307	1.9526	6.7781
222	57.55	0.01632	17.3110	4.4578	12.8532
321	65.60	0.01658	35.0776	7.0076	28.0700

[a]Data from Halder and Mitra [77].

Table 9-4 lists the experimental data together with the numerical quantities involved in the correction of the seven line profiles for instrumental broadening.

If we let

$$C_1 = \frac{K\lambda(\delta 2\theta)^*}{2\pi^2 \cos\theta}, \qquad C_2 = \frac{T_a^* \lambda^2}{4\pi^2 \cos^2\theta}, \qquad C_3 = \frac{4\tan^2\theta}{d^2},$$

equation 9-78c assumes the simplified form

$$W_{2\theta}^* = \frac{C_1}{p} - \frac{C_2}{p^2} + C_3\langle e^2 \rangle,$$

or

$$\frac{W_{2\theta}^*}{C_3} = \frac{C_1}{C_3}\frac{1}{p} - \frac{C_2}{C_3}\frac{1}{p^2} + \langle e^2 \rangle.$$

If now T_a^* is neglected, in conformity with the usual practice, $C_2 = 0$, and we are left with the expression:

$$\frac{W_{2\theta}^*}{C_3} = \frac{C_1}{C_3}\frac{1}{p} + \langle e^2 \rangle. \tag{9-100a}$$

For a second diffraction line,

$$\frac{(W_{2\theta}^{*\prime})}{C_3'} = \frac{C_1'}{C_3'}\frac{1}{p} + \langle e^2 \rangle, \tag{9-100b}$$

and elimination of $\langle e^2 \rangle$ from equations 9-100a and 9-100b yields

$$p = \frac{\Delta(C_1/C_3)}{\Delta(W_{2\theta}^*/C_3)}. \tag{9-101}$$

Thus from various pairs of lines several values of the "true" crystallite size p (see equation 9-62) can be determined, and substitution of these p's in equation 9-100a then provides values of the variance of the mean square strain $\langle e^2 \rangle$.

For the seven reflections concerned, Table 9-5 outlines the numerical evaluation of the constants C_1 and C_3 and of $W_{2\theta}^*/C_3$. Table 9-6 gives the calculation of p for four pairs of reflections, resulting in a mean value of 232 Å. When this value is substituted in equation 9-100a using successively the appropriate pairs of $W_{2\theta}^*/C_3$ and C_1/C_3 values from Table 9-5, an overall mean $\langle e^2 \rangle^{1/2}$ value of 0.0032 is obtained. These results may be compared with those of McKeehan and Warren[78] for 0.75 per cent thoriated tungsten filings by the method of Fourier analysis: $L = 200$ Å, $\langle \epsilon^2 \rangle^{1/2} = 0.005$. McKeehan and Warren interpret the size parameter, 200 Å, as a measure of the distance between layers of slip planes or layers of dislocations. In conclusion, it should be noted that in view of the differences between the definitions of the size parameters p and L and the strain parameters $\langle e^2 \rangle^{1/2}$ and $\langle \epsilon^2 \rangle^{1/2}$, there is no a priori

Table 9-5. Numerical Values of C_1, C_3 and $W_{2\theta}^*/C_3{}^a$

hkl	$C_1(10^{-3}$ Å$^{-1})$	$C_3($Å$^{-2})$	C_1/C_3 $(10^{-3}$ Å$)$	$W_{2\theta}^*/C_3$ (10^{-5})
110	0.87849	0.10857	8.09146	4.4229
200	1.38258	0.49468	2.79489	2.1583
211	1.67420	1.31102	1.27702	1.5266
220	1.87375	1.87200	1.00093	1.4431
310	2.25399	5.78013	0.38995	1.1726
222	2.87470	11.85011	0.24259	1.0846
321	3.79361	26.92883	0.14087	1.0424

^{a}Data from Halder and Mitra[77].

Table 9-6. Calculation of Variance Size Parameter p^a

$hkl - h'k'l'$	$\Delta(C_1/C_3)$ $(10^{-3}$ Å$)$	$\Delta(W_{2\theta}^*/C_3)$ (10^{-5})	$p($Å$)$
110 − 220	7.09053	2.9798	238
200 − 310	2.40494	0.9857	242
211 − 222	1.03443	0.4420	234
220 − 321	0.86006	0.4007	215
		$\bar{p} =$	232

^{a}Data from Halder and Mitra[77].

necessity for close numerical agreement between the outcomes of the two investigations.

C. Deformed Cubic Metals, Fourier and Integral-Breadth Methods, Compound Fault Probability Evaluated (Wagner and Aqua[54]).

Filings of SAP-aluminum, silver/10 per cent indium, silver/15 per cent indium, tungsten, and niobium were passed through a 150-mesh screen and compacted into briquets with Duco cement and acetone as binder. The line-broadening reference specimens were the same filings after annealing, further screening through 325-mesh, and compacting into briquets. The x-ray patterns were recorded diffractometrically with nickel-filtered $CuK\alpha$ radiation, and the measured reflection profiles were corrected for $K\alpha_1\alpha_2$ doubling by the method of Rachinger (see Section 9-1.1). Figure 9-22 shows the tungsten 220 reflections for (A) the annealed filings, (B) the unannealed filings, and (C) profile B after applying the Rachinger correction. It is seen that the severity of the cold-work distortion introduced by filing results in pronounced broadening of the $K\alpha_1$ profile relative to that of the annealed reference specimen. The Rachinger-corrected reflections from the cold-worked specimens were further corrected for instrumental broadening by the Fourier-transform method of Stokes (Section 9-1.1). Table 9-7 gives the integral breadths of eight reflections from annealed aluminum $(\delta s)_R$, as calculated (I) from the sums of the Fourier coefficients A_L of the Stokes Fourier-transform corrected profiles, and (II) from the integrated intensities above background divided by the corresponding peak heights (see Section

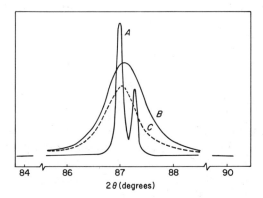

Fig. 9-22.   Profiles of the 220 reflection of tungsten filings. (A) Annealed, (B) unannealed, (C) profile B after application of the Rachinger correction for $K\alpha$-doublet broadening. $CuK\alpha$ radiation, 3° divergence slit, 0.05° receiving slit, medium-resolution Soller slits, 4° take-off angle. (Courtesy of C. N. J. Wagner and E. N. Aqua and Plenum Press, *Advances in X-Ray Analysis*, Vol. 7, 1964, p. 46.)

Table 9-7. Integral Breadths of Reflections from
Annealed Aluminum Determined in Two Ways[a,b]

hkl	(I) $1\bigg/\sum\limits_{-\infty}^{+\infty} A_L$ L	(II) Integrated intensity $\overline{}$ Peak height
111	1.75	1.67
200	1.64	1.56
220	1.35	1.34
311	1.27	1.23
222	1.27	1.31
400	1.17	1.17
331	1.14	1.19
420	1.14	1.19
Average of 311 $\cdots$ 420	1.19	1.22

[a]Data from Wagner and Aqua [54].
[b]Breaths $(\delta s)_R$ expressed in $\text{Å}^{-1} \times 10^3$.

5-2.2B). The figures on the bottom line of the table are the average
breadths for the five lines occurring at the largest Bragg angles.

For each of the cold-worked specimens, the Fourier coefficients A_L
were handled in the usual way. Thus the slopes of plots of $\ln A_L$ against
h_0^2 for $h_0^2 \to 0$ gave values of the rms strain components $\langle \epsilon_L^2 \rangle^{1/2}$ (see
equation 9-57), after which the size coefficients A_L^S were plotted as a func-
tion of L (refer to Fig. 9-15), the abscissa intercepts of the initial slopes
giving $\overline{D}$ or $\overline{D}_{\text{eff}}$ depending on the absence or presence of faulting, respectively. In extrapolating the initial slopes to the abscissa axis, due
allowance was made for a possible "hook effect," as has been discussed in
relation to Fig. 9-13 and equation 9-54. When size and faulting were
both present, application of equation 9-94 to several reflections from a
given specimen permitted the calculation of the coherently diffracting
domain size $\overline{D}$ and the compound fault probability $1.5\alpha + \beta$. The results
of this Warren-Averbach procedure are listed in the columns designated
method I in Tables 9-8 and 9-9. In method I the values of the rms strain
$\langle \epsilon^2 \rangle^{1/2}$ are averaged approximately over the effective crystallite size D_{eff},
that is, between $0.1 \overline{D}_{\text{eff}} < L < \overline{D}_{\text{eff}}$.

As an illustration of the extraction of $\overline{D}_{\text{eff}}$ (effective crystallite size)
values from plots of A_L^S versus L, Figure 9-23 shows separate plots for

Table 9-8. Effective Crystallite Sizes $\bar{D}_{eff}$, Domain Sizes $\bar{D}$ or $\bar{D}^2/D$, and Compound Fault Probabilities $1.5\alpha+\beta$ in Cold-Worked Metals and Alloys[a]

Metal		Method			
		I	II	III	IV
Tungsten	Domain size (Å)	220	400	850	375
Aluminum	Domain size (Å)	400	610	720	610
Niobium	$\bar{D}_{eff}$, [110] (Å)	215	425	4000	600
	$\bar{D}_{eff}$, [100] (Å)	140	250	540	225
	$1.5\,\alpha+\beta$	0.010	0.013		0.021
	Domain size (Å)	305	700		6000
Ag/10 per cent In	$\bar{D}_{eff}$, [111] (Å)	120	170	400	185
	$\bar{D}_{eff}$, [100] (Å)	70	120	270	120
	$1.5\,\alpha+\beta$	0.043	0.035	0.017	0.042
	Domain size (Å)	260	250	600	315
Ag/15 per cent In	$\bar{D}_{eff}$, [111] (Å)	80	150	185	155
	$\bar{D}_{eff}$, [100] (Å)	45	90	150	90
	$1.5\,\alpha+\beta$	0.070	0.064	0.018	0.067
	Domain size (Å)	200	305	225	345
Formula used for domain size		$\bar{D}$	$\bar{D}^2/\bar{D}$	$\bar{D}^2/\bar{D}$	$\bar{D}^2/\bar{D}$

[a]Data from Wagner and Aqua[54].

Table 9-9. Microstrains (Distortions) in Cold-Worked Metals and Alloys[a]

Metal	Strain	Method			
		I	II	III	IV
Tungsten	Isotropic	0.0028	0.0042	0.0063	0.0042
Aluminum	Isotropic	0.0007	0.0009	0.0006	0.0010
Niobium	[110]	0.0050	0.0067	0.0066	0.0063
	[100]	0.0025	0.0037	0.0032	0.0036
Ag/10 per cent In	[111]	0.0038	0.0060	0.0048	0.0051
	[100]	0.0050	0.0107	0.0090	0.0096
Ag/15 per cent In	[111]	0.0035	0.0068	0.0041	0.0068
	[100]	0.0055	0.0120	0.0092	0.0114

[a]Data from Wagner and Aqua[54].

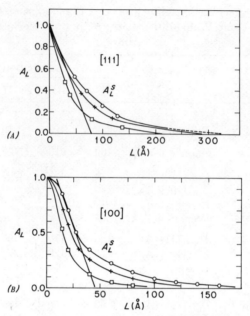

Fig. 9-23. Fourier coefficients A_L and $A_L{}^S$ of a cold-worked silver–15% indium alloy plotted as functions of L. (*A*) coefficients of 111 and 222 reflections: $\bigcirc$, $A_L{}^S$; +, 111; $\square$, 222. (*B*) coefficients of 200 and 400 reflections: $\bigcirc$, $A_L{}^S$; +, 200; $\square$, 400. (Courtesy of C. N. J. Wagner and E. N. Aqua and Plenum Press, *Advances in X-Ray Analysis*, Vol. 7, 1964, p. 46.)

the 111 and 222 reflections (part *A*) and 200 and 400 reflections (part *B*) of cold-worked silver–15 per cent indium alloy, which is anisotropic, from which respective crystallite-size values of 80 and 45 Å were obtained in the [111] and [100] directions from the abscissa intercepts (see also column I of Table 9-8). It should be pointed out that for isotropic metals, such as tungsten and aluminum, all the reflections are used to give one average strain-corrected curve of $A_L{}^S$ versus L, yielding a single mean crystallite, or domain, size.

The data in columns II, III, and IV of Tables 9-8 and 9-9 were obtained from integral breadths by the following three methods:

II. Stokes' Fourier-transform-corrected integral breadths: $(\delta s)_i = 1/\Sigma A_L$. Separation of size and strain by plotting $(\delta s)_i{}^2$ against s^2 (equation 9-91), that is, Gaussian size and strain broadening assumed. Strain defined as $e \simeq 1.25 \langle \epsilon^2 \rangle^{1/2}$ (equation 9-85).

III. Same as II, except separation of size and strain by plotting $(\delta s)_i$ against s (equation 9-90), that is, Cauchy size and strain broadening assumed.

IV. Integral breadths $(\delta s)_i{}^{pa}$ extracted from experimentally observed breadths $(\delta s)_{cw}$ by the parabolic-approximation correction, equation 9-89a, in the form

$$\frac{(\delta s)_i{}^{pa}}{(\delta s)_{cw}} = 1 - \frac{(\delta s)_R^2}{(\delta s)_{cw}^2}, \tag{9-89b}$$

where $(\delta s)_R$ is the breadth of the line of the reference specimen. Separation of size and strain by plotting $(\delta s)_i{}^2$ against s^2 (equation 9-91), that is, Gaussian size and strain broadening assumed. Strain defined as $e \approx 1.25 \langle \epsilon^2 \rangle^{1/2}$ (equation 9-85).

We illustrate the conversion of the observed (Rachinger-corrected) integral breadth $(\delta s)_{cw}$ of the 220 tungsten reflection to the breadth $(\delta s)_i{}^{pa}$ by means of equation 9-89b. The reader may refer to the data presented in Table 9-10 and Fig. 9-24.

$$\frac{\sin \theta}{\lambda} = \frac{(h^2 + k^2)^{1/2}}{2a} = \frac{8^{1/2}}{2 \times 3.165} = 0.477 \text{ Å}^{-1}.$$

$$\sin \theta = 0.447 \times 1.542 = 0.689 \quad \text{and} \quad \theta = 43.55°.$$

$$\cos \theta = 0.7248 \quad \text{and} \quad \frac{\cos \theta}{\lambda} = 0.470.$$

From measurements of the 220 reflections of annealed (R) and cold-worked (cw) filings followed by the Rachinger correction for $K\alpha$ doublet broadening:

$$(\delta 2\theta)_R = 0.165° = 0.00288 \text{ radians.}$$
$$(\delta 2\theta)_{cw} = 0.984° = 0.01715 \text{ radians.}$$

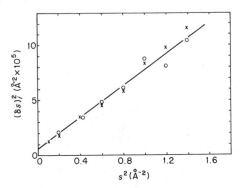

Fig. 9-24.   Plot of $(\delta s)_i{}^2$ against s^2 for tungsten filings: $\times$, $(\delta s)_i{}^{St}$, integral breadths corrected by Stokes' Fourier-transform method; $\bigcirc$, $(\delta s)_i{}^{pa}$, integral breadths corrected by the parabolic approximation. (Courtesy of C. N. J. Wagner and E. N. Aqua and Plenum Press, *Advances in X-Ray Analysis*, Vol. 7, 1964, p. 46.)

Table 9-10. Integral-Breadth Data for Tungsten[a,b,c]

	Corrected Integral Breadths of Reflections from Cold-Worked Specimens		
hkl	$(\delta s)_i^{St}$ ($\text{Å}^{-1} \times 10^3$)	$(\delta s)_i^{pa}$ ($\text{Å}^{-1} \times 10^3$)	$(\delta s)_R/(\delta s)_{cw}$
110	4.25	4.52	0.400
200	5.81	5.76	0.268
211	6.72	6.82	0.226
220	7.64	7.81	0.168
310	9.05	9.30	0.131
222	9.85	8.02	0.135
321	10.75	10.35	0.095

[a]Data from Wagner and Aqua[54].
[b]Body-centered cubic lattice, $a = 3.165$ Å.
[c]$(\delta s)_i^{St}$ corrected by Stokes' Fourier-transform method; $(\delta s)_i^{pa}$ corrected by parabolic approximation, equation 9-89b; $(\delta s)_R$ integral breadth of reflection from reference specimen; $(\delta s)_{cw}$ uncorrected integral breadth of reflection from cold-worked specimen.

Then

$$(\delta s)_R = \frac{\cos \theta}{\lambda} \times 0.00288 = 1.351 \times 10^{-3} \text{ Å}^{-1}.$$

$$(\delta s)_{cw} = \frac{\cos \theta}{\lambda} \times 0.01715 = 8.04 \times 10^{-3} \text{ Å}^{-1}.$$

$$\frac{(\delta s)_R}{(\delta s)_{cw}} = 0.168.$$

Substitution of numerical values in equation 9-89b gives:

$$(\delta s)_i^{pa} = (\delta s)_{cw}\left[1 - \frac{(\delta s)_R^2}{(\delta s)_{cw}^2}\right]$$

$$= 8.04[1 - (0.168)^2] \times 10^{-3}$$

$$= 7.81 \times 10^{-3} \text{ Å}^{-1}$$

$$[(\delta s)_i^{pa}]^2 = 6.1 \times 10^{-5} \text{ Å}^{-2}$$

In Fig. 9-24 the quantities $[(\delta s)_i^{pa}]^2$, as calculated in the manner just outlined, for the seven tungsten reflections along with the corresponding $[(\delta s)_i^{St}]^2$ values (see columns 3 and 2, respectively, of Table 9-10) have been plotted as a function of s^2 and the best straight line drawn through all the data points. Then from equation 9-91 the mean crystallite size L

and strain e are found to be:

$$\text{Ordinate intercept} = 0.62 \times 10^{-5}\ \text{Å}^{-2}.$$

$$L^2 = \frac{1}{0.62} \times 10^{-5} \simeq 16 \times 10^4\ \text{Å}^2.$$

$$L \simeq 400\ \text{Å}.$$

$$\text{Slope} = (2e)^2 = 0.715 \times 10^{-4}$$

$$e^2 = 0.1788 \times 10^{-4}$$

$$e = 0.0042.$$

Because of the isotropic character of tungsten and aluminum, the data from all the measured reflections may, in principle, at least, be used to determine mean size and strain parameters. However, in the case of the other three specimens the anisotropy requires that size and strain parameters be evaluated independently for different crystallographic directions. Thus in Fig. 9-25 reflections 200 and 400 are used to define a straight line providing size and strain parameters in the [100] direction,

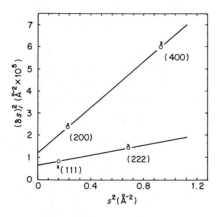

Fig. 9-25. Plots of $(\delta s)_i^2$ against s^2 for the 200 and 400 reflections and for the 111 and 222 reflections of silver–15% indium alloy: ×, $(\delta s)_i^{St}$, integral breadths corrected by Stokes' Fourier-transform method; ○, (δs_i^{pa}) integral breadths corrected by the parabolic approximation. (Courtesy of C. N. J. Wagner and E. N. Aqua and Plenum Press, *Advances in X-Ray Analysis*, Vol. 7, 1964, p. 46.)

while 111 and 222 are utilized similarly for the [111] direction. We may note in this connection that the third order of any reflection hkl from a cubic polycrystalline specimen is always blended with another reflection of higher multiplicity, thus 300 and 221, 330 and 411, 333 and 511, and so on; consequently when the anisotropy is appreciable, as is more commonly the case in cubic metals and alloys, only two resolved orders of a reflection of simple indices are available unless drastic preferred orientation of a particular mode is present[1].

Limitations of space dictate that the reader must refer to the original paper[54] for additional details. However, a few important general observations may be made here. The strains e calculated with equation 9-91 are generally larger than the rms values $\langle \epsilon_L{}^2 \rangle^{1/2}$ averaged over the dimensions of the crystallite, as is predicted by equation 9-85. Actually the e's tend to approach the rms values for $L \to 0$, but the rms strains are not very accurate for $L \to 0$ because of the "hook effect." This is a result of the Fourier coefficients at small L values being very sensitive to the accuracy with which the reflection tails can be measured.

The Cauchy formula for separation of size and strain (equation 9-90, method III) leads to size parameters $\bar{D}_{\text{eff}}$, $\bar{D}$, and $\bar{D}^2/\bar{D}$ that are usually much larger, and faulting probabilities that are much smaller, than the values found by the other three methods, which are in fair agreement among themselves. The Warren-Averbach Fourier method (I) is assuredly the most accurate and generally reliable way of analyzing the broadening of powder reflections. Integral breadths may also be used with considerable confidence when the Gaussian relationship between size and strain (equation 9-91, method II) is employed and when relatively large fault densities are to be assessed. For flat-specimen parafocusing geometry, as used in diffractometry, the parabolic approximation (equation 9-89a or 9-89b) for eliminating instrumental broadening gives integral breadths that agree well with those provided by the Stokes Fourier method of correction.

D. Comparison of Size and Strain Values Derived by Four Methods. The results of five representative studies, selected from a much larger number published to date, are summarized in Table 9-11. Of these investigations only that of Lewis and Northwood (IV) actually encompasses all four methods. A fifth method, integral breadths with both size and strain broadening assumed to be Cauchy, has been eliminated because (1) in practice it consistently leads to sizes that are much larger and faulting probabilities that are much smaller than those yielded by the other methods, as pointed out in Section 9-2.7C, and (2) *in principle* a model of size and strain distributions that would result in Cauchy/Cauchy broadening must be regarded as atypical of the great majority of actual polycrystalline specimens exhibiting both size and strain effects. This is, of course, not to rule out with finality the possibility of its appearance in specimens of exceptional histories. We must emphasize, further, that relative to the two methods based on integral breadths which are included in Table 9-11, their validity is conditioned by the acceptance of one of two arbitrary models of size and strain broadening, neither of which is apt to be a really faithful portrayal of any particular specimen. However, experimental and theoretical evidence tends to favor somewhat the applicability of the Cauchy-Gaussian model to the "typical unknown" specimen encountered in practice. The applicability of the Fourier and variance methods, unlike that of integral breadths, is not dependent upon the acceptance of specific models of size and strain broadening.

We must caution the reader to keep in mind that differences in the definitions of the parameters being measured are an underlying source of disagreement among certain classes of numerical results contained in Table 9-11. Thus the Fourier and integral-breadth methods, as ordinarily used, lead to the volume-average dimension perpendicular to the reflecting planes, while the variance method gives the cube root of the volume of the crystallite (coherently diffracting domain). The Fourier method provides the rms strain $\langle \epsilon_L^2 \rangle^{1/2}$ in a direction perpendicular to the reflecting planes and averaged over the distance L, whereas the integral-breadth method gives an approximate upper limit e such that $e \simeq 1.25 \langle \epsilon_L^2 \rangle^{1/2}$, and the variance gives $\langle e^2 \rangle^{1/2}$. Experience tends to show that in actual practice both the Fourier and variance methods are more reliable for the derivation of strain than of size parameters, a consequence of the sensitive dependence of size measurements upon the accuracy with which the tails of the reflections can be measured. Certainly, the more extreme differences in Table 9-11 are observed to exist among size parameters determined by different methods. Aside from the numerical results given in Table 9-11, the domain size obtained also depends, of course, on the extent of faulting and whether or not it is specifically allowed for.

Mitra and Misra[83] have described an interesting theoretical investigation of the differences to be anticipated among numerical sizes and and strains derived by line-shape (Fourier), integral-breadth, and variance analysis of reflection profiles in which the actual strain distribution is postulated to be either Cauchy or Gaussian. It was found that in general the three methods yield different values of both size and strain, but that the possibility of their being equal, though small, cannot be excluded.

E. Additional Literature. We now present a listing of some additional theoretical and experimental investigations arranged in four subject categories. It is regrettable that the existence of a vast body of publications in this field makes it impossible to present anything other than a more-or-less random selection, even though some effort has been made to avoid many significant omissions. Our hope is that the literature cited will prove useful to the reader who wishes to embark on a more comprehensive study of size, strain, and stacking-fault analysis by x-ray methods. Each item in the listing includes consecutively the reference

Table 9-11. Size and Strain Parameters Derived by Four Methods

Investigation	Ref.	Material		Preparation No.	Warren-Averbach (Fourier)	Variance	Integral Breadths	
							Cauchy-Gaussian[a]	Gaussian-Gaussian[b]
I. Halder and Wagner (1966)	[42]	Tungsten	Size (Å)		210	140	430	650
			Strain		0.0040[c] 0.0025[d]	0.0043	0.0037	0.0010
II. Aqua (1966)	[79]	Aluminum	Size (Å)		400	500		
			Strain		0.0007	0.0022		
III. Trömel and Urmann (1968)	[80]	Cadmium oxide	Size (Å)	6	150	170	180	
				1	370	440	460	
				2	440	570	560	
				3	760	1200	860	
				4	1100	1300	980	
			Strain	6	0.0074	0.0040	0.0056[e]	
				1	0.0045	0.0025	0.0031	
				2	0.0029	0.0019	0.0012	
				3	0.0027	0.0022	0.0020	
				4	0.0012	0.0008	0.0005	
IV. Lewis and Northwood (1969)	[81]	Lithium fluoride Ball-mill treatment A, B, or C and planes Size (Å)	A 100		690	7000	2000	1200
			111		900	7000	2000	1600

684

B 100	420	1900	1200	750
111	510	1200	830	750
C 100	300	360	520	440
111	410	610	560	500
Strain				
A 100	0.00066	0.0018	0.0013	0.0014
111	0.00039	0.00074	0.00094	0.00097
B 100	0.0014	0.0022	0.0020	0.0022
111	0.00087	0.0015	0.0014	0.0016
C 100	0.0023	0.0034	0.0033	0.0037
111	0.0016	0.0029	0.0023	0.0026

V. De and Sen Gupta (1970) [82]

Size (Å)

Ag–7.15% Zn	195	116		277
Ag–22.54% Zn	134	101		202
Ag–32.94% Zn	110	79		182
Cu–3.23% Sb	137	47		177
Cu–5.79% Sb	112	43		122
Ag–70.12% Pd	240	132		

Strain

Ag–7.15% Zn	0.00254	0.00437		0.00389
Ag–22.54% Zn	0.00268	0.00853		0.00548
Ag–32.94% Zn	0.00289	0.00772		0.00541
Cu–3.23% Sb	0.00261	0.00355		0.00548
Cu–5.79% Sb	0.00367	0.00500		0.00755
Ag–70.12% Pd	0.00218	0.00522		

[a]Cauchy size and Gaussian strain broadening (equation 9-92).

[b]Gaussian size and strain broadening (equation 9-91).

[c]At $L = 0$.

[d]Averaged over $L = 0$ to D_{eff}.

[e]Specified by Trömel and Urmann as η, the apparent strain (equation 9-81).

number, authors, year of publication, specimen studied (experimental work), and in a few cases additional explanatory matter in parentheses.

I. Size and strain separation (principally theoretical)
    [84] Kochendörfer, 1944
    [85] Tetzner, Schrader and Lang, 1968 (extension of Kochendörfer's theory)
    [86] Pines and Sirenko, 1962
    [87] Pines and Surovtsev, 1963
    [88] Kovalenok and Smushkov, 1963
    [88a]Harrison, 1966, 1967
    [88b]Ungar, 1967
    [89] La Fleur and Koopmans, 1968
    [90] La Fleur, 1969

II. Size and strain separation, size distributions (principally experimental)

[91] Quinn and Cherin, 1962	MgO (crystallite size distributions using equation 9-54)
[92] Pitts and Willets, 1961	AgBr/AgI (in photographic emulsions)
[93] Willets, 1965	AgBr/AgI (in photographic emulsions)
[94] Buchanan and Miller, 1966	Isotactic polystyrene (compared results from Fourier, integral-breadth, and paracrystalline [73] methods)
[95] Mitra, 1964	Pure aluminum
[96] Mitra and Misra, 1967	Magnesium
[97] Lihl, Ebel, and Stahl, 1968	Gold (thin films)

III. Stacking-fault analysis (mostly theoretical)
    [98] Warren and Averbach, 1952
    [99] Warren and Warekois, 1955
    [100] Williamson and Hall, 1953
    [101] Lele, 1969
    [102] Sato, 1969
    [103] Lele and Rama Rao, 1970
    [104] La Fleur, 1970

IV. Stacking-fault analysis (mostly experimental)
    [104a] Anantharaman and Christian, 1956

[105-1,2] Wagner, 1957	Copper and α-brass [105-1]; silver [105-2]
[106] Mitra and Halder, 1964	Cobalt (hexagonal)
[107] Ahlers and Vassamillet, 1967	Copper–germanium alloys
[108] De and Sen Gupta, 1968	Copper–antimony alloys
[109] De, 1969	Silver–palladium alloys

In conclusion, we wish to recommend as a modern general treatment in German Chapter 26 in the recent revision of Glocker[110].

9-3 DETERMINATION OF CRYSTALLITE SIZE – NO LATTICE IMPERFECTIONS

9-3.1 The Scherrer Equation

This topic is of special importance because the nature or history (including pretreatment) of many specimens precludes the presence of lattice strains or other imperfections in significant amounts. Under these circumstances the breadth β of the pure diffraction profile can be ascribed solely to small crystallite size, and appropriate theoretical relationships may be utilized to yield the mean crystallite dimension, and in favorable cases, limited information about the size distribution as well. In Section 9-2.3 Scherrer's equation[45] was given with β in 2θ units in the form

$$L = \frac{K\lambda}{\beta \cos \theta},$$

(9-60)

as well as with the breadth expressed in δs units by equation 9-61.

Bragg[111] has given a simplified derivation of the Scherrer equation employing only the ordinary principles of optical diffraction. Referring to Fig. 9-26, consider diffraction of x-rays of wavelength λ by a crystal in the form of a platelet. Suppose that the platelet consists of p atomic planes (hkl) of spacing d parallel to the surface, so that for moderately large values of p the thickness of the platelet is very close to pd. A relationship is to be found between $L_{hkl} = pd$ and the width at half-maximum intensity of the reflection (hkl). For this simplified discussion no account is taken of the lateral extent of the planes (hkl), which is to say, the dimensions of the crystallite in directions parallel to the planes.

If we employ the reflection analogy, the amplitude of a diffracted ray

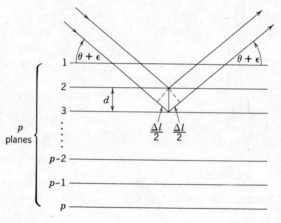

Fig. 9-26.

will be a maximum when the path difference Δl between rays reflected from successive planes is equal to a multiple of the wavelength:

$$\Delta l = 2d \sin \theta = n\lambda. \tag{9-102}$$

When the glancing angle differs from θ by a small amount ϵ, the path difference may be written:

$$\Delta l = 2d \sin (\theta + \epsilon)$$
$$= 2d \, [\sin \theta \cos \epsilon + \cos \theta \sin \epsilon]$$
$$= n\lambda \cos \epsilon + \sin \epsilon \, 2d \cos \theta.$$

Since the diffracted amplitude is appreciable only for very small values of ϵ, this may be written with good accuracy in the form

$$\Delta l = n\lambda + 2\epsilon d \cos \theta.$$

The corresponding phase difference is

$$\frac{2\pi}{\lambda} \Delta l = 2\pi n + \frac{4\pi}{\lambda} \epsilon d \cos \theta$$
$$= \frac{4\pi \epsilon d \cos \theta}{\lambda}.$$

By a well-known principle of optics [112], if n equal vectors of amplitude a differ in phase by successive uniform increments, the resultant amplitude is

$$an \frac{\sin \alpha}{\alpha}, \tag{9-103}$$

α being one-half the phase difference between the first and last vectors of the series. To return to the present problem, the phase difference of the first and pth planes is

$$\phi = \frac{4\pi p \epsilon d \cos \theta}{\lambda},$$

and by formula 9-103 the resultant amplitude of the reflected wave is

$$A = \frac{ap \sin \left(\dfrac{2\pi p \epsilon d \cos \theta}{\lambda}\right)}{\dfrac{2\pi p \epsilon d \cos \theta}{\lambda}}. \tag{9-104}$$

The amplitude of the reflected ray is a maximum when $\epsilon = 0$ and the reflected rays from all p planes are in phase:

$$A_0 = ap.$$

Bearing in mind that the intensity of an electromagnetic wave varies as the square of the amplitude, at half maximum intensity we can write:

$$\frac{A^2}{A_0^2} = \frac{1}{2} = \frac{\sin^2(\phi/2)}{(\phi/2)^2}. \tag{9-105}$$

From a plot of $[\sin^2(\phi/2)]/(\phi/2)^2$ as a function of $\phi/2$, it is found that equation 9-105 is satisfied when $\phi/2 = 1.40$. Then

$$\frac{2\pi p \epsilon_{1/2} d \cos \theta}{\lambda} = 1.40,$$

and the full angular width at half-maximum intensity of the reflection becomes

$$\beta_{hkl} = 4\epsilon_{1/2} = \frac{4 \times 1.40\lambda}{2\pi p d \cos \theta}$$

$$= \frac{0.89\lambda}{L_{hkl} \cos \theta}, \tag{9-106}$$

which is the Scherrer equation, with $K = 0.89$.

Scherrer's original derivation[45, 113] was based on the assumptions of Gaussian line profiles and small cubic crystals of uniform size, in which case K of equation 9-60 is 0.94 rather than 0.89 as in 9-106, β in both instances being $\beta_{1/2}$, the breadth at half-maximum intensity. Employing integral breadths β_i, Stokes and Wilson[46, 114] developed a more generalized treatment of crystallite-size broadening which is independent of the distribution of crystallite shape and symmetry. Their derivation leads to an *effective crystallite dimension* L_{hkl}, which is the volume average of the crystallite dimension T normal to the reflecting planes hkl,

$$\beta_i(2\theta) = \frac{\lambda}{L_{hkl}\cos\theta}. \tag{9-107}$$

L_{hkl} may be defined by

$$L_{hkl} = \frac{1}{V}\int T\,dV. \tag{9-108}$$

Furthermore, if A is the projected cross-sectional area of the crystal parallel to the reflecting planes,

$$V = A\langle T\rangle, \tag{9-109}$$

and L_{hkl} can also be expressed as

$$L_{hkl} = \frac{\langle T^2\rangle}{\langle T\rangle}. \tag{9-110}$$

We should note that the concept of the crystallite dimension presented by Stokes and Wilson leads directly to a formula identical with the Scherrer equation except that the constant K assumes a value of unity (compare equations 9-60 and 9-107).

Crystallites of Markedly Anisotropic Shapes. If lattice distortions may be presumed absent, the Scherrer equation with L defined as L_{hkl}, the volume average of the crystallite dimension in a direction normal to the reflecting planes, has a special value in interpreting an x-ray powder pattern consisting of lines varying conspicuously in breadth as a function of the indices hkl. This is illustrated by Fig. 9-27, which shows the powder diagram of a nickel hydroxide [Ni(OH)$_2$] specimen. It is immediately evident that the lines vary greatly in sharpness, and closer

Fig. 9-27.   Debye-Scherrer pattern of nickel hydroxide showing variations in line width.

inspection shows that lines of the type $00 \cdot l$ are very broad while those with zero l index are sharp. Lines of mixed indices tend to be intermediate in sharpness, those with relatively large values of l being more diffuse and those with relatively large values of h or k, or both, being sharper. X-ray diffractometer measurements of the half-maximum breadths lead to dimensions L_{hkl} of about 40 Å from the $00 \cdot 2$ line and about 200 Å from lines of the type $hk \cdot 0$. These results show the mean crystallite shape to be platelike, with the large dimension in the basal ab plane and the smaller dimension in the c direction. This is illustrated in Fig. 9-28. The approximately equal sharpness of all the $hk \cdot 0$ lines permits no conclusions to be drawn concerning the mean shape of the plates, this being indicated by the use of a circular profile in the drawing.

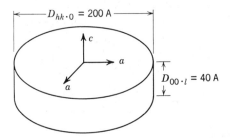

Fig. 9-28. Mean shape of a nickel hydroxide crystallite as deduced from line-breadth measurements.

Actually, any regular or irregular plane polygon approximating a circle in shape or a mixture of polygons of rather different shapes leads to the same diffraction effects, namely, relatively sharp $hk \cdot 0$ lines, just as long as the several dimensions in the ab plane are sufficiently large.

Certain other rather distinct crystal habits are revealed by their x-ray patterns as long as the dimensions are sufficiently small to produce line broadening. Thus, needlelike crystals with the needle axis coinciding with b show broad $h0l$ reflections and sharp $0k0$ reflections, whereas hkl lines are intermediate in breadth according to the relative sizes of the indices. Lathlike crystals in which b, a, and c are, respectively, the longest, intermediate, and shortest dimensions (see Fig. 9-29) generate wide $00l$ reflections, intermediate $h00$ widths, and $0k0$ lines of maximum

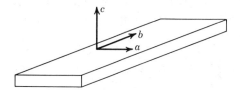

Fig. 9-29.   Lathlike crystal habit.

sharpness. Again of course this presupposes that the dimensions are within the line-broadening range. As was true for a platelike habit, measurements of the breadths of lines of different index types can yield the mean dimensions of needlelike and platelike crystals within limits. When the crystallites are regular polyhedra, such as cubes, tetrahedra, or octahedra, it is sometimes possible to determine the particular shape involved by means of a more rigorous analysis of the diffraction effects. In this connection the reader is referred to Section 9-2.3A, in particular Table 9-2 and related text.

9-3.2 The Variance Method

By reference to Section 9-2.3A, and in particular equation 9-73, it can be seen that in the absence of lattice distortions and with neglect of the taper parameter T_a, the line-profile variance is given as a function of the "true" crystallite size p by

$$W_{2\theta}^S = \frac{K\lambda\Delta2\theta}{2\pi^2 p \cos \theta},$$
(9-111)

in which $\Delta2\theta$ is the range of integration used in the calculation of $W_{2\theta}^S$. We may also remark in regard to the variance-range function (equation 9-59),

$$W = W_0 + k\sigma,$$

that in the absence of lattice distortions the ordinate intercept W_0 is determined only by instrumental aberrations[41]. Hence in a plot of W versus σ for the diffraction profile corrected for instrumental broadening, the intercept should be very close to zero.

9-3.3 Size Distributions[32, 33]

If lattice distortions and faulting may with good reason be presumed absent or negligibly small, we may employ an extension of the Warren-Averbach theory developed by Bienenstock[32, 33] to obtain the size distribution $p(n)$. We present herewith only Bienenstock's significant results, neglecting the intermediate derivational steps. The departure point is equation 9-40 of Warren and Averbach (Section 9-2.2A) for the experimentally observable power P' diffracted by a small distorted crystallite. Neglecting lattice distortions and summing over m_1' and m_2', we can modify equation 9-40 to the form

$$P'(h_3) \sin^2 (\pi h_3) = \frac{1}{2} K' \sum_{m_1} \sum_{m_2} \{1 - \cos 2\pi h_3 n (m_1, m_2)\},$$
(9-112)

which is the basic equation of the line profile resulting from finite crystallite size alone. In equation 9-112, K' is a slowly varying function of angle which can usually be considered constant over the range of interest.

With $p(n)$ defined as the number of columns $n-1$ cells in length, and N_c as the total number of columns, equation 9-112 can be rewritten as

$$\sum_{n=1}^{\infty} p(n) \cos (2\pi h_3 n) = N_c - \frac{2}{K'} P'(h_3) \sin^2 (\pi h_3),$$

and application of Fourier's integral theorem yields

$$p(n) = \frac{-4}{K'} \int_{-1/2}^{+1/2} P'(h_3) \sin^2 (\pi h_3) \cos (2\pi h_3 n) \, dh_3. \qquad (9\text{-}113)$$

Now applying the normalizing condition,

$$\frac{4}{K'} \int_{-1/2}^{+1/2} P'(h_3) \sin^2 (\pi h_3) \, dh_3 = N_c,$$

which simply states that the sum over columns of the partial contributions, which are independent of column length, is proportional to the total number of columns, we obtain as the normalized size distribution function:

$$\frac{p(n)}{N_c} = \frac{\int_{-1/2}^{+1/2} P'(h_3) \sin^2 (\pi h_3) \cos (2\pi h_3 n) \, dh_3}{\int_{-1/2}^{+1/2} P'(h_3) \sin^2 (\pi h_3) \, dh_3}. \qquad (9\text{-}114)$$

Bienenstock[32] shows that Warren and Averbach's result is equal to the first term in a series expansion of the right-hand side of equation 9-114. The Fourier size coefficients $A_n{}^S$ of the Fourier size expansion of $P'(h_3)$ may be defined by

$$P'(h_3) = \sum_{-\infty}^{+\infty} A_n{}^S \cos 2\pi n h_3, \qquad (9\text{-}115)$$

after which substitution of equation 9-115 in 9-114, reversal of the order of integration and summation, and performance of the integration lead to the result[33]

$$\frac{p(n)}{N_c} = \frac{2A_n - A_{n+1} + A_{n-1}}{A_0 - A_1}. \qquad (9\text{-}116)$$

This equation makes it possible to evaluate the column-length (size) distribution $p(n)$ directly from the Fourier coefficients calculated by Stokes' Fourier method for correction of the experimental line profiles for instrumental broadening.

By reference to equation 9-114, it can be seen that it is the weighted power P' (h_3) $\sin^2 \pi h_3$ rather than P' (h_3) itself that is important in Bienenstock's analytical procedure. Whereas the method of Warren and Averbach gives undue weight to the tails of the reflection because of the factor $(\pi h_3)^2$[31], Bienenstock's method reduces the importance of these regions, which are precisely the ones measurable with the least accuracy. Finally, it should be noted that for *very broad lines* it may be necessary to allow for the variation of K' with angle (see equation 9-112).

Using the variance approach, Wilson[115] has shown that when a distribution of sizes exists, the two values of the mean crystallite size derivable from the slope and intercept of the variance-range plot (equation 9-59) permit one to set an upper limit to the number-average crystallite size.

9-3.4 Some Practical Considerations

From the experimental point of view, the accuracy of a crystallite-size determination is mainly dependent upon the accuracy with which the pure diffraction breadth β can be measured. The choice of K is largely predetermined by the definitions of β and L that have been adopted, as well as by the indices of the reflecting planes in certain procedures. Consider for a moment the several curves of β/B as a function of b/B in relation to the accuracy and precision with which β can be deduced (see Figs. 9-6 and 9-9). Corresponding to a given experimental uncertainty in the determination of b/B, there is a related uncertainty in β/B that varies somewhat from one curve to another but which, in any case, is profoundly dependent upon the size of b/B. For illustration consider a large instrumental contribution b and a small pure diffraction breadth β. In this event B is only slightly larger than b. If $b/B = 0.9$ and the experimental error in measuring b and B is somewhat optimistically assumed to be ± 5 per cent, the error in b/B is

$$\pm \sqrt{5^2 + 5^2} = \pm 7 \text{ per cent.}$$

This means that b/B is known to fall in the range 0.84 to 0.96, and assuming the Gaussian–Gaussian correction curve (Fig. 9-6) to hold, this leads to a most likely β/B ratio of 0.44, with an uncertainty range extending from 0.26 to 0.55. The other limiting correction curve is the straight line, which applies to diffractometer measurements at high angles (see Fig. 9-9), and this leads to a range of β/B values extending from 0.16 to 0.40. Evidently, the precision is exceedingly poor at best for b/B as large as 0.9, and similar computations show that the precision is still unsatisfactory at $b/B = 0.8$. As the crystallite size becomes very large (and β very small), the ratio b/B inevitably increases into the unfavorable range above 0.50, in spite of the best possible adjustment of the experimental conditions. However, it can be seen from Figs. 9-6 and 9-9 that at small values of b/B the various correction curves lead to very similar β values. In fact, at very small ratios (small crystallite sizes) it is unnecessary to apply any correction, since the directly observed breadth B is equal to β within experimental accuracy.

The optimum experimental conditions are those making b as small as possible and B as large as possible. Since b is due mainly to instrumental broadening, except at large angles where the spectral breadth becomes significant, parafocusing geometries are superior to the Debye-Scherrer arrangement even when the sample diameter is made very small. Photographic focusing methods are equivalent to the high-resolution diffractometer technique, except for the greater labor involved in working up the intensity profile from film densities and the lower precision of such intensity measurements. When the crystallite size is smaller than about 300 Å, Debye-Scherrer cameras can be used satisfactorily, provided the sample diameter is made as small as practicable and the Bragg angle is properly selected. Similarly, methods of no special precision are effective up to 300 or 400 Å, but in the range 400 to 2000 Å greater refinements become mandatory as the size increases.

Another means of making the ratio b/B small is to increase θ, since the pure diffraction broadening varies as $1/\cos\theta$. At Bragg angles less than 30 or 40° the value of $\cos\theta$ does not fall greatly below unity, so that it becomes necessary to work in the back-reflection region to gain a real advantage from this effect. Thus at 80° the breadth β is increased by a factor of 6. For smaller crystallite sizes there is ordinarily no need to work at such a large angle because b/B will be sufficiently small for precise results even at low angles. However, when the crystallite dimension exceeds 500 Å, precise measurements are scarcely possible in the low-angle region.

In correcting for instrumental broadening with the aid of curves of the types given in Figs. 9-6 and 9-9, it is important that the geometrical conditions affecting the b and B profiles be as nearly identical as possible, particularly in the larger crystallite-size range wherein b may equal or exceed $\frac{1}{2}B$. Regardless of what experimental technique is used, this can be accomplished by observing two simple rules: (1) Intimately mix the "unknown" and standard powders and record the two lines needed from the pattern of the mixture. (2) Correct the unknown line in question by employing a standard line at about the same Bragg angle. These rules should be faithfully followed in all Debye-Scherrer work. When parafocusing techniques are used, the same rules should be observed when the absorption coefficients of the two materials are either small or differ considerably. If the linear absorption coefficients of both powders are high, for instance, $\mu = 40$ or more when the interstices are included in the calculations, good results can be obtained by mounting and analyzing the two materials separately. Fine quartz powder, which displays a typical packing behavior, assumes a state in which solid and interstices constitute about 40 and 60 per cent, respectively, of the total sample volume. Hence the linear absorption coefficient of the powder (μ_p) is only 0.40 as large as that of the solid material (μ_s):

$$\mu_p = \frac{\rho_p}{\rho_s} \times \mu_s \cong 0.4 \times 93 = 37.$$

Here ρ_p and ρ_s are the respective densities of powder and solid. The preceding discussion presupposes a sufficient sample thickness to give maximum absorption. If sufficient intensity can be obtained by employing a very thin film of powder mounted on a plane supporting surface, any absorptive differences between the two substances become much less important, and it is usually unnecessary to mix them.

The choice of a reference powder is not always simple. First, it should produce diffraction lines in the immediate neighborhood of the lines of the unknown to be measured. It should be a substance not subject to lattice distortions, which would produce line broadening, and the crystallite-size distribution should include no dimensions smaller than a few tenths of a micron and preferably none larger than about 20μ. Such a specimen gives sharp, smooth Debye-Scherrer lines at all angles. Appreciable amounts of material with dimensions of $40\,\mu$ or larger result in "granulated" or spotty lines which are difficult to microphotometer, and these sizes also produce undesirable fluctuations in diffractometrically recorded intensities. However, a size distribution extending to $0.1\,\mu$ or lower produces very appreciable broadening in the sensitive back-reflection region. We have had good results with a reference powder prepared by crushing and grinding quartz to pass a 325-mesh screen, then fractionating the fine powder into three portions with the size ranges <5, 5 to 20, and $>20\,\mu$ by sedimentation in methanol, and finally using the 5- to 20-μ fraction as the reference material. This fraction showed no crystallite-size broadening at large Bragg angles.

9-3.5 Illustrative Analyses

A. Crystallite Shape—Magnesium Oxide (MgO) Powder. To illustrate the possibility of deducing the mean crystallite shape from the dependence of line breadth on the indices of reflection, some experiments of Berry[116] on MgO powder will be described briefly. For two diffraction lines $h_1k_1l_1$ and $h_2k_2l_2$ produced by a specimen, the crystallite-size equation may be written

$$L = \frac{K_1\lambda}{\beta(h_1k_1l_1)\cos\theta_1},$$

$$L = \frac{K_2\lambda}{\beta(h_2k_2l_2)\cos\theta_2}.$$

The ratio R between the breadths of the two lines is then

$$R = \frac{\beta(h_1k_1l_1)}{\beta(h_2k_2l_2)} = \frac{K_1\cos\theta_2}{K_2\cos\theta_1}. \qquad (9\text{-}117)$$

By comparing the experimentally observed ratios $\beta(h_1k_1l_1)/\beta(h_2k_2l_2)$ with the theoretical ones calculated from the right-hand side of expression 9-117 upon insertion of the proper K's corresponding to the various crystallite shapes, it should be possible to select the best shape on the basis of the quality of the agreement observed. The absolute and relative breadths at half-maximum intensity of three pure diffraction profiles (deduced from the uncorrected experimental profiles by a Fourier integral analysis similar to that of Stokes[4]) were found to be as follows:

$$\beta_{200} = 0.0127 \text{ radian}, \qquad \beta_{200}/\beta_{220} = 0.993,$$
$$\beta_{220} = 0.0128 \text{ radian}, \qquad \beta_{222}/\beta_{200} = 1.244,$$
$$\beta_{222} = 0.0159 \text{ radian}, \qquad \beta_{222}/\beta_{220} = 1.236.$$

The ratios of these observed relative breadths to the theoretical ones for crystallites of various shapes are given in Table 9-12. It is seen that the best agreement as indicated by the closest average approach to a ratio of unity is given by the cube.

Table 9-12.

	Tetra-hedron	Octa-hedron	Dodeca-hedron	Sphere	Cube
$\dfrac{R_{200/220}\,(\text{obs})}{R_{200/220}\,(\text{calc})}$	0.820	1.220	1.170	1.155	1.090
$\dfrac{R_{222/200}\,(\text{obs})}{R_{222/200}\,(\text{calc})}$	1.050	0.790	0.870	0.905	0.940
$\dfrac{R_{222/220}\,(\text{obs})}{R_{222/220}\,(\text{calc})}$	0.855	0.965	1.005	1.050	1.020

B. MgO from Decomposition of MgCO$_3$ at 600°C. The profiles of the 200 MgO reflection at 42.7° 2θ and the 11·2 quartz reflection at 50.0° were manually counted with the aid of a *low-resolution* diffractometer, using CuKα radiation. The Fourier transform procedure of Stokes[4] was applied to the data in order to determine the pure diffraction profile of the MgO peak with the greatest possible accuracy. Figure 9-30 portrays the experimentally observed $g(\epsilon)$ (quartz) and $h(\epsilon)$(MgO) peaks and the pure diffraction profile of MgO, $f(\epsilon)$, as derived by the Fourier analysis. The crystallite-size equation 9-106 assumes the numerical value,

$$L_{200} = \frac{0.9\lambda}{\beta_{1/2}\cos\theta}$$

$$= \frac{0.9 \times 1.542 \times 57.3}{\beta_{1/2} \times 0.931} = \frac{85.4}{\beta_{1/2}},$$

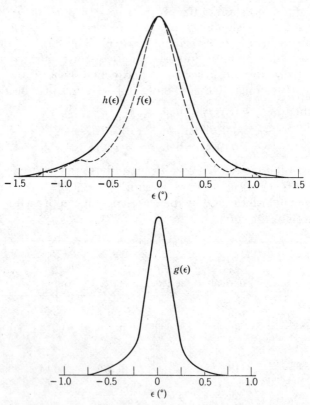

Fig. 9-30. Diffraction profiles involved in the Fourier analysis of the 200 maximum of MgO by the method of Stokes (x-ray diffractometer with broad x-ray source).

and substitution of 0.530° for $\beta_{1/2}$ gives

$$L_{200} = 161 \text{ Å.}$$

C. MgO from Decomposition of MgCO$_3$ at 900°C. (Fig. 9-31). This figure illustrates the application of high-resolution diffractometry to the measurement of large crystallite sizes at low angles. The instrument was adjusted for high resolution by employing a 1° x-ray divergence slit and a 0.025° receiving slit, and the counting rates with CuKα radiation were recorded manually. For the geometrical broadening profile the 11·2 quartz β line was used, the nickel filter being removed for this measurement. This added to the accuracy of the determination by removing the

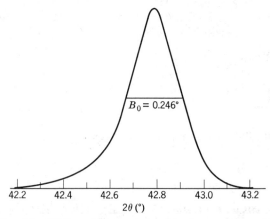

$B_0 = 0.246°$

42.2   42.4   42.6   42.8   43.0   43.2

2θ (°)

Fig. 9-31. Profile of the 200 line of MgO prepared by decomposition of MgCO$_3$ at 900°C (x-ray diffractometer adjusted for high resolution).

need for correcting the reference profile for $K\alpha$ doublet broadening, although the MgO peak still required this correction. The results of the calculations were as follows:

$L_{200} = 85.4/\beta_{1/2}$ (see Example B)
  $B_0 = 0.246°$ (± 0.010)
    $B = 0.186°$ (± 0.010) (corrected for α_{12} separation, using the curves in Figs. 9-7 and 9-8)
    $b = 0.115°$ (± 0.010)
$b/B = 0.618$ (± 0.070)
$\beta/B = 0.448$ (± 0.076) (using the low-angle curve in Fig. 9-9)
    $\beta = 0.083°$
$L_{200} = 1030 \text{ Å.}$
          Approximate precision limits in L_{200}: 900–1230 Å

This determination could have been done with appreciably greater precision in the high-angle region, using resolved α_1 lines.

D. Micronized Quartzite Powder, Fraction $< 5\,\mu$. The powder was sedimented from methanol to remove all particles larger than $5\,\mu$. The diffraction pattern with Cu$K\alpha$ radiation showed appreciable broadening only in the high-angle region, so this served as a good test of high-resolution diffractometry applied to a crystallite size near the the upper limit of the measurable range. The $24 \cdot 0\,\alpha_1$ peak at $2\theta = 146.6°$ was step-scanned, and the profile compared with a reference profile of the $24 \cdot 0$ α_1 peak of a 5- to 20-μ fraction of pure quartz powder (see Fig. 9-32). The latter peak is conspicuously broadened due to the natural spectral breadth of the radiation, as can be seen by comparing its breadth, 0.220°, with that of the $11 \cdot 2\,\beta$ line at 44.9°, 0.115°, (refer to example C). The following numerical results were obtained:

$$L_{24\cdot 0} = \frac{0.9 \times 1.540 \times 57.3}{\beta_{1/2} \times 0.2874} = 276/\beta_{1/2}$$

$$b = 0.220° \ (\pm 0.015)$$
$$B = 0.370° \ (\pm 0.015)$$
$$b/B = 0.595 \ (\pm 0.050)$$
$$\beta/B = 0.405 \ (\pm 0.050) \ (\text{using the high-angle curve of Fig. 9-9})$$
$$\beta = 0.150°$$
$$L_{24\cdot 0} = 1840 \ \text{Å}$$

Approximate precision limits in $L_{24\cdot 0}$: 1700–2100 Å

E. L_c Dimension of a Carbon Black. By L_c or L_{002} we refer to the thickness of a parallel-layer group in the direction perpendicular to the

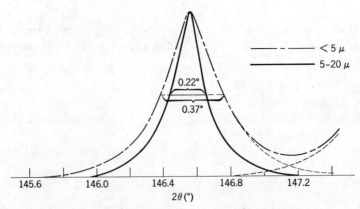

Fig. 9-32. Shapes of the $(24.0)\alpha_1$ peaks of $<$ 5- and 5-20-μ fractions of quartz powder (x-ray diffractometer adjusted for high resolution).

layers. In this connection the reader should again examine Section 9-2.5B. Figure 9-33 illustrates how a diffraction profile computed for a mixture of two discrete sizes fits the observed 00·2 line profile of a carbon black better than that of a single size. As pointed out in Section 9-2.5B, only reflection of types hk and $00l$ appear, and in the present instance the latter are very broad because of the few layers constituting the average parallel-layer group. The scattering profile of a $00l$ reflection is then of the form [56, 117, 118]

$$I = \frac{\sin^2 \left(\dfrac{2\pi N d \sin \theta}{\lambda} \right)}{N \sin^2 \left(\dfrac{2\pi d \sin \theta}{\lambda} \right)}, \qquad (9\text{-}118)$$

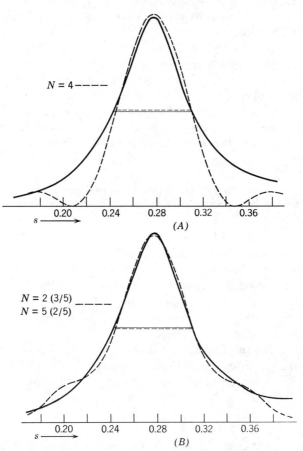

Fig. 9-33.   Comparison of theoretical profiles (---) with the observed profile (——) of the 00.2 diffraction band of a carbon-black specimen.

if each parallel-layer group consists of N layers; for a distribution of N's it can be written:

$$I = \sum_N \frac{P_N \sin^2\left(\frac{2\pi N d \sin\theta}{\lambda}\right)}{N \sin^2\left(\frac{2\pi d \sin\theta}{\lambda}\right)}. \tag{9-119}$$

The experimentally observable 002 profile is rather asymmetric because of (1) the asymmetric contribution of the (00) maximum and (2) the variation of the atomic scattering factor f with angle. In Fig. 9-33 these effects have been eliminated by multiplying the observed intensities by the factor $s^2/f(s)^2$, where $s = 2(\sin\theta)/\lambda$[117]. Finally these corrected intensities were normalized with an arbitrary factor to permit direct comparison of the shapes and breadths of the experimental and theoretical line profiles. In Fig. 9-33A it is seen that $N = 4$ gives good agreement with experiment as regards breadth at half-maximum intensity, but the observed profile is much broader at the base. Figure 9-33B shows the comparatively good agreement obtained on the assumption that three-fifths of the parallel-layer groups consist of two layers, and two-fifths of five layers. On the basis of an observed interlayer spacing of 3.60 Å for this sample, these N's correspond to L_{002} dimensions of 3.60 and 14.4 Å, respectively, if L_{002} for a parallel-layer group is defined as

$$L_{002} = (N-1)d_{002}.$$

The effective L_{002} for this carbon black is then about

$$(0.6 \times 3.60) + (0.4 \times 14.4) = 7.9\text{ Å}.$$

It must be remembered of course that the values $N = 2$ and 5 do not give a literal picture of the distribution of thicknesses of the parallel-layer groups, but they do serve to show that the actual distribution of N's, which must be continuous, extends from values as small as 2 to values greater than 5. Somewhat over half of the sample by weight consists of groups with two or three layers, and somewhat less than half of groups with four, five, or six layers.

The foregoing much oversimplified method of analysis has been superseded by a more sophisticated treatment[119–121] which fits the 00l profile to a relatively detailed stack-height distribution,

$$\sum_{i=1}^{i_{max}} f_i N_i,$$

including f_i, the volume fraction of unassociated single layers and D, the fraction of so-called disorganized material. The reader is cautioned that such a distribution analysis constitutes only one possible interpretation of the 00l line profile. The alternative interpretation of a very imperfect lattice characterized by a mean defect-free distance, advanced by Ergun [62], is possibly more realistic and deserves serious consideration.

F. L_a Dimension of a Carbon Black. With reference to Section 9-2.5B, L_A or L_{hk} may be used to denote the mean dimension in the plane of a two-dimensional-layer structure. The specimen was a carbon black possessing the random-layer lattice structure, as was demonstrated by the presence of only hk and 00l reflections in its diffraction pattern. In order to eliminate scattering due to general radiation and the air in the path of the x-ray beam, photographic patterns were prepared in an evacuated Debye-Scherrer camera, using crystal monochromatized Mo$K\alpha$ and Cu$K\alpha$ radiations. The sample was in the form of a thin cake of 0.6-mm thickness, through which the beam penetrated normally. The films were microphotometered and the transmission curves converted to intensity curves in the usual way by comparison with the transmission values of a graded density scale. The immediate object was to determine the mean layer dimension L_{hk} from the breadth of an hk profile, using the appropriate K value of 2.0 (see Section 9-2.5B). Because of the great breadth of the selected reflection 10 the geometrical broadening, in this case between 0.5 and 1.0°, could be neglected.

Two procedures were followed. In the first, the Mo$K\alpha$ data alone were used. The profile of the 10 peak, after correcting for the variation of absorption with angle, is shown in Fig. 9-34A. From the observed half-

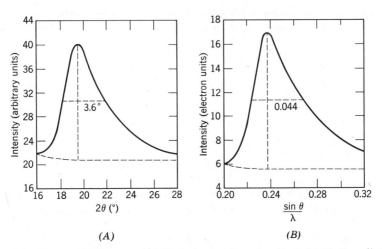

Fig. 9-34. Photographically recorded 10 maximum of a carbon black. (A) Mo$K\alpha$ radiation, uncorrected for polarization. (B) Mo$K\alpha$ and Cu$K\alpha$ radiations combined, corrected for polarization.

maximum breadth of 3.6° the mean layer dimension is found to be

$$L_{10} = \frac{2.0\lambda}{\beta_{1/2} \cos \theta}$$

$$= \frac{2.0 \times 0.711 \times 57.3}{3.6 \times 0.986} = 23.0 \text{ Å}.$$

The precision limits are approximately ± 3 Å.

The second procedure was based on a recognition of the error introduced into the foregoing calculation by the appreciable variation in the polarization factor over an angular range as large as that involved here. This time both the Mo and Cu data were utilized. The experimental intensity curves were separately corrected for absorption and polarization and then combined to give the profile shown in Fig. 9-34B as a function of $(\sin \theta)\lambda$. In a manner parallel to the derivation of equation 9-61 from 9-60, in terms of δs the crystallite-size equation for a two-dimensional structure becomes

$$L_{hk} = \frac{2.0}{\delta s}. \tag{9-120}$$

Substitution of the observed breadth, $\delta s = 2\delta \, [(\sin \theta)/\lambda] = 0.088 \text{ Å}^{-1}$, in this formula gives

$$L_{10} = 22.7 \text{ Å},$$

imperceptibly smaller than the value arrived at when the polarization correction was not taken into account.

Using the Debye equation, Diamond[58, 122] has shown how to derive the distribution of layer dimensions L_a from the hk profiles of carbons by least squares.

GENERAL REFERENCES

1. C. S. Barrett and T. B. Massalski, *Structure of Metals*, 3rd ed., McGraw-Hill, New York, 1966, Chapter 16.

2. D. R. Buchanan and R. L. Miller, "X-Ray Line Broadening in Isotactic Polypropylene," *J. Appl. Phys.*, **37**, 4003 (1966).

3. S. Ergun, "Direct Method for Unfolding Convolution Products—Its Application to X-Ray Scattering Intensities," *J. Appl. Crystallogr.*, **1**, 19 (1968).

4. S. Ergun, "X-Ray Scattering by Very Defective Lattices," *Phys. Rev.*, **B1**, 3371 (1970).

*5. R. Glocker, *Materialprüfung mit Röntgenstrahlen*, 5th ed., Springer, Berlin, 1971, Chapter 27.

6. A. Guinier, *X-Ray Diffraction in Crystals, Imperfect Crystals and Amorphous Bodies* (English translation by P. Lorrain and D. Sainte-Marie Lorrain), W. H. Freeman, San Francisco, 1963, Chapter 7.

7. A. Kochendörfer, "The Determination of Particle Size and Lattice Distortions in Crystalline Materials from the Breadths of X-Ray Lines," *Z. Kristallogr.*, **105**, 393 (1944).

8. J. I. Langford and A. J. C. Wilson, "On Variance as a Measure of Line Broadening in Diffractometry: Some Preliminary Measurements on Annealed Aluminum and Nickel and on Cold-Worked Nickel," in *Crystallography and Crystalline Perfection* (G. N. Ramachandran ed.), Academic Press, New York, 1963, pp. 207–222.

9. L. I. Mirkin, *Handbook of X-Ray Analysis of Polycrystalline Materials* (English translation by J. E. S. Bradley), Consultants Bureau, New York, 1964, Chapter 8.

10. A. R. Stokes, "A Numerical Fourier-Analysis Method for the Correction of Widths and Shapes of Lines on X-Ray Diffraction Photographs," *Proc. Phys. Soc.* (London), **A61**, 382 (1948).

11. A. Taylor, *X-Ray Metallography*, Wiley, New York, 1961, Chapters 14, 15.

12. C. N. J. Wagner and E. N. Aqua, "Analysis of the Broadening of Powder Pattern Peaks from Cold-Worked Face-Centered and Body-Centered Cubic Metals," *Advances in X-Ray Analysis*, Vol. 7, Plenum Press, New York, 1964, p. 46.

13. B. E. Warren, "X-Ray Studies of Deformed Metals," *Prog. Met. Phys.*, **8**, 147 (1959).

*14. B. E. Warren, *X-Ray Diffraction*, Addison-Wesley, Reading, Massachusetts, 1969, Chapter 13.

15. B. E. Warren and B. L. Averbach, "The Effect of Cold-Work Distortion on X-Ray Patterns," *J. Appl. Phys.*, **21**, 595 (1950).

16. A. J. C. Wilson, "On Variance as a Measure of Line Broadening in Diffractometry. General Theory and Small Particle Size," *Proc. Phys. Soc.* (London), **80**, 286 (1962).

17. A. J. C. Wilson, "On Variance as a Measure of Line Broadening in Diffractometry. II. Mistakes and Strains," *Proc. Phys. Soc.* (London), **81**, 41 (1963).

SPECIFIC REFERENCES

[1] B. E. Warren, *X-Ray Diffraction*, Addison-Wesley, Reading, Massachusetts, 1969.

[2] A. F. Jones and D. L. Misell, *J. Phys.*, **A3**, 462 (1970).

[3] L. P. Smith, *Phys. Rev.*, **46**, 343 (1934).

[4] A, R. Stokes, *Proc. Phys. Soc.* (London), **A61**, 382 (1948).

[5] H. C. Burger and P. H. van Cittert, *Z. Phys.*, **79**, 722 (1932).

[6] S. Ergun, *J. Appl. Crystallogr.*, **1**, 19 (1968).

[7] Reference 1, pp. 258–262.

[8] L. E. Alexander, *X-Ray Diffraction Methods in Polymer Science*, Wiley, New York, 1969, pp. 467–472.

[9] W. D. Hoff and W. J. Kitchingman, *J. Sci. Instrum.*, **43**, 654 (1966).

[10] Reference 8, p. 457.

[11] See, for example, E. C. Titchmarsh, *Introduction to the Theory of Fourier Integrals*, Clarendon Press, Oxford, 1948.

[12] W. A. Rachinger, *J. Sci. Instrum.*, **25**, 254 (1948).

[13] A. Papoulis, *Rev. Sci. Instrum.*, **26**, 423 (1955).

[14] D. T. Keating, *Rev. Sci. Instrum.*, **30**, 725 (1959).

[15] C. M. Mitchell and P. M. de Wolff, *Acta Crystallogr.*, **22**, 325 (1967).

[16] H. J. Edwards and K. Toman, *J. Appl. Crystallogr.*, **3**, 157 (1970).

[17] See, for example, I. V. Sokolnikoff and E. S. Sokolnikoff, *Higher Mathematics for Engineers and Physicists*, McGraw-Hill, New York, 1941, pp. 554–558.

[18] B. E. Warren and P. Bodenstein, *Acta Crystallogr.*, **18**, 282 (1965).

[19] B. E. Warren and P. Bodenstein, *Acta Crystallogr.*, **20**, 602 (1966).

[20] A. Taylor and H. Sinclair, *Proc. Phys. Soc.* (London), **57**, 108 (1945).

[21] F. W. Jones, *Proc. Roy. Soc.* (London), **166A**, 16 (1938).

[22] L. E. Alexander, *J. Appl. Phys.*, **19**, 1068 (1948).

[23] See, for example, *Internationale Tabellen zur Bestimmung von Kristallstrukturen*, Vol. II, Gebrüder Borntraeger, Berlin, 1935, pp. 588–608; A. Guinier, *Théorie et Technique de la Radiocristallographie*, 2nd ed., Dunod, Paris, 1956, pp. 709–724; A. Guinier, *X-Ray Crystallographic Technology* (English translation by T. L. Tippell), Hilger and Watts, London, 1952, pp. 306–320.

[24] W. Ruland, *Acta Crystallogr.*, **18**, 581 (1965).

[25] W. Ruland, *J. Appl. Crystallogr.*, **1**, 90 (1968).

[26] L. E. Alexander, *J. Appl. Phys.*, **25**, 155 (1954).

[27] A. H. Compton and S. K. Allison, *X-Rays in Theory and Experiment*, 2nd ed., Van Nostrand Reinhold, New York, 1935, Chapter 9.

[28] Reference 1, pp. 264–275.

[29] B. E. Warren and B. L. Averbach, *J. Appl. Phys.*, **21**, 595 (1950).

[30] F. Bertaut, *C. R. Acad. Sci.*, Paris, **228**, 492 (1949).

[31] B. E. Warren, *Prog. Met. Phys.*, **8**, 147 (1959).

[32] A. Bienenstock, *J. Appl. Phys.*, **32**, 187 (1961).

[33] A. Bienenstock, *J. Appl. Phys.*, **34**, 1391 (1963).

[34] F. Bertaut, *C. R. Acad. Sci.*, Paris, **228**, 187 (1949).

[35] A. J. C. Wilson, *Nature*, **193**, 568 (1962).

[36] A. J. C. Wilson, *Proc. Phys. Soc.* (London), **80**, 286 (1962).

[37] A. J. C. Wilson in *Advanced Methods of Crystallography* (G. N. Ramachandran, ed.), Academic Press, London, 1964.

[38] A. J. C. Wilson, *J. Appl. Crystallogr.*, **3**, 71 (1970).

[39] H. J. Edwards and K. Toman, *J. Appl. Crystallogr.*, **3**, 165 (1970).

[40] M. Tournarie, *C. R. Acad. Sci.*, Paris, **242**, 2161 (1956).

[41] J. I. Langford and A. J. C. Wilson in *Crystallography and Crystal Perfection* (G. N. Ramachandran, ed.), Academic Press, New York, 1963, pp. 207–222.

[42] N. C. Halder and C. N. J. Wagner, *Advances in X-Ray Analysis*, Vol. 9, Plenum Press, New York, 1966, p. 91.

[43] N. K. Misra and G. B. Mitra, *Acta Crystallogr.*, **23**, 867 (1967).

[43a] M. Tournarie, *C. R. Acad. Sci.*, Paris, **242**, 2016 (1956); A. J. C. Wilson, *Proc. Phys. Soc.* (London), **85**, 807 (1965); G. B. Mitra and N. K. Misra, *Brit. J. Appl. Phys.*, **17**, 1319 (1966); R. A. Young, R. J. Gerdes, and A. J. C. Wilson, *Acta Crystallogr.*, **22**, 155 (1967); J. I. Langford, *J. Appl. Crystallogr.*, **1**, 48 (1968).

[44] A. J. C. Wilson, *Proc. Phys. Soc.* (London), **81**, 41 (1963).

[45] P. Scherrer, *Gött. Nachr*, **2**, 98 (1918).

[46] A. R. Stokes and A. J. C. Wilson, *Proc. Camb. Phil. Soc.*, **38**, 313 (1942).

[47] A. J. C. Wilson, *X-Ray Optics*, Methuen, London, 1949, pp. 37–40.

[48] Reference 47, p. 5.

[49] A. R. Stokes and A. J. C. Wilson, *Proc. Phys. Soc.* (London), **56**, 174 (1944).

[50] D. R. Buchanan, R. L. McCullough, and R. L. Miller, *Acta Crystallogr.*, **20**, 922 (1966).

[51] F. R. L. Schoening, *Acta Crystallogr.*, **18**, 975 (1965).

[52] N. C. Halder and C. N. J. Wagner, *Acta Crystallogr.*, **20**, 312 (1966).

[53] M. Trömel and H. Hinkel, *Ber. Bunsenges. Phys. Chem.*, **69**, 725 (1965).

[54] C. N. J. Wagner and E. N. Aqua, *Advances in X-Ray Analysis*, Vol. 7, Plenum Press, New York, 1964, p. 46.

[55]   Reference 1, pp. 275–312.

[56]   B. E. Warren, *Phys. Rev.*, **59**, 693 (1941).

[57]   A. J. C. Wilson, *Acta Crystallogr.*, **2**, 245 (1949).

[58]   R. Diamond, *Acta Crystallogr.*, **10**, 359 (1957).

[59]   W. Ruland, *Acta Crystallogr.*, **22**, 615 (1967).

[60]   R. Perret and W. Ruland, *J. Appl. Crystallogr.*, **1**, 257 (1968).

[61]   S. Ergun, *J. Appl. Crystallogr.*, **3**, 153 (1970).

[62]   S. Ergun, *Phys. Rev.*, **B1**, 3371 (1970).

[63]   F. Zernike and J. A. Prins, *Z. Phys.*, **41**, 184 (1927).

[64]   O. Kratky, *Phys. Z.*, **34**, 482 (1933).

[65]   J. J. Hermans, *Rec. Trav. Chim. Pays-Bas*, **63**, 5 (1944).

[66]   R. Hosemann, *Z. Phys.*, **128**, 1, 464 (1950).

[67]   G. Porod, *Acta Phys. Austriaca*, **3**, 66 (1949).

[68]   G. Porod, *Kolloid-Z.*, **125**, 51 (1952).

[69]   R. Hosemann, *Acta Crystallogr.*, **4**, 520 (1951).

[70]   R. Bonart, *Z. Kristallogr.*, **109**, 296 (1957).

[71]   R. Bonart and R. Hosemann, *Kolloid-Z.*, *Z. Polymere*, **186**, 16 (1962).

[72]   R. Hosemann and S. N. Bagchi, *Direct Analysis of Diffraction by Matter*, North Holland, Amsterdam, 1962.

[73]   R. Bonart, R. Hosemann and R. L. McCullough, *Polymer*, **4**, 199 (1963).

[74]   R. Hosemann and W. Wilke, *Faserforsch. Textiltech.*, **15**, 521 (1964).

[75]   Reference 8, pp. 424–436.

[76]   B. E. Warren and B. L. Averbach, *J. Appl. Phys.*, **23**, 497 (1952).

[77]   N. C. Halder and G. B. Mitra, *Proc. Phys. Soc.* (London), **82**, 557 (1963).

[78]   M. McKeehan and B. E. Warren, *J. Appl. Phys.*, **24**, 52 (1953).

[79]   E. N. Aqua, *Acta Crystallogr.*, **20**, 560 (1966).

[80]   M. Trömel and E. Urmann, *Ber. Bunsenges. Phys. Chem.*, **72**, 580 (1968).

[81]   D. Lewis and D. O. Northwood, *J. Phys.*, **D2**, 21 (1969).

[82]   M. De and S. P. Sen Gupta, *J. Phys.*, **D3**, 33 (1970).

[83]   G. B. Mitra and N. K. Misra, *J. Phys.*, **D1**, 495 (1968).

[84]   A. Kochendörfer, *Z. Kristallogr.*, **A105**, 393, 438 (1944).

[85]   G. Tetzner, R. Schrader and A. Lang, *Z. Chem. Deut.*, **8**, 352 (1968).

[86]   B. Ya. Pines and A. F. Sirenko, *Soviet Phys. – Crystallogr.*, **7**, 15 (1962).

[87]   B. Ya. Pines and I. Ya. Surovtsev, *Soviet Phys. – Crystallogr.*, **8**, 390 (1963).

[88]   R. V. Kovalenok and I. V. Smushkov, *Soviet Phys. – Crystallogr.*, **8**, 494 (1963).

[88a]   J. W. Harrison, *Acta Crystallogr.*, **20**, 390 (1966); **23**, 175 (1967).

[88b]   T. H. Ungar, *Acta Crystallogr.*, **23**, 174 (1967).

[89]   P. L. G. M. La Fleur and K. Koopmans, *Acta Crystallogr.*, **A24**, 311 (1968).

[90]   P. L. G. M. La Fleur, *Acta Crystallogr.*, **A25**, 643 (1969).

[91]   H. F. Quinn and P. Cherin, *Advances in X-Ray Analysis*, Vol. 5, Plenum Press, New York, 1962, p. 94.

[92]   E. Pitts and F. W. Willets, *Acta Crystallogr.*, **14**, 1302 (1961).

[93]   F. W. Willets, *Brit. J. Appl. Phys.*, **16**, 323 (1965).

[94]   D. R. Buchanan and R. L. Miller, *J. Appl. Phys.*, **37**, 4003 (1966).

[95]   G. B. Mitra, *Acta Crystallogr.*, **17**, 765 (1964).

[96]   G. B. Mitra and N. K. Misra, *Acta Crystallogr.*, **22**, 454 (1967).

[97]   F. Lihl, H. Ebel, and G. Stahl, *Acta Phys. Austriaca*, **28**, 279 (1968).

[98]   B. E. Warren and B. L. Averbach, *J. Appl. Phys.*, **23**, 1059 (1952).

[99]   B. E. Warren and E. P. Warekois, *Acta Met.*, **3**, 473 (1955).

[100]   G. K. Williamson and W. H. Hall, *Acta Met.*, **1**, 22 (1953).

[101]   S. Lele, *Acta Crystallogr.*, **A25**, 351, 551 (1969).

[102]   R. Sato, *Acta Crystallogr.*, **A25**, 309, 387 (1969).

[103]   S. Lele and P. Rama Rao, *Scripta Met.*, **4**, 799 (1970).

[104]   P. L. La Fleur, *Acta Crystallogr.*, **A26**, 431 (1970).

[104a] T. R. Anantharaman and J. W. Christian, *Acta Crystallogr.*, **9**, 479 (1956).

[105]   C. N. J. Wagner, *Acta Met.*, **5**, 427, 477 (1957).

[106]   G. B. Mitra and N. C. Halder, *Acta Crystallogr.*, **17**, 817 (1964).

[107]   M. Ahlers and L. F. Vassamillet, *Advances in X-Ray Analysis*, Vol. 10, Plenum Press, New York, 1967, p. 265.

[108]   M. De and S. P. Sen Gupta, *Acta Crystallogr.*, **A24**, 269 (1968).

[109]   M. De, *J. Appl. Crystallogr.*, **2**, 82 (1969).

[110]   R. Glocker, *Materialprüfung mit Röntgenstrahlen*, 5th ed., Springer, Berlin, 1971, Chapter 26.

[111]   W. L. Bragg, *The Crystalline State*, Vol. I, *A General Survey*, G. Bell, London, 1949, p. 189.

[112]   See, for example, A. Schuster and J. W. Nicholson, *An Introduction to the Theory of Optics*, 3rd ed., Edward Arnold, London, 1924, p. 11.

[113]   Reference 1, pp. 251–254.

[114]   Reference 1, pp. 254–257.

[115]   A. J. C. Wilson, *J. Appl. Crystallogr.*, **1**, 194 (1968).

[116]   C. R. Berry, *Phys. Rev.*, **72**, 942 (1947).

[117]   R. E. Franklin, *Acta Crystallogr.*, **3**, 109 (1950).

[118]   R. W. James, *The Crystalline State*, Vol. II, *The Optical Principles of the Diffraction of X-rays*, G. Bell, London, 1948, p. 4.

[119]   L. E. Alexander and S. R. Darin, *J. Chem. Phys.*, **23**, 594 (1955).

[120]   L. E. Alexander and E. C. Sommer, *J. Phys., Chem.*, **60**, 1646 (1956).

[121]   S. S. Pollack and L. E. Alexander, *J. Chem. Eng. Data*, **5**, 88 (1960).

[122]   R. Diamond, *Acta Crystallogr.*, **11**, 129 (1958).

C H A P T E R 10

INVESTIGATION OF PREFERRED
ORIENTATION AND TEXTURE

10-1 ORIENTATION AND TEXTURE IN MATERIALS

Previous considerations of Debye-Scherrer and back-reflection powder patterns (see Chapter 4) assumed a specimen whose crystallites were randomly oriented. Such a sample with suitable crystallite size gives a diffraction pattern whose rings are smooth and of uniform intensity around their entire circumference. Highly asymmetric crystallites, however, such as platelike or needle-shaped ones, tend to orient preferentially in a mechanically prepared sample. The resulting polycrystalline aggregates may deviate widely from complete randomness, and are said to possess *texture*. Their powder diffraction patterns in turn exhibit smooth rings with nonuniform intensities around their circumference, and this aspect of a powder diffraction ring is conclusive evidence of preferential crystallite orientation rather than a purely random one. During powder-sample preparation the possibility of such *shape textures* must be kept in mind.

Ideally random polycrystalline aggregates are rare in both nature and commerce. Crystalline rocks and minerals commonly develop *orientation texture* in crystallizing from the melt or during metamorphosis. Natural and artificial fibers show orientation texture as a result of long chainlike molecules orienting during growth or manufacture (Fig. 10-1). Mechanical operations, such as drawing, stretching, and rolling, performed on fibers and polymeric materials may increase these orientation effects. Cold-drawn wires and rolled metal sheet show strong preferred orientation and texture (Figs. 10-2 and 10-3). If the metal is heated during such forming operations, annealing and recrystallization textures may result. Cast metals develop oriented grains perpendicular to the walls of the mold, and metal films from electroplating, vacuum evaporation, and sputtering show special textures. All such texturing gives the materials

709

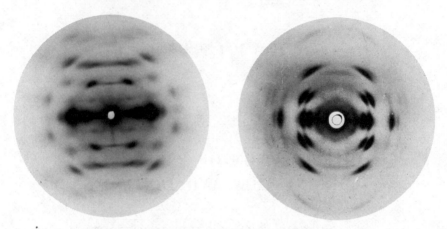

Fig. 10-1.   Fiber texture patterns. Fiber axis vertical. (*A*) Italian hemp, native cellulose. (Courtesy of W. T. Astbury.) (*B*) β-poly-L-alanine. [Courtesy of C. Bamford et al., *Nature*, **173**, 27 (1954)].

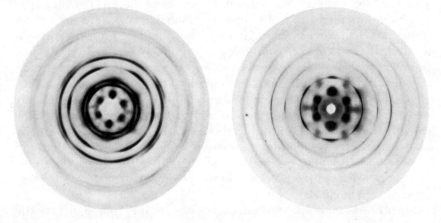

Fig. 10-2.   Fiber patterns of wires. Fiber axis vertical. (*A*) aluminum, face-centered cubic. (*B*) iron, body-centered cubic.

physical and mechanical properties vastly different from those of the corresponding random aggregate. Indeed, the properties of the oriented materials may even be objectionable. It is important therefore to be able to determine the nature of these textures, and x-ray diffraction is the principal method for their complete characterization.

In the process of drawing a wire or rolling a sheet of metal, the dimensional change in the metal is accompanied by plastic deformation of the metal through cleavage and slippage of the metal crystals along certain

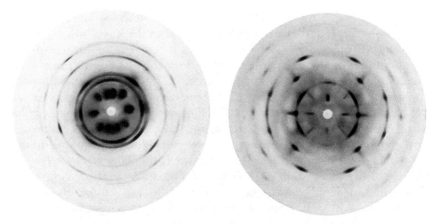

Fig. 10-3. Texture patterns of cold-rolled metal sheet. Rolling direction vertical. (*A*) copper, face-centered cubic. (*B*) iron–cobalt alloy (35 per cent cobalt), body-centered cubic.

crystallographic directions. This gliding usually occurs along those crystal planes most densely populated with atoms. Face-centered cubic metals, such as aluminum and copper, usually glide along the {111} planes; body-centered cubic metals, like iron and molybdenum, slip along {110} or {112}; and hexagonal elements, magnesium, zinc, and so on, glide very readily along {0001}. Starting with randomly oriented material, these gliding processes not only fragment the metal crystals but tend to rotate them so that they take up preferred orientations with respect to the drawing or rolling directions.

Wire drawing offers the simplest kind of preferred orientation and texture, because essentially a single direction, the wire axis, is involved. In forming an aluminum wire, slip occurs along the {111} planes, and the aluminum crystallites are left with their [111] directions parallel to the wire axis. The crystallites, however, have all possible azimuthal orientations about the [111] direction. In this respect the orientation in a wire is similar to that in fibers, such as asbestos, silk, and ramie, and this grain orientation process is often referred to as "fibering," the wire axis being known as the "fiber axis." The fibering in an aluminum wire (Fig. 10-2*A*) is practically 100 per cent parallel to [111]. Other face-centered metal wires usually have a double fiber texture with [111] of some crystals, and [100] of others parallel to the wire axis; for example, copper is 60 per cent [111] and 40 per cent [100], gold is 50-50, and silver 75-25, respectively. Body-centered cubic metals, however, usually have a simple [110] texture (Fig. 10-2*B*), whereas wire textures of hexagonal metals are less simply interpreted.

The situation in rolled sheet is more complicated. In the rolling process the sheet is lengthened at the expense of its thickness, but the width is practically unchanged. Two types of texturing arise in this treatment, an alignment of some crystallographic direction parallel to the rolling direction, and an orientation of a set of crystal planes parallel to the surface of the sheet. The particular planes and directions thus oriented are characteristic of the metal, and the degree of orientation is a function of the number of passes, diameter of the rolls, and the per cent reduction. The principal orientation texture for cold-rolled face-centered cubic metals is one (Fig. 10-3A) in which the (110) plane is parallel to the surface of the sheet and the [$\bar{1}$12] direction is parallel to the rolling direction, which description is usually written (110)[$\bar{1}$12]. Among the metals characterized by this principal texture are copper, aluminum, gold, silver, nickel, and platinum, as well as several face-centered alloys. A number of these have superposed on the foregoing orientation a second one designated (112)[11$\bar{1}$], and in rare instances even a trace of a third texture. Iron and steel have had by far the most study among body-centered cubic metals, and their cold-rolled texture is one (Fig. 10-3B) that is ideally represented by the symbol (001)[110]. Actually, there is a fair amount of deviation from the ideal alignment in each of its aspects. Molybdenum, likewise, has had some study, and its principal orientations are similar to those of iron. The predominating texture of rolled sheet in hexagonal close-packed metals is the anticipated one with the basal plane parallel to the rolling plane, provided the metal has approximately the ideal axial ratio for close packing of spheres ($c/a = 1.633$). The metals zinc and cadmium, which differ widely from this ratio (1.856 and 1.885, respectively), show almost no orientation of (0001) in the plane of the sheet.

It should be noted too that the preferred orientations in rolled sheet may differ considerably between the surface and the interior, the inhomogeneity of texture being to some extent a function of the degree of reduction. This effect is much less after a drastic reduction (such as 99 per cent) than for intermediate degrees. Hot rolling is not necessarily free from orienting effects, and in some instances these are similar to the effects of cold rolling. Cross-rolling, in which the material is rolled in two directions 90° apart, introduces complicated and unexpected effects in the sheet texture. The drawing operations in the production of tubes and cups from sheet material cause a flow of metal leading to strong orientation effects. Frequently, these are such as to require an anneal between some stages of the draw to relieve the orientation and permit further deformation. For further data on the numerous studies in this field, the reader is referred to the excellent summary in Barrett and Massalski[1].

Mineral silicate fibers and natural and synthetic organic fibers are similarly made up of bundles of crystals, with one axis parallel to the fiber axis and with random orientation about this axis. Certain synthetic polymers, such as polyethylene, polyesters, and polyamides, on extrusion or drawing through a die become oriented. Likewise, rubber, although amorphous in the unstretched state, becomes crystalline on stretching, the crystals all being aligned with one axis parallel to the stretching direction. Since organic fiber and polymer patterns have been treated in great detail in Alexander's recent book, *X-ray Diffraction Methods in Polymer Science* [2], the treatment of preferred orientation and texture in this chapter is devoted chiefly to fibered metals, with only brief treatment of mineral and organic fibers.

10-2 GEOMETRY OF FIBER PATTERNS

10-2.1 Ideal Fiber Patterns

The fiber texture in a wire involves essentially a single direction, the wire axis, and is therefore the simplest example of texturing. As a result of plastic slippage and rotation, the crystallites ideally end up after the drawing operation with a single crystallographic direction parallel to the wire axis. The crystallites, however, have all possible azimuthal orientations about the wire axis. By specifying the crystallographic direction [uvw] parallel to the wire axis, the wire texture can be specified completely.

It is useful to consider an ideal case of wire texture for a face-centered cubic metal wire with the [111] axis as the fiber direction. On reviewing reciprocal lattice concepts (Sections 1-3.7 and 3-2.8) and the opening paragraphs of Chapter 4, it is seen that in a random powder aggregate the reciprocal-lattice vectors of the various crystal planes lie on reciprocal-lattice spheres of radius ρ_{hkl}. The ideal fiber is intermediate, however, between a single crystal and the random aggregate. If its [111] fiber direction is made the vertical axis of the reciprocal lattice sphere, the locus of the 111 reciprocal-lattice points is at the north and south poles of the sphere and along the $\pm 19°28'$ latitude circles (Fig. 10-4A). The intersection of the Ewald sphere (sphere of reflection) with these circles during diffraction gives rise to four diffraction spots on what would have been the 111 diffraction ring for the unoriented metal. From a table listing the angular inclination of plane normals to several zone axes, it is possible to construct ideal fiber diagrams for various fiber textures. Table 10-1 lists inclination angles* for plane normals of cubic crystals.

---

*These angles ρ are not to be confused with ρ_{hkl}, the reciprocal lattice spacings.

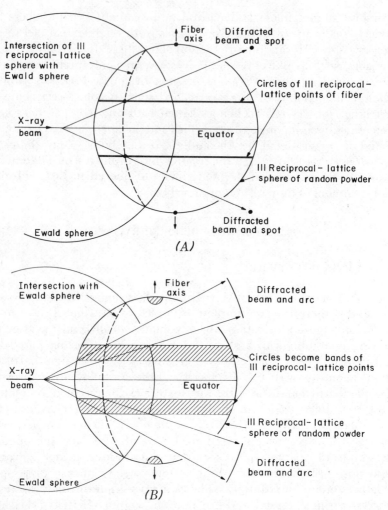

Fig. 10-4. Reciprocal-lattice construction of diffraction by a wire or fiber (schematic). (*A*) Ideal fiber. (*B*) Real fiber.

Figure 10-5 illustrates ideal fiber diagrams prepared from data in Table 10-1.

These diagrams are easily prepared by bearing in mind the index triplets possible on powder patterns of the lattice type under consideration (Fig. 6-11), and drawing a series of concentric circles with radii proportional to $r = D \tan 2\theta$ in Fig. 6-9A for the rings of a transmission monochromatic-pinhole powder pattern of the cubic lattice type involved. The

Table 10-1. Inclination Angles ρ between Plane Normals in Cubic Crystals[3]

Lattice Plane	Zone or Fiber Axis Direction[a]			
	[100]	[110]	[111]	[112]
(100)	0°, 90°	45°, 90°	54° 44′	35° 16′, 65° 54′
(110)	45°, 90°	0°, 60°, 90°	35° 16′, 90°	30° 1′, 54° 44′, 73° 13′, 90°
(111)	54° 44′	35° 16′, 90°	0°, 70° 32′	19° 28′, 61° 52′, 90°
(210)	26° 34′, 63° 26′, 90°	18° 26′, 50° 46′, 71° 34′	39° 14′, 75° 2′	24° 6′, 43° 5′, 56° 47′, 79° 29′, 90°
(211)	35° 16′, 65° 54′	30° 1′, 54° 44′, 73° 13′, 90°	19° 28′, 61° 52′, 90°	0°, 33° 33′, 48° 11′, 60°, 70° 32′, 80° 24′
(221)	48° 11′, 70° 32′	19° 28′, 45° 76° 22′, 90°	15° 48′, 54° 44′, 78° 54′	17° 43′, 35° 16′, 47° 7′, 65° 54′, 74° 12′, 82° 12′
(310)	18° 27′, 71° 34′, 90°	26° 34′, 47° 52′, 63° 26′, 77° 5′	43° 6′, 68° 35′	25° 21′, 39° 14′, 49° 48′, 58° 55′, 75° 2 ′, 82° 35 ′
(311)	25° 14′, 70° 27′	31° 29′, 64° 47′, 90°	29° 30′, 58° 30′, 79° 58′	10°, 42° 24′, 60° 30′, 75° 45′, 90°

[a]It is recalled that the direction [hkl] is the normal to the plane (hkl), so that these values are the usual normal angles between the corresponding pairs of planes.

trace of the fiber axis is next drawn in vertically on the diagram. Then, for the given fiber-axis direction, the angular positions ρ of the intensity maxima for each plane are plotted on the corresponding powder diffraction ring, using the data of Table 10-1. These spots, in general, occur as groups of four, symmetrically related, except when $\rho = 90°$ there are only two spots at opposite ends of a horizontal diameter of the ring. The resulting fiber diagram is indeed analogous to the rotation photograph of a single crystal in which the crystal is completely rotated about an axis perpendicular to the incident beam. Although the fiber was not rotated, the same effect is obtained because the individual crystals have all possible orientations about the fiber axis. In Fig. 10-5 the layer lines have been sketched in to show the relationship to the layer lines of single-crystal photographs (Fig. 10-6). The vertical row lines are also evident

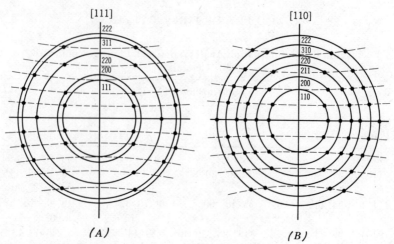

Fig. 10-5.　Ideal fiber diagrams. (*A*) Face-centered cubic metal, [111] fiber axis. (*B*) Body-centered cubic metal, [110] fiber axis.

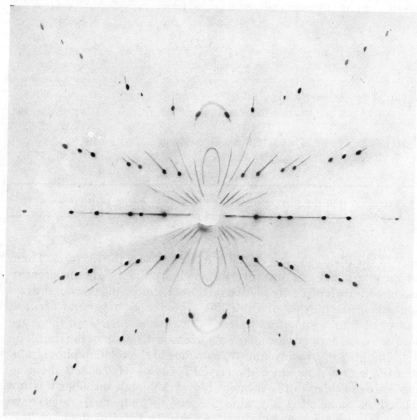

Fig. 10-6.　Flat-film rotation photograph of a single crystal of urea about the *c* axis. Note the layer lines and vertical row lines.

in Fig. 10-5A. Ideally, no spot should occur at positions corresponding to $\rho = 0$ and 180°, since here the plane normal is parallel to the fiber axis, and the plane itself is parallel to the x-ray beam.

10-2.2 Bragg Geometry of Fiber Patterns

The Bragg reflection geometry used by Glocker[4] is convenient for discussing the origin of a diffraction ring on a wire pattern (Fig. 10-7). The beam is perpendicularly incident on the vertical wire at O. Diffraction occurs from the plane (hkl) (exaggerated in the diagram), preferred orientation producing an intensity maximum on the Debye-Scherrer ring at S. The wire axis on the film is the direction UW, and U is the center of the diffraction ring and the position of the undeviated beam on the film. ON is the ideal normal to the set of planes with preferred orientation, and ρ is the angle between their normal and the wire axis. The film angle δ between UW and US is then related to ρ by the expression

$$\cos \delta = \frac{\cos \rho}{\cos \theta}, \tag{10-1}$$

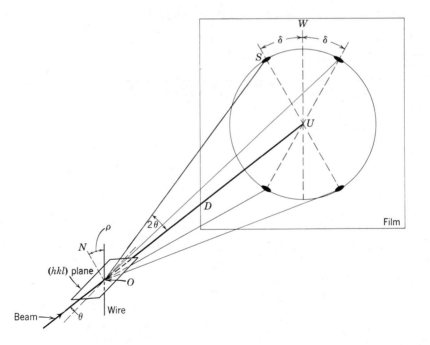

Fig. 10-7.   Geometry of the intensity maxima on rings of a fiber pattern (rather idealized).

as demonstrated by Polanyi[5] for single-crystal rotation patterns. Indeed, spot S occurs four times on the powder ring, as depicted in Fig. 10-7, since the plane (hkl) has no azimuthal preferred orientation about the wire axis. The four spots are strictly analogous to the four reflections of the plane (hkl) on a single-crystal rotation photograph, and the fiber pattern has other aspects of the rotation pattern.

If angle θ were zero, equation 10-1 indicates that angles δ and ρ would be equal. Actually, for the first few diffraction rings on a fiber pattern, θ is small enough that $\cos \theta$ is very nearly unity, and the expression

$$\angle \delta = \angle \rho \qquad (10\text{-}2)$$

is a useful approximation of equation 10-1, since it permits easy preparation of ideal fiber diagrams for various fiber-axis directions.

When the fiber axis is inclined to the x-ray beam at an angle β less than 90°, the orientation intensity maxima still occur in groups of four, but are no longer so symmetrically arranged. As depicted schematically in Fig. 10-8, they are crowded together in that area of the diagram toward which the axis leans, the spread apart in opposite areas. Under this circumstance two values for the angle δ arise as expressed by the following more complicated versions of relation 10-1:

$$\cos \delta = \frac{\cos \rho - \cos \beta \sin \theta}{\sin \beta \cos \theta},$$

and $\qquad\qquad\qquad\qquad\qquad\qquad\qquad\qquad\qquad\qquad (10\text{-}3)$

$$\cos \delta' = \frac{\cos \rho - \cos (180 - \beta) \sin \theta}{\sin (180 - \beta) \cos \theta}.$$

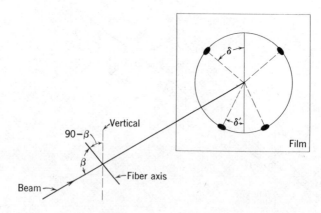

Fig. 10-8.   Effect of tilting fiber axis on orientation of intensity maxima.

10-2.3 Real Fiber Patterns

Real fibering is never perfect. The inclination of the planes to the fiber direction always deviates slightly from the ideal angle ρ. This results in the reciprocal-lattice-point circles of Fig. 10-4A widening into bands (Fig. 10-4B). The intersections of these bands with the Ewald sphere are arc segments, and the diffraction patterns of real fibers show arcs along the powder rings (Figs. 10-1 to 10-3) instead of the sharp spots of the ideal pattern. The arcs are generally most intense at their centers, indicative of a tendency for a symmetrical distribution of inclinations about the ideal angle ρ. The lengths of the arcs vary from a few degrees to practically continuous rings, and are thus a measure of the amount of preferential orientation or fibering, and in turn of the cold work done on the specimen. The crystal grains are misaligned to such an extent that diffracted beams from a plane come forth in a cone around the ideal position. The semiapex angle of this cone is half the angular arc length of the intensity maximum, and gives a measure of the deviation of the plane normals from the ideal. It is entirely possible for the plane normals to deviate sufficiently for a forbidden spot to occur at $\delta = 0$. For example, the pattern of aluminum wire with MoKα radiation (Fig. 10-2A) shows a strong forbidden 111 spot at $\delta = 0$. Under the conditions $\theta_{(111)} = 8°45'$, and it is necessary only for these planes to deviate from the ideally perfect alignment by this amount to give a reflection. The presence of such forbidden spots and their relative intensity can thus give evidence as to the perfection of the fibering.

Real fiber diffraction patterns usually differ from the ideal in still another respect. Inside the smallest diffraction ring may be found radial streaks or bands extending from the center toward each spot (Figs. 10-2 and 10-3). These originate from diffraction of the white radiation that is always present except in crystal-monochromatized beams. They have their inner edge determined by the low-wavelength limit of the beam, and tend to terminate at the wavelength 0.485 Å, the absorption edge of the silver in the film emulsion. The same intensity maxima are represented in these bands or streaks as appear on the diffraction rings.

10-3 PREPARATION OF FIBER PATTERNS

Fiber diffraction patterns are very simple to prepare. Two considerations enter into the choice of radiation to be used; it must not excite fluorescence in the specimen, and it must have sufficient penetrating power for transmission patterns in the study of metal sheet. Accordingly, MoKα

radiation is the choice for most studies, but Cu$K\alpha$ can be used for investigations of aluminum and magnesium. The pattern will be simplified if the beam is well filtered to remove $K\beta$ radiation. Fine wires and thin sheet or foil (a thickness of 0.005 to 0.008 in. is satisfactory for steel) make the best specimens. Orientation in the surface of a thick specimen may be studied, however, by reflection of the x-ray beam from its surface, in which case only the part of the pattern is obtained that is not obstructed by the specimen. When sheet patterns are taken for pole-figure work (Section 10-6.1), a series of photographs is prepared with the beam incident on the specimen at different angles. This necessitates an absorption correction[6,7] if any quantitative consideration of intensities is to be made. Bakarian[8] has suggested a convenient means for avoiding this correction. A T-shaped piece is cut from the sheet with a jeweler's saw, as in Fig. 10-9. The stem is then ground to cylindrical shape and lightly etched to remove surface cold work due to the grinding. Absorption is constant for all orientations about the stem of such specimens. Obviously, the transverse direction may be made the stem axis when it is so desired.

A very suitable holder for wire and small sheet specimens is a goniometer head with cross slides and arcs (Fig. 10-10). The slides permit translations in two directions at right angles to each other and normal to the holder axis, and the arcs allow limited angular adjustment about lateral axes parallel to the cross slides. Angular positions about the axis of the goniometer head are determined from a horizontal circle mounted in conjunction with it. The wire or sheet specimen is mounted with a low-melting wax (such as picein) or modeling clay on a brass pin which fits the hole in the holder on the upper arc. Adjustment of the specimen

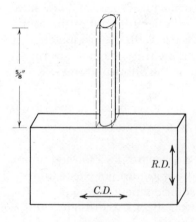

$\frac{5}{8}''$

R.D.

C.D.

Fig. 10-9. Constant-absorption specimen of sheet material for preferred-orientation studies. [Courtesy of P. Bakarian, *Trans. AIME*, **147**, 266 (1942).]

Fig. 10-10. Goniometer head for mounting and orienting specimen. (Courtesy of O. von der Heyde.)

may be carried out rather simply with the aid of a low-power microscope or telemicroscope. The fiber axis, or rolling direction of a sheet specimen, is preliminarily adjusted in the wax or clay to be approximately parallel to the goniometer axis, which by construction is accurately perpendicular to the beam-collimating system. With sheet specimens it is well to orient them so that the plane of the sheet is parallel to one of the holder arcs.

Final adjustment proceeds, after the microscope has been placed with its optical axis normal to the goniometer axis. The specimen holder is rotated to a position with one arc parallel to the microscope axis, and it is observed whether the fiber axis (rolling direction, or plane of the sheet) is accurately parallel to the vertical cross hair in the microscope. This observation is repeated with the specimen holder rotated 180° from the first position. There rarely is parallelism at this stage, but adjustment of the appropriate arc by means of its screw, and of the cross hair too (unless it is known in advance that the cross hair is accurately adjusted to the vertical axis of the holder), will improve the alignment between the fiber axis and the vertical cross hair. This process of adjustment with observations 180° apart must be repeated until there is perfect parallelism between the fiber axis and the microscope cross hair for the two positions 180° apart. The crystal holder is next rotated 90° to bring the other arc parallel to the microscope axis, and a similar adjustment procedure carried out for this position. At this point it may be discovered that the fiber axis, although parallel to the holder axis, is still not on the holder axis. This position is achieved by appropriate movement of the cross slides by means of their adjusting screws. When sheet patterns are desired at other than normal incidence, various angular rotations about the rolling direction can be obtained by proper settings on the horizontal circle. If similar angular settings about the transverse axis are desired, they are best obtained by reorienting the specimen with the transverse axis on the goniometer vertical axis.

Pinholes of 0.5 to 1.0 mm diameter (rarely larger) are suitable for fiber patterns. The flat film, either rectangular or circular, is placed at a convenient distance, 4 to 6 cm, from the specimen, and normal to the beam. This will be recognized as essentially an aspect of the monochromatic-pinhole technique discussed in Section 4-3, to which description the reader is directed for calibration and other details. So many factors, such as pinhole size, nature and thickness of the specimen, choice of radiation, and so on, enter into the exposure time that the investigator is urged to make a preliminary series of trial exposures as a basis for estimating the optimum exposure time for his studies.

Eventually, in preparing preferred-orientation patterns, the investigator encounters material so coarse-grained as to make it impossible to detect the intensity maxima on the diffraction rings because of their spotty character. This troublesome problem, however, is readily solved by some variety of scanning or integrating camera[9–16] which permits the x-ray beam to scan a large portion of the surface of the specimen during a single exposure. For instance, Smith and Hinde's[13] device has an upper scanning limit of 10 mm, and Underwood's[12] equipment permits scans of 1 in.² or 1 cm² of the sample by the use of separate cams. In the resulting patterns line spottiness is either eliminated or greatly reduced. Hay[10] has described a simple integrating camera which permits the production of both transmission and back-reflection patterns.

The powder diffractometer is very important in studies of orientation and texture because of its high accuracy in intensity measurements. Normally, the diffractometer detector measures intensities only along the equator of photographs, such as Figs. 10-1 to 10-3. From Fig. 10-4A, however, it is evident that, if the detector is held stationary at the 2θ angular position for a diffraction ring and the sample is rotated slowly about the incident x-ray beam direction, a scan around the powder diffraction ring is possible. The result is a point-by-point measure of the intensity distribution along the diffraction ring, and it can be carried out on oriented samples to measure rings like those in Figs. 10-2 and 10-3. The amount of rotation about the x-ray beam provides the angle δ along the ring. When a transmission arrangement is used, an absorption correction for the diffracted beam is required. The diffractometer (usually automated) is thus the basic instrument for detailed studies of orientation and texture through the preparation of pole figures (Section 10-6.2).

10-4 ANALYSIS OF SIMPLE FIBER PATTERNS

The analysis of an experimental fiber pattern proceeds in several steps. The lattice type and the indices of the powder rings appearing on the pattern are first identified. If the composition of the alloy is known, its lattice type will usually also be known, and the ring indices can then be identified by mere inspection and comparison with the line sequences of Fig. 6-11. When nothing is known about the lattice type, it is possible to identify the various simple lattices from the ratios of the $\sin \theta$ values, passing from the innermost ring outward. Since $\sin \theta \propto \sqrt{h^2 + k^2 + l^2}$ for the cubic lattices, the ratio sequences for the face-centered and body-centered cubic lattices are:

Face-centered cubic: $1.73 : 2.00 : 2.83 : 3.32 : 3.46 : 4.00$,
Body-centered cubic: $1.41 : 2.00 : 2.45 : 2.83 : 3.16 : 3.46$.

A similar application to the identification of the hexagonal close-packed lattice is possible, but is more complicated.

The lattice type and the diffraction rings having been identified, the next step is to measure the angles δ from the 12-o'clock position on the rings to the midpoints of the various intensity maxima on each ring. Comparison of these δ values with a table of angles ρ between various plane normals (Table 10-1) should yield a single zone axis consistent with all the data. This zone axis is the fiber axis. If only a part of the data is consistent with one crystallographic direction, it may indicate a double fiber texture, in which instance the remaining data will be found consistent with a second zone axis, and both these directions are parallel to the fiber axis. A hint as to the fiber axis may be had from the rings that have intensity maxima at the 12-o'clock position. Such maxima arise from planes that ideally would be normal to the fiber axis and thus not appear. Their presence indicates that alignment of the plane normals with the fiber axis is not perfect, but deviates by as much as angle θ, or that the fiber axis is inclined to the x-ray beam by an angle $90 - \theta$. Where no such spot appears on any ring, one can frequently be made to appear (and thus identify the fiber axis) by tipping the fiber by various angles, $\beta = 90 - \theta$, in the vertical plane containing the fiber and the beam.

A third step might be to determine the angular lengths of the intensity maxima as a measure of the extent of deviation from perfect fibering.

In analyzing the pattern of rolled sheet, the procedure is analogous to that for a wire pattern. The lattice type and rings are identified, and the direction parallel to the rolling direction, which corresponds to the fiber axis of the wire, is determined from measurement of the δ angles and reference to Table 10-1. This identification may account for all the maxima observed, but the ideal fiber pattern may call for maxima not present on the pattern for the sheet material. This is evidence that the orientation in sheet material is limited as compared to wire. Inclination of the fiber axis to the plane of the sheet can be detected by the appearance of spots at the 6- and 12-o'clock positions on the rings, as in wire patterns.

Identification of the face or plane parallel to the sheet surface is less simple. The investigator usually assumes the identity of this plane and then calculates the amount of angular rotation about either the rolling direction or the transverse direction required to bring specific planes into reflecting position. The presence or absence of maxima for these planes on the film is an indication of the correctness of the identity of the plane assumed parallel to the sheet. These calculations are complicated for most instances, but Glocker[4] has presented a series of curves which provide a complete solution for the cubic lattice rotating about the [110] direction. The most complete analysis of texture patterns is obtained by the preparation of pole figures (Section 10-6).

A final brief consideration is due mineral- silicate and organic fibers at this point. These materials derive their fibrous properties as a result of being formed from long chainlike molecules which lie either closely parallel to the fiber axis or along spirals about the fiber axis. Long, thin bundles of these chain molecules make up the fiber crystallites, which in turn have all possible orientations about their long axis which is parallel to the fiber axis. The perfection of fibering in these materials can vary all the way from the slight orientation leading to a pair of broad intensity maxima on a diffraction halo, to the high degree of ordering that produces a pattern (Fig. 10-1) closely akin to a single-crystal rotation pattern. Moreover, oriented organic fibers and textured polymers are usually only partially crystalline, the remainder being amorphous, a situation that further complicates their study. Alexander has treated the analysis of these patterns, in great detail in a separate book [2], so it seems appropriate to refer the serious worker in polymers directly to this publication. Certain characteristic information, however, which can be obtained in very simple fashion, is detailed in the following paragraphs.

The distance at which the molecular pattern repeats in the direction of the fiber axis is obtained from a measurement of the height of the first layer line at its apex directly above the undeviated beam. If no diffraction maximum occurs at this point, we must estimate where the layer line lies by sketching it in from the positions of other maxima on it. Then, if T is the repeat distance along the fiber axis,

$$T = \frac{\lambda}{\sin \mu},\tag{10-4}$$

where μ, the angle subtended by the layer line, is

$$\mu = \tan^{-1} \frac{S}{D},\tag{10-5}$$

S being the layer-line height, and D the specimen-to-film distance. The fiber-axis repeat distance for hemp (native cellulose) (Fig. 10-1A) is about 10.3 Å, corresponding to the length of two glucose residues in the chain.

Having determined T, the vertical dimension of the unit cell, it may be possible to determine the other cell dimensions from a consideration of the equatorial reflections. We first consider the simplest possibility, a rectangular cell projection, and attempt to match the values of $\log d$ for the spots with points on a $\log d$ chart such as Fig. 6-12. Failing in this simple technique, we may resort to more powerful methods based on the reciprocal lattice [2, 17]. It is to be noted that these equatorial spacings are related to the distances between chains in the fiber.

10-5 REPRESENTATION OF PREFERRED ORIENTATION

10-5.1 Pole Figures

It is seen that the nature and degree of preferred orientation of the crystallites in textured materials is difficult to describe and represent clearly. To provide a convenient means for presenting such results in the study of metals, Wever [18] made an adaptation of the single-crystal stereographic projection referred to as a *pole figure*. The pole figure is a map of the statistical distribution of the normals to given {*hkl*} planes of a polycrystalline sample. It depicts the direction of the preferred orientation and the extent of angular deviation from the ideal in a form easily comprehended, and thereby provides a very complete picture of the texture of a metal or polymer. For instance, Fig. 10-16 presents a simple pole-figure diagram of rolled silicon–iron sheet. The double-hatched areas represent areas with a high concentration of the (100) plane normals or *poles*. The single-hatched areas have a smaller concentration of (100) poles. As a basis for understanding the plotting of pole figures, an introduction to stereographic projection is in order.

10-5.2 The Stereographic Projection

A stereographic projection, widely used for description and analysis in geometrical crystallography, is derived from a *spherical projection*. To prepare a spherical projection, the crystal is assumed to be placed at the center of a hollow, transparent sphere. With the centers of the crystal and sphere coincident, normals are drawn from the common center to

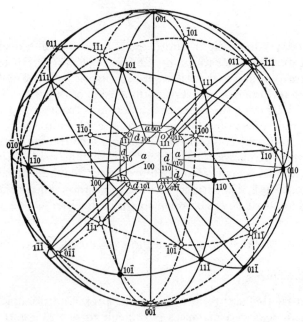

Fig. 10-11.   The spherical projection of a crystal. (From E. Dana, *A Textbook of Mineralogy*, 1932; courtesy of Wiley.)

each face and extended until they intersect the surface of the sphere (Fig. 10-11). The intersections of the face normals with the reference sphere are known as the *poles* of the faces, and they constitute the spherical projection of the crystal. One of the most useful features of the spherical projection is the ease with which angular measurements can be made on it by a system of angular coordinates analogous to the lines of latitude and longitude used on the globe of the earth. The angle between two planes is the angle between their face normals, that is, between their poles, and is measured on the spherical projection as the number of degrees on a great circle passing through the poles.

In spite of the ease of depicting such angular relationships on a sphere, it is much more convenient in practice to work on flat sheets of paper. This has led to various schemes for mapping the spherical projection without destroying the useful angular relations between the poles. Most important of these is the stereographic projection, derived from the spherical projection by considering the equatorial plane of the sphere (or a plane tangent to the sphere at the north pole) as a plane of projection. Lines are then imagined as drawn through the poles of the upper hemisphere to the south pole of the sphere as in Fig. 10-12. The intersections of these lines with the plane of projection (equatorial or tangent) locate the poles on the stereographic projection. It is evident that the only difference in the projection when the tangent plane is used rather than the equatorial plane is one of size, distances on the former being twice those on the latter, since the tangent projection plane is twice the distance of the equatorial plane from the south pole, the center of projection. This projection scheme places all the poles of the northern hemisphere within a circle (the *primitive* or *basic* circle) defined by the projections of the poles lying on the equatorial plane of the spherical projection. If the poles of the southern hemisphere are similarly projected their projections lie outside the primitive circle and, indeed, extend the diagram to infinity. The inconvenience of such an unduly large diagram can be overcome by an alternative method of projection of the southern poles. Their projection is accomplished by shifting the center of projection to the north pole (and the tangent projection plane, if used, to the south pole) and projecting in the same manner as before. This scheme restricts the entire projection to the area within the primitive circle. Projections of poles of the lower half in this instance may be distinguished from those of the upper half by using open rings to represent the former and dots for the latter.

Some of the properties of the stereographic projection may now be observed. Reference to Figs. 10-11 and 10-12 reveals that the poles of planes parallel to the north-south axis of the reference sphere project

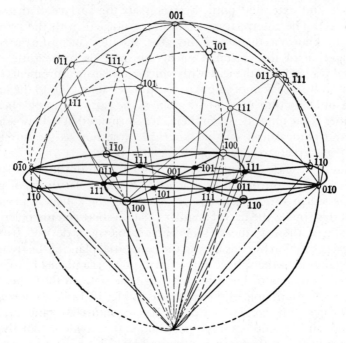

Fig. 10-12.   Derivation of the stereographic projection from the spherical projection. (From E. Dana, *A Textbook of Mineralogy*, 1932; courtesy of Wiley.)

onto the primitive circle. In fact, the poles of the faces in a zone always lie on a great circle of the reference sphere. If the plane of this great circle contains the north-south axis, the zone projects on the stereographic projection as a straight line passing through the center of the primitive circle. Zones represented by other great circles project as symmetrical pairs of curved arcs about a diameter of the primitive. Another property readily demonstrated is that any small circle on the spherical projection is projected as a circle on the stereographic projection. This and many other useful constructions and measurements on the stereographic projection can be carried out in simple fashion with compass, ruler, and protractor[19].

The entire angular net of latitude and longitude, so convenient on the sphereical projection, projects stereographically in the same fashion as just described for great and small circles. The great Russian crystallographer, Wulff, prepared such stereographic nets for use in measuring angles on stereographic projections. The meridional type (Fig. 10-13) was prepared with the north-south axis of the ruled globe parallel to the

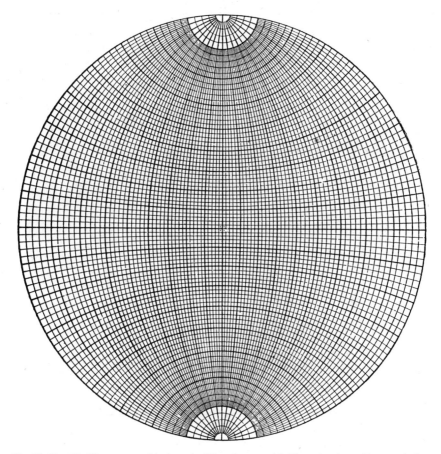

Fig. 10-13.   Wulff stereographic net, meridional type, with 2° graduations. (By permission from *Structure of Metals*, by C. Barrett, 1st ed. Copyright 1943, McGraw-Hill Book Company.)

plane of projection, and is especially useful for angular measurements on the curved zonal lines of a stereographic projection, and for angular rotations about an axis lying in the plane of the basic circle. The polar or equatorial net (Fig. 10-14) is convenient for angular measurements on the straight zonal lines, and for rotations about an axis perpendicular to the basic circle. The nets are usually used in connection with a stereographic projection or diagram prepared to the same scale on tracing paper. When the diagram is placed over the net, the transparency of the paper permits orientation and alignment, and observation of the direction which poles will move on rotation about an axis.

The angle between two poles is equal to their difference in latitude

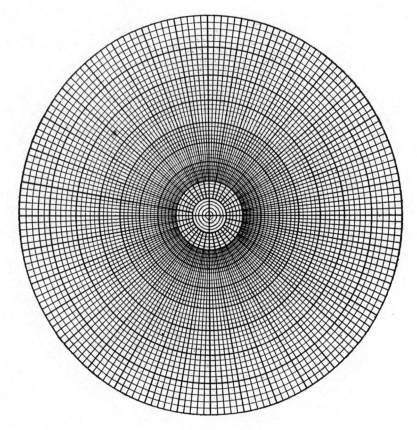

Fig. 10-14. Stereographic net, polar type, with 2° graduations. (By permission from *Structure of Metals*, by C. Barrett, 1st ed. Copyright 1943, McGraw-Hill Book Company.)

when they lie on the same meridian. To make such a measurement their zone line is brought onto a meridian, and the angular difference in latitude read off. If the zone line is diametral, it may be aligned with the diametral meridian of Fig. 10-13, or with any diametral meridian of the polar net (Fig. 10-14). Curved zonal lines must be aligned with the meridian of the same curvature on the net in Fig. 10-13 before the difference in latitude is measured. Similarly, for straight zone lines which can be brought into coincidence with the equator of Fig. 10-13, angular distances between poles are the differences in their longitudes. To rotate a pole about an axis represented by a diameter of the basic circle, the vertical meridian of Fig. 10-13 is aligned with the axial direction. The pole is then moved the required number of degrees along the proper curve of constant latitude. Rotations about an axial direction normal to the basic circle are carried out with the polar net, the point again being moved the proper amount along a circle of constant latitude. From angular data such as those of Table 10-1, and a consideration of the relationship between faces in a zone, it is a simple matter to prepare a standard projection for a cubic crystal. In an excellent chapter on the stereographic projection, Barrett and Massalski[20] present additional problems readily handled with the Wulff net. A simple stereographic plotting table has been described by Dyson[21], and the computation of stereographic projections for noncubic crystals has been treated by Palm[22].

10-5.3 Inverse Pole Figures

The pole figures we have been discussing are direct pole figures, in that the x-ray data are plotted directly on the projections. An inverse pole figure is obtained by mathematically "inverting" the x-ray data for a normal pole figure. This may give a density distribution of the axial direction in a polycrystalline sample (such as a wire) on a stereographic projection of the crystal lattice in some chosen standard orientation. Inverse pole figures offer some advantage over direct pole figures for fiber textures, and are frequently applied to sheet textures also. Precise intensity data from a considerable number of reflections are required, so that the method may be laborious if automatic methods are not available. It is our purpose (to conserve space) to treat the preparation of direct pole figures, and to refer the reader to references [23–34] on inverse pole figures.

10-6 PREPARATION OF POLE FIGURES

10-6.1 Photographic Methods

Pole figures were first prepared from orientation photographs. A series of patterns is required to determine the extent of the areas delimiting the poles. This series might be prepared by first taking a pattern with the beam normal to the surface of a sheet specimen (rolling direction vertical). Then photographs are taken with the specimen rotated about the rolling direction in successive steps of 5 or 10°. Small areas near the top and bottom of the pole figure are not covered by this series of patterns, but are readily evaluated from a short series of photographs prepared by rotation about the cross direction. The gathering and plotting of the data from such a series of patterns for a (100) pole figure of a rolled alloy was treated in detail in the first edition of this text[35].

The very considerable labor involved in the preparation of pole figures has been the most unattractive aspect of preferred-orientation studies. Attention therefore was given to devices that would reduce the number of photographs required and the labor of plotting. As early as 1930, Kratky[36] described a texture goniometer with which a single exposure on an oscillating film gives the information ordinarily obtained from a series of exposures. In these devices the beam traverses a thin flat sample P inclined at a given angle. A segment of the desired diffraction ring is allowed to pass through a ring slot S in a stationary shield A placed in front of the film F (Fig. 10-15). The film and shield may be provided either as concentric cylinders parallel to the beam[36] or to the rotation axis D, or as a flat, parallel arrangement normal to the beam direction[37, 38a,b]. In each instance the film is coupled to the sample-rotation axis by gears, so that a rotational oscillation of the sample causes the film to describe a translational motion. With this arrangement each angular position α between the sheet normal and the incident beam corresponds to a unique position on the film. Then, by slowly oscillating through an angular range (0 to 45°, 0 to 60°), the section of the Debye ring is spread out into an elongated shape. Both angles α and β (positions of the intensity maxima on the Debye rings) are readily measured on the film, or a chart may be prepared to convert positions on the film directly to positions on the pole figure. These continuous pole-figure cameras have not been widely used because of the efficiency of the counter diffractometer for such studies. It is possible, however, to combine a scanning mechanism with the sample mount and thus study coarse-grained as well as fine-grained materials with them.

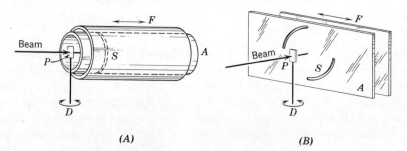

(A)   *(B)*

Fig. 10-15. Principle of the continuous pole-figure camera or texture goniometer. (*A*) Cylindrical film and shield. (*B*) Flat film and shield. (By permission from *Structure of Metals*, by C. Barrett, 1st ed. Copyright 1943, McGraw-Hill Book Company.)

10-6.2 Diffractometric Techniques

A. Introduction. Pole-figure intensities are conveniently and accurately obtained with a counter diffractometer. The counter is fixed at the angle 2θ corresponding to the chosen planes [for example, (200)], and the specimen is rotated about two perpendicular axes so as to permit the measurement of the intensities diffracted by these selected planes oriented at various angles within the specimen. Established procedures for preparing quantitative pole figures of metals[39] and polymers[40] are well documented. Both transmission and reflection techniques are important for metal studies, whereas the transmission technique is the more important for polymers. The complete pole figure, however, requires the use of both techniques. The portion of the latitude range from 0 to 60 or 70° in α is accessible by the transmission technique, and the zone from $\alpha = 30$ to 90° yields readily to the reflection technique. The angular range in which the two techniques overlap, $\alpha = 30$ to 60°, provides data for placing the intensities on a common basis when the specimen has been studied by both techniques. The data are usually presented as a polar stereographic projection (Figs. 10-14 and 10-16) with the machine or rolling (R) and transverse (T) directions on the

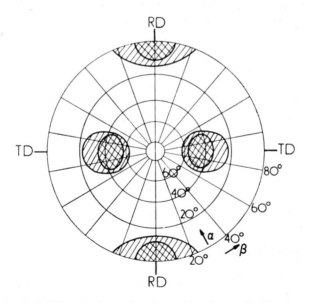

Fig. 10-16. Typical (100) pole figure for silicon–iron sheet. *TD*, Transverse direction; *RD*, rolling direction. [Courtesy of C. Milner and J. James, *J. Sci. Instrum.*, **30**, 77 (1953). Copyright, The Institute of Physics.]

equator and the normal direction (N) at $\alpha = 90°$. Some investigators use the angle, $90 - \alpha = \phi$, measured from the center of the polar chart and referred to as the colatitude.

The intensity required at any particular polar coordinate position α, β (Fig. 10-16) is the integrated intensity of the reflection (hkl) above background I_i. In referring to Fig. 10-17, this quantity can be measured for a fixed α and β by scanning the reflection over the 2θ interval between two background positions $2\theta_{b1}$ and $2\theta_{b2}$, and then subtracting from the integrated counts the background counts (hatched area). Often it is assumed that I_i is proportional to the overall peak intensity I_p minus the background intensity at $2\theta_p$ (calculated as the average of I_{b1} and I_{b2}). A more reliable approximate measure of I_i can be had by multiplying the peak height above background by the width at half-maximum intensity $\omega_{1/2}$.

If one uses highly monochromatic x-rays rather than β-filtered radiation, the background correction is greatly reduced and may even become negligibly small relative to I_p. Should this be the case, and if it can also be shown that absorption corrections vary only slightly over the pertinent ranges of α and β, then the directly measured peak intensities I_p may be considered as roughly proportional to the true pole densities and plotted directly on the pole figure. The nature and quantitative evaluation of the corrections for x-ray absorption by the specimen are

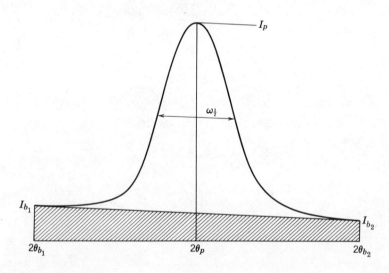

Fig. 10-17. Profile of a reflection as a function of 2θ; α and β constant. (From L. Alexander, *Diffraction Methods in Polymer Science*, 1969, Wiley.)

treated below in connection with discussions of transmission and reflection geometries. The investigator must always bear in mind that any intensities he may use other than the integrated intensity above background I_i are only approximate intensities and, accordingly, introduce a proportionate error into the pole figure.

B. Transmission Technique; Sheet Specimen. Most diffractometric transmission technique is based on the outstanding work of Decker, Asp, and Harker[7]. The simple specimen holder (Fig. 10-18) of Decker, Asp, and Harker is now usually replaced by a more sophisticated one (Fig. 10-21). In either instance the geometry of the technique, relative to Cartesian axes, is shown in Fig. 10-19. The Z coordinate coincides with the diffractometer axis, and the XY plane contains the incident (s_0) and diffracted (s) x-ray beams and the center of the specimen. The plane of the specimen coincides with the YZ plane when $\alpha = 0°$, and the arrangement is the symmetrical-transmission arrangement (Section 5-4.3)[41]. Usually, the machine (rolling) direction R is oriented in the Z direction. Then, with $\alpha = 0°$, the diffracting-plane poles coincident with Y lie in the plane of the specimen. A rotation of the specimen then about its normal N (β rotation), which coincides with X, constitutes a scan of the periphery of the pole figure at $0°$ latitude. By turning the specimen about the Z axis to various desired values of α, and in each case rotating the specimen about its normal N (β rotation), circles at higher α latitudes can be scanned. Note that when $\alpha \neq 0°$ the transmission geometry is asymmetrical. Also, from Fig. 10-19 it is evident that the transmission method fails when α approaches $90° - \theta$, because the diffracted beam and the specimen surface become parallel at this angle. Thus for a Bragg angle $\theta = 15°$ the *limiting* value of α is $75°$. The *largest practicable* value, however, is about $65°$.

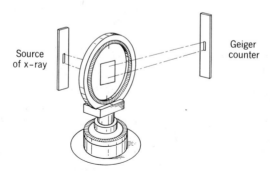

Fig. 10-18. Simple diffractometer specimen holder for preferred-orientation studies. [Courtesy of B. Decker, E. Asp, and D. Harker, *J. Appl. Phys.*, **19**, 388 (1948).]

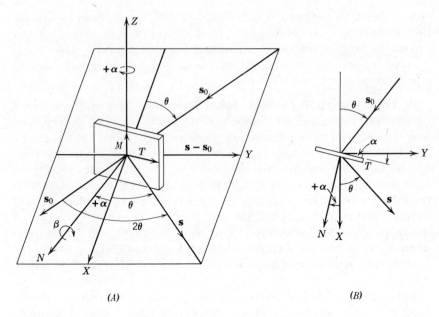

(A)   (B)

Fig. 10-19. Geometry of the transmission technique for pole-figure measurements. (From L. Alexander, *Diffraction Methods in Polymer Science*, 1969, Wiley.)

With symmetrical-transmission geometry the optimum specimen thickness to yield maximum diffracted intensity is given by equation 5-63 of Section 5-4.3:

$$t_m = \frac{1}{\mu \sec \theta}. \tag{5-63}$$

Thus for a light-element polymer (carbon, oxygen, nitrogen, and hydrogen only) the linear absorption coefficient μ for $CuK\alpha$ radiation is in the range 5 to 8 cm^{-1}. The optimal thickness by 5-63 is from 1.0 to 2.0 mm for moderate and small values of θ. For aluminum $\mu = 131$ cm^{-1} for $CuK\alpha$ x-rays, and aluminum sheet must be only about 0.07 mm thick for transmission studies with $CuK\alpha$ radiation ($\theta = 25°$). With $MoK\alpha$ radiation $\mu = 13.9$ cm^{-1} for aluminum, and aluminum sheet of approximately 0.7 mm thickness may be studied by transmission. But sheet steel specimens ($\mu = \sim 303$ cm^{-1}) must be ≤ 0.03 mm in thickness for transmission studies, even with the penetrating $MoK\alpha$ radiation, if optimum intensities are to be achieved. It will be seen that pole-figure studies of heavy metals are simpler when done with the reflection technique (Section 10-6.2C).

In addition to the use of a specimen of optimal thickness, it is necessary to keep geometrical broadening of the diffracted ray to a practicable minimum. The equatorial divergence (in the XY plane) of the incident beam must be limited so as to provide satisfactory resolution of the pattern. The attainable resolution depends on the widths of the reflections, and two reflections cannot be resolved unless their individual widths at half-maximum intensity are less than their angular separation. Years ago the factors contributing to the observed reflection profile in conventional powder diffractometry were analyzed[42] (also see Sections 5-2.1 and 5-2.2) in detail. Alexander, however, found that a simpler semiquantitative approach[43], involving some approximations, could be applied to pole-figure intensity profiles and sufficed to determine the choice of experimental conditions. The findings of this study[43] may be summarized as follows:

1. Considering resolution and effective accessible angular range, the $+\alpha$ region is superior to the $-\alpha$ region for measurement. Since the $+\alpha$ and $-\alpha$ hemispheres are symmetrically equivalent in most textures, one need make measurements only in the $+\alpha$ region.*

2. The calculation yields a good approximation of the experimental profile width or, in some instances, its lower limit. Accordingly, the anticipated resolution might be somewhat poorer, but never better, than that corresponding to the calculated receiving-aperture width ω_s.

3. Variations in ω_s (and therefore in the overall line width, ω) with α are significant, thereby demonstrating that accurate studies demand integrated rather than peak intensity measurements.

4. Smaller aperture widths than those indicated provide little if any improvement in resolution, and at the same time reduce the intensity. Of course, if the circumstances do not require optimal resolution, wider apertures than indicated may be used to provide greater intensities.

5. A line source is effective in improving resolution only when the other factors, particularly the half-maximum width of the pure diffraction profile $(\omega_0)_{1/2}$, are also small.

These conclusions are based on an assumed goniometer radius of 145 mm, values of $(\omega_0)_{1/2} = 0$ to $0.50°$, $\theta = 15$ to $30°$, specimen thickness $t = 1$ mm usually, and an incident beam divergence $\epsilon = 0.25$ to $0.50°$.

A "line" source is necessary for highest resolution. Such a source has a length of approximately 10 mm in the axial, or Z, direction, and Soller-slit collimators (Section 5-1.2) are necessary to restrict the axial

---

*The reader should note that the sense of positive α rotation adopted in this book is the simplest mathematical interpretation for actual measurements, and is *opposite* to the convention of Decker, Asp, and Harker[7].

divergence of both the direct and diffracted beams. To achieve good intensities the axial dimension of the specimen should equal or exceed the source length. The receiving aperture, for a line source, should be somewhat larger in the axial (Z) direction than the source length, usually 10 to 15 mm. When a "spot" focus is used, a reasonable compromise between resolution and intensity is achieved through the use of the following relation[42],

$$\omega'_s = [(\omega'_\epsilon)^2_{1/2} + (\omega'_x)^2_{1/2}]^{1/2}, \qquad (10\text{-}6)$$

where ω'_s is the height of the receiving aperture in angular units, ω'_ϵ is the axial divergence of the primary beam, and ω'_x is the Z dimension of the focal spot in angular units. Representative values of $(\omega'_\epsilon)_{1/2} = 0.25°$ and $(\omega'_x)_{1/2} = 0.40°$ lead to the value: $\omega'_s = 0.47°$.

When $\alpha \neq 0°$, the resulting asymmetrical-transmission geometry causes the x-ray path length through the specimen to increase over that for $\alpha = 0°$. Decker, Asp, and Harker[7] derived a formula for correcting intensities measured at nonzero values of α. Corrected for the change in sense of positive α adopted in this book, this formula is

$$\frac{I_{0°}}{I_\alpha} = \frac{\mu t \exp(-\mu t/\cos \theta)}{\cos \theta} \times \frac{[\cos(\theta - \alpha)/\cos(\theta + \alpha)] - 1}{\exp[-\mu t/\cos(\theta - \alpha)] - \exp[-\mu t/\cos(\theta + \alpha)]}$$

$$\text{(transmission technique).} \qquad (10\text{-}7)$$

A knowledge of μt for the specimen under investigation is required before equation 10-7 can be used. One can calculate an approximate value of μt from the measured thickness t of the specimen and a knowledge of its chemical composition. A more accurate value can be obtained by experimentally measuring μt. The diminution in intensity of a monochromatic beam of x-rays is determined when passed perpendicularly through the specimen. The monochromatic beam used may be a reflection from a single crystal or a strong beam diffracted from a crystalline powder. For optimal thickness specimens $\mu t \cong 1.00$. Equation 10-7 then gives for $+\alpha = 30°$ correction factors for $\theta = 15.0$, 30.0, and 45.0° of 1.196, 1.564, and 2.811 respectively.

A common practice which circumvents the absorption problem is to normalize the intensities from the textured specimen to the intensities at the same angles by a randomly oriented specimen of the same material. A pole figure of relative pole densities can be plotted directly from these normalized intensities. A value greater than unity denotes a greater than average density of poles, whereas numbers smaller than unity indicate smaller than average pole densities. A severe problem with this normalization method is that it may be difficult to procure a truly random

specimen of the same thickness and absorption coefficient as the textured sample. In the case of polymers, the attainment of a random
sample has been discussed by Jones[44], Wilchinsky[45], Lindenmeyer
and Lustig[46], and Desper and Stein[47]. In metal work, Grewen,
Sauer, and Wahl[48] and Bragg and Packer[49] describe means for
calculating the random intensity without the use of a random sample.
Wilchinsky[50] describes a pole-figure integrator which carrys out a
time-averaging process during a diffractometer scan, and provides
relative intensities which are the same as those of the random specimen.
Holland, Engler, and Powers[51] used a random powder compact for a
random sample, and Klenck[52] used fine annealed filings sprinkled on
a glass slide coated with vacuum grease.

C. Reflection Technique; Sheet Specimen. Schulz[53] first described
the reflection technique with a counter diffractometer. Reflection geometry (Fig. 10-20) places the α axis along the Cartesian axis X. The
specimen plane lies in the equatorial XY plane when $\alpha = 0°$, and the
specimen normal N is coincident with Z. The machine (rolling) direction
is normally aligned with X. It is evident that the reflection geometry is
most favorable when $\alpha = 90°$ and N is coincident with Y. Suppose we

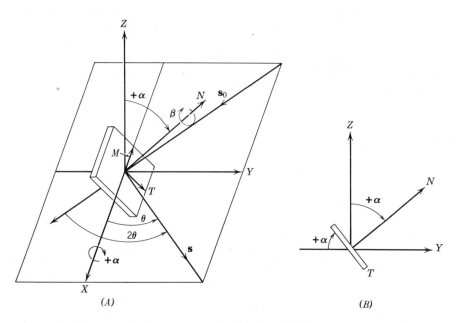

Fig. 10-20. Geometry of the reflection technique for pole-figure measurements. (From
L. Alexander, *Diffraction Methods in Polymer Science*, 1969, Wiley.)

have obtained the outer region, $\alpha = 0$ to $60°$, of a pole figure by the transmission technique. Then the remaining central zone of the figure can be completed by reflection measurements. A convenient experimental technique is to set the longitude β to successive fixed values, scanning along each β longitude line over the range $\alpha = 30$ to $90°$ by means of the α motion. Intensity data from the overlap region, $\alpha = 30$ to $60°$, may be used to place the transmission and reflection intensities on a common basis.

In the reflection technique the x-ray path through the specimen is a minimum for $\alpha = 90°$ and increases with decrease in α. The minimum specimen thickness to yield maximum diffraction intensities when $\alpha = 90°$ is given by the relation

$$t \geq \frac{3.2\rho}{\mu\rho'} \sin \theta. \tag{5-11}$$

For a "light" element polymer with μ about 5 to 8 cm^{-1} and a representative Bragg angle $\theta = 18°$, equation 5-11 indicates a minimum thickness of 1.5 to 2.0 mm. For $\alpha \ll 90°$ thinner samples may be used. In the interest of good resolution, however, it is useful to restrict t for polymer samples to about 1 mm, although there is some loss of intensity for $\alpha \sim 90°$. For most oriented metal samples, ρ/ρ' essentially equals unity and t is readily calculated. Indeed, nearly all flat, fine-grained metal specimens make excellent diffractometer reflection samples from the standpoint of x-ray beam penetration because the thickness required by equation 5-11 is small except for the lightest metals. Schulz[53] has provided the following absorption correction formula to be applied to intensities measured at values of $\alpha < 90°$:

$$\frac{I_{90°}}{I_\alpha} = \frac{1 - \exp(-2\mu t/\sin \theta)}{1 - \exp(-2\mu t/\sin \theta \sin \alpha)} \quad \text{(reflection technique).} \tag{10-8}$$

For typical experimental conditions, $\theta = 0$ to $45°$ and $\mu t \geq 1.00$, the correction is close to unity and may be neglected unless exceptionally accurate results are required. It is common practice, as in the transmission technique, to normalize measured intensities against corresponding intensities from a "random" specimen and thus avoid applying these corrections.

Alexander[43] applied the same semiquantitative analysis to reflection profiles as in the transmission case. Because of the appreciable axial (Z) height of the incident beam, defocusing occurs when $\alpha < 90°$, and broadening of the diffracted beam at the receiver results. Indeed, a line source cannot be used in the reflection method except at α close to $90°$.

Fig. 10-21.   Siemens pole-figure goniometer with transmission-specimen holder. 1, Transmission-specimen holder; 2, vertical (β) circle; 3, toothed outer rim which houses vertical circle; 4, azimuthal circle; 5, direct-beam collimator; 6, diffracted-beam collimator; 7, drive motor. (Courtesy of Siemens Aktiengesellschaft, and L. Alexander, *Diffraction Methods in Polymer Science*, 1969, Wiley.)

When a spot focus up to $(\omega_0)_{1/2} = 0.50°$ is used, the optimal receiving-aperture width falls between 0.3 and 0.4°, a value comparable to that for the transmission method. Equation 10-6 is also applicable to the reflection method for calculating the minimum axial (Z) dimension of the receiving aperture. The aperture's height may be increased (but not its width) to enhance intensities, but with some sacrifice of resolution.

D. Special Instrumentation. Pole-figure (texture) goniometers with counter recording are available from several purveyors of diffraction equipment.* A commercial goniometer is shown in Fig. 10-21 mounted with a transmission specimen holder (1). In the transmission technique

*For instance: The Diano Corporation, Industrial X-ray Division, 2 Lowell Avenue, Winchester, Massachusetts (successor to General Electric Company for diffraction equipment); Philips Electronic Instruments, 750 South Fulton Avenue, Mt. Vernon, New York; Siemens Aktiengesellschaft, Schöneberger Strasse 2-4, Berlin, Germany (available in the United States through Siemens Corporation, 186 Wood Avenue South, Iselin, New Jersey).

Fig. 10-22.   Siemens pole-figure goniometer with reflection-specimen holder. 1', Reflection-specimen holder in $\alpha = 90°$ position; 2, vertical circle which provides α motion; 5, direct-beam collimator; 6, diffracted-beam collimator; 7, drive motor. (Courtesy of Siemens Aktiengesellschaft, and L. Alexander, *Diffraction Methods in Polymer Science*, 1969, Wiley.)

the large vertical circle (2) bearing the specimen holder (1) rotates about a horizontal axis in the outer, stationary, toothed ring (3). This movement provides the β rotation about the specimen normal N. Measurements at higher latitudes are achieved by turning the vertical-circle assembly about its vertical (Z) axis to the desired α value, whereupon the β scan is performed as before. Gear trains with synchronous motor drives permit motion individually about either the α or β circles.

During reflection technique, α motion is provided by the vertical circle, and β motion by the small horizontal azimuth circle. Figure 10-22 depicts the specimen (1') in the reflecting position with $\alpha = 90°$. The direct- and diffracted-beam collimating systems, (5) and (6), respectively, and the drive motor (7) are also displayed. It is possible to scan the pole figure along a spiral path (Fig. 10-23) by simultaneous motion about the α and β circles, a technique first used by Holden[54] to study metal textures. To suit the precision of counting or degree of resolution desired, adjustments on the Siemens goniometer provide different scanning speeds or spirals of different pitches. A spiral pitch of 5° (5° increase in α per 360° rotation of β) may be combined with a suitable

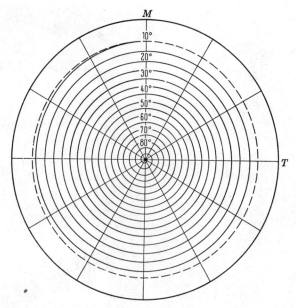

Fig. 10-23. Spiral pole-figure pattern corresponding to a simultaneous scan of α and β. Five-degree pitch. M, Machine (rolling) direction; T, transverse direction. (From L. Alexander, *Diffraction Methods in Polymer Science*, 1969, Wiley.)

scanning speed to map the range $\alpha = 0$ to $75°$ in either 90 or 180 min. A pole figure of greater detail results from a pitch of $2.5°$ and a 180-min scan of the range $\alpha = 0$ to $75°$. It is to be noted that Holden's α is actually the colatitude ϕ. The tedious plotting of strip-chart intensities on the spiral stereographic polar chart was eliminated by Chirer[55] by the "texturograph," a direct registering device for use with the Siemens goniometer.

Geisler[56] used a turntable-type recorder directly coupled to a pole figure goniometer to produce pole figures automatically. The device synchronizes the specimen-orientation coordinates α and β with the α coordinate of the printer arm, and the rotation angle β of the turntable with its attached pole-figure chart. The General Electric Company marketed an improved version (Fig. 10-24) which prints out the pole figure on an X-Y recorder. The printer head provides the X motion, and the chart paper the Y motion. The specimen's β-angle rotation is first resolved into $\sin^2 \beta$ and $\cos^2 \beta$ signals, which, respectively, actuate the X and Y motions and thereby produce one circular traverse of the printing head per $360°$ rotation of the specimen. The sine- and cosine-signal amplitudes are related to the latitude setting of the specimen so as to

Fig. 10-24. Automatic *XY* pole-figure recorder (*A*) coupled to a diffractometer with a pole-figure goniometer (*B*). Also visible are counter (*C*), x-ray tube (*D*), cam for absorption compensation (*E*), mode of operation control (F), and counting and scaling units (G). (Courtesy of General Electric Company, and I. Cohen, Westinghouse Bettis Atomic Power Laboratory. From L. Alexander, *Diffraction Methods in Polymer Science*, 1969, Wiley.)

provide β-circle diameters on the chart corresponding to the colatitude ϕ of the specimen.

The numerical printer of the General Electric unit is of the multipoint type, with inks of 10 different colors, permitting the registration of 10 levels of diffracted intensity. It is possible to apply the absorption corrections required by the transmission or reflection technique (see equations 10-7 and 10-8) by the use of an aluminum cam of proper shape which modulates the voltage applied to the printer. The resulting colored numbers on the pole-figure chart are the pole densities on an arbitrary scale, except for background corrections. A rather similar device was described by Baro and Ruer[57], and used to produce the (110) pole figure of low carbon steel sheet shown in Fig. 10-25. Eichhorn [58] also has used an *X-Y* recorder to plot spiral-scanned pole-figure data automatically.

Modern four-circle single-crystal diffractometers are ideal for pole-figure intensity measurements. Any instrument modifications required

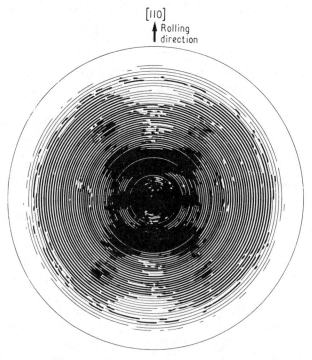

Fig. 10-25. Pole figure (110) of a low carbon steel sheet drawn with a spiral step of 1.25°. Isointensity lines were drawn with three black pens of different thicknesses. [Courtesy of R. Baro and D. Ruer, *J. Phys.*, **E3**, 541 (1970). Copyright the Institute of Physics.]

pertain to construction of beam collimators of sufficiently large aperture, and suitable specimen holders for transmission and reflection techniques. The β and α motions for the transmission technique are provided by the **X** and ω circles, respectively, while for reflection technique the α and β motions are provided by the **X** and Φ circles, respectively. The **X** circle for transmission measurements should be of the 360° (full-circle) rather than the 90° type.

Single-crystal diffractometers capable of automatic control by means of computers, punched cards, or paper tape are the most valuable for pole-figure measurements. Alty[59] has written and described a computer program for producing pole figures from an x-ray texture goniometer. The data are corrected for background and absorption and stored on magnetic tape in a form suitable for quantitative calculations. Alty used the program for calculating Hall coefficients of materials from their pole figures. Computer time on the E.E.L.M. KDF9 for one complete pole-figure output and Hall calculation is 30 sec, and the storage required is under 16K. Desper[60] used on-line control of the Picker Four-Angle Computer System (FACS-1) by a PDP-8S computer with teletype input and output. Among several operations programed was the determination of pole figures of sheet specimens in reflection or transmission, or of fibers or small particles. Chao[61] has developed a method for automatically producing pole figures by direct printout from a digital computer. The pole-figure diffraction intensities are recorded on punched tape or punched cards and fed into the computer (IBM 360/50) which applies the necessary corrections and prints out the pole figure. Whereas a manual plot of the data requires 8 to 12 hours, and computer-directed plotting by the VP622 drafting machine requires 65 min, direct printout plotting requires only 20 to 45 min based on the complexity of the plotting array.

Segmüller and Angilello[62] have described an automatic pole-figure operation. They use a commercial pole-figure goniometer (Philips) controlled by an IBM 1800 computer. A digital voltmeter is used to digitize the rate meter output. The pole-density data are placed in disk storage, and after processing are made into a pole-figure map by a 1627 plotter attached to the computer. Segmüller and Angilello are restricted to the inner part of the pole figure, since the Philips pole-figure goniometer is not suited to the automation of the transmission technique. It is noted too that their angles α are actually the colatitude ϕ.* In another study Segmüller[64] attached a Tektronix 611 storage oscilloscope to an IBM 1800 computer. This unit permits the display of the pole figures on the oscilloscope in much less time (about one-eight) than that needed to plot them on the plotter. Once a pole-figure trace has been written on the oscilloscope screen, it stays until the whole screen is erased, thus allowing time for the figure to be studied or photographed.

The reader will appreciate that it is inappropriate to discuss these computer-controlled operations in detail. Almost no two laboratory's needs or facilities are the same. Hence we see as our greatest service to the reader in the present instance the brief mention of several computer-controlled systems with additional references to others. By reference to these as a nucleus, the interested reader can begin to evaluate the complex problem of having his laboratory automated for pole-figure studies. Other computer systems and programs to be consulted in pole-figure technology are those of Love[65], Montgomery[66], Elias and Heckler [67], Morris and Heckler[68], and Slane and Hultgren[69].

---

*Huggins and Green[63] suggest that the use of a radiation shield is desirable with the Philips texture goniometer, and describe an easily demountable shield which reduces random scattered radiation 500-fold.

E. The Specimen and its Alignment. The reflection method is the major technique for texture studies in metals. Unfortunately, the useful colatitude angle ϕ range is limited by defocusing effects[70]. These effects, however, can be minimized by precise alignment of the specimen surface. Chernock et al.[71] designed an automatic specimen holder for rapid precise alignment of the specimen. With it a single flat specimen gives the center portion of a pole figure, and no absorption correction is required. The holder features a removable jig which fits into the inner ring and permits fast alignment of the specimen. Two specimen-scanning devices are also described which allow large numbers of grains in the specimen surface to be successively illuminated by the incident beam during an intensity measurement. The alignment procedure of Chernock et al. requires backing off the sample, in order to remove the jig, and then replacing the specimen surface in its aligned position. Although Chernock and co-workers[71] worked out a satisfactory check for the accuracy of their specimen alignment, Singer[72] felt it desirable that the jig be removable from the goniometer ring without disturbing the aligned specimen. In a jig[72], based on that of Chernock et al., he provides this feature along with interchangeability between the Schulz head and the standard diffractometer specimen mount. His device is claimed to yield high precision of alignment with a minimum of skill.

The transmission method is the desired technique for obtaining the outer or peripheral part of a pole figure. But for metals and materials that are quite opaque to x-rays, only the reflection method is available. Hence workers have sought schemes by which the entire pole figure could be obtained by the reflection technique. Mueller and Knott[73] achieved this end by the use of seven different sections of the sample to obtain the data for one quadrant of the pole figure. A single piece of the sheet provides the section for data at colatitude $\phi = 0°$. Six other sections are made up of a number of small platelets cut from the sheet and mounted in a jig so as to give a plane surface at an angle of 45° or 90° to the sheet surface. The assembled blocks of platelets in their jigs are surface ground, hand-polished, and etched to form the array of sheet edges into a plane surface suitable for diffraction measurements. Figure 10-26 depicts the orientation of Mueller and Knotts' seven sheet sections to cover one quadrant of the pole figure.

Others quickly took up the use of the *composite* sample prepared by stacking and clamping together sections cut from the sheet material [51, 67, 74–76], although the work in preparing such specimens might seem formidable enough to discourage their routine use. Meieran[74] and Lopata and Kula[75] showed that a complete quadrant of a pole

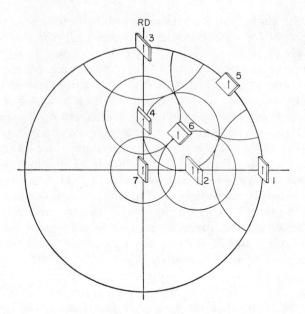

Fig. 10-26.   Quadrant of a pole figure, showing area covered by the various sheet specimens of Mueller and Knott. [Courtesy of M. Mueller and H. Knott, *Rev. Sci. Instrum.*, **25**, 1115 (1954).]

figure can be obtained from a single composite sample. Pieces of the sheet were stacked to form a cube with the normal, transverse, and rolling directions of the pieces in common alignment. The pieces were bonded together with epoxy cement, and a cut was made at 54.7° to each of these principal reference directions, as shown in Fig. 10-27. The normal to the cut face projects at equal angles into one quadrant of a stereographic projection with the usual polar orientation of these directions. One has merely to make spiral diffraction intensity measurements about this normal with increasing angular rotation (corresponding to $\phi = 0$ to 55°) about an axis lying in the plane of the surface. When these data are plotted on a polar stereographic net offset 54.7° from the center, one quadrant of a pole figure is obtained (Fig. 10-28).

Leber[76] improved the sampling technique by cutting the platelets 45° to the rolling direction. These he stacked and bonded at an angle of 54.7° to a reference plane, and ground and lapped a flat-specimen surface parallel to this plane. Elias and Heckler[67] wishing to avoid the presence of the diffracting power of the bonding medium, an epoxy resin, prepared composite specimens without any bonding material.

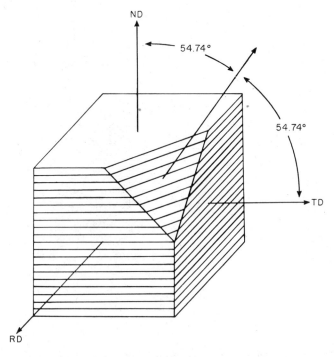

Fig. 10-27.   Composite specimen of Meieran and Lopata and Kula (schematic). [Courtesy of J. Elias and A. Heckler, *Trans. TMS-AIME*, **239**, 1237 (1967).]

Believing also that prolonged etching might lead to significant inter-lamellar attack of the platelet faces, they used a minimum of etching in preparing the specimen surfaces.

In addition to the great gain of obtaining a complete quadrant of a pole figure from a single sample, the composite sample has other ad-vantages. Such specimens require little or no correction for absorption or defocusing. The texture is averaged through the thickness of the sheet, and the average is obtained from a larger than usual area of the sheet. Preparation of the specimen is, however, costly, often requiring large amounts of material and extensive machining. Leber[76] believed his technique minimized these cost factors somewhat.

A sample technique particularly applicable to thicker sheet material is that of Norton[77]. Here the sample takes the form of small, cylindrical rods cut from the sheet in such a manner that the rod axes are parallel to to the plane of the sheet and make various angles with the rolling direc-tion. With a series of such rods of constant diameter, the absorption is

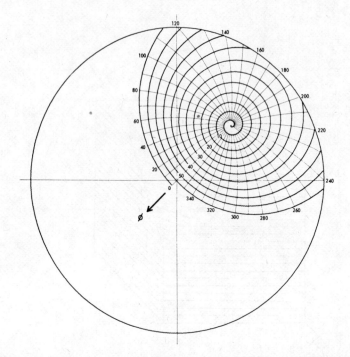

Fig. 10-28.   Spiral plotting net for a single quadrant of a pole figure. [Courtesy of J. Elias and A. Heckler, *Trans. TMS-AIME*, **239**, 1237 (1967).]

constant, and the troublesome problem of absorption corrections is avoided. Preparation of the specimen rods offers no particular difficulty, and other experimental details are simple.

10-7   MISCELLANEOUS

Leber[78] has pointed out that wires fabricated from body-centered cubic metals frequently show textures that are not the full[110] texture normally associated with such metals. Rather they show an intermediate texture which, from its similarity to the texture obtained by forming a cylinder from rolled sheet metal (rolling direction parallel to the cylinder axis), is called a cylindrical texture. Extensive deformation converts the cylindrical texture into a normal fiber texture. The cylindrical texture is most pronounced at the surface of a wire, and decreases toward the normal fiber texture as the center of the wire is approached. As indicated by Leber, Fig. 10-2B is not a pure normal fiber pattern but is indeed partly a cylindrical texture. Leber, in his study[78] of tungsten wires, showed how cylindrical textures can be distinguished.

van Someren and Sprenkle[79] were interested in studying metal films prepared by chemical vapor deposition. Such films are characterized by a tendency to have one class of planes parallel to the surface. Tungsten vapor-deposited by hydrogen reduction of WF_6 at 500 to 800°C preferentially has (100) planes parallel to the deposit surface. Thus the material is essentially isotropic in the plane of the deposit, but is highly oriented perpendicular to the surface. van Someren and Sprenkle describe an accessory for the Philips diffractometer which permits the spatial investigation of these oriented planes when the metal is deposited on a cylindrical surface. The situation with these specimens is similar to the cylindrical textures described by Leber[78], except that there is no fiber direction in the film parallel to the cylinder axis. The method gives information only about the spatial distribution of the one preferentially oriented plane, and it cannot give information about the relative orientation of several planes to one another.

Microtechniques (Section 4-4) can be extremely useful for examination of fibers (Fig. 4-42). Early studies of natural and synthetic fibers led to micromethods for obtaining the diffraction effects from single fibers less than 0.001 in. in diameter[80, 81]. Such methods permit selected portions, such as the "core" and "skin," of larger fibers to be investigated, and were employed by Fankuchen and Mark[82] in a study of the mechanism of drawing of nylon.

GENERAL REFERENCES

*1. C. S. Barrett and T. B. Massalski, *Structure of Metals*, 3rd ed., McGraw-Hill, New York, 1966, Chapter 2, 9, 19, 20, 21.

2. A. Taylor, *X-ray Metallography*, Wiley, New York, 1961, Chapter 13.

3. L. V. Azároff, *Elements of Crystallography*, McGraw-Hill, New York, 1968, pp. 531–548.

4. G. Wassermann and J. Grewen, *Texturen metallischer Werkstoffe*, Springer, Berlin, 1962.

5. F. A., Underwood, *Textures in Metal Sheets*, Macdonald, London, 1961.

6. E. Bocke, *Die Anwendung der stereographische Projektion bei kristallographischen Untersuchungen*, Gebrüder Borntraeger, Berlin, 1911.

7. J. K. Wood, "An Analytical Discussion of the Construction of Pole Figures," *J. Appl. Phys.*, **19**, 784 (1948).

8. M. V. King, "The Precession Method in Fiber Diffraction Photography," *Acta Crystallogr.*, **21**, 629 (1966).

9. U. Mukhopadhyay and C. A. Taylor, "Application of Optical Transform Methods to Interpretation of X-ray Fiber Photographs," *J. Appl. Crystallogr.* **4**, 20 (1971).

10. J. L. Pentecost and C. H. Wright, "Preferred Orientation in Ceramic Materials Due

to Forming Techniques," *Advances in X-ray Analysis*, Vol. 7, Plenum Press, New York, 1964, p. 174.

11. R. L. Miller, "Volume Fraction Analysis of Phases in Textured Alloys," *Trans. ASM*, **61**, 592 (1968).

12. K. Little, "Investigation of Nylon 'Texture' by X-ray Diffraction," *Brit. J. Appl. Phys.*, **10**, 225(1959).

13. R. O. Williams, "Analytical Methods for Representing Complex Textures by Biaxial Pole Figures," *J. Appl. Phys.*, **39**, 4329 (1968).

SPECIFIC REFERENCES

[1]   C. S. Barrett and T. B. Massalski, *Structure of Metals*, 3rd ed., McGraw-Hill, New York, 1966, Chapters 9, 19, 20, 21.

[2]   L. E. Alexander, *X-ray Diffraction Methods in Polymer Science*, Wiley, New York, 1969.

[3]   R. M. Bozorth, *Phys. Rev.*, **26**, 390 (1925).

[4]   R. Glocker, *Materialprüfung mit Röntgenstrahlen*, 5th ed., Springer, Berlin, 1971, p. 467.

[5]   M. Polanyi, *Z. Phys.*, **7**, 149 (1921).

[6]   B. F. Decker, *Am. Soc. Test. Mater., Proc.*, **43**, 785 (1943). R. Smoluchowski and R. W. Turner, *Rev. Sci. Instrum.*, **20**, 173 (1949).

[7]   B. F. Decker, E. T. Asp, and D. Harker, *J. Appl. Phys.*, **19**, 388 (1948).

[8]   P. W. Bakarian, *Trans. AIME*, **147**, 266 (1942).

[9]   W. L. Fink and D. W. Smith, *Symposium on Radiography and X-ray Diffraction Methods*, American Society for Testing Materials, Philadelphia, 1937, p. 200.

[10] R. H. Hay, *Rev. Sci. Instrum.*, **18** 801 (1947).

[11] J. W. Hickman and A. G. Kleinknecht, *Rev. Sci. Instrum.*, **20**, 573 (1949).

[12] F. A. Underwood, *J. Sci. Instrum.*, **29**, 128 (1952).

[13] G. W. Smith and R. M. Hinde, *J. Sci. Instrum.*, **33**, 391 (1956).

[14] D. A. Northrop, *Rev. Sci. Instrum.*, **31**, 1160 (1960).

[15] G. Gandolfi, *Mineral Petrog. Acta*, **13**, 67 (1967).

[16] E. J. Graeber and D. A. Jelinek, *Norelco Rep.*, **13**, 91 (1966).

[17] C. W. Bunn, *Chemical Crystallography*, 2nd ed., Clarendon Press, Oxford, 1961, pp. 188–193, 348–359.

[18] F. Wever, *Z. Physik*, **28**, 69 (1924); *Trans. AIME*, **93**, 51 (1931).

[19] See particularly F. C. Phillips, *An Introduction to Crystallography*, 3rd ed., Longmans, Green, London, 1963; Wiley, New York, Chapter 2.

[20] Reference 1, Chapter 2.

[21] D. J. Dyson, *Z. Kristallogr.*, **122**, 307 (1965).

[22] J. H. Palm, *Z. Kristallogr.*, **123**, 388 (1966).

[23] G. B. Harris, *Phil. Mag.*, **43**, 113 (1952).

[24] P. R. Morris, *U.S. AEC Report FMPC-310* (1953).

[25] P. R. Morris, *J. Appl. Phys.*, **30**, 595 (1959).

[26] M. H. Mueller, W. P. Chernock, and P. A. Beck, *Trans. TMS–AIME*, **212**, 39 (1958).

[27] L. K. Jetter, C. J. McHargue, and R. O. Williams, *J. Appl. Phys.*, **27**, 368 (1956).

[28] R. O. Williams, *Trans. TMS–AIME*, **215**, 646 (1959).

[29] C. M. Mitchell and J. F. Rowland, *Acta Met.*, **2**, 559 (1954).

[30] R. J. Roe and W. R. Krigbaum, *J. Chem. Phys.*, **40**, 2608 (1964).

[31] A. J. Heckler, J. A. Elias, and A. P. Woods, *Trans. TMS–AIME*, **239**, 1241 (1967).

[32] R. M. S. B. Horta, W. T. Roberts, and D. V. Wilson, *Trans. TMS–AIME*, **245**, 2525 (1969).

[33] H. J. Bunge and W. T. Roberts, *J. Appl. Crystallogr.*, **2**, 116 (1969).

[34] M. J. Nasir and H. J. Bray, *Acta Crystallogr.*, **23**, 555 (1967).

[35] H. P. Klug and L. E. Alexander, *X-ray Diffraction Procedures*, 1st ed., Wiley, New York, 1954, pp. 574–579.

[36] O. Kratky, *Z. Kristallogr.*, **72**, 529 (1930).

[37] C. S. Barrett, *Trans. AIME*, **93**, 75 (1931).

[38a] R. Smoluchowski and R. W. Turner, *Physica*, **16**, 397 (1950).

[38b] C. J. Milner and J. A. James, *J. Sci. Instrum.*, **30**, 77 (1953).

[39] *1965 Book of ASTM Standards*, American Society for Testing and Materials, Philadelphia, 1965, Part 31, pp. 151–167, Standard Method E81-63.

[40] Reference 2, pp. 209–240.

[41] Reference 2, Section 2-2.1.

[42] L. E. Alexander, *J. Appl. Phys.*, **21**, 126 (1950); **25**, 155 (1954).

[43] Reference 2, pp. 217–221, 224–226.

[44] J. W. Jones, *Advances in X-ray Analysis*, Vol. 6, Plenum Press, New York, 1963, p. 223.

[45] Z. W. Wilchinsky, *J. Appl. Polymer Sci.*, **7**, 923 (1963).

[46] P. H. Lindenmeyer and S. Lustig, *J. Appl. Polymer Sci.*, **9**, 227 (1965).

[47] C. R. Desper and R. S. Stein, *J. Appl. Phys.*, **37**, 3990 (1966).

[48] J. Grewen, D. Sauer, and H. P. Wahl, *Scripta Met.*, **3**, 53 (1969).

[49] R. H. Bragg and C. M. Packer, *J. Appl. Phys.*, **35**, 1322 (1964).

[50] Z. W. Wilchinsky, *Rev. Sci. Instrum.*, **40**, 592 (1969).

[51] J. R. Holland, N. Engler, and W. Powers, *Advances in X-ray Analysis*, Vol. 4, Plenum Press, New York, 1961, p. 74.

[52] M. M. Klenck, *Advances in X-ray Analysis*, Vol. 11, Plenum Press, New York, 1968, p. 447.

[53] L. G. Schulz, *J. Appl. Phys.*, **20**, 1030 (1949).

[54] A. N. Holden, *Rev. Sci. Instrum.*, **24**, 10 (1953).

[55] E. G. Chirer, *J. Sci. Instrum.*, **44**, 225 (1967).

[56] A. H. Geisler, *Trans. ASM*, **45A**, 131 (1953).

[57] R. Baro and D. Ruer, *J. Phys.*, **E3**, 541 (1970).

[58] R. M. Eichhorn, *Rev. Sci. Instrum.*, **36**, 997 (1965).

[59] J. L. Alty, *J. Appl. Phys.*, **39**, 4189 (1968).

[60] C. R. Desper, *Advances in X-ray Analysis*, Vol. 12, Plenum Press, New York, 1969, p. 404.

[61] H. C. Chao, *Advances in X-ray Analysis*, Vol. 12, Plenum Press, New York, 1969, p. 391.

[62] A. Segmüller and J. Angilello, *J. Appl. Crystallogr.*, **2**, 76 (1969).

[63] F. G. Huggins and G. J. Green, *J. Phys.*, **E1**, 668 (1968).

[64] A. Segmüller, *J. Appl. Crystallogr.*, **2**, 259 (1969).

[65] G. R. Love, *Trans. TMS–AIME*, **242**, 746 (1968).

[66] G. L. Montgomery, *Trans. TMS–AIME*, **242**, 762 (1968).

[67] J. A. Elias and A. J. Heckler, *Trans. TMS–AIME*, **239**, 1237 (1967).

[68] P. R. Morris and A. J. Heckler, *Trans. TMS–AIME*, **245**, 1877 (1969).

[69] J. A. Slane and F. Hultgren, *Advances in X-ray Analysis*, Vol. 14, Plenum Press, New York, 1971, p. 231.

[70] W. P. Chernock and P. A. Beck, *J. Appl. Phys.*, **23**, 341 (1952).

[71] W. P. Chernock, M. H. Mueller, H. R. Fish, and P. A. Beck, *Rev. Sci. Instrum.*, **24**, 925 (1953).

[72] J. Singer, *Rev. Sci. Instrum.*, **26**, 963 (1955).

[73] M. H. Mueller and H. W. Knott, *Rev. Sci. Instrum.*, **25**, 1115 (1954).

[74] E. S. Meieran, *Rev. Sci. Instrum.*, **33**, 319 (1962).

[75] S. L. Lopata and E. B. Kula, *Trans. TMS–AIME*, **224**, 865 (1962).

[76] S. Leber, *Rev. Sci. Instrum.*, **36**, 1747 (1965).

[77] J. T. Norton, *J. Appl. Phys.*, **19**, 1176 (1948).

[78] S. Leber, *Technical Information Series Report No. 60-LMC-146*, General Electric Company, Engineering No. 731, Refractory Metals Laboratory, 1331 Chardon Road, Cleveland, Ohio, 1960, 17 pp.

[79] L. van Someren and E. Sprenkle, *Rev. Sci. Instrum.*, **40**, 56 (1969).

[80] O. Kratky and H. Mark, *Fortschritte auf dem Gebeite der Hochpolymern*, Springer, Berlin, 1938.

[81] I. Fankuchen and H. Mark, *Rec. Chem. Prog. (Kresge-Hooker Sci. Libr.)*, **4**, 54 (1943).

[82] I. Fankuchen and H. Mark, *J. Appl. Phys.*, **15**, 364 (1944).

CHAPTER 11

STRESS MEASUREMENT IN METALS

Stress-strain phenomena are conveniently illustrated by the dimensional changes of a uniform metal bar when a tension is applied along its axis. Under such stress the bar is lengthened, and its area of cross section decreased in direct proportion to the applied stress, provided the bar's elastic limit has not been exceeded. The resultant effect on each individual crystallite in the bar is an extension in the direction parallel to the bar axis (Fig. 11-1) and a compression in directions normal thereto. Crystal planes perpendicular to these extensive or compressive forces have their interplanar spacings changed by amounts $\pm \Delta d$, and measurement of this change yields a measure of the elastic strain and thereby of the stress as well. For crystals lying in the surface of the bar, such measurements can be made with high precision by the back-reflection method (described in detail in Section 4-3.2).

Stress measurement by x-ray diffraction actually dates from the pioneer work of Lester and Aborn[1] in 1925, but it did not compete favorably with other types of strain gages until 1930, when Sachs and Weerts[2] introduced the back-reflection technique to the problem. In its simplest and earliest application, it yielded only the sum of the principal stresses at the surface. Barrett and Gensamer[3], however, soon demonstrated that the principal stresses could be determined individually, and shortly thereafter it was shown that d for the unstressed material need not be obtained (merely two measurements of the stressed specimen)[4], a most useful finding since often it is impossible to attain the unstressed condition. This method of two exposures may be replaced by a single-exposure technique[5, 6] which is simpler and less time-consuming, although the precision attainable is somewhat less than that provided by the two-exposure procedure[7]. In the last two decades diffractometric techniques have come into general use because of their superiority in the registration of line profiles.

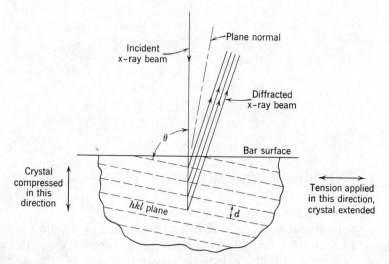

Fig. 11-1. Stress–strain phenomena and back-reflection of x-rays in a surface crystal of a metal bar.

11-1 ADVANTAGES AND DISADVANTAGES OF DIFFRACTION METHODS

Stress measurement by x-ray diffraction has certain very definite advantages. To begin with, it is the only nondestructive method for determining initial, residual, or "locked-in" stresses in a specimen without cutting it up to relieve the stresses. This is possible because it is not necessary to make measurements of the material in the unstressed condition, a universal requirement of other methods. Whereas other strain gages measure both elastic and plastic strains, *diffraction procedures determine only elastic stresses*. Another useful feature is the fact that the method measures the strain at a point usually not over 1 to 2 mm in diameter. Studies of localized stresses and of steep stress gradients thereby become possible.

Against the foregoing advantages is the fact that poorer precision is obtained unless the grain size in the metal is right, that is, neither too large nor too small. With a suitably fine-grained annealed steel free from microscopic stresses, an accuracy of 2000 to 3000 psi is easily attained. Severely cold-worked or hardened steels, however, are usually too fine-grained and give diffuse lines which increase the error to several times these values when photographic recording is employed. Nevertheless, since the advent of the use of x-ray diffractometry in stress stress analysis[8], it has been found possible with this technique to make measurements of hardened steels with acceptable accuracy. Exceedingly coarse-grained structures produce spotty lines, with a resulting adverse effect on the reliability of measurement; however, oscillation of the film (or specimen) through a limited angular range about the direct-beam axis greatly ameliorates this difficulty. Another disadvantage of both photographic and diffractometric methods is that only surface stresses can be detected and measured because the x-ray beam is incapable of penetrating the metal to a depth greater than about 0.001 in. Finally, the method requires more costly and cumbersome equipment than the mechanical and electrical devices commonly used, and it is more time-consuming because of the photographic recording.

11-2 ELASTIC STRESS–STRAIN RELATIONSHIPS

The classical theory of elasticity[9] assumes the materials under consideration to be perfectly elastic, homogeneous, and isotropic. It is further assumed that their elastic constants are the same in compression and combined stress systems as in simple tension, and that they adhere faithfully to Hooke's law. No actual material satisfies all these requirements, but metals approximately satisfy them. Another limitation is that the strains be small, so that the object is not appreciably changed dimensionally or in shape.

A strain e is defined as

$$e = \Delta l / l, \qquad (11\text{-}1)$$

where Δl is the change in length of a stressed body whose initial length was l. If it is assumed that this strain was produced by a stress σ acting in

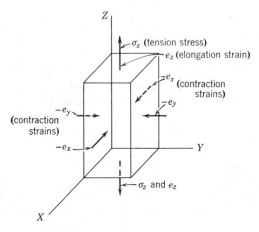

Fig. 11-2.   Strain directions during a uniaxial tension stress in a tetragonal prism.

a single direction, Hooke's law* requires that

$$e = \frac{\sigma}{E}, \tag{11-2}$$

where E is Young's modulus. If a tension σ_z is applied along the Z axis of the tetragonal prism in Fig. 11-2, the body is elongated in the Z direction and the strain is

$$e_z = \frac{\sigma_z}{E}. \tag{11-3}$$

At the same time, the body contracts equally along its X and Y axes, and these strains are related to e_z through Poisson's ratio ν as follows†:

$$-e_x = -e_y = \nu e_z = \frac{\nu \sigma_z}{E}, \tag{11-4}$$

where the negative signs denote contraction. The simple stress system just discussed is one-dimensional with respect to the direction of stress. The strains for two- and three-dimensional stress systems, however, are readily obtained from expressions 11-3 and 11-4 by the principle of superposition, giving

$$e_x = \frac{1}{E} [\sigma_x - \nu(\sigma_y + \sigma_z)],$$

$$e_y = \frac{1}{E} [\sigma_y - \nu(\sigma_z + \sigma_x)], \tag{11-5}$$

$$e_z = \frac{1}{E} [\sigma_z - \nu(\sigma_x + \sigma_y)].$$

The strains considered above are designated *normal strains* since they are produced by stresses normal to a surface. Usually, such normal strains are accompanied by additional strains, *shear strains*, in a plane normal to the stress direction. The shearing stress causes parallel planes within the body to slide past one another as in Fig. 11-3, and the shear strain γ is defined as the relative displacement of parallel planes at unit distance:

$$\gamma = \frac{d}{l} = \tan \alpha. \tag{11-6}$$

Hooke's law: Within the elastic limit of a body, the strain produced is directly proportional to the applied stress. The proportionality constant, the stress required to produce unit strain, is known as Young's modulus. Typical values are given in Table 11-1.

†Poisson's ratio is the ratio (per unit length) of the transverse contraction to the elongation for a uniform bar during tensile stress. Its value is approximately 0.3 for most metals.

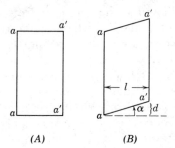

Fig. 11-3.   Parallelogram (B) resulting from a vertical shear stress on rectangle (A).

(A)   *(B)*

The shear strain and the shear stress τ are related through the modulus of elasticity in shear G by the relation

$$\gamma = \frac{\tau}{G}. \tag{11-7}$$

Figure 11-4 illustrates the relation between the normal and shear stresses in an element of a two-dimensional stress system. The symbol τ_{zy} denotes a shear stress perpendicular to the Z axis and acting in the direction of the Y axis. It is readily shown that under conditions of equilibrium

$$\tau_{zy} = \tau_{yz}, \tag{11-8}$$

so that only the three quantities σ_y, σ_z, and τ_{yz} are needed to define the system. Three-dimensional stress systems obviously involve three such two-dimensional systems, the complete analysis of which reveals that no more than six of the various stress components (σ_x, σ_y, σ_z, τ_{xy}, τ_{yz}, and τ_{zx}) are required to define completely the state of stress in a solid.

The normal stresses just considered are not necessarily the maximum normal stresses within the body. The latter, however, are of such im-

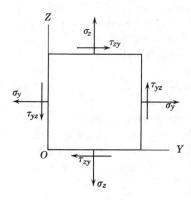

Fig. 11-4.   Normal and shear stresses in an element of a two-dimensional stress system.

portance in considerations of elastic behavior that they are called the *principal stresses*. These principal stresses σ_1, σ_2, and σ_3, which may always be made parallel to the three rectangular coordinate axes by a suitable axial choice[9], are further characterized by the fact that they act over sections of zero shear stress. A set of equations analogous to equation 11-5 relates the principal stresses to the principal strains e_1, e_2, and e_3.

An elegant approach to the three-dimensional stress–strain problem in an isotropic medium is to observe that under such stress a spherical element of volume is deformed into an ellipsoid. Then, if the principal stresses are taken parallel to X, Y, and Z of a set of rectangular coordinate axes, the equation,

$$\frac{X^2}{\sigma_1{}^2}+\frac{Y^2}{\sigma_2{}^2}+\frac{Z^2}{\sigma_3{}^2}=1, \tag{11-9}$$

for the *stress ellipsoid* may be written. Any point X_n, Y_n, Z_n on the surface of this ellipsoid represents the components of a normal stress σ_n whose direction and magnitude are close to, but not precisely, those of the radius vector to this point. More exactly, σ_n is given by the relation (which approximates the stress ellipsoid)

$$\sigma_n=\sigma_1\alpha_1{}^2+\sigma_2\alpha_2{}^2+\sigma_3\alpha_3{}^2, \tag{11-10}$$

where α_1, α_2, and α_3 are the direction cosines of the normal stress σ_n relative to the coordinate axes and principal strains. A similar expression,

$$e_n=e_1\alpha_1{}^2+e_2\alpha_2{}^2+e_3\alpha_3{}^2, \tag{11-11}$$

holds for the approximate ellipsoid of strain.

11-3 SUM OF THE PRINCIPAL STRESSES IN A SURFACE

Equations 11-5 are readily used to determine the sum of the principal stresses in a surface. If a steel specimen (Fig. 11-5) is subject to principal stresses σ_1 and σ_2 in its surface plane, the stress σ_3 normal to its surface is zero. The normal strain e_3 perpendicular to the surface is then

$$e_3=-\frac{\nu}{E}\,(\sigma_1+\sigma_2), \tag{11-12}$$

and represents a contraction if σ_1 and σ_2 are tensile stresses. To determine e_3 it is necessary merely to measure the change in d spacing of planes parallel to the surface, by taking back-reflection diffraction patterns in the stressed and unstressed condition[2, 10]. Precision measurements of the patterns yield d_s (stressed) and d_U (unstressed) for

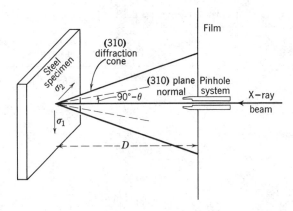

Fig. 11-5.   Back-reflection from the (310) planes in steel (Co$K\alpha$ radiation).

use in the relation

$$\sigma_1 + \sigma_2 = -\frac{E}{\nu}\frac{\Delta d}{d} = -\frac{E}{\nu}\left(\frac{d_s - d_U}{d_U}\right). \qquad (11\text{-}13)$$

An important advantage of the back-reflection method is the ease and precision with which small changes in d can be measured as θ approaches 90°. For measuring $\sigma_1 + \sigma_2$ we select a plane (hkl) and a radiation, such that the combination yields a diffraction ring at the highest possible angle θ. For iron and steel the best combinations are (a) the (310) planes and cobalt $K\alpha$ radiation (Fig. 11-5), which produces a reflection at $\theta = 80°37.5'$, and (b) the (211) planes and chromium $K\alpha$ radiation, yielding a reflection at $\theta = 77°30'$. With Co$K\alpha$ radiation the strain thus measured is along the (310) plane normals, which make an angle of 9°22.5' with the normal to the specimen surface, and, since these reflecting planes are not exactly parallel to the surface, the resulting value of the stress $\sigma_1 + \sigma_2$ is slightly smaller than its true value by approximately 7 per cent. Useful plane and radiation combinations for the study of other metals are listed in Table 11-1.

When the photographic method is employed, the precision of the results depends to a considerable extent upon the quality of the diffraction patterns yielded by the specimen. If they are sharp, it will be possible to measure the d values to about ± 0.0001 Å. This error corresponds to a stress of about ± 3000 psi in an iron or steel specimen. With diffuse lines the error can be several times this much. The unstressed value of d may be obtained by removing the load or, with locked-in stress, by cutting out a small piece of the metal. If the latter is undesirable, the d value for unstressed material of similar composition may be used.

Table 11-1.   Plane, Radiation, and E Data for Stress Measurements

Metal	Radiation, $K\alpha$	Plane	Reflection Angle, θ	Typical E (10^6 psi)
Iron	Co	(310)	80°37.5′	12–14
	Cr	(311)	77°30′	
Steel	Co	(310)	80°37.5′	30
Martensitic	Cr	(211)	77°30′	
Austenitic	Cr	(220)	64°	
Aluminum	Cu	(511)/(333)	81°	9–10
	Cu($K\beta$)	(444)	76°30′	
	Co	(420)	81°18′	
	Cr	(311)	69°45′	
Copper	Co	(400)	81°46.5′	15–18
Brass (68 per cent Cu)	Co	(400)	75°30′	13
Brass (cartridge)	Ni	(331)	79°	13
Magnesium	Fe	(105)	83°	6
Titanium (cubic alloys)	Cu	(213)	~ 60°40′	16–18

11-4   COMPONENT OF STRESS IN ANY DESIRED DIRECTION IN A SURFACE

The sum of the stresses $\sigma_1 + \sigma_2$ in a surface is not of too great utility. A more generally useful quantity is the surface stress in a given direction, which may be determined from two photographs of the stressed material, one with the beam perpendicular to the surface and another with it inclined at a known angle to the surface and lying in the vertical plane fixed by the surface direction of interest[4]. Consider that the stress σ_ϕ is desired at the point O in the ϕ direction of the metal plate of Fig. 11-6. It can be obtained from photographs taken along the Z direction and the ψ direction. The principal stresses σ_1, σ_2, and σ_3 are taken parallel to X, Y, and Z, respectively, and α_1, α_2, and α_3 are the direction cosines of the ψ direction relative to these axes. In terms of the experimental angles ψ and ϕ, the direction cosines may be written

$$\alpha_1 = \sin\psi\cos\phi,$$
$$\alpha_2 = \sin\psi\sin\phi, \tag{11-14}$$
$$\alpha_3 = \cos\psi = (1 - \sin^2\psi)^{1/2}.$$

By relation 11-10 the stress in the ψ direction is

$$\sigma_\psi = \sigma_1(\sin\psi\cos\phi)^2 + \sigma_2(\sin\psi\sin\phi)^2 + \sigma_3\cos^2\psi. \tag{11-15}$$

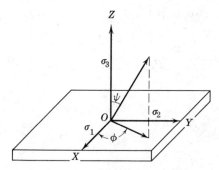

Fig. 11-6. Angular relations for determination of the stress component in the ϕ direction.

This in turn reduces to σ_ϕ when $\psi = 90°$:

$$\sigma_\phi = \sigma_1 \cos^2 \phi + \sigma_2 \sin^2 \phi. \tag{11-16}$$

To evaluate equation 11-16 from x-ray strain measurements, the strain in the ψ direction is expressed in terms of expression 11-11:

$$e_\psi = e_1 (\sin \psi \cos \phi)^2 + e_2 (\sin \psi \sin \phi)^2 + e_3 \cos^2 \psi. \tag{11-17}$$

Remembering that $\sigma_3 = 0$, and substituting e_1 and e_2 from equation 11-5, we may write

$$e_\psi = \frac{1}{E} \left[(\sigma_1 - \nu\sigma_2) \cos^2 \phi \sin^2 \psi + (\sigma_2 - \nu\sigma_1) \sin^2 \phi \sin^2 \psi \right] + e_3 \cos^2 \psi \tag{11-18}$$

which, on rearranging and collecting terms, becomes

$$e_\psi = \frac{1+\nu}{E} (\sigma_1 \cos^2 \phi + \sigma_2 \sin^2 \phi) \sin^2 \psi + e_3. \tag{11-19}$$

Substitution of equation 11-19 in equation 11-16 thus leads to

$$\sigma_\phi = (e_\psi - e_3) \frac{E}{(1+\nu) \sin^2 \psi}. \tag{11-20}$$

The quantity $e_\psi - e_3$ is determined by measuring the d spacing of the appropriate high-θ plane (Table 11-1) when oriented essentially perpendicular to the ψ direction and when nearly normal to Z. Thus, in terms of d_ψ, d_Z, and the unstressed spacing d_U,

$$e_\psi - e_3 = \frac{d_\psi - d_U}{d_U} - \frac{d_Z - d_U}{d_U} = \frac{d_\psi - d_Z}{d_U}. \tag{11-21}$$

Since a measurement of d_U may be difficult, or even impossible, the problem is usually avoided by writing equation 11-21 as the close approximation

$$e_\psi - e_3 = \frac{d_\psi - d_Z}{d_Z}. \tag{11-22}$$

It is then only necessary to substitute equation 11-22 in equation 11-20 to obtain as a working relation

$$\sigma_\phi = \frac{d_\psi - d_Z}{d_Z} \cdot \frac{E}{(1+\nu)\ \sin^2 \psi}, \tag{11-23}$$

which with the further substitution,

$$\frac{d_\psi - d_Z}{d_Z} = \frac{\Delta d}{d} \simeq \cot \theta \Delta \theta,$$

may be written in the alternative form

$$\sigma_\phi = \frac{E}{1+\nu}\ \frac{\cot \theta \Delta \theta}{\sin^2 \phi}. \tag{11-24}$$

The numerical value of E is independent of ψ for metals that are isotropic or nearly so. Even though this is not true for anisotropic specimens, equations 11-23 and 11-24 are still generally applicable in stress measurement. As long as two values of the interplanar spacing are determined at the same angles ψ_1 and ψ_2, their difference $d_{\psi_2} - d_{\psi_1}$ will always be proportional to the stress in spite of possible differences in E at ψ_1 and ψ_2, so that[8]

$$\sigma_\phi = K(d_{\psi_2} - d_{\psi_1}). \tag{11-25}$$

K, known as the stress constant, can easily be evaluated for any given material by stressing the specimen in known increments and measuring the *difference* in d values at the two selected ψ angles, commonly $0°$ (d_Z) and $45°$.

For the evaluation of any given experimental technique, Norton[7] has emphasized the importance of dividing the stress constant K (equation 11-25) into two factors,

$$K = K_E \cdot K_G, \tag{11-26}$$

in which

$$K_E = \frac{E}{1+\nu}, \tag{11-27}$$

the elastic constant of the material, and K_G depends only on the geometrical properties inherent in the technique.

11-4.1 Photographic Techniques

Experimental details of the back-reflection method, including calibration of the specimen-to-film distance, are adequately treated in Section 4-3.2. The specimen support and orienting device varies greatly with the nature of the specimen. Figure 11-7 displays a simple adjustable support

Fig. 11-7.   Simple specimen holder for stress measurements in small metal plates.

we have used in studying surface stresses in metal plates resulting from grinding operations. It is frequently necessary, however, to take the diffraction equipment to the specimen rather than the specimen to the equipment. Readily portable and mobile diffraction units for this purpose, with the back-reflection camera attached to the x-ray tube (Fig. 11-8), have been described[11]. Exposure times vary with the experimental conditions from 10 min to several hours. Rotation or oscillation of the film about its center during the exposure is usually necessary to prevent spottiness of the diffraction lines.

Unless one wishes to study the surface effects of grinding, shot-peening, or machining, it is important to prepare the surface properly by removing any surface layer not typical of the condition being studied. The surface scale is first removed, and the area flattened by careful grinding or filing. After this preparation the surface is vigorously etched

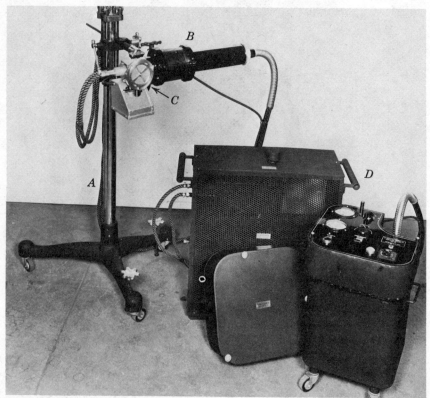

Fig. 11-8. Portable x-ray equipment for back-reflection method of stress analysis. (*A*) Adjustable stand for x-ray tube; (*B*) x-ray tube; (*C*) back-reflection camera directly attached to x-ray tube; (*D*) transformer and controls. (Courtesy of H. R. Isenburger, *Machinery*, July 1947).

(30 per cent nitric acid solution for steel) to remove the cold-worked layer. The etching leaves a fairly rough surface, which is smoothed with fine abrasive paper and then lightly etched with a 10 per cent alcoholic solution of nitric acid. Surfaces prepared in this way should give very reproducible results.

A. Double-Exposure Technique (DET). The experimental procedure in preparing the photograph at normal incidence follows that suggested in Section 11-3. For the inclined photograph the value of $\psi_0 = 45°$ is commonly taken, leading to the reflection geometry depicted in Fig. 11-9. Since the sets of planes P_1 and P_2 contributing to the diffraction halo make different angles, $\psi = \psi_0 \pm \eta$, with the surface normal, the strains in the two sets are different, and their Bragg angles are slightly different. Accordingly, the angles η_1 and η_2 are nearly, but not quite,

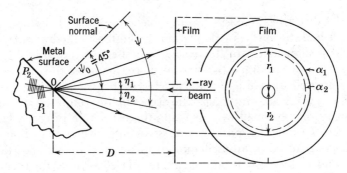

Fig. 11-9.   Reflection geometry for the photograph at inclined incidence ($\psi_0 = 45°$).

equal. The diffraction ring in turn is not a perfect circle, and measurement of r_1 leads to d_1 for the P_1 planes at angle $\psi_0 - \eta_1$, whereas r_2 gives d_2 for the P_2 planes at angle $\psi_0 + \eta_2$, the latter being the more sensitive. Calibration of the distance D may be done by either of the suggested methods (Section 4-3.2). Because of the distortion of the diffraction halo, it is not possible to remove ring spottiness by complete rotation of the film

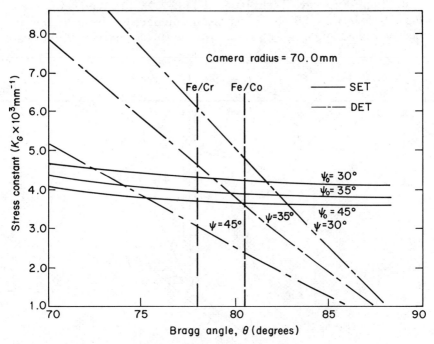

Fig. 11-10.   Stress constant K_G as a function of θ. (Courtesy of J. T. Norton, *Advances in X-ray Analysis*, Vol. 11, Plenum Press, 1968, p. 401.)

about the direct-beam axis. Considerable smoothing of the lines can be
accomplished, however, by oscillation through a limited angular range.
This has a slight effect of decreasing the effective ψ on the side of the
film next to the specimen (r_2), and of increasing ψ on the opposite side.
This same oscillation also introduces a bit of uncertainty in the azimuthal
direction ϕ by an amount $\pm \Delta \phi$. The errors in the computed stresses may
be kept negligibly small by restricting the oscillation range to less than
$\pm 30°$[12]. Norton[7] reports that a great improvement in line quality
can also be realized by an oscillation of the angle ψ by $\pm 2°$. For the DET
the geometrical stress constant of equation 11-26 assumes the form[7]

$$K_G = \frac{\cot \theta_0}{2R \sin^2 \psi_0},$$ (11-28)

in which R is the specimen-to-film distance and θ_0 is the mean value of θ
corresponding to the reflection profile. In Fig. 11-10 the variation of K_G
with θ is plotted for ψ or ψ_0 equal to 30, 35, and 45°. It is seen that K_G,
which is a measure of the sensitivity of the technique employed, de-
creases with increasing θ or ψ_0, or both, for the DET.

B. Single-Exposure Technique (SET) [5-7, 12, 13]. With reference to
Fig. 11-11, in the SET the direct beam impinges on the specimen surface at a fixed angle ψ_0 with respect to the surface normal N, the beam

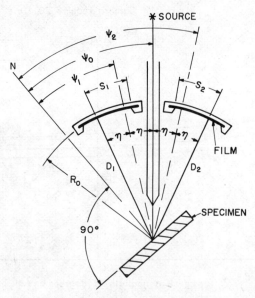

Fig. 11-11. Schematic diagram of single-exposure technique. (Courtesy of J. T. Norton,
Advances in X-ray Analysis, Vol. 11, Plenum Press, 1968, p. 401.)

being inclined toward the direction in which the component σ_ϕ is to be determined. Two diffracted beams (D_1 and D_2) are measured from planes of a given index triplet (*hkl*) whose normals make angles ψ_1 and ψ_2, respectively, with the surface normal. The beams D_1 and D_2 are registered simultaneously on separate films or on two sides of one film, and the position of the diffraction ring of the specimen is measured with respect to that of a calibrating substance at the top ($\psi_1 = \psi_0 - \eta$) and at the bottom ($\psi_2 = \psi_0 + \eta$) of the ring, from which spacings d_1 and d_2 are derived. Then application of equation 11-20 to this pair of measurements yields two simultaneous equations, the solution of which is[12]

$$\sigma_\phi = \frac{E}{1+\nu} \frac{d_1 - d_2}{d_\perp} \frac{1}{\sin^2 \psi_0 \sin^2 \eta}. \tag{11-29}$$

If ψ_0 is 45°, the factor $\sin^2 \psi_0$ in this equation becomes unity.

In the application of equation 11-29, it is $d_1 - d_2$ rather than $d_\perp$ that must be measured with great accuracy; for $d_\perp$ it is quite satisfactory to use the accepted interplanar spacing of the metal concerned in its unstressed state. With reference again to equation 11-26, for the SET the value of K_G is[7]

$$K_G = \frac{1}{4R \sin^2 \theta_0 \sin 2\psi_0}, \tag{11-30}$$

which may be compared with equation 11-28 for the DET. By reference to Fig. 11-10, it is seen that the sensitivity of the SET, unlike that of the DET, is virtually independent of θ and does not vary markedly with changes in ψ_0. Hence with the SET appreciable errors in the diffraction angle or inclination angle have but negligible effects upon the derived stress values.

C. General Considerations [7]. To paraphrase Norton[7], for either the SET or the DET, the most serious problem encountered in the determination of *absolute* stress values is the proper evaluation of the elastic constant K_E in equation 11-26. Extensive investigations to date have still not disclosed a satisfactory theoretical basis for its numerical calculation, and if nominal, mechanically determined bulk values of E and ν for a given metal are used, errors as large as 30 per cent may be incurred. The "true" value of K_E evidently depends on several factors including the composition of the specimen, its structure, its prior heat treatment, and the x-radiation and diffracting planes utilized. Hence at present the only practical course of action seems to be to determine the stress constant K experimentally with a specimen and conditions identical to those of the actual x-ray measurements.

Finally, it is perhaps worthwhile to summarize the relative merits of The DET and SET as follows[7]: (1) Because of the DET's lower K_G value, the precision of its results is better than that of the SET by 10 to 30 per cent. However, this apparent advantage is actually attainable only if the alignment is very precise and only with the expenditure of double or triple the time for each individual measurement. (2) Measurements by the SET can be made more quickly and simply and with less demand upon the skill of the worker. (3) Duplicate measurements are a practical necessity in the SET; at the same time they can be performed at least as quickly as a single measurement by the DET. (4) The experience of many years has demonstrated that the SET is very useful and well adapted to many practical stress-measurement problems.

11-4.2 Diffractometric Techniques

The pioneering diffractometric technique described by Christenson and Rowland[8] in 1953 was substantially improved in several respects by Koistinen and Marburger and reported in another significant paper in 1959[14]. The basic features of their method are embodied in most current diffractometric techniques. The Koistenen-Marburger technique, besides proving generally applicable to all metal systems, achieved notable success in the measurement of stresses in hardened and very high-strength steels, a class of materials that at best yield only poor analytical results by the photographic method.

The double-exposure technique has been used in most of the diffractometric investigations thus far reported. The sample-supporting device must be very precisely constructed so as to permit the specimen surface to be rotated through the desired angle ψ_0* about an axis coincident with the goniometer axis. Because of the considerable breadth of most back-reflection lines employed in stress analysis (as much as 8° at half-maximum intensity in some steels), it is not necessary that the collimation be sharp except under certain special conditions. Therefore the direct-beam divergence in the equatorial plane is commonly set at 3 or 4°, and the angular width of the receiving slit 0.2 to 0.5°.

As shown in Fig. 11-12, if the parafocusing condition (and therefore optimal definition of the reflection) is to be preserved at ψ values other than 0°, the normally constant specimen-to-receiver distance, or goniometer radius R, must be reduced by an amount D to a radius R' [12, 15]

---

*In its usage in this discussion, the angle ψ is to be carefully distinguished from ψ_0: ψ = angle between the normal to the specimen surface and the normal to the diffracting planes; ψ_0 = angle between the normal to the specimen surface and the incident beam.

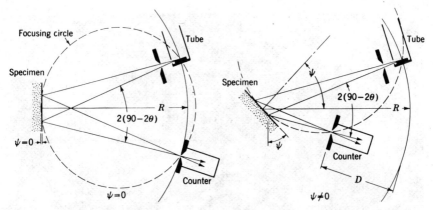

Fig. 11-12. Diffractometric parafocusing geometry for zero and nonzero inclination angles ψ. (From *Structure of Metals*, 3rd ed., by C. S. Barrett and T. B. Massalski. Copyright 1966 by McGraw-Hill Book Company. Used with permission of the copyright owner.)

such that

$$R' = R - D = R\frac{\cos\left[\psi + (90° - \theta)\right]}{\cos\left[\psi - (90° - \theta)\right]}. \tag{11-31}$$

The achievement of this condition requires a modification of a standard diffractometer providing for very precise movement of the receiver carriage along a radial track, or ways, until the receiving slit is situated at R'. Figure 11-13 shows a diffractometer, as modified by Wallace and Terada[16], which provides this required radial adjustment of the receiver and which also includes a tensometer for subjecting reference specimens to known stresses. With respect to the theoretical need for a reduction in R for $\psi \neq 0°$, the critical nature of the alignment required has led several investigators[15, 17, 18] to maintain a fixed R in actual practice despite the loss of good focusing. This practice does not result in unacceptable distortion and displacement of the diffracted-beam profile provided the direct-beam divergence is limited to about 1° and the receiving-slit width made as small as possible without excessive loss of intensity. Ogilvie[19] adopted a compromise solution with the counter at its standard fixed distance but the receiving slit of fixed width (0.060 in.) moved radially to the calculated parafocusing position R'.

Zantopulos and Jatczak[17] investigated the systematic errors in measured stresses caused by specimen geometry and beam divergence when the centroid of the reflection is used as the criterion of its position. They found that at $2\theta = 156°$ the systematic errors resulting from the fixed-R (nonfocusing) technique were only about one-third as large as

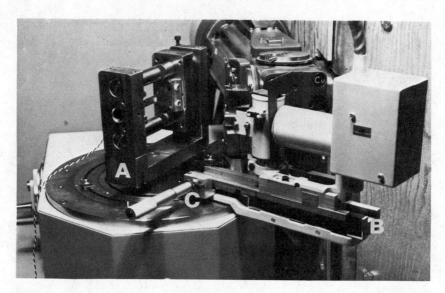

Fig. 11-13. Diffractometer as modified by Wallace and Terada for residual stress measurements. (A) Tensometer, (B) sliding track for the counter, (C) micrometer for adjust-ysis, Vol. 14, Plenum Press, 1971, p. 389.)

the errors caused by aberrations of the parafocusing technique and, furthermore, that the latter errors increased as 2θ increased above $156°$ and, conversely, decreased as 2θ decreased. For either technique the magnitude of the errors was not large ($< \pm 7500$ psi), even with a beam divergence as large as $3°$ and convex specimen surfaces; moreover, for a divergence of $1°$, the systematic errors amounted to only ± 2300 psi or less. It is of course true that any of the above changes in the goniometer radius or in the widths of the divergence and receiving slits introduce appreciable changes in the intensity of the diffracted beam; however, for the measurement of stresses it is not the intensity of the reflection, but its position, that is significant and must be precisely determined.

The Lorentz (L) and polarization (P) factors vary appreciably over the angular ranges occupied by the broadened back reflections from stressed metals. Furthermore, the diffraction profiles are considerably distorted as a consequence of the absorption of the x-ray beams in the specimen for inclinations ψ deviating much from $0°$. Hence all "raw" reflection profiles must be corrected for the $L \cdot P$ factor by application of the factor

$$\frac{1}{L \cdot P} = \frac{\sin^2 \theta \cos \theta}{1 + \cos^2 2\theta},$$

(11-32)

and, when $\psi \neq 0°$, for absorption[14] by means of the factor

$$\frac{1}{A(\psi, \theta)} = \frac{1}{1 - \tan \psi \cot \theta}, \tag{11-33}$$

before their peak positions are determined.

Diffractometric recording makes possible a more objective and precise determination of the peak position than is possible with photographic reflections. Following a method first employed by Ogilvie[19], it has become the usual practice to fit the corrected intensity profile to a parabola using either three or five points. Ogilvie used five points and determined the best fit by least squares. For reasons of simplicity and speed, the three-point parabolic fit has been most generally used. Figure 11-14 illustrates the procedure followed in fitting a parabola to three points distributed at uniform angular intervals about the apex of the corrected intensity profile. If $2\theta_1$, $2\theta_2$, and $2\theta_3$ are three sequential data points separated by the angular interval $\Delta 2\theta$, it can be shown that the location of the vertex $2\theta_0$ of the vertical-axis parabola is given by[14]

$$2\theta_0 = 2\theta_1 + \frac{\Delta 2\theta}{2}\left[\frac{3a+b}{a+b}\right]. \tag{11-34}$$

Figure 11-15, from the work of Koistinen and Marburger[14], shows the "as-measured" data points over the central portion of a back-reflection at $\psi = 0°$, including uniformly spaced points (solid circles) together with their corrected counterparts (solid squares) and the parabola fitted to them (solid line). With $\Delta 2\theta = 1.0°$, $2\theta_1 = 154.5°$, $a = 4.5$ sec, and $b =$

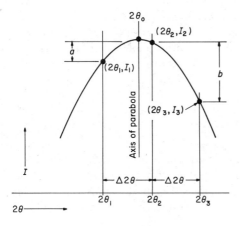

Fig. 11-14.   A vertical-axis parabola fitted to three points near the peak of a reflection profile. (Courtesy of D. P. Koistinen and R. E. Marburger and the American Society for Metals – Metallurgical Society of AIME[14].)

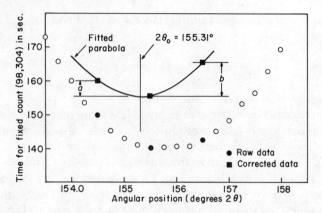

Fig. 11-15. Peak position (155.31° 2θ) as determined from a parabola fitted to corrected x-ray intensity data for $\psi = 0°$. (Courtesy of D. P. Koistinen and R. E. Marburger and the American Society for Metals — Metallurgical Society of AIME [14].)

10.0 sec, the numerical solution of equation 11-34 is:

$$2\theta_0 = 154.5° + \frac{1}{2}\frac{3 \times 4.5 + 10.0}{4.5 + 10.0}$$
$$= 154.5° + 0.81° = 155.31°,$$

This angle, then, designates the location of the vertex of the fitted parabola, which becomes the criterion of the position of the corrected reflection profile. Note that the peak and parabolic profiles of Fig. 11-15 are inverted with respect to those of Fig. 11-14, which results from the plotted data being seconds for a fixed count rather than counts per second.

Kelly and Short[20], on the basis of the three-point parabola-fitting procedure, derived expressions for calculating the standard deviation σ_S in a given stress measurement due to random counting statistics alone. They also showed how to estimate the number of counts required at each point to attain a desired standard deviation in the calculated stress. For $\psi = 45°$ and with a number of simplifying assumptions, Kelly and Short arrived at the expression

$$\sigma_S = \frac{K \Delta 2\theta}{2(1-p)N^{1/2}}, \tag{11-35}$$

in which K is the stress constant (see equation 11-25), $\Delta 2\theta$ is the fixed angular increment between the three points of measurement across the reflection (which are assumed to be symmetrically distributed about the maximum point), N is the accumulated count at the maximum points of the 0 and 45° reflection profiles (assumed to be of equal intensity), and $1-p$ represents the *relative difference* in accumulated counts between the central and outer points of the parabola, it being further assumed that $p \geqslant 0.85$. Figure 11-16 is a graph of σ in ksi (psi $\times 10^{-3}$) as a function of $\Delta 2\theta/(1-p)$ for $\psi = 45°$ and various typical values of N as calculated with equation 11-35[20].

In regard to the possible need for making background corrections prior to determination of the reflection position, it can be said that in nearly all instances the $2\theta_0$ values derived from an experimental reflection profile before and after subtraction of the background are found to differ from each other only negligibly. Furthermore, the extreme breadths of the α_1 reflection profiles of hardened metals together with the close proximity of the α_2 profiles actually render it impracticable to attempt meaningful background corrections in such circumstances. Finally, since in nearly all stress measurements it is the *shift* in peak position rather than the absolute position that is important (Δd rather than d, as has been pointed out in connection with the stress equation 11-29),

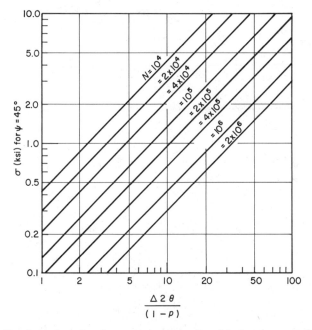

Fig. 11-16. Standard deviation in measurements of residual stress as a function of $\Delta 2\theta/(1-p)$ for various values of accumulated counts N. Plots calculated using equation 11-35 with $\psi = 45°$. (Courtesy of C. J. Kelly and M. A. Short, *Advances in X-ray Analysis*, Vol. 14, Plenum Press, 1971, p. 377.)

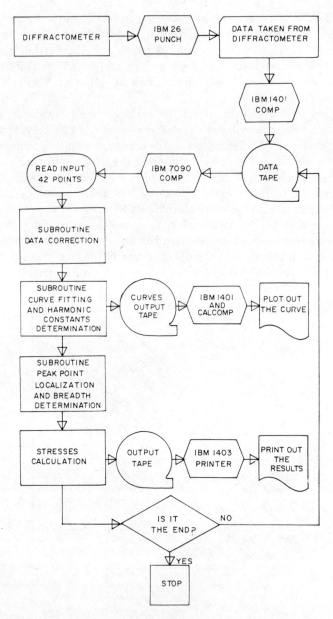

Fig. 11-17.   Flow chart for automatic recording and processing of diffractometric residual-stress data. (Courtesy of G. Koves and C. Y. Ho and Philips Electronic Instruments [21].)

background corrections tend to be irrelevant, especially as they are virtually invariant for a series of specimens of the same type.

A special potential of diffractometric techniques not afforded by photographic methods is that of automatic data registration and processing. For the study of magnetic iron alloys using $CrK\alpha$ radiation and the 211 reflection, Koves and Ho[21] modified a Norelco high-angle goniometer to permit convenient setting of the inclination angle ψ to 0 or 45°, and adjustment of the goniometer radius to the corresponding values of R and R' for parafocusing. A step-scan mode of counting was employed (Section 5-4.5), and the numerical output impulses, instead of actuating the customary printer, were modified so as to operate a key punch. Then the only manual step in the entire operation was the transferral of the output data card deck from the key punch to the computer card reader. Figure 11-17 is a flow chart of the data processing and computing operations performed by an IBM 7090 computer. In order to permit the evaluation of both microstrains and microstresses (Chapter 9) as well as macrostresses Koves and Ho measured 21 discrete points over the entire reflection profile. These "raw" intensity data are corrected for the Lorentz-polarization and absorption factors, after which the profile is subjected to Fourier harmonic analysis, yielding a least-squares fitted analytic equation. The microstress is calculated from the reflection breadth, and the macrostress from the peak displacement for $\psi = 0°$ and 45° by equation 11-24. Koves and Ho[21] wrote a special Fortran program for processing the data and calculating the micro- and macro-stresses. From Fig. 11-17 it can be seen that in addition to a print-out of the output, the option is also available of obtaining automatic plots of the line profiles by means of a Calcomp plotter. These investigators found that repeated scans of a given specimen showed very satisfactory reproducibility, amounting to a maximum deviation in the difference between the two peak positions ($\psi = 0$ and 45°) of 0.01° 2θ.

Braski and Royster[22] utilized a General Electric XRD-5 diffractometer to study residual stresses in cubic titanium alloy sheet with $CuK\alpha$ radiation and the (213) planes. They wrote a computer program to make the $L \cdot P$ and $A(\theta, \psi)$ corrections, fit three-point parabolas to the 0° and 45° ψ reflections, and calculate the stress values.

11-4.3 Selected Investigations

A. Hardened Steel: Comparison of X-ray and Mechanical Stress Measurements[14, 23]. In the course of studies of the effects of grinding at various depths below the surface, Letner[23] devised a very precise

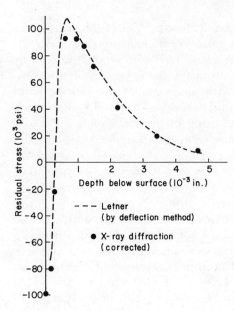

Fig. 11-18. Comparison of stresses measured by x-ray and mechanical-deflection methods. Residual stress as a function of depth below a ground surface on hardened steel (RC 59). (Courtesy of D. P. Koistinen and R. E. Marburger and the American Society for Metals—Metallurgical Society of AIME[14].)

mechanical-deflection method for measuring residual stresses with the aid of a comparator. Figure 11-18 compares some of Letner's measurements with x-ray stress values on the same specimens obtained by Koistinen and Marburger using the diffractometric technique described in their article[14], which includes the $L \cdot P$ and $A(\psi, \theta)$ corrections, a correction for the known penetration of the x-ray beam, and the three-point parabolic fit of the reflection profile. It is seen that very good agreement was obtained over the entire stress range, $-100,000$ to $+100,000$ psi.

B. Aluminum Alloy 2024 and Ingot Iron: Determination of Elastic Constants[24].

Barrett[25] showed that for uniaxial stresses applied to an isotropic solid under homogeneous deformation:

$$\epsilon_{\psi, L} = \frac{\sigma_L}{E} [(1 + \nu) \sin^2 \psi - \nu]. \tag{11-36}$$

In this equation E and ν have their usual meanings, $\epsilon_{\psi, L}$ is the lattice strain at the inclination angle ψ lying in the plane defined by the surface normal and the longitudinal direction, and σ_L is the microstress in the longitudinal direction. The object of the study (by Donachie and Norton [24]) was to confirm the applicability of equation 11-36 to the two metals and to compare the numerical values of ν and E determined by x-ray diffraction with the mechanically determined, or nominal bulk, values.

For both metals the specimens were $\frac{1}{16}$-in. sheet, heat-treated after machining to make them as strain-free as possible in their initial condition. The x-ray measurements were performed with a General Electric XRD-3 diffractometer modified to permit adjustment of the specimen-to-receiver distance in order to preserve the parafocusing condition for an inclination angle ψ. For each metal a series of specimens was measured at stresses up to 30,000 psi and for $\psi = 0$, 15, 32.5, 44.5, and 61.5°, and the interplanar spacings of two reflections were determined by employing two wavelengths. The numerical results obtained for the mean elastic constants E, ν, and $K_E = E/(1+\nu)$ are listed in Table 11-2 and compared with the nominal bulk values for each metal.

Figure 11-19 is a schematic representation of equation 11-36 showing that the theory predicts $\epsilon_{\psi,L}$ to be a linear function of $\sin^2 \psi$ with

$$\epsilon = -\frac{\nu\sigma_L}{E} \qquad \text{for} \qquad \sin^2 \psi = 0,$$

$$\epsilon = \frac{\sigma_L}{E} \qquad \text{for} \qquad \sin^2 \psi = 1,$$

$$\epsilon = 0 \qquad \text{for} \qquad \sin^2 \psi = \frac{\nu}{1+\nu}.$$

By way of illustration Fig. 11-20 shows plots of ϵ versus $\sin^2 \psi$ for one set of experimental measurements of the (420) planes of 2024-aluminum alloy. It is evident that the three plots are linear and that they intersect rather well at $\epsilon = 0$, leading to well-defined values of the elastic con-

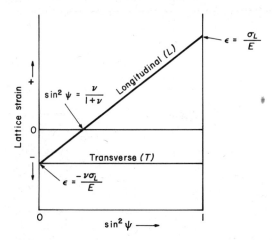

Fig. 11-19.   Schematic representation of lattice strain as function of $\sin^2\psi$ for a uniaxial stress (equation 11-36). (Courtesy of M. J. Donachie, Jr. and J. T. Norton and the American Society for Metals — Metallurgical Society of AIME [24].)

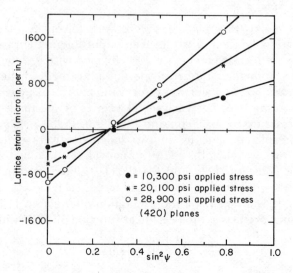

Fig. 11-20. Longitudinal lattice strain versus $\sin^2 \psi$ in 2024 = aluminum alloy for stresses in the elastic range. (Courtesy of M. J. Donachie, Jr. and J. T. Norton and the American Society for Metals — Metallurgical Society of AIME [24].)

stants ν and E. Likewise, a representative set of experimental data for the (310) planes of ingot iron provided linear plots of quality similar to that of Fig. 11-20 for aluminum.

An examination of the data of Table 11-2 shows that for the aluminum alloy the "x-ray" elastic constants for the (511)/(333) and (420) lattice planes agree with each other and differ from the bulk elastic constants by not more than 6.5 per cent. Contrariwise, the "x-ray" elastic constants

Table 11-2. X-Ray and Bulk Values of the Elastic Constants of Aluminum Alloy and Ingot Iron[a]

Number of Specimens	Radiation	Planes	$\bar{E}(10^6\,\text{psi})$	$\bar{\nu}$	$K_E = \bar{E}/(1+\nu)$
202Y-Aluminum alloy:					
5	CuKα	(511)/(333)	11.47	0.349	8.51
2	CoKα	(420)	11.46	0.345	8.51
Nominal bulk values			10.8	0.33	8.13
Armco ingot iron:					
4	CrKα	(211)	36.36	0.304	27.9
3	CoKα	(310)	27.42	0.356	20.2
Nominal bulk values			29.0	0.28	22.6

[a]Data from Donachie and Norton [24].

for the (211) and (310) planes of ingot iron differ very considerably from each other, and the K_E values for the (211) and (310) planes are, respectively, 24 per cent higher and 11 per cent less than the nominal bulk value of 22.6. Donachie and Norton[24] conclude that the most likely cause of these differences lies in the anisotropy of iron.

In general, efforts to calculate theoretical values of the elastic constants for different planes of various materials have been unsuccessful. Thus, as suggested in Section 11-4, the conversion of "x-ray" lattice strains to macrostresses requires the prior empirical establishment of the appropriate elastic constants for each material studied. Of course, in actual practice the numerical value of the working stress constant, $K = K_E \cdot K_G$, rather than K_E, is evaluated directly, thus including the proportionality constant of the geometrical arrangement.

C. High-Strength Aluminum Alloys: Residual Stress Measurements

[26]. In research on fatigue in high-strength aluminum alloys, Hilley, Wert, and Goodrich[26] employed the diffractometric technique of Koistinen and Marburger ([14] and Section 11-4.2) to (1) determine the elastic constants for several 5083-aluminum alloys, (2) measure the residual macrostress as a function of degree of cold work induced by rolling, and (3) evaluate the extent of relaxation realized by the removal of layers of different thickness. The experimental conditions were selected with a view to minimizing sources of error in the measurements, especially by the selection of maximum practicable values of 2θ and by effecting high precision in the calibration of the elastic constants for the crystallographic planes involved.

The 5083-aluminum alloy contained 4.58 per cent magnesium, 0.70 per cent manganese, and slight traces of iron and silicon. The as-received H-323 temper alloy, in the form of 0.094-in. sheet, was cold-rolled in the course of the investigation to yield reductions of 10, 20, and 30 per cent in thickness. The x-ray measurements were made with a General Electric XRD-5 diffractometer having a CA-8L high-intensity copper tube and proportional counter. The (511)/(333) reflection was measured with a 3° primary-beam divergence slit and a 0.5° detector slit for $\psi = 0°$ and a 0.2° slit for $\psi = 45°$. The target take-off angle was 4°, and the preamplifier housing was modified so as to permit measurements to be made as high as 164° 2θ for $\psi = 45°$. For $\psi = 45°$ the sample-to-receiver distance R was reduced as required to maintain the parafocusing condition. The recorded intensities were corrected for $L \cdot P$ and $A(\psi, \theta)$ and fitted to a vertical-axis parabola by the three-point procedure. Best results were achieved when the three measurements were made at 2θ values corresponding to intensities 85 per cent or more above background, and when the difference in the corrected intensities on either side of the central data point did not vary by more than a factor of 2. The specimens were subjected to known uniaxial tensile strains in a straining jig mounted on the goniometer to permit calibration of the elastic constants following the procedure prescribed in Society of Automotive Engineers Technical Report 182[27].

For the determination of the stress distribution as a function of depth, successive layers of the specimen were removed by electropolishing in a 25 volumetric per cent solution of perchloric acid in absolute alcohol. Figure 11-21 shows the variation of longitudinal and transverse residual surface stresses with per cent reduction in thickness performed by single-pass rolling at room temperature. With increasing reduction in thickness, the stresses are seen to increase to maximum values of about 10 ksi at 20 per cent reduction in the longitudinal direction, and 8.8 ksi near 14 per cent reduction in the transverse direction. The marked deformation-induced recovery in the transverse direction above 14 per cent reduction can be accounted for, in part at least, by the production of edge cracks.

Figure 11-22 shows the distribution of longitudinal and transverse stresses with depth below the rolling surface for the as-received H-323 temper alloy. It is noteworthy that a second stress maximum occurs about 0.003 in. below the surface, and a minimum at a depth of about 0.0005 to 0.001 in. Figure 11-23 indicates that, except for a marked increase in magnitude, a similar stress distribution with depth in the

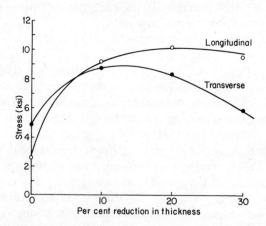

Fig. 11-21. Variation of longitudinal and transverse residual surface stresses in 5083 — aluminum alloy with rolling reduction at room temperature. (Courtesy of M. E. Hilley, J. J. Wert, and R. S. Goodrich, *Advances in X-ray Analysis*, Vol. 10, Plenum Press, 1967, p. 284.)

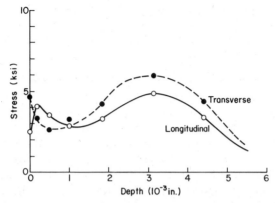

Fig. 11-22. Distribution of longitudinal and transverse residual stresses with depth for H323-temper 5083-aluminum alloy. (Courtesy of M. E. Hilley, J. J. Wert, and R. S. Goodrich, *Advances in X-ray Analysis*, Vol. 10, Plenum Press, 1967, p. 284.)

longitudinal direction persists after rolling to a 10 per cent reduction in thickness. Likewise, the transverse measurements (not shown) gave a stress distribution resembling the original curve except for an increase in magnitude to a maximum value of about 12 ksi. Hilley, Wert, and Goodrich[26] were unable to offer an explanation for the minimum at 0.0005 to 0.001 in. below the rolling surface. This investigation was also

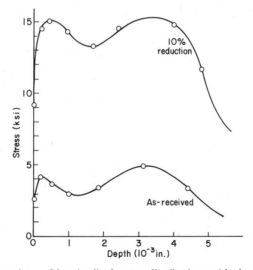

Fig. 11-23. Comparison of longitudinal stress distributions with depth for as-received and 10 per cent cold-rolled 5083-aluminum alloy. (Courtesy of M. E. Hilley, J. J. Wert, and R. S. Goodrich, *Advances in X-ray Analysis*, Vol. 10, Plenum Press, 1967, p. 284.)

continued to greater depths, where it was anticipated that compressive stresses would be encountered. It was felt certain that the high residual tensile stresses found at and below the surface would be detrimental to the fatigue behavior of the alloy.

D. Measurement of a Triaxial Residual Stress[28]. The difficult problem of measuring a triaxial residual stress in the interior of a body has been studied by Rosenthal and Norton[28]. Earliest investigations of this problem evolved the well-known Sachs boring-out method[29], which is limited to cylindrical bars and tubes. Later, Stablein[30] applied similar relations to rectangular and flat bars. The method proposed by Rosenthal and Norton, however, can be applied to more general types of stress, and is particularly suited to the study of stress in thick welded plates. Although all these techniques are based on some means of relaxation of the stress whereby the change produced in the interior is determined from the change observed on the surface, Rosenthal and Norton have applied this principle in a different manner by dealing with a small element that has been cut free from the body, rather than with the body itself. The procedure is best understood by considering the examination of a weld in butt-welded steel plates.

If it is desired to determine the residual stress along the weld axis, a narrow rectangular block is removed from the weld (by sawing) as depicted in Fig. 11-24A. The block is made narrow enough so that its removal has relieved most of the stress in a transverse direction through it, but has relieved only a part of the longitudinal or axial stress. If the length

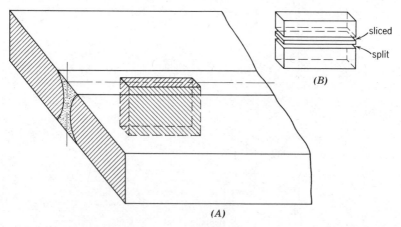

(A)

Fig. 11-24. (A) Removal of block from weld for study of triaxial residual stress. (B) Splitting and slicing of block for progressive relaxation of stress. (Courtesy of D. Rosenthal and J. T. Norton, *Welding J.* (Research Supplement), **24**, 295-s (1945).)

of the block is twice the thickness or more, an approximately linear law of relaxation is obtained through the thickness at all stages of relaxation. From the value of the stress on the top and bottom faces of the plate, the values of the stress relieved along the thickness direction can be calculated. The strain measurements required can be made by means of wire strain gages or, preferably, by x-ray diffraction. First, the relaxation of each block removed is measured in the longitudinal direction. Next, the value of the stress left in the block must be determined. This is done by splitting the block in half (Fig. 11-24B) and successively cutting slices about ⅛ in. thick from the midsurface of each half, progressing outward to the bottom and top faces of the original block. At each stage the amount of stress left is measured on the two outer faces. It is in the last stages of this progressive relaxation of the block that x-ray measurements are especially advised to avoid certain extrapolations and the use of correction factors required with wire gages. The amount of stress relieved in the interior of the specimen is then calculated from the progressive relaxation of the two outer surfaces, the total stress at various levels within the

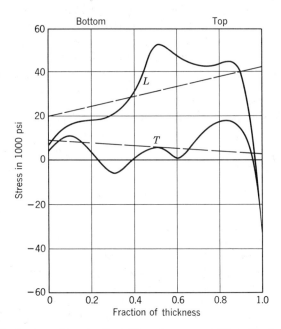

Fig. 11-25. Distribution of longitudinal (L) and transverse (T) residual stress across the thickness of a weld in a 1-in. welded plate. ---, Relief of stress due to removal of block; ——, total relief of stress. (Courtesy of D. Rosenthal and J. T. Norton, *Welding J.* (Research Supplement), **24**, 295-s (1945).)

body being simply the sum of the values obtained in these two relaxation steps. It is to be noted that this procedure measures the transverse and longitudinal stresses only, and that from them the stresses acting in the direction of the thickness are calculated by means of the differential equations of equilibrium. Complete details with actual examples are given by Rosenthal and Norton. Figure 11-25 illustrates the distribution of longitudinal and transverse stress across the thickness of a weld in butt-welded plates of 1-in. mild steel as determined by the method.

E. Other Experimental Work and Information. The intended scope of this chapter and space limitations preclude a detailed treatment of further important studies. We only briefly summarize a few and cite some additional references to guide the reader who is interested in pursuing this subject further.

For specimens deformed plastically (beyond the elastic limit) Wagner, Boisseau, and Aqua[31] have developed an alternative approach to the measurement of residual stress which utilizes Bragg-Brentano para-focusing ($\psi = 0°$) throughout and permits the simultaneous determination of a stacking-fault probability, α (see Section 9-2.5). In a noteworthy diffractometric investigation of surface residual stresses in cold-rolled α brass, Wallace and Terada[16] compared the numerical stress values obtained by the conventional two-exposure method with values arrived at following the theory of Wagner, Boisseau, and Aqua[31]. They concluded that stress data obtained by the latter method may at times be subject to error as a consequence of parameter changes induced by deformation.

MacDonald[32], using a Siemens diffractometer with $CuK\alpha$ radiation and the two-exposure technique, investigated stress distributions in carburized steel. Carburization introduces a constant residual stress σ_R into the specimen such that the overall observed stress σ_ϕ of equations 11-23 and 11-24 must be replaced by the sum

$$\sigma_\phi = \sigma_A + \sigma_R,$$

σ_A being the applied elastic stress. This results in a minor modification of the standard equations.

Diffractometric attachments for applying known tensile or bending stresses to the specimen *in situ* have been described by a number of investigators[19, 26, 32–37]. McCune[38] enclosed the goniometer of a Norelco diffractometer in a helium-filled polyethylene bag in order to improve the peak-to-background ratio for measurements of retained austenite and residual stress in steel with a very soft radiation such as Cr $K\alpha$.

The reader is referred to a bibliography[39] published in 1949 which lists 240 references on x-ray stress analysis. In 1968, Norton[40] published an excellent review of methods of x-ray stress measurement, including a good treatment of the practical aspects. In addition, Barrett[13] and Barrett and Massalski[12] have summarized much of the significant work in stress analysis published prior to the publication dates of their books.

11-5 PROBLEMS RAISED BY PLASTIC DEFORMATION

It is well known that when a metal specimen is stressed, by tension or bending, beyond its elastic limit and then released, the shifts in the diffraction lines indicate the existence of residual macroscopic compressive stresses adjacent to the surface even though these are not revealed by mechanical measurements[12]. Despite the fact that this problem has occupied the attention of theoretical investigators for many years, the various factors contributing to this phenomenon are still not well understood.

Studies by Donachie and Norton[41] and by Wood and Dewsnap[42] have shown that the residual lattice strains resulting from plastic deformation are not to any major extent caused by an intergranular stress system, as had been proposed by Greenough[43–45]. In this connection it must be remembered that the grains contributing to a given diffraction line change as the angle of inclination ψ varies; hence the lattice strain measured is the mean of the strains in all the crystallites (grains) whose reflecting normals are in the directions specified by ϕ and ψ[46]. There is also the problem of the dependence of the elastic coefficients upon the crystallographic direction of measurement in anisotropic substances (that is to say, dependence on the indices of the diffracting planes), even when the distortion does not exceed the elastic limit.

With more direct reference to plastic deformation, it is possible that surface grains may have a lower yield point than interior grains relative to stresses, with the result that the former might be plastically deformed while the latter are still within the elastic limit[12]. It may also well be true that the least distorted grains contribute most strongly to the diffraction peaks and thus exert a stronger influence on their measured positions than the more highly strained grains, which would tend to generate more diffuse reflections. Likewise the portions of a grain close to a boundary may differ in the same manner from the interior portions [12, 13, 46, 47].

Fortunately, the foregoing worrisome but largely indeterminate factors do not invalidate the x-ray method of stress measurement when it is circumspectly applied. All the evidence tends to show that the coherently diffracting regions of the grains are restrained by their environments in such a way that the average grain conforms to the theory of elasticity, and that consequently the diffracting regions are subjected to a stress which, on the average, is constant throughout the specimen [41]. Barrett and Massalski[12], as well as others, suggest that when uncertainty exists as to whether or not plastic deformation has taken place, data should be recorded using several wavelengths on the premise that the average of such a set of values is apt to be closer to the true macroscopic stress than a value derived from a single wavelength. Cullity [48] concludes that the x-ray method is valid whether or not plastic flow has occurred, provided only that such plastic flow has not taken place mainly in one direction. He feels, furthermore, that nonhomogeneous plastic flow does not invalidate the x-ray results, even when stacking faults are present, provided the stress is determined from two x-ray measurements (at normal incidence and at the angle ψ, or at angles ψ_1 and ψ_2) on the same stressed specimen, which is to say, the single- or double-exposure method as described in Section 11-4.1. For treatments of these matters in greater depth the reader is referred to the literature cited in this section as well as to the General References.

GENERAL REFERENCES

1. C. S. Barrett, *Structure of Metals,* 2nd ed., McGraw-Hill, New York, 1952, Chapter 14.

*2. C. S. Barrett and T. B. Massalski, *Structure of Metals,* 3rd ed., McGraw-Hill, New York, 1966,, Chapter 17.

3. A. L. Christenson and E. S. Rowland, "X-ray Measurement of Residual Stress in Hardened High Carbon Steel," *Trans. ASM,* **45**, 638 (1953).

4. B. D. Cullity, *Elements of X-ray Diffraction*, Addison-Wesley, Reading, Massachusetts, 1956, Chapter 17.

*5. R. Glocker, *Materialprüfung mit Röntgenstrahlen*, 5th ed., Springer, Berlin, 1971, Chapter 27.

6. G. B. Greenough in *X-ray Diffraction of Polycrystalline Materials* (H. S. Peiser, H. P. Rooksby, and A. J. C. Wilson, ed.), Institute of Physics, London, 1955, Chapter 30.

7. D. P. Koistinen and R. E. Marburger, "A Simplified Procedure for Calculating Peak Position in X-ray Residual Stress Measurements on Hardened Steel," *Trans. ASM,* **51**, 537 (1959).

8. M. H. Miller in *Handbook of X-rays* (E. F. Kaelble, ed.), McGraw-Hill, New York, 1967, Chapter 19.

9. L. I. Mirkin, *Handbook of X-ray Analysis of Polycrystalline Materials* (English translation by J. E. S. Bradley), Consultants Bureau, New York, 1964, Chapter 7.

10.   J. T. Norton, "Review of Methods of X-ray Stress Measurement", *Norelco Rep.*, **15**, 50 (1968).

*11.   A. Taylor, *X-ray Metallography*, Wiley, New York, 1961, Chapter 15.

SPECIFIC REFERENCES

[1]   H. H. Lester and R. H. Aborn, *Army Ordanance*, **6**, 120, 200, 283, 364 (1925–1926).

[2]   G. Sachs and J. Weerts, *Z. Phys.*, **64**, 344 (1930).

[3]   C. S. Barrett and M. Gensamer, *Phys. Rev.*, **45**, 563 (1934); *Physics* **7**, 1 (1936); also R. Glocker and E. Osswald, *Z. tech. Phys.*, **16**, 237 (1935).

[4]   F. Gisen, R. Glocker, and E. Osswald, *Z. tech. Phys.*, **17**, 145 (1936).

[5]   R. Glocker, B. Hess, and O. Schaaber, *Z. tech. Phys.*, **19**, 194 (1938).

[6]   D. E. Thomas, *J. Appl. Phys.*, **19**, 190 (1948).

[7]   J. T. Norton, *Advances in X-ray Analysis*, Vol. 11, Plenum Press, New York, 1968, p. 401.

[8]   A. L. Christenson and E. S. Rowland, *Trans. ASM*, **45**, 638 (1953).

[9]   For a detailed treatment the reader is referred to treatises such as the following: G. H. Lee, *An Introduction to Experimental Stress Analysis*, Wiley, New York, 1950, Chapters 1 and 2; S. Timoshenko, *Theory of Elasticity*, McGraw-Hill, New York, 1934. Very comprehensive treatment.

[10]   F. Wever and H. Möller, *Arch. Eisenhüttenw.*, **5**, 215 (1931–1932).

[11]   D. E. Thomas, *J. Sci. Instrum.*, **18**, 135 (1941); J. T. Norton and D. Rosenthal, *Proceedings of the Society for Experimental Stress Analysis*, Vol. 5, 1947, p. 71; H. R. Isenburger, *Machinery*, p. 167 (July 1947); also see reference 7.

[12]   C. S. Barrett and T. B. Massalski, *Structure of Metals*, 3rd ed., McGraw-Hill, New York, 1966, Chapter 17.

[13]   C. S. Barrett, *Structure of Metals*, 2nd ed., McGraw-Hill, New York, 1952, Chapter 14.

[14]   D. P. Koistinen and R. E. Marburger, *Trans. ASM*, **51**, 537 (1959).

[15]   H. R. Woehrle, F. P. Reilly III, W. J. Barkley III, L. A. Jackman, and W. R. Clough, *Advances in X-ray Analysis*, Vol. 8, Plenum Press, New York, 1965, p. 38.

[16]   W. Wallace and T. Terada, *Advances in X-ray Analysis*, Vol. 14, Plenum Press, New York, 1971, p. 389.

[17]   H. Zantopulos and C. F. Jatczak, *Advances in X-ray Analysis*, Vol. 14, Plenum Press, New York, 1971, p. 360.

[18]   R. C. Larson, *Advances in X-ray Analysis*, Vol. 7, Plenum Press, New York, 1964, p. 31.

[19]   R. E. Ogilvie, *Stress Measurement with the X-ray Spectrometer*, M.S. Thesis, Massachusetts Institute of Technology, 1952.

[20]   C. J. Kelly and M. A. Short, *Advances in X-ray Analysis*, Vol. 14, Plenum Press, New York, 1971, p. 377.

[21]   G. Koves and C. Y. Ho, *Norelco Rep.*, **11**, 99 (1964).

[22]   D. N. Braski and D. M. Royster, *Advances in X-ray Analysis*, Vol. 10, Plenum Press, New York, 1967, p. 295.

[23]   H. R. Letner, *Trans. ASME*, **77**, 1089 (1955).

[24]   M. J. Donachie, Jr., and J. T. Norton, *Trans. ASM*, **55**, 51 (1962).

[25]   Reference 13, p. 316.

[26]   M. E. Hilley, J. J. Wert, and R. S. Goodrich, *Advances in X-ray Analysis*, Vol. 10, Plenum Press, New York, 1967, p. 284.

[27]   A. L. Christenson (ed.), D. P. Koistinen, R. E. Marburger, M. Senchyshen, and W. P.

Evans, *Measurement of Stress by X-ray*, SAE TR-182, Society of Automotive Engineers, 1960.

[28]   D. Rosenthal and J. T. Norton, *Welding J.* (Research Supplement), **24**, 295-s (1945).

[29]   G. Sachs, *Z. Metallk.*, **19**, 352 (1927).

[30]   F. Stablein, *Krupp. Monatsh.*, **12**, 93 (1931).

[31]   C. N. J. Wagner, J. P. Boisseau, and E. N. Aqua, *Trans. AIME*, **233**, 1280 (1965).

[32]   B. A. MacDonald, *Advances in X-ray Analysis*, Vol. 13, Plenum Press, New York, 1970, p. 487.

[33]   S. R. Maloof and H. R. Erhard, *Rev. Sci. Instrum.*, **23**, 687 (1952).

[34]   M. J. Donachie, Sc.D. Thesis Massachusetts Institute of Technology, 1958.

[35]   L. S. Birks, *Rev. Sci. Instrum.*, **25**, 963 (1954).

[36]   E. L. Bartholomew, Jr., and R. R. Biederman, *Rev. Sci. Instrum.*, **37**, 77 (1966).

[37]   R. A. Coyle, *J. Physics*, **E3**, 930 (1970).

[38]   R. A. McCune, *Advances in X-ray Analysis*, Vol. 6, Plenum Press, New York, 1963, p. 85.

[39]   H. R. Isenburger, *Bibliography on X-ray Stress Analysis*, St John X-Ray Laboratory, Califon, New Jersey, 1949.

[40]   J. T. Norton, *Norelco Rep.*, **15**, 50 (1968).

[41]   M. J. Donachie, Jr., and J. T. Norton, *Trans. MS AIME*, **221**, 962 (1961).

[42]   W. A. Wood and N. Dewsnap, *J. Inst. Met.*, **77**, 65 (1950).

[43]   G. B. Greenough, *Prog. Metal Phys.*, **3**, 176 (1952).

[44]   G. B. Greenough, *Proc. Roy. Soc.* (London), **197A**, 556 (1949).

[45]   G. B. Greenough, *J. Iron Steel Inst.*, **169**, 235 (1951).

[46]   G. B. Greenough in *X-ray Diffraction of Polycrystalline Materials* (H. S. Peiser, H. P. Rooksby, and A. J. C. Wilson, ed.), Institute of Physics, London, 1955, Chapter 30.

[47]   R. I. Garrod and G. A. Hawkes, *Brit. J. Appl. Phys.*, **14**, 422 (1963).

[48]   B. D. Cullity, *J. Appl. Phys.*, **35**, 1915 (1964).

CHAPTER 12

RADIAL-DISTRIBUTION STUDIES OF
NONCRYSTALLINE MATERIALS

This chapter deals with those substances in which the degree of regularity of the atomic positions is very small. Although a regular crystalline arrangement is not required for the production of diffraction effects, as was first pointed out by Debye[1], in contrast to the sharp diffraction effects of crystalline materials, it is found that liquids, glasses, resins, unoriented polymers, and so on, generate only one or more broad diffuse halos.

In monatomic liquids and gases, the atomic environment about any reference atom is constantly changing; nevertheless, a small degree of local order results from the fact that two atoms cannot be separated by a distance smaller than the sum of two atomic radii. In molecular gases and liquids, additional fixed intramolecular distances are introduced which are determined by the lengths of the bonds and any characteristic angles between them. A new structural feature appears in glasses, resins, and unoriented solid polymers; that is, each atom possesses permanent neighbors at definite distances and in definite directions, although in general these vectorial properties relating an atom to its atomic environment are not the same for any two atoms in the assemblage. Figure 12-1 is a two-dimensional illustration of the difference between the arrangement of atoms in a hypothetical crystal A and glass B of the same chemical composition, A_2O_3[2]. In both cases the arrangement of O atoms about A atoms is the same, but the crystal possesses an additional ordering of the AO_3 groups into a repetitive pattern which the glass lacks.

Although there is no sharp dividing line between crystalline and so-called amorphous materials, for clarity in this discussion we somewhat arbitrarily designate as *crystalline* those materials characterized by three-dimensional periodicity over appreciable distances, say, of the order of six or more unit translations. Conversely, materials possessing only one-

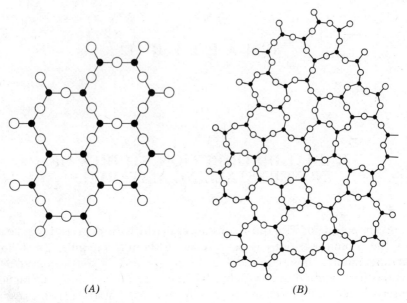

(A)   (B)

Fig. 12-1. Two-dimensional representation of the difference between a crystal A and glass B of the same chemical composition A_2O_3. (From W. H. Zachariasen, *J. Am. Chem. Soc.*, **54**, 3841 (1932), copyright by the American Chemical Society. Reprinted by permission of the copyright owner.)

or two-dimensional, or lesser, degrees of order are referred to as *noncrystalline*. Moreover, this term also embraces materials possessing such severe three-dimensional lattice distortions as to generate relatively diffuse diffraction patterns. As noted above, a degree of local order prevails even in liquids, glasses, resins, and so on, as well as in the most minute fragments of a crystalline material. As used in this chapter, the term amorphous is to be considered synonymous with *noncrystalline*. Figure 12-2 shows the considerable resemblance among patterns produced by some very different noncrystalline materials.

In this chapter we treat almost exclusively randomly oriented noncrystalline materials. The lack of any overall structural regularity removes from the patterns of such specimens differentiation of the scattering in a directional sense, and this has the direct consequence that the available intensity information permits the determination of the magnitudes of the interatomic vectors, but not their directions. The results can be portrayed as a radial-distribution function (RDF), which specifies the density of atoms or electrons as a function of the radial distance from any reference atom or electron in the system. Although the intensity scattered by (randomly oriented) noncrystalline materials can be

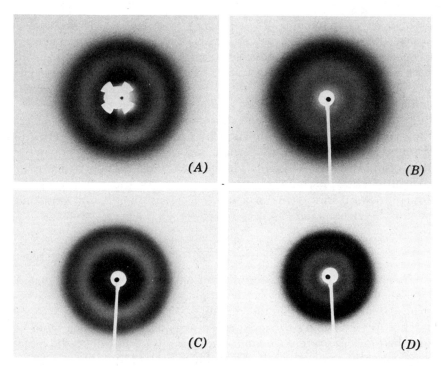

Fig. 12-2. Monochromatic-pinhole patterns of several noncrystalline substances. (*A*) Pyrex glass; (*B*) urea formaldehyde resin; (*C*) ethocel resin; (*D*) rosin.

expressed as a function of any angular variable such as θ, 2θ, or $\sin\theta$, it is especially appropriate and useful to specify its dependence on $(\sin\theta)/\lambda$, $s = 2(\sin\theta)/\lambda$ or $S = 4\pi(\sin\theta)/\lambda$.

12-1 THEORY

The theoretical groundwork for radial-distribution analysis was laid by Debye[1], who showed that the intensity in electron units scattered by a noncrystalline array of atoms at the angle θ is given by

$$I = \sum_m \sum_n f_m f_n \frac{\sin S r_{mn}}{S r_{mn}}, \tag{12-1}$$

wherein f_m and f_n are the respective atomic scattering factors of the mth and nth atoms, and r_{mn} is the magnitude of the vector separating these two atoms. The double summation is taken over all pairs of atoms in the assemblage. The validity of equation 12-1 is also contingent upon the atomic array's assuming all orientations in space, which in effect is realized for noncrystalline materials without rotation of the specimen.

Equation 12-1 makes it possible to establish the correct atomic configuration by comparing the experimental intensity function with theoretical functions computed for various models. This trial-and-error procedure was followed in the earlier studies with varying degrees of success, but more recently it has become the almost universal practice to invert the experimental intensity function by means of the Fourier integral theorem as first suggested by Zernike and Prins[3], and so to obtain the RDF of the specimen directly without any a priori assumptions as to its structure. The application of the Fourier integral theorem is perfectly straightforward in the case of a substance consisting of only one kind of atom. Then equation 12-1 becomes

$$I = Nf^2 \sum_m \frac{\sin Sr_{mn}}{Sr_{mn}} \tag{12-2}$$

if it is assumed that the environment of one atom is the same as that of any other atom. The following development follows closely the mathematical treatments in the literature[3–6].

Since in performing the summation of equation 12-2 each atom in turn becomes the reference atom, there are N terms due to the interaction of each atom with itself. The value of each of these terms is unity, since in the limit as $r_{mn} \to 0$, $(\sin Sr_{mn})/Sr_{mn} \to 1$. So equation 12-2 may be written

$$I = Nf^2 \left(1 + \sum_{m'} \frac{\sin Sr_{mn}}{Sr_{mn}}\right), \tag{12-3}$$

if it be understood that the summation excludes the origin atom. The distribution of atoms about any reference atom may now be regarded as a continuous function and the summation replaced by an integral,

$$I = Nf^2 \left[1 + \int_0^\infty 4\pi r^2 \rho(r) \frac{\sin Sr}{Sr} dr\right]. \tag{12-4}$$

Here $\rho(r)$ is the number of atoms per unit volume at a distance r from the reference atom, and $4\pi r^2 \rho(r)\, dr$ is the number of atoms contained in a spherical shell of radius r and thickness dr. Letting ρ_0 be the average density of atoms in the sample, equation 12-4 may be rewritten as

$$I = Nf^2 \Big\{1 + \int_0^\infty 4\pi r^2 [\rho(r) - \rho_0] \frac{\sin Sr}{Sr} dr$$

$$+ \int_0^\infty 4\pi r^2 \rho_0 \frac{\sin Sr}{Sr} dr \Big\}. \tag{12-5}$$

The integral of the last term of equation 12-5 represents the scattering by a hypothetical object of the same form as the specimen but of rigorously uniform electron density. This is the central scattering, which occurs at such small angles as to be unresolvable from the direct beam. Hence, if attention is limited to experimentally observable intensities, equation 12-5 can be simplified to the form

$$\frac{I}{Nf^2} - 1 = \int_0^\infty 4\pi r^2 [\rho(r) - \rho_0] \frac{\sin Sr}{Sr} \, dr. \tag{12-6}$$

By means of the Fourier integral theorem[7], this expression can be transformed to

$$r[\rho(r) - \rho_0] = \frac{1}{2\pi^2} \int_0^\infty Si(S) \sin rS \, dS,$$

or

$$4\pi r^2 \rho(r) = 4\pi r^2 \rho_0 + \frac{2r}{\pi} \int_0^\infty Si(S) \sin rS \, dS, \tag{12-7}$$

where

$$i(S) = \frac{I}{Nf^2} - 1. \tag{12-8}$$

Randall has used the Fourier integral theorem to develop equivalent equations, but in terms of a different atomic-density function[8]. Debye and Menke[9] were the first to apply Fourier integral analysis to the study of a noncrystalline substance consisting of one kind of atom, namely, liquid mercury.

It is now necessary to consider the changes in the mathematical formulary introduced by the presence of more than one kind of atom [6, 10]. For this case the general formula for the intensity of scattering (equation 12-1) can be written

$$I = N \sum_p f_p^2 + \sum_m^{m \neq n} \sum_n f_m f_n \frac{\sin Sr_{mn}}{Sr_{mn}}, \tag{12-9}$$

where $m, n, \ldots$ denote the kinds of atoms constituting some appropriate unit of structure (such as one molecule) of which the entire specimen is regarded as being composed, and N is the number of such units. The first summation is to be taken over all the atoms in a unit; the second is to be taken over every pair of atoms in the specimen regardless of which units they belong to.

Once again regarding the distribution of atoms about any reference atom as being continuous, let an atom of type m be the reference atom and suppose that the average numbers of atoms of types $m, n, \ldots$, lying in a spherical shell of radius r and thickness dr are $a_m, a_n, \ldots$. A weighted density function $\rho_m(r)$ can then be defined such that

$$4\pi r^2 \rho_m(r)\, dr = \sum_m a_m f_m, \tag{12-10}$$

the summation being taken over all the atoms in the unit of structure previously selected. Expression 12-9 can now be transformed to

$$I = N\left[\sum_m f_m^2 + \sum_m f_m \int 4\pi r^2 \rho_m(r)\, \frac{\sin Sr}{Sr}\, dr\right]. \tag{12-11}$$

Unlike the parallel formula, equation 12-5, this expression cannot be directly inverted by using the Fourier integral theorem, since both f_m and $\rho_m(r)$ are functions of S. In order to accomplish this desired result, it is necessary to resort to an approximation which detracts somewhat from the exactness of the treatment but nevertheless gives useful results.

Suppose that the scattering factor of atom m can be expressed with the necessary accuracy in terms of the scattering factor f_e of a single electron in the form

$$f_m = K_m f_e, \tag{12-12}$$

which confers upon K_m the significance of an effective number of electrons per atom of type m. This assumes that the angular dependence of f is the same for all atoms, which is a fair approximation for atoms of not too different atomic numbers. The atomic-density function $\rho_m(r)$ of equation 12-10 can now be expressed in terms of f_e as

$$\rho_m(r) = f_e g_m(r), \tag{12-13}$$

$g_m(r)$ being the electron-density function, which must be employed in the new treatment. Substitution of equations 12-12 and 12-13 in equation 12-11 gives

$$I = N\left\{\sum_m f_m^2 + 4\pi f_e^2 \int \left[\sum_m K_m g_m(r)\right] r^2 \frac{\sin Sr}{Sr}\, dr\right\}. \tag{12-14}$$

It should be noted that K_m and $g_m(r)$ are not functions of S as was true of $\rho_m(r)$ in equation 12-11. By proceeding in the same way as in deriving equation 12-6 from 12-4, equation 12-14 can be transformed to the expression

$$\frac{I}{N} - \sum_m f_m^2 = 4\pi f_e^2 \int_0^\infty \sum_m K_m[g_m(r) - g_0] r^2 \frac{\sin Sr}{Sr}\, dr. \tag{12-15}$$

Application of the Fourier integral theorem yields

$$4\pi r^2 \sum_m K_m g_m(r) = 4\pi r^2 g_0 \sum_m K_m + \frac{2r}{\pi} \int_0^\infty S i(S) \sin rS\, dS, \tag{12-16}$$

in which

$$i(S) = \frac{[(I/N) - \sum\limits_{m} f_m{}^2]}{f_e{}^2} = \sum_m K_m{}^2 \left(\frac{I}{N \sum\limits_m f_m{}^2} - 1 \right). \qquad (12\text{-}17)$$

The determination of a radial-distribution function with the aid of expression 12-7 or 12-16 comprises two main steps: first, the numerical evaluation of the function $i(S)$ from experimental scattering data and, second, numerical computation of the integral

$$\int_0^\infty Si(S) \sin rS \, dS.$$

Very refined experimental techniques are required as detailed below.

12-2 EXPERIMENTAL REQUIREMENTS

The formulas just developed apply to the *coherent*, or *unmodified*, scattering by the specimen (see Section 2-3.4). Therefore a way must be found to separate the coherent scatter of wavelength λ from other types of scatter contributing to the observed intensity, such as *incoherent* (*modified*) scatter and scatter due to the general radiation, air, and possibly materials used in supporting the specimen. Prior to the development of high-precision diffractometric techniques, it was the general practice to employ a Debye-Scherrer camera in conjunction with a crystal monochromator to eliminate undesired wavelengths. The camera was either evacuated or filled with hydrogen or helium to exclude air scatter. This photographic technique is still acceptable in situations in which highly discriminative structural information is not required. However, the almost universal availability of diffractometers has rendered the Debye-Scherrer method largely obsolete, and indeed counter registration is regarded as mandatory for all investigations of an exacting nature. Monochromatization is accomplished by means of a crystal in the direct or diffracted beam or, alternatively, with the use of Ross balanced filters. In connection with either of these techniques, it is recommended that pulse-height discrimination be employed in order to exclude as far as possible all residual wavelengths differing considerably from that of the desired $K\alpha$ radiation. A correction may be applied for air scattering[11], or it may be largely eliminated by providing a helium-filled path for the primary and scattered x-rays[12]. The scattering power f^2 of helium is only about 6 per cent of the scattering power of oxygen. In general, the incoherent scatter cannot be removed experimentally, but it can be computed theoretically and subtracted from the total intensity curve. Unlike crystalline substances, noncrystalline materials scatter coherently at all angles, so that the intensities must be recorded over a continuous range of angles, and the usual notion of the intensity falling to a background level between discrete maxima is not applicable.

In principle the integrals of equations 12-7 and 12-16 require that the intensity measurements extend to very large S values and, indeed, in all refined experimental work the upper limit of S ($\sim 4\pi/\lambda$) should be as large as practicable, about $17\,\text{Å}^{-1}$ and $22\,\text{Å}^{-1}$ for MoKα and AgKα, respectively, provided that discernible interference effects extend to these limits, of course. This demand is not always as rigorous as might be expected, for the integral of equation 12-4 approaches zero as S becomes sufficiently large, and hence the experimental coherent intensity I approaches the independent coherent scattering Nf^2 at large values of S. This is illustrated in Fig. 12-3, in which the experimental (A) and independent (B) scattering curves of a synthetic polyisoprene sample are compared. Independent scattering is defined as the hypothetical scattered intensity from an assemblage of atoms when each one scatters independently of the others so that no interference effects are produced. Evidently, the total independent coherent scattered intensity

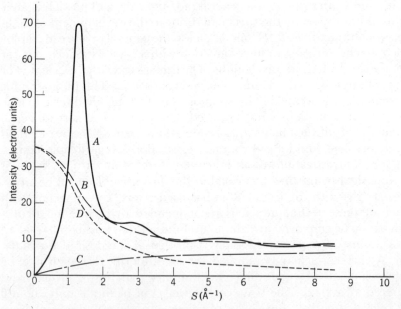

Fig. 12-3. Experimental and independent scattering curves for a synthetic polyisoprene sample. (A) Experimental scattering; (B) total independent scattering; (C) incoherent scattering; (D) independent coherent scattering.

from N atoms of scattering factor f_m is $Nf_m{}^2$ in electron units (see curve D in Fig. 12-3, which is plotted for $N = 1$). For numerical values of the atomic scattering factors of the elements, the reader is referred to Appendix VII or the literature[13, 14]. Addition of the coherent (D) and incoherent (C) independent scattering curves gives the total independent scattering (B).

In practice, then, it suffices to measure the scattering to an angle large enough to insure that $i(S)$ has sensibly reached a steady value of zero (see equation 12-8). For many noncrystalline substances this state is reached for S between 8 and 10, although often the range must be extended. Generally speaking, the greater the degree of atomic order the larger the value of S that must be reached. A radiation of short wavelength such as MoKα is commonly employed to supply intensity data at larger S values; for smaller values of S, a radiation of longer wavelength is perferable in order to resolve the intensity detail to best advantage.

By inspecting equations 12-8 and 12-17, it is seen that I must be expressed in the same units as Nf^2 or $N \sum_m f_m{}^2$. Traditionally, this has been accomplished by scaling the properly corrected experimental scattering curve (including both coherent and incoherent scattering) to the theoretical total-independent-scattering curve at large values of S, where the two functions are assumed to agree closely. The theoretical incoherent scattering is then subtracted from the fitted experimental curve, leaving the desired coherent scatter I. In applying diffractometric techniques in radial-distribution analysis Ergun, Bayer, and Van Buren [15] have developed a sensitive method by means of which the overall experimental intensity data are utilized so as to yield simultaneously (1) μT, the optimum value of the absorption exponent of the specimen, (2) the scaling factor K required to reduce I to electron units (in which Nf^2 or $N \sum_m f_m{}^2$ is expressed), and (3) the required $i(S)$ curve. We describe this method in greater detail in Section 12-4.

12-3 CORRECTION AND SCALING OF EXPERIMENTAL INTENSITIES TO ABSOLUTE (ELECTRON) UNITS

12-3.1 Correction for Air Scatter

The necessity for making this correction, or for experimentally eliminating air scatter, has been pointed out above. In diffractometry its experimental elimination is not ordinarily convenient, so that a geometrical correction must be applied. The two most useful geometrical arrangements in diffractometry are symmetrical reflection and symmetrical transmission (see Fig. 5-52*A* and *C*, respectively). We now summarize a treatment of this correction given by Ergun[11]. In Fig. 12-4*A* and *B* the quadrilaterals CDEF represent the cross sections of the air volume the scattering from which is received by the detector in the absence of a specimen. When a flat specimen of thickness T is inserted, this area is reduced by GHIJ and in addition the sample attenuates part of the intensity scattered by the remaining air volume. If we denote the *ratios* of the air-scattered intensity received with and without the sample in place by a_r and a_t, respectively, for the symmetrical-reflection and symmetrical-transmission geometries, it can be shown that

$$a_r = \frac{1}{2} + \left(\frac{1}{2} - \frac{T\cos\theta}{R\beta}\right)\exp\left(\frac{-2\mu T}{\sin\theta}\right), \qquad (12\text{-}18)$$

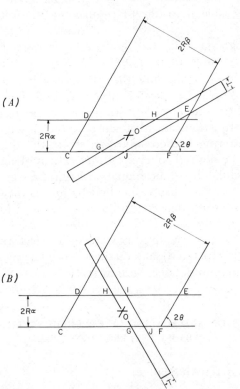

Fig. 12-4. Cross section of the scattering air volume and the sample. (*A*) Symmetrical-reflection and (*B*) symmetrical-transmission geometries. (S. Ergun[11]. Reprinted from *Chemistry and Physics of Carbon*, Vol. 3 (P. L. Walker, Jr., ed.), p. 225, by courtesy of Marcel Dekker, Inc.)

$$a_t = \left(1 - \frac{T \sin \theta}{R\beta}\right) \exp\left(\frac{-\mu T}{\cos \theta}\right).$$ (12-19)

In these equations R is the goniometer radius, and β is the equatorial angle subtended at the specimen by the detector slit. The magnitude of the air scatter with the specimen in place is obtained by measuring the air scattering in the absence of the specimen and multiplying the result by equation 12-18 or 12-19. Under typical experimental conditions the air-scattered intensity at low angles may amount to as much as 10 per cent of the total scatter measured from the sample.

12-3.2 Correction for Absorption by the Sample

In the Debye-Scherrer technique a cylindrical specimen is used. The correction for absorption in a cylindrical specimen is a complicated operation, but the literature contains tabulated factors and curves for applying this correction, computed by graphical integration, which embrace a large range of values of the sample radius and absorption coefficient[16]. For organic substances and penetrating radiations like $MoK\alpha$ and $AgK\alpha$, the correction is likely to show negligible variation with angle, in which case it can be ignored.

For the two symmetrical diffractometric geometries, the following corrections for absorption by the specimen are appropriate. In symmetrical reflection, when the specimen is too thin to give maximum diffracted intensity (see equation 5-11), the correction to be used is

$$\frac{I_\infty}{I_T} = [1 - \exp(-2\mu T \cos \theta)]^{-1} \quad \text{(symmetrical reflection)},$$ (12-20)

while for symmetrical transmission the thickness required to give maximum diffracted intensity is given by equation 5-63, and no matter what the thickness, the following correction is to be applied:

$$\frac{I_{0°}}{I_{2\theta}} = \frac{\exp\left[-\mu T(1 - \sec \theta)\right]}{\sec \theta} \quad \text{(symmetrical transmission)}.$$ (12-21)

Occasionally, circumstances require that a portion or all of the pattern be recorded by transmission with the direct beam normally incident upon the surface of the flat specimen, in which case the absorption correction to be applied is

$$\frac{I_{0°}}{I_{2\theta}} = \frac{\mu T(1 - \sec 2\theta)}{\exp\left[\mu T(1 - \sec 2\theta)\right] - 1}.$$ (12-22)

As explained at length in Sections 5-4.2C and 5-4.3, the symmetrical-reflection arrangement has serious deficiencies at small Bragg angles, especially for relatively thick and low-absorbing specimens. Conversely, the symmetrical-transmission technique is unsuitable at very large Bragg angles for corresponding reasons. Therefore in precise experimental studies it is strongly recommended that transmission be employed for $2\theta < 60°$ and reflection for $2\theta > 60°$, the specific angle of 60° being only suggestive. To provide for normalization of the reflection and transmission curves to a common level, so as to provide one continuous scattering curve over the entire 2θ range, an appreciable angular overlap amounting to, say, 30° should be allowed. A special additional warning must be sounded: Serious losses of diffracted intensity may occur if the inappropriate technique is used at very small or very large Bragg angles, because under these conditions diffraction by certain portions of the irradiated quadrilaterals GHIJ (see Fig. 12-4) may elude the detector slit. Ergun[11] has discussed this problem in some detail. Errors of this kind are more easily avoided by employing the correct experimental geometry than by correcting faulty data.

It is worthwhile to mention that, when the diffraction pattern is highly diffuse, broader collimation may be employed, even in transmission geometry, without loss of detail of the interference pattern and with a valuable increase in intensity. We also reiterate that, whether photographic or diffractometric techniques are employed, it is helpful to use a longer wavelength at low Bragg angles in order to provide the greatest possible dispersion of this portion of the pattern.

12-3.3 Correction for Polarization

For crystal-monochromatized radiation a ray scattered by the sample is diminished in intensity by the factor

$$P = \frac{1 + \cos^2 2\theta' \cos^2 2\theta}{1 + \cos^2 2\theta'}, \tag{12-23}$$

in which θ' is the Bragg angle for the reflecting planes of the monochromatizing crystal. Thus, for the (200) planes of sodium chloride, $\cos^2 2\theta'$ is 0.937 for Mo$K\alpha$ and 0.723 for Cu$K\alpha$ radiation[17]. Hence the observed intensities should be normalized to the unpolarized reference level by multiplying them by the reciprocal of equation 12-23.

12-3.4 Correction for Incoherent Scattering

After the experimental intensity curve has been corrected for the above three factors (12-3.1, 12-3.2, and 12-3.3), it is fitted to the theoretical total-independent-scattering curve at large S values (traditional method) or scaled to absolute (electron) units by some other method, after which the incoherent scattering is subtracted off. To clarify the method of correcting for the incoherent scattering, the following brief summary of the theory is given. The total intensity, coherent and incoherent, scattered by an atom of atomic number Z is[18]

$$I_t = I_e[f^2 + R(Z - \sum f_n^2)], \tag{12-24}$$

in which I_e is the scattering for a single electron, f_n is the scattering factor of the nth electron in the atom, and R is the Breit-Dirac recoil factor. Formerly, R was usually set equal to unity, which is still an acceptable approximation for elements of high atomic number. For light elements and for present-day high-precision diffractometric measurements, however, it is essential that R be numerically evaluated if the maximum amount of information inherent in the experimental data is to be extracted. In equation 12-24 the terms $I_e f^2$ and $I_e R(Z - \sum f_n^2)$ represent, respectively, the coherent and incoherent scattering by the atom. Numerical values of $\sum f_n^2$ have been tabulated[19]. By letting I_e equal unity, the scattering of both kinds is expressed in terms of the scattering per electron (electron units). If anomalous dispersion is appreciable, the coherent scattering factor f should be corrected for it (see Section 3-3.4D).

Keating and Vineyard[20] have recalculated more exactly the incoherent scattering function $Z - f_n^2$ of carbon, obtaining for $(\sin \theta)/\lambda = 0.6$ a value about 7 per cent smaller than the earlier result of Compton and Allison[17]. Rodriguez and Pings[21] have suggested semiempirical equations for closely approximating the quantum-mechanical results for various elements calculated both with and without the effect of exchange terms. Under very favorable, and exceptional, conditions, it has been found possible to measure experimentally the incoherent scattering from certain substances[22–24], but an experimental solution of the problem is not as yet generally applicable in radial-distribution studies, or in any other studies necessitating the measurement of continuous scattering over large angular ranges.* Thus the standard present-day

---

*It has been called to our attention by Ergun that rhodium, ruthenium, and technetium are of special value as balanced filters for AgKα radiation[24a]. Ergun, Braun, and Fitzer [24] showed that the incoherent (Compton) scattering can be removed experimentally by employing balanced ruthenium–rhodium filters in connection with a monochromator to isolate the Kα doublet. Ergun[24b] subsequently found that the rhodium filter can be dispensed with.

If A_{α_2} is the absorption exponent μT of the ruthenium filter for AgKα$_2$ radiation, and if two sets of data $I_1(S)$ and $I_2(S)$ are taken with the primary beam, respectively, unfiltered and filtered with ruthenium, it can be shown that $I_1(S) - [\exp(-A_{\alpha_2})]I_2(S)$ is the intensity free from Kα$_2$. If a third set of data $I_3(S)$ is now taken with the diffracted beam filtered with ruthenium, the incoherent scattering can be evaluated because $I_2(S) - I_3(S)$ is simply the incoherent intensity attenuated by the sample and the filter. Thus the three sets of data permit the evaluation of the pure coherent intensity due to AgKα$_1$.

practice is to calculate the incoherent scattering, including allowance for the Breit-Dirac factor and, in the most accurate work, allowance as well for the differences in counter response and in the absorption factors of the specimen, filters, and air path for the coherent and incoherent scattering[25, 26].

Equation 2-16 for the change in wavelength of the incoherent (λ') relative to the coherent (λ) scattering may be written

$$\lambda' - \lambda = \Delta\lambda = \frac{2h}{mc} \sin^2 \theta,$$

from which the Breit-Dirac recoil factor is

$$R = \left(\frac{\lambda'}{\lambda}\right)^3 = \left(1 + \frac{2h}{mc} \frac{\sin^2 \theta}{\lambda}\right)^3. \tag{12-25}$$

According to Ergun[11], equation 12-25 applies when *intensities* are measured (as when ionization chambers are used), whereas when the *number of photons per unit area per unit time* is measured, as with counters and scalers, the correction factor should be

$$\left(\frac{\lambda'}{\lambda}\right)^2 = \left(1 + \frac{2h}{mc} \frac{\sin^2 \theta}{\lambda}\right)^2, \tag{12-26}$$

or

$$\left(\frac{\lambda'}{\lambda}\right)^2 = \left(1 + 0.0486 \frac{\sin^2 \theta}{\lambda}\right)^2$$

upon substitution of the numerical values of h, m, and c.

Analytical expressions for the scaling factor K have been given by Hultgren, Gingrich, and Warren[27], Krogh-Moe[28], and Norman[29]. However, inasmuch as their expressions involve integrals in S extending from 0 to ∞, the imposition of an arbitrary upper limit S_{max} by actual data necessitates the exercise of careful judgment in the selection of K, just as in fitting the experimental curve $I_{coh} + I_{inc}$ to the theoretical curve $f + I_{inc}$, so that the former oscillates smoothly about the latter at large values of S.

12-4 UNIFIED DETERMINATION OF μT, $i(S)$, AND SCALING FACTOR K[15]

If μ is the linear absorption coefficient of the coherent x-rays, x is the path length in the sample of the direct ray to its point of interception by an infinitesimal volume element dV, and y is the path length of the corresponding diffracted ray, the net attenuation suffered is exp $[-\mu(x+y)]$. Because of its longer wavelength the incoherent scattered radiation has a somewhat different absorption coefficient $\mu B(S)$; therefore its attenuation is $\exp[-\mu B(S)]$. Over a limited wavelength range, μ is proportional to λ^3, so that B may be taken as the reciprocal of the Breit-Dirac recoil factor R. Then the total scattered intensity is given by[15]

$$I(S) = \frac{I_0 N \rho}{P(S)M} \int \left[\langle F^2(S) \rangle \exp[-\mu(x+y)] \right.$$
$$\left. + \sum_p C_p(S) \exp\{-\mu[x+yB(S)]\} \right] dV, \qquad (12\text{-}27)$$

in which I_0 is the incident-beam intensity, N is Avogadro's number, ρ is the specimen density in grams per cm³, M is the formula weight, $P(S)$ is the polarization factor, $C_p(S)$ is the incoherent scattering factor corrected for the Breit-Dirac recoil effect, and $\langle F^2(S) \rangle$ is the Debye interference function (equation 12-9). Equation 12-9 may be reexpressed as

$$\langle F^2(S) \rangle = \left[\sum_p f_p^2(S) \right][1 + i(S)], \qquad (12\text{-}28)$$

$i(S)$ being the oscillatory part of the interference function (second term of equation 12-9) in atomic units,

$$\frac{1}{\sum_p f_p^2} \sum_m^{m \neq n} \sum_n f_m f_n \frac{\sin S r_{mn}}{S r_{mn}}. \qquad (12\text{-}29)$$

It is now expedient to generalize equation 12-27 as follows:

$$\Phi(S, \mu T) = K[1 + g(S, \mu T)] + Ki(S). \qquad (12\text{-}30)$$

For the symmetrical-reflection geometry, integration of equation 12-27 and comparison with equation 12-30 gives for Φ and g:

$$\Phi(S, \mu T) = \frac{I(S)P(S)2\mu T}{\left[\sum_p f_p^2(S) \right][1 - \exp(-2\mu T \csc \theta)]},$$

$$(12\text{-}31)$$

$$g(S, \mu T) = \frac{\sum_p C_p(S) 2\left(1 - \exp\{-[B(S)+1]\mu T \csc \theta\}\right)}{\sum_p f_p^2(S)[B(S)+1][1 - \exp(-2\mu T \csc \theta)]}.$$

If $I(S)$ of equation 12-31 is in electron (Thomson) units, the normalization factor K of equation 12-30 is given by

$$K = \frac{N \rho A T}{M},$$

A being the cross-sectional area of the beam at the specimen, and T the thickness of the specimen.

For the symmetrical-transmission geometry, equations 12-31 are replaced by:

$$\Phi(S,\mu T) = \frac{I(S)P(S)}{\left[\sum_p f_p^2(S)\right] \exp\left(-\mu T \sec\theta\right) \sec\theta},$$

$$g(S,\mu T) = \frac{\sum C_p(S)\left(1 - \exp\left\{-\left[B(S)-1\right]\mu T \sec\theta\right\}\right)}{\sum_p f_p^2(S)[B(S)-1]\mu T \sec\theta}. \tag{12-32}$$

We may observe in connection with equations 12-32 and 12-30 that K is again the normalization factor, possessing the same value as in the reflection method.

The essential item to be evaluated in the solution of equations 12-30 and 12-31 or of 12-30 and 12-32 is, of course, the normalized intensity function $i(S)$, but optimum numerical values of μT and K are found as well. If some arbitrary value is assigned to μT, equation 12-30 may be used to test its validity and at the same time provide values of K and $i(S)$. If the assigned value of μT is correct, a plot of Φ versus g gives a straight line [modulated only by the oscillations of $i(S)$] with slope and ordinate intercept both equal to K. If the selected value of μT is too small, the Φ plot curves downward at higher values of g, and the slope and intercept yield different K values. Contrariwise, if μT is too large, the Φ plot curves upward and again the slope and intercept do not give the same value of K. The function Φ is very sensitive to the choice of μT at high angles, so that visual inspection of plots corresponding to different μT values permits the selection of the best μT. If the solution is to be obtained by computer, the program should be written so as to begin the calculations with an initial *low* value of μT, which is then increased by increments until a least-squares correlation is obtained with the slope and intercept having the same value. The application of this unified method for evaluating μT, $i(S)$, and K to a paracrystalline carbon is illustrated in Section 12-8.2.

12-5 REPRESENTATIVE EXPERIMENTAL PROCEDURE

A clearer understanding of the method of radial-distribution analysis can best be gained by following through an actual example in which only moderately sophisticated techniques, including photographic registration of the pattern, were employed. The subject is a specimen of commercial carbon black, which can be treated with the theory applicable to substances consisting of one kind of atom (see equations 12-2 to 12-8). The structure of carbon black has long been of great interest to workers in both pure and applied chemistry. Fourier integral analysis was first applied to a carbon black by Warren[30], who concluded from the study that the specimen examined contained single graphite layers and that furthermore it was not possible to specify the relative orientations of the layers from the radial-distribution curve.

The x-ray specimen was prepared by molding the pure powder into a small disk approximately 0.8 mm thick in a hydraulic spectrographic sample-pelleting press. From the weight (0.0693 gram) of the disk and area (0.90 cm²) of one face, the quantity μT in equation 12-22 for the absorption correction can be computed:

$$\mu T = \frac{\mu}{\rho}\rho T = \frac{w}{A}\frac{\mu}{\rho} = 0.077\frac{\mu}{\rho}.$$

From standard tables of the mass absorption coefficients[31]:

For CuKα, $\mu T = 0.077 \times 5.50 = 0.424.$
For MoKα, $\mu T = 0.077 \times 0.70 = 0.054.$

Photographic exposures were made with both CuKα and MoKα radiations monochromatized sufficiently by reflection from the (200) planes of a flat pentaerythritol crystal. The patterns, shown in Fig. 12-5, were prepared on double-coated nonscreen x-ray film in an evacuated Debye-

Fig. 12-5. Diffraction patterns of a commercial carbon black prepared in a 57.3-mm cylindrical camera. (A) MoKα and (B) CuKα radiation monochromatized with pentaerythritol.

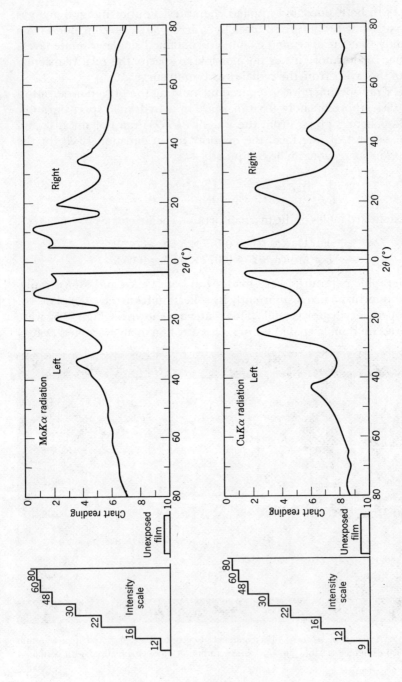

Fig. 12-6. Microphotometer records of the carbon-black patterns shown in Fig. 12-5. Minor irregularities due to film grain have been eliminated in reproducing the curves.

Scherrer camera 57.3 mm in diameter, the beam penetrating through the flat specimen at normal incidence. The exposure conditions were as follows:

Radiation	Cu Kα	Mo Kα
Exposure (hr)	3	12
Peak voltage (kV)	35	40
Tube current (mA)	20	20
Maximum (sin θ)/λ recorded	0.40	0.90
Collimator diameter (mm)	1	1

With the experimental arrangement employed, the absorption correction becomes unreliable for angles approaching 90°, which prescribes the upper limit of $(\sin \theta)/\lambda$. An unexposed portion of the film was used for the zero of photographic density.

The patterns were microphotometered with a Leeds and Northrup recording microphotometer, the precautions outlined in Chapter 6 being observed. The recorder patterns, including the reference scale of intensity steps, are shown in Fig. 12-6. Table 12-1 illustrates the method of working up the data. Readings from both sides of the undeviated beam were averaged and converted to overall intensities by comparison with a plot of the reference intensities. The intensity corresponding to the film fog level, 9.9 arbitrary units, was next subtracted to give the net intensity, which was then corrected for polarization and absorption by means of equations 12-23 and 12-22, respectively. The application of these corrections was expedited by first plotting these functions against 2θ, using the proper numerical values of $\cos^2 2\theta'$ in equation 12-23 and μT in equation 12-22, after which the numerical values of the corrections for any angle could be read off by inspection. The data of Table 12-1 are for that portion of the CuKα microphotometer curve lying between $2\theta = 20$ and 30°.

The corrected MoKα intensities were then plotted as a function of $(\sin \theta)/\lambda$ over the range 0.06 to 0.90, and comparison with the total independent-scattering curve for carbon (see curve B of Fig. 12-8) indicated that the experimental intensity curve agreed closely with the form of the independent-scattering curve in the range 0.67 to 0.85. The small portion above 0.85 was rejected because of the poorer reliability of the absorption correction at the diffraction angles concerned, namely, $2\theta = 75$ to 80°. A factor for normalizing the entire molybdenum intensity curve to electron units was then deduced by comparing the experimental and B curves at regular intervals over the range $(\sin \theta)/\lambda = 0.675$ to 0.85, as shown in Table 12-2.

Table 12-1. Calculations Illustrating the Conversion of Microphotometer Readings to Corrected Intensities in Arbitrary Units[a]

$2\theta(°)$	Microphotometer Readings			Total Intensity	Net Intensity	Correction Factors		Corrected Intensity (I_{Cu})	$(\sin \theta)/\lambda$
	Left	Right	Average			Polarization	Absorption		
20	4.18	4.16	4.17	23.4	13.5	1.125	1.014	15.4	0.113
21	3.61	3.59	3.60	26.4	16.5	1.131	1.016	19.0	0.118
22	3.10	2.94	3.02	29.8	19.9	1.138	1.017	23.0	0.124
23	2.56	2.53	2.54	33.4	23.5	1.145	1.019	27.4	0.129
24	2.25	2.22	2.23	36.1	26.2	1.152	1.020	30.8	0.135
25	2.51	2.39	2.45	34.3	24.4	1.161	1.022	29.0	0.140
26	3.22	3.00	3.11	29.2	19.3	1.168	1.024	23.1	0.146
27	4.25	3.96	4.10	23.8	13.9	1.176	1.026	16.8	0.151
28	5.26	5.09	5.17	19.9	10.0	1.186	1.028	12.2	0.157
29	6.19	5.91	6.05	17.4	7.5	1.195	1.030	9.2	0.162
30	6.72	6.51	6.61	15.9	6.0	1.204	1.033	7.5	0.168

[a]Film background = 9.9 intensity units.

Table 12-2. Deduction of a Factor to Convert the Experimental Molybdenum Intensity Curve to Electron Units

$\dfrac{\sin\theta}{\lambda}$	I_{Mo}	B (eu)	B/I_{Mo}
0.675	12.4	7.0	0.565
0.700	13.0	6.9	0.531
0.725	13.1	6.8	0.519
0.750	12.5	6.7	0.536
0.775	12.3	6.7	0.545
0.800	12.6	6.8	0.540
0.825	12.6	6.7	0.532
0.850	12.4	6.6	0.532
		Average factor	0.538

The molybdenum intensity data were next multiplied by the factor 0.538, replotted, and compared with the $CuK\alpha$ curve to determine a second factor for the purpose of normalizing the copper data to electron units. Both radiations were found to yield good intensity data in the range $(\sin\theta)/\lambda = 0.16$ to 0.40, so that it was decided to compare the molybdenum and copper intensities at small intervals over this range in order to arrive at a best value of the normalizing factor. The copper intensities before and after normalization to electron units with a factor of 2.68 are plotted in Fig. 12-7. In Fig. 12-8 is shown the final best intensity curve (A) in electron units based on both radiations. Compared with it are the theoretical independent scattering curves for carbon, coherent (D), incoherent (C), and coherent plus incoherent (B).

Before the integral of equation 12-7 can be evaluated, it is necessary to compute the amplitude function $Si(S)$ for the complete range of S covered experimentally. The numerical data required can be easily understood by expressing equation 12-8 in terms of the independent scattering functions B, C, and D in Figs. 12-3 and 12-8:

$$i(S) = \frac{I - Nf^2}{Nf^2}$$

$$= \frac{(A-C)-D}{D} = \frac{A-B}{D}. \tag{12-33}$$

Table 12-3 illustrates an appropriate way of tabulating the data in the process of computing $Si(S)$ from the experimental intensities, $I = A - C$. In this table only limited portions of the intensity data at low and high

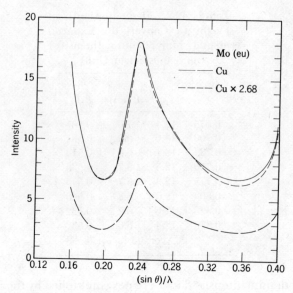

Fig. 12-7. Normalization of the scattering curve obtained with $CuK\alpha$ radiation to that obtained with $MoK\alpha$ in electron units.

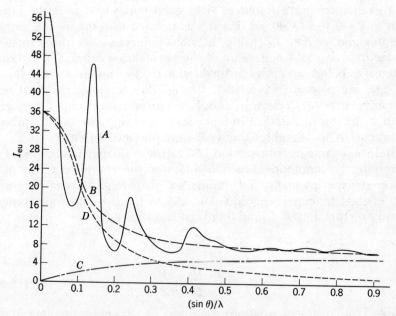

Fig. 12-8. Final scattering curve of carbon black (A) in electron units compared with theoretical independent scattering curves: coherent (D), incoherent (C), and total independent scattering (B).

Table 12-3. **Scheme for Computing the Function $Si(S)$ from the Experimental Intensity Curve (Carbon Black)**

$\dfrac{\sin\theta}{\lambda}$	$S=$ $4\pi\dfrac{\sin\theta}{\lambda}$	A (eu)	B	$A-B$	D	$i(S)=$ $\dfrac{A-B}{D}$	$Si(S)$
0.10	1.256	18.4	23.7	-5.3	21.2	-0.250	-0.314
0.12	1.508	31.2	19.6	$+11.6$	17.5	$+0.663$	$+1.000$
0.14	1.759	46.1	16.8	$+29.3$	14.6	$+2.001$	$+3.520$
0.16	2.010	16.1	14.9	$+1.2$	12.4	$+0.097$	$+1.195$
0.18	2.262	8.2	13.1	-4.9	10.5	-0.467	-1.056
0.20	2.513	6.6	12.3	-5.7	8.9	-0.640	-1.608
0.22	2.765	10.3	11.5	-1.2	7.6	-0.158	-0.437
0.24	3.016	18.1	10.8	$+7.3$	6.6	$+1.106$	$+3.336$
...	...	...	...	...	...	...	...
...	...	...	...	...	...	...	...
...	...	...	...	...	...	...	...
0.64	8.042	7.3	7.33	-0.03	2.27	-0.013	-0.105
0.66	8.294	7.3	7.29	$+0.01$	2.19	$+0.005$	$+0.038$
0.68	8.545	7.4	7.24	$+0.16$	2.10	$+0.076$	$+0.649$
0.70	8.796	7.6	7.19	$+0.41$	2.02	$+0.203$	$+1.786$
0.72	9.048	7.7	7.15	$+0.55$	1.95	$+0.282$	$+2.552$
0.74	9.299	7.6	7.11	$+0.49$	1.87	$+0.262$	$+2.436$
0.76	9.550	7.2	7.06	$+0.14$	1.78	$+0.079$	$+0.754$
0.78	9.802	7.3	7.02	$+0.28$	1.70	$+0.165$	$+1.617$
0.80	10.053	7.5	6.98	$+0.52$	1.63	$+0.319$	$+3.207$
0.82	10.304	7.5	6.94	$+0.56$	1.56	$+0.359$	$+3.699$
0.84	10.556	7.4	6.90	$+0.50$	1.49	$+0.336$	$+3.547$
0.86	10.807	7.2	6.86	$+0.34$	1.43	$+0.238$	$+2.572$
0.88	11.058	6.7	6.82	-0.12	1.37	-0.088	-0.973
0.90	11.310	6.8	6.78	$+0.02$	1.31	$+0.015$	$+0.170$

values of $(\sin\theta)/\lambda$ have been transformed to $Si(S)$ values for special illustrative purposes that will be apparent later. Figure 12-9 shows a plot of the complete $Si(S)$ curve for carbon black as a function of S (solid line).

It remains to compute the integral of equation 12-7. A modern electronic computer can of course be readily programmed for this purpose. In their early radial-distribution studies, Warren and associates employed a Coradi harmonic analyzer effectively[4, 5, 10, 30]. It is also possible to perform the computations graphically[32] or analytically with the aid of a desk calculator, although tediously. We employed a

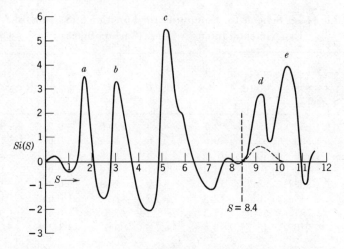

Fig. 12-9.   Plot of the experimental amplitude function $Si(S)$ for carbon black.

chart-and-strip technique for reducing the labor of an analytical evaluation of the integral[33].

The mean atomic density ρ_0 of equation 12-7 can be computed from the observed density of the solid material by means of the relationship,

$$\rho_0 = \frac{Nd}{A \times 10^{24}},$$

d being the density in grams per cubic centimeter, N Avogadro's number, and A the atomic weight. Assuming the density to be the same as that of graphite, 2.25 grams/cm³, we obtain the result,

$$\rho_0 = \frac{2.25 \times 6.02 \times 10^{23}}{12 \times 10^{24}} = 0.113,$$

and equation 12-7 may be written for the carbon black calculations in the form

$$4\pi r^2 \rho(r) = 1.420r^2 + 0.0666r \sum Si(S) \sin rS, \qquad (12\text{-}34)$$

in which ΔS of the summation has been included in the coefficient of the last term.

The solid line of Fig. 12-10A is the RDF of the carbon black computed as indicated in equation 12-34, using all the amplitude data of Fig. 12-9. The parabolic curve (broken line) represents the average atomic distribution in the sample. Figure 12-10B is the differential RDF, which gives the difference between the observed and the average distributions. The solid curve of Fig. 12-11 is obtained by dropping that portion of the

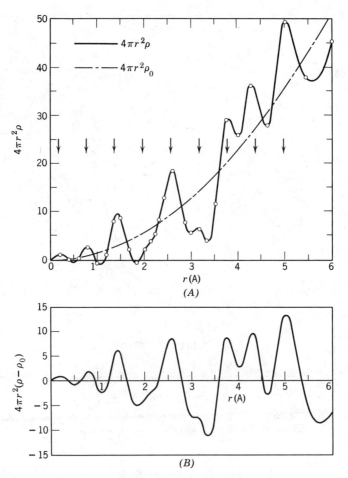

Fig. 12-10.   (A) RDF and (B) differential RDF of carbon black based on the complete $Si(S)$ curve of Fig. 12-9. Arrows denote locations of expected ripple maxima due to a discrete spurious peak in $Si(S)$ at $S = 10.4$.

$Si(S)$ curve above $S = 8.4$, and the broken-line curve results when the portion above $S = 8.4$ is replaced by the broken-line curve of Fig. 12-9 extending from 8.4 to 10.0. The significance of the differences among these three curves is considered in the following section on sources of error. It is pertinent to point out here, however, that since the experimental intensities were reduced to electron units, the areas under the peaks of the RDF give directly the respective numbers of carbon atom neighbors at the several radial distances r.

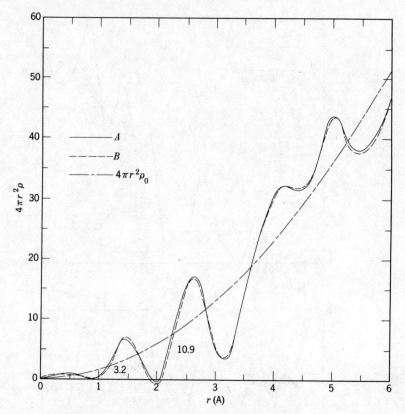

Fig. 12-11.   RDF of carbon black as modified by (A) eliminating that portion of the $Si(S)$ curve above $S = 8.4$ and (B) replacing it with the broken-line curve of Fig. 12-9.

12-6 SOURCES OF ERROR

12-6.1 Choice of Increment ΔS in the Computation of $\Sigma\, Si(S) \sin rS\, \Delta S$

The numerical evaluation of the integral of equation 12-7 or 12-16 is most commonly performed by computing the equivalent summation

$$\sum_{S_{min}}^{S_{max}} Si(S) \sin rS\, \Delta S.$$

Practical considerations require that the increment ΔS be a fixed one, which means that it must be small enough so that $r\Delta S$ will also be small relative to the period in $\sin rS$ (i.e., $2\pi/r$) at the largest value of r considered significant in the analysis. For example, for $\Delta S = 0.1$ Å^{-1} there would be only about five increments per half cycle in sin rS for $r = 6.0$ Å. A more generally acceptable value of ΔS would be about 0.05 Å^{-1}, although in some recent highly precise studies of carbon blacks extending to $r = 50$ Å or more[34] a much smaller increment was of course required. Commonly, however, for simplicity in the calculations ΔS can be set equal to $2\pi/120$, for example, which permits evaluation of the integral of equation 12-7 for any value of r expressible in the form $120/N$, where N is an integer. Figure 12-12 from the excellent review paper by Kruh[35] shows the effect on the inversion of an arbitrary intensity function resulting from the replacement of a small increment in S of 0.05 (curve A) with a large one, 0.50 (curve B). The latter curve illustrates dramatically how when the size of the increment approaches that of the period in rS, as is true for $r \simeq 10$ Å, the error in the RDF can become very large indeed.

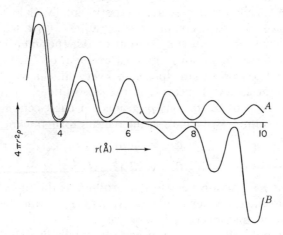

Fig. 12-12. Inverted curves of an arbitrary $Si(S)$ function for (A) $\Delta S = 0.05$ Å^{-1}, and (B) $\Delta S = 0.5$ Å^{-1}. (From R. F. Kruh, *Chem. Rev.*, **62**, 319 (1962), copyright by the American Chemical Society. Reprinted by permission of the copyright owner.)

12-6.2 Scaling of the Experimental Intensity Curve; Absorption Corrections

It goes without saying that inaccurate normalization of the experimental curve to electron units results in deficiencies in the RDF. Even with the most accurate intensity measurements and very precise calculations, a limiting factor is the accuracy with which the atomic scattering factors are known. When the scaling is performed by fitting the experimental curves to the theoretical curve at large values of S, the accuracy with which the incoherent scattering is evaluated may have a profound effect upon the results when the sample is composed of light elements.

The effect of an incorrect scaling factor is to displace the $Si(S)$ curve up or down, often attended by a noticeable tilt effect about a hinge at the point of matching of the experimental and theoretical curves[4, 36]. Errors in the absorption correction exert somewhat similar effects on the $Si(S)$ curve. Such long-period variations affect mainly the low-r region of the RDF[4], modifying both the areas and positions of the first few peaks[37, 38]. The presence of pronounced distortions due to these two causes can be detected by the appearance of spurious diffraction ripples between 0 and the first legitimate maximum in the function $4\pi r^2 \rho(r)$.

The absorption exponent μT of equations 12-20 to 12-22 must be accurately known for the specimen and radiation in question if the correction for absorption by the (flat) specimen is to be reliable. Now μT may be either calculated or measured experimentally. Calculation of μT actually involves a measurement of T, together with calculation of μ from the known elemental composition of the specimen (see Section 2-3.3). Of these two operations, the measurement of T with due allowance for possible porosity of the specimen is very difficult to perform with the necessary accuracy. However, if the absorption exponent is re-expressed in terms of the density ρ (grams per cubic centimeter), mass m, area A, and volume V of a flat specimen of uniform thickness T,

$$\mu T = \frac{\mu}{\rho}\rho T = \frac{\mu}{\rho}\frac{\rho V}{A} = \frac{\mu}{\rho}\frac{m}{A}, \qquad (12\text{-}35)$$

it is seen that μT is equivalent to the product of the mass absorption coefficient and the mass per unit area, the lattice quantity being subject to rather accurate measurement in most cases.

More commonly μT is measured experimentally by determining the attenuation suffered by a monochromatic x-ray beam on penetrating the flat specimen perpendicularly,

$$\mu T = \log_e \frac{I_0}{I}.$$

This method is generally reliable except for inhomogeneous solids that give rise to appreciable small-angle scattering, in which case the transmitted intensity is thereby excessively reduced, resulting in μT being estimated too high[39]. Ergun and Tiensuu[40] have described a method for overcoming this deficiency that involves comparison of the reflected and transmitted intensities from the same specimen at various scattering angles.

With respect to a given fractional error in the scale factor, Bienen-stock[38] has shown that a similar fractional error is to be expected in the area of the legitimate peaks. He suggests a method for determining the correction required to rescale correctly an improperly scaled independent scattering curve, namely, $I' = (1 + a)I$, instead of I as de-fined by the theoretical equations 12-1 to 12-17. Bienenstock performs the Fourier transformation of the *calculated independent scattering curve*, after which comparison with the previously determined RDF permits the estimation of a on the basis of the magnitude of the initial maxima. The data can then be rescaled, and a more accurate RDF obtained. For details the reader is referred to Bienenstock's article[38].

12-6.3 Discrete Errors in $Si(S)$; Termination-of-Series Errors

It is easy to show that the Fourier transform of a discrete peak at S is a general ripple extending throughout the $4\pi r^2 \rho$ diagram with a period of $\Delta r = 2\pi/S$, the first maximum occurring at $r = \pi/2S$. Similarly, a discrete minimum in $Si(S)$ produces a ripple with minima at the same sites. Broader maxima produce well-resolved ripples at small values of r, which diminish in sharpness and become imperceptible as r increases. Hence in a radial-distribution curve a discrete error of considerable mangitude in $Si(S)$ causes the superposition of such ripples on an other-wise normal diagram. The ease of recognition of these spurious effects depends upon their sharpness and, as indicated above, they can be detected most readily at small r's where no valid peaks can occur. If several orders can be identified, it may be possible to locate the offending maximum or minimum in the $Si(S)$ function from the relation $S = 2\pi/\Delta r$.

It is now profitable to return to the carbon-black analysis, which was left in an incomplete state in Section 12-5. The functions of Fig. 12-10 are seen to contain two clear-cut maxima in the low-r region besides a number of other minor sharp peaks throughout the diagram. The first peak occurs at about $r = 0.15$, and the succeeding sharp peaks display a rather consistent period of $\Delta r = 0.60$. The offending maximum in $Si(S)$ should therefore be expected at $S = 2\pi/\Delta r = 6.28/0.60 \cong 10.5$. Indeed, Fig. 12-9 does show a large peak (e) at 10.4, and when this is eliminated by replacing the original solid-line curve by the smoothed-off broken-line curve above $S = 8.4$, the resulting RDF (broken line in Fig. 12-11) is seen to be free of the entire ripple sequence. The same result is achieved by terminating the $Si(S)$ function at 8.4 (solid line). The only obvious abnormality still perceptible is a low maximum at $r = 0.5$ Å. It is seen that these ways of correcting the distribution curve implicitly assume that peaks d and e in Fig. 12-9 are either partially or wholly fictitious.

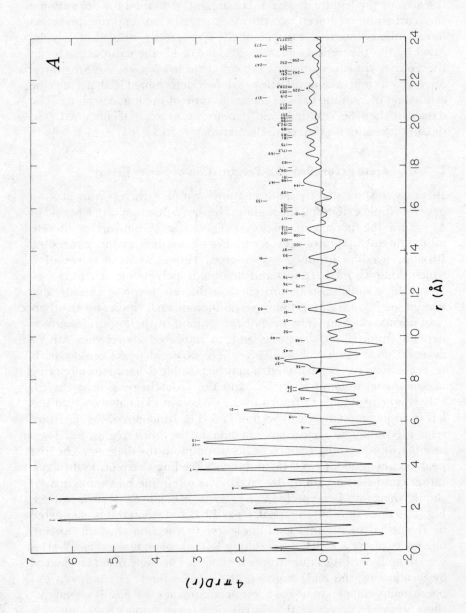

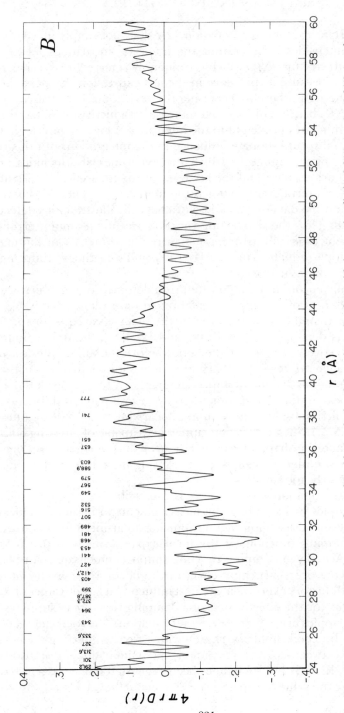

Fig. 12-13. Atomic RDF $4\pi r \rho_\Delta(r)$ of a carbon black calculated from intensity data with integration limits $0.05(2\pi) < S < 2.64(2\pi)$ and no damping factor. (A) $r = 0$ to 24 Å; (B) $r = 24$ to 60 Å. The peaks are identified by integers $n = p^2 + q^2 + pq$, which correspond to all interatomic distances within a graphite layer; p and q assume all positive integral values including zero. (Reprinted with permission from J. R. Townsend and S. Ergun, *Carbon*, **6**, 19 (1968), Pergamon Press.)

It is important to note that equation 12-7 does not simply invert the normalized intensities I, but rather the intensity fluctuations $I - Nf^2$, modified by the factor $1/Nf^2$, which is commonly referred to as a sharpening function because its presence in the $i(S)$ expression sharpens the features of the RDF, thereby increasing the resolution. Unfortunately, the factor $1/Nf^2$, because of its rapid increase with increasing S, has the effect of assigning undue weight to the intensities at high angles, so that they assume a disproportionate influence in the integration. But this is precisely the angular range in which the experimental difficulties of measurement are greatest and the accuracy of the theoretical values of the coherent and incoherent scatter is the most critical. The $Si(S)$ curve of Fig. 12-9 and the data of Table 12-3 dramatically illustrate this defect in the method. Thus the sharpening function accentuates minor errors of measurement, especially in the high-angle region, thereby contributing to fictitious ripple sequences in the RDF $4\pi r^2 \rho$ and sometimes producing false negative troughs in it as well.

A universally encountered, and often troublesome, source of error is the incompleteness of the experimental integration range, that is, S_{min} to S_{max} rather than 0 to ∞ as required by the theory (see equations 12-7 and 12-16). As demonstrated by Bragg and West[41], the upper finite limit S_{max} effectively terminates the series before the amplitudes have necessarily fallen to zero. Furthermore, they showed that when the intensity fluctuations are still of appreciable size at S_{max}, the result is that each major peak in the RDF is bracketed by several pairs of diffraction ripples. The first subsidiary maximum appears on either side at a distance $\Delta r = \frac{4}{3}(S_{max})$. Such diffraction ripples are capable of affecting both the position and the shape of a bona fide peak and, furthermore, they reduce the area under the central peak, causing the estimate of atomic concentration to be too small.

When, as is commonly the case, it is impossible to suppress such diffraction ripples by extending the measurements to higher S values, the same end can be accomplished by applying an artificial temperature factor, or damping function, of the form $\exp(-A \sin^2 \theta)$ to the $Si(S)$ amplitudes. Actually, a damping function simply compensates more or less fully for the oversharpening effect of the factor $1/Nf^2$ in the $Si(S)$ function. Both factors exert their greatest influence at larger values of S, which accounts for the effectiveness of a damping function in suppressing spurious ripples arising from series termination. The coefficient A of the damping function should be chosen with a view to effecting the best compromise between loss of resolution and elimination of spurious ripples in the RDF. It has sometimes been suggested that A be selected so as to reduce the value of $\exp(-A \sin^2 \theta)$ to about 0.10 at $S = S_{max}$.

The lower finite limit of integration S_{min} excludes the intensity information in the region $S = 0$ to S_{min}, in particular what is technically referred to as small-angle scattering, and results in long-period oscillations in the RDF. Figure 12-13 clearly illustrates this phenomenon in the RDF of a carbon black which extends over the exceptionally large r range from 0 to 60 Å[34]. The strongly inhomogeneous character of carbon black generates strong scattering at small angles, with a resulting pronounced long-period oscillation. For $S_{min} = 0.05$ (2π) the period is $1/0.05 = 20$ Å. We may mention at this point that the $4\pi r^2 \rho(r)$ curve of many noncrystalline substances assumes an average negative value in the lower-r region, an artifact resulting from the exclusion of significant amounts of scattering at small angles.

12-7 SPECIFIC PROCEDURES FOR MINIMIZING ERRORS

12-7.1 Application of a Damping Factor

A brief discussion of the value of this technique for dealing with termination-of-series errors has been given in Section 12-6.3. The multiplication of the $Si(S)$ function by a damping factor of the form $\exp(-A'S^2)$ actually has a more general usefulness in reducing most sources of error that produce short-period ripples in the RDF, for example, improper scaling and "oversharpening." The indicated procedure, then, is to apply the $\exp(-A'S^2)$ damping factor to the $Si(S)$ curve and note whether or not all spurious peaks are removed at low r's and, if not, to recheck carefully the curve-fitting procedure and absorption corrections for systematic errors. Figure 12-14 shows the experimental $Si(S)$ function of a natural-rubber specimen before and after multiplication by a damping factor, $\exp(-A'S^2)$, with $A' = 0.025$. The reduction in the amplitudes at higher S values is conspicuous. Figure 12-15 shows the corresponding RDFs. The uncorrected function (A) exhibits several symptoms of errors, including ripple maxima at 0.65 and 3.20 Å and a deep negative trough at 2.0 Å. In addition, the number of nearest neighbors at 1.5 Å must be 2 on the basis of our chemical knowledge of the rubber structure, whereas the area under the 1.5 Å peak is found to be only about 1.5. Curve B in Fig. 12-15 shows that the use of a temperature factor has eliminated both the ripples at 0.65 and 3.20 Å and the negative trough, but a spurious peak remains at about 0.75 Å. This must therefore denote a further error in curve fitting or in the absorption correction, which in all probability is also responsible for the deficiency in the area of the maximum at 1.5 Å.

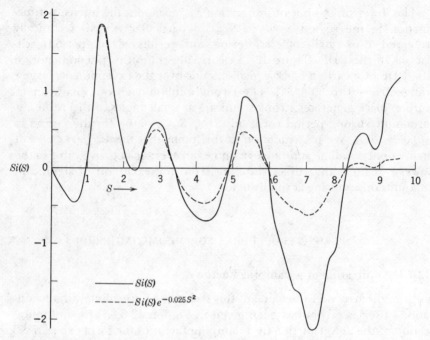

Fig. 12-14.   Effect on the $Si(S)$ function for natural rubber of multiplication by a damping factor, $\exp(-0.025\,S^2)$.

12-7.2 Use of an Electronic Distribution Function

Finbak [42, 43] eliminated the oversharpening effect and magnification of $Si(S)$ errors at high angles by employing an electronic RDF, $\sigma(r)$, which is defined in terms of the probability $\sigma(r)\,dr$ of one electron in an atomic or molecular system being found at a distance between r and $r + dr$ from another electron. The difference between the actual distribution of electrons in the sample and the mean density for substances composed of one kind of atom is shown to be

$$\sum_m \sigma_m(r) + \sigma_l(r) = \left(\frac{4R^2 m^2 c^4}{\pi e^4 I_0}\right) r \int_0^\infty S(I - Nf^2) \sin rS\, dS, \quad (12\text{-}36)$$

$$= Cr \int_0^\infty S(I - Nf^2) \sin rS\, dS,$$

wherein the subscripts m and I refer, respectively, to electron-electron distances within a molecule and between different molecules, C is a constant, and I is the observed intensity of the coherent scattering. This expression may be compared with the corresponding atomic RDF,

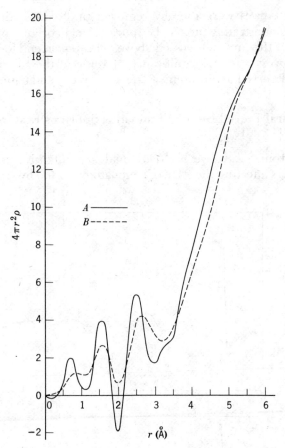

Fig. 12-15.   RDFs of a natural rubber specimen (A) before and (B) after inclusion of a damping coefficient exp $(-0.025\,S^2)$ with the $Si(S)$ function.

which from equations 12-7 and 12-8 can be written

$$4\pi r^2[\rho(r) - \rho_0] = \frac{2r}{\pi}\int_0^\infty S\left(\frac{I - Nf^2}{Nf^2}\right)\sin rS\,dS. \qquad (12\text{-}37)$$

It is seen that the basic point of difference in the two functions is the inclusion in the integral of equation 12-37 of an f^2 in the denominator which is absent from equation 12-36. Consequently, the electronic RDF is not unduly sensitive to minor intensity errors at large S values as is true of the atomic function. Using the electronic RDF, Finbak recalculated the published data for nine molten metals and liquid noble gases and showed that subsidiary maxima present in the original atomic

distribution curves were thereby removed or greatly reduced in size
[43], thus demonstrating them to be spurious detail originating from one
or another of the causes discussed above. This put an end to long-stand-
ing conjectures as to the significance of some of the subsidiary peaks
previously observed in the atomic RDFs of mercury and liquid tin.

12-7.3 General Procedure for Eliminating Spurious Features from the RDF [44, 45]

Kaplow, Strong, and Averbach [44] made an intensive study of the
various errors affecting the RDF of a monatomic liquid by starting with a

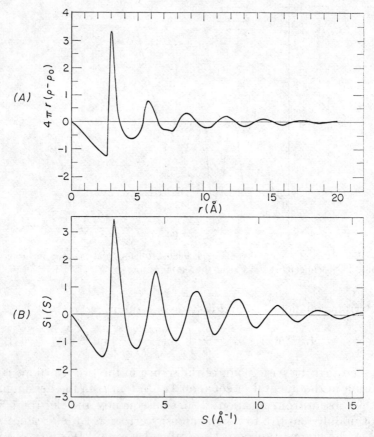

Fig. 12-16. (A) Differential RDF $G(r) = 4\pi r(\rho - \rho_0)$, of liquid mercury and (B) corresponding $Si(S)$ curve obtained by inversion of (A). [Courtesy of R. Kaplow, S. L. Strong,
and B. L. Averbach, *Phys. Rev.*, **138**, A 1336 (1965).]

known differential atomic RDF, in this case

$$G(r) = 4\pi r(\rho - \rho_0) \qquad (12\text{-}38)$$

for mercury (Fig. 12-16A), Fourier-inverting it to obtain the corresponding $Si(S)$ curve (Fig. 12-16B), and then subjecting the latter curve to
various specific errors of the types discussed in Section 12-6 and noting
the resulting distortions of the $G(r)$ curve. The initial $G(r)$ curve had itself
been refined by the corrective procedure to be described in this section;
hence both the $G(r)$ and $Si(S)$ curves were believed to be representative
of liquid mercury to a high degree of reliability.

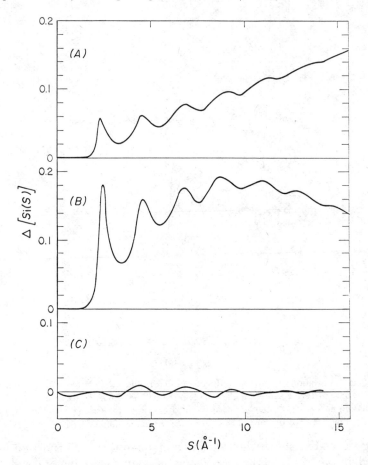

Fig. 12-17.   Errors in $Si(S)$ curve in Fig. 12-16B: (A) caused by 1 per cent error in scale
factor; (B) caused by error in scattering factor of $\epsilon = 0.03 \exp(-0.005\ S^2)$ followed by
rescaling; (C) residual error after using correction procedure. [Courtesy of R. Kaplow,
S. L. Strong, and B. L. Averbach, $Phys.\ Rev.$, $\mathbf{138}$, A 1336 (1965).]

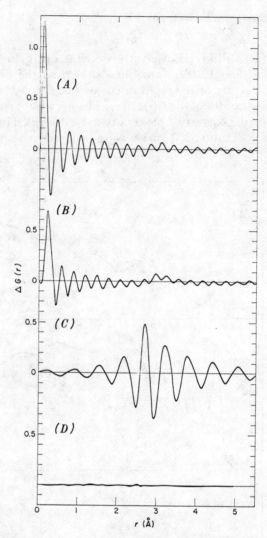

Fig. 12-18.   Variations in $G(r)$ resulting from errors in $Si(S)$: (A) from 1 per cent error in scale factor; (B) from an error $\epsilon = 0.03 \exp(-0.005\,S^2)$ followed by rescaling; (C) from termination at $S = 10\,\text{Å}^{-1}$. (D) Residual error after using correction procedure. [Courtesy of R. Kaplow, S. L. Strong, and B. L. Averbach, *Phys. Rev.*, **138**, A 1336 (1965).]

Figure 12-17A shows the calculated error in this $Si(S)$ function introduced by an artificial 1 per cent error in the scale factor, while Fig. 12-18A shows the equivalent oscillatory error in the $G(r)$ function arrived at by Fourier transformation of the analytically expressed scaling error in $Si(S)$. It is seen that the $\Delta G(r)$ error function in Fig. 12-18A exhibits sharp, large-amplitude oscillations as r approaches zero. The elimination of these pronounced oscillations is a valuable criterion for verifying that a scaling error has been corrected.

Kaplow and co-workers postulate that errors in the scattering factors along with errors in the absorption correction and instrumental alignment assume the form of a slowly varying function of S. They assume, furthermore, that such a slowly varying function is largely, or at least partially, compensated by the correction of the scaling error as described above. Figure 12-17B shows the remaining error in $Si(S)$ after the imposition of a slowly varying error of the form $\epsilon = 0.03 \exp(-0.005 S^2)$ and rescaling as just indicated. The resulting residual error $\Delta[Si(S)]$ after applying the correction procedure is given in Fig. 12-17C. From Fig. 12-18B it is noted that the spurious oscillations of $G(r)$ corresponding to the $\Delta[Si(S)]$ curve in Fig. 12-17B have been considerably reduced relative to those in Fig. 12-18A.

Figure 12-18C shows the errors in the $G(r)$ curve resulting from terminating the $Si(S)$ data at $S = 10 \, \text{Å}^{-1}$. It is important to observe that the spurious ripples of greatest amplitude occur in the region of the first bona fide maximum at about 3 Å, where they are capable of causing serious interpretative difficulties. Figure 12-19A shows the effect on $G(r)$ of terminating the $Si(S)$ series at 6, 10, and 14 Å^{-1}, from which the direction required for the corrections can be deduced.

Lack of space precludes a detailed description of the corrective procedure of Kaplow et al.; however, we may present a qualitative summary [46]. First, the experimental $Si(S)$ curve is analytically extended beyond the limit S_{max} according to the following criteria: (1) the transform of the final $G(r)$ function must of course reproduce the measured $Si(S)$ curve at all values of S below S_{max}; (2) at r's below the first legitimate peak, the $G(r)$ must be linear. Second, the effect on $G(r)$ of limiting the experimental $Si(S)$ function to three different S_{max} values is noted, and an estimate of the correct $G(r)$ is made. This corrected $G(r)$ curve is then transformed to yield a new $Si(S)$ curve.

The foregoing two-stage process is repeated until a self-consistent *extension* of $Si(S)$ subject to the above two criteria is obtained. Kaplow, Rowe, and Averbach [46] state that the results of experience show it to be unnecessary to extend the $Si(S)$ function in this manner beyond a value of $S = 2S_{max}$. Furthermore, the final experimental initial slope of the smoothed $G(r)$ function, theoretically $4\pi\rho_0$ (see equation 12-38), was invariably found to yield a value of ρ_0 in good agreement with the measured bulk density of the specimen. Figure 12-19B shows the final $G(r)$ function after correction for termination errors.

In actual practice Kaplow, Strong, and Averbach [44] found it most satisfactory to first correct for termination of series and then analyze the remaining oscillations at small values of r for the other errors. They also

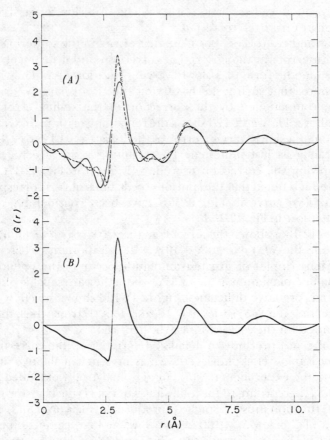

Fig. 12-19. (A) Effects on $G(r) = 4\pi r(\rho - \rho_0)$ of terminating the $Si(S)$ data at $6,\ldots$; 10, $---$; and $14\,\text{Å}^{-1}$, ——. (B) $G(r)$ function after correction for termination errors. [Courtesy of R. Kaplow, S. L. Strong, and B. L. Averbach, *Phys. Rev.*, **138**, A 1336 (1965).]

found that several iterations may be required before a convergent (invariant) $G(r)$ function is obtained. Although this procedure requires the calculation of many transforms, the ready accessibility of fast electronic computers renders the operations quite practicable.* Although, strictly speaking, one cannot be certain that all errors have been removed from the final RDF, the criteria of (a) vanishingly small oscillations at small r values and (b) agreement between ρ_0 from the initial

*Information on the computing techniques may be obtained by addressing R. Kaplow, Department of Metallurgy, Massachusetts Institute of Technology, Cambridge, Massachusetts.

slope of $G(r)$ and the measured bulk density constitute effective measures of the efficacy of the procedure.

Cargill[47] has discussed in some detail the nature of the distortion of the RDF resulting from neglect of the x-ray scattering at small angles, which is to say, from commencing the integration in $i(S)$ at S_{min} rather than at $S = 0$. He shows that the net effect is concentrated in the region of small r and that the resulting $G(r)$ function corresponds to a material of greater average density than that of the sample under investigation.

12-7.4 Method for Correcting the RDF for Termination-of-Series Errors Only[34]

Townsend and Ergun[34] have derived a rather complicated formula which expresses the quantitative relationship between $\rho(r)$, the radial atomic density distribution for integration limits of 0 to ∞ in S, and $\rho_\Delta(r)$, the corresponding distribution for finite limits M_1 and M_2 in S. The equation can be solved to yield a RDF fully corrected for termination-of-series errors on condition that numerical values of the following quantities are known for the specimen under study:

$\delta_0{}^2 =$ a single discrete value of the mean square deviation in the inter-atomic distances throughout the structure,

$\tilde{n}_k =$ average number per atom of all interatomic distances that have the magnitude r_k.

In connection with these two conditions, it must be realized that $\delta_0{}^2$ is actually an artificial concept with only limited physical significance, inasmuch as there must actually exist a distribution of δ_k values, any given δ_k being the most appropriate value of δ for a peak in the RDF corresponding to the particular radial distance r_k.

Although at the time of publication of their article Townsend and Ergun had not generalized their method to encompass the more realistic case of variable δ_k, they convincingly demonstrated the effectiveness of their method when applied to a hypothetical aromatic model, for which $\tilde{n}_k$ could be precisely evaluated and with δ_0 set equal to 0.026 Å in order to simplify the results. M_1 and M_2 were assigned the values 2 and 16 Å^{-1}, respectively, approximately simulating the lower and upper integration limits in S for MoKα x-rays. Figure 12-20 shows the RDF $4\pi r\rho_\Delta(r)$ calculated for the finite data limits, while Fig. 12-21 shows the $4\pi r\rho_\infty(r)$ curve corrected to the integration limits 0 and ∞ in S.

Townsend and Ergun remark that in performing the solution for $\rho_\infty(r)$ the quantities $\tilde{n}_k$ and δ_0 must be regarded as variables to be assigned numerical values such as to give self-consistent results. Self-consistency

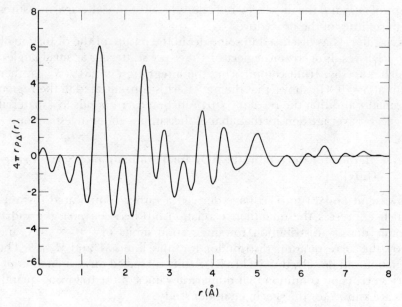

Fig. 12-20. Atomic RDF $4\pi r \rho_\Delta(r)$ of a hypothetical aromatic molecule calculated for the finite integration limits $2 < S < 16$ Å^{-1}. (Reprinted with permission from J. R. Townsend and S. Ergun, *Carbon*, **6**, 19 (1968), Pergamon Press.)

is to be recognized on the basis of the criteria: (a) the RDF is free of small ripples near the peaks, (b) the peaks are essentially Gaussian in shape with a mean square width of $\delta_0{}^2$, (c) the number of atoms given by the kth peak is correct, which is to say,

$$\int 4\pi r^2 \rho_k(r) \, dr = \tilde{n}_k.$$

Townsend and Ergun[34] suggest that the generalization of the method might involve the selection of an initial value of δ_0 appropriate for the most prominent maxima of the RDF followed by the carrying-out of refinements to yield best values of $\tilde{n}_k$ and ρ_k for the various individual peaks.

Because of the complexity of the concepts and mathematical expressions involved in the method of Townsend and Ergun, it is necessary to refer the interested reader to their article[34] for full details of the theory in its present state. The excellent results achieved in its illustrative application to a known molecule, as described above, seem to offer great promise for its future utility once the theory has been further generalized along the lines indicated.

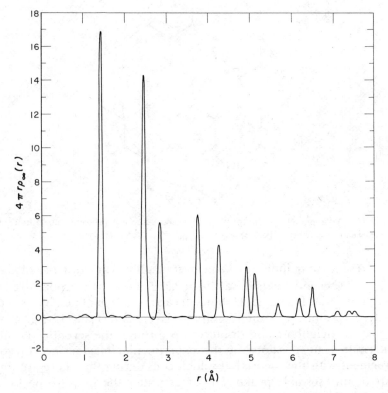

Fig. 12-21. Atomic RDF $4\pi r \rho_\infty(r)$, from Fig. 12-20 after correction to the integration limits $0 < S < \infty$. (Reprinted with permission from J. R. Townsend and S. Ergun, *Carbon*, **6**, 19(1968), Pergamon Press.)

12-8 PRACTICAL EXAMPLES

12-8.1 Carbon Black

The derivation of the radial-distribution curve and its correction for spurious detail have been described in previous sections. The following interpretative remarks are based on the corrected function of Fig. 12-11, which now contains only the low maximum at 0.50 Å as evidence of persisting abnormalities. These are most likely to be due to curve fitting or errors in the absorption correction and are probably inconsequential, judging from the weakness of this maximum.

As Warren[30] has shown for another carbon black, the distribution curve agrees with the existence in the sample of single graphite layers.

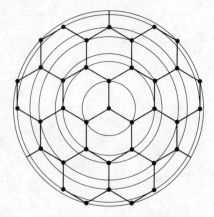

Fig. 12-22.   Arrangement of carbon atom neighbors about a reference atom in a single graphite layer. [Courtesy of B. E. Warren, *J. Chem. Phys.*, **2**, 551 (1934).]

The present curve indicates 3.2 nearest neighbors at a distance of about 1.45 Å, 10.9 second neighbors at about 2.55 Å, and further concentrations at 4.1 and 5.0 Å. Figure 12-22 shows the arrangement of neighbors about a reference atom in a single graphite layer, and Table 12-4 gives the number of neighbors and distances up through the seventh coordination circle. The positions of the experimental maxima are in good agreement with the theoretical values, considering that there are three pairs of unresolvable peaks. The areas under the first two peaks, 3.2 and 10.9, are somewhat larger than the theoretical. A partial explanation is that the density of carbon black assumed in computing the $4\pi r^2 \rho_0$ curve, 2.25 grams/cm^3 is probably too large (see Section 12-5). Minor systematic errors in curve fitting or the absorption correction may

Table 12-4.   Numbers of Neighbors and Distances in a Single Layer of the Graphite Structure

Number of Neighbors	Distance (Å)	Average Distance (Å)
3	1.42	1.42
6	2.46	2.60
3	2.84	
6	3.75	4.00
6	4.25	
6	4.92	5.00
6	5.11	

account for the remaining discrepancy. The radial-distribution findings do not permit us to say whether or not the single graphite layers are associated in groups which display the truly crystalline interlayer arrangement characteristic of graphite. The reader is reminded, however, that diffraction evidence of a different kind does answer this question by showing that the arrangement is of the random-layer-lattice type[48].

12-8.2 Carbon Black: Unified Determination of $\mu T, i(S)$, and Scaling Factor K

The underlying theory has been presented in Section 12-4 and here we only briefly describe its application to a carbon black by Ergun, Bayer,

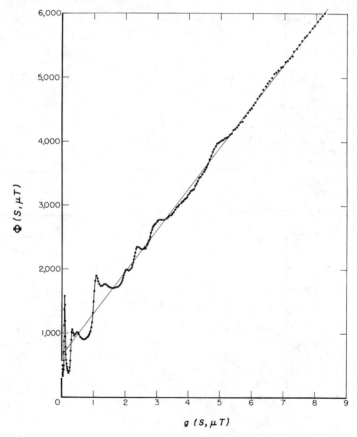

Fig. 12-23. Plot of $\Phi(S,\mu T)$ versus $g(S,\mu T)$ for the determination of K and μT by equation 12-32. Carbon-black data, symmetrical-transmission technique. [Courtesy of S. Ergun, J. Bayer, and W. Van Buren, *J. Appl., Phys.*, **38**, 3540 (1967).]

and Van Buren[15]. The scattered intensities in symmetrical transmission were obtained diffractometrically with MoKα radiation, using a
1° divergence slit with medium-resolution Soller slits. The flat specimen
was 2.34 mm thick and 50 mm long and pressed to a density of 1.18
grams/cm³. Figure 12-23 is a plot of Φ versus g as per equation 12-32.
From the best linear plot through the data points so as to give equal
slope and ordinate intercept, these quantities were found to have the
numerical value 651 (the scale factor K of equation 12-30), and μT
was 0.167.

With MoKα radiation and a 1° divergence slit, the same specimen was
also measured in symmetrical reflection, except that intensity measure-

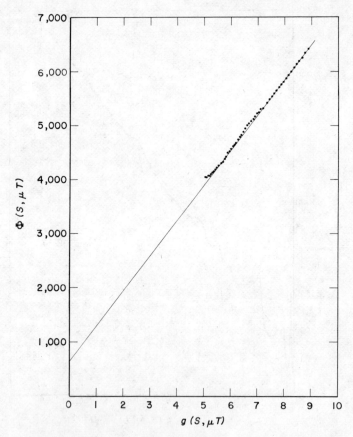

Fig. 12-24.   Plot of $\Phi(S, \mu T)$ versus $g(S, \mu T)$ for the determination of K and μT by equation 12-31. Carbon-black data, symmetrical-reflection technique. [Courtesy of S. Ergun,
J. Bayer, and W. Van Buren, *J. Appl. Phys.*, **38**, 3540 (1967).]

ments in the region $2\theta < 40°$ were not included because of the inability of the detector to intercept the full diffracted beam at the lower angles (see Section 12-3.2). The plot of Φ versus g for the usable higher-angle region is shown in Fig. 12-24. The least-squares line to give equal slope and intercept was calculated and extrapolated to $g = 0$ as shown, yielding $\mu T = 0.167$, the same result as from transmission, and slope and intercept equal to 650. The identity of these numerical values with those derived from the transmission data may be regarded as a useful check of their validity. For the final interference function $1 + i(S)$ derived by this procedure, the reader is referred to the original article[15].

An examination of Fig. 12-13 shows that the coherently diffracting domains extend over regions at least as large as 60 Å. Furthermore, out to large r values the peaks can be unambiguously assigned $00l$ and hk indices that are characteristic of graphite-like layers. The existence of domains of this order of size and even much larger has also been confirmed by electron microscopy[49] and small-angle x-ray scattering[50], a finding in conspicuous disagreement with the typically much smaller values of the L_a and L_c dimensions of the parallel-layer groups pictured as comprising carbon blacks on the basis of the simple theory of line broadening (see Sections 9-2.5B and 9-3.5E and F). These contradictory results induced Ergun to carry out theoretical studies leading to two alternative interpretations of the experimental observations: (a) large domains characterized by high densities of lattice defects[51, 52], and (b) large domains affected by pronounced lattice strains[51, 53]. The latter interpretation, in particular, appears to explain very satisfactorily the different experimental findings for carbon blacks described above [54].

12-8.3 Silica Glass

Warren and co-workers have studied SiO_2 and B_2O_3 in the vitreous state[10, 30, 55]. They prepared patterns of SiO_2 in an evacuated cylindrical camera, using crystal-monochromatized $CuK\alpha$ and $MoK\alpha$ radiations and employing a cylindrical specimen 1 mm in diameter. The patterns were microphotometered and the intensities evaluated as described earlier in this chapter. An electronic RDF, shown in Fig. 12-25, was derived using equations 12-9 to 12-17 for substances composed of more than one kind of atom. The first peak, at 1.62 Å, is clearly resolved and encloses an area of 970 electrons. Now the area of a peak due to the interaction of an atom of type M and one of type N in a compound of the composition M_aN_b is

$$A = 2 \times aK_M \times n_N K_N, \tag{12-39}$$

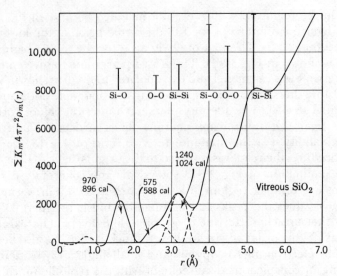

Fig. 12-25. Radial-distribution curve for vitreous SiO_2. [Courtesy of B. E. Warren, H. Krutter, and O. Morningstar, *J. Am. Ceram. Soc.*, **19**, 202 (1936).]

where n_N is the number of atoms of type N surrounding each M atom at the specified distance r. The curves of F_{Si}/f_e and f_O/f_e over the angular range of the experiment show the best values of K_{Si} and K_O to be 16.0 and 7.0, respectively, from which

$$n_O = \frac{970}{2 \times 1 \times 16 \times 7} = 4.3.$$

This result shows that each silicon atom is surrounded by four oxygen atoms at a distance of 1.62 Å, and substitution of 4 in equation 12-39 gives 896 for the calculated area. If the coordination is tetrahedral and the center-to-apex distance is 1.62 Å, geometrical calculations yield an edge length of $1.62 \sqrt{8/3} = 2.65$ Å, which means that an oxygen-oxygen peak is to be anticipated at 2.65 Å with an area of $2 \times 6 \times 7 \times 7 = 588$ electrons. Indeed, this peak is seen to be present although only partially resolved from a larger peak at 3.2 Å. That this latter maximum is due to Si–Si interactions is readily shown. Assuming the Si–O–Si bond system to be linear, each Si will be tetrahedrally surrounded by four others at a distance of 3.24 Å, the resulting area per SiO_2 unit of structure being $4 \times 16 \times 16 = 1024$ electrons. The experimental maximum shown resolved in Fig. 12-25 is in satisfactory agreement with this model.

Thus the radial-distribution analysis has clearly confirmed the presence of tetrahedral coordination of Si by O in vitreous SiO_2. The lengths of the vectors connecting Si and O, and O and O, in neighboring tetrahedra vary with the mutual orientations of these tetrahedra about the Si–O–Si axes connecting their centers. Hence the vertical lines in the figure indicating the positions and relative areas of the peaks due to these larger interatomic distances are in the nature of averages, whereas the positions and heights of the first three peaks are perfectly definite.

Much more recently, Cartz[56] reinvestigated the structure of silica glass in an effort to find evidence for longer-range local ordering. His procedure was to compare the experimental radial intensity function with intensity functions calculated for various structure models. Even though the method was incapable of providing a definitive solution, it was found that the observed scattering is compatible with the existence of certain preferred intertetrahedral linkages of the types known to be present in cristobalite and tridymite. Nonplanar rings consisting principally of five, six, or seven tetrahedra and a most probable Si–O–Si angle of about 150° are indicated by the investigation. This value of the angle agrees well with earlier results obtained by Zarzycki[57] from radial-distribution studies. For vitreous germanium dioxide (GeO_2) he found that the Ge–O–Ge angle approaches 180°.

12-8.4 Liquid Argon

The Fourier analysis of liquid argon is cited as an example of liquid studies. The present remarks are only a brief commentary on the work of Eisenstein and Gingrich[58], although numerous investigators have studied this element in the solid and liquid states. Very refined experimental conditions were required, since argon is crystalline at liquid-nitrogen temperature and gaseous at liquid-oxygen temperature under atmospheric pressure. In the first studies the gas was liquefied by cooling it to 90°K with old liquid air in a thin-walled Pyrex capillary at a pressure of 50 psi, after which an evacuated camera of 9.53-cm radius was used to obtain a pattern that would be as complete as possible. Exposures of 140 hr were required, using Mo$K\alpha$ radiation reflected from the (200) planes of rock salt. In the later experiments argon was examined at various temperatures and pressures near the triple point. Figure 12-26 shows the intensity and radial-distribution curves obtained by increasing the temperature along the saturated-vapor curve, and compared with these results is the distribution for argon gas at 149.3°K and 46.8 atm. Noticeable effects are the progressive smearing out of the pattern, shift of the first peak toward lower angles, and increasing small-angle scatter. Compared with crystalline argon, in which there

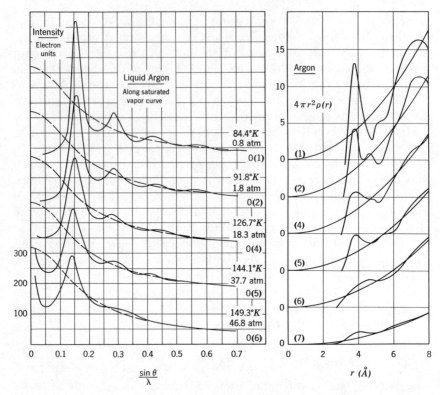

Fig. 12-26. Intensity curves and radial-distribution curves for liquid argon under five experimental conditions ranging from 84.4°K and 0.8 atm to 149.3°K and 46.8 atm. [Courtesy of N. S. Gingrich, *Rev. Mod. Phys.*, **15**, 90 (1943).]

are 12 nearest neighbors at 3.82 Å, liquid argon at 84.4°K and 0.8 atm shows 10.2 to 10.9 neighbors at 3.79 Å.

For an excellent review of the more important radial-distribution studies of elements in the liquid state previous to 1942, the reader is referred to an article by Gingrich [5].

12-8.5 Vitreous Selenium

Kaplow, Rowe, and Averbach [46] investigated the atomic arrangement in vitreous selenium following the general procedure described in Section 12-7.3 for correcting the experimental RDF. There are three polymorphic forms of crystalline selenium based on two molecular configurations. In the hexagonal form, stable at room temperature, the atoms are arranged in three linear chains parallel to the c axis in such a manner as to be characterizable as a threefold helix. In the α and β

monoclinic modifications, the atoms comprise identical, but differently arranged, eight-atom puckered rings. The vitreous form is obtained when the liquid is quickly cooled, the glass transition occurring at about 10°C.

The amorphous specimens studied by Kaplow and co-workers were a quenched and cast sample and a vapor-deposited layer-sample about $500\,\mu$ thick. The samples were ground to <325-mesh and compressed into $1 \times \frac{1}{2} \times \frac{1}{8}$ in. briquettes. At low angles ($0.6 < S < 6\,\text{Å}^{-1}$), Co$K\alpha$ radiation was used with a lithium fluoride (LiF) monochromator in the direct beam, the diffracted beam being received by a proportional counter capable of eliminating the $\lambda/2$ harmonic. For $2 < S < 15\,\text{Å}^{-1}$, Rh$K\alpha$ radiation was employed in conjunction with a LiF monochromator in the diffracted beam and a scintillation counter. The latter set of experimental conditions had the advantage of virtually excluding the incoherent scattering at the larger angles. Intensity data for the vitreous specimens were obtained by means of two scans over the low-angle region and four over the high-angle region, successive scans being made in opposite directions. For reference purposes intensity data from sintered, powdered polycrystalline hexagonal samples were also recorded. Figure 12-27 shows the reduced-intensity curves $Si(S)$ for the cast amorphous

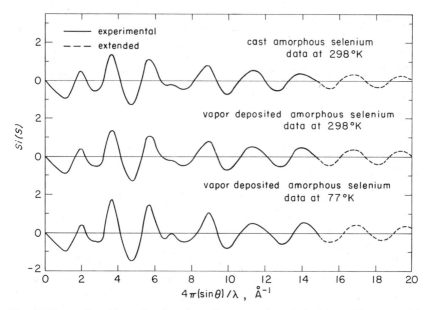

Fig. 12-27. Reduced-intensity functions for amorphous selenium. [Courtesy of R. Kaplow, T. A. Rowe, and B. L. Averbach, *Phys. Rev.*, **168**, 1068 (1968).]

specimen and for the vapor-deposited specimen as measured at 298 and 77°K. Figure 12-28 gives the differential RDFs derived from the three intensity functions of Fig. 12-27 by Fourier transformation and correction according to the procedure of Section 12-7.3.

Kaplow, Rowe, and Averbach[46] calculated the *theoretical* RDFs for the three crystalline forms and then endeavored to introduce perturba-

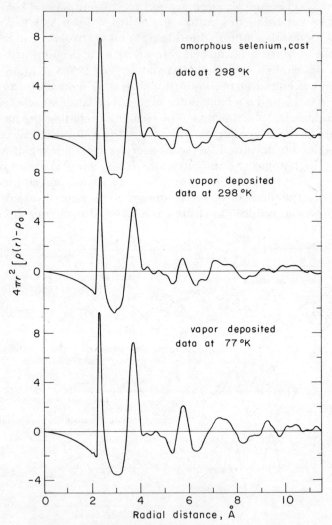

Fig. 12-28. Differential radial-distribution functions for amorphous selenium derived from the three intensity functions in Fig. 12-27. [Courtesy of R. Kaplow, T. A. Rowe, and B. L. Averbach, *Phys. Rev.*, **168**, 1068 (1968).]

tions into these ordered structures so as to arrive at the best possible match of the resulting modified RDFs with the experimental RDFs of amorphous selenium. For the calculation of the crystalline RDFs, the numbers of atomic neighbors in successive coordination shells at distances $r_1, r_2, \ldots, r_i$ about a given reference atom were determined from the known structures, the discrete peaks in the RDFs being appropriately broadened by using a Gaussian distribution function:

$$4\pi^2\rho(r) = \sum_{i=1}^{\infty} \frac{C_i}{(2\pi\sigma_i^2)^{1/2}} \exp \frac{(r-r_i)^2}{2\sigma_i^2}. \tag{12-40}$$

In equation 12-40 C_i is the number of atoms in the ith shell at distance r_i from a given origin (reference) atom, and σ_i^2 is the mean-square amplitude of vibration of an atom at distance r_i with respect to the origin atom. The parameters σ_i were treated as variables and determined by a fitting procedure with coupling coefficients defined by

$$\gamma_i = \frac{\sigma_i^2}{\sigma_\infty^2}, \tag{12-41}$$

where σ_∞ is the vibrational amplitude of two atoms at infinite separation relative to each other (independent vibrations). Thus γ_i equal to 1 and 0

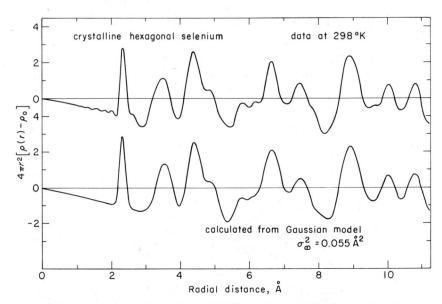

Fig. 12-29. Comparison of experimental differential radial-distribution function and calculated function from Gaussian model of crystalline hexagonal selenium. [Courtesy of R. Kaplow, T. A. Rowe, and B. L. Averbach, *Phys. Rev.*, **168**, 1068 (1968).]

correspond, respectively, to completely uncoupled and completely coupled vibrations.

The coupling coefficients calculated in this way for hexagonal selenium at room temperature are given in Table 12-5, and Fig. 12-29 compares the experimental and best calculated differential RDFs for crystalline hexagonal selenium. It may be remarked that there are two sets of coupling coefficients in Table 12-5, the first (denoted by a superscript *b*) being for atoms along a chain and the second set for atoms in different chains. The former set commences with a very low value of 0.13 for adjacent atoms and then increases with distance along the chain to unity, whereas the intermolecular values of γ_i start with a largely uncoupled value of 0.70 for adjacent atoms in neighboring chains and increase to unity for more distant chains.

Table 12-5. **Coupling Coefficients for Hexagonal Selenium at Room Temperature.** $\sigma_\infty^2 = 0.055 \text{ Å}^2$ [a]

Interatomic Separation, r_i (Å)	Coupling Coefficients, γ_i
2.32[b]	0.13
3.47	0.70
3.68[b]	0.55
4.36	0.65
4.49	0.75
4.93	0.85
4.95[b]	0.65
5.70	0.70
6.06	0.90
6.13	0.75
6.59	0.95
6.70	1.00
6.77	1.00
6.80[b]	1.00
7.06	1.00
7.27	0.90
7.52	0.90
7.54	0.95

[a]Kaplow, Rowe, and Averbach[46], p. 1071.
[b]Intrachain distance; the others are interchain distance.

Encouraged by the excellent correspondence obtained between the experimental and calculated curves for crystalline hexagonal selenium by optimization of the γ_i and σ_∞ parameters, Kaplow and co-workers proceeded to seek a reasonable fit for the experimental RDF of amorphous selenium. Initially, they tried to achieve acceptable agreement with various randomly oriented microcrystalline models without success, although agreement with the α-monoclinic microcrystalline model was superior to the other two, suggesting the vitreous form to be largely composed of rings rather than chains. Finally, a Monte Carlo optimization procedure was adopted, commencing with 100 atoms in a spherical array arranged precisely as in one of the three crystalline structures and then displacing individual atoms various distances and directions relative to the ensemble. In the course of the latter process, the new position of any given atom was retained if the agreement between the calculated and experimental RDFs was improved, but its initial position was kept if improvement did not result. After the testing of approximately 10^5 atomic movements, the model converged to the point where no additional atomic displacement improved the fit.

Figure 12-30 compares the final best fit of the experimental RDF with the calculated RDFs for perturbed α-monoclinic and β-monoclinic structures. On the whole the agreement is seen to be quite good. Figure

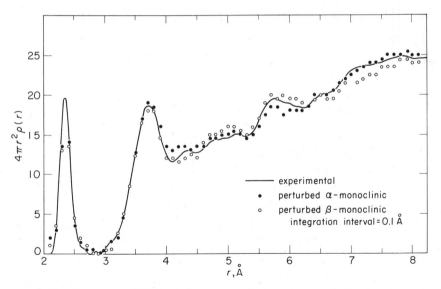

Fig. 12-30. Experimental RDF of amorphous selenium compared with functions calculated from perturbations of α- and β-monoclinic ring structures. [Courtesy of R. Kaplow, T. A. Rowe, and B. L. Averbach, *Phys. Rev.*, **168**, 1068 (1968).]

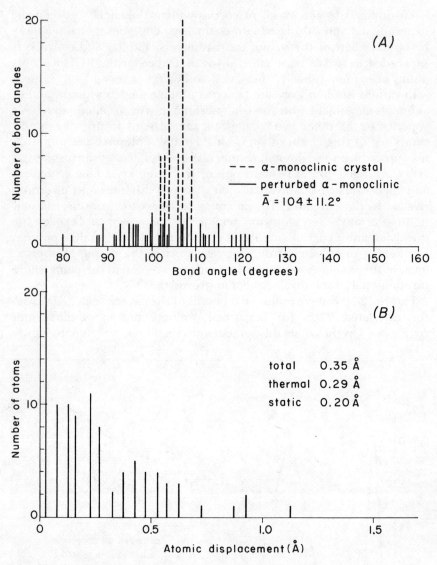

Fig. 12-31. Perturbed α-monoclinic ring structure. (*A*) Distribution of nearest-neighbor bond angles; (*B*) rms atomic displacements in ångström units. [Courtesy of R. Kaplow, T. A. Rowe, and B. L. Averbach, *Phys. Rev.*, **168**, 1068 (1968).]

12-31 portrays the perturbations of the final amorphous structure with respect to the α-monoclinic structure used as the starting point, A being the distributions of the nearest-neighbor bond angles for the α-monoclinic structure and the perturbed model, and B the rms atomic displacements. Corresponding results were calculated for the perturbed β-monoclinic and hexagonal structures, but they are not reproduced here because of space limitations. Much larger perturbations of the hexagonal chain structure were required to achieve a good fit of the RDFs than for either of the monoclinic ring structures.

It was concluded that the structure of vitreous selenium is comprised of approximately 95 per cent ring symmetry and 5 per cent weak trigonal symmetry, the latter presumed due to occasional opening of the eight-membered rings so that the atoms in the vicinity of the openings simulate the nearest-neighbor trigonal symmetry of the chain form rather than the symmetry of the ring. A few greatly deformed chains could produce the same effect. These conclusions from the radial-distribution investigation are supported by optical and Raman spectra, all of which constitutes strong evidence for the presence of Se_8 rings but much weaker evidence for near-neighbor trigonal symmetry.

12-8.6 Identification of Noncrystalline Patterns

The simplicity and diffuseness of noncrystalline patterns preclude the setting up of any general scheme of identification such as has been done with crystalline powder patterns. However, when the number of possibilities is limited and only pure species are encountered, the positions and intensities of the two or three discernible maxima are often uniquely characteristic of the several compounds. This is illustrated in Table 12-6 with the positions and intensities of the characteristic maxima of a number of synthetic resins and plastics. It should be pointed out that the positions of the peaks are computed from the observed angles in ångström units, using the Bragg equation as if reflection took place from sets of planes in a crystalline lattice. Evidently, this is an incorrect procedure, but the equivalent crystalline d spacings nevertheless are very useful for earmarking a noncrystalline pattern without the need for specifying the wavelength of the radiation used. Havriliak[59] has published the equivalent d spacings for the two strongest maxima of various polyacrylates and polymethacrylates.

Noncrystalline maxima, of course, denote the frequent occurrence of particular interatomic distances in a largely disordered substance. These distances are given in only an approximate sense by the Bragg equation. More accurate values can be obtained by analyzing the interference

Table 12-6. Maxima Observed in Patterns of Typical Resins and Plastics[a]

Sample	First Peak(Å)	Second Peak(Å)	Third Peak(Å)
Polystyrene	9.2 s	4.6 vs	
Phenolic resin A	9.8 w	4.7 m	
Phenolic resin B	9.9 w	4.8 s	
Phenolic resin C	10.8 w	4.8 s	
Methyl methacrylate	11.6 w	6.6 s	3.1 w
Ethocel resin	12.5 s	4.5 m	
Unsaturated ketone resin	12.0 w	6.2 s	3.0 w
Vinyl acetate AYAF	7.0 s	4.0 m	
Methyl acrylate	7.3 mw	4.1 m	
Glyptal resin	8.0 w	4.5 s	
Urea-formaldehyde resin A	8.7 w	4.4 s	
Urea-formaldehyde resin B	9.7 w	4.6 m	
Cumarone-indene resin	9.3 s	5.1 s	
Phenol-acetaldehyde resin	9.4 ms	4.8 ms	

[a]Equivalent Bragg spacings and relative intensities are tabulated.

function $(\sin Sr)/Sr$ of expression 12-2. Consider for a moment just those interatomic vectors with some particular frequently occurring magnitude R. That portion of the scattered intensity due to these numerous equal distances can then be written

$$I_R(S) = N_R f^2(S)\left(1 + \frac{\sin SR}{SR}\right). \tag{12-42}$$

The form of the function $1 + [(\sin SR)/SR]$ is shown in Fig. 12-32. Following the nonobservable maximum at $SR = 0$ (equivalent to $2\theta = 0$), successive maxima of decreasing amplitude occur at $SR = (1+4n)\,\pi/2$, where $n = 1,2,3,\ldots$. The strongest observable maximum occurs when

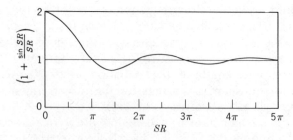

Fig. 12-32. The scattering function $1 + (\sin SR)/SR$.

$n = 1$, from which

$$SR = \frac{5}{2}\pi = R \times \frac{4\pi \sin \theta}{\lambda}$$

and

$$R = \frac{5}{8} \times \frac{\lambda}{\sin \theta}. \tag{12-43}$$

Also

$$R = \frac{5}{4} \times \frac{\lambda}{2 \sin \theta} = 1.25 \times d_{\text{Bragg}}. \tag{12-44}$$

The meaning of this expression is that the frequently occurring inter-atomic distance R responsible for a strong maximum in the diffraction pattern at angle θ is equal to 1.25 times the d spacing calculated with the aid of the Bragg equation. Because of the rapid decrease of $f^2(S)$ with S in equation 12-42, the actual factor to be used in equation 12-44 is some-what less than 1.25, a value of about 1.22 being more appropriate[60].

Compared with equation 12-42 for a truly random assemblage of atoms, the following equation (12-45) gives the intensity function in the equatorial plane for assemblages of parallel cylinders[61] or lines[62] with a most frequently occurring distance R' between their centers in the same plane:

$$I_R(S) = F^2[1 + J_0(SR')]. \tag{12-45}$$

In this equation F is the scattering factor of an isolated cylinder or line, and J_0 is a Bessel function of zero order, for which the first and principal maximum occurs at $SR' = 7$, so that

$$R' = \frac{7}{2\pi} \times \frac{\lambda}{2 \sin \theta} = 1.11 \times d_{\text{Bragg}}. \tag{12-46}$$

Equation 12-46 states that when an amorphous specimen is composed of bundles of parallel chain molecules, a hypothetical Bragg spacing calculated from the position of the principal amorphous peak should be multiplied by 1.11 to obtain the interchain distance.

In long-chain polymers, both natural and synthetic, the innermost peak is nearly always the strongest and is commonly considered to arise from interatomic vectors between adjacent chains. On the basis of this assumption, equation 12-44 or 12-46 can be used to obtain an approximate value of the interchain separation from the position of the first diffraction maximum. Table 12-7 illustrates the variation in the spacing R or R' for a number of linear polymers. The influence of the side groups upon the separation of the chains is very noticeable.

It is of course difficult to say for any given specimen whether equation

12-44 or 12-46 is preferable. For highly oriented but amorphous linear polymers, equation 12-46 is probably more suitable, whereas for randomly oriented polymers a value intermediate between R and R' may be more realistic. It is also true that these equations are valid only insofar as the actual structures correspond to the models on which the equations are based. For example, equation 12-46 takes no account of details of structure within the molecular chains, such as bulky side groups or substituent atoms of high scattering power, which may profoundly modify the calculated scattering patterns[63].

12-8.7 Other Representative Studies

A. Liquid Hydrocarbons. Mendel[64] has applied radial-distribution analysis to the study of liquid cyclohexane and benzene using a helium-filled cylindrical camera and selective filters to protect the film from fluorescence radiation, and Otvos and Mendel[65] have employed a similar technique in an investigation of the short-range structures of mesitylene, cyclohexane, and benzene. In both studies the diffraction specimen was in the form of a liquid jet irradiated by a monochromatic beam.

B. Binary Alloys. Wagner and associates[66] have studied the structures of binary liquid alloys of silver–tin, gold–tin, and copper–tin, and vapor-quenched alloys of silver–copper and silver–germanium. For the liquid-state measurements, the alloy specimens were contained in a graphite crucible and heated by means of a pyrolytic-graphite resistance

Table 12-7.   Position of Most Intense Peak in the Amorphous Patterns of Various Linear Polymers[a]

Substance	$d = \dfrac{\lambda}{2 \sin \theta}$	$R = \dfrac{5}{8} \dfrac{\lambda}{\sin \theta}$	$R' = \dfrac{7}{4\pi} \dfrac{\lambda}{\sin \theta}$
Polyethylene (27°C)	4.5	5.5	5.0
Polyethylene (170°C)	4.9	6.0	5.4
Polytrifluorochlorethylene	5.5	6.8	6.1
Cellulose	~5.0	6.2	5.5
Polyisobutylene	6.3	7.8	6.9
Natural rubber	4.8	5.9	5.3
Polyisoprene	5.0	6.2	5.5
Polymethylmethacrylate	6.6	8.1	7.3
Octacosane (molten at 150°C)	4.8	5.9	5.3

[a]Positions are given in ångström units.

ribbon to temperatures as high as 1200°C[67]. Structural features of the specimens were ascertained by comparing the experimental $i(S)$ curves with intensity curves calculated for proposed models.

C. Aggregates of Oriented Linear and Planar Molecules. Ruland[68] has explored the factors involved in spherical Fourier-transform analysis of substances consisting of randomly oriented elements possessing two types of noncrystalline internal ordering: (1) bundles of linear molecules, and (2) stacks of planar molecules, both devoid of positional correlations other than the parallelism of the principal axes. He points out that the accumulated evidence to date is becoming convincing that the short-range order in amorphous linear macromolecular systems is anisotropic, as a result of the tendency of neighboring chains to align themselves in bundles, and that the consequent nature of the order in these structures is essentially the same as that in the nematic phase of liquid crystals except for differences in the sizes of the anisotropic domains. Ruland expresses the view that information on the intramolecular structure is more accessible in the RDF, whereas the intermolecular arrangement is more clearly expressed by the intensity curve. He presents illustrations of the two types of internal ordering as follows: type 1, nylons 6, 7, 66 and 11[26, 69]; type 2, (a) maceral vitrinite and extraction products [70], and (b) polyaromatic compounds carbonized at low temperatures[71].

D. Helical Molecules in Solution. Kirste[72] and Kirste and Wunderlich[73] have calculated the scattering curves of helical structure elements of polymers in solution, with particular attention to the small-angle-scattering region. Rather good agreement was achieved between the experimental intensity function of polymethylmethacrylate and a curve calculated for folded helical elements possessing syndiotactic stereoregularity[73].

E. Biological Systems. Among studies of interest in molecular biology we may cite the helical characterization of proteins by Arndt and Riley[36, 74], which was performed by comparison of the average experimental $i(S)$ curve for six α proteins with $i(S)$ functions calculated for a number of helical conformations. It was concluded that the α_1-helix model displayed the best agreement. According to Riley[36], experience has shown that differentiation of closely related molecular models on the basis of the calculated RDFs tends to be uncertain, whereas the differences among the $i(S)$ curves are often sufficiently clear to render the distinctions unambiguous. This apparently anomalous phenomenon appears to result from the fact that small structural modifications tend to produce differences localized in one part of the intensity curve, which makes them more obvious.

Balyuzi and Burge[75] have used electronic RDFs to analyze the structural features of two amorphous cross-linked polymers with the object of ascertaining the efficacy of the method when applied to amorphous polymers of biological importance. They conclude that the method is capable of providing structural information about the *rigid regions* within the polymer systems concerned.

F. Oriented Systems. Although this topic is beyond the scope of the present chapter, we cite a few significant studies for the guidance of the interested reader. Among these are the application of the cylindrically symmetrical distribution function[76, 77] to the study of cellulose and other fibers by Norman[78–80], and the use of a simplified *difference* cylindrical-distribution function by Milberg and Daly[81–84] in structural investigations of linear synthetic organic and inorganic polymers exhibiting low degrees of preferred orientation. A summary of the foregoing programs of investigation has been given by Alexander[85].

Heyn[86, 87] studied the transverse structure of cellulose fibers by treating them as assemblages of parallel, solid, rodlike fibrils and Fourier-inverting the equatorial intensity function. Ruland and Dewaelheyns[88] invented an electronic apparatus for spherically averaging the intensity data from fibers, thus permitting standard radial-distribution methods to be applied.

12-9　FURTHER REMARKS ON EXPERIMENTAL TECHNIQUES

More needs to be said about handling liquid samples. With reference first to photographic techniques, noncorrosive liquids of low vapor pressure are most commonly photographed in the form of open jets of small diameter. Harvey, in studies of methyl[89] and ethyl[90] alcohol, circulated the liquid through a "closed" system with a small Fountainette pump, adjusting the flow until turbulence was eliminated and the stream resembled a smooth, stationary cylinder when observed through a microscope. Katzoff[91] and Morgan and Warren[92] used similar circulatory systems in producing sample jets for the study of water. By enclosing the main reservoir in a Dewar vessel, it was possible to prepare diffraction patterns at below room temperature. Morgan and Warren obtained photographs at a series of higher temperatures extending to 83°C by insulating the reservoir from the camera, protecting the film from the high temperatures, and designing the camera so as to suppress the vapor scattering. In their photographic studies of liquid hydrocarbons (Section 12-8.7A), Mendel and Otvos[64, 65] also irradiated the specimens in liquid-jet form with the liquid circulating in a closed system of small volume, making it possible to obtain long exposures with only 30 ml of liquid. Paalman and Pings[93] employed a cylindrical beryllium sample cell of 0.035-in. inside diameter for extensive diffraction measurements of liquid nitrogen at 100 psi and $-193°C$.

Rodriguez and Pings[94] constructed a Lucite cell for measuring liquid scattering with a Norelco vertical goniometer. A more complete description of this diffractometric apparatus and its application to the study of liquids confined under their own vapor pressure or in inert gases at temperatures from 0 to 50°C has been given by Rodriguez, Caputi, and Pings[95]. Another liquid-diffraction unit which fits a Norelco vertical goniometer and permits diffraction measurements of confined fluids at temperatures between 80 and 300°K and pressures up to 100 atm, has been described by Honeywell et al.[96]. Hallock[97] has built a liquid cell for diffractometry which is relatively sophisticated but nevertheless mechanically rugged. This cell, which respresents a refinement of a basic design by Moss, Kellers, and Bearden[98], is primarily intended for low-temperature measurements but can also be used at pressures up to 60 mm Hg. In this connection special mention must again be made of the parafocusing diffractometer designed specifically for measurements of scattering by liquids, including molten metals and alloys, by Kaplow and Averbach[99] (see Section 5-4.2F).

In regard to the need for highly monochromatic x-rays in radial-distribution analysis, the reader's attention is directed to the treatment of balanced filters in Section 2-3.6B. Doubly bent crystal monochromators especially designed for the measurement of diffusely scattered x-rays have been described by Chipman[100] and by Reid and Smith [101]. A useful treatment of the design of balanced filters for low-intensity measurements has been given by Soules, Gordon, and Shaw [102].

With a more general view, we conclude this section with a reiteration of the experimental factors that are of prime importance in radial-distribution analysis. Relative to the potential structural information procurable, it is fair to say that the quantity and quality of the results are a direct reflection of the pains that have been exercised in measuring and processing the experimental intensities. We reemphasize that strictest attention must be paid to the following matters: (1) accuracy of measurement of the "raw" intensities, including the use of highly monochromatic radiation, (2) maximization of the $(\sin \theta)/\lambda$ range of measurement, (3) correction for systematic errors, including correction for instrumental broadening if the interferences are relatively sharp, (4) quantitative allowance for incoherent scattering, especially for samples composed of light elements, (5) scaling of the corrected intensity function to absolute (electron, or Thomson) units, (6) correction for the finite angular range of measurement (termination-of-series errors). A specific diffractometric technique of exceptional power for optimizing the accuracy of intensity measurements is the *continuous*, or *repetitive*, averaging of counting rates over long periods of time using a multichannel analyzer (refer to Section 5-4.5). Ergun and associates have applied this technique with outstanding success to the study of carbons by extending the period of measurements over several weeks so as to yield high statistical accuracy of counting at all points on the experimental curve. Typically, they accumulated at least 10^5 counts at each point, resulting in the statistical accuracy of counting being 0.3 per cent or better.

12-10 CHARACTERIZATION OF ORDERING IN POLYMERS

This subject, though not strictly relevant to radial-distribution analysis, is of such great practical importance to investigators working in the general area of poorly ordered and noncrystalline technological polymers that we consider it worthwhile to include a brief synopsis together with a number of literature references.

The traditional and artificial concept of crystalline and amorphous regions in polymers has undergone a radical transformation in recent years[103]. It is now generally recognized that lattice imperfections of various kinds are so prevalent and of such large magnitude that there can be only very limited physical meaning in attempting to differentiate crystalline from amorphous phases. It is probably more realistic to regard a given polymer *in toto* as an imperfectly ordered continuum or as a highly defective, or highly strained, structure[51]. It has also gradually become acceptable to call such structures *paracrystalline*[68], even though the type of ordering does not correspond to the precise mathematical definition of the coordination statistics as proposed by Hosemann[104].

Despite this inherent inapplicability of the terms *crystalline* and *amorphous* to polymers, there is undoubted utility in employing the terms for empirically categorizing a given polymer specimen relative to others on an arbitrary (and quantitatively uncertain) numerical scale of ordering. This is particularly apropos in the characterization of polymers for many technological applications. Accordingly, we herewith refer the reader to some key literature that may hopefully be pertinent to his needs in the matter of categorizing polymers with respect to their degrees of ordering:

 I. General treatment and overview [105]

 II. "Classical" analysis of crystalline and amorphous fractions [106–111]

 III. Analysis of crystalline and amorphous fractions with allowance for lattice distortions [25, 69]

 IV. Determination of a crystallinity index [112, 113]

  In an effort to represent more satisfactorily the low degrees of ordering prevailing in many technological polymers, Lippert[114] has proposed the concept of *significant structure*, defined by

$$\int_{S_{\min}}^{S_{\max}} S |i(S)| \, dS,$$

which is evidently an empirical measure of the degree to which the fully corrected and reduced intensity function of the polymer in question deviates from the coherent intensity curve that would be generated by all the atoms of the polymer scattering independently of one another.

GENERAL REFERENCES

1. J. Bouman (ed.), *Selected Topics in X-ray Crystallography*, Interscience, New York, 1951, pp. 191–210, 263–296.

2. S. Ergun and M. Berman, "X-ray Diffraction Profiles of Defective Layered Lattices Showing Preferred Orientation," *Acta Crystallogr.*, **A29**, 12 (1973).

3. C. Finbak, "The Structure of Liquids," *Acta Chem. Scand.*, **3**, 1279, 1293 (1949).

4. N. S. Gingrich, "The Diffraction of X-rays by Liquid Elements," *Rev. Mod. Phys.*, **15**, 90 (1943).

*5. A. Guinier, *X-ray Diffraction in Crystals, Imperfect Crystals and Amorphous Bodies* (translated by P. Lorrain and D. Sainte-Marie Lorrain), W. H. Freeman, San Francisco, 1963, Chapter 3.

*6. R. W. James, *The Crystalline State*, Vol. II. *The Optical Principles of the Diffraction of X-rays*, G. Bell, London, 1948, Chapter 9.

7. R. F. Kruh, "Diffraction Studies of the Structure of Liquids," *Chem. Rev.*, **62**, 319 (1962). Excellent bibliography.

8. R. F. Kruh in *Handbook of X-rays* (E. F. Kaelble, ed.), McGraw-Hill, New York, 1967, Chapter 22.

9. M. E. Milberg, "A Simplified Cylindrical Distribution Function," *J. Appl. Phys.*, **33**, 1766 (1962).

10. D. P. Riley, "Spherically Symmetric Fourier Transforms and Medium-Range Radial-Distribution Functions in the X-ray Determination of Complex Molecular Structures," in *Non-crystalline Solids* (V. D. Frechette, ed.), Wiley, New York, 1960, Chapter 2.

11. W. Ruland, "The Structure of Amorphous Solids," *Pure Appl. Chem.*, **18**, 489 (1969).

12. J. R. Townsend and S. Ergun, "Analysis of Termination Effects on Atomic Radial Density Curves," *Carbon*, **6**, 19 (1968).

13. B. K. Vainshtein, "On the Theory of the Radial Distribution Method," *Sov. Crystallogr.*, **2**, 24 (1957).

14. C. N. J. Wagner, "Diffraction Analysis of Liquid and Amorphous Alloys," *Advances in X-ray Analysis*, Vol. 12, Plenum Press, New York, 1969, pp. 50–71.

15. B. E. Warren and N. S. Gingrich, "Fourier Integral Analysis of X-ray Powder Patterns," *Phys. Rev.*, **46**, 368 (1934).

16. J. Waser and V. Schomaker, "The Fourier Inversion of Diffraction Data," *Rev. Mod. Phys.*, **25**, 671 (1953).

SPECIFIC REFERENCES

[1] P. Debye, *Ann. Phys.*, **46**, 809 (1915).

[2] W. H. Zachariasen, *J. Am. Chem. Soc.*, **54**, 3841 (1932).

[3] F. Zernike and J. A. Prins, *Z. Phys.*, **41**, 184 (1927).

[4] B. E. Warren and N. S. Gingrich, *Phys. Rev.*, **46**, 368 (1934).

[5] N. S. Gingrich, *Rev. Mod. Phys.*, **15**, 90 (1943).

[6] R. W. James, *The Crystalline State*, Vol. II. *The Optical Principles of the Diffraction of X-rays*, G. Bell, London, 1948, Chapter 9.

[7] E. C. Titchmarsh, *Introduction to the Theory of Fourier Integrals*, Oxford University Press, New York, 1937.

[8] J. T. Randall, *The Diffraction of X-rays and Electrons by Amorphous Solids, Liquids, and Gases*, Wiley, New York, 1934, pp. 107–121.

[9] P. Debye and H. Menke, *Ergeb. Tech. Röntgenk.*, **2**, 1 (1931).

[10] B. E. Warren, H. Krutter, and O. Morningstar, *J. Am. Ceram. Soc.*, **19**, 202 (1936).

[11] S. Ergun in *Chemistry and Physics of Carbon*, Vol. 3 (P. L. Walker, Jr., ed.), Marcel Dekker, New York, 1968, pp. 211–288.

[12] R. A. McCune, *Advances in X-ray Analysis*, Vol. 6, Plenum Press, New York, 1963, p. 85.

[13] A. H. Compton and S. K. Allison, *X-rays in Theory and Experiment*, Van Nostrand Reinhold, New York, 1935, p. 781.

[14] *International Tables for X-ray Crystallography*, Vol. III, Kynoch Press, Birmingham, England, 1963, pp. 202–207.

[15] S. Ergun, J. Bayer, and W. Van Buren, *J. Appl. Phys.*, **38**, 3540 (1967).

[16] See Section 3-3.4G together with references noted; F. C. Blake, *Rev. Mod. Phys.*, **5**, 180 (1933); A. Taylor and H. Sinclair, *Proc. Phys. Soc.* (London), **57**, 108 (1945).

[17] J. Morgan and B. E. Warren, *J. Chem. Phys.*, **6**, 666 (1938).

[18] Reference 13, p. 140.

[19] Reference 13, p. 782.

[20] D. T. Keating and G. H. Vineyard, *Acta Crystallogr.*, **9**, 895 (1956).

[21] S. E. Rodriguez and C. J. Pings, *Acta Crystallogr.*, **18**, 979 (1965).

[22] W. Ruland, *Brit. J. Appl. Phys.*, **15**, 1301 (1964).

[23] B. E. Warren and G. Mavel, *Rev. Sci. Instrum.*, **36**, 196 (1965).

[24] S. Ergun, W. Braun, and E. Fitzer, *Rev. Sci. Instrum.*, **41**, 1133 (1970).

[24a] M. Berman and S. Ergun, *Rev. Sci. Instrum.*, **41**, 870 (1970).

[24b] S. Ergun, manuscript in preparation.

[25] W. Ruland, *Acta Crystallogr.*, **14**, 1180 (1961).

[26] W. Ruland, *Norelco Rep.*, **14**, 12 (1967).

[27] R. Hultgren, N. S. Gingrich, and B. E. Warren, *J. Chem. Phys.*, **3**, 351 (1935).

[28] J. Krogh-Moe, *Acta Crystallogr.*, **9**, 951 (1956).

[29] N. Norman, *Acta Crystallogr.*, **10**, 370 (1957).

[30] B. E. Warren, *J. Chem. Phys.*, **2**, 551 (1934).

[31] *International Tables for X-ray Crystallography,* Vol. III, Kynoch Press, Birmingham, England, 1963, pp. 162–165; also Appendix V of this book.

[32] Reference 5, p. 98.

[33] H. P. Klug and L. E. Alexander, *X-ray Diffraction Procedures*, 1st ed., Wiley, New York, 1954, pp. 603–605.

[34] J. R. Townsend and S. Ergun, *Carbon*, **6**, 19 (1968).

[35] R. F. Kruh, *Chem. Rev.*, **62**, 319 (1962).

[36] D. P. Riley in *Non-crystalline Solids* (V. D. Frechette, ed.), Wiley, New York, 1960, pp. 26–52.

[37] R. F. Kruh in *Handbook of X-rays* (E. F. Kaelble, ed.), McGraw-Hill, New York, 1967, Chapter 22.

[38] A. Bienenstock, *J. Chem. Phys.*, **31**, 570 (1959).

[39] D. R. Chipman, *J. Appl. Phys.*, **26**, 1387 (1955).

[40] S. Ergun and V. H. Tiensuu, *J. Appl. Phys.*, **29**, 946 (1958).

[41] W. L. Bragg and J. West, *Phil. Mag.*, (7) **10**, 823 (1930).

[42] C. Finbak, *Acta Chem. Scand.*, **3**, 1279 (1949).

[43] C. Finbak, *Acta Chem. Scand.*, **3**, 1293 (1949).

[44] R. Kaplow, S. L. Strong, and B. L. Averbach, *Phys. Rev.*, **138**, A1336 (1965).

[45] R. R. Fessler, R. Kaplow, and B. L. Averbach, *Phys. Rev.*, **150**, 34 (1966).

[46] R. Kaplow, T. A. Rowe, and B. L. Averbach, *Phys. Rev.*, **168**, 1068 (1968).

[47] G. S. Cargill III, *J. Appl. Crystallogr.*, **4**, 277 (1971).

[48] J. Biscoe and B. E. Warren, *J. Appl. Phys.*, **13**, 364 (1942); also see Sections 6-10.2D and 9-3.5F.

[49] L. L. Ban in *Surface and Defect Properties of Solids*, Vol. 1, Chemical Society, London, 1972, pp. 54–94.

[50] H. Kuroda, *J. Colloid Sci.*, **12**, 497 (1957).

[51] S. Ergun, *Phys. Rev.*, **B1**, 3371 (1970).

[52] S. Ergun and M. Berman, *Acta Crystallogr.*, **A29**, 12 (1973).

[53] S. Ergun, *Carbon,* in press.

[54] S. Ergun, private communication.

[55] B. E. Warren, *J. Appl. Phys.*, **8**, 645 (1937).

[56] L. Cartz, *Z. Kristallogr.*, **120**, 241 (1964).

[57] J. Zarzycki, *Verres Réfractaires*, No. 1, 1957, p. 3.

[58] A. Eisenstein and N. S. Gingrich, *Phys. Rev.*, **58**, 307 (1940); **62**, 261 (1942).

[59] S. Havriliak, Jr., *Polymer*, **9**, 289 (1968).

[60] See also A. Guinier, *X-ray Diffraction in Crystals, Imperfect Crystals and Amorphous Bodies*, W. H. Freeman, San Francisco, 1963, pp. 72–74.

[61] G. Oster and D. P. Riley, *Acta Crystallogr.*, **5**, 272 (1952).

[62] L. E. Alexander and E. R. Michalik, *Acta Crystallogr.*, **12**, 105 (1959).

[63] L. E. Alexander, *X-ray Diffraction Methods in Polymer Science*, Wiley, New York, 1969, pp. 379–381.

[64] H. Mendel, *Acta Crystallogr.*, **15**, 113 (1962).

[65] J. W. Otvos and H. Mendel, *Acta Crystallogr.*, **15**, 657 (1962).

[66] C. N. J. Wagner, *Advances in X-ray Analysis*, Vol. 12, Plenum Press, New York, 1969, p. 50.

[67] M. L. Joshi, *Rev. Sci. Instrum.*, **36**, 678 (1965).

[68] W. Ruland, *Pure Appl. Chem.*, **18**, 489 (1969).

[69] W. Ruland, *Polymer*, **5**, 89 (1964).

[70] W. Ruland, *Proceedings of the Fifth Conference on Carbon*, Vol. 1, Pergamon Press, New York, 1962, p. 429.

[71] W. Ruland, *Carbon*, **2**, 365 (1965).

[72] R. G. Kirste, *Z. phys. Chem.*, N.F., **42**, 358 (1964).

[73] R. G. Kirste and W. Wunderlich, *Macromol. Chem.*, **73**, 240 (1964).

[74] U. W. Arndt and D. P. Riley, *Phil. Trans. Roy. Soc.* (London), **A247**, 409 (1955).

[75] H. H. M. Balyuzi and R. E. Burge, *Biopolymers*, **10**, 777 (1971).

[76] D. Wrinch, *Fourier Transforms and Structure Factors*, ASXRED Monograph No. 2, 1946.

[77] G. H. Vineyard, *Acta Crystallogr.*, **4**, 281 (1951).

[78] N. Norman, *On the Cylindrically Symmetrical Distribution Method in X-ray Analysis*, Ph.D. Thesis, University of Oslo, 1954.

[79] N. Norman, *Acta Crystallogr.*, **7**, 462 (1954).

[80] N. Norman in *Selected Topics in Structure Chemistry*, Universitetsforlaget, Oslo, 1967, p. 195.

[81] M. E. Milberg, *J. Appl. Phys.*, **33**, 1766 (1962).

[82] M. E. Milberg, *J. Appl. Phys.*, **34**, 722 (1963).

[83] M. E. Milberg and M. C. Daly, *J. Chem. Phys.*, **39**, 2966 (1963).

[84] M. E. Milberg, *J. Polymer Sci., Part A-1*, **4**, 801 (1966).

[85] Reference 63, pp. 384–389.

[86] A. N. J. Heyn, *J. Appl. Phys.*, **26**, 519 (1955).

[87] A. N. J. Heyn, *J. Appl. Phys.*, **26**, 1113 (1955).

[88] W. Ruland and A. Dewaelheyns, *J. Sci. Instrum.*, **44**, 236 (1966).

[89] G. G. Harvey, *J. Chem. Phys.*, **6**, 111 (1938).

[90] G. G. Harvey, *J. Chem. Phys.*, **7**, 878 (1939).

[91] S. Katzoff, *J. Chem. Phys.*, **2**, 841 (1934).

[92] J. Morgan and B. E. Warren, *J. Chem. Phys.*, **6**, 666 (1938).

[93] H. H. Paalman and C. J. Pings, *Rev. Sci. Instrum.*, **33**, 496 (1962).

[94] S. E. Rodriguez and C. J. Pings, *Rev. Sci. Instrum.*, **33**, 1469 (1962).

[95] S. E. Rodriguez, R. W. Caputi, and C. J. Pings, *Rev. Sci. Instrum.*, **36**, 449 (1965).

[96] W. I. Honeywell, C. M. Knobler, B. L. Smith, and C. J. Pings, *Rev. Sci. Instrum.*, **35**, 1216 (1964).

[97] R. H. Hallock, *Rev. Sci. Instrum.*, **41**, 1107 (1970).

[98] T. H. Moss, C. F. Kellers, and A. J. Bearden, *Rev. Sci. Instrum.*, **34**, 1267 (1964).

[99] R. Kaplow and B. L. Averbach, *Rev. Sci. Instrum.*, **34**, 579 (1963).

[100] D. R. Chipman, *Rev. Sci. Instrum.*, **27**, 164 (1956).

[101] J. S. Reid and T. Smith, *J. Phys.*, **E2**, 601 (1969).

[102] J. A. Soules, W. L. Gordon, and C. H. Shaw, *Rev. Sci. Instrum.*, **27**, 12 (1956).

[103] W. O. Statton, *J. Polymer Sci., Part C*, **18**, 33 (1966).

[104] For example, see R. Hosemann and S. N. Bagchi, *Direct Analysis of Diffraction by Matter*, North Holland, Amsterdam, 1962.

[105] Reference 63, Chapter 3.

[106] J. M. Goppel, *Appl. Sci. Res.*, **A1**, 3, 18 (1947); J. J. Arlman, *Appl. Sci. Res.*, **A1**, 347 (1949); J. M. Goppel and J. J. Arlman, *Appl. Sci. Res.*, **A1**, 462 (1949).

[107]   P. H. Hermans and A. Weidinger, *J. Appl. Phys.*, **19**, 491 (1948); *J. Polymer Sci.*, **4**, 135 (1949); **5**, 565 (1950).

[108]   P. H. Hermans and A. Weidinger, *J. Polymer Sci.*, **4**, 709 (1949); **5**, 269 (1950).

[109]   P. H. Hermans and A. Weidinger, *Makromol. Chem.*, **44-46**, 24 (1961); **50**, 98 (1961).

[110]   L. E. Alexander, S. Ohlberg, and G. R. Taylor, *J. Appl. Phys.*, **26**, 1068 (1955).

[111]   G. Challa, P. H. Hermans, and A. Weidinger, *Makromol. Chem.*, **56**, 169 (1962).

[112]   J. H. Wakelin, H. S. Virgin, and E. Crystal, *J. Appl. Phys.*, **30**, 1654 (1959).

[113]   W. O. Statton, *J. Appl. Polymer Sci.*, **7**, 803 (1963).

[114]   E. L. Lippert, Jr., "Significant Structure in Polymers, Measurement by X-ray Diffractometry," presented at the American Physical Society Meeting, San Diego, March 19–22, 1973.

APPENDIX I

LAYOUT FOR A DIFFRACTION LABORATORY

Planners of an x-ray diffraction laboratory, as well as the staffs of such laboratories, should study and be familiar with every aspect of x-ray safety as presented in the general references at the end of Chapter 2. Unlike radiographic installations, which may be placed within a permanent total enclosure to which no person has access while the x-rays are being generated, diffraction equipment requires varying amounts of an operator's attention during the diffraction exposure or measurements, depending upon the type of study under way. The diffraction unit itself therefore must meet the requirements of a class A "totally protective installation."* This is normally achieved by the protective and safety features built into a commercial diffraction unit. Otherwise, it must be provided by mounting the tube in a "totally protective housing" which (except for the direction of the useful beam) reduces the dosage rate at contact with the tube housing to less than 2.5 mR/hr†. In addition, each port or window of the x-ray tube must be provided with an automatic beam shutter so arranged that it remains closed except when a collimating system is in position. The beam exit of the camera in turn should be provided with a beam stop of lead glass provided with a fluoroscopic screen for observing the beam alignment. All equipment and accessories must be so designed that operating personnel any place in the laboratory are never subjected to more than the permissible daily dose of 20 mR. Some slight amount of exposure to radiation cannot be avoided in a diffraction laboratory, but it is not difficult to keep it below one-tenth of the permissible exposure.

Uninformed laboratory designers have occasionally in the past gone to needless expense in their efforts to provide a safe diffraction laboratory. Some have actually sheathed the entire laboratory in sheet lead to

*See, for instance, *Safety Code for the Industrial Use of X-rays*, American Standards Association, New York, 1946.

†Note that the permissible daily dose of x-radiation has been revised to a total of 0.02 R.

protect the personnel of adjoining laboratories. This is obviously unnecessary, for if radiation of such intensity as to be dangerous to occupants of neighboring laboratories were escaping, how could it be safe for the diffraction laboratory's own staff? Some safety expert has suggested that the diffraction laboratory should be free enough from escaping radiation that one might store the supply of unexposed x-ray film in the room without additional protection, under which conditions the laboratory is entirely safe for the operating personnel, and of course for all persons in adjoining laboratories.

The planner of an x-ray diffraction laboratory will either have space in a new building to arrange in the most convenient manner, or space in an older building the utilization of which may impose certain limitations. Modern laboratory construction makes wide use of the module idea, the module being a repetitive structural unit of space between the corridor and the outside wall. It is the unit of work space required by the individual worker, the typical module being approximately 8×25 ft. A laboratory is planned to occupy one, two, or more such modules by erection of movable partitions at the proper positions. Services for each module are usually provided in vertical utility stacks in the outside walls or in the corridor walls. In remodeling older space, existing partitions and services may dictate to some extent the space available, and in turn the convenience of the final layout.

Because of the hazards of continued exposure to x-rays, good planning segregates the diffraction equipment from the desk or office space, and makes necessary a little larger floor area per man for x-ray laboratories than for general laboratories. For a one-man laboratory, for instance, a minimum of 300 ft^2 is required, but a two-man laboratory can be conveniently arranged in 500 ft^2. Figure I-1 suggests a layout for a two-man diffraction laboratory. It occupies three average-size modules, and has an area of slightly over 500 ft^2. The spaces and facilities as depicted are regarded as a minimum for a modern diffraction laboratory for research and/or control. It is noted that the arrangement provides convenient circulation between the preparation space, the x-ray units, and the darkroom. Likewise, the film-measuring equipment, calculating facilities, and files are handy to the operators' desks. The room for film viewing, goniometric studies, and microscopic work has daylight but may be darkened at will by means of the dark shade in channel.

Adequate storage is provided in this plan, by means of suitable cupboards and shelves, except that there is need for storage space for bulky items of equipment used only infrequently. For instance, most laboratories have occasional use for a dry- or controlled-atmosphere box in mounting samples, for a high-vacuum bench mounted on casters

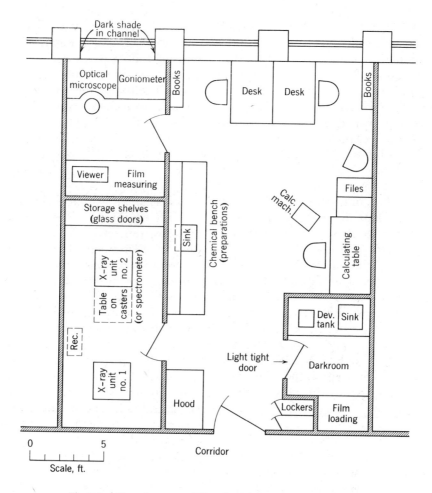

Fig. I-1.   Plan of an x-ray diffraction laboratory for two workers.

so that it may be wheeled into the laboratory for certain preparatory work, for thermostats for constant-temperature diffraction studies, and so on. Suitable storage for such equipment, in less costly space, should be planned for on the same floor of the building. Modifications of the floor plan to suit modules of other dimensions, or irregular floor areas, will present themselves on a little study.

Most domestic commercial diffraction units have a power demand of 2 to 6 kVA, and usually operate on either a 100/130-V or 200/250-V,

50/60-cycle, ac line. Some foreign-made units, however, have power demands up to 10 kVA. The number and make of the units determines the power requirements of the room housing the x-ray equipment. In addition, a 120-V ac circuit for auxiliary equipment is desirable. A water supply (preferably filtered and pressure regulated) together with drain facilities is needed for cooling tubes and transformers. Vacuum for vacuum-back cassettes, and compressed air should be supplied in this room. The hood and chemical preparation bench in the main laboratory should be supplied with the usual utilities of a chemical laboratory. The microscope and goniometer bench should have 120-V alternating current, gas, air, vacuum, water, and drain.

Darkroom facilities are best provided as a small individual darkroom conveniently located with respect to the x-ray equipment. The central community darkroom used by several groups is inconvenient and unsatisfactory because of the different safelight requirements for various photographic materials, the need for different processing solutions for the various materials, the divided responsibility for keeping the equipment clean and in order, and the time wasted in going to and from a distant room. Properly ventilated inside rooms are preferred for darkrooms, since outside rooms are more difficult to maintain at constant temperature, and, if they have windows, they are more difficult to make completely light-tight. Their general lighting fixtures should be of the incandescent type rather than fluorescent because the frequent on-off switching greatly shortens the life of fluorescent lamps. Another serious objection to the latter is the very appreciable and rather persistent afterglow which some show, and which is sufficient to fog some photographic materials. The photographic illumination is provided by overall indirect illumination with the proper safelight (Wratten Series 6B, indirect) together with spot illumination (Series 6B, direct) over the film-loading and processing areas. Chilled water is a necessity for keeping processing solutions at the proper temperature. If chilled water is not piped throughout the building, it can be furnished by a special darkroom refrigeration unit, or a drinking-fountain unit may serve as the supply.

A well-designed combination film-processing and darkroom-sink unit is very desirable. Commercially available units may be used here, but they frequently require larger volumes of processing solutions than are economical for the quantity of film being developed. Figure I-2 depicts a stainless-steel processing and sink unit that is readily fabricated.* Section *A* is a waterbath in which three tanks of processing solutions are

---

*The Calumet Manufacturing Company, 6550 N. Clark St., Chicago, Illinois specializes in the fabrication of darkroom processing units.

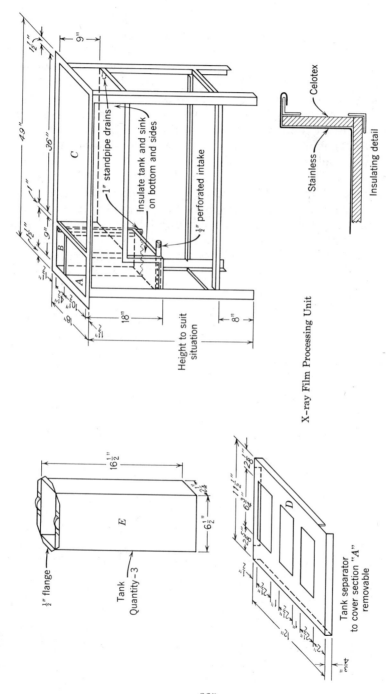

X-ray Film Processing Unit

Fig. I-2. Design for a stainless-steel film-processing unit and sink.

865

suspended by means of the separator cover D. Each processing tank E has a capacity of 1 gal of solution, and is supplied with a floating lid to prevent evaporation and/or oxidation of solutions between periods of use. Section B is the film-washing tank, and C is the darkroom sink of such dimensions that it may be used (when desired) for holding trays of processing solution. The entire unit, except for the angle-iron support, is fabricated of 20-gage stainless steel. Since the cooled water and solutions might cause condensation on the outside of the unit under excessively humid conditions, the sink, water jacket, and wash compartment have been insulated with 0.5 in. of Celotex. Proper temperature is maintained in the processing tanks by admitting water of controlled temperature through the perforated intake pipe at the bottom of A. This water circulates around the tanks, then passes into the wash compartment B through openings at the bottom of the partition between A and B, and finally leaves B through the overflow standpipe drain at the rear corner. The desired water temperature is maintained to $\pm 0.5°$ by blending warm and cold streams of water with a thermostatic mixing valve.*

*Such valves are manufactured by the Jordan Valve Division, Richards Industries, Inc., Helen and Blade Sts., Cincinnati, Ohio, and by the Powers Regulator Company, 3436 Oakton St., Skokie, Illinois.

A P P E N D I X II

THE HANDLING AND PROCESSING
OF X-RAY FILM

It is assumed at the start that the reader is familiar with the rudiments of photographic theory and the handling of photographic materials in general. If such is not the case, reference may be made to any of the well-known books on photographic principles and techniques.* The diffraction worker is urged to discipline himself in his darkroom technique, since slovenly darkroom habits can lead to more disappointing diffraction films than almost any other cause.

The Handling of X-ray Film

Modern x-ray films for diffraction studies are composed of a transparent, blue-tinted, cellulose acetate base coated on both sides with a layer of sensitive silver emulsion about 0.001 in. thick. Double coating provides maximum speed and contrast in the film, and allows developing, fixing, and drying in the shortest possible time.† The emulsion is especially sensitive to x-rays, but it is also sensitive to ordinary light and must be handled only in the light of a special x-ray safelight.

X-ray film must always be handled with care. Physical strains, such as pressure, creasing, buckling, friction, and abrasion, may produce sensitized areas which lead to black markings on the developed film. Avoid drawing the film rapidly from the carton, camera, or cassette.

---

*C. B. Neblette, *Photography, Its Principles and Practice*, 4th ed., Van Nostrand, Reinhold, New York, 1942; C. E. K. Mees, *The Theory of the Photographic Process*, 3rd ed., (T. H. James, ed.), Macmillan, New York, 1966; T. H. James and G. C. Higgins, *Fundamentals of Photographic Theory*, Wiley, New York, 1948.

†Agfa Non-Screen x-ray film and Eastman No-Screen x-ray film, because of their high contrast and relatively high speed, are very satisfactory for diffraction work. Kodak industrial x-ray film, type K, is even faster than the latter, but has somewhat less contrast. Table 4-3 lists x-ray films available throughout the world.

Large films should be handled by the corners; small films are conveniently held by the edges. *All contact with fingers that are moist, or contaminated with chemicals, must be avoided!* Use a *clean* towel freely to assure that the hands are dry and clean. The film is always handled in its individual black-paper folder until the moment it is to be loaded into the camera. In many instances the black-paper folder is not removed for loading in the camera.

All cutting and punching of film is done with the black paper around it. Best results are obtained in cutting film if a straightedge is placed on the cutting board at the proper position to act as a stop against which the left-hand edge of the film, in its folder, is held while shearing with the knife. The straightedge is first adjusted with the lights on, and should be held in place with weights. Then, by the light of the safelight, the film is put in place and held securely while being cut.

Safe illumination for handling and processing film is most simply obtained by using the proper safelight filters in lamps with bulbs of the correct wattage. Wratten safelights, series 6B, are the standard for use with x-ray film. In this light unexposed film may be handled at a distance of 3 ft for 1 min; at 2 ft for $\frac{1}{2}$ min. After exposure the film is more sensitive to the light of the safelight than before. This must be kept in mind while loading into hangers for development. The emulsions, however, are less sensitive when wet, and the films may be exposed to the light for longer periods after development has been started.

The Processing of X-ray Film

Standardized time-temperature processing procedures in tanks is recommended. Such methods permit a reliable exposure technique to be quickly established, and prevent overdevelopment. A water-jacketed tank development unit (Appendix I) very conveniently controls the temperature of the solutions through the circulation of water at the proper temperature. The usual condition in a darkroom is that the solutions are too warm and must be cooled. Tap water, blended with warm water, is cold enough for cooling except in the summer when it is necessary to blend tap water with ice water to achieve desired temperatures. An accurate darkroom timer,* preferably with a safelight illuminated dial, is a necessity for timing development.

Development. Films should be processed in the developer solution recommended by the manufacturer for the time and at the temperature

---

*The Eastman electric x-ray timer, made by Eastman Kodak Company, Rochester New York, is very satisfactory.

(normally 68°F) recommended. The temperature of the developer is checked immediately before the films are immersed in it. When it is not feasible to cool the solution to 68°F, it is usually possible to process at a higher temperature for a shorter time. Most manufacturers provide a development-time table, such as the following:

Temperature		Development Time (min)	
F°	C°	Normal	Maximum
60	15	$8\frac{1}{2}$	16
65	18	6	10
68	20	5	8
70	21	$4\frac{1}{2}$	7
75	24	$3\frac{1}{4}$	$5\frac{1}{2}$

Suitable stainless-steel film-developing hangers should be provided on which the film is placed for processing. With the developer at the proper temperature, the film is immersed in the developer solution and the timer simultaneously actuated. The film should be moved up and down in the solution a few times after immersion to insure that no air bubbles are clinging to it. Then, at $\frac{1}{2}$-min intervals during development, the film should be agitated or moved about in the solution to prevent uneven development. When the timer sounds, indicating that development is complete, the film is removed from the developer, rinsed in running water or an acid stop bath for 30 sec, and then placed in the fixing bath.

In tank processing of films, each film carries from the tank a certain amount of developer when removed at the end of the process. Thus, after several films have been developed the level of the developer will have fallen in the tank, and fresh developer must be added to maintain the level of the solution. Although the addition of fresh developer slows up exhaustion of the solution, its activity gradually diminishes. With such maintenance we can develop the equivalent of 110 5×7 in. films in 1 gal of developer at the normal time of 5 min. Then an additional equivalent of films may be developed by increasing the development time 1 min. Oxidation by the air and deterioration with age are also factors in depletion of the developer solution. Evaporation and, to some extent, air oxidation may be retarded through a close-fitting floating lid when the solution is not in use. Many diffraction laboratories, however, find that they do not develop enough films to exhaust the developer through use before it deteriorates from age. In such instances it is probably desirable to renew the entire solution at the end of a fixed time, regardless of the number of films processed. When exhaustion and deterioration are allowed to go too far, stain, dichroic fog, loss of contrast, and lack of detail may result. Obviously, since these solutions are used over and over, great care must be taken to avoid contaminating them.

Stop Bath. When films are removed from the developer solution, a considerable amount of the solution clings to the emulsion. If the films are placed directly in the fixer solution, the alkali from the developer solution retained by the emulsion will quickly neutralize the acid of the fixer solution and impair its efficiency. This can be prevented by a 30-sec rinse in running water or, better, by the use of an acid stop bath. The stop bath consists of 6 oz (180 ml) of 28 per cent acetic acid in 1 gal of water. A rinse of 30 sec in this bath effectively removes the alkaline developing solution, and prolongs the life of the fixing bath. The stop bath should be replaced with fresh solution as soon as it fails to show an acid reaction with litmus paper.

Fixing. Fixing removes the unchanged silver salts left in the emulsion, thereby permanently fixing the image on the film. When the films are first placed in the fixer solution, and once or twice during fixation, the hangers should be agitated to assure uniform action of the chemicals. The temperature of the solution should not be greatly different from that of the developer. To fix the film completely and harden the emulsion thoroughly, it should be left in the solution at least 15 min. The solution has become exhausted when it has lost its acidity or when it takes an unusually long time to remove the unexposed silver halide from the emulsion. The volume of the fixing bath should be kept constant by the addition of small amounts of fresh solution if necessary. When the developer solution is discarded for fresh solution, the fixer solution should also be replaced by an entirely fresh one.

Washing and Drying. After fixing, the films should be washed in running water, which is circulating freely enough so that the entire emulsion area receives frequent changes. The hourly flow of water should be such that there is a complete change of water every 5 min, and the time of washing should be at least 20 min. In warm weather the films must be removed from the tank as soon as washing is completed, to prevent softening of the emulsion. When washing is completed, the film should be sponged carefully with a photographic cellulose sponge to remove any sediment that may have settled on it. Finally the sponge should be squeezed as dry as possible and the excess water wiped from the film. The film is then hung up to dry in its hanger in a place where the air is as free from dust as possible. Films dry best in warm, circulating, dry air.

APPENDIX III

MISCELLANEOUS CONSTANTS AND NUMERICAL DATA*

Avogadro's number $N = 6.02257 \times 10^{23}$ mole^{-1}

Velocity of light $c = 2.997925 \times 10^{10}$ cm sec^{-1}

Electronic charge $e = 4.80296 \times 10^{-10}$ esu

Electronic rest mass $m_0 = 9.10904 \times 10^{-28}$ gram

Planck's constant $h = 6.62554 \times 10^{-27}$ erg sec

Boltzmann's constant $k = 1.38053 \times 10^{-16}$ erg deg^{-1}

Mass of atom of unit atomic weight $1/N = 1.66042 \times 10^{-24}$ gram

1 electron volt $= 1.602095 \times 10^{-12}$ erg

Zero degrees Centigrade $= 273.15$ degrees Kelvin

Base of natural logarithms $e = 2.7182818$

1 radian $= 57.29578$ degrees

1 degree $= 0.017453$ radian

$\pi = 3.14159265$

1 cm $= 10^8$ Å $= 10^4\,\mu = 0.39370$ in.

1 in. $= 2.540005$ cm

$\mathrm{Log}_e x = 2.302585\,\log_{10} x$

*The numerical values of the first nine constants above are from S. P. Clark, Jr. (ed.), *Handbook of Physical Constants*, rev. ed., Geological Society of America, Memoir 97, 1966.

APPENDIX IV

INTERNATIONAL ATOMIC WEIGHTS – 1969*

Element	Symbol	Atomic Number	Atomic Weight	Element	Symbol	Atomic Number	Atomic Weight
Actinium	Ac	89	227	Gadolinium	Gd	64	157.25
Aluminum	Al	13	26.98	Gallium	Ga	31	69.72
Americium	Am	95	[243]	Germanium	Ge	32	72.59
Antimony	Sb	51	121.75	Gold	Au	79	196.97
Argon	A	18	39.948	Hafnium	Hf	72	178.49
Arsenic	As	33	74.92	Helium	He	2	4.003
Astatine	At	85	[210]	Holmium	Ho	67	164.93
Barium	Ba	56	137.34	Hydrogen	H	1	1.0080
Berkelium	Bk	97	[247]	Indium	In	49	114.82
Beryllium	Be	4	9.012	Iodine	I	53	126.90
Bismuth	Bi	83	208.98	Iridium	Ir	77	192.2
Boron	B	5	10.81	Iron	Fe	26	55.85
Bromine	Br	35	79.904	Krypton	Kr	36	83.80
Cadmium	Cd	48	112.40	Lanthanum	La	57	138.90
Calcium	Ca	20	40.08	Lead	Pb	82	207.2
Californium	Cf	98	[251]	Lithium	Li	3	6.941
Carbon	C	6	12.011	Lutetium	Lu	71	174.97
Cerium	Ce	58	140.12	Magnesium	Mg	12	24.305
Cesium	Cs	55	132.91	Manganese	Mn	25	54.938
Chlorine	Cl	17	35.453	Mercury	Hg	80	200.59
Chromium	Cr	24	51.996	Molybdenum	Mo	42	95.94
Cobalt	Co	27	58.93	Neodymium	Nd	60	144.24
Columbium (see Niobium)				Neptunium	Np	93	[237]
				Neon	Ne	10	20.179
Copper	Cu	29	63.54	Nickel	Ni	28	58.71
Curium	Cm	96	[247]	Niobium	Nb	41	92.91
Dysprosium	Dy	66	162.50	Nitrogen	N	7	14.007
Erbium	Er	68	167.26	Osmium	Os	76	190.2
Europium	Eu	63	151.96	Oxygen	O	8	15.9994
Fluorine	F	9	18.998	Palladium	Pd	46	106.4
Francium	Fr	87	[223]	Phosphorus	P	15	30.974

*A value given in brackets denotes the mass number of the most stable known isotope.

Atomic Weights (continued)

Element	Symbol	Atomic Number	Atomic Weight	Element	Symbol	Atomic Number	Atomic Weight
Platinum	Pt	78	195.09	Strontium	Sr	38	87.62
Plutonium	Pu	94	[244]	Sulfur	S	16	32.06
Polonium	Po	84	209	Tantalum	Ta	73	180.95
Potassium	K	19	39.102	Technetium	Tc	43	[97]
Praseodymium	Pr	59	140.91	Tellurium	Te	52	127.60
Promethium	Pm	61	[145]	Terbium	Tb	65	158.93
Protactinium	Pa	91	231	Thallium	Tl	81	204.37
Radium	Ra	88	226.03	Thorium	Th	90	232.04
Radon	Rn	86	222	Thulium	Tm	69	168.93
Rhenium	Re	75	186.2	Tin	Sn	50	118.69
Rhodium	Rh	45	102.91	Titanium	Ti	22	47.90
Rubidium	Rb	37	85.468	Tungsten	W	74	183.85
Ruthenium	Ru	44	101.07	Uranium	U	92	238.03
Samarium	Sm	62	150.4	Vanadium	V	23	50.94
Scandium	Sc	21	44.96	Xenon	Xe	54	131.30
Selenium	Se	34	78.96	Ytterbium	Yb	70	173.04
Silicon	Si	14	28.09	Yttrium	Y	39	88.91
Silver	Ag	47	107.868	Zinc	Zn	30	65.37
Sodium	Na	11	22.990	Zirconium	Zr	40	91.22

MASS ABSORPTION COEFFICIENTS μ/ρ OF THE ELEMENTS ($Z = 1$ TO 83) FOR A SELECTION OF WAVELENGTHS*

Element	Atomic Number	Ag$K\alpha$ 0.5609 Å	Mo$K\alpha$ 0.7107 Å	Cu$K\alpha$ 1.5418 Å	Co$K\alpha$ 1.7902 Å	Fe$K\alpha$ 1.9373 Å	Cr$K\alpha$ 2.2909 Å
H	1	0.371	0.380	0.435	0.464	0.483	0.545
He	2	0.195	0.207	0.383	0.491	0.569	0.813
Li	3	0.187	0.217	0.716	1.03	1.25	1.96
Be	4	0.229	0.298	1.50	2.25	2.80	4.50
B	5	0.279	0.392	2.39	3.63	4.55	7.38
C	6	0.400	0.625	4.60	7.07	8.90	14.5
N	7	0.544	0.916	7.52	11.6	14.6	23.9
O	8	0.740	1.31	11.5	17.8	22.4	36.6
F	9	0.976	1.80	16.4	25.4	32.1	52.4
Ne	10	1.31	2.47	22.0	35.4	44.6	72.8
Na	11	1.67	3.21	30.1	46.5	58.6	95.3
Mg	12	2.12	4.11	38.6	59.5	74.8	121
Al	13	2.65	5.16	48.6	74.8	93.9	152
Si	14	3.28	6.44	60.6	93.3	117	189
P	15	4.01	7.89	74.1	114	142	229
S	16	4.84	9.55	89.1	136	170	272
Cl	17	5.77	11.4	106	161	200	318
A	18	6.81	13.5	123	187	232	366
K	19	8.00	15.8	143	215	266	417

*Prepared by special permission from *International Tables for X-ray Crystallography*, Vol. III, Kynoch Press, Birmingham, England, 1972, pp. 162–165. The numerical values are expressed in units of square centimeters per gram. Values in italics are of low accuracy. An intervening absorption edge is denoted by a bar separating two successive numerical values. Additional mass absorption coefficient data at other wavelengths are presented in the above reference.

Mass Absorption Coefficients (*continued*)

Element	Atomic Number	AgKα 0.5609 Å	MoKα 0.7107 Å	CuKα 1.5418 Å	CoKα 1.7902 Å	FeKα 1.9373 Å	CrKα 2.2909 Å
Ca	20	9.28	18.3	162	243	299	463
Sc	21	10.7	21.1	184	273	336	*513*
Ti	22	12.3	24.2	208	308	377	<u>*571*</u>
V	23	14.0	27.5	233	343	419	68.4
Cr	24	15.8	31.1	260	381	<u>463</u>	79.8
Mn	25	17.7	34.7	285	<u>414</u>	57.2	93.0
Fe	26	19.7	38.5	308	52.8	66.4	108
Co	27	21.8	42.5	<u>313</u>	61.1	76.8	125
Ni	28	24.1	46.6	45.7	70.5	88.6	144
Cu	29	26.4	50.9	52.9	81.6	103	166
Zn	30	28.8	55.4	60.3	93.0	117	189
Ga	31	31.4	60.1	67.9	105	131	212
Ge	32	34.1	64.8	75.6	116	146	235
As	33	36.9	69.7	83.4	128	160	258
Se	34	39.8	74.7	91.4	140	175	281
Br	35	42.7	79.8	99.6	152	190	305
Kr	36	45.8	84.9	108	165	206	327
Rb	37	48.9	90.0	117	177	221	351
Sr	38	52.1	95.0	125	190	236	373
Y	39	55.3	<u>100</u>	134	203	252	396
Zr	40	58.5	15.9	143	216	268	419
Nb	41	61.7	17.1	153	230	284	441
Mo	42	64.8	18.4	162	243	300	463
Tc	43	<u>67.9</u>	19.7	172	257	316	485
Ru	44	10.7	21.1	183	272	334	*509*
Rh	45	11.5	22.6	194	288	352	*534*
Pd	46	12.3	24.1	206	304	371	*559*
Ag	47	13.1	25.8	218	321	391	*586*
Cd	48	14.0	27.5	231	338	412	*613*
In	49	14.9	29.3	243	356	432	*638*
Sn	50	15.9	31.1	256	373	451	*662*
Sb	51	16.9	33.1	270	391	472	*688*
Te	52	17.9	35.0	282	407	490	707
I	53	19.0	37.1	294	422	*506*	722
Xe	54	20.1	39.2	306	436	*521*	<u>763</u>
Cs	55	21.3	41.3	318	450	*534*	*793*
Ba	56	22.5	43.5	330	463	*546*	<u>*461*</u>
La	57	23.7	45.8	341	475	<u>*557*</u>	202
Ce	58	25.0	48.2	352	486	<u>*601*</u>	219

Mass Absorption Coefficients (*continued*)

Element	Atomic Number	AgKα 0.5609 Å	MoKα 0.7107 Å	CuKα 1.5418 Å	CoKα 1.7902 Å	FeKα 1.9373 Å	CrKα 2.2909 Å
Pr	59	26.3	50.7	363	_497_	_359_	236
Nd	60	27.7	53.2	374	_543_	_379_	252
Pm	61	29.1	55.9	386	_327_	172	268
Sm	62	30.6	58.6	_397_	_344_	182	284
Eu	63	32.2	61.5	_425_	156	193	299
Gd	64	33.8	64.4	_439_	165	203	314
Tb	65	35.5	67.5	_273_	173	214	329
Dy	66	37.2	70.6	_286_	182	224	344
Ho	67	39.0	73.9	128	191	234	359
Er	68	40.8	77.3	134	199	245	373
Tm	69	42.8	80.8	140	208	255	387
Yb	70	44.8	84.5	146	217	265	401
Lu	71	46.8	88.2	153	226	276	416
Hf	72	48.8	91.7	159	235	286	430
Ta	73	50.9	95.4	166	244	297	444
W	74	53.0	99.1	172	253	308	458
Re	75	55.2	103	179	262	319	473
Os	76	57.3	106	186	272	330	
Ir	77	59.4	110	193	282	341	_502_
Pt	78	61.4	113	200	291	353	_517_
Au	79	63.1	115	208	302	365	_532_
Hg	80	64.7	117	216	312	377	_547_
Tl	81	66.2	119	224	323	389	_563_
Pb	82	67.7	120	232	334	402	_579_
Bi	83	69.1	120	240	346	415	_596_

APPENDIX VI

QUADRATIC FORMS FOR THE CUBIC SYSTEM*

$h^2+k^2+l^2$	$\sqrt{h^2+k^2+l^2}$	$\log_{10}(h^2+k^2+l^2)$	Lattice	hkl
1	1.00000	0.00000		100
2	1.41421	0.30103	B	110
3	1.73205	0.47712	FD	111
4	2.00000	0.60206	BF	200
5	2.23607	0.69897		210
6	2.44949	0.77815	B	211
7				
8	2.82843	0.90309	BFD	220
9	3.00000	0.95424		300, 221
10	3.16228	1.00000	B	310
11	3.31662	1.04139	FD	311
12	3.46410	1.07918	BF	222
13	3.60555	1.11394		320
14	3.74166	1.14613	B	321
15				
16	4.00000	1.20412	BFD	400
17	4.12311	1.23045		410, 322
18	4.24264	1.25527	B	411, 330
19	4.35890	1.27875	FD	331
20	4.47214	1.30103	BF	420
21	4.58258	1.32222		421
22	4.69042	1.34242	B	332
23				
24	4.89898	1.38021	BFD	422
25	5.00000	1.39794		500, 430
26	5.09902	1.41497	B	510, 431
27	5.19615	1.43136	FD	511, 333
28				

*All the entries are possible reflections from a simple cubic lattice. The letters F, B, and D represent, respectively, the possible reflections from face-centered, body-centered, and diamond cubic lattices.

Cubic Quadratic Forms (*continued*)

$h^2+k^2+l^2$	$\sqrt{h^2+k^2+l^2}$	$\log_{10}(h^2+k^2+l^2)$	Lattice	hkl
29	5.38516	1.46240		520, 432
30	5.47723	1.47712	B	521
31				
32	5.65685	1.50515	BFD	440
33	5.74456	1.51851		522, 441
34	5.83095	1.53148	B	530, 433
35	5.91608	1.54407	FD	531
36	6.00000	1.55630	BF	600, 442
37	6.08276	1.56820		610
38	6.16441	1.57978	B	611, 532
39				
40	6.32455	1.60206	BFD	620
41	6.40312	1.61278		621, 540, 443
42	6.48074	1.62325	B	541
43	6.55742	1.63347	FD	533
44	6.63325	1.65321	BF	622
45	6.70820	1.66276		630, 542
46	6.78233	1.67210	B	631
47				
48	6.92820	1.68124	BFD	444
49	7.00000	1.69020		700, 632
50	7.07107	1.69897	B	710, 550, 543
51	7.14143	1.70757	FD	711, 551
52	7.21110	1.71600	BF	640
53	7.28011	1.72428		720, 641
54	7.34847	1.73239	B	721, 633, 552
55				
56	7.48331	1.74819	BFD	642
57	7.54983	1.75587		722, 544
58	7.61577	1.76343	B	730
59	7.68115	1.77084	FD	731, 553

A P P E N D I X   V I I

ATOMIC AND IONIC SCATTERING FACTORS

Table 1. Mean Atomic Scattering Factors, in Electrons, from Self-Consistent or Variational Wave Functions[a]

Element and Ionic Charge	Z	$(\sin\theta)/\lambda$ (Å^{-1})											
		0.00	0.10	0.20	0.30	0.40	0.50	0.60	0.70	0.80	0.90	1.00	1.10
H	1	1.00	0.811	0.481	0.251	0.130	0.071	0.040	0.024	0.015	0.010	0.007	0.005
H^{-1}	1	2.000	1.064	0.519	0.255	0.130	0.070	0.040	0.024	0.015	0.010	0.007	0.005
He	2	2.000	1.832	1.452	1.058	0.742	0.515	0.358	0.251	0.179	0.129	0.095	0.071
He^{-1}	2	3.000											
Li	3	3.000	2.215	1.741	1.512	1.269	1.032	0.823	0.650	0.513	0.404	0.320	0.255
Li^{+1}	3	2.000	1.935	1.760	1.521	1.265	1.025	0.818	0.647	0.510	0.403	0.319	0.254
Li^{-1}	3	4.000	2.176	1.743	1.514	1.269	1.033	0.826	0.654	0.516	0.408	0.323	0.257
Be	4	4.000	3.067	2.067	1.705	1.531	1.367	1.201	1.031	0.878	0.738	0.620	0.519
Be^{+1}	4	3.000	2.583	2.017	1.721	1.535	1.362	1.188	1.022	0.870	0.735	0.618	0.520
Be^{+2}	5	2.000	1.966	1.869	1.724	1.550	1.363	1.180	1.009	0.855	0.721	0.606	0.508

	Z												
B	5	5.000	4.066	2.711	1.993	1.692	1.534	1.406	1.276	1.147	1.016	0.895	0.783
B+1	5	4.000	3.471	2.551	1.962	1.688	1.536	1.410	1.283	1.154	1.028	0.908	0.798
B+2	5	3.000	2.757	2.290	1.928	1.707	1.552	1.414	1.278	1.144	1.016	0.896	0.786
B+3	5	2.000	1.979	1.919	1.824	1.703	1.566	1.420	1.274	1.132	0.999	0.877	0.767
C	6	6.000	5.126	3.581	2.502	1.950	1.685	1.536	1.426	1.322	1.218	1.114	1.012
C (valence)	6	6.000	(5.093)	(3.561)	(2.506)	(1.975)	(1.712)	(1.553)	(1.434)	(1.322)	(1.207)	(1.096)	(0.993)
C+2	6	4.000	3.686	2.992	2.338	1.910	1.672	1.533	1.429	1.332	1.233	1.131	1.030
C+3	6	3.000	2.842	2.487	2.133	1.874	1.697	1.564	1.447	1.335	1.225	1.116	1.012
C+4	6	2.000	1.986	1.945	1.880	1.794	1.692	1.579	1.459	1.338	1.219	1.104	0.994
N	7	7.000	6.203	4.600	3.241	2.397	1.944	1.698	1.550	1.444	1.350	1.263	1.175
N+3	7	4.000	3.772	3.227	2.635	2.172	1.869	1.682	1.558	1.461	1.373	1.287	1.199
N+4	7	3.000	2.890	2.619	2.306	2.038	1.837	1.690	1.573	1.472	1.375	1.281	1.188
N-1	7	8.000	6.688	4.631	3.186	2.364	1.929	1.694	1.551	1.446	1.352	1.263	1.170
O	8	8.000	7.250	5.634	4.094	3.010	2.338	1.944	1.714	1.566	1.462	1.374	1.296
O+1	8	7.000	6.493	5.298	4.017	3.016	2.356	1.956	1.717	1.567	1.461	1.374	1.296
O+2	8	6.000	5.647	4.776	3.771	2.924	2.327	1.948	1.716	1.568	1.463	1.378	1.301
O+3	8	5.000	4.760	4.151	3.410	2.745	2.246	1.913	1.701	1.562	1.463	1.382	1.308
O-1	8	9.000	7.836	5.756	4.068	2.968	2.313	1.934	1.710	1.566	1.462	1.373	1.294
F	9	9.000	8.293	6.691	5.044	3.760	2.878	2.312	1.958	1.735	1.587	1.481	1.396
F-1	9	10.000	9.108	7.126	5.188	3.786	2.885	2.323	1.972	1.747	1.596	1.486	1.399
Ne	10	10.000	9.363	7.824	6.087	4.617	3.536	2.794	2.300	1.976	1.760	1.612	1.504
Na	11	11.00	9.76	8.34	6.89	5.47	4.29	3.40	2.76	2.31	2.00	1.78	1.63
Na+1	11	10.000	9.551	8.390	6.925	5.510	4.328	3.424	2.771	2.314	2.001	1.785	1.634

Table 1. Atomic Scattering Factors (continued)

Element and Ionic Charge	Z	$(\sin\theta)/\lambda(\text{Å}^{-1})$											
		0.00	0.10	0.20	0.30	0.40	0.50	0.60	0.70	0.80	0.90	1.00	1.10
Mg	12	12.00	10.50	8.75	7.46	6.20	5.01	4.06	3.30	2.72	2.30	2.01	1.81
Mg^{+2}	12	10.00	9.66	8.75	7.51	6.20	4.99	4.03	3.28	2.71	2.30	2.01	1.81
Al	13	13.00	11.23	9.16	7.88	6.77	5.69	4.71	3.88	3.21	2.71	2.32	2.05
Al^{+1}	13	12.00	10.94	9.22	7.90	6.77	5.70	4.71	3.88	3.22	2.70	2.32	2.04
Al^{+2}	13	11.00	10.40	9.17	7.95	6.79	5.70	4.71	3.88	3.22	2.71	2.33	2.05
Al^{+3}	13	10.00	9.74	9.01	7.98	6.82	5.69	4.69	3.86	3.20	2.70	2.32	2.04
Si	14	14.00	12.16	9.67	8.22	7.20	6.24	5.31	4.47	3.75	3.16	2.69	2.35
Si^{+3}	14	11.00	10.53	9.48	8.34	7.27	6.25	5.30	4.44	3.73	3.14	2.67	2.34
Si^{+4}	14	10.00	9.79	9.20	8.33	7.31	6.26	5.28	4.42	3.71	3.13	2.68	2.33
P	15	15.00	13.17	10.34	8.59	7.54	6.67	5.83	5.02	4.28	3.64	3.11	2.69
S	16	16.00	14.33	11.21	8.99	7.83	7.05	6.31	5.56	4.82	4.15	3.56	3.07
S^{-1}	16	17.00	15.00	11.36	8.95	7.79	7.05	6.32	5.57	4.83	4.16	3.57	3.08
S^{-2}	16	18.00	(15.16)	(10.74)	(8.66)	(7.89)	(7.22)	(6.47)	(5.69)	(4.93)	(4.23)	(3.62)	(3.13)
Cl	17	17.00	15.33	12.00	9.44	8.07	7.29	6.64	5.96	5.27	4.60	4.00	3.47
Cl^{-1}	17	18.00	16.02	12.20	9.40	8.03	7.28	6.64	5.97	5.27	4.61	4.00	3.47
A	18	18.00	16.30	12.93	10.20	8.54	7.56	6.86	6.23	5.61	5.01	4.43	3.90

Element	Ion	Z												
K		19	19.00	16.73	13.73	10.97	9.05	7.87	7.11	6.51	5.95	5.39	4.84	4.32
	K^{+1}	19	18.00	16.68	13.76	10.96	9.04	7.86	7.11	6.51	5.94			
Ca		20	20.00	17.33	14.32	11.71	9.64	8.26	7.38	6.75	6.21	5.70	5.19	4.69
	Ca^{+1}	20	19.00	17.21	14.35	11.70	9.63	8.26	7.38	6.75	6.21	5.70	5.19	4.68
	Ca^{+2}	20	18.00	16.93	14.40	11.70	9.61	8.25	7.38	6.75	6.22	5.70	5.18	4.68
Sc		21	21.00	18.72	15.39	12.39	10.12	8.60	7.64	6.98	6.45	5.96	5.48	5.00
	Sc^{+1}	21	20.00	18.50	15.43	12.43	10.13	8.61	7.64	6.98	6.45	5.96	5.48	5.00
	Sc^{+2}	21	19.00	17.88	15.27	12.44	10.18	8.64	7.65	6.98	6.45	5.96	5.48	5.01
	Sc^{+3}	21	18.00	17.11	14.92	12.38	10.22	8.68	7.67	6.98	6.44	5.96	5.49	5.02
Ti		22	22.00	19.41	16.07	13.20	10.83	9.12	7.98	7.22	6.65	6.19	5.72	5.29
	Ti^{+1}	22	21.00	19.52	16.39	13.25	10.77	9.06	7.95	7.21	6.66	6.18	5.73	5.28
	Ti^{+2}	22	20.00	18.86	16.19	13.25	10.82	9.10	7.96	7.21	6.66	6.18	5.73	5.28
	Ti^{+3}	22	19.00	18.09	15.82	13.16	10.84	9.14	7.99	7.22	6.65	6.18	5.73	5.29
V		23	23.00	20.47	17.03	14.03	11.51	9.63	8.34	7.48	6.86	6.39	5.94	5.53
	V^{+1}	23	22.00	20.54	17.37	14.11	11.46	9.57	8.31	7.47	6.87	6.39	5.95	5.52
	V^{+2}	23	21.00	19.86	17.14	14.10	11.51	9.61	8.32	7.47	6.86	6.38	5.95	5.52
	V^{+3}	23	20.00	19.07	16.76	13.99	11.52	9.65	8.36	7.48	6.87	6.38	5.95	5.53
Cr		24	24.00	21.93	18.37	15.01	12.22	10.14	8.72	7.75	7.09	6.58	6.14	5.74
	Cr^{+1}	24	23.00	21.58	18.40	15.03	12.21	10.13	8.71	7.75	7.09	6.58	6.14	5.74
	Cr^{+2}	24	22.00	20.87	18.13	15.00	12.26	10.18	8.74	7.76	7.09	6.58	6.14	5.72
	Cr^{+3}	24	21.00	20.07	17.72	14.87	12.26	10.22	8.77	7.78	7.09	6.58	6.14	5.74

Table 1. Atomic Scattering Factors (*continued*)

Element and Ionic Charge	Z	$(\sin\theta)/\lambda(\text{Å}^{-1})$											
		0.00	0.10	0.20	0.30	0.40	0.50	0.60	0.70	0.80	0.90	1.00	1.10
Mn	25	25.00	22.61	19.06	15.84	13.02	10.80	9.20	8.09	7.32	6.77	6.32	5.93
Mn^{+1}	25	24.00	22.60	19.42	15.96	13.00	10.75	9.17	8.08	7.33	6.78	6.33	5.93
Mn^{+2}	25	23.00	21.89	19.16	15.94	13.05	10.80	9.19	8.09	7.33	6.77	6.33	5.93
Mn^{+3}	25	22.00	21.07	18.71	15.78	13.04	10.84	9.23	8.12	7.34	6.78	6.32	5.93
Mn^{+4}	25	21.00	20.22	18.18	15.55	12.98	10.84	9.25	8.13	7.35	6.79	6.34	5.94
Fe	26	26.00	23.68	20.09	16.77	13.84	11.47	9.71	8.47	7.60	6.99	6.51	6.12
Fe^{+1}	26	25.00	23.63	20.45	16.92	13.82	11.41	9.67	8.45	7.60	6.99	6.52	6.11
Fe^{+2}	26	24.00	22.89	20.15	16.87	13.86	11.46	9.69	8.46	7.60	6.99	6.51	6.11
Fe^{+3}	26	23.00	22.09	19.72	16.74	13.87	11.50	9.73	8.48	7.61	6.99	6.52	6.12
Fe^{+4}	26	22.00	21.22	19.15	16.46	13.78	11.51	9.77	8.52	7.64	7.00	6.52	6.11
Co	27	27.00	24.74	21.13	17.74	14.68	12.17	10.26	8.88	7.91	7.22	6.70	6.29
Co^{+1}	27	26.00	24.66	21.49	17.89	14.67	12.11	10.21	8.85	7.91	7.22	6.71	6.29
Co^{+2}	27	25.00	23.91	21.17	17.84	14.72	12.17	10.25	8.87	7.91	7.22	6.71	6.29
Co^{+3}	27	24.00	23.09	20.71	17.68	14.71	12.21	10.29	8.90	7.92	7.22	6.70	6.28
Ni	28	28.00	25.80	22.19	18.73	15.56	12.91	10.85	9.33	8.25	7.48	6.90	6.47
Ni^{+1}	28	27.00	25.69	22.55	18.90	15.57	12.86	10.80	9.31	8.24	7.48	6.91	6.46
Ni^{+2}	28	26.00	24.93	22.21	18.83	15.61	12.91	10.84	9.32	8.25	7.48	6.91	6.46
Ni^{+3}	28	25.00	24.10	21.72	18.65	15.58	12.95	10.88	9.36	8.26	7.48	6.90	6.46
Cu	29	29.00	27.19	23.63	19.90	16.48	13.65	11.44	9.80	8.61	7.76	7.13	6.65
Cu^{+1}	29	28.00	26.71	23.59	19.92	16.50	13.66	11.45	9.80	8.61	7.75	7.12	6.64

884

	Z												
Cu+2	29	27.00	25.95	23.24	19.84	16.52	13.70	11.47	9.82	8.62	7.76	7.13	6.65
Cu+3	29	26.00	25.11	22.75	19.65	16.50	13.74	11.53	9.86	8.64	7.77	7.13	6.64
Zn	30	30.00	27.92	24.33	20.77	17.42	14.51	12.16	10.37	9.04	8.08	7.37	6.84
Zn+2	30	28.00	26.96	24.27	20.86	17.48	14.54	12.18	10.37	9.04	8.07	7.36	6.83
Ga	31	31.00	28.65	24.92	21.47	18.26	15.38	12.95	11.02	9.54	8.46	7.64	7.05
Ga+1	31	30.00	28.35	24.98	21.50	18.26	15.37	12.94	11.02	9.54	8.45	7.65	7.05
Ga+3	31	28.00	27.12	24.78	21.65	18.38	15.41	12.94	11.00	9.53	8.44	7.64	7.05
Ge	32	32.00	29.52	25.53	22.11	19.02	16.19	13.72	11.68	10.08	8.87	7.96	7.29
Ge+2	32	30.00	28.64	25.58	22.22	19.05	16.18	13.71	11.68	10.08	8.87	7.97	7.29
Ge+4	32	28.00	27.25	25.19	22.32	19.21	16.26	13.72	11.66	10.06	8.85	7.96	7.29
As	33	33.00	30.47	26.20	22.69	19.69	16.95	14.48	12.37	10.67	9.34	8.32	7.57
As+1	33	32.00	29.79	25.85	22.38	19.33	16.56	14.11	12.07	10.44	9.18	8.23	7.52
As+2	33	31.00	29.33	25.86	22.43	19.33	16.54	14.09	12.06	10.43	9.17	8.23	7.52
As+3	33	30.00	28.74	25.79	22.47	19.34	16.52	14.07	12.04	10.42	9.17	8.23	7.52
As+5	33	28.00	27.34	25.51	22.89	19.95	17.07	14.49	12.34	10.63	9.31	8.31	7.56
Se	34	34.00	31.43	26.91	23.24	20.28	17.63	15.20	13.06	11.27	9.83	8.71	7.86
Se+6	34	28.00	27.42	25.78	23.38	20.62	17.82	15.25	13.04	11.23	9.80	8.70	7.87
Br	35	35.00	32.43	27.70	23.82	20.84	18.27	15.91	13.78	11.93	10.41	9.19	8.24
Br+7	35	28.00	27.48	26.00	23.80	21.21	18.53	15.98	13.74	11.85	10.33	9.13	8.21
Br-1	35	36.00	32.81	27.65	23.76	20.82	18.27	15.91	13.77	11.92	10.40	9.18	8.24

[a]Prepared by special permission from *International Tables for X-ray Crystallography*, Vol. III, Kynoch Press, Birmingham, England, 1962, pp 202–207. Values in parentheses have been obtained by interpolation.

Table 2. Mean Atomic Scattering Factors, in Electrons, from the Thomas-Fermi-Dirac Statistical Model[a]

Element	Z	$(\sin\theta)/\lambda$ (Å⁻¹)											
		0.00	0.10	0.20	0.30	0.40	0.50	0.60	0.70	0.80	0.90	1.00	1.10
Ca	20	20.00	18.09	14.77	12.26	10.29	8.79	7.61	6.66	5.88	5.26	4.73	4.28
Ca⁺¹	20	19.00	17.73	14.91	12.23	10.28	8.80	7.60	6.66				
Ca⁺²	20	18.00	17.09	14.86	12.34	10.27	8.77	7.62	6.67				
Sc	21	21.00	19.03	15.59	12.97	10.91	9.33	8.08	7.09	6.27	5.60	5.05	4.57
Sc⁺¹	21	20.00	18.67	15.73	12.93	10.90	9.34	8.08	7.08				
Sc⁺²	21	19.00	18.04	15.69	13.04	10.88	9.31	8.10	7.09				
Sc⁺³	21	18.00	17.28	15.43	13.11	10.96	9.29	8.06	7.10				
Ti	22	22.00	19.96	16.41	13.68	11.53	9.88	8.57	7.52	6.65	5.95	5.36	4.86
Ti⁺¹	22	21.00	19.61	16.55	13.64	11.52	9.89	8.56	7.52				
Ti⁺²	22	20.00	18.99	16.52	13.75	11.50	9.86	8.58	7.52				
Ti⁺³	22	19.00	18.24	16.27	13.82	11.58	9.84	8.55	7.53				
V	23	23.00	20.90	17.23	14.39	12.15	10.43	9.05	7.95	7.05	6.31	5.69	5.15
V⁺¹	23	22.00	20.56	17.37	14.36	12.15	10.44	9.05	7.95				
V⁺²	23	21.00	19.94	17.35	14.46	12.12	10.41	9.07	7.96				
V⁺³	23	20.00	19.19	17.11	14.54	12.19	10.38	9.04	7.97				
Cr	24	24.00	21.84	18.05	15.11	12.78	10.98	9.55	8.39	7.44	6.67	6.01	5.45
Cr⁺¹	24	23.00	21.50	18.20	15.07	12.78	10.99	9.54	8.40				
Cr⁺²	24	22.00	20.89	18.18	15.18	12.75	10.97	9.56	8.40				
Cr⁺³	24	21.00	20.15	17.96	15.26	12.82	10.94	9.53	8.41				

Mn	25	25.00	22.77	18.88	15.84	13.41	11.54	10.04	8.84	7.85	7.03	6.34	5.75
Mn^{+1}	25	24.00	22.44	19.02	15.79	13.42	11.55	10.04	8.84				
Mn^{+2}	25	23.00	21.84	19.01	15.90	13.38	11.53	10.06	8.84				
Mn^{+3}	25	22.00	21.10	18.80	15.99	13.45	11.50	10.03	8.85				
Mn^{+4}	25	21.00	20.30	18.42	15.97	13.54	11.53	10.00	8.82				
Fe	26	26.00	23.71	19.71	16.56	14.05	12.11	10.54	9.29	8.25	7.39	6.67	6.06
Fe^{+1}	26	25.00	23.39	19.85	16.52	14.05	12.12	10.54	9.29				
Fe^{+2}	26	24.00	22.79	19.85	16.62	14.02	12.09	10.56	9.29				
Fe^{+3}	26	23.00	22.06	19.65	16.71	14.08	12.06	10.54	9.30				
Fe^{+4}	26	22.00	21.26	19.28	16.71	14.18	12.09	10.50	9.28				
Co	27	27.00	24.65	20.54	17.29	14.69	12.67	11.05	9.74	8.66	7.77	7.01	6.37
Co^{+1}	27	26.00	24.33	20.68	17.25	14.70	12.68	11.04	9.74				
Co^{+2}	27	25.00	23.74	20.69	17.35	14.66	12.66	11.07	9.74				
Co^{+3}	27	24.00	23.01	20.50	17.44	14.72	12.63	11.04	9.76				
Ni	28	28.00	25.60	21.37	18.03	15.34	13.25	11.56	10.20	9.08	8.14	7.35	6.68
Ni^{+1}	28	27.00	25.28	21.52	17.98	15.34	13.25	11.55	10.20				
Ni^{+2}	28	26.00	24.69	21.53	18.08	15.30	13.24	11.58	10.19				
Ni^{+3}	28	25.00	23.97	21.35	18.18	15.36	13.20	11.56	10.21				
Cu	29	29.00	26.54	22.21	18.76	15.98	13.82	12.07	10.66	9.49	8.52	7.70	7.00
Cu^{+1}	29	28.00	26.22	22.35	18.71	15.99	13.83	12.07	10.66				
Cu^{+2}	29	27.00	25.64	22.37	18.81	15.95	13.81	12.09	10.65				
Cu^{+3}	29	26.00	24.93	22.20	18.91	16.00	13.77	12.07	10.68				

887

Table 2.   Atomic Scattering Factors (*continued*)

Element	Z	$(\sin\theta)/\lambda(\text{Å}^{-1})$											
		0.00	0.10	0.20	0.30	0.40	0.50	0.60	0.70	0.80	0.90	1.00	1.10
Zn	30	30.00	27.48	23.05	19.50	16.64	14.40	12.59	11.12	9.91	8.90	8.05	7.32
Zn^{+2}	30	28.00	26.59	23.32	19.55	16.60	14.40	12.61	11.12				
Ga	31	31.00	28.43	23.89	20.25	17.29	14.98	13.11	11.59	10.33	9.29	8.40	7.64
Ga^{+1}	31	30.00	28.12	24.03	20.19	17.30	14.99	13.10	11.60				
Ga^{+3}	31	28.00	26.84	23.90	20.39	17.30	14.93	13.11	11.61				
Ge	32	32.00	29.37	24.73	20.99	17.95	15.57	13.63	12.06	10.76	9.68	8.76	7.97
Ge^{+2}	32	30.00	28.50	24.91	21.03	17.92	15.57	13.65	12.06				
Ge^{+4}	32	28.00	27.02	24.45	21.18	18.05	15.53	13.60	12.07				
As	33	33.00	30.32	25.58	21.74	18.61	16.16	14.16	12.54	11.19	10.07	9.11	8.30
As^{+1}	33	32.00	30.02	25.72	21.68	18.63	16.16	14.16	12.54				
As^{+2}	33	31.00	29.45	25.76	21.77	18.58	16.16	14.18	12.53				
As^{+3}	33	30.00	28.75	25.62	21.88	18.61	16.11	14.17	12.55				
Se	34	34.00	31.26	26.42	22.49	19.28	16.75	14.69	13.02	11.62	10.46	9.47	8.63
Br	35	35.00	32.21	27.27	23.24	19.95	17.35	15.22	13.50	12.06	10.86	9.84	8.97
Kr	36	36.00	33.16	28.12	24.00	20.62	17.95	15.76	13.98	12.50	11.26	10.21	9.31
Rb	37	37.00	34.11	28.97	24.75	21.29	18.55	16.30	14.47	12.94	11.66	10.58	9.65
Rb^{+1}	37	36.00	33.82	29.11	24.70	21.31	18.55	16.30	14.48				
Sr	38	38.00	35.06	29.83	25.51	21.96	19.15	16.84	14.96	13.39	12.07	10.95	9.99
Y	39	39.00	36.01	30.68	26.28	22.64	19.76	17.39	15.46	13.84	12.48	11.32	10.34
Zr	40	40.00	36.96	31.54	27.04	23.32	20.37	17.94	15.95	14.29	12.89	11.70	10.68
Zr^{+4}	40	36.00	34.72	31.39	27.25	23.39	20.31	17.92	15.97				

	Z												
Nb	41	41.00	37.91	32.40	27.81	24.01	20.98	18.49	16.45	14.74	13.31	12.08	11.04
Mo	42	42.00	38.86	33.25	28.57	24.69	21.60	19.04	16.95	15.20	13.73	12.46	11.39
Mo⁺¹	42	41.00	38.59	33.39	28.51	24.72	21.59	19.04	16.96				
Tc	43	43.00	39.81	34.12	29.34	25.38	22.21	19.60	17.46	15.65	14.15	12.85	11.74
Ru	44	44.00	40.76	34.98	30.12	26.07	22.83	20.16	17.96	16.12	14.57	13.24	12.10
Rh	45	45.00	41.72	35.84	30.89	26.76	23.46	20.72	18.47	16.58	14.99	13.63	12.46
Pd	46	46.00	42.67	36.70	31.67	27.46	24.08	21.28	18.98	17.05	15.42	14.02	12.82
Ag	47	47.00	43.63	37.57	32.44	28.16	24.71	21.85	19.50	17.52	15.85	14.42	13.19
Ag⁺¹	47	46.00	43.37	37.71	32.38	28.18	24.70	21.85	19.50				
Cd	48	48.00	44.58	38.44	33.22	28.85	25.34	22.42	20.02	17.99	16.28	14.81	13.56
In	49	49.00	45.53	39.31	34.00	29.56	25.97	22.99	20.53	18.46	16.71	15.21	13.93
Sn	50	50.00	46.49	40.17	34.78	30.26	26.60	23.56	21.05	18.93	17.15	15.61	14.30
Sb	51	51.00	47.45	41.05	35.57	30.96	27.24	24.14	21.58	19.41	17.59	16.02	14.67
Te	52	52.00	48.40	41.92	36.35	31.67	27.87	24.71	22.10	19.89	18.03	16.42	15.05
I	53	53.00	49.36	42.79	37.14	32.38	28.51	25.29	22.63	20.37	18.47	16.83	15.42
Xe	54	54.00	50.32	43.66	37.93	33.09	29.16	25.87	23.16	20.86	18.92	17.24	15.80
Cs	55	55.00	51.27	44.54	38.72	33.80	29.80	26.46	23.69	21.34	19.36	17.65	16.18
Ba	56	56.00	52.23	45.41	39.51	34.51	30.44	27.04	24.22	21.83	19.81	18.07	16.57
La	57	57.00	53.19	46.29	40.30	35.23	31.09	27.63	24.76	22.32	20.26	18.48	16.95
Ce	58	58.00	54.15	47.16	41.09	35.94	31.74	28.22	25.30	22.81	20.71	18.90	17.34
Pr	59	59.00	55.11	48.04	41.89	36.66	32.39	28.81	25.84	23.31	21.17	19.32	17.72
Nd	60	60.00	56.07	48.92	42.69	37.38	33.04	29.40	26.38	23.80	21.62	19.74	18.11
Pm	61	61.00	57.02	49.80	43.48	38.10	33.69	29.99	26.92	24.30	22.08	20.16	18.51

Table 2. Atomic Scattering Factors (continued)

Element	Z	(sin θ)/λ(Å⁻¹)											
		0.00	0.10	0.20	0.30	0.40	0.50	0.60	0.70	0.80	0.90	1.00	1.10
Sm	62	62.00	57.98	50.68	44.28	38.82	34.35	30.59	27.46	24.80	22.54	20.58	18.90
Eu	63	63.00	58.94	51.56	45.08	39.55	35.01	31.19	28.01	25.30	23.00	21.01	19.29
Gd	64	64.00	59.91	52.45	45.88	40.27	35.66	31.79	28.56	25.80	23.46	21.44	19.69
Tb	65	65.00	60.87	53.33	46.68	41.00	36.33	32.39	29.11	26.31	23.93	21.87	20.09
Dy	66	66.00	61.83	54.21	47.49	41.73	36.99	32.99	29.66	26.81	24.39	22.30	20.49
Ho	67	67.00	62.79	55.10	48.29	42.46	37.65	33.59	30.21	27.32	24.86	22.73	20.89
Er	68	68.00	63.75	55.98	49.10	43.19	38.31	34.20	30.76	27.83	25.33	23.17	21.29
Tm	69	69.00	64.71	56.87	49.90	43.92	38.98	34.81	31.32	28.34	25.80	23.60	21.70
Yb	70	70.00	65.67	57.75	50.71	44.66	39.65	35.42	31.88	28.85	26.28	24.04	22.11
Lu	71	71.00	66.64	58.64	51.52	45.39	40.32	36.03	32.44	29.37	26.75	24.48	22.51
Hf	72	72.00	67.60	59.53	52.33	46.13	40.99	36.64	33.00	29.88	27.23	24.92	22.92
Ta	73	73.00	68.56	60.42	53.14	46.86	41.66	37.25	33.56	30.40	27.70	25.36	23.33
W	74	74.00	69.52	61.31	53.95	47.60	42.33	37.87	34.12	30.92	28.18	25.80	23.74
Re	75	75.00	70.49	62.20	54.76	48.34	43.01	38.48	34.69	31.44	28.66	26.25	24.16
Os	76	76.00	71.45	63.09	55.58	49.08	43.68	39.10	35.26	31.96	29.14	26.70	24.57
Ir	77	77.00	72.42	63.98	56.39	49.83	44.36	39.72	35.82	32.48	29.63	27.14	24.99
Pt	78	78.00	73.38	64.87	57.21	50.57	45.04	40.34	36.39	33.01	30.11	27.59	25.41
Au	79	79.00	74.35	65.77	58.02	51.31	45.72	40.96	36.96	35.53	30.60	28.04	25.83
Au⁺¹	79	78.00	74.14	65.88	57.96	51.35	45.70	40.97	36.97				
Hg	80	80.00	75.31	66.66	58.84	52.06	46.40	41.59	37.54	34.06	31.08	28.50	26.25
Hg⁺²	80	78.00	74.65	66.90	58.79	52.05	46.41	41.59	37.53				
Tl	81	81.00	76.27	67.55	59.66	52.81	47.08	42.22	38.12	34.60	31.59	28.96	26.68
Tl⁺¹	81	80.00	76.07	67.67	59.59	52.84	47.06	42.23	38.12				
Tl⁺³	81	78.00	75.03	67.82	59.71	52.76	47.08	42.24	38.10				

	Z	82.00	77.24	68.45	60.48	53.56	47.77	42.85	38.69	35.13	32.08	29.42	27.11
Pb	82	82.00	77.24	68.45	60.48	53.56	47.77	42.85	38.69	35.13	32.08	29.42	27.11
Pb^{+3}	82	79.00	76.00	68.71	60.53	53.50	47.76	42.87	38.68				
Bi	83	83.00	78.20	69.34	61.30	54.30	48.45	43.47	39.27	35.66	32.57	29.87	27.53
Po	84	84.00	79.17	70.24	62.12	55.05	49.14	44.10	39.85	36.19	33.06	30.33	27.96
At	85	85.00	80.13	71.13	62.94	55.80	49.82	44.73	40.43	36.73	33.55	30.79	28.38
Rn	86	86.00	81.10	72.03	63.76	56.56	50.51	45.36	41.01	37.26	34.05	31.25	28.81
Fr	87	87.00	82.07	72.93	64.58	57.31	51.20	45.99	41.59	37.80	34.55	31.71	29.24
Ra	88	88.00	83.03	73.82	65.41	58.06	51.89	46.63	42.17	38.34	35.04	32.17	29.67
Ac	89	89.00	84.00	74.72	66.23	58.82	52.58	47.26	42.75	38.88	35.54	32.64	30.10
Th	90	90.00	84.97	75.62	67.06	59.57	53.27	47.90	43.34	39.42	36.05	33.10	30.54
Pa	91	91.00	85.93	76.52	67.88	60.33	53.97	48.53	43.93	39.96	36.55	33.57	30.97
U	92	92.00	86.90	77.42	68.71	61.09	54.66	49.17	44.51	40.50	37.05	34.04	31.41
Np	93	93.00	87.87	78.32	69.54	61.84	55.35	49.81	45.10	41.04	37.55	34.50	31.84
Pu	94	94.00	88.83	79.22	70.37	62.60	56.05	50.45	45.69	41.59	38.06	34.97	32.28
Am	95	95.00	89.80	80.12	71.20	63.36	56.75	51.09	46.28	42.14	38.56	35.45	32.72
Cm	96	96.00	90.77	81.02	72.03	64.13	57.45	51.73	46.87	42.68	39.07	35.92	33.16
Bk	97	97.00	91.74	81.92	72.86	64.89	58.15	52.37	47.47	43.23	39.58	36.39	33.60
Cf	98	98.00	92.70	82.82	73.69	65.65	58.85	53.02	48.06	43.78	40.09	36.86	34.04
Es	99	99.00	93.67	83.73	74.52	66.41	59.55	53.66	48.66	44.33	40.60	37.34	34.49
Fm	100	100.00	94.64	84.63	75.35	67.18	60.25	54.31	49.25	44.88	41.11	37.82	34.93
Md	101	101.00	95.61	85.53	76.18	67.94	60.95	54.96	49.85	45.44	41.63	38.29	35.38
No	102	102.00	96.58	86.44	77.02	68.71	61.66	55.61	50.45	45.99	42.14	38.77	35.83
—	103	103.00	97.55	87.34	77.85	69.48	62.36	56.26	51.05	46.55	42.66	39.25	36.27
—	104	104.00	98.52	88.25	78.69	70.25	63.07	56.91	51.65	47.10	43.17	39.74	36.72

[a]Reprinted by special permission with abridgement from *International Tables for X-ray Crystallography*, Vol. III, Kynoch Press, Birmingham, England, 1962, pp. 210–212.

APPENDIX VIII

LORENTZ AND POLARIZATION FACTORS

Table 1. The Polarization Factor $(1 + \cos^2 2\theta)/2$ as a Function of Sin θ

$\sin \theta$	$\dfrac{1 + \cos^2 2\theta}{2}$	$\sin \theta$	$\dfrac{1 + \cos^2 2\theta}{2}$
0.00	1.000	0.50	0.625
0.02	0.999	0.52	0.605
0.04	0.997	0.54	0.587
0.06	0.993	0.56	0.570
0.08	0.987	0.58	0.553
0.10	0.980	0.60	0.539
0.12	0.972	0.62	0.527
0.14	0.962	0.64	0.516
0.16	0.950	0.66	0.508
0.18	0.937	0.68	0.503
0.20	0.923	0.70	0.500
0.22	0.908	0.72	0.501
0.24	0.891	0.74	0.504
0.26	0.874	0.76	0.512
0.28	0.856	0.78	0.524
0.30	0.836	0.80	0.539
0.32	0.816	0.82	0.559
0.34	0.795	0.84	0.584
0.36	0.774	0.86	0.615
0.38	0.753	0.88	0.651
0.40	0.731	0.90	0.692
0.42	0.709	0.92	0.740
0.44	0.688	0.94	0.794
0.46	0.666	0.96	0.856
0.48	0.645	0.98	0.924

Table 2.   The Combined Lorentz and Polarization Factor as a Function of Sin θ^a

sin θ	Debye-Scherrer Method[b] $\left(\dfrac{1+\cos^2 2\theta}{\sin^2 \theta \cos \theta}\right)$	Oscillating- or Rotating-Crystal Method $\left(\dfrac{1+\cos^2 2\theta}{\sin 2\theta}\right)$
0.00	∞	∞
0.025	3197	39.962
0.050	797.0	19.925
0.075	352.6	13.221
0.100	197.0	9.851
0.125	125.0	7.815
0.150	85.95	6.446
0.20	47.11	4.711
0.25	29.17	3.647
0.30	19.48	2.922
0.35	13.68	2.394
0.40	9.973	1.995
0.45	7.487	1.685
0.50	5.774	1.443
0.55	4.576	1.258
0.60	3.744	1.123
0.65	3.189	1.037
0.70	2.869	1.004
0.75	2.730	1.024
0.80	2.808	1.123
0.85	3.148	1.338
0.90	3.921	1.764
0.95	5.848	2.778
1.00	∞	∞

[a](The values in this table are reprinted from *Internationale Tabellen zur Bestimmung von Kristallstrukturen*, Gebrüder Borntraeger, Berlin, 1935, Vol. 2, pages 567–568.)

[b]For more elaborate tables of these values see the *International Tables for X-ray Crystallography*, Vol. II, Kynoch Press, Birmingham, England, 1959; Debye-Scherrer method, pp. 270, 271; oscillating- or rotating-crystal method, pp. 268, 269.

Table 3. The Combined Lorentz and Polarization Factor as a Function of θ^a

θ (°)	Debye-Scherrer Method $\left(\dfrac{1 + \cos^2 2\theta}{\sin^2 \theta \cos \theta}\right)$	Oscillating- or Rotating-Crystal Method $\left(\dfrac{1 + \cos^2 2\theta}{\sin 2\theta}\right)$
0.0	∞	∞
1.0	6563	57.272
1.5	2916	38.162
2.0	1639	28.601
2.5	1048	22.860
3.0	727.2	19.029
3.5	533.6	16.289
4.0	408.0	14.231
4.5	321.9	12.628
5.0	260.3	11.344
6.0	180.06	9.411
7.0	131.70	8.025
8.0	100.31	6.980
9.0	78.80	6.163
10.0	63.41	5.506
12.0	43.39	4.510
14.0	31.34	3.791
16.0	23.54	3.244
18.0	18.22	2.815
20.0	14.44	2.469
22.5	11.086	2.121
25.0	8.730	1.845
27.5	7.027	1.622
30.0	5.774	1.443
32.5	4.841	1.300
35.0	4.123	1.189
37.5	3.629	1.105
40.0	3.255	1.046
42.5	2.994	1.0115
45.0	2.828	1.000
47.5	2.744	1.0115
50.0	2.731	1.046
52.5	2.785	1.105
55.0	2.902	1.189
57.5	3.084	1.300
60.0	3.333	1.443
62.5	3.658	1.622

Table 3. (*continued*)

$\theta(°)$	Debye-Scherrer Method $\left(\dfrac{1+\cos^2 2\theta}{\sin^2 \theta \cos \theta}\right)$	Oscillating- or Rotating- Crystal Method $\left(\dfrac{1+\cos^2 2\theta}{\sin 2\theta}\right)$
65.0	4.071	1.845
67.5	4.592	2.121
70.0	5.255	2.469
72.0	5.920	2.815
74.0	6.749	3.244
76.0	7.814	3.791
78.0	9.221	4.510
80.0	11.182	5.506
81.0	12.480	6.163
82.0	14.097	6.980
83.0	16.17	8.025
84.0	18.93	9.411
85.0	22.78	11.344
85.5	25.34	12.628
86.0	28.53	14.231
86.5	32.64	16.289
87.0	38.11	19.029
87.5	45.76	22.860
88.0	57.24	28.601
88.5	76.35	38.162
89.0	114.56	57.272
90.0	∞	∞

[a]The values in this table are reprinted from *Internationale Tabellen zur Bestimmung von Kristallstrukturen*, Gebrüder Borntraeger, Berlin, 1935, Vol. 2, pages 567–568.

APPENDIX IX

THE DEBYE-WALLER TEMPERATURE FACTOR $e^{-(B\sin^2\theta)/\lambda^2}$ [a]*

$B \times 10^{16}$ \ $\dfrac{\sin\theta}{\lambda}\times10^{-8}=$	0.0	0.1	0.2	0.3	0.4	0.5	0.6	0.7	0.8	0.9	1.0	1.1	1.2
0.0	1.000	1.000	1.000	1.000	1.000	1.000	1.000	1.000	1.000	1.000	1.000	1.000	1.000
0.1	1.000	0.999	0.996	0.991	0.984	0.975	0.964	0.952	0.938	0.923	0.905	0.886	0.866
0.2	1.000	0.998	0.992	0.982	0.968	0.951	0.931	0.906	0.880	0.850	0.819	0.785	0.750
0.3	1.000	0.997	0.988	0.973	0.953	0.928	0.898	0.863	0.826	0.784	0.741	0.695	0.649
0.4	1.000	0.996	0.984	0.964	0.938	0.905	0.866	0.821	0.774	0.724	0.670	0.616	0.562
0.5	1.000	0.995	0.980	0.955	0.924	0.882	0.834	0.782	0.726	0.667	0.607	0.548	0.487
0.6	1.000	0.994	0.976	0.947	0.909	0.860	0.804	0.745	0.681	0.615	0.549	0.484	0.421
0.7	1.000	0.993	0.972	0.939	0.894	0.839	0.776	0.710	0.639	0.567	0.497	0.429	0.365
0.8	1.000	0.992	0.968	0.931	0.880	0.818	0.750	0.676	0.599	0.523	0.449	0.380	0.314
0.9	1.000	0.991	0.964	0.923	0.866	0.798	0.724	0.644	0.561	0.482	0.406	0.336	0.273
1.0	1.000	0.990	0.960	0.915	0.852	0.779	0.698	0.613	0.527	0.445	0.368	0.298	0.236
1.1	1.000	0.989	0.957	0.907	0.839	0.759	0.672	0.584	0.494	0.410	0.333	0.264	0.205
1.2	1.000	0.988	0.953	0.898	0.826	0.740	0.649	0.556	0.464	0.378	0.301	0.234	0.178
1.3	1.000	0.987	0.950	0.890	0.813	0.722	0.626	0.529	0.435	0.349	0.273	0.207	0.154
1.4	1.000	0.986	0.946	0.882	0.800	0.704	0.604	0.503	0.408	0.332	0.247	0.184	0.133
1.5	1.000	0.985	0.942	0.874	0.787	0.687	0.582	0.479	0.383	0.297	0.223	0.167	0.116
1.6	1.000	0.984	0.938	0.866	0.774	0.670	0.562	0.458	0.359	0.274	0.202	0.144	0.100
1.7	1.000	0.983	0.935	0.858	0.762	0.654	0.543	0.436	0.337	0.252	0.183	0.128	0.086
1.8	1.000	0.982	0.931	0.850	0.750	0.638	0.523	0.414	0.316	0.233	0.165	0.113	0.075
1.9	1.000	0.981	0.927	0.842	0.739	0.622	0.505	0.394	0.296	0.215	0.149	0.100	0.065
2.0	1.000	0.980	0.924	0.834	0.727	0.607	0.487	0.375	0.278	0.198	0.135	0.089	0.056

[a]Values are from *Internationale Tabellen zur Bestimmung von Kristallstrukturen*, Gebrüder Borntraeger, Berlin, 1935, Vol. 2, pages 574–575.

*A more elaborate table of temperature factor data is presented in the *International Tables for X-ray Crystallography*, Vol. II, Kynoch Press, Birmingham, England, 1959, pp. 242–264.

APPENDIX X

WARREN'S POWDER PATTERN POWER THEOREM*

Problems involving diffraction in powder samples of imperfect crystals are often simplified by the use of the powder pattern power theorem. The intensity from a small parallelopipedon crystal can be expressed as

$$I = I_e F^2 \frac{\sin^2(\pi/\lambda)(\mathbf{s} - \mathbf{s}_0) \cdot N_1 \mathbf{a}_1}{\sin^2(\pi/\lambda)(\mathbf{s} - \mathbf{s}_0) \cdot \mathbf{a}_1} \frac{\sin^2(\pi/\lambda)(\mathbf{s} - \mathbf{s}_0) \cdot N_2 \mathbf{a}_2}{\sin^2(\pi/\lambda)(\mathbf{s} - \mathbf{s}_0) \cdot \mathbf{a}_2}$$

$$\times \frac{\sin^2(\pi/\lambda)(\mathbf{s} - \mathbf{s}_0) \cdot N_3 \mathbf{a}_3}{\sin^2(\pi/\lambda)(\mathbf{s} - \mathbf{s}_0) \cdot \mathbf{a}_3}, \tag{X-1}$$

where N_1, N_2, N_3 are the numbers of unit cells along the $\mathbf{a}_1$, $\mathbf{a}_2$, $\mathbf{a}_3$ directions. If the diffraction vector is represented in terms of the continuous variables h_1, h_2, h_3 by the relation

$$(\mathbf{s} - \mathbf{s}_0)/\lambda = h_1 \mathbf{b}_1 + h_2 \mathbf{b}_2 + h_3 \mathbf{b}_3$$

and this expression is used in equation X-1, the intensity in electron units from a single parallelopipedon crystal takes the form

$$I_{eu}(h_1 h_2 h_3) = F^2 \frac{\sin^2 \pi N_1 h_1}{\sin^2 \pi h_1} \frac{\sin^2 \pi N_2 h_2}{\sin^2 \pi h_2} \frac{\sin^2 \pi N_3 h_3}{\sin^2 \pi h_3}. \tag{X-2}$$

The quantity $I_{eu}(h_1 h_2 h_3)$ is an example of what we shall call an "interference function," the intensity in electron units from one crystal, expressed in terms of the coordinates in reciprocal space of the diffraction vector.

We first imagine a sample of crystals in which the interference function is zero everywhere in reciprocal space except close to the position $(h_1' h_2' h_3')$. Let $90 - \theta$ be the angle between $-\mathbf{s}_0$ and $h_1' \mathbf{b}_1 + h_2' \mathbf{b}_2 + h_3' \mathbf{b}_3$ when $(\mathbf{s} - \mathbf{s}_0)/\lambda = h_1' \mathbf{b}_1 + h_2' \mathbf{b}_2 + h_3' \mathbf{b}_3$. If there are M crystals in the sample, the number of crystals dM, whose vectors $h_1' \mathbf{b}_1 + h_2' \mathbf{b}_2 + h_3' \mathbf{b}_3$ make angles

*Reproduced by special permission in slightly modified form from B. E. Warren, *X-Ray Diffraction*, Addison-Wesley, Reading, Massachusetts, 1969, Chapter 13.

between $90-(\theta+\alpha)$ and $90-(\theta+\alpha+d\alpha)$ with the direction $-\mathbf{s}_0$, is given by

$$\frac{dM}{M} = \frac{2\pi r^2 \sin\left[90-(\theta+\alpha)\right] d\alpha}{4\pi r^2},$$

and since we are interested only in small values of α,

$$dM = \frac{M}{2} \cos\theta \, d\alpha.$$

Figure X-1 illustrates one crystal in the sample whose $h_1'\mathbf{b}_1 + h_2'\mathbf{b}_2 + h_3'\mathbf{b}_3$ vector is related to the primary direction $\mathbf{s}_0/\lambda$ and the diffraction direction $\mathbf{s}/\lambda$ so as to satisfy exactly the relation $(\mathbf{s}-\mathbf{s}_0)/\lambda = h_1'\mathbf{b}_1 + h_2'\mathbf{b}_2 + h_3'\mathbf{b}_3$. By means of small angles β perpendicular to the paper and γ in the plane of the paper, we represent a slightly different direction $\mathbf{s}'/\lambda$ for the diffracted beam. On a receiving surface which is normal to $\mathbf{s}/\lambda$ and at a distance R from the sample, we produce a small area $dA = R^2 d\beta \, d\gamma$ by means of small changes $d\beta$ and $d\gamma$. We are interested in the contribution from crystals whose $h_1'\mathbf{b}_1 + h_2'\mathbf{b}_2 + h_3'\mathbf{b}_3$ direction makes angles between $90-(\theta+\alpha)$ and $90-(\theta+\alpha+d\alpha)$ with the primary beam direction $\mathbf{s}_0/\lambda$. It is simplest to keep the $h_1'\mathbf{b}_1 + h_2'\mathbf{b}_2 + h_3'\mathbf{b}_3$ vector fixed, and let the primary beam change to the direction $\mathbf{s}_0'/\lambda$ which makes an angle $90-(\theta+\alpha)$ with $h_1'\mathbf{b}_1 + h_2'\mathbf{b}_2 + h_3'\mathbf{b}_3$. The total power arriving at the receiving surface is then given by integrating the intensity from one crystal over the area of the receiving surface, and then integrating over the number of crystals in each range of orientation with respect to the primary beam:

$$P = I_e \iiint I_{eu}(h_1 h_2 h_3) R^2 \, d\beta \, d\gamma \frac{M}{2} \cos\theta \, d\alpha. \tag{X-3}$$

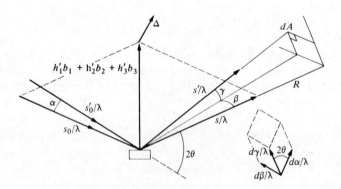

Fig. X-1.   Relations involved in the derivation of the powder pattern power theorem.

The interference function $I_{eu}(h_1h_2h_3)$ refers to the position in reciprocal space

$$h_1\mathbf{b}_1 + h_2\mathbf{b}_2 + h_3\mathbf{b}_3 = h_1'\mathbf{b}_1 + h_2'\mathbf{b}_2 + h_3'\mathbf{b}_3 + \mathbf{\Delta},$$

where

$$\mathbf{\Delta} = \frac{\mathbf{s}'}{\lambda} - \frac{\mathbf{s}}{\lambda} - \left(\frac{\mathbf{s}_0'}{\lambda} - \frac{\mathbf{s}_0}{\lambda}\right).$$

Changes $d\alpha\,d\beta\,d\gamma$ cause changes in the tip of the $\mathbf{\Delta}$ vector of amounts $d\boldsymbol{\alpha}/\lambda$ normal to $\mathbf{s}_0'/\lambda$ and $d\boldsymbol{\beta}/\lambda$ and $d\boldsymbol{\gamma}/\lambda$ normal to $\mathbf{s}/\lambda$. The three small vector changes are shown separately in Fig. X-1, with $d\boldsymbol{\alpha}/\lambda$ and $d\boldsymbol{\gamma}/\lambda$ in the plane of the paper and $d\boldsymbol{\beta}/\lambda$ perpendicular to the paper. The variation $d\alpha\,d\beta\,d\gamma$ causes the tip of the $\mathbf{\Delta}$ vector to sweep through a small volume in reciprocal space:

$$dV(RS) = \frac{d\alpha}{\lambda}\frac{d\gamma}{\lambda}\sin 2\theta\frac{d\beta}{\lambda} = \frac{\sin 2\theta}{\lambda^3}\,d\alpha\,d\beta\,d\gamma. \qquad \text{(X-4)}$$

From equation X-4 the product $d\alpha\,d\beta\,d\gamma$ can be expressed in terms of a volume element $dV(RS)$, and equation X-3 becomes a volume integral in reciprocal space:

$$P = \frac{I_e R^2 M \lambda^3}{4}\iiint \frac{I_{eu}(h_1h_2h_3)}{\sin\theta}\,dV(RS). \qquad \text{(X-5)}$$

The derivation has been carried through with the restriction that the interference function differs from zero only in the immediate vicinity of $h_1'h_2'h_3'$. If $I_{eu}(h_1h_2h_3)$ differs from zero over an extended region, the result applies to each small volume element in the region, and equation X-5 becomes a completely general theorem. For a powder sample the power on the receiving surface due to any region in reciprocal space is obtained from equation X-5 by integrating $I_{eu}(h_1h_2h_3)/\sin\theta$ over the desired volume in reciprocal space. Since it is a powder sample, the power expressed by equation X-5 is the total power in a hollow cone whose axis is the direction of the primary beam.

It is often convenient to express the volume element in terms of the variables $h_1h_2h_3$:

$$dV(RS) = \mathbf{b}_1\,dh_1 \cdot \mathbf{b}_2\,dh_2 \times \mathbf{b}_3\,dh_3 = v_b\,dh_1\,dh_2\,dh_3 = \frac{1}{v_a}\,dh_1\,dh_2\,dh_3.$$

Equation X-5 then takes the form

$$P = \frac{I_e R^2 M \lambda^3}{4v_a}\iiint \frac{I_{eu}(h_1h_2h_3)}{\sin\theta}\,dh_1\,dh_2\,dh_3. \qquad \text{(X-6)}$$

The powder pattern power theorem is represented by either equation X-5 or equation X-6. For imperfect structures it is usually simplest to express an interference function in terms of the variables $h_1h_2h_3$ in reciprocal space, and the observable quantity P is then obtained from $I_{eu}(h_1h_2h_3)$ by means of the theorem.

AUTHOR INDEX

Numbers in parentheses following page numbers denote specific references that appear on the given pages but without the author's name. Numbers in *italics* indicate pages on which the complete references are given.

Abell, J.S., 383 (198), 387, *417*

Aborn, R.H., 755, *789*

Abrahams, S.C., 152 (59), *173*, 254, (137), *269*, 382, 382 (183), *417*

Adams, C.E., 521, *563*

Ahlers, M., 687, *708*

Aitken, D.W., 330 (94), 333, 333 (94), 335 (98), 412, *414*, *415*

Aka, E.Z., 203 (20), *267*

Alexander, E., 478 (88), *503*

Alexander, L.E., 70 (34), 107, *118*, 176, 179, 182, 185, 188, 208 (28), 209, 229, 240, (5, 174), 241, 242, *266*, *267*, *270*, 290 (26-28), 291 (27,28), 292, 294 (27, 28, 36), 297, 319 (88), 335 (88), 336, 340 (88), 341, 342 (88), 343 (88), 360 (143), 365 (145), 366 (145), 375 (160), 389 (210), 390, 391 (210), 412 (231), *413*, *414*, *416*, *418*, 439, *501*, 532, 535 (33), 535-539, 554 (46), 556 (46), 561, 564, *565*, 583, 605 (54), 610 (69), 611, (70, 72), 612 (74, 76), *616*, *617*, 619 (8), 620 (10), 635, 641 (26), 670, 702 (119, 120, 121), *705*, *706*, *707*, *708*, 713, 725 (2), 732 (35), 733 (40), 734, 735 (41), 736, 737, 737 (42), 738 (42), 739-744, *752*, *753*, 814 (33), 849 (62), 850 (63), 852, 855 (105, 110), *857*, *858*, *859*

Alexander, T.G., 412 (231), *418*

Allison, S.K., 88, 117, *118*, 137, *172*, 642, 706, 799 (13), 803 (18, 19), *856*

Alty, J.L., 745, *753*

Amorós, J.L., 148 (91), *174*

Anantharaman, T.R., 495 (123), *503*, 686, *708*

Anderson, H.C., 526 (16), *563*

Angilello, J., 746, *753*

Antia, D.P., 558, *565*

Antonoff, G., 3, *56*

Appel, A., 382 (184), *417*

Aqua, E.N., 664, 675-681, 681 (54), 684, 705, *706*, *707*, 786, *790*

Arlman, J.J., 855 (106), *858*

Armstrong, W.E., 208 (29), *267*, 424, 476, *500*, *502*

Arndt, U.W., 313 (83), 315, 316, 321, 322 (83), 323 (83), 327, 327 (83), 328 (83), 335 (83), 340 (83), 342, 342 (83, 113), 343, 348 (122), 349, 350 (122), 412, *414*, *415*, 851, *858*

Arnell, R.D., 238 (100), *269*, 543, 558, *564*

Arrington, J.S., 352 (127), 354, *415*

Aruja, E., 248, *269*, 424 (6), *500*

Ashworth, W.J., 116

Asp, E.T., 720 (7), 735, 737, 738, *752*

Astbury, W.T., 65 (13), *117*, 710

Atlee, Z.J., 66 (25), *118*

Auld, J.H., 485 (102), *503*

901

SUBJECT INDEX

gas amplification factor of,
315
monochromatizing effect of,
317, 318, 324, 325
quenching agents for, 314,
315, 321, 322
Geiger-Müller, 271, 313-322
noise pulses of, 321, 324-326
nonlinearity of response of,
320, 323, 335-344
plateau of, 316, 317, 324-326,
330
proportional, 311, 314-316,
321-326
pulse-amplitude distribution
of, 322, 324, 328, 345
pulses of, 314-316, 319-329
quantum-counting efficiency
of, 305, 317-319, 324,
326, 329
recovery time of, 319, 320
resolving time of, 320
response curves of, 316, 317,
324, 325, 329
scintillation, 311, 318, 326-
330
semiconductor, 330-335
solid-state, 330-335
statistical accuracy of intensity measurements with,
360-364
threshold voltage of, 316,
317, 319, 324
use of, for monitoring direct
beam, 343, 344, 410, 412
with pulse-height discrimination, 323
voltage stabilization for,
324
Counting modes, fixed-time
and fixed-count, 313, 403
Counting-rate meter, 312, 313
Cristobalite (α), mixtures
of, with quartz, 535
occurrence of, in dust samples, 556, 557
Crookes tubes, 61
Cryostats, for low-temperature

diffractometric techniques,
381-387
Crystal face or plane, 4
Crystal forms, 19-22
Crystalline-amorphous ratio
in polymers, 854
Crystalline fine structure,
early concepts of, Haüy's,
28, 29
Huygens', 28
Crystalline powders, see Powders
Crystalline materials, 791
Crystalline state, 1-4
Crystallite size, accuracy in
determinations of, 624,
642, 694-696
apparent, 657
determination of, by variance
method, 657-659, 692
in absence of lattice imperfections, 687-704
shape factor K for, 656, 657,
659, 667, 668, 689, 690
effective, 666, 676-678, 689,
690
effect of, on diffraction
pattern, 202-206, 224,
225, 494-496, 539-542
on diffractometric intensities, 365-367
on line breadths, 495, 496
on line profiles, 290
illustrative determinations of,
696-704
number-average, 694
of reference substances in size
determinations, 696
practical considerations in
determination of, 694-696
true, 657
volume-average, 658
Crystallite size and lattice
imperfections (strains),
comparison of values of,
obtained by four methods,
682-685
dependence of numerical results
on definitions of, 682-685